SLURRY HANDLING
DESIGN OF SOLID–LIQUID SYSTEMS

Elsevier Handling and Processing of Solids Series

Advisory Editor: R. Clift, Guildford, UK

Other titles in series:

K. Rietema—The Dynamics of Fine Powders

SLURRY HANDLING DESIGN OF SOLID–LIQUID SYSTEMS

Edited by

Nigel P. Brown

The British Petroleum Company plc, Britannic House, Moor Lane, London, EC2Y 9BU, UK

and

Nigel I. Heywood

Warren Spring Laboratory, Gunnels Wood Road, Stevenage, Hertfordshire, SG1 2BX, UK

ELSEVIER APPLIED SCIENCE
LONDON and NEW YORK

ELSEVIER SCIENCE PUBLISHERS LTD
Crown House, Linton Road, Barking, Essex 1G11 8JU, England

Sole distributor in the USA and Canada
ELSEVIER SCIENCE PUBLISHING CO., INC
655 Avenue of the Americas, New York, NY 10010 USA

WITH 74 TABLES AND 295 ILLUSTRATIONS

British Library Cataloguing in Publication Data

Slurry handling: design of solid–liquid systems
I. Brown, N. P. II. Heywood, N. I.
621.8'672—dc20

ISBN 1-85166-645-1

Library of Congress Cataloging-in-Publication Data

Slurry handling: design of solid–liquid systems / edited by N.P. Brown and N.I. Heywood.
p. cm. — (Elsevier handling and processing of solids series)
Includes bibliographical references and index.
ISBN 1-85166-645-1
1. Hydraulic conveying. 2. Slurry—Pipe lines.
I. Brown, N. P. (Nigel P.) II. Heywood, N. I. (Nigel I.) III. Series.
TJ898.S549 1991
621.8'672—dc20 91-15406
CIP

Photoset by Alden Multimedia Ltd
Printed in Great Britain at the Universities Press Cambridge

Dedication

To our contributing authors and other engineers who strive to create systems of practical use to mankind.

Preface

This design handbook has been conceived to provide guidance on the design of systems to handle and transport slurries and comprehensive information on slurry properties and flow behaviour.

This handbook is focused on system design and does not attempt to provide guidance on the design of specific items of process equipment. It is recognised that considerable knowledge of the design and application of equipment resides with specialist manufacturers. The handbook strives to provide sufficient background information on equipment to enable informed enquiries to be made of vendors and to assist in the location of more specific documentation. Transport of slurries is considered in both pipes and flumes. Where possible, physically based treatments are presented to enable the steady-state behaviour to be analysed and predicted. Transient pipe flow is also considered. As appropriate, distinction is made between the long-distance commodity pipeline and short in-plant applications. The integrated system is considered in terms of economic evaluation and commissioned projects.

The complex nature of solid–liquid mixtures often dictates that experimental studies have to be undertaken in order to provide both data and confidence for predictive approaches. Bench-scale and pilot-plant studies are discussed in relation to experimental measurement of slurry properties and behaviour.

The design of slurry systems requires the application of expertise from many diverse areas of engineering. The aim has been to present practical information that has been gained from experience with slurries and slurry systems complemented with theoretical treatments and interpretive reviews of information available in the literature.

Contributions to this handbook have been invited from engineers that are active in the design of slurry systems. Editorial intervention has largely been restricted to developing the content and ensuring continuity and a consistent approach to terminology and nomenclature.

SI units have been used throughout. However, it is recognised that Imperial and US customary units are still widely used both for design calculations and equipment specification. Appendix 1 provides tables to assist in the conversion to SI units of these and other systems of units that are routinely encountered. In instances where equipment is described using non-SI units both systems are given. Where possible, variables used in equations are defined using the base dimensions of length–mass–time (LMT) enabling any coherent system of units to be used. This approach is discussed briefly in Appendix 2.

The editors would like to thank their contributing authors and their employers without whose assistance this handbook would not have been possible.

Nigel P. Brown & Nigel I. Heywood
London, February 1991

Contents

1

Classification and Characterisation of Slurries

Nigel P. Brown
The British Petroleum Company plc, London, UK

&

Nigel I. Heywood
Warren Spring Laboratory, Stevenage, UK

1. INTRODUCTION

This chapter attempts to provide background to the rather inexact comprehension of the description *slurry*. Classification and characterisation schemes are proposed based on physical properties and flow behaviour. These schemes are applied to slurry technologies that offer potential commercial importance. Sources of information are provided on general aspects of slurries including flow behaviour, slurry handling equipment and large projects.

2. WHAT IS A SLURRY?

Slurries are encountered in many industries where they are referred to by various descriptions, including

- *broths*, fibrous particle–liquid mixtures in biotechnological industries,
- *dispersions* (of colloidal particles) in fine chemical industries,
- *muds* in the oil drilling industry,
- *pastes* in pharmaceutical and food industries,
- *pulps*, of wood in the paper industry, and of ground minerals in the mineral processing industries,
- *silts* when dredged from water courses,
- *slimes*, *concentrates* and *tailings* in mineral processing,
- *slips* in the ceramics industry,
- *sludges* when associated with sewage and waste, and
- *suspensions* across a wide range of industries.

The common aspect of all these mixtures is that they consist of solids mixed with a liquid and can therefore be regarded as two-phase, solid–liquid mixtures. Slurries are fluids and can be transported in pipes and flumes, in contrast to other wet solids that cannot readily be pumped and are not perceived to flow under the influence of gravity. Such solid–liquid mixtures are better transported by other means that are more appropriate for bulk solids such as conveyer belts and screw feeders. The flow behaviour of solid–liquid mixtures extends from those which are similar to single-phase liquids to those in which it is appropriate to consider the solids and liquid as distinctly separate phases.

Other examples of slurries include printing inks, paints, cosmetics, pharmaceuticals and foodstuffs.

3. WHY ATTEMPT TO CLASSIFY A SLURRY?

The enormous range of mixtures described as slurries is encountered in a wide range of industrial sectors that include chemical engineering, civil engineering (soil mechanics, waste water, concrete pumping), mineral processing and the food industry. Classification schemes attempt to provide a rational basis for describing the physical appearance and flow behaviour of solid–liquid mixtures.

Numerous schemes are possible; the two considered here are based on *physical properties* and *rheology*, or flow behaviour.

4. CLASSIFICATION BASED ON PHYSICAL PROPERTIES

Attributes that are commonly used to characterise a slurry are the basic physical properties of the constituents, in particular those of the solids:

- *densities* of the constituent phases,
- *concentration* of solids,
- characteristic *particle size* (or, more appropriately, particle size distribution), and
- characteristic *particle shape.*

4.1. Particle Size and Shape

Terminology associated with differently sized particles has developed within many disciplines; the scheme reproduced in Table 1.1 originates in civil engineering (BSI, 1975).

Chapter 2 considers various schemes that can be used to describe mathematically the shape of irregular particles.

4.2. Settling Behaviour

Descriptions such as *coarse-particle*, *granular* and *non-cohesive* are usually associated with slurries that settle and *fine-particle* with those of a low settling tendency. Quantification of these subjective descriptions has to be made with regard to the *time frame* over which settling is deemed to be important. The majority of particles contained in most industrially important slurries will settle appreciably, given sufficient time, in the absence of supporting forces originating within the flow structure.

TABLE 1.1
Descriptive Terminology Applied to Rounded and Sub-angular Mineral Particles[a]

Size range, d/(mm)	*Sub-description*	*Main description*
$d <$ 0·002		Clay
0·002 $\leqq d \leqq$ 0·006	Fine	
0·006 $\leqq d \leqq$ 0·02	Medium	Silt
0·02 $\leqq d \leqq$ 0·06	Coarse	
0·06 $\leqq d \leqq$ 0·2	Fine	
0·2 $\leqq d \leqq$ 0·6	Medium	Sand
0·6 $\leqq d \leqq$ 2	Coarse	
2 $\leqq d \leqq$ 6	Fine	
6 $\leqq d \leqq$ 20	Medium	Gravel
20 $\leqq d \leqq$ 60	Coarse	
60 $\leqq d \leqq$ 200		Cobbles

[a]Adapted from BSI (1975).

An apparently settling slurry can sometimes, for the purposes of design, be treated as *non-settling* if, for example, the mean residence time of a slurry in a pipeline is small compared to the particle settling rate. Similarly, what appears to be a relatively slowly settling slurry could cause potential problems in a long-distance, near-horizontal pipeline. For instance, the residence time of a slurry with a mean flow velocity of 1 $m\,s^{-1}$ in a 100 km long pipe is some 30 h. This time is sufficient for particles with a 0·001 $mm\,s^{-1}$ settling velocity to double in concentration in the lower half of a 200 mm pipe. The effect of the settled solids would be to increase the frictional resistance of the slurry; this could precipitate a blockage due to excessive pumping requirements. The techniques discussed in Chapter 2 can be applied to predict settling behaviour under static conditions.

Slow settling tendency is a requirement for slurries that are to be rheologically characterised using rotational viscometric techniques. This constraint on the nature of the slurry arises from the practicality of making the measurement and also from a consideration of the appropriateness of the measurement technique to predict the pipeline flow behaviour.

There is considerable temptation to classify slurries rigidly between the limiting cases of *settling* and *non-settling* as a result of the behaviour of a static sample of the slurry. Such a scheme is undoubtedly useful for a wide range of slurries but, like any broad scheme, it must be applied with caution. Whilst physical attributes are useful, the rheological or flow behaviour of the slurry is of utmost importance because it reflects the resistance of the slurry to flow.

5. CHARACTERISATION BY FLOW BEHAVIOUR

5.1. Introduction

A classification scheme for single- and two-phase fluids based on various attributes of flow behaviour is shown in Fig. 1.1. This figure attempts to show the interrelation between the various attributes. Broken vertical lines are used to indicate that a distinct boundary does not exist.

5.2. Flow Regime

The flow regime provides an indication of the distribution of solids in the cross-section of the pipe under flowing conditions. The difficulties encountered in predicting the distribution of solids under dynamic conditions is discussed in Chapter 3. Homogeneous flow is an attribute that can be associated only with a single-phase liquid or slurries composed of particles of colloidal dimensions. Colloidal materials (characterised by primary particle diameters of, typically, less than 2 μm) are maintained in suspension by molecular movement within the liquid (Brownian motion). Larger diameter particles are maintained in suspension by other means that include electrical effects, support from structure-building mol-

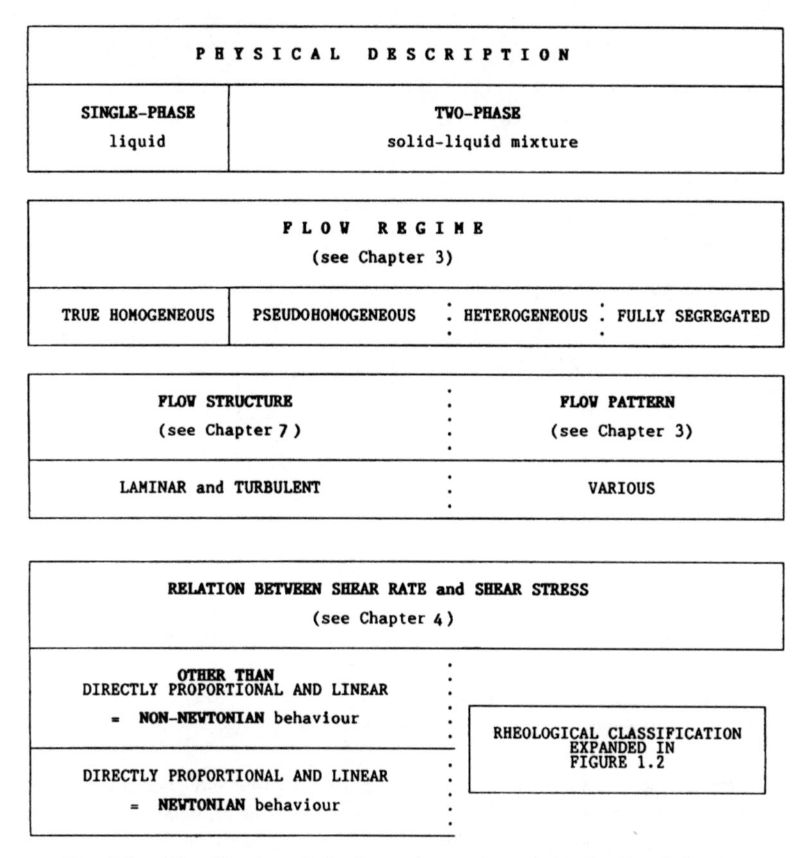

Fig. 1.1. Classification of single- and two-phase fluids by flow behaviour.

ecules (such as polymers), and at high concentrations of solids, direct mechanical support. Slurries in which the solids do not readily settle under conditions of no shear are frequently classified as pseudohomogeneous.

5.3. Flow Structure and Flow Pattern

The flow structure of pseudohomogeneous slurries is considered to range from laminar through transitional to turbulent. The flow structure in each of these regimes is considered for pipe flows in Chapter 7. Mixtures that show a marked tendency for the solids to settle under conditions of no-flow are exhibiting a distinct two-phase characteristic. The flow patterns exhibited by such slurries are considered in Chapter 3.

<table>
<tr><td colspan="4">NON - NEWTONIAN FLUID</td><td>NEWTONIAN FLUID</td></tr>
<tr><td colspan="2"></td><td>TIME DEPENDENT</td><td colspan="2">TIME INDEPENDENT</td></tr>
<tr><td rowspan="2">PURELY VISCOUS</td><td>SHEAR THINNING</td><td>THIXOTROPIC</td><td>PSEUDOPLASTIC and YIELD PSEUDOPLASTIC</td><td rowspan="2">PURELY VISCOUS</td></tr>
<tr><td>SHEAR THICKENING</td><td>RHEOPECTIC</td><td>DILATANT and YIELD DILATANT</td></tr>
<tr><td>VISCO ELASTIC</td><td colspan="3">MANY FORMS</td><td></td></tr>
</table>

Fig. 1.2. Simplified rheological classification of homogeneous and pseudohomogeneous fluids.

5.4. Relation between Shear Rate and Shear Stress

Although the flow behaviour of slurries containing dispersed (non-flocculated) particles can often be approximated to Newtonian behaviour, flocculated particle slurries result generally in non-Newtonian rheological properties. Newtonian behaviour defines the fluids behaviour when the rate at which the fluid is sheared is directly proportional to the shear stress exhibited by the fluid at a plane boundary; the coefficient of proportionality is the (dynamic) viscosity. Non-Newtonian behaviour is characterised by a viscosity that is dependent upon the rate of shear.

Figure 1.2 shows a simplified rheological classification scheme applicable to homogeneous and pseudohomogeneous slurries. In order to simplify interpretation and analyses it is frequently assumed, at least initially, that a slurry exhibits purely viscous, time-independent behaviour. In practice this assumption is often satisfactory. However, the possibility of elastic response (indicating a tendency for the fluid to return to its original shape after shear) and time-dependent behaviour should not be excluded. Viscoelasticity requires considerably more experimental measurement effort in the laboratory than purely viscous non-Newtonian fluids. The majority of industrially important slurries are Newtonian or shear thinning. If shear thinning, a proportion of these slurries may, in addition, show thixotropic behaviour. Thixotropic behaviour is shown by a progressive decrease in viscosity under conditions of either constant shear stress or shear rate with a recovery in viscosity when the shear is removed.

Shear-thinning behaviour is the progressive reduction in viscosity with increasing shear stress or shear rate (often realised as increasing flow or agitation). The flow property is time independent. Shear thickening, or dilatancy, is less commonly encountered in pseudohomogeneous slurries and tends to occur in high concentration non-flocculated slurries. The response of slurries at low shear rates is often interpreted as solid-like behaviour, as a minimum shear stress apparently is required to be applied before the mixture shears and becomes fluid like. The minimum shear stress to shear the fluid is referred to as a yield stress. Slurries that apparently display this behaviour are described as yield-pseudoplastic or yield-dilatant. However, yield-dilatancy is relatively uncommon. In general, the complex, shear rate dependent rheological behaviour is described by empirical models. This frequently results in a slurry flow property being associated with a particular model. For example, slurries consisting of fine-particle coal are often described as Bingham-like, i.e. those exhibiting a yield stress, but with a linear relationship between shear stress and shear rate after yield.

The physical and chemical bases for the rheological behaviour of *non-settling* slurries are discussed in Chapter 4.

6. SLURRY TECHNOLOGIES

6.1. Introduction

In addition to conventional dilute-phase transport several slurry technologies have received continued attention in the technical literature for the conveyance of commodities. Within each of the technologies discussed below there are many proprietary techniques.

6.2. Dense-Phase Slurries

Dense-phase slurry technology is typified by high solids contents, with concentrations at, or slightly above, the freely settled solids concentration. The freely settled concentration is that which can be attained by allowing the solids to settle freely in the liquid without compaction. Under this condition the solids can be visualised as being in the most open packing configuration, which for spheres is referred to as open square packing displaying a theoretical volumetric concentration of 52·4%. Rounded, narrowly graded, sand particles closely approach this ideal concentration. Higher concentrations are possible with wide particle size distributions or with angular broken materials such as limestone or coal because even when freely settled they tend to interlock (Furnas, 1931; White & Walton, 1937). Dense-phase conveying is frequently considered to be an attractive technology because the amount of fluid required is less than for conventional dilute-phase transport. Under conditions of no-flow, a clear fluid layer tends to form at the top of the solids; it is usually appropriate to classify such slurries as *settling* (see, for example, Brown, 1981).

6.3. Stabilised Slurries

Slurries containing coarse particles that would normally be expected to settle, but which do not produce supernatant liquid on standing are referred to as *stabilised.* Stabilised slurries have also been referred to as *mixed slurries* because a common practice of creating stability is by mixing solids of two sizes to form a bi-modal (or dumbbell-shaped) size distribution. The overall concentration of the solids in the slurry and the particle size distribution of the fines is manipulated to reduce the viscosity of the slurry (and hence reduce pumping costs) yet maintain the ability of the slurry to support coarse particles when stationary, or pumped at low velocities.

Elliott and Gliddon (1970) identified that particle size distributions which approached the closest particle packing produced a stabilised slurry that could be characterised as a Bingham plastic. Duckworth and Pullam (1981) have suggested that it is not necessarily the particle size distribution *per se* that is responsible for stability but that it is also attributable to the rheological properties of the fine slurry formed from particles that pass a 500 μm screen. The pioneering work, carried out by Elliott and Gliddon, used volumetric concentrations up to 51%, with a top particle size of 12 mm. Stabilised coal slurries containing material with a top size of 50 mm have been shown to be stable (see, for example, Brookes & Snoek, 1988); the nature of these slurries has been investigated using three conceptual flow models (Brown, 1988). At volumetric concentrations between 56 and 73% these slurries could be characterised as pseudohomogeneous although models that assumed a two-phase nature also provided reasonable predictions of flow behaviour.

Proprietary particle size distributions used in stabilised slurry technology enable slurries to be pumped at concentrations up to 40% higher than the freely poured concentration of the run-of-mine coal. At this concentration the slurry may form a stable cargo and be suitable for direct transportation by ship. Pumping slurries at such high concentration can be achieved only at the expense of high frictional pressure drops and, as such, appear to be attractive only for short-distance conveying. Lower concentration slurries can be pumped economically under flow conditions that may be regarded as laminar.

6.4. Energy Slurries

As a consequence of the oil crises in the 1970s coal was once again looked at as a principle source of energy. Transportation and convenience of use posed major obstacles to large-scale commercial use. Although coal had been transported hydraulically in sizes ranging from run-of-mine down to specially prepared mixtures (with a top-size of the order 1–2 mm) commercial combustion of these slurries requires drying and pulverising. The production of coal–liquid mixtures, with the handling convenience of oil and burning characteristics that promised direct-firing in existing boilers using oil or pulverised fuel burners, resulted in the emergence of several energy slurry technologies.

A wide range of liquids has been investigated as the vehicle including heavy fuel oil, crude oil, water, methanol, liquid gases (such as LNG and LPG) and liquid carbon dioxide, each offering project-specific advantages. The maximum particle size of the coal used is typically less than 300 μm with between 70 and 80% less than 74 μm; this distribution is typically similar to, or finer than, conventional pulverised coal.

Coal–oil mixtures consist of crude or fuel oils loaded with pulverised coal at concentrations of typically 40–50% (by mass). Depending on the relative heating values, approximately 40% of the energy in the slurry is obtained from the coal. Finer grinding of the coal, down to average particle sizes in the range 10–30 μm have reportedly produced slurries that exhibit stability without the use of additives. The rapid rise in slurry viscosity with coal concentration frequently makes more highly loaded slurries less attractive.

The concentration of solids in coal–water mixtures is typically 60–70% (by mass), although proprietary technologies report the use of concentrations up to 75%. To promote long-term stability dispersants are commonly added to coal–water mixtures, at typical concentrations between 0·5 and 1% (by mass).

7. SOURCES OF INFORMATION

7.1. Abstracts

Heat Transfer and Fluid Flow Service (*HTFS*) *Digest*, HTFS (Harwell, National Engineering Laboratory and Chalk River) in co-operation with Lavis Marketing, UK. Published every two months since 1967.
(Abstracts covering a wide topic base; those particularly relevant include instrumentation and measuring techniques, and non-Newtonian systems.)

Solid–Liquid Flow Abstracts, BHRA, The Fluid Engineering Centre, Cranfield, UK, 1968–1991. (Now published by Elsevier Advanced Technology under the title *Fluid Abstracts: Process Engineering*.)
(Abstracts of highly relevant literature commencing 1968; see Thornton (1970) for a biography covering earlier literature.)

7.2. Journals

Bulk Solids Handling, Trans Tech Publications, Clausthal-Zellerfeld, Germany. Published every two months.
(Occasional articles on slurry transport.)

International Journal of Multiphase Flow, Pergamon Press, Oxford, UK. Published every two months.
(Predominantly mathematical treatment of multiphase flows.)

Journal of Pipelines, Elsevier. Vol 1 (1981)–Vol 7 (ceased publication 1988).
(Contributions covered all aspects of slurry pipeline design and operation.)

Powder Technology, Elsevier Sequoia, Lucerne, Switzerland. Two or three issues per volume, three volumes per year.
(All aspects of particle technology and systems containing particulate solids.)

Solid–Liquid Flow, GEPS (Groupe d'etudes et de publications scientifiques), France. Four issues per year, commencing 1989.
(Interdisciplinary journal including work on sediment transport (rivers and maritime), underground flow, industrial flows (filtration, mixing, pipeline transport) and mathematical modelling of two-phase flow. Diary of forthcoming conferences and meetings. Information on research laboratories.)

7.3. Books

Bain, A.G. & Bonnington, S.T. (1970). *The Hydraulic Transport of Solids by Pipeline*. Pergamon Press, Oxford, UK.
(Most aspects of slurry system design.)

Baker, P.J. & Jacobs, B.E.A. (1979). *A Guide to Slurry Pipeline Systems*. BHRA, The Fluid Engineering Centre, Cranfield, UK.

Cheremisinoff, N.P., (ed.) (1986). *Slurry Flow* (Vol. 5 of *Encyclopedia of Fluid Mechanics*). Gulf Publishing Co., Houston, TX, USA.

Cheremisinoff, N.P., (ed.) (1988). *Rheology and Non-Newtonian Flows* (Vol. 7 of *Encyclopedia of Fluid Mechanics*). Gulf Publishing Co., Houston, TX, USA.

Cheremisinoff, N.P. & Gupta, R. (eds) (1983). *Handbook of Fluids in Motion*. Ann Arbor Science, The Butterworth Group.
(Five contributions on solid–liquid flow).

Fayed, M.E. & Otten, L. (eds) (1984). *Handbook of Powder Science and Technology*. Van Nostrand Reinhold Co., New York, USA.
(Storage of particulate solids, transport of solids in pipelines, size reduction of solids, sedimentation and filtration.)

Govier, G.W. & Aziz, K. (1972). *The Flow of Complex Mixtures in Pipes*. Van Nostrand Reinhold Co. New York, USA.
(Useful review of predictive techniques.)

Neville, A.M. (1975). *Properties of Concrete*. Pitman International, London, UK.

Shamlou, P.A. (1988). *Handling of Bulk Solids: Theory and Practice*. Butterworths, London, UK.

Skelland, A.H.P. (1967). *Non-Newtonian Flow and Heat Transfer*. John Wiley & Sons, New York, USA.
(Practical reference for non-Newtonian fluid mechanics.)

Tattersall, G.H. & Banfill, P.F.G. (1983). *The Rheology of Fresh Concrete*. Pitman Advanced Publishing Programme, London, UK.
(Modelling and experimental techniques.)

Thornton, W.A. (1970). *Hydraulic Transport of Solids in Pipes: A Bibliography.* BHRA, The Fluid Engineering Centre, Cranfield, UK.
(Bibliography (predating *Solid–Liquid Flow Abstracts*) covering the period 1950–1969.)

Wallis, G.B. (1969), *One-Dimensional Two-Phase Flow*, McGraw-Hill, New York, USA.
(Mathematical modelling of two-phase slurries and other systems.)

Wasp, E.J., Kenny, J.P. & Gandhi, R.L., (1977), *Solid–Liquid Flow Slurry Pipeline Transportation.* Trans Tech Publications, Clausthal-Zellerfeld, Germany.
(Design of slurry systems.)

Wöhlbier, R.H. (ed.) (1986), *Hydraulic Conveying and Slurry Pipeline Technology.* Trans Tech Publications, Clausthal-Zellerfeld, Germany.
(Reprints of selected articles from *Bulk Solids Handling*, 1981–85.)

Zandi, I. (ed.) (1971). *Advances in Solid–Liquid Flow in Pipes and its Application.* Pergamon Press, Oxford, UK.
(Contributions cover a wide range of topics.)

7.4. Conference Proceedings

Advances in Particulate Technology (Symposium and Exhibition), Institution of Chemical Engineers, UMIST, Manchester, UK, 6–8 Apr. 1987 (published in *Chem. Eng. Sci.*); University of Surrey, UK, 15–17 Apr. 1991 (published in *Powder Technol.*).

Dredging Technology Conferences (International Symposium on Dredging Technology), BHRA Fluid Engineering, Cranfield, UK.
1st, Canterbury, UK, 17–19 Sept. 1975.
2nd, College Station, TX, USA, 2–4 Nov. 1977.
3rd, Bordeaux, France, 5–7 Mar. 1980.
4th, Singapore, 19–22 Apr. 1983.
(Pipeline systems and pumps, with particular reference to dredging operations.)

European Conferences on Coal–Liquid Mixtures, Institution of Chemical Engineers, Rugby, UK.
1st, Cheltenham, UK, 5–6 Oct. 1983.
2nd, London, UK, 16–18 Sept. 1985.
3rd, Malmö, Sweden, 14–15 Oct. 1987.
(Slurry preparation, stabilisation and on-site handling, demonstration projects, commercial applications, atomisation, burner design, heat.)

Hydraulic Transport of Solids (Conference and Workshop), Institution of Mechanical Engineers, London, UK, 2 Dec. 1980.

Hydrotransport Conferences (International Conferences on the Hydraulic Transport of Solids in Pipes), BHRA Fluid Engineering, Cranfield, UK.
1st, Coventry, UK, 1–4 Sept. 1970.

2nd, Warwick, UK, 20–22 Sept. 1972.
3rd, Golden, CO, USA, 15–17 May 1974.
4th, Banff, Alberta, Canada, 18–21 May 1976.
5th, Hannover, Germany, 8–11 May 1978.
6th, Canterbury, UK, 26–28 Sept. 1979.
7th, Sendai, Japan, 4–6 Nov. 1980.
8th, Johannesburg, Republic of South Africa, 25–27 Aug. 1982.
9th, Rome, Italy, 17–19 Oct. 1984.
10th, Innsbruck, Austria, 29–31 Oct. 1986.
11th, Stratford-upon-Avon, UK, 19–21 Oct, 1988.
(All aspects of slurry transport using pipes: measurement and instrumentation, fluid mechanics, mathematical modelling, coal, minerals, pumps, wear, operational experience, pumping systems.)

IADC/SPE Drilling Conferences, International Association of Drilling Contractors/Society of Petroleum Engineers, Dallas, TX, USA.
New Orleans, LA, USA, 5–8 Mar. 1985.
Dallas, TX, USA, 9–12 Feb. 1986.
New Orleans, LA, USA, 15–18 Mar. 1987.
Dallas, TX, USA, 28 Feb.–2 Mar. 1988.
(All aspects of drilling, including reciprocating pumps, separation equipment, and transport of drilled cuttings in non-Newtonian drilling fluids in annular well-bore geometrics.)

Interpipe (International Pipeline Technology Exhibition and Conference), Houston, TX, USA.
8th, 5–7 Feb. 1980. (2 vols)
9th, 24–26 Feb. 1981.
10th, 23–25 Feb. 1982.
(Pipeline design, construction techniques, pipeline protection, instrumentation and controls, pumps, and specialised applications, including slurries.)

International Symposium on Freight Pipelines.
Selected papers from the 2nd (1977), 3rd (1981), 4th (1982) and 5th (1987) conferences appear in the *Journal of Pipelines.*
6th, Colombia, MO, USA, 10–12 May 1989. (Hemisphere Publishing Corp., New York, USA)
(Slurry technology, capsule and pneumatic systems.)

Powtech (International Exhibition and Conference on Powder and Bulk Solids Technology) Institution of Chemical Engineers, London, UK.
8th, Birmingham, UK, 10–13 Mar. 1981.
9th, Birmingham, UK, 8–11 Mar. 1983.
10th, Nürnberg Messe, Germany, 9–11 May 1984.
11th, Birmingham, UK, 5–8 Mar. 1985.
(Recent conferences contained contributions on solid–liquid interaction, gas–solid interaction, and bulk solids handling.)

Slurry Transport Association (International Technical Conference on Slurry Transportation), Batelle Memorial Institute, Columbus, OH, USA.
1st, Columbus, OH, USA, 3–4 Feb. 1976.
Slurry Transport Association, Washington, DC, USA.
2nd, Las Vegas, NV, USA, 2–4 Mar. 1977.
3rd, Las Vegas, NV, USA, 29–31 Mar. 1978.
4th, Las Vegas, NV, USA, 28–30 Mar. 1979.
5th, Lake Tahoe, NV, USA, 26–28 Mar. 1980.
6th, Las Vegas, NV, USA, 24–27 Mar. 1981.
7th, Lake Tahoe, NV, USA, 23–26 Mar. 1982.
8th, San Francisco, CA, USA, 15–18 Mar. 1983.
9th, Lake Tahoe, NV, USA, 21–22 Mar. 1984.
10th, Lake Tahoe, NV, USA, 26–28 Mar. 1985.
11th, Hilton Head, SC, USA, 16–18 Mar. 1986.
(International Conference on Slurry Technology), Slurry Technology Association, Washington, DC, USA.
12th, New Orleans, LA, USA, 31 Mar.–3 Apr. 1987.
(International Conference on Coal and Slurry Technology), Coal and Slurry Technology Association, Washington, DC, USA.
13th, Washington, DC, USA, 13–15 Apr. 1988.
14th, Clear Water, FL, USA, 24–27 Apr. 1989.
(Current position of pipeline projects worldwide, operational reviews, coal, direct fired slurries, market evaluations and economics, rheology and characterisation, slurry combustion, projects and applications, materials and equipment, design, planning, pipeline equipment.)

SPE Conference (Annual Technical Conference and Exhibition), Society of Petroleum Engineers, Dallas, TX, USA.
Dallas, TX, USA, 27–30 Sept. 1987.
Houston, TX, USA, 2–5 Oct. 1988.
(Drilling volume contains papers similar to those in the IADC/SPE Conference series.)

The Winter Annual Meeting (ASME)
Slurry Transport and Pneumatic Handling, Phoenix, AZ, USA, 14–19 Nov. 1982.
International Symposium on Slurry Flows, Anaheim, CA, USA, 7–12 Dec. 1986.
3rd International Symposium on Liquid–Solid Flows, Chicago, IL, USA, 28 Nov.–12 Dec. 1988.
(Mathematical modelling, experimental, pumps and instrumentation.)

7.5. Saskatchewan Research Council Reports

Available from Saskatchewan Research Council, 15 Innovation Blvd, Saskatoon, Saskatchewan, Canada, S7N 2X8.

Gillies, R.G., Haas, D.B., Small, M.H. & Husband, W.H.W. (1981). *Hydraulic testing of coarse coal slurries in pipelines up to 20 in. in diameter*. Report number E 725-10-C-82.

(Pipeline flow characteristics of coal in water slurries were measured in pipes of nominal diameter: 50, 100, 150, 250 and 500 mm. The coal particle sizes ranged from < 1 mm to < 50 mm.)

Gillies, R.G., Haas, D.B., Small, M.H. & Husband, W.H.W. (1981). *Coarse coal in water slurries—full scale pipeline tests.* Report number E 725-7-C-81.
(Four coals were tested: two thermal and two metallurgical, at three nominal size distributions: < 2 mm in 50, 150, 250 and 500 mm pipes, < 9·5 mm in 100 and 150 mm pipes, < 50 mm in 150, 250 and 500 mm pipes.)

Gillies, R.G., Husband, W.H.W. & Small, M.H. (1985). *Hydraulic testing of coarse particle slurries* (Phase 1: Sand slurries in a 250 mm pipeline). Report number R 833-2-C-85.
(The pipeline behaviour of sand/gravel in water slurries was extensively tested in a project designed to provide a better understanding of the flow characteristics of coarse particle slurries. Sand with particle sizes ranging from < 1 mm to > 2 × < 6 mm were tested. The results were interpreted using a two-layer model.)

Gillies, R.G., Husband, W.H.W., Small, M.H. & Shook, C.A. (1986). *Hydraulic testing of coarse particle slurries* (Phase 2: Coal slurry tests in a 250 mm pipeline). Report number R 833-2-C-85.
(The pipeline behaviour of slurries consisting of < 50 mm coal in water was studied. The results were interpreted using a two-layer model.)

Haas, D.B., Gillies, R., Small, M. & Husband, W.H.W. (1980). *Study of the hydraulic properties of coarse particles of metallurgical coal when transported in slurry form through pipelines of various diameter.* Report number E 835-1-C-80.
(Three bituminous coals were tested: < 20 mm in a 100 mm pipe and < 50 mm in 150 mm and 250 mm pipes at volumetric concentrations up to 50%.)

Husband, W.H.W. (1983). *Slurry pipeline development centre in Saskatoon.* Report number E 725-4-C-83.
(Description of the test equipment and services provided by the SRC's pipeline facility at the year of publication.)

Postlethwaite, J. & Tinker, E.B. (1973). *Erosion–corrosion studies in slurry pipelines.* Final report to the Saskatchewan Research Council from Corrosion Labs, Faculty of Engineering, University of Saskatchewan, Regina Campus. Report number E 73–22.
(Sand, iron ore, iron concentrate, potash, limestone and coal tested in 50 mm steel pipes.)

Schriek, W., Smith, L.G., Haas, D.B. & Husband, W.H.W. (1973). *Experimental studies on solids pipelining of Canadian commodities for The Transportation Development Agency* (Report 2: Experimental studies on the hydraulic transport of limestone). Report number E 73-10.
(Coarse and fine limestone, typically < 2 mm, tested at volumetric concentrations up to 30%, in 50, 100 and 150 mm pipes.)

Schriek, W., Smith, L.G., Haas, D.B & Husband, W.H.W. (1973). *Experimental studies on solids pipelining of Canadian commodities for The Canadian Transport Commission and The Transportation Development Agency* (Report 3: Experimental studies on the hydraulic transport of iron ore). Report number E 73-12.
(Behaviour of water based slurries tested in a tube viscometer and in 50, 100, 150, 200, 250 and 300 mm pipes. Two ores (< 0·3 and < 0·6 mm) were tested at volumetric concentrations up to 40%.)

Schriek, W., Smith, L.G., Haas, D.B. & Husband, W.H.W. (1973). *Experimental studies on solids pipelining of Canadian commodities for The Canadian Transport Commission and The Transportation Development Agency* (Report 4: Hydraulic studies on the hydraulic transport of potash). Report number E 73-16.
(< 3 mm, nominal size potash in brine slurries tested in 50, 100, 150 and 250 mm pipes at volumetric concentrations up to 50%.)

Schriek, W., Smith, L.G., Haas, D.B. & Husband, W.H.W. (1973). *Experimental studies on solids pipelining of Canadian commodities for The Canadian Transport Commission and The Transportation Development Agency* (Report 5: Experimental studies on the hydraulic transport of coal). Report number E 73-17.
(Lignite, thermal and metallurgical coals, typically < 3·5 mm, tested in 50, 100, 150, 200, 250 and 300 mm pipes at volumetric concentrations up to 50%.)

Schriek, W., Smith, L.G., Haas, D.B. & Husband, W.H.W. (1973). *Experimental studies on solids pipelining of Canadian commodities for The Canadian Transport Commission and The Transportation Development Agency.* (Report 7: Experimental studies on the hydraulic transport of two different sands in water in 2, 4, 6, 8, 10 and 12 in. pipelines). Report number E 73-21.
(2 mm sand used in Report E73-20 (see below) extensively studied at volumetric concentrations up to 35%.)

Shook, C.A. (1973). *Behaviour of inclined slurry pipelines at shutdown.* Report number E 75-12.
(The experimental study was conducted with slurries of iron ore concentrates in a long 50 mm (nominal bore) pipe loop. It was found that the length of the inclined pipe did not significantly affect the ultimate settled condition.)

Shook, C.A. (1973). *Transient flow: 1—Sliding in inclined flows at shutdown; 2—Transient flow of slurries.* Report Number E 73-23.
(The phenomena were investigated experimentally using slurries of narrowly sized sand in 50 mm and 150 mm pipes. The density current phenomenon was found to be the likely cause of blockage in pipelines of moderate slope and was studied in the 50 mm pipe loop using slurries consisting of limestone, iron concentrate, and coal.)

Shook, C.A., Schriek, W., Smith, L.G., Haas, D.B. & Husband, W.H.W. (1973). *Experimental studies on solids pipelining of Canadian commodities for The Canadian Transport Commission and The Transportation Development Agency* (Report 6: Experimental studies on the hydraulic transport of sands in liquids of varying properties in 2 and 4 inch pipelines). Report number E 73-20.

(Fluid density and viscosity were varied to obtain laminar and turbulent flow conditions, slurries composed of two narrowly graded sands (with mean sizes of 2 and 5 mm) and mixtures of the two, at volumetric concentrations up to 42%.)

Smith, L.G., Husband, W.H.W., Haas, D. & Richardson, A. (1976). *Experimental studies on the pipelining of coal-in-oil slurries.* Report number E 76-3.
(Pressure drops for powdered coal in crude oil mixtures conducted in 50 mm (nominal diameter) pipe test loop. Also reported are results of separation tests using a continuous centrifuge.)

8. REFERENCES

BSI (1975). BS 1377: Methods of test for soils for civil engineering purposes. British Standards Institution, London, UK.

Brookes, D.A. & Snoek, P.E. (1988). The potential for Stabflow coal slurry pipelines—an economic study. In *Proc. HT 11*, paper A3, pp. 33–48.*

Brown, N.P. (1981). Friction mechanisms in hydraulic conveying at high solids concentration. PhD thesis, University of London, UK.

Brown, N.P. (1988). Three scale-up techniques for stabilised coal–water slurries. In *Proc. HT 11*, paper F1, pp. 267–83.*

Duckworth, R.A. & Pullam, L. (1981). The hydraulic transport of coarse coal. Paper presented at 2nd National Conference on Rheology, British Society of Rheology (Australian Branch), May.

Elliott, D.E. & Gliddon, B.J. (1970). Hydraulic transport of coal at high concentrations. In *Proc. HT 1*, paper G2, pp. 25–56.*

Furnas, C.C. (1931). Grading aggregates. 1—Mathematical relations for beds of broken solids of maximum density. *Ind. Eng. Chem.*, **23**(9), 1052–8.

Thornton, W.A. (1970). *Hydraulic Transport of Solids in Pipes: A Bibliography*. BHRA, The Fluid Engineering Centre, Cranfield, UK.

White, H.E. & Walton, S.F. (1937). Particle packing and particle shape. *J. Amer. Ceram. Soc.*, **20**, 155–66.

*Full details of the Hydrotransport (HT) series of conferences can be found on pp. 11–12 of this chapter.

2

The Settling Behaviour of Particles in Fluids

Nigel P. Brown
The British Petroleum Company plc, London, UK

1. INTRODUCTION

Prediction of the fall behaviour of particles finds application in many aspects of the design of slurry systems, including

- the design of slow settling slurries,
- prediction of the behaviour of slurries in hoisting applications, and
- the design and operation of separation equipment.

This chapter presents techniques that enable the fall velocity of both regular and irregular shaped particles to be predicted in both Newtonian and non-Newtonian fluids. Figure 2.1 shows the extent of the calculation procedures covered by this chapter.

The terminal settling velocity of a sphere in a large extent of quiescent, or

PARTICLE ENVIRONMENT	NATURE OF FLUID		
	NEWTONIAN	NON-NEWTONIAN	
		WITHOUT A YIELD STRESS	WITH A YIELD STRESS
SINGLE PARTICLE IN AN UNBOUNDED, QUIESCENT FLUID	SPHERE: Section 2.1 REGULAR, NON-SPHERICAL: Section 2.2 IRREGULAR: Section 2.2	SPHERE: Section 3.2.1 REGULAR, NON-SPHERICAL: Section 3.2.2	SPHERE: Section 3.3.2
OTHER PARTICLES	SPHERE: Section 2.3 REGULAR, NON-SPHERICAL: Section 2.3		
OTHER PARTICLES AND CYLINDRICAL CONFINEMENT	SPHERE: Section 2.3		
CYLINDRICAL CONFINEMENT	SPHERE: Section 2.4		

Fig. 2.1. The extent of calculation schemes for settling velocity covered in this chapter.

stagnant, Newtonian fluid has been extensively studied. Such ideal conditions provide a confident bench-mark against which predictions for more complex situations can be compared. Most particles of practical interest are irregular and settling conditions frequently depart from those of a single particle settling in a large extent of fluid. Where departures from idealised conditions are substantial, or where the application is different to that in the reference from where the technique was reported, recourse should always be made to experimental verification before detailed design is carried out.

2. TERMINAL VELOCITY IN A NEWTONIAN FLUID

2.1. Spheres

The motion of particles under the influence of gravity is characterised by initial acceleration after which descent continues at the free-fall or terminal velocity. Under these conditions the gravitational force on the particle is balanced by frictional drag consisting of viscous and inertial components. Conventionally, this complex physical situation is overcome by defining a drag coefficient, in the same manner as the Fanning friction factor for flow of fluids in pipes, as the ratio of a shear stress experienced at the boundary to the kinetic energy per unit volume of fluid (or dynamic head).

Under free-fall the drag coefficient is evaluated by equating the forces acting on

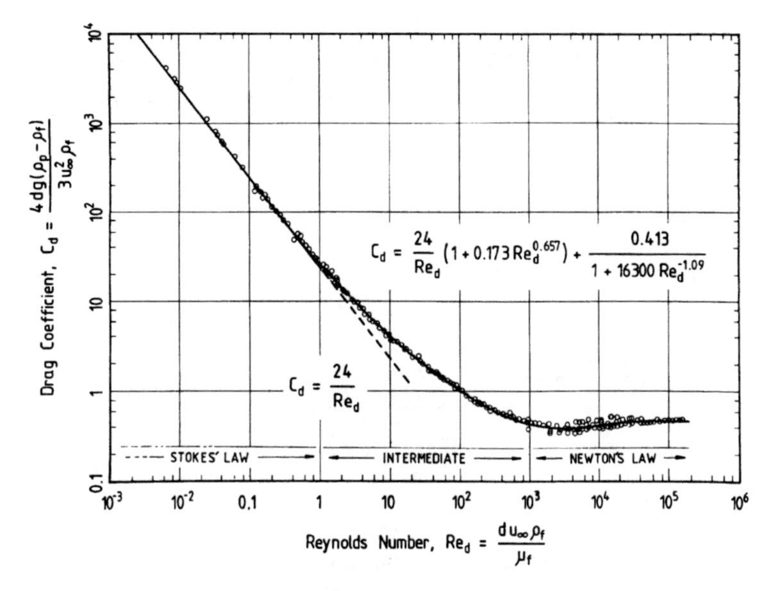

Fig. 2.2. Relation between Reynolds number and drag coefficient for spheres falling in a Newtonian fluid.

a single sphere in a semi-infinite expanse of fluid.

$$C_d = \frac{4dg(\rho_p - \rho_f)}{3u_\infty^2 \rho_f} \tag{2.1}$$

where C_d is the drag coefficient [1], d is the diameter of sphere [L], g is the gravitational acceleration [L T^{-2}], u_∞ is the terminal velocity [L T^{-1}], ρ_f is the density of fluid [M L^{-3}], and ρ_p is the density of sphere [M L^{-3}].

In an analogous manner to the use of the friction factor in pipe flow the drag coefficient is correlated with the Reynolds number for a sphere evaluated at the terminal velocity

$$Re_d = du_\infty \rho_f / \mu_f \tag{2.2}$$

Re_d is the Reynolds number for a sphere [1], u_∞ is the terminal velocity [L T^{-1}] and μ_f is the dynamic viscosity of the fluid [M L^{-1}T^{-1}].

Figure 2.2 is commonly referred to as the *standard* drag curve for spheres. At low values of the Reynolds number, below approximately unity, where viscous forces predominate, the flow *creeps* around the surface of the sphere. This flow regime is amenable to analysis (see, for example, Bird *et al.*, 1960) which was first carried out by Stokes (1851) and in recognition of this, it is often referred to as the *Stokes' law* regime. Two-thirds of the total drag is associated with viscous forces, the remaining one-third is attributable to inertia. The inertial contribution increases with Reynolds number; analytic expressions become less reliable and recourse has to be

TABLE 2.1
Drag Relationships for Spheres

Fall regime and range of Reynolds number (Re_d)	*Relation for drag coefficient* (C_d)
Stokes' law $Re_d \leqslant 1$	$C_d = 24Re_d^{-1}$
Intermediate $1 < Re_d \leqslant 1000$	$C_d = 24Re_d^{-1}(1 + 0.15Re_d^{0.687})$ (Schiller & Naumann, 1933)
Newton's law $1000 < Re_d \leqslant 2 \times 10^5$	$C_d = 0{\cdot}44$

made to fitting curves to the experimental data. For design purposes two regimes are commonly defined: the intermediate regime covering Reynolds numbers in the interval $1 < Re_d \leqslant 1\,000$ and the turbulent, or more correctly termed, *Newton's law* regime covering Reynolds numbers in the range $1\,000 < Re_d \leqslant 2 \times 10^5$. Expressions for calculating the drag coefficient in each range are shown in Table 2.1. Where greater precision is required Clift *et al.* (1978, p. 111, *et seq.*) provide a comprehensive review of drag relations.

Equations that cover discrete ranges of Reynolds numbers present an unnecessary complication in computation schemes. Turton and Levenspiel (1986) have fitted a single equation to the available experimental data that covers Reynolds numbers of practical interest up to values of 2×10^5.

$$C_d = \frac{24}{Re_d}(1 + 0{\cdot}173Re_d^{0{\cdot}657}) + \frac{0{\cdot}413}{1 + 16\,300Re_d^{-1{\cdot}09}} \tag{2.3}$$

where C_d is the drag coefficient, defined by eqn (2.1) [1], and Re_d is the particle Reynolds number, defined by eqn (2.2) [1].

This curve is plotted on experimental data in Fig. 2.2. The authors claim that over 90% of the experimental drag data lies within 10% of the correlation, which is claimed to be comparable to the capabilities of the discrete relations reviewed by Clift *et al.* (1978, p. 111, *et seq.*).

ITERATIVE CALCULATION SCHEME

Except in the Newton's law regime, the terminal velocity appears in both the equations for the drag coefficient and the Reynolds number; a trial-and-error procedure has to be adopted. The scheme below can be used to calculate the terminal velocity of a single sphere of known diameter and physical properties falling through a fluid of known properties. It can equally be applied to calculate the diameter of a sphere of known terminal velocity.

(1) Assume a value for the Reynolds number.

(2) Calculate the drag coefficient using eqn (2.3) (or alternatively the appropriate relation given in Table 2.1).
(3) From a knowledge of the particle and fluid properties calculate the terminal settling velocity from the expression for the drag coefficient (eqn (2.1)).
(4) Calculate the Reynolds number, using eqn (2.2).
(5) If the agreement between the calculated and assumed value of the Reynolds number provides an unacceptable error in the terminal velocity the calculated value can be refined by reintroducing the Reynolds number into step (2).

EXPLICIT CALCULATION SCHEME

Turton and Clark (1987) reported an equation that is explicit in terminal velocity.

$$u_\infty = \left(\frac{g\mu_f(\rho_p - \rho_f)}{\rho_f^2}\right)^{1/3}\left(\left(\frac{18}{d_*^2}\right)^{0\cdot824} + \left(\frac{0\cdot321}{d_*}\right)^{1\cdot214}\right)^{-1\cdot214} \quad (2.4)$$

where d_* is the dimensionless diameter of sphere [L]

$$d_* = d\left(\frac{g\rho_f(\rho_p - \rho_f)}{\mu_f^2}\right)^{1/3} \quad (2.5)$$

where d is the diameter of sphere [1], g is the gravitational acceleration [LT^{-2}], u_∞ is the terminal velocity [LT^{-1}], ρ_f is the density of fluid [ML^{-3}], ρ_p is the density of sphere [ML^{-3}], and μ_f is the dynamic viscosity of the fluid [$ML^{-1}T^{-1}$].

Equation (2.4) is the result of fitting a continuous curve to the 408 data points that Turton and Levenspiel (1986) used to generate eqn (2.3). Although there is systematic under-prediction of the terminal velocity in the range $20 \leqslant d_* \leqslant 200$, the equation should provide adequate quality predictions for preliminary design purposes.

2.2. Non-spherical Particles

2.2.1. Introduction

Particles encountered in most industrial applications rarely have perfect spherical form, but on occasion can be approximated by regular shapes. Prediction of the fall behaviour of non-spherical particles is complicated by the need to describe the shape of the particles to enable appropriate studies to be located. A further complication arises as a consequence of shape, namely the fall attitude of the particles. This is particularly important for oblate particles (those in which one dimension is much smaller than the other two). A number of experimentalists (including Peden & Luo, 1987) report that oblate particles tend to present their maximum area to the direction of motion. However, the stability of fall path appears to depend upon the fall regime (Clift *et al.*, 1978).

Empirical shape factors provide a means of formalising the description of the shape of irregular particles. Clift *et al.* (1978, pp. 16–22) provide a comprehensive

TABLE 2.2
Values of the Volumetric Shape Factor for Various Particles[a]

Shape of particle	*Shape factor*
Equi-dimensional particles, ϕ_v	
Sphere	0·524 ($=\pi/6$)
Cube	0·696
Tetrahedron	0·328
Irregular particles, ϕ'_v	
Angular, tending to tetrahedral form	0·38
Angular, tending to prisimoidal form	0·47
Sub-angular	0·51
Rounded	0·54

[a]Adapted from Heywood (1962).

review of shape factors. Two correlations for drag of non-spherical particles have attracted much practical attention; the volumetric shape factor and sphericity.

2.2.2. Volumetric Shape Factor

The volumetric shape factor is defined as the volume of the particle divided by a volume term obtained from the projected area diameter of the particle.

$$\phi_v = V_p / d_a^3 \tag{2.6}$$

where d_a is the projected area diameter of particle [L], and also the diameter of a circle with the same projected area as the particle

$$d_a = (4A_{p,p}/\pi)^{0.5} \tag{2.7}$$

in which $A_{p,p}$ is the projected area of particle [L^2], V_p is the volume of particle [L^3], and ϕ_v is the volumetric shape factor [1].

Clift *et al.* (1978, pp. 18–19) examine some techniques that have been used to estimate the projected diameter. Without the use of sophisticated image analysis equipment the measurements tend to be very tedious.

Extensive experimental studies by Heywood (1962) suggest that the value of the shape factor for mineral particles averages 0·20–0·25 although very oblate particles may have values of 0·1 or less. Values of the volumetric shape factor for regularly shaped particles are given in Table 2.2. The volumetric shape factor may also be estimated for irregular particles (Heywood, 1962, pp. 3–4) from the shape factor for a characteristic particle and the dimensions of the actual particle.

$$\phi_v = \frac{\phi'_v}{(B/T)(L/B)^{0.5}} \tag{2.8}$$

where B is the breadth of particle [L], L is the length of particle [L], T is the thickness of particle perpendicular to the plane of greatest static stability [L], ϕ_v is the volumetric shape factor [1], and ϕ'_v is the volumetric shape factor for a characteristic particle [1].

The dimensions used in the above expression are selected such that $T < B < L$.

Heywood (1938, pp. 268–9) provides a useful table and graph that relate the volumetric shape factor to the ratio of the mean projected diameter to screen size for different crushed minerals in the size range 50 μm to 60 mm with ratios of length to breadth between 1·3 and 1·7.

Heywood (1938, p. 273) correlated the fall behaviour of particles using the volumetric shape factor by modifying the definition of particle Reynolds number and the drag coefficient.

$$C_{da} = \frac{8\phi_v d_a g(\rho_p - \rho_f)}{\pi u_\infty^2 \rho_f} \tag{2.9}$$

where C_{da} is the drag coefficient based on the projected area diameter of the particle [1], d_a is the projected area diameter of particle [L], g is the gravitational acceleration [L T^{-2}], u_∞ is the terminal velocity [L T^{-1}], ϕ_v is the volumetric shape factor, defined by eqn (2.6) [1], ρ_f is the density of fluid [M L^{-3}], and ρ_p is the density of sphere [M L^{-3}].

$$\mathrm{Re}_{da} = d_a u_\infty \rho_f / \mu_f \tag{2.10}$$

where the remaining undefined variables are as follows: Re_{da} is the particle Reynolds number based on the projected area diameter [1], and μ_f is the dynamic viscosity of the fluid [M L^{-1}T^{-1}].

Heywood (1938) reports curves, from which experimental data points are omitted, in the range of Reynolds number between $0.02 \leqslant \mathrm{Re}_{da} \leqslant 1\,000$.

CALCULATION SCHEME FOR REGULARLY SHAPED PARTICLES

(1) Calculate the projected area diameter from eqn (2.7).
(2) Calculate the shape factor from eqn (2.6), or Table 2.2.
(3) The terminal velocity is calculated using the same scheme used for spheres provided that the Reynolds number and drag coefficient take the definitions given above.

CALCULATION SCHEME FOR IRREGULAR PARTICLES

(1) Estimate the projected area diameter (see, for example, Clift *et al.*, 1978; Heywood, 1938, pp. 268–9).
(2) From a knowledge of the volume of the particle calculate the shape factor from eqn (2.6). Alternatively, from the dimensions and shape factor for a characteristic particle estimate the shape factor from eqn (2.8).
(3) Refer to step (3) in the preceding scheme.

2.2.3. Sphericity

Sphericity is defined as the surface area of a sphere with the same volume as the particle divided by the surface area of the particle.

$$\psi = 4\pi \left(\frac{3V_p}{4\pi}\right)^{2/3} \Big/ A_p \tag{2.11}$$

where A_p is the surface area of the particle [L^2], V_p is the volume of the particle [L^3], ψ is sphericity [1].

Although the volume of the particle can be readily estimated (for example, using fluid displacement) the surface area requires more complex experimental procedure. Brown *et al.* (1950, p. 77) showed that the sphericity can be estimated from a screen analysis of the particles.

$$\psi = \bar{d}/(\bar{d}_v n) \tag{2.12}$$

where $\bar{d}$ is the average particle diameter in a size interval from a screen analysis [L], and $\bar{d}_v$ is the average volume equivalent diameter [L]; $\bar{d}_v$ is also the diameter of a sphere with the same volume as that of the particle [L].

$$\bar{d}_v = (6\bar{V}_p/\pi)^{1/3} \tag{2.13}$$

in which $\bar{V}_p$ is the average volume of the particles [L^3], n is the ratio of the specific surfaces [1], which equals the ratio of the surface area per unit mass of the particles to that of spheres of the same diameter.

Values of the ratio of the specific surfaces as a function of average particle size for several crushed minerals are given in Fig. 2.3.

Wadell (1934) correlated experimental measurements of terminal velocity for non-spherical particles using sphericity and the volume equivalent diameter as the length scale in the Reynolds number and drag coefficient. A family of curves is also presented by Brown *et al.* (1950, Fig. 70), that appear to be based on the data originally plotted by Wadell (1934, Fig. 3). The data used by Wadell are plotted on the curves published by Brown *et al.* in Fig. 2.4. Considerable caution is advised when using this figure outside the range covered by experimental data. Brown *et al.* (1950) provide a useful source of values of sphericity reproduced here in Table 2.3.

TABLE 2.3
Sphericity and the Ratio of $\bar{d}_v/\bar{d}$ for Regularly Shaped Particles[a]

Shape	*Sphericity* (ψ)	$\bar{d}_v/\bar{d}$
Sphere	1·0	1·0
Octahedron	0·847	0·965
Cube	0·806	1·24
Prism (length of side, *a*)		
$a \times a \times 2a$	0·767	1·564
$a \times 2a \times 2a$	0·761	0·985
$a \times 2a \times 3a$	0·725	1·127
Cylinders (radius, *r* and height, *h*)		
$h = r/15$	0·254	0·368
$h = r/10$	0·323	0·422
$h = r/3$	0·594	0·630
$h = r$	0·827	0·909
$h = 3r$	0·860	1·31
$h = 10r$	0·691	1·96
$h = 20r$	0·580	2·592

[a]Adapted from Brown *et al.* (1950, Table 18).

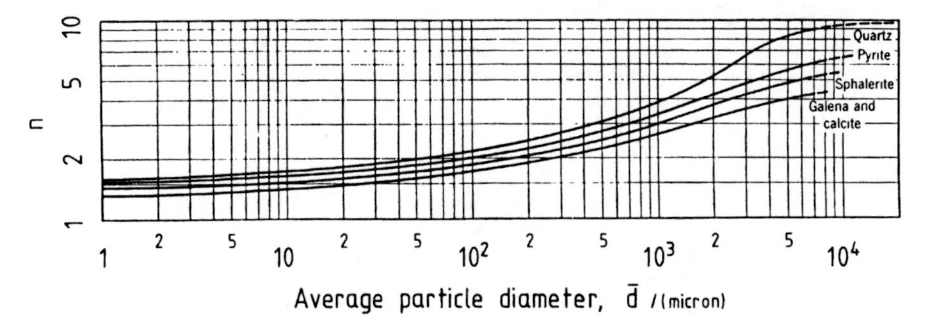

Fig. 2.3. Ratio of specific surfaces, n, as a function of average particle diameter for various crushed minerals. (Adapted from Brown *et al.*, 1950, Fig. 17.)

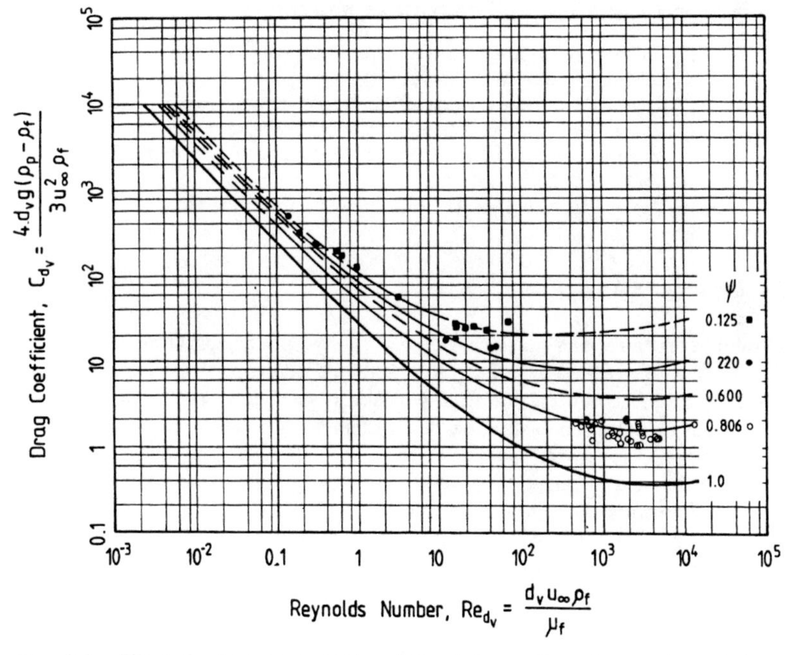

Fig. 2.4. Effect of sphericity, ψ, on the relation between Reynolds number and drag coefficient. Solid lines from Brown *et al.* (1950, Fig. 70), data points from Wadell (1934, Fig. 3).

CALCULATION SCHEME FOR REGULARLY SHAPED PARTICLES

(1) Use eqn (2.11) or Table 2.3 to obtain the sphericity.
(2) Calculate the volume equivalent diameter from eqn (2.13). Alternatively, estimate the value from Table 2.3 by multiplying the ratio given in the final column by the average particle diameter.
(3) Assume a value for the Reynolds number.
(4) Obtain the drag coefficient for the sphericity from Fig. 2.4.
(5) Calculate the terminal velocity from eqn (2.1), with the drag coefficient evaluated using the volume equivalent diameter.
(6) Calculate the Reynolds number from eqn (2.2), using the volume equivalent diameter.
(7) If the agreement between the calculated and assumed value of the Reynolds number provides an unacceptable error in the terminal velocity the calculated value can be refined by reintroducing the Reynolds number into step (3).

CALCULATION SCHEME FOR IRREGULAR PARTICLES

(1) Estimate the average particle diameter from an appropriate size interval on the grading curve.
(2) Obtain the value of the ratio of the specific surfaces from Fig. 2.3.
(3) Obtain the volume equivalent diameter either from experiment or by interpolating data for the appropriate shape from Table 2.3.
(4) Calculate the sphericity from eqn (2.12).
(5) Proceed to step (3) in the preceding calculation scheme.

2.3. Effect of Other Particles

At dilute concentrations (in the order of 1%, by volume) particles move individually. The upward motion of the displaced fluid on the particles becomes more pronounced as the concentration of particles is increased. The result of these interactions is that a mono-sized particle system tends to settle *en masse* with all the particles moving at the same velocity. When this type of settling occurs it is customary to denote the settling velocity of the slurry by the velocity of the interface with which the term *sedimenting velocity* is often associated. The physical phenomenon is often referred to, somewhat ambiguously, as *hindered settling*. In terms of particle concentration the upper limit to the hindered settling regime is that at which mechanical support of the particles takes place. At such concentrations it is more appropriate to predict the relative motion of the fluid by techniques applicable to porous media (see, for example, Scheidegger, 1960). While the settling characteristics of very low concentrations of spheres at low Reynolds numbers has been theoretically studied (see, for example, Happel & Brenner, 1973) for most practical situations resort has to be made to empirical correlations.

The settling behaviour of mono-sized particles has been systematically studied by Richardson and Zaki (1954*a*, *b*). They concluded that the effect of particle concentration on the velocity of the sedimenting particles can be described by a

TABLE 2.4
Exponents in the Richardson–Zaki Equation[a]

Range of Reynolds number (Re_d)	*Expression for exponent* (z)
$0.002 < Re_d \leqslant 0.2$	$z = 4.65 + 19.5(d/D)$
$0.2 < Re_d \leqslant 1$	$z = (4.35 + 17.5(d/D))Re_d^{-0.03}$
$1 < Re_d \leqslant 200$	$z = (4.45 + 18(d/D))Re_d^{-0.1}$
$200 < Re_d \leqslant 500$	$z = 4.45Re_d^{-0.1}$
$500 < Re_d \leqslant 7000$	$z = 2.39$

[a]Adapted from Richardson & Zaki (1954*a*).

relation of the form

$$u = u_\infty(1 - C)^z \qquad (2.14)$$

where C is the volumetric concentration of particles [1], z is the exponent given in Table 2.4 [1], u is the velocity of sedimenting spheres [L T^{-1}] and u_∞ is the terminal velocity of a sphere [L T^{-1}].

The values of the exponent in the Richardson–Zaki equation were obtained from experiments carried out on both the sedimentation and fluidisation of particles. In the latter experiments the particles were maintained in suspension by an upward flow of liquid. The Reynolds numbers of the particles in the experiments comfortably overlapped which enabled the investigators to conclude that the data collected by the two techniques could be represented in the same form. The experimental data were obtained with particles greater than 100 μm in diameter at concentrations up to about 45% (by volume). In the experimental equipment the ratio of the particle diameter to that of the container, d/D, was in the range $0.002 \leqslant d/D \leqslant 0.1$.

At low Reynolds numbers, the value of the exponent is similar to that reported by Lewis *et al.* (1949) and Fair and Geyer (1954, p. 677). In situations where wall effects are insignificant the effect of concentration on the settling velocity of a slurry can be estimated from Fig. 2.5.

In the fluidisation experiments, in addition to spheres regular non-spherical particles were studied. Cylinders, hexagonal prisms, cubes and plates were studied with a typical dimension of approximately 6 mm. The best correlation for non-spherical particles was obtained using the volumetric shape factor.

$$z = 2.7\phi_v^{0.16}; \qquad \text{for } 0.174 \leqslant \phi_v \leqslant 0.71 \qquad (2.15)$$

where z is the exponent in the Richardson–Zaki equation (eqn (2.14)) [1], and ϕ_v is the volumetric shape factor, defined by eqn (2.6) [1].

Sedimentation of binary mixtures of spheres at low Reynolds number ($\leqslant 0.2$) has been studied by Mirza and Richardson (1979). They concluded that the velocity of the interfaces could be predicted by the correlation for mono-sized spheres (eqn (2.14)) by the addition of 0·4 to the value of the exponent. As a result of the work with binary mixtures a model is proposed for multi-sized particle

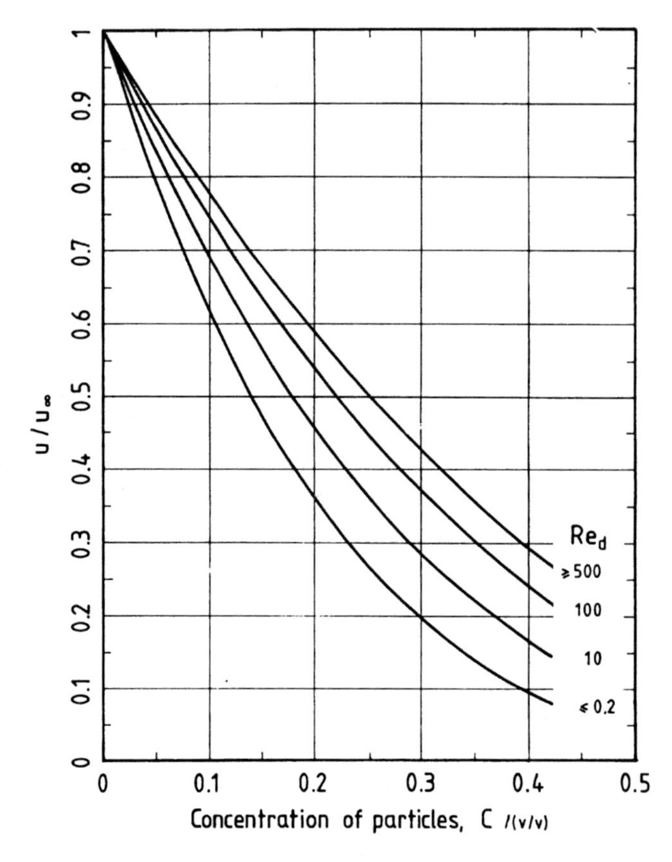

Fig. 2.5. Effect of particle concentration on relative terminal velocity.

systems. For the same range of Reynolds number, Scott (1984) proposes a modification to the diameter (Scott & Mandersloot, 1979) term in the Richardson–Zaki equation to allow for multi-sized spheres. In making the modification to the diameter the applicability is necessarily restricted to situations where wall effects are under consideration. Scott (1984) provides a useful review of 28 settling models; a summary is to be found in Mandersloot *et al.* (1987).

2.4. Effect of Confinement

Interaction of the fluid displaced by a falling particle with the wall of the container slows down the descent of the sphere. In practice, it has been found that the wall effect is dependent upon the fall regime of the sphere and the ratio of the diameter

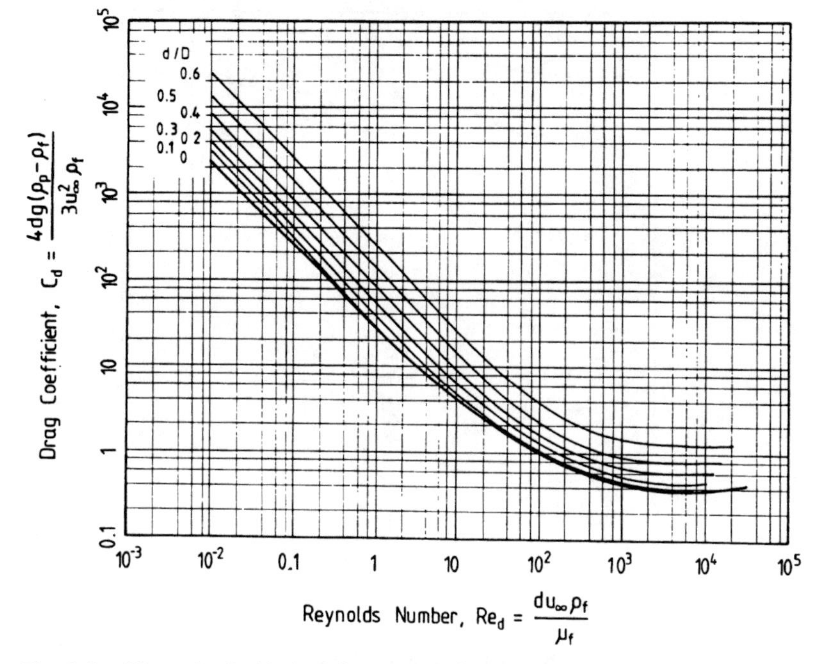

Fig. 2.6. Effect of cylindrical confinement on the relation between Reynolds number and relative terminal velocity. (Adapted from Fidleris & Whitmore (1961, Fig. 2).)

of the sphere to the internal diameter of the cylindrical container in which the sphere is falling.

An extensive review of the motion of spheres in bounded fluids at low Reynolds numbers has been made by Happel and Brenner (1973) where the extent of the studies reflects the amenability of this regime to analysis and the historic importance placed upon the measurement of viscosity using a falling sphere. At higher Reynolds numbers wall effects are described by correlations.

Fidleris and Whitmore (1961) have compared correlations that predict the effect of confinement on data collected in an extensive experimental study. Their experiments investigated the axial fall of spheres at Reynolds numbers in the range $0{\cdot}05 \leqslant Re_d \leqslant 2 \times 10^4$ in cylindrical geometries with ratios of the sphere to tube diameter, d/D, in the range $0{\cdot}05 \leqslant d/D \leqslant 0{\cdot}6$.

The effect of the diameter ratio on the dependence of drag coefficient on Reynolds number that they reported is shown in Fig. 2.6. These data are almost identical to those obtained by McNown *et al.* (1948, Fig. 4) and those reported by Clift *et al.* (1978, Fig. 9.4). Fidleris and Whitmore (1961) report that 95% of their data are within 3% of the solid lines for a particular diameter ratio. From Fig. 2.6 it can be seen that for a given diameter ratio the reduction in terminal velocity, due

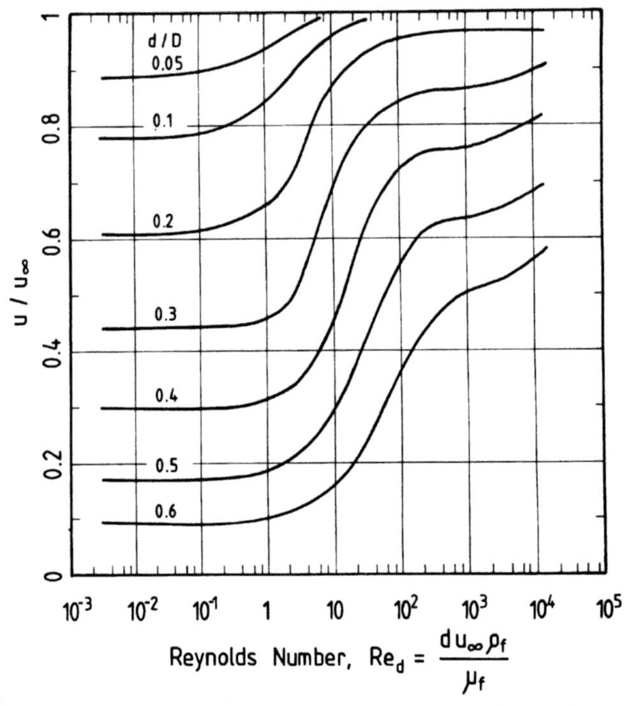

Fig. 2.7. Effect of diameter ratio on the relation between Reynolds number and relative terminal velocity. (Adapted from Fidleris & Whitmore, 1961, Fig. 4.)

to the presence of a cylindrical confining wall, decreases upon transition from the Stokes' law to Newton's law fall regime.

Fidleris and Whitmore (1961) also present graphs, based on their experimental results, that enable terminal velocities known under conditions of either bounded or infinite extents of fluid to be converted to the unknown condition. They present the corrected velocity as a function of the appropriate Reynolds number. One of their figures is redrawn as Fig. 2.7. The information in this figure enables the terminal velocity in a bounded fluid to be estimated from that predicted at infinite dilution.

The authors compare their experimental data with predictions from formulae that are commonly used to correct terminal velocities for confinement. In the Stokes' law regime, (for $Re_d \leqslant 0{\cdot}2$) the equation proposed by Francis (1933) provides predictions that are within $\pm 1\%$ of their experimental data for diameter ratios less than 0·6.

$$u = u_\infty \left(\frac{1 - (d/D)}{1 - 0{\cdot}475(d/D)} \right)^4 \tag{2.16}$$

where d is the diameter of sphere [L], D is the diameter of cylindrical container [L], u is the velocity of a sphere in cylinder of diameter, D [$L T^{-1}$], and u_∞ is the terminal velocity of sphere [$L T^{-1}$].

In the Newton's law regime the authors concluded that the most satisfactory predictions were obtained from the relation suggested by Munroe (1888).

$$u = u_\infty(1 - (d/D)^{1 \cdot 5}) \tag{2.17}$$

Where the variables are as defined for eqn (2.16). When eqn (2.17) is used at Reynolds numbers in the range $1\,000 \leqslant Re_d \leqslant 3\,000$ for diameter ratios up to 0·6, predictions are reported to be in error by a maximum of $\pm 2.5\%$. In the intermediate regime Fig. 2.7 can be used to correct predictions of terminal velocity in an unbounded fluid to those expected under bounded conditions.

3. THE BEHAVIOUR OF PARTICLES IN NON-NEWTONIAN FLUIDS

3.1. Introduction

Non-Newtonian fluids are distinguished from Newtonian fluids because the viscosity is dependent upon the rate at which the fluid is sheared; hence the use of a single viscosity is no longer appropriate. Instead, an empirical relation is fitted to the rheological measurements. The size of parameters in the model are used as a means of describing the complex behaviour of the fluid. The flow curve is often highly dependent upon the nature of the fluid and as such it is inappropriate to provide values of the parameters for models for different fluids, in the same way that viscosities are tabulated for Newtonian fluids.

Caution should be exercised regarding the accuracy of predictions for particles settling in non-Newtonian fluids. It is recommended that particular attention should be addressed to the measurement techniques used to establish the flow curve and, in particular, ensure that the relevant part of the flow curve is used. As a guide, the shear rate experienced by the fluid at the surface of a sphere of diameter d, settling at a velocity u, is of the order u/d; the maximum occurs at the equator and is $3u/d$. The applicability of a particular technique should preferably be confirmed through experiment.

Studies that have considered the motion of particles in non-Newtonian fluids have been confined to isolated particles falling in two relatively simple classes of fluid; those that appear to possess a yield stress and those that do not. The more complex effects of viscoelasticity and time dependency on particle drag do not appear to have been studied systematically. A comprehensive review of the literature on the motion of particles in non-Newtonian fluids is given by Chhabra (1986).

3.2. Fluids without a Yield Stress

The simplest non-Newtonian behaviour is pseudoplastic. This behaviour is frequently referred to as shear-thinning since the viscosity decreases with increasing shear rate.

3.2.1. Terminal Velocity of Spheres

Slattery and Bird (1961) report the fall of spheres in solutions of carboxymethyl cellulose at five concentrations in which the effects of confinement were assessed as negligible. Flow curves for the fluids were measured in a tube viscometer with due consideration being given to the shear rates over which measurements were made. They found that the fluids were Newtonian in character at low shear rates whilst at higher values the fluid was described by the power law model. They concluded that the Ellis model adequately described the flow curve.

$$\tau = (A + B\tau^{\alpha-1})^{-1}\dot{\gamma} \tag{2.18}$$

where A is a parameter [L M^{-1} T], B is a parameter [L$^{\alpha}$ M$^{-\alpha}$ T$^{2\alpha-1}$], α is a parameter [1], $\dot{\gamma}$ is the shear rate [T^{-1}], and τ is the shear stress [M L^{-1} T^{-2}].

Using dimensional analysis, the general description of the drag coefficient in terms of fluid and particle properties was established by a relation of the form

$$C_d = f(\text{Re}_{d,2}, E, \alpha) \tag{2.19}$$

where $\text{Re}_{d,2}$ is the particle Reynolds number [1],

$$\text{Re}_{d,2} = Bdu_{\infty}^{(2\alpha-1)}\rho_f^{\alpha} \tag{2.20}$$

in which B, α are parameters in the Ellis model (eqn (2.18)), d is the diameter of sphere [L], u_{∞} is the terminal velocity [L T^{-1}], and ρ_f is the density of fluid [M L^{-3}].

$$E = A^{(2\alpha-1)}d^{2(\alpha-1)}\rho_f^{(\alpha-1)}/B \tag{2.21}$$

in which the remaining undefined variable is A, the parameter in the Ellis model (eqn (2.18)).

The formulation of the parameter E in the above equation is reported here as modified by Skelland (1967, p. 145). At low particle Reynolds numbers ($\text{Re}_{d,2} < 0.1$), Slattery and Bird (1961) established that, on logarithmic coordinates, the Reynolds number was linearly correlated with a modified drag coefficient, $C_{d,2}$, as shown in Fig. 2.8. In this figure the equation of the solid line is given by

$$C_{d,2} = 24/\text{Re}_{d,2} \tag{2.22}$$

where

$$\begin{aligned}\log_{10}(C_{d,2}) = {} & -2{\cdot}1013(\alpha-1) - 2{\cdot}0303(\alpha-1)^2 \\ & + \log_{10}(C_d)(1 + 1{\cdot}0342(\alpha-1) + 3{\cdot}5017(\alpha-1)(1/(E+1)) \\ & - 3{\cdot}7789(\alpha-1)(1/(E+1))^2 + 1{\cdot}0502(\alpha-1)^2)\end{aligned} \tag{2.23}$$

in which E is a parameter, defined by eqn (2.21), C_d is the drag coefficient, defined by eqn (2.1) [1], $C_{d,2}$ is the modified drag coefficient [1], α is a parameter in the Ellis model (eqn (2.18)), and $\text{Re}_{d,2}$ is the particle Reynolds number, defined by eqn (2.20) [1].

At higher particle Reynolds numbers ($0{\cdot}1 \leqslant \text{Re}_{d,2} \leqslant 5000$), the value of the modified drag coefficient, $C_{d,2}$, was best described by the curves reproduced in Fig. 2.9 (from which experimental data have been excluded to aid clarity). The relations

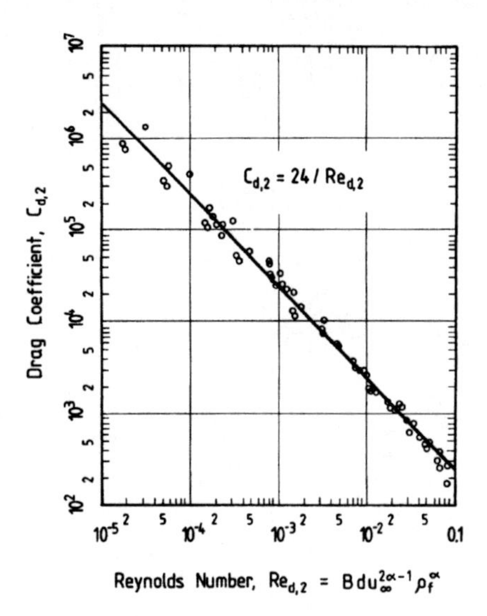

Fig. 2.8. Relation between Reynolds number and drag coefficient for spheres falling in an Ellis fluid at low Reynolds number. (Adapted from Bird & Slattery, 1961, Fig. 3.)

between the particle Reynolds number and the modified drag coefficient were obtained from a total of 294 data points with an average error, reported by the authors, slightly greater than $\pm 10\%$.

Although this approach was derived for a fluid that can be characterised by the Ellis model, the equations may be adapted for use with the simpler power law model (eqn (2.24)) by setting $A = 0$, $\alpha = 1/n$ and $B = K^{-1/n}$.

CALCULATION SCHEME

(1) Estimate a value for the terminal velocity, u_∞.
(2) Calculate the particle Reynolds number, $Re_{d,2}$, from eqn (2.20).
(3) If $Re_{d,2} < 0.1$ calculate the modified drag coefficient from eqn (2.22), move to step (5).
(4) If $0{\cdot}1 \leqslant Re_{d,2} \leqslant 5000$ estimate the modified drag coefficient, at the appropriate value of α, from Fig. 2.9.
(5) Calculate the drag coefficient from eqn (2.23).
(6) Calculate the terminal velocity from the definition of the drag coefficient given in eqn (2.1).
(7) If the estimated and calculated values of the terminal velocity are not in sufficient agreement the calculated terminal velocity can be used to evaluate a new estimate of the Reynolds number, in step (2), and the trial-and-error procedure repeated.

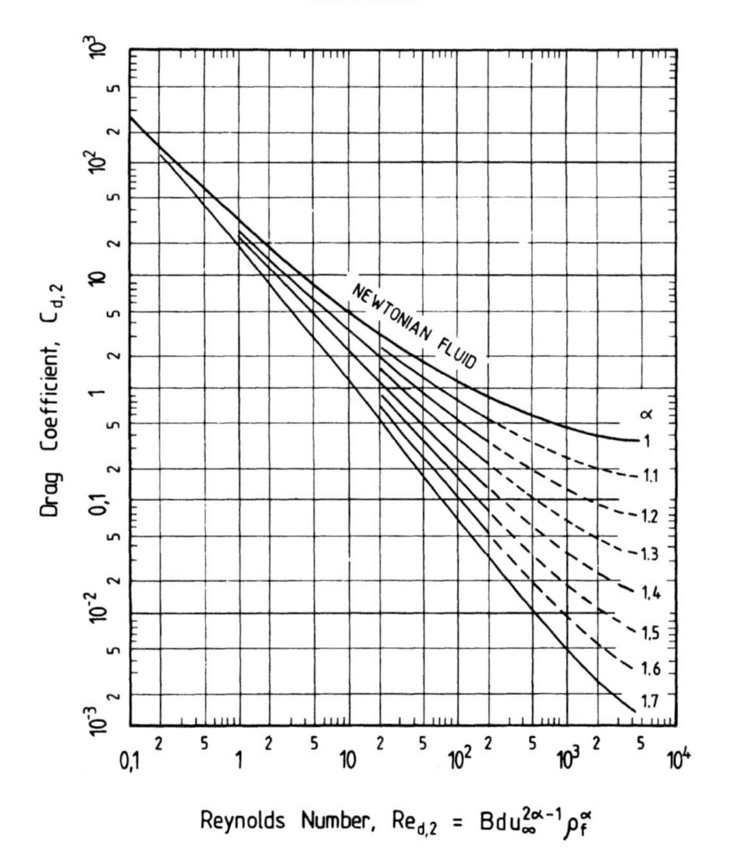

Fig. 2.9. Relation between Reynolds number and drag coefficient for spheres falling in an Ellis fluid at high Reynolds number. (Adapted from Bird & Slattery, 1961, Fig. 4.)

3.2.2. Terminal Velocity of Non-spherical Particles

Peden and Luo (1987) reported data, corrected for wall effects, on the settling of spheres, discs and rectangular plates in idealised fluids that approximate the behaviour of drilling fluids used in the oil industry. Flow curves were established over a range of shear rates, $\dot{\gamma}$, in the range $5 \leqslant \dot{\gamma} \leqslant 1021\ s^{-1}$ using a concentric cylinder viscometer. The Ostwald–de Waele model (power law model) was fitted to the flow curve.

$$\tau = K\dot{\gamma}^n \qquad (2.24)$$

where K is the power law consistency index $[M\,L^{-1}\,T^{n-2}]$, n is the power law flow index ($n \leqslant 1$) [1], $\dot{\gamma}$ is the shear rate $[T^{-1}]$, and τ is the shear stress $[M\,L^{-1}\,T^{-2}]$.

Sphericity (defined by eqn (2.12)) was used to describe the shape of the plate-like particles. Experimental data were correlated using a drag coefficient and particle Reynolds number for a power law fluid; in both instances the length-scale is volume equivalent diameter (defined by eqn (2.13)).

$$\mathrm{Re}_{\mathrm{dv,l}} = d_v^n u_\infty^{2-n} \rho_f / K \tag{2.25}$$

where d_v is the volume equivalent diameter, defined by eqn (2.13) [L], K is the power law consistency index [M L^{-1} T^{n-2}], $\mathrm{Re}_{\mathrm{dv,l}}$ is the particle Reynolds number for a power law fluid based on the volume equivalent diameter [1], u_∞ is the terminal velocity of particle [L T^{-1}], and ρ_f is the fluid density [M L^{-3}].

$$C_{\mathrm{dv}} = \psi' a \mathrm{Re}_{\mathrm{dv,l}}^{-b} \tag{2.26}$$

where

a = $39{\cdot}8 - 9n$; for $\mathrm{Re}_{\mathrm{dv,l}} < 5$
= $42{\cdot}9 - 23{\cdot}9n$; for $1 < \mathrm{Re}_{\mathrm{dv,l}} < 200$

b = $1{\cdot}2 - 0{\cdot}47n$; for $\mathrm{Re}_{\mathrm{dv,l}} < 5$ and $n > 0{\cdot}45$
= $1{\cdot}0 - 0{\cdot}33n$; for $1 < \mathrm{Re}_{\mathrm{dv,l}} < 200$

C_{dv} = drag coefficient (defined by eqn (2.1)) but based on the volume equivalent diameter, defined by eqn (2.13) [1]

$\mathrm{Re}_{\mathrm{dv,l}}$ = particle Reynolds number for a power law fluid, defined by eqn (2.25) [1]

ψ' = $1{\cdot}5 - 0{\cdot}5\psi$; for $\mathrm{Re}_{\mathrm{dv,l}} < 1$
= $2{\cdot}6 - 1{\cdot}6\psi$; for $\mathrm{Re}_{\mathrm{dv,l}} > 1$
ψ is defined by eqn (2.11) or (2.12)

The authors report that the deviation of the drag coefficients for the irregular particles was within approximately $\pm 10\%$ of the corresponding data for spheres.

CALCULATION SCHEME

> The terminal velocity is calculated in the same manner as that for a Newtonian fluid except that eqn (2.25) is used to obtain the Reynolds number and eqn (2.26) the drag coefficient.

3.3. Fluids with a Yield Stress

Fluids that do not appear to shear at low levels of applied stress are commonly attributed with a characteristic referred to as a *yield stress* and are often described as viscoplastic. The simplest model that has been used to characterise viscoplastic fluids is attributed to Bingham.

$$\tau = \tau_{\mathrm{y,B}} + k_{\mathrm{B}} \dot{\gamma} \tag{2.27}$$

where k_{B} is the Bingham plastic viscosity [M L^{-1} T^{-1}], $\dot{\gamma}$ is the shear rate [T^{-1}], τ is the shear stress [M L^{-1} T^{-2}], and $\tau_{\mathrm{y,B}}$ is the Bingham yield stress [M L^{-1} T^{-2}].

A generalised form of the Bingham model is attributed to Herschel–Bulkley and described by

$$\tau = \tau_{\mathrm{y,HB}} + k_{\mathrm{HB}} \dot{\gamma}^m \tag{2.28}$$

where k_{HB} is the Herschel–Bulkley consistency parameter [$M L^{-1} T^{m-2}$], m is the Herschel–Bulkley flow behaviour index [1], $\dot{\gamma}$ is the shear rate [T^{-1}], τ is the shear stress [$M L^{-1} T^{-2}$], and $\tau_{y,HB}$ is the Herschel–Bulkley yield stress [$M L^{-1} T^{-2}$].

3.3.1. The Limit of Static Equilibrium

Viscoplastic fluids are capable of preventing solids from settling, at least at a perceptible rate. A comprehensive review of published investigations into the motion of particles in viscoplastic fluids has been undertaken by Atapattu *et al.* (1988). They concluded that the limit of static equilibrium of a particle in an unsheared Bingham fluid could be described in terms of a yield–gravity parameter obtained from a force balance on a supported particle.

$$Y = \tau_{y,B}/(dg(\rho_p - \rho_f)) \quad (2.29)$$

where Y is the yield–gravity parameter [1], d is the particle diameter [L], g is the gravitational acceleration [$L T^{-2}$], ρ_f is the density of fluid [$M L^{-3}$], ρ_p is the density of sphere [$M L^{-3}$], and $\tau_{y,B}$ is the Bingham yield stress defined by eqn (2.27) [$M L^{-1} T^{-2}$].

Published data indicated that the parameter fell into one of two groups, taking values of between 0·04 and 0·08, or values close to 0·2. The yield stress predicted from this wide range of values indicates the need for fluids of very different properties. Atapattu *et al.* (1988) identified that the two groups of values appear to be associated with the manner in which the experiments are carried out. One factor that is evident in all work associated with viscoplastic fluids is the certainty to which the yield stress is measured. Indeed, the actual existence of a true yield stress has been questioned (Barnes & Walters, 1985) and problems associated with measuring yield stress can be large (Cheng, 1986). An alternative approach to designing fluid systems to support solids is based on Stokes' law and the zero-shear viscosity of the suspending medium (Cheng, 1989).

3.3.2. Terminal Velocity of Spheres

Ansley and Smith (1967) measured the terminal velocity of silver spheres in tomato sauce. The spheres were detected in the opaque field by irradiating them with neutrons and detecting the gamma activity. The flow curve of the fluid was measured using a tube viscometer and although they noted non-linear behaviour at low shear rates, the fluid was characterised as a Bingham fluid. The parameters in the Bingham model were obtained from the tangent to the flow curve at a shear rate of u_∞/d.

Experimental data were correlated using the drag coefficient and a non-dimensional *dynamic* parameter in a manner similar to that used by Hedström (1952) to describe the flow of Bingham fluids in pipes. Although the use of the dynamic parameter was successful in correlating their experimental data, it did not provide the appropriate relation for creeping flow where the Bingham model was taken in the Newtonian limit ($k_B = \mu$ and $\tau_{y,B} = 0$). By reanalysing existing data and performing new experiments at small values of the dynamic parameter Atapattu *et al.* (1988) conclude that this was due to the technique used to obtain the parameters for the Bingham model.

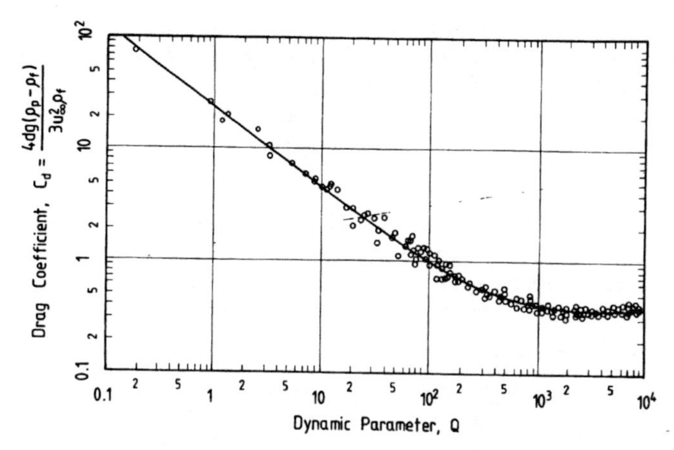

Fig. 2.10. Relation between dynamic parameter and drag coefficient for spheres falling in a Herschel–Bulkley fluid. (Adapted from Hanks & Sen, 1983, Fig. 2.)

Hanks and Sen (1983) extended the work of Ansley and Smith (1967) to investigate the fall of spheres in yield-pseudoplastic fluids. The terminal velocity of spheres up to 25 mm in diameter was measured in a 150 mm tube using aqueous solutions of laponite at 10 different concentrations. The rheological properties of the fluid were measured in a tube viscometer, under conditions that were anticipated in the particle fall experiments. The flow curves were characterised by the Herschel–Bulkley model (eqn (2.28)).

Using an analogous derivation to the earlier investigators, a particle Reynolds number and Hedström number were defined for the fluid. The resulting dynamic parameter, given by eqn (2.30), is identical to that obtained by Ansley and Smith in the Bingham limit ($m = 1$ and $k_{\mathrm{HB}} = k_{\mathrm{B}}$).

$$Q = \frac{2^{6-2m} d^{2m-2} u_\infty^{4-2m} k_{\mathrm{HB}}^{2(1-m)/m} \rho_f \left(\dfrac{m}{(1+3m)}\right)^{2m}}{\dfrac{7\pi}{24} \tau_{\mathrm{y,HB}}^{(2-m)/m} + 2^{3-m} d^{m-2} u_\infty^{2-m} k_{\mathrm{HB}}^{(2-m)/m} \left(\dfrac{m}{(1+3m)}\right)^{m}} \tag{2.30}$$

where d is the diameter of sphere [L], k_{HB} is the Herschel–Bulkley consistency parameter [M L^{-1} T^{m-2}], m is the Herschel–Bulkley flow behaviour index [1], Q is the dynamic parameter for a Herschel–Bulkley fluid [1], u_∞ is the terminal velocity [L T^{-1}], ρ_f is the density of fluid [M L^{-3}], and $\tau_{\mathrm{y,HB}}$ is the Herschel–Bulkley yield stress [M L^{-1} T^{-2}].

Hanks and Sen (1983) correlated their experimental terminal velocity data in terms of a drag coefficient and dynamic parameter. Their experimental data are reproduced in Fig. 2.10, where the Newtonian drag curve for spheres is shown by the solid line. Although the correlation is good and scatter of the data comparable to that for a Newtonian fluid, it should be noticed that data are scarce at small values of the dynamic parameter. Importantly, the data points superimpose those for a Newtonian fluid as would be expected in the Newtonian limit for the dynamic parameter (when $k_{HB} = \mu_f$, $m = 1$ and $\tau_{HB} = 0$).

Hanks and Sen (1983) investigated the potential discrepancy between their data and those of Ansley and Smith (1967). In contrast to Atapattu *et al.* (1988) they concluded that systematic experimental errors present in the determination of the physical properties used by Ansley and Smith (1967) to calculate the drag coefficient may have been responsible for the poor agreement with the Stokes' law regime. The evidence from both investigators appears plausible, indicating that the measurement of the fluid and particle properties require careful attention and that the most appropriate model be fitted to the rheological data.

CALCULATION SCHEME

> Calculation of the terminal velocity is analogous to that for a Newtonian fluid with the dynamic parameter (defined by eqn (2.30)) used in place of the Reynolds number.

4. SETTLING IN A SHEARED FLUID

Settling of particles in a sheared flow field is sometimes referred to as *dynamic settling*. The motion of a particle falling under gravity is not well understood and cannot be reliably predicted. The structure of the fluid has the potential to affect not only the drag experienced by the particle but also its position in the flow field and its orientation.

A particle situated in a simple sheared flow field will experience a force normal to the direction of shear. This *lift*, or *Magnus*, force arises from the pressure difference across the particle and causes it to move in the direction of increasing velocity. Radial migration of particles away from the pipe wall in the upward flow of granular materials transported in water has been observed by numerous investigators including Englemann (1978) and Newitt *et al.* (1961). In both of these studies the effect was noticed at high flow rates. Radial particle migration in laminar flow fields is considered by Lawler and Lu (1971).

Literature relating to dynamic settling is sparse and often contradictory. This possibly reflects the complexity of fluid–solid interaction and the strong effect of experimental conditions on the observations. Considerable experimental effort has been devoted to the subject area by the drilling industry (see, for example, Sample & Bourgoyne (1977) and Sifferman *et al.* (1974)), where the upward transport of drilled cuttings in vertical annuli is of importance. The effect of fluid turbulence on a single particle suspended centrally in a pipe up which water is flowing under turbulent conditions has been studied by Richardson and Meikle (1961).

5. REFERENCES

Ansley, R.W. & Smith, T.N. (1967). Motion of spheres in a Bingham plastic. *AIChE J.*, **13**, 1193–6.

Atapattu, D.D., Chhabra, R.P. & Uhlherr, P.H.T. (1988). Particle drag and equilibrium in viscoplastic fluids. In *Proceedings of the International Symposium on the Hydraulic Transport of Coal and Other Minerals*. Bhubaneswar, India, pp. 253–60.

Barnes, H.A. & Walters, K. (1985). The yield stress myth? *Rheol. Acta*, **24**, 323–6.

Bird, R.B., Stewart, W.E. & Lightfoot, E.N. (1960). *Transport Phenomena*. John Wiley & Sons, New York, USA.

Brown, G.G. *et al.* (1950). *Unit Operations*. John Wiley & Sons Inc., New York, USA.

Cheng, D.C-H. (1986). Yield stress: a time dependent property and how to measure it. *Rheol. Acta*, **25**, 542–54.

Cheng, D.C-H. (1989). Rheological design for suspending solids. Report LR 711 (MP/BM), Warren Spring Laboratory, Stevenage, UK.

Chhabra, R.P. (1986). Steady non-Newtonian flow about a rigid sphere. In *Flow Phenomena and Measurement* (Vol. 1 of *Encyclopedia of Fluid Mechanics*), ed. N.P. Cheremisinoff. Gulf Publishing Co., Houston, TX, USA, pp. 983–1033.

Clift, R., Grace, J.R. & Weber, M.E. (1978). *Bubbles, Drops and Particles*. Academic Press, London, UK.

Engelmann, H.E. (1978). Vertical hydraulic lifting of large solids—a contribution to marine mining. Paper presented at 10th Annual Offshore Tech. Conference, Houston, TX, USA, 8–11 May, pp. 731–40.

Fair, G.M. & Geyer, J.C. (1954). *Water Supply and Waste-Water Disposal*. John Wiley & Sons Inc., New York, USA.

Fidleris, V. & Whitmore, R.L. (1961). Experimental determination of the wall effect for spheres falling axially in cylindrical vessels. *Brit. J. Appl. Phys.*, **12**, 490–4.

Francis, A.W. (1933). Wall effect in the falling-ball method for viscosity. *Physics*, **4**, 403–6.

Hanks, R.W. & Sen, S. (1983). The influence of yield stresses and fluid rheology on particle drag coefficients. In *Proc. STA 8*, pp. 71–80.*

Happel, J. & Brenner, H. (1973). *Low Reynolds Number Hydrodynamics* (2nd edn). Noordhoff, Leyden, The Netherlands.

Hedström, B.O.A. (1952). Flow of plastics materials in pipes. *Ind. Eng. Chem.*, **44**(3), 651–6.

Heywood, H. (1938). Measurement of the fineness of powdered materials. *Proc. Inst. Mech. Engrs.*, **140**, 257–347.

Heywood, H. (1962). Uniform and non-uniform motion of particles in fluids. In *Interaction between Fluids and Particles*. Institution of Chemical Engineers, London, UK, pp. 1–8.

Lawler, M.T. & Lu, P-C. (1971). The role of lift in the radial migration of particles in pipe flow. In *Advances in Solid–Liquid Flow in Pipes and its Application*, ed. I. Zandi. Pergamon Press Inc., New York, USA, pp. 39–57.

Lewis, W.K., Gilliland, E.R. & Bauer, W.C. (1949). Characteristics of fluidised particles. *Ind. Eng. Chem.*, **41**, 1104–17.

Mandersloot, W.G.B., Scott, K.J. & Geyer, C.P. (1987). Sedimentation in the hindered settling regime. In *Advances in Solid–Liquid Separation*, ed. H.S. Muaralidhara. Battelle Press, Columbus, OH, USA, pp. 63–77.

McNown, J.S., Lee, H.M., McPherson, M.B. & Engez, S.M. (1948). Influence of boundary proximity on the drag of spheres. In *Proceedings of the 7th International Congress of Applied Mechanics*, London, **2**(1), pp. 17–29.

Mirza, S. & Richardson, J.F. (1979). Sedimentation of suspensions of particles of two or more sizes. *Chem. Eng. Sci.*, **34**, 447–54.

Munroe, H.S. (1888). The English vs the Continental system of jigging—is close sizing advantageous? *Trans. Amer. Inst. Mining Met. Eng.*, **17**, 637–59.

Newitt, D.M., Richardson, J.F. & Gliddon, B.J. (1961). Hydraulic conveying of solids in vertical pipes. *Trans. Inst. Chem. Engrs.*, **39**, 93–100.

Peden, J.M. & Luo, Y. (1987). Settling velocity of variously shaped particles in drilling and fracturing fluids. *SPE Drilling Eng.*, Dec, **2**(4), 337–43.

Richardson, J.F. & Zaki, W.N. (1954*a*). Sedimentation and fluidisation: Part 1. *Trans. Inst. Chem. Engrs.*, **32**, 35–53.

Richardson, J.F. & Zaki, W.N. (1954*b*). The sedimentation of a suspension of uniform spheres under conditions of viscous flow. *Chem. Eng. Sci.*, **3**, 65–73.
Richardson, J.F. & Meikle, R.A. (1961). Sedimentation and fluidisation: Part 4. *Trans. Inst. Chem. Eng.*, **39**, 357–62.
Sample, K.J. & Bourgoyne, A.T. (1977). An experimental evaluation of correlations used to predict cutting slip velocity. *52nd Annual Fall Technical Conference of the Society of Petroleum Engineers.* AIME, Denver, CO, USA, paper SPE 6645.
Scheidegger, A.E. (1960). *The Physics of Flow through Porous Media* (2nd edn). University of Toronto Press, Canada.
Schiller, L. & Naumann, A. (1933). Über die grundlegenden Berechnungen bei der Schwerkraftaufbereitung. *Z. Ver. deut. Ing.*, **77**, 318. (In German.)
Scott, K.J. (1984). Hindered settling of a suspension of spheres. Critical evaluation of equations relating settling rate to mean particle diameter and suspension concentration. Report CENG 497, CSIR, Pretoria, Republic of South Africa.
Scott, K.J. & Mandersloot, W.G.B. (1979). The mean particle size in hindered settling of multi-sized spheres. *Powder Technol.*, **24**, 99–101.
Sifferman, T.R., Myers, G.M., Haden, E.L. & Wahl, H.A. (1974). Drill-cutting transport in full-scale vertical annuli. *J. Petr. Tech.*, **XXVI** (Nov), 1295–302.
Skelland, A.H.P. (1967). *Non-Newtonian Flow and Heat Transfer*. John Wiley & Sons, New York, USA.
Slattery, J.C. & Bird, R.B. (1961). Non-Newtonian flow past a sphere. *Chem. Eng. Sci.*, **16**, 231–41.
Stokes, G.G. (1851). On the effect of internal friction of fluids on the motion of pendulums. *Trans. Cambridge Phil. Soc.*, **9**(2), 8–106.
Turton, R. & Clark, N.N. (1987). An explicit relationship to predict spherical particle terminal velocity. *Powder Technol.*, **53**, 127–9.
Turton, R. & Levenspiel, O. (1986). A short note on the drag correlation for spheres. *Powder Technol.*, **47**, 83–6.
Wadell, H. (1934). The coefficient of resistance as a function of Reynolds number for solids of various shapes. *J. Franklin Inst.*, **217**, 459–90.

* Full details of the Slurry Transport Association (STA) series of conferences can be found on p. 13 of Chapter 1.

3

Flow Regimes of Settling Slurries in Pipes

Nigel P. Brown

The British Petroleum Company plc, London, UK

1. FLOW REGIMES AND FLOW PATTERNS

Slurries that show a marked tendency for the solids to settle under conditions of no-flow exhibit two-phase behaviour. For horizontal and near-horizontal flows the distribution of particles over the cross-section of the pipe is often described in terms of flow regimes. Gravity-dominated flows also exhibit a wide range of flow patterns which are used to describe the motion of the particles along a straight section of pipeline. Although both terms are used to aid the description of non-vertical flows, in practice their use is limited because of the highly subjective nature of the terminology.

The flow features exhibited by settling slurries result from the complex processes of momentum interchange between the two phases. If correctly interpreted, flow features provide considerable guidance for the rational selection of techniques to predict hydraulic behaviour and for suitable operating conditions for pipelines.

2. CHARACTERISATION OF SLURRIES

Characterisation schemes for slurries were introduced in Chapter 1. When considering flow features it is convenient to classify slurries as belonging to one of two broad classes; as either a *non-settling* slurry, in which the solid phase shows little or no tendency to settle under conditions of no-flow, or as a *settling* slurry. Typical slurries that show little tendency to settle are sewage sludge, drilling muds, fine coal slurries and concentrated suspensions of fine limestone (such as material fed to cement kilns). Slurries that show a marked tendency to settle include slurries consisting of coarse coal, potash, rock and aggregates.

In practice, this classification scheme is often realised from a consideration of the experimental techniques that are available to provide estimates of the energy requirements to pump slurries through pipes. Non-settling slurries can often be treated as single-phase systems. Viscometric techniques can frequently be used to provide appropriate data for non-settling slurries in much the same manner as applied to single-phase liquids. However, phase separation under shear, not associated with gravitational forces, can occur and may have to be assessed. Viscometric techniques developed for single-phase fluids are inapplicable to a slurry in which the solids settle out when the mixture is not subject to shear. These slurries have to be tested in more elaborate test loops. From a mechanistic viewpoint mixtures that show a marked tendency to settle are showing distinct two-phase behaviour and the mechanisms by which momentum is transferred through the material can be considered separately for each phase.

3. HYDRAULIC CHARACTERISTICS OF SLURRIES

Flow features are a visual manifestation of the processes by which momentum is transferred from the bulk of the slurry to the pipe wall where it is dissipated and realised as a pressure drop. Engineers concerned with pipeline flows of slurries are principally interested in the amount of energy that has to be expended to pump unit mass of the solids along unit length of pipeline. The data that are used to provide this information are usually expressed in the form of a graph, at a fixed solids concentration, of pressure drop against throughput, or superficial velocity. Such a graph is often referred to as the hydraulic characteristic of the slurry. Figure 3.1 shows examples of the hydraulic characteristic exhibited by both settling and non-settling slurries. In this figure both axes are logarithmic; the ordinate is hydraulic gradient, a convenient means of describing pressure drop measured over a length of straight pipe, and the abscissa is superficial velocity, defined as volumetric throughput divided by the cross-sectional area of the pipe. The lowest line is included to illustrate the behaviour of a Newtonian fluid, such as water in turbulent flow in a smooth pipe. Many slurries show a marked deviation from Newtonian behaviour, especially at low rates of shear.

Terminology associated with single-phase fluids is frequently used to describe the flow regime of non-settling slurries; application to slurries that show settling

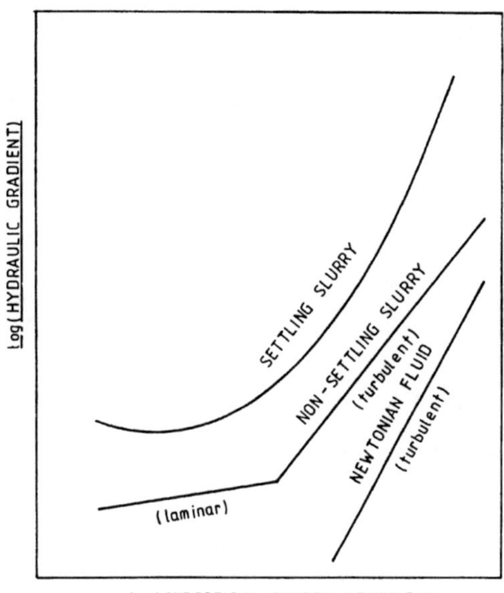

Fig. 3.1. Schematic examples of hydraulic characteristics.

tendency is inappropriate. The hydraulic characteristic of a typical non-settling slurry is similar to that of a single-phase liquid except that the slope of the characteristic may change within the region of commercial operating conditions. This change reflects the transition from laminar to turbulent flow as the velocity is increased. In general, the transition is very sensitive to the rheology of the slurry as are the friction losses in laminar flow.

The upper curve in Fig. 3.1 represents the behaviour of a typical settling slurry in non-vertical flow. Settling slurries do not tend to show an abrupt change in flow behaviour with increasing throughput. Changes in flow behaviour are often less well defined than those exhibited by a non-settling slurry. These mixtures can exhibit a complex range of flow patterns, depending upon the physical properties of the carrier fluid and the transported solid, and the superficial velocity and concentration of the slurry. Slurries in which the solids do not tend to settle to a marked extent do not exhibit visually recognisable flow features.

4. FLOW REGIMES EXHIBITED BY SETTLING SLURRIES

The flow regimes exhibited by settling slurries are closely related to the distribution of solids within the cross-section of the pipeline. Figure 3.2 provides a schematic representation of the distribution of solids in a horizontal or near-horizontal pipe that are associated with features of the hydraulic characteristic. The upper curve

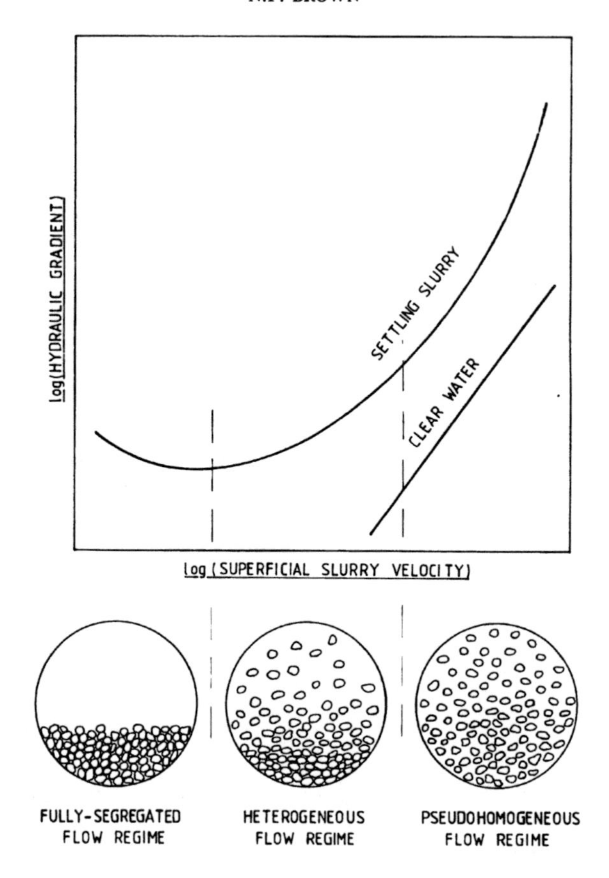

Fig. 3.2. Comparison of the distribution of solids in a pipeline to the shape of the hydraulic characteristic.

represents the hydraulic characteristic of a settling water-based slurry and the lower straight line that for the clear carrier fluid. The axes in this figure are logarithmic. The three circles represent the distribution of the particles within the pipeline. In this illustration the solids are considered to fall into a narrow size distribution. As a broad generalisation, to the left of the *knee* exhibited by the characteristic the solids are present as a gravity bed. Such a flow regime is referred to as fully segregated or two-layer. At very high velocities (usually well above those that would be considered commercially viable for such a slurry) and with favourable solid and liquid properties, the solids may approach an even distribution throughout the pipe. This flow regime is often referred to as pseudo- or quasi-homogeneous. True homogeneous flow, in which the solids are evenly distributed throughout the pipe, rarely occurs in settling slurries; it is, however, often considered a characteristic of a non-settling slurry. Between the two extreme cases lies the heterogeneous flow regime. This regime is by far the most complex as the characteristics of the extreme flow regimes are embodied in the flow; some particles may be present in the form of a bed, others supported by fluid turbulence.

As the superficial velocity of a settling slurry is reduced, a minimum in the hydraulic gradient occurs which is closely associated with the formation of a bed of solids in the pipe. The velocity at the minimum is sometimes referred to as the bedding velocity (as the throughput is reduced), or the limit of deposition (as the throughput is increased). The velocity occurring at the minimum is often casually referred to as the critical velocity. The number of different terms used in the literature to describe phenomena occurring at, or in the vicinity of, the minimum in the hydraulic characteristic can cause confusion. This confusion is only heightened by several terms being defined in different manners. Carleton and Cheng (1974) present an excellent review of terminology used in the literature.

5. EXPERIMENTAL EVIDENCE FOR FLOW REGIMES

Axisymmetric concentration and velocity profiles are associated with well-developed flows of settling slurries in vertical flows. At high velocities, photographs of vertical counter-gravity flow of a coarse-particle (Newitt *et al.*, 1961) slurry clearly show radial particle migration. A particle-rich core of solids is formed at the axis of the pipe surrounded by a particle-lean annulus.

Experimental measurements of both particle velocity profiles and chord averaged concentration profiles are useful to illustrate the flow regimes that occur at extreme flow rates. Measurements performed on two slurries consisting of closely graded sands with different particle diameters provide a convenient illustration of the flow regimes in horizontal flow. The two sands are characterised by the d_{50} particle diameter; the diameter at which 50% of the material is undersized on the grading curve. Sand A has a particle diameter approximately equal to 0·15 mm (with 90% of the solids between 0·1 and 0·25 mm) and sand B a diameter of 0·53 mm (with 90% of the solids between 0·3 and 1·0 mm). Figure 3.3 shows the hydraulic characteristic obtained from the two sand–water slurries flowing in a 50 mm pipeline at an in-situ solids concentration of 20% (by volume). The hydraulic data are plotted using linear axes. The head loss data for the flow of clear water, in the same pipe, are included to provide a datum for comparison purposes.

Figures 3.4 and 3.5 show the distribution of the sand in the cross-section of a 50 mm pipe. The data points are measurements of chord-averaged solids concentration profiles made using a traversing gamma-photon densitometer. The ordinate of these figures is the relative position from the top of the pipe.

Figure 3.4 shows measurements of the solids concentration at a superficial slurry velocity just above that at which deposition occurs. The superficial velocity for sand A is 1·7 ms^{-1} and for sand B, 2·1 ms^{-1}. The data in the figure show that sand B, with the larger particle size, is concentrated in the lower part of the pipe. The finer sand A is also concentrated in the lower half of the pipe, although the velocity

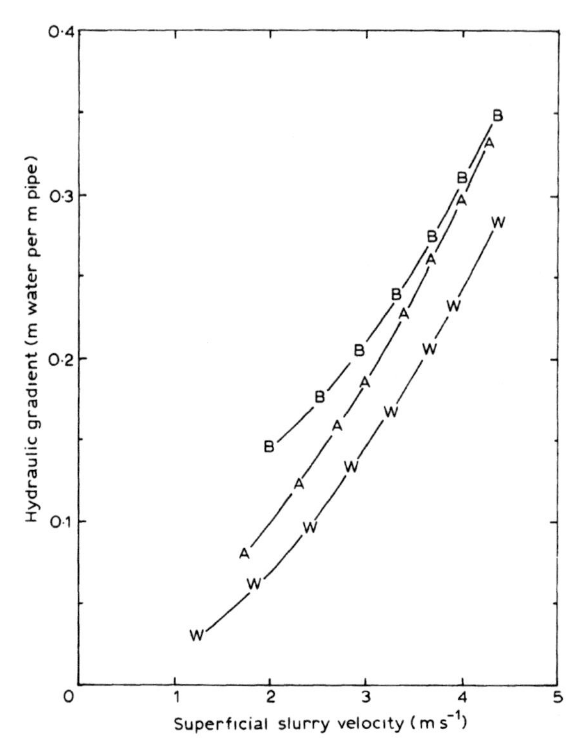

Fig. 3.3. Hydraulic characteristic for water (W) and the water-based slurries of sands A and B at an in-situ concentration of 20% (by volume).

gradient in this lower layer has caused the gravity bed to dilate. Under conditions of no-flow a 20% sand–water slurry would normally occupy the lower one-third of the pipe.

The data in Fig. 3.5 show similar measurements of chord-averaged solids concentration but at a much higher superficial velocity of $4{\cdot}2\,\mathrm{ms}^{-1}$ for sand A and $4{\cdot}3\,\mathrm{ms}^{-1}$ for sand B. At this higher velocity the finer sand (A) is more evenly distributed throughout the pipe than the coarser sand (B). Sand B exhibits a distinctly skew solids profile even at this high velocity.

The heterogeneous behaviour exhibited by the coarser sand can often be identified on the hydraulic characteristic (shown in Fig. 3.3) by frictional losses that tend to converge toward the head loss line for the clear fluid as the flow rate is increased and become parallel at high flow rate. In comparison the pseudohomogeneous behaviour, exhibited by the finer sand at the higher velocity, is associated with a hydraulic characteristic that tends to pull away from the clear fluid losses with a greater slope as the superficial slurry velocity is increased above that associated with bed formation.

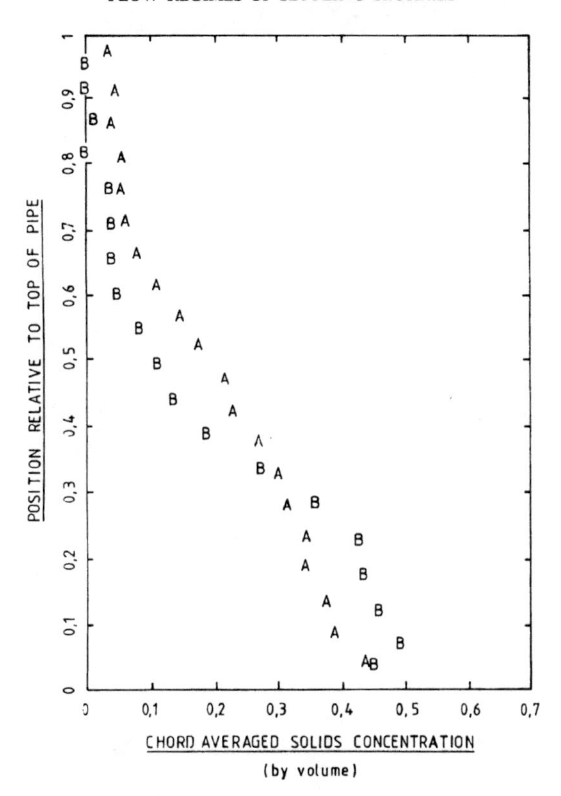

Fig. 3.4. Chord-averaged concentration profiles for sands A and B measured just above the bedding velocity (sand A = $1{\cdot}7\,\mathrm{m\,s^{-1}}$, sand B = $2{\cdot}1\,\mathrm{m\,s^{-1}}$) at an in-situ concentration of 20% (by volume). (Measurements for sand A adapted from Brown *et al.*, 1983.)

For sand A local particle velocity measurements were also made. They were measured using a small probe resembling a pitot tube (Brown *et al.*, 1983). These measurements were made at the same superficial velocities as the solids concentration profiles presented in the two previous figures. The measurements are shown in Fig. 3.6. In this figure the open symbols represent the profile measured in the vertical plane from top (= 1) to bottom (= 0) in the pipe, while the closed symbols show measurements made in the horizontal plane. Both profiles pass through the centre of the pipe and were made in a section of pipe 100 pipe diameters from an upstream bend.

At the lower superficial velocity the vertical profile is skewed indicating the presence of a bed of sheared solids in the pipe. The highest velocities occur above the bed. The horizontal profiles, made across the centre of the pipe (and plotted

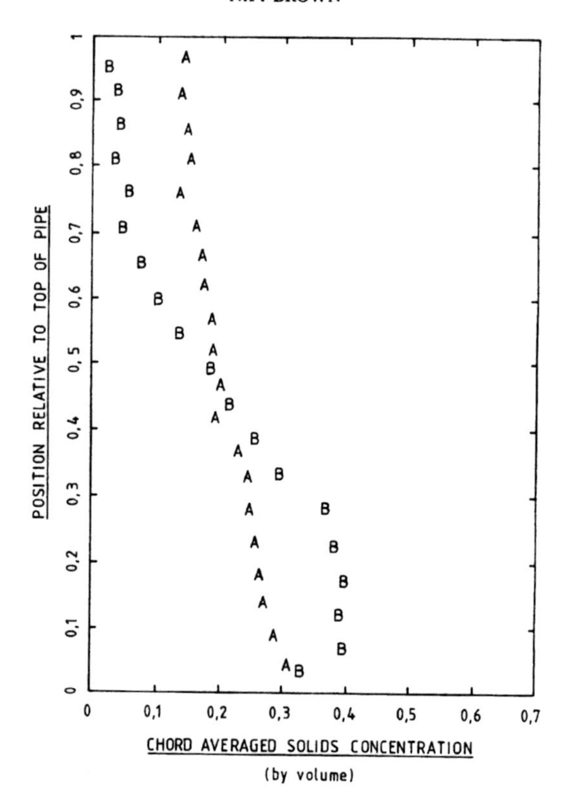

Fig. 3.5. Chord-averaged profiles of particle concentration for sands A and B (sand A = $4{\cdot}2\,\mathrm{m\,s^{-1}}$, sand B = $4{\cdot}3\,\mathrm{m\,s^{-1}}$) measured at an in-situ concentration of 20% (by volume). (Adapted from Brown *et al.*, 1983.)

using closed symbols), are very similar to those predicted by the $\frac{1}{7}$th power law model frequently used for single-phase flows. The profile predicted using the power law model is plotted on the figure using a solid line. At the higher superficial velocity the profiles in both the horizontal and vertical planes are similar, a feature associated with pseudohomogeneous flow. The solid line again represents the profile predicted by the $\frac{1}{7}$th power law model based on the maximum velocity in the horizontal plane.

6. THE PHENOMENA OF FLOW PATTERNS

Flow patterns are used to describe the motion of the particles in non-vertical flow of the slurry. They are often visually intriguing and have been the subject of numerous laboratory studies.

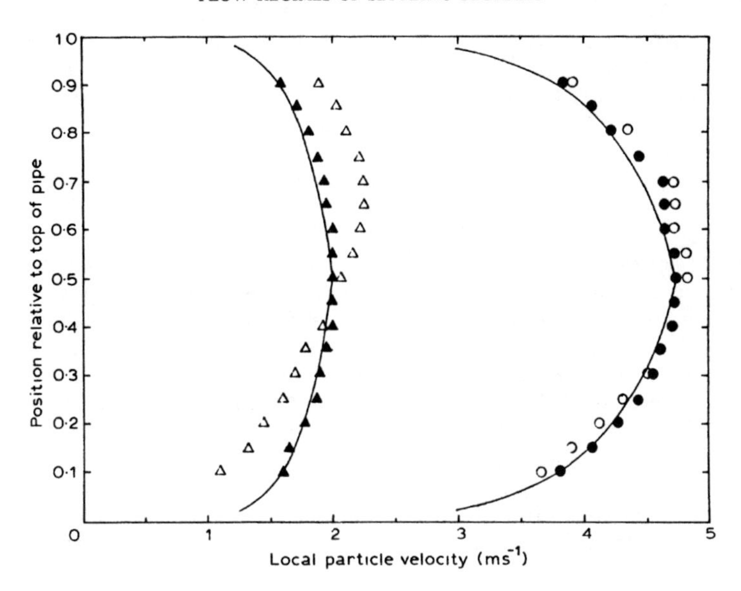

Fig. 3.6. Velocity profiles for sand A. Open symbols—vertical plane, closed symbols—horizontal plane. Superficial velocity: △, 1·7 m s^{-1}; ○, 4·2 m s^{-1}. (Adapted from Brown *et al.*, 1983.)

The motion of solid particles over horizontal, stationary and moving beds of the same granular material has been extensively investigated by Bagnold (1973) and Francis (1973). As the velocity is increased, the bed grains can be observed to jump up into the moving fluid stream as if they were acted upon by some momentary impulse at the bed surface. The particle then follows a smooth downward trajectory to the bed under the influence of gravity. By careful experimentation, Francis (1973) has shown that this phenomenon of particle saltation can exist in the absence of fluid turbulence and must, therefore, be due to a separate dynamic process from that of transport in suspension by the internal eddy motions of a turbulent fluid. It has been further shown that the forward motion of saltating grains is opposed by a frictional force of the same order as the immersed weight of solids, the friction coefficient being approximated by the angle of internal friction of the solids in the granular bed. This is perhaps not surprising since the saltation arises solely from momentary particle–particle contact. Saltating particles are usually associated with the gravity bed of solids in the pipe.

Increasing the superficial velocity causes the shear stress on the bed surface to increase, entering a range of stress where elongated dunes or ripples form at the interface. The downstream face of these dunes have a greater slope than the face opposing the flow. Particles are lifted from the bed and since they cannot be supported by fluid turbulence, are redeposited on the bed. The effect of these mobile grains is to increase the height of the dunes to such a point that the wall stresses overcome the driving force on the bed; this results in the bed gradually decelerating to a momentary halt. The restriction in the waterway above the stationary dunes causes the fluid shear stress to increase to such a value as to reinstate movement of the bed. Conditions at the surface are potentially unstable, and the bed as a whole moves in a series of unsteady jerks. This represents a wholly unsatisfactory operating condition for a pipeline as such behaviour is often associated with an increasing solids hold-up in the line that is likely to precipitate a blockage.

As the superficial velocity is increased, particles begin to be supported when the random succession of upward impulses imparted upon the particles by the eddies of fluid turbulence exceed gravitational attraction. Suspended solids can, therefore, be distinguished from bed-load as the particles which remain out of contact with the bed for an indefinite time depending upon the essentially random nature of turbulence. Bagnold (1973) noted that a critical bed shear stress is necessary to initiate turbulent suspension, its value being dependent upon grain size. A distinctly asymmetric solids concentration profile exists across the pipe.

Upon further increasing the superficial velocity the enhanced intensity of fluid turbulence, and hence lift forces imparted on the particles, enables an increasing proportion of the solids to become suspended. The vertical solids concentration profile tends to become less pronounced, while the velocity profiles of fluid and particles tend to resemble those of single-phase fluids in both the horizontal and vertical planes.

The preceding description can be applied only to materials with a rounded shape that fall into a narrow size distribution. The features in the flow patterns become less well defined as the size distribution is broadened to include both fine and coarse material. The larger particles tend to remain in the bed as the smaller particles become suspended in the faster moving fluid above the bed. Such a description of a succession of flow patterns is inappropriate for slurries composed of a mixture of materials of different physical properties such as density and shape.

7. EXPERIMENTAL TECHNIQUES FOR DISTINGUISHING BETWEEN FLOW FEATURES

Most slurries of commercial interest are often extremely murky with the result that precise definitions and limits are difficult, if not impossible, to establish. Frequently, the only visual observations that are possible are whether or not particles can be seen on the bottom of the pipe and whether or not they are moving. Even these somewhat trivial observations can be difficult to make in an objective manner. Considerable success has been achieved in removing the subjectivity associated with observations of pipeline flows through the application of measurement techniques, such as those discussed in Chapter 11.

8. PREDICTING FLOW REGIMES

In the design of pipelines conveying settling slurries information is required on the flow behaviour. The use of experimental data is highly advisable. In preference the data should be obtained with the slurry that is the subject of the study in operational-sized pipes. Where this is not practicable two courses of action are available. The behaviour can either be predicted or interpreted from published experimental data. While the use of experimental data is always preferable, the wide variety of slurries that are encountered often results in unrewarded searches of the literature.

Many of the correlations reported in the literature were developed from experiments observed in small diameter pipes using ideal slurries. Early investigators (for example, Newitt *et al.*, 1955; Durand, 1953) expended considerable effort exploring the detailed features of the flows. The results were often given in the form of detailed maps showing the flow regime under different conditions. Carleton and Cheng (1974) have made a comprehensive review of published correlations for calculating design velocities for settling slurries. They concluded that there is considerable variation in the predictions and they recommended that the correlations should be applied only within the range of experimental conditions on which they were based. More recently, Turian *et al.* (1987) presented a useful review of published correlations for critical velocity, which they define as the minimum velocity demarcating flows in which solids form a bed at the bottom of the pipe from fully suspended flows (*sic*). Their review forms a useful basis for the selection of appropriate published data as the liquid and particle properties of the slurry, pipe diameter and extent of experimental data is tabulated for each of the 34 correlations studied. The majority of experiments were conducted in pipes smaller than 150 mm.

Due principally to the questionable worth of these correlations for applications other than those for which they were developed, the recent trend, for general design studies, has turned away from correlation towards physically based techniques. A physically based design technique is described by Wilson in Chapter 6.

9. REFERENCES

Bagnold, R.A. (1973). The nature of saltation and of 'bed-load' transport in water. *Proc. Royal Soc.*, **A332**, 473–504.

Brown, N.P., Shook, C.A., Peters, J. & Eyre, D. (1983). A probe for point velocities in slurry flows, *Can. J. Chem. Engng.*, **61**, 597–602.

Carleton, A.J. & Cheng, D.C-H. (1974). Design velocities for hydraulic conveying of settling slurries. In *Proc. HT3*, paper E5, pp. 57–74.*

Durand, D. (1953). Ecoulements de mixture en conduites verticules-influence de la densite des materiaux sur les characteristiques de refoulement en conduite horizontale, *La Houille Blanche*, **8**, 124. (In French)

Francis, J.R.D. (1973). Experiments on the motion of solitary grains along the bed of a water stream. *Proc. Royal. Soc.*, **A332**, 443–71.

Newitt, D.M., Richardson, J.F., Abbott, M. & Turtle, R.B. (1955). Hydraulic conveying of solids in horizontal pipes. *Trans. Inst. Chem. Engrs*, **33**, 93–113.

Newitt, D.M., Richardson, J.F. & Gliddon, B.J. (1961). Hydraulic conveying of solids in vertical pipes. *Trans. Inst. Chem. Engrs*, **39**, 93–100.

Turian, R.M., Hsu, F-L. & Ma, T-W. (1987). Estimation of the critical velocity in pipeline flow of slurries. *Powder Technol.*, **51**, 35–47.

*Full details of the Hydrotransport (HT) series of conferences can be found on pp. 11–12 of Chapter 1.

4

Rheological Characterisation of Non-settling Slurries

Nigel I. Heywood

Warren Spring Laboratory, Stevenage, UK

1. PHYSICAL AND CHEMICAL BASES FOR RHEOLOGICAL PROPERTIES

1.1. Introduction

If the solid particles in a slurry settle sufficiently slowly under gravity, the slurry may, for a number of applications, be considered *non-settling* (see Chaper 1). Solids

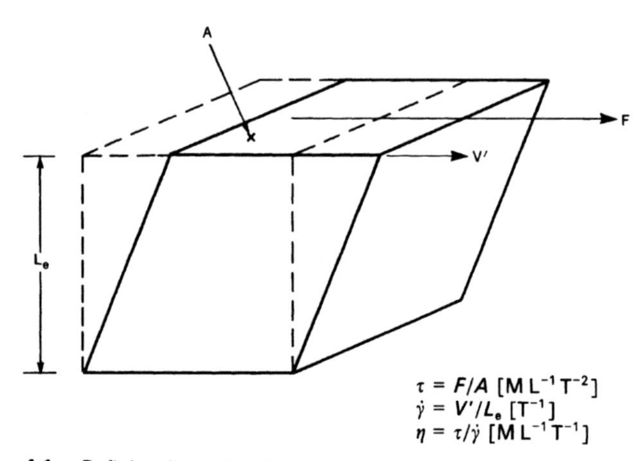

Fig. 4.1. Defining figure for shear stress, τ, shear rate, $\dot{\gamma}$, and viscosity, η.

are then assumed to be distributed uniformly throughout the liquid phase and, for the purposes of assessing its rheological properties, the slurry can be considered as pseudohomogeneous. The presence of particles can then be unimportant in some circumstances, although when a shear stress is applied to the slurry to cause it to flow, the particles may move away from the shearing surface such as that in a viscometer or the wall of a pipeline. This can result in a lower solids concentration at the wall compared with the bulk solids concentration giving rise to *wall-slip*. In addition, because most particles do not deform when shear is applied, particle jamming can occur in some flow applications and in some viscometric geometries (e.g. cone-and-plate viscometer).

Particles less than a few micrometres exhibit Brownian motion due to their thermal energy. This causes collision between two or more particles, and, if their combined momentum is sufficient to overcome the energy barrier (see section 1.2) between particles as a result of the particle surface charge (zeta potential), these particles can flocculate. This process leads to the entrapment of suspending medium. Instead of individual particles being the primary determinant of the slurry's rheological properties, it is these flocs which act as the relevant *flow units*. Even a slurry where only a few percent of the solids are sub-micrometre can be in a highly flocculated state (see section 1.3).

The viscosity of a slurry is defined as the ratio of shear stress to shear rate (see Fig. 4.1). In general, slurry viscosity will be a function of shear rate and temperature and is determined essentially by three main types of interaction (Cheng, 1980):

- hydrodynamic interaction between the liquid and the particles, which gives rise to viscous dissipation in the liquid;
- interparticle attraction, which promotes the formation of flocs, aggregates, agglomerates (structure);

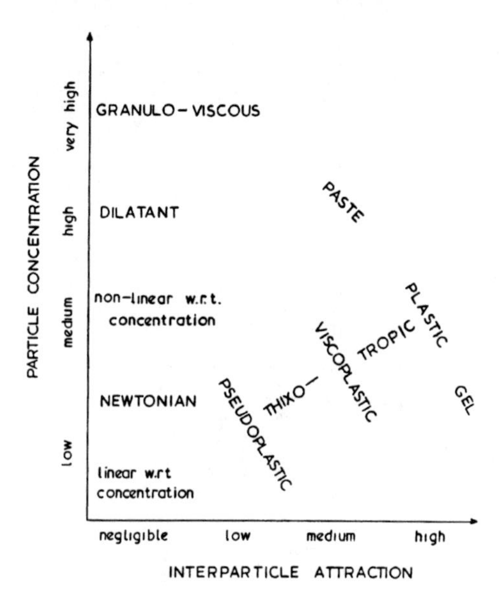

Fig. 4.2. Relationship between the three categories of interaction in a slurry and the resulting rheological behaviour. (Adapted from Cheng (1980).)

- particle–particle interaction causing surface frictional energy dissipation.

The relative importance of these types of interaction determine the class of rheological property according to Fig. 4.2 (Cheng, 1980). The various flow properties on the left of this figure occur for deflocculated slurries or slurries which have a low level of flocculation, with the degree of flocculation increasing from left to right. At any level of interparticle attraction (and therefore degree of flocculation), viscosity will rise with increasing solids concentration. In all cases, viscosity rises relatively slowly with increases in solids content at low concentrations, but can rise rapidly with concentration when the maximum particle packing is approached. For randomly packed spheres this occurs at about 62% solids by volume.

1.2. Non-flocculated (Deflocculated or Dispersed) Slurries

Non-flocculated, or fully dispersed, slurries either occur naturally as a result of the constituents of an industrial slurry, or are created by the addition of a chemical deflocculent. Sometimes the slurry may be only partially deflocculated if an incorrect concentration of deflocculent (too low or too high) has been used. The individual particles in the slurry act as the *flow units* and this often results in Newtonian flow behaviour (constant viscosity) at low to medium solids concentration and medium to high shear rates, but shear-thickening behaviour (viscosity

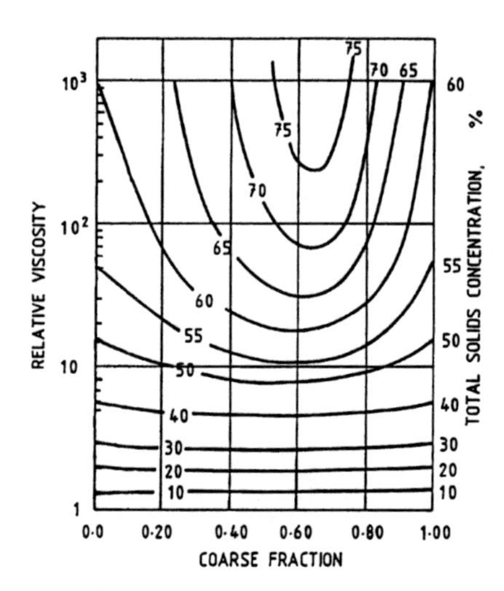

Fig. 4.3. Predictions of relative viscosity of a slurry with a bimodal particle size distribution of non-flocculated particles. (Adapted from Farris, 1968.)

increases with shear rate) at medium to high solids concentrations and/or medium to high shear rates.

At any given solids concentration exceeding about 40% (by volume), the viscosity of a non-flocculated slurry will be greatest for mono-sized solids and will reduce progressively as the particle size distribution is progressively widened (Farris, 1968). Theoretical predictions of this effect are illustrated in Fig. 4.3 for a slurry with a bimodal particle size distribution where relative slurry viscosity (slurry viscosity divided by the suspending medium viscosity) is plotted as a function of the fraction of coarse solids. Triangular diagrams can also be constructed for trimodal slurries (Cheng *et al.*, 1990). Such behaviour obviously has important implications if viscosity is to be minimised when slurry is handled at a given concentration. Conversely, by appropriate adjustment of the particle size distribution, it is possible to maximise the solids content to meet a specified maximum viscosity.

Particles comprising a non-flocculated slurry will settle more slowly under gravity compared with those in the equivalent flocculated slurry, but the resulting sediment is almost always much more cohesive and can often be very difficult to resuspend. On the other hand, sediment formed from the settling of a deflocculated slurry after some considerable time is normally highly voluminous and is readily resuspended. The supernatant liquid formed during the settling of a non-flocculated slurry is usually cloudy, whereas that of a flocculated slurry is generally clear.

1.3. Flocculated Slurries

The potential energy between two particles in a slurry is the result of the attractive van der Waals forces acting between them and the Born repulsive forces, caused by the interaction of the electrical double layers on the particle surfaces at small particle–particle separation. The resulting zeta potential determines whether particles adhere on collision or whether they remain in a non-flocculated or dispersed state. When flocs form, the slurry develops an internal structure which usually imparts non-Newtonian, shear-thinning flow behaviour and sometimes also significant time dependent, thixotropic behaviour.

When shear stresses applied to a flocculated slurry are progressively raised, the average floc size reduces and the flocs trap progressively less suspending medium. This releases suspending medium which is now available to assist the relative motion between flocs. This process is observed on a macroscopic scale as a progressive reduction in viscosity. At high shear stresses, all the flocs have been reduced in size to almost the individual particle sizes. As a result, the flow behaviour of the flocculated slurry at high shear stresses becomes almost Newtonian and similar to that of the same slurry if the particles had been deflocculated chemically.

Experimental studies on the effect of changing the zeta potential on the rheological properties of a number of industrially important slurries have been undertaken by Horsley and Reizes (1980), Ishihara *et al.* (1987), Round and Hessari (1985) and Sikorski *et al.* (1982). Figure 4.4 shows some typical hydraulic gradient/slurry velocity data for the pipe flow of a 43% (by volume) sand at different pH levels (and therefore at different zeta potential in mV). The changes in pressure loss in the laminar regime are greater for higher sand concentrations. This illustrates that large zeta potentials (i.e. high negative charge on the particle surfaces) lead to low hydraulic gradients and can cause a transition from laminar to turbulent flow at constant slurry concentration and flow rate.

1.4. Fibrous Slurries

Examples of fibrous slurries include fermentation broths, crushed meal animal feed, various foodstuffs such as tomato purée and fruit pulps, as well as various types of sewage sludge and paper pulp. Many of these materials are characterised by long, flexible fibres which intertwine and entrap much of the suspending medium. The fibres may be flocculated but even in the absence of attractive interaction between the fibre molecules, the fibres can form flocs with an open structure.

The flow properties of such suspensions are determined by the overall volume of the flocs or flow units, which can often approach the close packing condition. Thus, even when the slurry contains only a few percent of dry solids, the suspension behaves as a dense suspension (Cheng, 1984) exhibiting high viscosities and its flow curve a function of the flow geometry. In addition, because the flocs are highly compressible they can readily deform according to the flow. The suspending medium can easily be expressed under pressure from the fibre flocs and so *wall-slip* is very common.

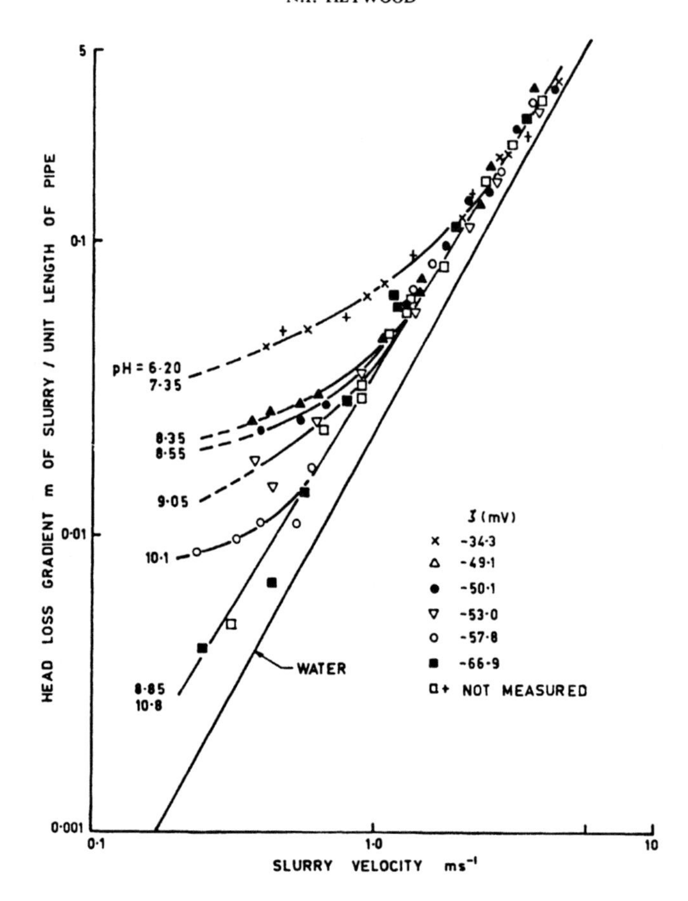

Fig. 4.4. Effect of pH and zeta potential of a sand slurry (43% by volume concentration; average particle size of 17 μm) in pipe flow. (Reproduced from Horsley & Reizes, 1980.)

2. TYPES OF RHEOLOGICAL PROPERTY

Much work has been undertaken to assist in our understanding of why slurries have certain rheological properties. Despite this, it is impossible to predict, with any reasonable degree of accuracy, the rheological properties of a given slurry no matter how well the slurry's physical and chemical properties may have been specified. This is partly because of the complex chemical and physical nature of industrial slurries and partly because the rheological properties of any multiphase fluid is inherently more complex than those of a single phase fluid, such as a liquid formed from one or more miscible *pure* liquids. Laboratory or pilot-plant measurements must therefore always be undertaken.

2.1. The Flow Curve

Slurry viscosity is just one of a number of rheological properties which can be used for slurry characterisation and engineering application purposes. Other rheological properties include extensional viscosity, normal stress and dynamic or elastic properties which are particularly important for polymer solutions and melts. Significant normal stresses can be developed in some high concentration slurries (Umeya & Otsubo, 1980) but most slurry rheological characterisation is based on the flow curve, i.e. a plot of shear stress against shear rate.

Slurry viscosity may be a constant over a wide range of shear rate (but not necessarily all relevant shear rates) in which case the slurry exhibits Newtonian flow behaviour. A large number of industrially important slurries exhibit a variable viscosity with shear rate and so the single parameter of viscosity is quite inadequate to describe the slurry shear property, even if the viscosity value is quoted at a specified shear rate. Here, for any end-use of the viscosity data, the relevant shear rate must be assessed and used in determining the measurement conditions.

Figure 4.5 shows the main classes of flow curve which slurries can exhibit. These are idealised representations because slurries can show at least one flow curve classification over the full shear rate range of some practical applications.

2.1.1. Time-Independent and Time-Dependent Properties

The simplest situation is one in which the flow curve obtained by viscometric test work does not depend on the time period over which the slurry sample is sheared. Thus, if a constant shear rate (or conversely shear stress) is applied, the resultant shear stress (or conversely shear rate) attains a constant value after a few seconds.

It is important to know whether the flow properties are time-independent in a number of applications. For instance, in steady pipe flow, a time-independent slurry will create a constant pressure gradient at any position along the pipeline, whereas a slurry exhibiting time-dependent flow properties will develop a variable pressure gradient along the pipe until the slurry structural state has reached an equilibrium level.

Time-dependent steady shear properties include thixotropy, and unsteady dynamic properties such as viscoelasticity. The latter is normally not detected under steady-state test conditions of practical importance in slurry handling. Thixotropic property is fairly common but is often overlooked. Slurries which can be thinned by subjecting them to shear energy can subsequently be handled more easily due to their reduced viscosity. This can be achieved through agitation prior to pumping through a pipeline (Boger *et al.*, 1979; Want *et al.*, 1982; Nguyen & Boger, 1985).

Slurries showing reverse properties to thixotropic slurries are often referred to as anti-thixotropic. Sometimes the term *rheopectic* is used, but it can be argued that

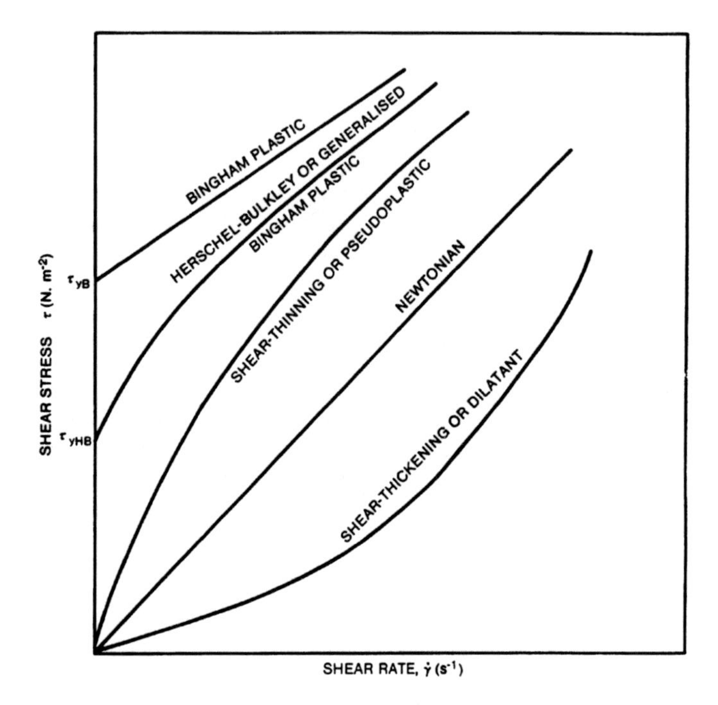

Fig. 4.5. Types of laminar flow behaviour. (τ_{yB} = Bingham plastic yield stress; τ_{yHB} = Herschel–Bulkley yield stress.)

rheopexy is in fact a special case of the thixotropy phenomenon. Fortunately, anti-thixotropic slurries are very rare.

2.1.2. Yield Stress

The yield stress of a slurry is the minimum shear stress required to initiate and sustain flow. For practical applications, the resultant sustainable flow must be at a suitable rate. Because the presence of a yield stress in a slurry is normally the result of structure due to the flocculated particles, the yield stress is not a constant but will vary according to the degree of structure in the slurry. Thus, a low yield stress will occur when the slurry structure is undeveloped or broken down, while a high yield stress results when the structural level is high.

The constant appearing in common flow models is normally referred to as a *yield stress*, but this quantity when estimated by curve-fitting shear stress/shear rate data (see section 4) is often merely a convenient curve-fitting parameter and may bear little relationship to the *true* yield stress. There are several reasons for this. First, the curve-fitting parameter is obtained by extrapolation of viscometric data to zero shear rate when the lowest shear rate used may be well above zero. Consequently, errors may be introduced depending on what extrapolation method is used.

Secondly, viscometric measurements performed on a slurry exhibiting thixotropic properties, or where phase separation under shear occurs (see section 3), will provide an equilibrium flow curve. The extrapolated yield stress (sometimes referred to as the *dynamic* yield stress) will generally be less than if it had been measured directly. Direct measurement gives the *static* yield stress and is the more reliable property for predicting important engineering design parameters such as the pressure gradient required for pipeline start-up (see section 3.7). An analogy may be drawn between this and the static and dynamic angles of wall friction (considered in Chapter 11).

Depending on the application and hence the time frame involved, there has been much argument over whether the yield stress concept is appropriate at all for fluids. Some argue it never exists in a fluid while others view it as a useful engineering tool (Hartnett & Hu, 1989; Astarita, 1990). As an aid to predicting pipeline flow behaviour there is no doubt of its use, but as an aid to assessing or formulating a slurry for suspending coarse particles it is probably a misleading concept (Cheng, 1989).

2.2. Common Flow Models

There are a number of model equations to choose from which relate shear stress to shear rate but only a few common models regularly act as the basis for engineering design calculations. These models are summarised in Chapter 7 and include the Newtonian model (constant viscosity), the power law model (describing pseudoplastic and shear-thickening or dilatant behaviour) and the three models which incorporate a yield stress parameter: the Bingham plastic, the generalised Bingham plastic (or yield-pseudoplastic) and the Casson models. They generally assume time-independent flow properties but estimates of the parameters appearing in them can be allowed to vary according to the structural state of the slurry. Approaches to parameter estimation for these flow models using experimental viscometric data are described in section 4.

3. MEASUREMENT OF SLURRY FLOW CURVE

Standard texts are available on flow curve and viscosity measurement (Van Wazer *et al.*, 1963; Whorlow, 1980; Collyer & Clegg, 1988; Barnes *et al.*, 1989) and on applications for the viscometric data (Walters, 1980), but slurries can cause a number of testing problems, such as *wall-slip*, owing to their two-phase composition (Cheng, 1984). These problems will be discussed where relevant throughout this section.

Flow curve measurements are made for laminar flow conditions only so checks need to be made that the data collected do indeed all correspond to primary laminar flow in the viscometer. Double check if the data appear to be showing some shear-thickening behaviour (Barnes, 1989; Boersma *et al.*, 1990); this may be the consequence of some data being obtained in the secondary laminar or turbulent flow regions in the viscometer. These data should be rejected if found to be affected in this way. Check also that the shear rate range used does cover that appropriate to flow in the pipe sizes of interest (see section 3.2.1 and Chapter 7).

Ideally, a number of flow curves should be generated to cover the typical variability which may occur in the slurry. Variables which can have a very marked effect on the flow curve include solids concentration, pH and particle size distribution. It is often best to go for a conservative design. For the prediction of pipe flow behaviour this will be based on the highest slurry viscosities which are likely to occur. These will tend to occur at the highest anticipated solids concentration (rather than the mean or most probable concentration), the pH which causes the highest degree of flocculation of the fines, and the narrowest particle size distribution if the slurry is non-flocculated and has a high concentration. A very conservative design would use the flow curve based on a slurry sample where all three of these factors had their greatest influence on maximising the slurry viscosity.

Wall-slip. One problem which regularly occurs when measuring the rheological properties of slurries is the phenomenon known as *wall-slip*. When a slurry is sheared either in a pipe, viscometer or in a piece of process equipment, a reduction in the solids concentration can occur at the shearing surface compared with the initial, uniform bulk concentration. This can happen for a variety of reasons. First, particle rotation in a velocity gradient results in particle motion away from the shearing surface (Magnus effect). Secondly, expression of the suspending phase can occur through the solids adjacent to the wall and, thirdly, it is geometrically impossible to have the same solids concentration at the wall as in the bulk of the flow. This last effect becomes more important as the particle size increases.

The overall effect of wall-slip can be large with some slurries and of minor importance with others; it is impossible to predict its significance in any flow situation without experimental tests. Because this phenomenon lowers the shear stress at a shearing surface compared with that which would have occurred if the solids concentration at the surface were the same as in the bulk, the apparent rheological slurry properties can be greatly modified; a lower slurry viscosity is apparent than is actually the case. Wall-slip can often occur in a viscometer and there are techniques to correct for it so that the flow properties corresponding to the bulk concentration of the slurry are obtained (see sections 3.3.2 and 3.4.2). These techniques apply only to laminar flow conditions, however, and very little is known about slip in turbulent flow.

If slip occurs in a viscometer, it will also probably occur to the same extent under the same shear conditions in pipelines and in process equipment. The degree of slip may need to be quantified (for instance, as a function of wall shear stress and characteristic dimensions) and applied to the design so that over-conservative designs are not made. In pipeline flow, for instance, the effect is normally advantageous as a reduced frictional pressure loss results for a given flow rate, or, conversely, for a given discharge pressure developed by a pump an enhanced flow rate is possible.

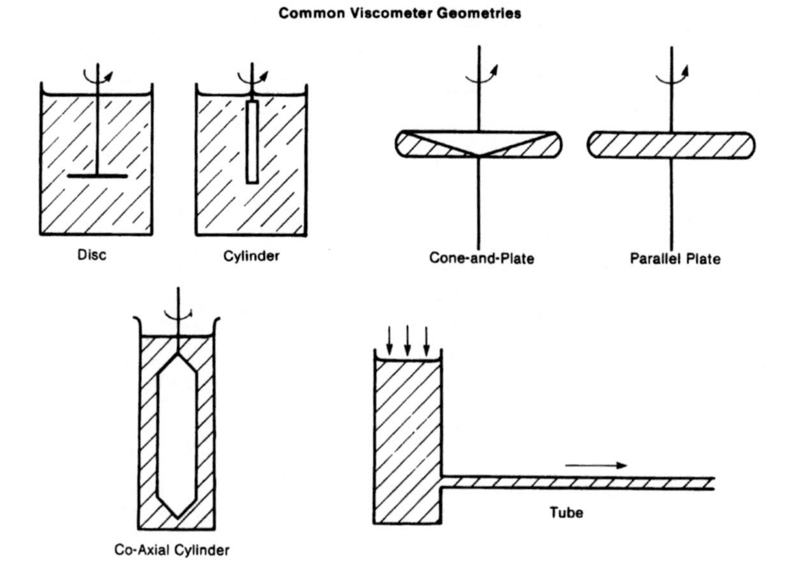

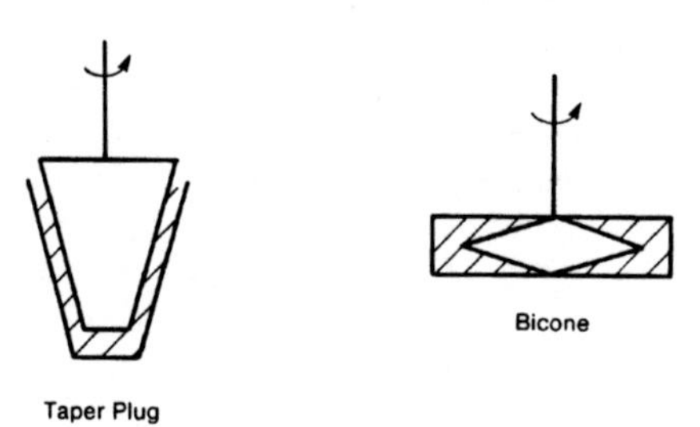

Fig. 4.6. Various viscometer geometries.

3.1. Viscometer Geometries

Commercial viscometers (Heywood, 1985) employ a wide range of geometries for slurry viscosity and flow curve measurement (Sarmiento *et al.*, 1979; Darby, 1984; Hanks & Bowman, 1985). Figure 4.6 gives some of the more common geometries. Flow curve measurements can be made using all these geometries but there are several drawbacks with many of the options. For instance, *particle-jamming* effects can frequently occur using the cone-and-plate geometry and

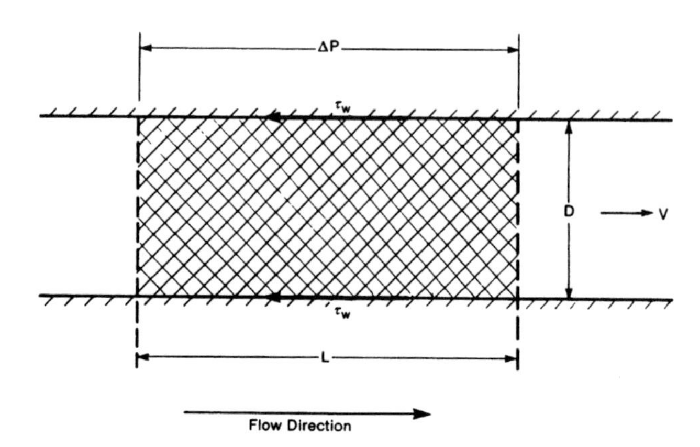

Fig. 4.7. Force balance on a length of pipe.

even the use of specially truncated cones often fails to remove the problem. Wall-slip cannot readily be identified in the cone-and-plate, parallel-plate and disc geometries.

Because of these considerations, the two most useful viscometer geometries for slurry flow curve measurements are the tube (Heywood & Richardson, 1978) and the coaxial cylinder (Hanks, 1988).

3.2. Specification of Shear Rate Range

The first step is to identify the appropriate shear rate (and/or shear stress) range over which the slurry flow curve should be evaluated. Typical wall shear rates for full-scale pipe flow are in the range $10–200\,s^{-1}$, while much lower shear rates ($10^{-3}\,s^{-1}$ and lower) are relevant when a slurry is to be checked or formulated for suspending the coarsest particles in the slurry during storage or pipeflow.

3.2.1. Viscometric Tests for Prediction of Pipe Flow Behaviour

Frictional pressure drop prediction in either the laminar or the turbulent flow regimes is generally based on the laminar flow curve measured using one or more suitable viscometers. Both shear stress (see Fig. 4.7) and shear rate values range from zero at the pipe axis to a maximum value at the pipe wall. From a force balance on a cylindrical slug of slurry,

$$\tau_w \pi D L = \Delta P \pi D^2/4 \tag{4.1}$$

which, when rearranged, gives the wall shear stress

$$\tau_w = D\Delta P/(4L) \tag{4.2}$$

The nominal wall shear rate is given by

$$\dot{\gamma}_{nom} = 8V/D \tag{4.3}$$

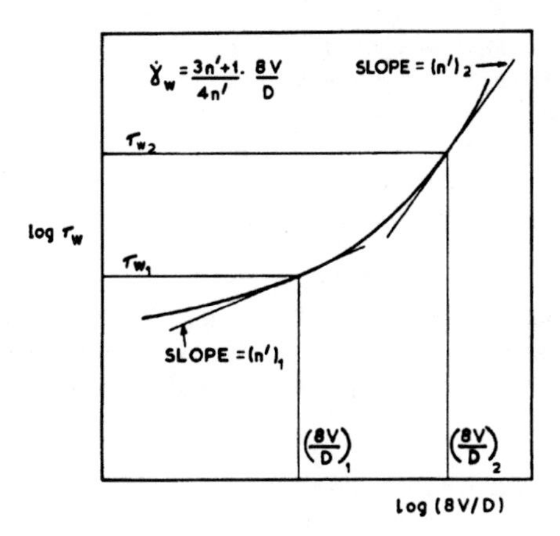

Fig. 4.8. Determination of true wall shear rate for pipe flow.

(this is the actual wall shear rate for a Newtonian slurry only) where the mean slurry (superficial) flow velocity

$$V = 4Q/(\pi D^2) \tag{4.4}$$

The true wall shear rate for a non-Newtonian slurry is given by the Rabinowitsch–Mooney equation:

$$\dot{\gamma}_w = 8V/D\,((1 + 3n')/(4n')) \tag{4.5}$$

where

$$n' = \frac{\mathrm{d}\,(\ln \tau_w)}{\mathrm{d}\,(\ln (8V/D))} \tag{4.6}$$

The true wall shear rate can therefore be up to twice the nominal wall shear rate for a highly shear-thinning slurry, e.g. $n' = 0{\cdot}3$, but the nominal value is usually a sufficient rule of thumb. In general, estimates of n' are required at each $8V/D$ as shown in Fig. 4.8, although if the slurry can be described well by the power law model, n' is a constant and is equal to the power law index, n (see Chapter 7).

More than one viscometer may be needed to cover the range of $8V/D$ down to at least $0{\cdot}1 \times 8V/D$ so that the relevant part of the slurry's flow curve is measured. Ideally, the flow curve should be measured over a shear rate range of zero to the maximum value given approximately by $8V/D$. However, in practice 0·1 of the maximum shear rate is a reasonable working lower limit for viscometric measurements.

If the viscometric data are to be used for estimating frictional pressure loss in pipe flow and a number of standard pipe sizes are to be included in the predictions, the lower limit should be $0{\cdot}1\ (8V/D_l)$ and the upper limit $8V/D_s$, where D_l and D_s are the largest and smallest pipe diameters under consideration, respectively.

3.2.2. Stabilised Slurry Formulation

A *non-settling* slurry can be formulated by incorporating a fine particle carrier medium for coarse particle hydraulic conveying. The carrier medium needs to be formulated so that it can support the coarse particles over considerable pipeline distances. The usual approach has often been to try to ensure that not only does the slurry exhibit a yield stress but that it is sufficiently large to support the coarsest particles.

Reliable values for yield stress of a slurry are notoriously difficult to measure and are usually dependent on the method of measurement (Cheng, 1986). Actual yield stress values are seldom related to the yield stress curve-fitting parameters appearing in the Bingham plastic, generalised Bingham plastic and Casson models given in Chapter 7.

Zero shear viscosity. An alternative approach (Cheng, 1989) to the formulation of carrier media for coarse particle support involves the measurement of the zero shear viscosity. Most slurries, even highly non-Newtonian slurries, will exhibit Newtonian behaviour at sufficiently low shear rates. These may be as low as 10^{-4}–$10^{-6}\,s^{-1}$. The zero shear viscosity is defined as this Newtonian viscosity.

The assumption is made that no slurry can support coarse particles indefinitely —there will always be some settling. The acceptable maximum allowable settling rate of coarse particles during the residence time of the slurry in the pipeline must be estimated. This will then specify the shear rate at which the zero shear viscosity must be measured because the maximum shear rate on a spherical particle (at the equator) when falling at its terminal velocity, u_∞, is given by

$$\dot{\gamma}_{pmax} = 3u_\infty/d \tag{4.7}$$

where d = particle diameter [L].

An average value is sometimes quoted as

$$|\dot{\gamma}_p| = 0{\cdot}6\,u_\infty/d \tag{4.8}$$

It can be seen that either of these shear rates will be very small for low values of u_∞, say of the order of mm h^{-1}. A different viscometer to that used for the relevant pipe velocity shear rate range will probably be needed, although some recent designs may be able to cover the full range of typically about 10^{-6}–$200\,s^{-1}$. Normally, however, two shear rate windows of approximately 10^{-4}–$10^{-6}\,s^{-1}$ and 1–$200\,s^{-1}$ need to be covered.

If the zero shear viscosity, η_0, is used in the Stoke's equation

$$u_\infty = \frac{1}{18}\frac{(\rho_m - \rho_p)\,gd^2}{\eta_0} \tag{4.9}$$

it is possible to determine whether the estimated terminal settling velocity from eqn (4.9) is less than the maximum required. If it is not, the carrier medium will require reformulating, perhaps by increasing the fine solids loading to raise the viscosity, or alternatively by changing the carrier pH while maintaining the fine particle concentration.

3.3. Tube or Pipeline Viscometer

A tube or pipeline viscometer measures the volumetric flow rate/pressure drop relationship and is generally preferred for characterisation of slurry flow properties if the end-use for the data is pipeline design. This is because any wall-slip effects present can be scaled-up to larger pipe sizes. Even when wall-slip effects are not identified, the tube geometry is preferred when high slurry concentrations are characterised because the flow curves in these cases have been found to be dependent on the geometry.

Laboratory tube viscometers are generally once-through batch devices whereas pilot-scale pipeline viscometers allow recirculation of the slurry through a length of pipe across which the frictional pressure loss is measured. Thus, in the former case, end-effect errors normally occur and usually require a correction to be made (see section 3.3.2). In the latter case, pressure tappings are placed at two or more positions on the pipe, together with the associated differential pressure transducer, and no end-effect occurs.

Laboratory tube viscometers may be mounted horizontally or vertically and typically range in diameter from a few millimetres up to 30–50 mm, as shown in Fig. 4.9. The type A tube viscometer forces the slurry through a tube at a constant flow rate using a piston or ram as a pump and the resultant pressure drop is measured. The type B tube viscometer creates a near-constant pressure drop through a combination of applied compressed air pressure and slurry head, and the resultant volumetric flow rate is measured. Type C relies on the head of the slurry alone. It is therefore not possible to keep the pressure drop constant as the slurry level in the reservoir falls but for wide diameter reservoirs and small bore viscometer tubes, a near constant pressure drop can sometimes be assumed.

A balanced beam tube viscometer has been developed (Lazarus & Sive, 1984; Lazarus & Slatter, 1986) and used on industrial slurries such as fly ash. The viscometer (Fig. 4.10) consists of two pressure vessels mounted at each end of a steel joist which is supported by a knife-edge fulcrum. A precision load cell is located at one extremity under the centre of gravity of the vessel. Slurry is forced back and forth, at different flow rates, from one vessel to the other by air pressure. When the load cell reading is zero there is no resulting slurry head and the pressure differential is due to the applied air pressure only. At this point, the slope of the load versus time curve (giving the instantaneous volume flow rate) and the pressure differential are logged and processed by computer. This viscometer can therefore expedite slurry test work and reduce evaporation problems.

3.3.1. Calculation of Shear Stress and Shear Rate

Shear stress and shear rate are calculated at the wall of the tube using the expressions given in section 3.2.1 for determining the relevant shear rate ranges. The wall shear stress is calculated from the measured pressure differential using eqn (4.2) while wall shear rate is calculated from eqn (4.5).

3.3.2. Errors in Tube Viscometry

A number of checks must be made to ensure the validity of the viscometric data.

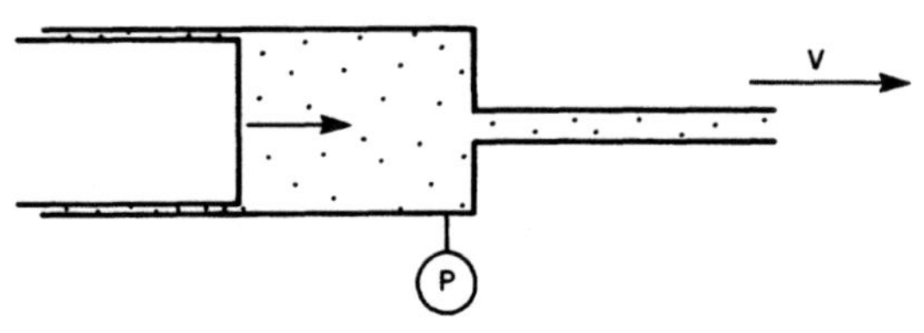

Controlled Flowrate. Type A

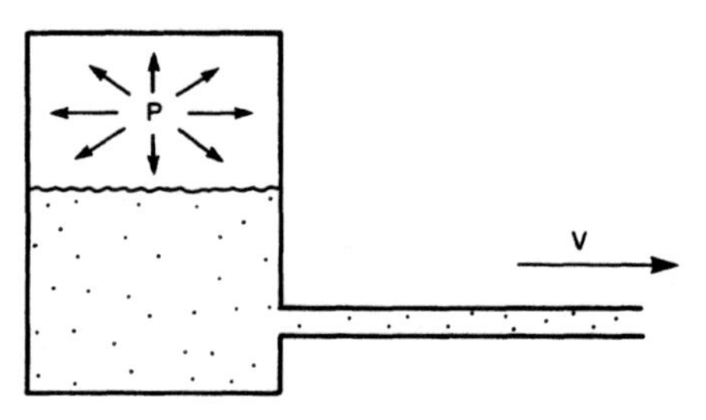

Controlled Pressure. Type B

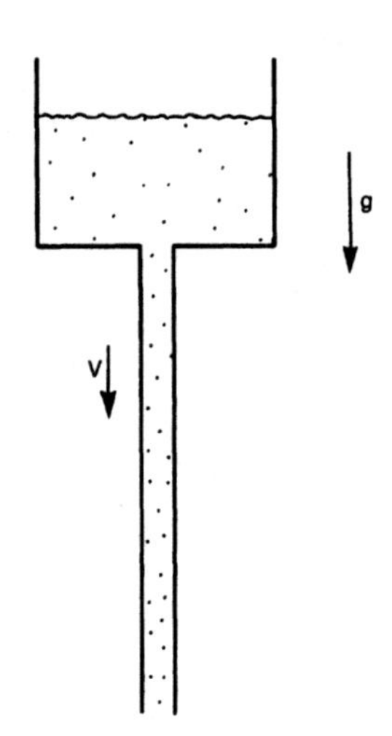

Gravity Driven. Type C

Fig. 4.9. Types of tube viscometer.

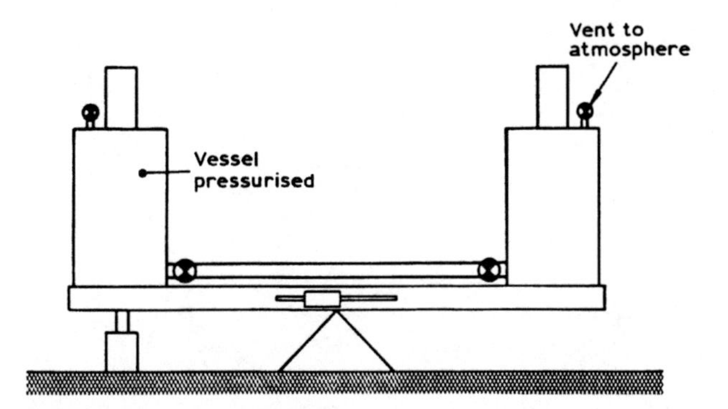

Fig. 4.10. Balance beam tube viscometer. (Reproduced from Lazarus & Sive, 1984.)

When these checks indicate that the data may be subjected to significant error sources, experimental procedures allow the original data to be corrected. First, a check is made to ensure the flow is laminar. All turbulent flow data must be rejected as they can in no way be used to predict the laminar flow curve of the slurry. Secondly, the data should be examined to see if they are affected by the two main error sources: end-effect and wall-slip.

Checking for laminar/turbulent flow. The laminar limit for Newtonian slurries is given by the well-known Reynolds number criterion:

$$\mathrm{Re} = \frac{\rho_\mathrm{m} VD}{\eta_\mathrm{m}} < 2100 \tag{4.10}$$

For a power law non-Newtonian slurry, the limit is given by (Ryan & Johnson, 1959)

$$\mathrm{Re}' = \frac{\rho_\mathrm{m}\, VD}{K\left(\dfrac{8V}{D}\right)^{n-1}\left(\dfrac{3n+1}{4n}\right)^{n}} < \frac{404}{n}\,(n+2)^{(n+2)/(n+1)}\left(\frac{4n}{3n+1}\right)^{2} \tag{4.11}$$

Equation (4.11) can be rearranged to give the critical wall shear stress in the viscometer for laminar flow breakdown:

$$\tau_\mathrm{wc} = \frac{\rho_\mathrm{m} D^2}{8}\left[\frac{n}{404}\,\frac{\left(\dfrac{3n+1}{4n}\right)^{2}}{(n+2)^{((n+2)/(n+1))}}\right]\left(\frac{8V}{D}\right)^{2} \tag{4.12}$$

Equation (4.12) can be plotted as log (τ_w) against log ($8V/D$) to give the laminar limit line; data above the line correspond to laminar flow, while data lying on it or below it must be rejected. Figure 4.11 gives an example of data obtained using a single size. The assumption has been made that the τ_w versus $8V/D$ data have been corrected for end-effect and wall-slip effect.

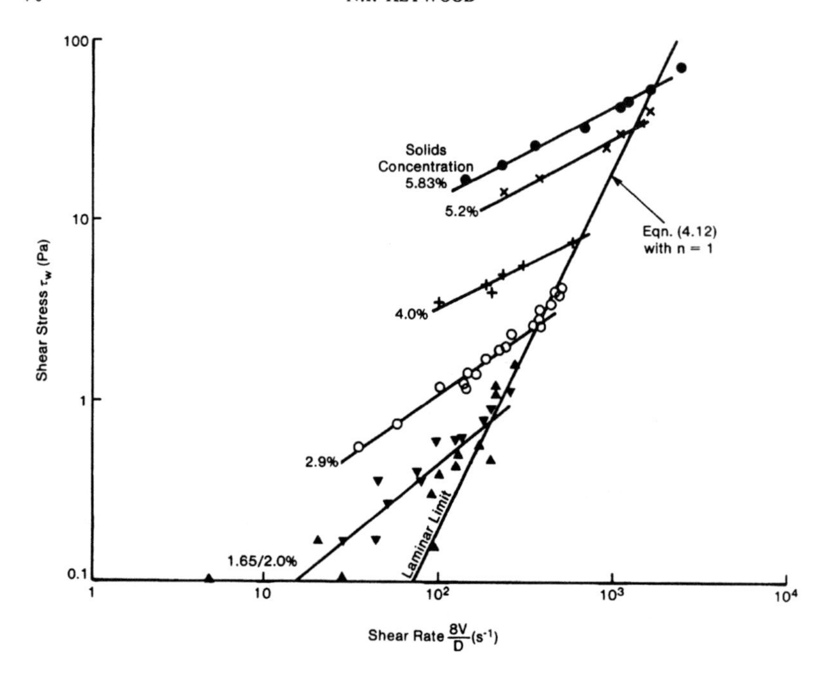

Fig. 4.11. Onset of turbulent flow in tubes of different diameter.

End-effect correction procedure. This error source can be minimised by using long tubes and/or large tube length-to-diameter ratios. It arises in the form of additional pressure losses at the entrance to and exit from the tube. These losses result from inertia effects and losses in the kinetic energy of the slurry brought about by rearrangement of the slurry streamlines on entry and exit.

To correct for the error, the volumetric flow rate/pressure drop relationship is established for a number of tubes of the same diameter but different length. The straight lines produced for each tube diameter when total pressure drop is plotted against tube length with flow rate as a parameter (see Fig. 4.12) are then extrapolated to zero tube length and the intercept for each flow rate is the end-effect pressure loss, P_e. The corrected pressure loss is then

$$P = P_t - P_e \tag{4.13}$$

From eqn (4.2) the corrected wall shear stress is

$$\tau_w = \frac{D\,(P_t - P_e)}{4\,L} \tag{4.14}$$

Identification and characterisation of any wall-slip effects. Wall shear stress versus nominal wall shear rate data should superpose when different tube diameters are used, provided that flow is in the laminar regime with all tube diameters.

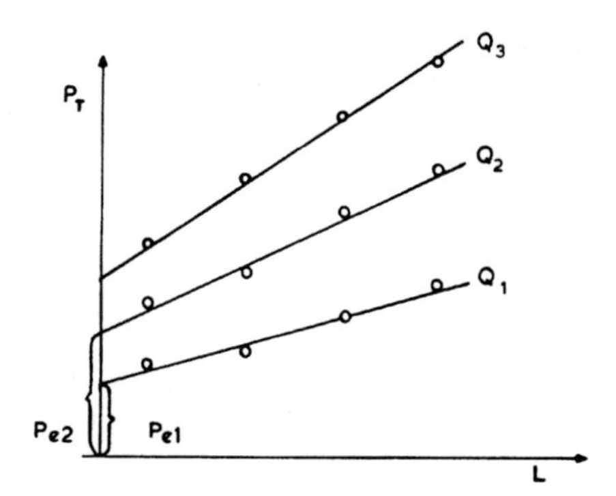

Fig. 4.12. Determination of end-effect in capillary tube viscometry.

If they do not superpose, wall-slip is present and can be characterised through a slip velocity (Jastrzebski, 1967) in order that the data are corrected and the slurry viscosity is not underestimated for frictional pressure loss prediction purposes. The wall slip velocity, V_s, is defined such that the flow rate for laminar flow when no slip is present is enhanced by an amount proportional to this velocity.

If wall-slip has been detected in viscometric tests, values for V_s can be estimated (Jastrzebski, 1967) as follows.

PROCEDURE

(1) Using plots of τ_w versus $8V/D$ obtained for different tube diameters, obtain by interpolation between data points values of $8V/D$ at fixed values of τ_w for the different tube sizes.
(2) Plot $8V/D$ against $1/D$ for the various values of τ_w. Example data are shown in Fig. 4.13.
(3) Draw a straight line through each set of constant shear stress data and determine the slope of each line. This slope is $8V_s$. The resulting V_s data can often be correlated by a *slip* coefficient, b, equal to V_s/τ_w. This coefficient appears to vary inversely with tube diameter, D, raised to some power which normally lies in the range 0–3.
(4) With V_s correlated as a function of both τ_w and D, predictions of pressure loss/flow rate in the laminar regime may be adjusted (see Chapter 7) for a slurry having any laminar flow property. Boger *et al.* (1979) have used the method for the correction of capillary tube data obtained for red mud waste generated in bauxite processing.

3.4. Coaxial Cylinder Viscometer

This geometry (Fig. 4.14) comprises an outer cup or cylinder which is filled with the slurry sample. An inner cylinder or *bob* is placed in the slurry such that the two cylinders are coaxial. Normally the inner cylinder is rotated at fixed speeds while the torque generated on the inner cylinder surface is measured. With more modern

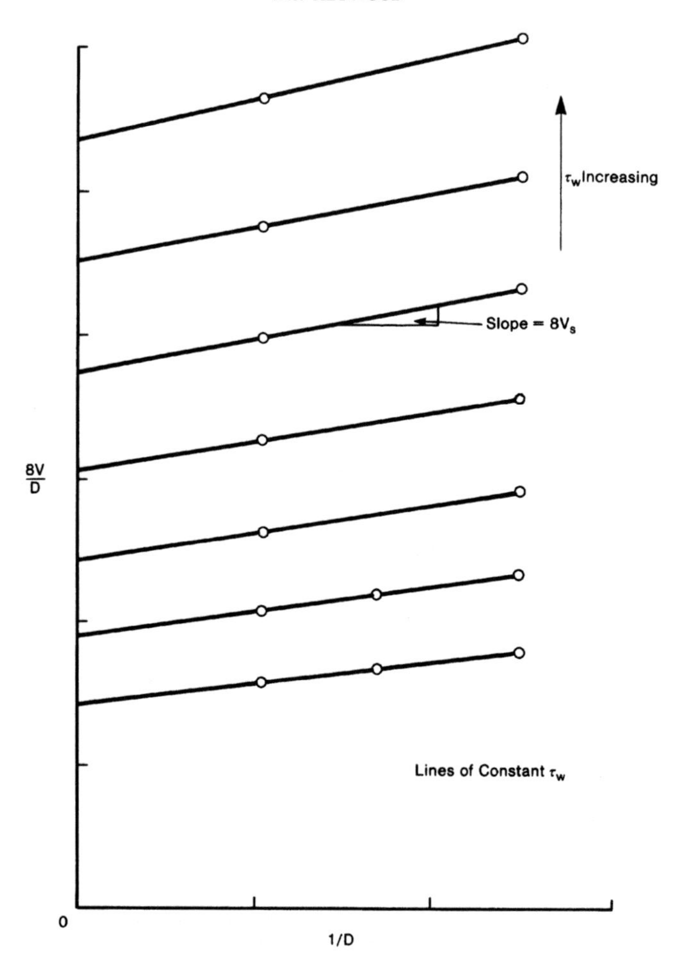

Fig. 4.13. Determination of wall-slip in a tube/pipe viscometer.

designs, often referred to as controlled stress rheometers, the torque level to rotate the cylinder is specified and the resultant rotational speed of the inner cylinder measured.

A coaxial cylinder viscometer is easier to set up than a tube viscometer and data collection is more rapid. Commercial viscometers now come with software for programming and controlling experiments as well as logging and interpreting data. Large particles can be tolerated, especially when the instrument is used in the *infinite sea* mode, i.e. when a bob or the inner cylinder is placed in a large expanse of slurry (Brown, 1988). It is the only rotational geometry

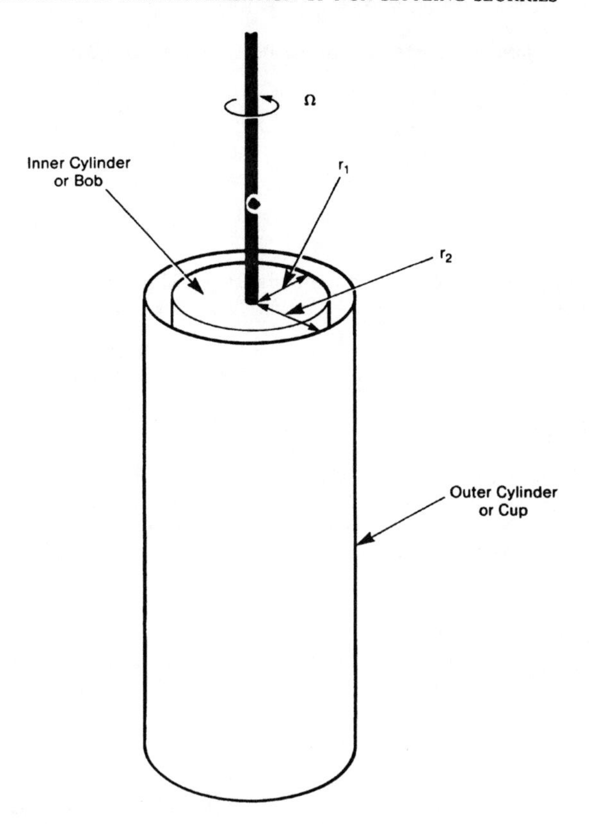

Fig. 4.14. The coaxial cylinder viscometric geometry.

that is appropriate for most slurry types, but in common with all rotational geometries, it is unsuitable for high viscosity slurries or pastes which may exhibit a yield stress (see section 3.7) as the inner cylinder tends to cut a hole in the slurry or paste sample and the tube geometry is then preferable. However, although the presence of wall-slip can be identified with the coaxial cylinder geometry, its quantification and application to pipe flows is limited because the two geometries are so dissimilar.

3.4.1. Calculation of Shear Stress and Shear Rate

The measured (or imposed) torque, T, on the inner cylinder surface can be related to the wall shear stress by

$$\tau_1 = \frac{T}{2\pi r_1^2 L_c} \tag{4.15}$$

where r_1 is the radius of the inner cylinder [L], and L_c is its length [L].

The corresponding shear rate on the inner cylinder surface is

$$\dot{\gamma}_1 = \frac{\Omega}{k}\left[1 + \frac{k}{a'} + \frac{1}{3}\left(\frac{k}{a'}\right)^2 (1 - a'') + \ldots\ldots\right] \tag{4.16}$$

where

$$k = \ln\left(\frac{r_2}{r_1}\right)$$

in which r_2 is the radius of the outer cylinder [L], Ω is the rotational speed of the inner cylinder in rad s^{-1},

$$a' = \frac{d\,(\ln T)}{d\,(\ln \Omega)} \quad \text{and } a'' = \frac{d\,(\ln a')}{d\,(\ln \Omega)} \tag{4.17}$$

PROCEDURE

To obtain the shear rate values at different Ω levels

(1) plot ln T versus ln Ω to obtain a' at different Ω values,
(2) plot ln a' as a function of ln Ω to obtain a'', and
(3) insert a' and a'' into eqn (4.16) for a given cylinder speed.

Use of 'infinite sea' approximation. When the inner cylinder is immersed in a large quantity of slurry contained in a vessel such that when the cylinder is rotated at its highest speed, the slurry adjacent to the vessel wall remains unsheared, the equation for shear rate on the cylinder surface may be simplified to

$$\dot{\gamma}_1 = \frac{2\Omega}{a'} \tag{4.18}$$

The advantage of this operation is that slurries containing large particles can often be successfully tested provided that the ratio of the largest particle size to the distance between the cylinder surface and the vessel wall is no more than about 1/6. Brown (1988) reports the use of this geometry for stabilised coal slurries with a top-size of 50 mm.

3.4.2. Errors in Coaxial Cylinder Viscometry

Checking for laminar flow breakdown. The use of low concentration slurries and/or high rotational cylinder speeds may create secondary flows which then invalidate the equations given above for shear rate at the cylinder's surface. For experimental data to be unaffected by secondary flow, the Taylor number must not exceed a critical value which is a function of the cylinder radii ratio (Sparrow *et al.*, 1964). The critical shear stress on the cylinder surface can be calculated for any speed and if any shear stress is below this critical value, the data set should be rejected. Critical Taylor numbers are available (Sparrow *et al.*, 1964) as a function of the cylinder radius ratio for the case of an inner cylinder rotating in a stationary outer cylinder. This is the most common situation for many commercially available

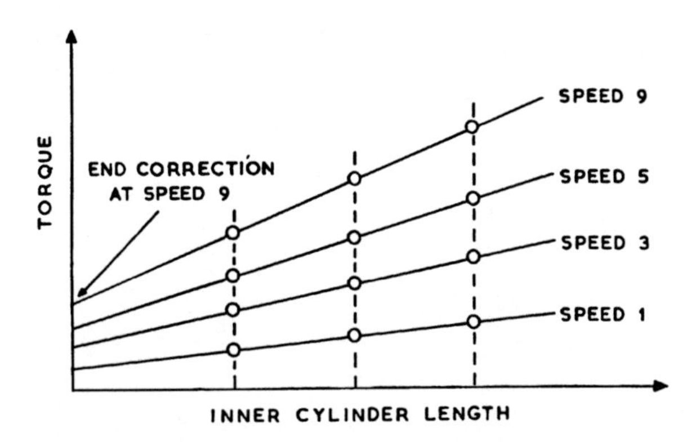

Fig. 4.15. End-effect determination in coaxial cylinder viscometry.

coaxial cylinder viscometers, but for rotation of the outer cylinder with a stationary inner cylinder (e.g. the Fann viscometer) critical values of the Taylor number are about 10 times larger.

End-effect correction. The greatest source of overprediction of viscosity arises from the additional torque contribution which arises from the two ends of the inner cylinder. Errors of 30% or more are possible. This end-effect can be minimised by using a long cylinder compared with its diameter or by trapping an air bubble in the slurry in the lower end of the cylinder if it has a concave surface. Alternatively, end-effect can be predicted theoretically or evaluated experimentally so that the data can be corrected. The latter method involves the use of the multicylinder method where several cylinders of the same diameter but different lengths are used and the results compared by plotting measured torque against cylinder length at different fixed speeds as shown in Fig. 4.15. The end-effect contribution at each speed, T_e, is estimated by extrapolation of the line to zero cylinder length and this contribution is subtracted from the original torque measurement to calculate shear stress from eqn (4.15).

The end-effect for a Newtonian slurry can be estimated from

$$\frac{T'}{T_e} = 1 + \frac{\pi L_c}{4A' r_1^3 (r_1^{-2} - r_2^{-2})} \tag{4.19}$$

where A' is a constant depending on the shape of the ends of the cylinder. For example, for the conical ends of the Contraves Rheomat Model 30 cylinders, A' is 2π and for an inner cylinder with flat ends (see Fig. 4.16), A' is $\sqrt{2\pi}$.

In the case of a power law slurry,

$$\frac{T'}{T_e} = 1 + \frac{\pi L_c}{4 A'(n) r_1} \left[\frac{2/n}{(1 - (r_1^2/r_2^2)^{2/n}}\right]^n \tag{4.20}$$

where the parameter A' is now a function of the power law index and the shape of the cylinders. Figure 4.16 shows how A' varies with power law index. For

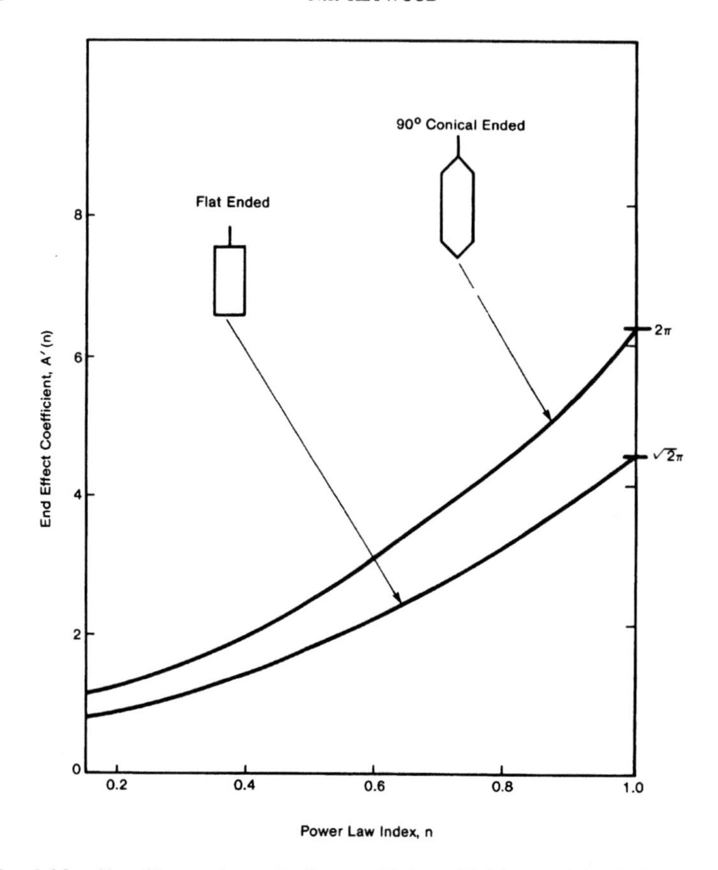

Fig. 4.16. Non-Newtonian end-effect coefficient $A'(n)$ in coaxial cylinder viscometry (Source: Warren Spring Laboratory data.)

other shapes of cylinder ends, values of A' are not generally available for either Newtonian or non-Newtonian slurries and hence must be estimated through experiment.

The corrected torque values (T) for use in eqn (4.15) are obtained by subtracting T_e from T'.

Wall-slip correction procedure. Wall-slip can be identified by using inner cylinders of different radius but the same length. If data from different cylinder sizes do not superpose then wall-slip is present. There are four procedures for wall-slip correction of the data (Cheng & Parker, 1976). If slip is present it is advisable to undertake experiments using a tube viscometer with tubes of different diameter to assess its significance for frictional pressure loss prediction in laminar flow. Slip velocity data obtained using the coaxial cylinder geometry should, in general, not be applied to pipe flow prediction as slip velocity correlations for one geometry are unlikely to be applicable to another.

3.5. Metzner–Otto Method for Slowly Settling Slurries

The concept of viscosity for rapidly settling slurries has little meaning and represents a grey area, but when the slurry can be considered as either settling or non-settling, depending on the residence time of the slurry in process equipment or a pipeline, it can be useful to estimate viscosity under certain conditions. The method can also be useful for slowly settling slurries. Measurements using either a tube or rotational viscometer are inappropriate because the degree of settling is normally significant during the course of the viscosity measurement in these situations.

In the Metzner–Otto method, an agitator can serve a dual purpose in that the torque generated by its rotation in the slurry sample can be measured at given rotational speeds and the flow generated maintains the solids in suspension. Such an approach has been derived from a method originally developed to assess the shear rate in agitated vessels containing non-Newtonian fluids (Metzner & Otto, 1957). Any agitator design can be used and attached to a commercially available rotational viscometer or to a vertical mixing shaft which incorporates a torque transducer. Because the flow field during agitation is poorly defined and shear rates vary with both position and time in the slurry, averaged values of shear rate and shear stress are used based on torque measurements. The method has been applied successfully to a wide range of slurries including thixotropic slurries (Godfrey *et al.*, 1974). The method is very useful but rarely, if ever, outlined in detail in standard rheology text books. Full details are therefore given here.

Before an agitator can be used for the viscosity measurement of a slurry, the agitator must be calibrated. This requires the estimation of two calibration constants, K_L and k_s. A Newtonian liquid of known viscosity is used to estimate K_L. This calibration is expressed as a plot of power number versus Reynolds number, defined respectively as

$$P_0 = \frac{2\pi T}{\rho_m N^2 D_a^5} \tag{4.21}$$

$$\mathrm{Re}_a = \frac{\rho_m N D_a^2}{\eta_m} \tag{4.22}$$

In the laminar regime ($\mathrm{Re}_a < 10$), the product of P_0 and Re_a is a constant K_L:

$$P_0 \, \mathrm{Re}_a = K_L \tag{4.23}$$

The non-Newtonian calibration gives the constant k_s relating torque-averaged shear rate to agitator speed:

$$\dot{\gamma}_a = k_s N \tag{4.24}$$

Three procedures are now outlined. The first two procedures are used to estimate the agitator calibration constants, K_L and k_s respectively, while the third procedure is used to measure the slurry flow curve.

PROCEDURE FOR ESTIMATION OF CALIBRATION CONSTANT, K_L

(1) Using a Newtonian liquid of known viscosity, measure agitator torque, T, for a range of agitator speeds, N.
(2) Calculate power number, P_0, and agitator Reynolds number, Re_a, from eqns 4.21 and 4.22 respectively.
(3) Plot P_0 versus Re_a on double logarithmic paper and either draw straight line by eye through all data for $Re_a < 10$ or use linear regression to obtain K_L according to eqn. (4.23).

PROCEDURE FOR ESTIMATION OF CALIBRATION CONSTANT, k_s

(1) Using a non-Newtonian fluid with known flow curve, obtain torque data, T, for a range of agitator speed, N.
(2) Calculate power number values, P_0, from eqn (4.21).
(3) Calculate agitator Reynolds numbers, Re_a, from eqn (4.23) using K_L value previously obtained with a Newtonian calibration fluid.
(4) Calculate viscosity values η_a, from Re_a values.
(5) Using the flow curve of the calibration non-Newtonian fluid, calculate $\dot{\gamma}_a$ from the η_a values using $\dot{\gamma}_a = \eta_a/\tau_a$.
(6) Plot $\dot{\gamma}_a$ against agitator speed, N, using linear axes.
(7) Estimate k_s by drawing a straight line through the origin on the $\dot{\gamma}_a$ versus N plot.

Once k_s has been determined for one agitator geometry, it has been found to remain relatively insensitive to the type of non-Newtonian calibration fluid. Values of k_s for paddles, six-blade turbines, curve-blade turbines and propellers typically lie in the range 7–13, when N is expressed in rev s^{-1}.

PROCEDURE TO MEASURE THE SLURRY FLOW CURVE

(1) Measure torque, T, for a range of agitator speeds, N, with the slurry.
(2) For each N value, estimate the slurry viscosity from

$$\eta'_m = \frac{2\pi}{K_L}\frac{T}{N\,D_a^3} \qquad (4.25)$$

(3) Calculate average shear rate values, $\dot{\gamma}_a$, from eqn (4.24).
(4) Calculate average shear stress values using

$$\tau_a = \eta'_m\,\dot{\gamma}_a = \left(\frac{2\,\pi\,k_s}{K_L}\right)\left(\frac{T}{D_a^3}\right) \qquad (4.26)$$

(5) Plot average shear stress, τ_a, against average shear rate, $\dot{\gamma}_a$, to obtain the slurry flow curve.

3.6. Fibrous Slurry Flow Curve

Fibrous materials, such as paper pulp, fermentation broths, various sewage sludge

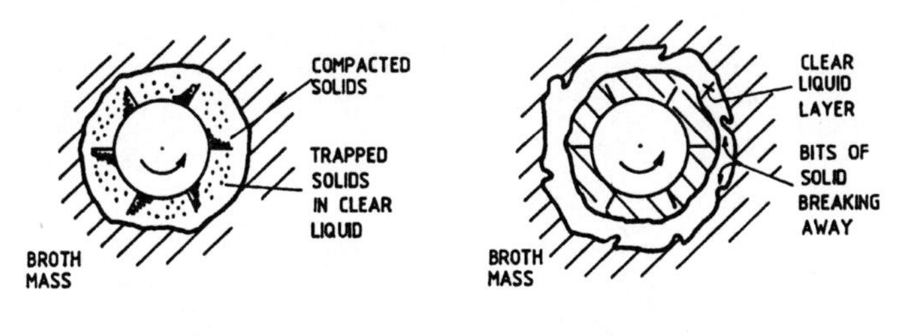

Fig. 4.17. Formation of clear liquid layer during agitation of a fibrous fermentation broth.

types (Frost & Ovens, 1982; Cheng & Heywood, 1984) and a variety of foodstuffs can pose severe problems in viscosity measurement because of their physical form (see section 1.4). As with slurries of hard particles at high concentrations, the flow curves of fibrous suspensions are sensitive to the characteristic gap used in a viscometer, such as the diameter in a tube/pipe viscometer or the cylinder separation in a coaxial cylinder viscometer. This effect is sometimes interpreted as wall-slip but the usual correction procedures often do not work. This is partly because the slip can occur at a position some distance from an agitator (if the Metzner–Otto method is used, see section 3.5) as shown in Fig. 4.17 or from the cylinder surface if a coaxial cylinder viscometer is used. The shaded area in Fig. 4.17 indicates the unsheared regions. Sometimes the only practical geometry to use is a comparatively large tube (12 mm or greater).

Figure 4.18 shows the results obtained for a digested sewage sludge at several solids concentrations. Various viscometer geometries were used: Brookfield discoidal spindles, Contraves coaxial cylinder systems using both rough and smooth inner cylinder surfaces, twisted-blade paddles, and 12·5 and 19 mm bore tube viscometers. Because the test sample does not retain a uniform solids concentration when subjected to shear, the usual equations for calculating non-Newtonian shear rate do not apply, but in the absence of an alternative procedure they were still used. There is very little agreement between data from different viscometer types so it is important to use at least two viscometer geometries and to construct an upper bound to the flow curve data for a conservative assessment of viscosity levels.

3.7. Thixotropic Behaviour

Time-dependent slurry properties are both difficult to detect and evaluate in a tube or pipeline viscometer. This is unfortunate because this is the preferred geometry for pipeline design. Methods using the coaxial cylinder geometry or the cone-and-plate geometry are much more advanced (Cheng & Evans, 1965; Nguyen & Boger, 1985) but resulting data cannot readily be applied to the prediction of pipe flow behaviour.

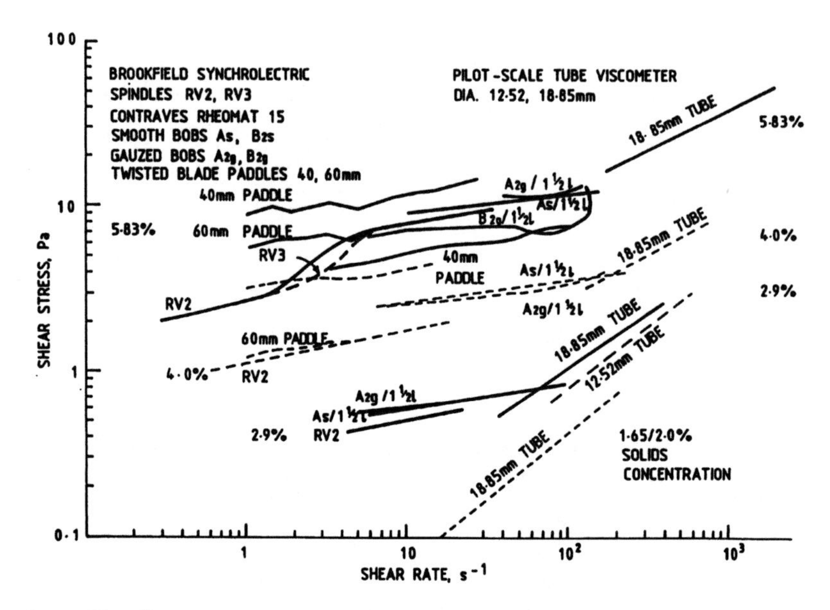

Fig. 4.18. Flow curves obtained using several viscometer geometries for digested sewage sludge at several solids concentrations. (Source: Warren Spring Laboratory data.)

Initial detection of thixotropic (or other time dependent) flow properties can be made using slurry which has been left undisturbed for a few hours or overnight, and loading a sample into a suitable rotational viscometer. Alternatively, the slurry sample can be left undisturbed in the viscometer before commencing test work. Repeated cycling by applying a continuously increasing shear rate (or, alternatively, shear stress) to the sample up to a maximum value, followed by a continuous decrease to zero, will result in positive (i.e. clockwise) hysteresis loops when shear stress (or the original torque) is plotted against shear rate (or rotational speed). Large hysteresis loops obtained using relatively low rates of increase in shear rate (i.e. instrumental inertia effects are minimised) are indicative of significant time-dependent flow property.

Continuous recirculation of thixotropic slurry in a pilot plant rig will reduce the structure in the slurry to some steady-state value and hence steady values of frictional pressure gradient as a function of flow rate. These values will provide equilibrium wall shear stress/wall shear rate data which can then be used to generate an equilibrium flow curve. Each time a new fixed flow rate is selected, time is required for a steady (equilibrium) pressure drop to develop. Because the rate of structural breakdown is normally much greater than the rate of build-up (see section 2.1.2), it is advisable to start at the lowest flow rate of interest and progressively increase the flow rate, recording the steady pressure loss each time.

This approach is valid provided the pump employed does not impart a significant degree of structural breakdown in the slurry. If a low shear recirculating method is not possible, it may be more realistic to measure the equilibrium flow curve using a rotational viscometer.

A suitable model can be fitted to the equilibrium flow curve and frictional pressure loss estimations made. This approach, however, does not provide any information on how the pressure gradient will progressively decrease (or increase) down the length of pipe as the structure moves towards an equilibrium value on pipeline start-up, or if slurry of a different structure is being continuously fed from a holding tank to the pump suction.

One of the areas of greatest significance of thixotropic property is in the start-up of a slurry pipeline. Because the slurry structure has been broken down through shear in the pump and in the pipeline to some steady-state level, the equilibrium pressure gradient during steady flow can often be significantly less than the start-up pressure gradient for a slurry which has been left undisturbed in the pipe for hours or even days. This start-up pressure requirement is usually estimated by measuring how any yield stress in the slurry is likely to develop (usually increase) with time under quiescent conditions. The yield stress normally arises as a result of distinct structure in the slurry and therefore is linked to thixotropic properties which may or may not be significant in any particular application.

Measurement of yield stress. The significance of the concept of a yield stress has been discussed in section 2.1.2. Estimates of the *dynamic* yield stress may be made by extrapolation of the flow curve to zero shear rate. Direct measurements of the *static* yield stress can be made using a number of techniques:

- using step functions of shear stress in a controlled stress rheometer;
- using step functions of applied pressure of increasing amplitude in a tube viscometer (sometimes referred to as a gun rheometer, see Fig. 4.19); and
- using a vane tester (Want *et al.*, 1982; Nguyen & Boger, 1983; Dzuy & Boger, 1985; Keating & Hannant, 1989).

Any of these methods can also be used to determine the development of yield stress with time (important, for instance, when a pipeline carrying thixotropic slurry is shutdown) by testing a number of slurry samples of different ages. However, all three methods listed above require judgement from the experimenter. The point at which a sample first starts to flow under a progressively increasing stress and sustains that flow needs to be judged and will sometimes depend upon the time-scale of the measurement. Hence, yield stress can be thought of as a time-dependent property (Cheng, 1986) and the yield point selected will be influenced by its end-use.

4. ESTIMATION OF FLOW CURVE PARAMETERS BY REGRESSION ANALYSIS

The measured flow curve data may be amenable to the construction of a single

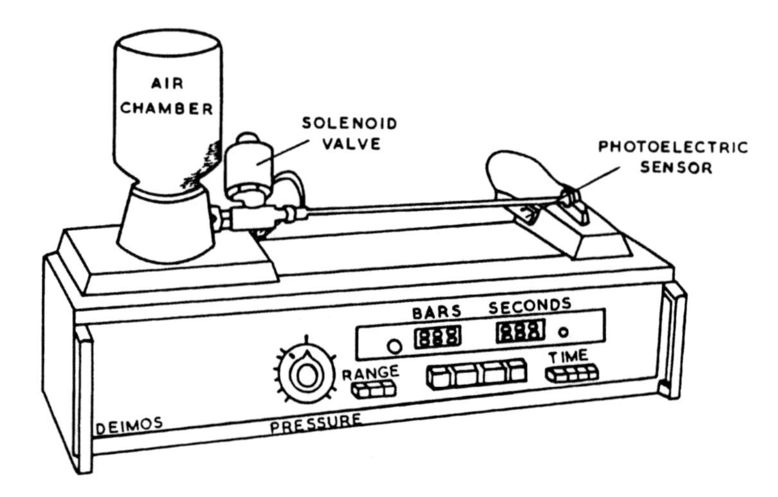

Fig. 4.19. Design features of the gun rheometer for direct yield stress measurement.

curve through them. Sometimes, however, there may be considerable scatter in the data. In this situation, it may be more appropriate to construct at least two curves: a *mean* curve constructed using all the data in a regression analysis and an *upper bound* curve. The latter curve is initially drawn by eye but is described using flow model parameters obtained through regression by selecting a number of $(\tau,\dot{\gamma})$ data sets from the drawn curve. The upper bound flow curve would normally represent the worst case for pipeline design and would lead to a conservative estimate of pump discharge pressure requirement. The curve would normally be able to take into account possible variations in slurry parameters such as solids concentration, particle size distribution, particle shape and pH.

Further factors can cause difficulties in attempting to draw a single flow curve through the data. These factors include the use of two or more different viscometric geometries (this can give differing degrees of phase separation during shear), sample variability taken from the same main batch of slurry sample, and uncorrected errors associated with the use of any viscometric geometry.

It is frequently not immediately obvious from the collated data which of the main flow models (summarised in Chapter 7) should be selected for further design usage. The following is suggested:

PROCEDURE

(1) Plot all the data using linear axes and, separately, on double logarithmic axes.
(2) If there is considerable scatter in the data, decide by eye or using simple linear regression, whether a straight line through the linear, or, alternatively, through the log, plot gives the better representation. If a regression analysis is undertaken by computer, cal-

culate the correlation coefficient to assess which is the better fit. Similarly decide for the *upper bound* curve. It is more convenient if a straight line representation on either linear or log coordinates is used for both the *mean* curve and the *upper bound* curve. If one of these two alternatives is acceptable, the added sophistication of using the three-parameter generalised Bingham model (Herschel–Bulkley model) is probably not warranted.

(3) If neither approach appears satisfactory because there is significant curvature of the data on both types of plot then the following situations can arise.

   (a) If there is data curvature on the double log plot with the curvature convex to the shear rate axis, and if a linear plot of the data does not produce a straight line, then it is worth exploring if the generalised Bingham model will adequately describe the data.

   (b) If there is data curvature on the double log plot and the curvature is concave to the shear rate axis, then the use of the generalised Bingham model is inappropriate. However, a curve fit is often possible and would result in a negative estimate for the yield stress parameter. Although this negative value has no physical meaning, it is still possible under some circumstances to provide meaningful frictional pressure loss estimates in the laminar or turbulent flow regimes. These circumstances have yet to be defined so no guidelines can be given here.

Estimates need to be made for the parameters appearing in the common flow model equations. As the Herschel–Bulkley, or generalised Bingham plastic, model can be reduced to the Newtonian, power law or Bingham plastic models, a least squares regression analysis can first be performed on the data to obtain all three parameters appearing in the model. It may then be possible to simplify the model by setting the yield stress parameter to zero if the estimate is close to zero and/or setting the flow behaviour index to unity if its estimate is close to unity.

Two methods are commonly used when carrying out a regression analysis on viscometric data (Heywood & Cheng, 1984):

- a non-linear least-squares regression on unweighted data, and
- a non-linear least-squares regression on weighted data.

Standard linear regression software packages can be used in either case. Both methods will provide sets of model parameters which will usually give predictions of the viscometric data to within $\pm 2\%$ of the original data when a comparison is made within the original shear rate range. Outside this shear rate range, the agreement can be poor and extrapolation to shear rates much above the maximum experimental value should be avoided. Extrapolation to shear rates below the experimental lowest value is unavoidable when predicting pressure for laminar flow in a pipeline. As both regression methods are equally valid, it is important to be aware of shear stress predictions below the lowest experimental shear rate.

5. NOTATION

a' Local slope of log–log plot of T against Ω [1]
a'' Local slope of log–log plot of slope a' against Ω [1]
A Area of edge of slurry element over which shear force is applied $[L^2]$
A' Parameter in eqn (4.20) [1]
d Particle diameter [L]
D Tube or pipe diameter [L]
D_a Diameter of agitator [L]
D_l Largest pipe diameter under consideration [L]
D_s Smallest pipe diameter under consideration [L]
F Shear force applied to the edge of a slurry element $[M\ L\ T^{-2}]$
g Gravitational acceleration $[L\ T^{-2}]$
k $\ln (r_2/r_1)$ [1]
k_s Proportionality constant between torque-averaged shear rate and agitator speed in eqn (4.24) [1]
K_L Product of agitator Reynolds number and power number defined by eqn (4.23) [1]
L Length of tube [L]
L_c Length of cylindrical part of inner bob of coaxial cylinder viscometer [L]
L_e Width of a slurry element sheared by a force, F [L]
n Flow behaviour index in power law model [1]
n' Local slope of log–log plot of wall shear stress versus nominal wall shear rate for tube flow [1]
N Rotational speed of agitator (rev s^{-1}) $[T^{-1}]$
P Pressure loss in tube viscometer corrected for end-effects $[M\ L^{-1}\ T^{-2}]$
P_e Pressure loss due to end effects in a tube viscometer $[M\ L^{-1}\ T^{-2}]$
P_o Power number for agitator, defined by eqn (4.21) [1]
P_t Total pressure loss measured using a tube viscometer $[M\ L^{-1}\ T^{-2}]$
ΔP Pressure drop across a cylindrical slug of slurry in pipe flow $[M\ L^{-1}\ T^{-2}]$
Q Volume flow rate $[L^3\ T^{-1}]$
r_1 Radius of inner cylinder in coaxial cylinder viscometer [L]
r_2 Radius of outer cylinder in coaxial cylinder viscometer [L]
Re Newtonian Reynolds number for pipe flow of slurry [1]
Re_a Agitator Reynolds number defined by eqn (4.22) [1]
Re′ Power law model Reynolds number for pipe flow of slurry [1]
T Torque on rotating agitator or inner cylinder surface of coaxial cylinder viscometer corrected for end-effect $[M\ L^2\ T^{-2}]$
T_e Torque from end-effect in coaxial cylinder viscometer $[M\ L^2\ T^{-2}]$
T' Uncorrected torque on inner bob of coaxial cylinder viscometer $[M\ L^2\ T^{-2}]$
u_∞ Terminal settling velocity of coarsest particle in a slurry $[L\ T^{-1}]$
V Superficial tube or pipe velocity $[L\ T^{-1}]$
V_s Wall-slip velocity in tube or pipe flow $[L\ T^{-1}]$
V' Velocity of edge of slurry element due to imposed shear force $[L\ T^{-1}]$

$\dot{\gamma}$ Shear rate [T^{-1}]
$\dot{\gamma}_a$ Torque-averaged shear rate using an agitator [T^{-1}]
$\dot{\gamma}_i$ Shear rate on inner bob surface in coaxial cylinder viscometer [T^{-1}]
$\dot{\gamma}_{nom}$ Nominal wall shear rate in pipe flow [T^{-1}]
$|\dot{\gamma}_p|$ Average shear rate on spherical particle surface [T^{-1}]
$\dot{\gamma}_w$ Wall shear rate in pipe flow corrected for non-Newtonian flow [T^{-1}]
$\dot{\gamma}_{pmax}$ Maximum shear rate on spherical particle surface [T^{-1}]
η_a Viscosity obtained on calibration using the Metzner–Otto method [$M\ L^{-1}\ T^{-1}$]
η_m Newtonian slurry viscosity [$M\ L^{-1}\ T^{-1}$]
η_0 Zero shear viscosity [$M\ L^{-1}\ T^{-1}$]
η'_m Slurry viscosity calculated using Metzner–Otto method [$M\ L^{-1}\ T^{-1}$]
ρ_m Slurry density [$M\ L^{-3}$]
ρ_p Particle density [$M\ L^{-3}$]
τ Shear stress [$M\ L^{-1}\ T^{-2}$]
τ_a Torque-averaged shear stress on an agitator [$M\ L^{-1}\ T^{-2}$]
τ_i Shear stress on inner bob surface in coaxial cylinder viscometer [$M\ L^{-1}\ T^{-2}$]
τ_w Wall shear stress in tube or pipe flow [$M\ L^{-1}\ T^{-2}$]
τ_{wc} Critical wall shear stress for laminar flow breakdown in tube or pipe flow [$M\ L^{-1}\ T^{-2}$]
Ω Rotational speed of inner bob of coaxial cylinder viscometer (rad s^{-1}) [T^{-1}]

6. REFERENCES

Astarita, G. (1990). Letter to the editor: the engineering reality of the yield stress. *J. Rheol.*, **34,** 275–7.

Barnes, H.A. (1989). Shear-thickening (dilatancy) in suspensions of non-aggregating solid particles dispersed in Newtonian liquids. *J. Rheol.*, **33**(2), 329–66.

Barnes, H.A., Hutton, J.F. & Walters, K. (1989). *An Introduction to Rheology*. Elsevier Science Publishers, Amsterdam, The Netherlands.

Boersma, W.H., Laven, J. & Stein, H.N. (1990). Shear-thickening (dilatancy) in concentrated dispersion. *AIChE J.*, **36**(3), 321–32.

Boger, D.V., Sarmiento, G. & Uhlherr, P.H.T. (1979). Flow properties of mineral slurries: red mud. In *Proceedings of Chemeca '79, the 7th Australian Conference on Chemical Engineering*, Australia, 202–7.

Brown, N.P. (1988). Three scale-up techniques for stabilised coal–water slurries. In *Proc. HT 11*, paper F1, pp. 267–84.*

Cheng, D.C.-H. (1980). Viscosity-concentration equations and flow curves for suspensions. *Chem. Ind.*, 17 May, **10,** 403–6.

Cheng, D.C.-H. (1984). Further observations on the rheological behaviour of dense suspensions. *Powder Technol.*, **37,** 255–73.

Cheng, D.C.-H. (1986). Yield stress: a time-dependent property and how to measure it. *Rheol. Acta*, **25**(5), 542–54.

Cheng, D.C.-H. (1989). Rheological design for suspending solids. Report LR 711 (MPBM), Warren Spring Laboratory, Stevenage, UK.

Cheng, D.C.-H. & Evans, F. (1965). Phenomenological characterisation of the rheological behaviour of inelastic reversible thixotropic and antithixotropic fluids. *Brit. J. Appl. Phys.*, **16,** 1599–617.

Cheng, D.C.-H. & Heywood, N.I. (1984). Viscometric testing and head loss correlations for pipeline design for fibrous suspensions. Report LR 502 (MH), Warren Spring Laboratory, Stevenage, UK.

Cheng, D.C.-H. & Parker, B.R. (1976). The determination of wall-slip velocity in the coaxial cylinder viscometer. In *Proceedings of the 7th International Congress on Rheology*, ed. C. Klason & J. Kubat, Swedish Society of Rheology, Gothenburg, Sweden, pp. 518–19.

Cheng, D.C.-H., Kruszewski, A.P., Senior, J.R. & Roberts, T.A. (1990). The effect of particle size distribution on the rheology of an industrial suspension. *J. Mater. Sci.*, **25**, 353–73.

Collyer, A.A. & Clegg, D.W. (eds) (1988) *Rheological Measurement*. Elsevier Applied Science Publishers, Barking, UK.

Darby, R. (1984). Determination and utilisation of rheological properties for prediction of flow behaviour of pseudohomogeneous slurries. In *Proc. STA 9*, pp. 107–16.

Dzuy, N.O. & Boger, D. (1985). Direct yield stress measurement with the vane method. *J. Rheol.*, **29**(3), 335–47.

Farris, R.J. (1968). Prediction of the viscosity of multimodal suspensions from unimodal viscosity data. *Trans. Soc. Rheol.*, **12,** 281–301.

Frost, R.C. & Ovens, J.A. (1982). A method of estimating viscosity and designing pumping systems for thickened heterogeneous sludges. In *Proc. HT 8*, paper P1, pp. 485–502.*

Godfrey, J.C., Yuen, T.H. & Edwards, M.F. (1974). Mixing of thixotropic fluids. In *Proceedings of the 1st European Mixing Conference*, ed. N.G. Coles. BHRA Fluid Engineering, Cranfield, UK, paper C3.

Hanks, R.W. (1988). Rotational couette viscometry of yield-power-law slurries. In *Proc. STA 13*, pp. 777–86.*

Hanks, R.W. & Bowman, D.M. (1985). The rheology of fine limestone slurries. In *Proc. STA 10*. pp. 145–50.*

Hartnett, J.P. & Hu, R.Y.Z. (1989). The yield stress: an engineering reality. *J. Rheol.*, **33**(4), 671–9.

Heywood, N.I. (1985). Selecting a viscometer. *Chem. Engr.*, Jun. **415,** 16–23.

Heywood, N.I. & Cheng, D.C.-H. (1984). Comparison of methods for predicting head loss in turbulent pipe flow of non-Newtonian fluids. *Trans. Inst. Meas. Contr.*, **6**(1), 33–45.

Heywood, N.I. & Richardson, J.F. (1978). Rheological behaviour of flocculated and dispersed kaolin suspensions in pipe flow. *J. Rheol.*, **22**(6), 599–613.

Horsley, R.R. & Reizes, J.A. (1980). The effect of zeta potential on the head loss gradient for slurry pipelines with varying slurry concentrations. In *Proc. HT 7*, paper D3, pp. 163–72.*

Ishihara, T., Katsuki, H. & Kuno, H. (1987). An experimental study on effects of solid concentration and zeta-potential on pseudoplastic flow of suspensions. *Rheol. Acta*, **26,** 172–81.

Jastrzebski, Z.D. (1967). Entrance effects and wall effects in an extrusion rheometer during flow of concentrated suspensions. *Ind. Eng. Chem. Fundam.*, **6,** 445–54.

Keating, J. & Hannant, D.J. (1989). The effect of rotation rate on gel strength and dynamic yield strength of thixotropic well cements measured using a shear vane. *J. Rheol.*, **33**(7), 1011–20.

Lazarus, J.H. & Sive, A.W. (1984). A novel balanced beam tube viscometer and the rheological characterisation of high concentration fly ash slurries. In *Proc. HT 9*, paper E1, pp. 207–26.*

Lazarus, J.H. & Slatter, P.T. (1986). Comparative rheological characterisation using a balanced beam tube viscometer and rotary viscometer. In *Proc. HT 10*, paper J2, pp. 291–302.*

Metzner, A.B. & Otto, R.E. (1957). Agitation of non-Newtonian fluids. *AIChE J.*, **3,** 3–10.

Nguyen, Q.D. & Boger, D.V. (1983). Yield stress measurement for concentrated suspension. *J. Rheol.*, **27**(4), 321–49.*

Nguyen, Q.D. & Boger, D.V. (1985). Thixotropic behaviour of concentrated bauxite residue suspensions. *Rheol. Acta*, **24**, 427–37.
Round, G.F. & Hessari, A.R. (1985). The effect of size distribution and pH on the rheology of coal slurries. In *Proc. STA 10*, pp. 151–6.*
Ryan, N.W. & Johnson, M.M. (1959). Transition from laminar to turbulent flow in pipes. *AIChE J.*, **5**, 433–5.
Sarmiento, G., Crabbe, P.G., Boger, D.V. & Uhlherr, P.H.T. (1979). Measurement of the rheological characteristics of slowly settling flocculated suspensions. *Ind. Eng. Chem. PDD*, **18**, 746–51.
Sikorski, C.F., Lehman, R.L. & Shepherd, J.A. (1982). The effect of viscosity reducing chemical additives on slurry rheology and pipeline transport performance for various mineral slurries. In *Proc. STA 7*, pp. 163–74.*
Sparrow, E.H., Munro, W.D. & Jonsson, V.K. (1964). Instability of the flow between rotating cylinders: the wide gap problem. *J. Fluid Mech.*, **20**, 35–46.
Umeya, K. & Otsubo, Y. (1980). Time-dependent behaviour of viscoelastic suspensions. *J. Rheol.*, **24**(2), 239–52.
Van Wazer, J.R., Lyons, J.W., Kim, K.Y. & Colwell, R.E. (1963) *Viscosity and Flow Measurement: A Laboratory Handbook of Rheology*. Interscience, New York, USA.
Walters, K. (ed.) (1980). *Rheometry: Industrial Applications*. Research Studies Press (John Wiley & Sons Ltd), Letchworth, UK.
Want, F.M., Colombera, P.M. & Boger, D.V. (1982). Pipeline design for the transport of high density bauxite residue slurries. In *Proc. HT 8*, paper E2, pp. 249–62.*
Whorlow, R.W. (1980). *Rheological Techniques*. Ellis Horwood Ltd, Chichester, UK.

*Full details of the Hydrotransport (HT) and the Slurry Transport Association (STA) series of conferences can be found on pp. 11–13 of Chapter 1.

5

Introduction to Predictive Techniques for Pipe Flow

Nigel P. Brown
The British Petroleum Company plc, London, UK

&

Nigel I. Heywood
Warren Spring Laboratory, Stevenage, UK

1. OVERVIEW

Predictive techniques complement many aspects of the design process. In this part of the handbook procedures are presented to aid the specification and prediction of

- pipeline operating velocity,
- steady-state pumping requirements, and
- transient pressure surges at start-up and shutdown.

Chapters 6–8 consider techniques that are applicable to the prediction of steady-state flow behaviour, specifically the operating velocity and frictional pressure gradient. Chapters 6 and 7 address pressurised pipe flow; Chapter 6 is directed towards slurries that tend to settle and Chapter 7 to slurries with a low settling tendency. Chapter 8 addresses unpressurised flows that exhibit a free surface. Chapter 9 considers transient flow behaviour. This chapter provides background to the techniques covered in Chapters 6–9 for predicting steady-state flow behaviour and relevant information not covered in the main chapters.

2. THE ROLE OF EXPERIMENTAL DATA

Flow behaviour of typical industrial slurries can vary widely. Data for slurries are highly specific to the slurry for which it has been measured and is dependent on a large number of physical variables including relative density of the phases, solids concentration and particle shape, and size distribution. In addition, the chemical environment of the particles and surface charge (or zeta potential) can be particularly important to flocculation in fine-particle systems and fluid–particle and particle–wall friction in coarse-particle systems.

There is no substitute for good quality experimental data. Testing the slurry in pipe loops (such as those described in Chapter 10) is often considered essential for novel or large-scale operations. Testing in flow loops generally requires substantial quantities of slurry and experiments are often costly to execute. Cheaper means of obtaining the data are frequently sought. Measurement of the rheological properties, using bench-scale viscometers, may be an appropriate technique for slow-settling slurries. Small-scale testing can also provide important frictional data for coarse-particle slurries. Experimental data are used with a flow model to predict pipeline behaviour. The majority of information that has been published on flow behaviour in pipes of commercially realistic diameters tends to be restricted to project-specific applications. A notable exception is data published by the Saskatchewan Research Council (Chapter 1 contains comprehensive references to reports in the public domain, see pp. 13–16). Data on a wider range of slurries are confined to research that has often been conducted using small diameter pipes.

3. SETTLING SLURRIES

3.1. Estimation of Operating Velocity and Pressure Drop

Estimation of the pressure drop is an essential precursor to pump sizing. In pressurised systems the total pressure change is the sum of hydrostatic (or potentially recoverable) pressure and frictional (irrecoverable) contributions. The relative importance of the two components is highly dependent on the application. The pressure drop exhibited by the flowing slurry is often very sensitive to the flow rate; hence the combination of pipe diameter and superficial velocity are usually determined early in the design.

3.2. Overview of Predictive Techniques

Traditionally, scaling data (most usually to larger pipe sizes) has been achieved using empirical curve fitting techniques, or semi-empirical correlation guided by dimensional analysis. While there is merit in using such approaches to interpolate data economically, prediction outside the range of experimental verification is unsatisfactory because rudimentary modelling techniques do not consider physical processes that occur within the slurry. Under these conditions the use of a physically based approach is recommended. Predictive techniques developed for gravity-dominated flows in horizontal and inclined pipes are not applicable to vertical transport. There are few data presented in the literature on frictional losses in pipe fittings and no general prediction techniques are available. Experience gained with water flows provides the only guidance.

3.3. Horizontal and Near-Horizontal Transport

3.3.1. Fluid–Particle Interaction and the Operating Velocity

Settling slurries exhibit a wide range of flow patterns that are discussed in Chapter 3. Pipelines transporting slurries that settle under conditions of no-flow are usually operated without a deposit. The formation of a slow-moving, or stationary, deposit is frequently associated with an unstable operating regime. Under these conditions a small decrease in flow can result in increased hold-up of solids which may precipitate a blockage if sufficient pumping capacity is not available to overcome the increase in pressure head. Prediction of the minimum velocity that will sustain transport is an essential starting point from which an operating velocity can be assigned and the associated energy requirements estimated.

The lack of standardisation has created much confusion over the definition of transition velocities between flow patterns and that of the operating velocity. Carleton and Cheng (1974) discuss in detail many of the descriptions commonly encountered in the literature. Although attempts have been made to formalise definitions of transitions in flow behaviour little success has been achieved until the comparatively recent development of techniques to detect flow patterns (such as that reported by Ercolani *et al.*, 1979).

3.3.2. Semi-empirical Techniques to Predict Flow Behaviour

Curve fitting and techniques based on correlating groups of dimensionless parameters were among the earliest techniques employed to enable predictions of operating velocity and pressure gradient to be made from experimental data. Encouraged by the success of such techniques in describing Newtonian fluid flow in pipes, early investigators of slurry flows (such as Durand *et al.*, 1953*a*, *b*; Newitt *et al.*, 1955) adopted correlation techniques for the more complex two-phase flows. Durand's publications reported an empirical relation, based on much experimental work, that can be used to determine the flow conditions under which a deposit would not be encountered in the pipe. Following the pioneering work, the technical literature contained much discussion on the numerical value of the coefficient used in this empirical relation. Many different correlations have been proposed, each reportedly providing an improved description of existing or new data. Comprehensive reviews of correlations are given by Carleton and Cheng (1974) and more recently by Turian *et al.* (1987). The wide variation in the indices used to correlate data serves to caution against the indiscriminate use of empirical techniques outside the application and range of variables for which they were developed.

3.3.3. Physically Based Approaches to Predict Flow Behaviour

In complete contrast to correlation techniques, distributed-parameter models have been developed (such as those reported by Roco & Shook, 1985) that attempt to model the physical processes that occur within the flow. The differential equations of motion that describe the flow behaviour in a small volume require complex numerical techniques to enable efficient solution over the entire flow domain. The complexity of two-phase flows precludes description solely in terms of fundamental equations. Intuitive, or empirical, descriptions of local flow structure have to be developed. These developments have become possible through greater understanding of the interactions between the fluid and particles. The detailed predictions from such models provide an objective basis for predicting the operating conditions of slurry pipelines. However, the models require careful calibration which reduces their generality and utility for preliminary design. At present, such techniques are the subject of research papers, where the emphasis is on understanding mechanisms through different modelling strategies. However, it would be reasonable to assume that the same predictive successes, currently enjoyed by analogous techniques used with single-phase fluids, will be gained with solid–liquid, two-phase systems.

Considerable success has been achieved in predicting the behaviour of slurries using less rigorous mechanistic approaches. Physically based models have been proposed that do not attempt to model the detailed flow structure. One of these approaches is the two-layer model. The name is derived from the form in which the model was conceived; the cross-section of the pipe is divided horizontally into two segmental flow areas. Particles existing in the lower segment form a gravity bed and are considered to slide along the bottom of the pipe. Solids in the upper segment are treated as a suspension. The model was originally applied to determining the limiting case where a stationary bed of solids first starts to slide (Wilson, 1970), thereby providing a prediction of the minimum transport velocity. The clear distinction between the two layers has gradually been eroded as research has enabled a greater insight into the mechanisms by which particles become suspended. The approach presented in Chapter 6 is the result of exhaustive and continual development of the two-layer model by one of its earliest proponents.

Physically based approaches enable the in-situ concentration of the slurry to be predicted, a quantity that is essential to the estimation of hydrostatic pressure contribution.

3.4. Inclined Transport

Inclined flows open the opportunity for complex flow behaviour (Wilson & Tse,

1984). As inclined flows are comparatively unimportant, very little analysis has been attempted. The majority of the studies that are reported (e.g. Shook & McLeod, 1975; Ferrini & Pareschi, 1980; Okada *et al.*, 1982) consider the motion of the particles on pipeline shutdown, in particular from the point of view of restart behaviour. Studies that have addressed the effect of inclination on deposit velocity include Kao & Hwang (1979), Noda *et al.* (1980), and Wilson & Tse (1984).

3.5. Vertical Transport

3.5.1. Fluid–Particle Interaction

Hoisting is important to many mining operations, whereas counter-gravity flow has found recent large-scale application in mine back-filling operations. The interaction between the fluid and solids in vertical, counter-gravity, pipeline transport of particulate slurries shares many common features with a moving fluidised bed. Fluidised beds are widely used for fluid–particle contact in chemical engineering operations. An unrestrained gravity bed of particles is described as *incipiently fluidised* once the upward passage of a fluid through the bed results in the particles no longer being supported by particle–particle contact. While the particles are in physical contact the resistance to the passage of fluid through the bed can be described by relations applicable to porous media. At fluid velocities above the condition of incipient fluidisation the solids are supported by fluid drag. Increasing the fluid velocity causes the bed to expand in a predictable manner (see, for example, Goddard & Richardson, 1969). Further increase in the fluid velocity causes the solids to be transported from the bed. Under these conditions the analogy between fluidisation and vertical pipeline transport is complete if solids were fed into the system above the fluid distributor.

Hoisting particles that are more dense than the fluid is characterised by relative motion between the particles and the fluid which results from the tendency of the solids to settle under the influence of gravity. The fall velocity of closely sized particles, relative to the fluid, can be estimated using the techniques discussed in Chapter 2. The effect of the slip-velocity is reflected in the in-situ concentration of the solids being greater than the delivered value. Gravity induced segregation of particles is most pronounced under conditions where the solids are broadly sized, of different density and where the transport velocity is low. The use of techniques to feed the solids into a hoist that operates in an intermittent or cyclic manner (such as those discussed in Chapter 17) can result in the particles being distributed in the form of a *slug* of solids separated by liquid. Redistribution of the particles can occur within the slugs as the coarse, or more dense, material settles through fine, or less dense particles. This can result in plugs of high-concentration material being formed that may choke the line. The effect of temporal and spatial variation of concentration in the inlet mixture to a hoist has been studied theoretically by Shook (1988). A similar approach is reported by Iyoho *et al.* (1987) in the study of the transport of drilled cuttings in an annular well-bore geometry.

Gravitational effects do not tend to cause a wide variety of flow patterns, such as those that may be experienced in the horizontal flow of the same material.

Steady operation is characterised by symmetrical radial concentration and velocity distributions (for a collection of data, see, for example, Govier & Aziz (1972, pp. 456–61)). Photographic evidence reported by Newitt *et al.* (1961) supports theoretical predictions of radial particle migration as the presence of a central core of solids is shown in the pipe surrounded by almost clear water. Similar observations are reported by Engelmann (1978) from studies on the hoisting of large diameter (13–52 mm) manganese nodules. The formation of slugs, characterised by axial concentration gradients, is associated with unsteady operation that can be traced to temporal changes in the flow rate of the fluid, feed rate of the solids, or changes in particle size distribution.

3.5.2. Selection of Operating Velocity

The operating velocity of systems that are intended to hoist closely graded materials can be selected from a consideration of the static behaviour under fluidisation. For the purposes of preliminary design, it is commonly assumed that the superficial fluid velocity should be twice the settling velocity of the particles (see, amongst others, Govier & Aziz, 1972, p. 468; Sifferman *et al.*, 1974). If the operating velocity falls below the minimum value required to fluidise the solids high axial stresses are generated that result in a plugged line.

3.5.3. Prediction of Pressure Gradient

At low concentrations of solids (in the region of 2–5% by volume) the frictional pressure drop at high flow rates is frequently reported to be almost indistinguishable from that of the flow of clear liquid. At higher concentrations, Einstein and Graf (1966) report that the frictional pressure drop can be approximated by a mixture-density correction to the clear fluid frictional loss. This semi-empirical approach has been verified by Televantos (1977) for sand at concentrations approaching the freely settled concentration of the solids. The pseudohomogeneous correction to the clear fluid loss is a commonly accepted approach for preliminary design purposes. A mechanistic approach to the prediction of frictional pressure gradients based on a radial force-balance has been reported by Wilson *et al.* (1979).

4. NON-SETTLING SLURRIES

4.1. Introduction

Non-settling slurries can often be considered as pseudohomogeneous mixtures of solids in a liquid carrier. As a consequence, frictional pressure loss in pipe flow is unaffected by the pipe orientation. This is therefore in marked contrast to settling slurry flows. Assessment of the flow properties of non-settling slurries is covered in Chapter 4, the prediction of pressure drop in Chapter 7 and the use of pressure drop information for pump selection and sizing in Chapters 13–18.

4.2. Prediction of Steady-State Behaviour

4.2.1. Information from Measurement of the Flow Curve

The important flow property to measure is the slurry's flow curve, i.e. a plot of shear stress versus shear rate. It cannot be emphasised too strongly that a relevant shear rate range must be selected for this; the range must be applicable to the range of mean slurry flow velocities in a range of potentially suitable pipe sizes as part of the initial hydraulic analysis. A different, much lower shear rate range needs to be used to determine the slurry's zero shear viscosity if particle settling rates are to be predicted (Cheng, 1989).

These measurements are undertaken using an appropriate viscometer or rheometer. There are many basic types available and commercial designs on the market. Heywood (1985) has discussed viscometer selection according to the material under test and the application for the data. A particularly useful viscometric geometry is a small diameter capillary tube (e.g. 2–6 mm) or a larger pipe size (e.g. 10–100 mm), operated on either a once-through or recirculating basis.

The slurry flow curve assists in providing the following information:

- the pressure loss incurred for flow in circular (and other geometry, e.g. annular) pipe (Chapter 7);
- the pressure loss incurred for flow through various types of pipe fitting, e.g. valves, bends, expansions and contractions, and some designs of flowmeter (Chapter 7);
- the potential for reducing the pressure loss through gas injection, pipe vibration, flow rate oscillation or some other means (Heywood, 1986; and Chapter 7);
- the mean flow velocity of the slurry corresponding to the limit of the laminar flow regime (Chapter 7);
- the settling rate of the various particle size fractions in slurry transported in a horizontal pipeline (Cheng, 1989);
- prediction of the velocity profile across a pipe diameter without the necessity to measure this directly. This can be useful for flowmeter selection and prediction of the flowmeter's performance (Chapter 23);
- selection and sizing of pump or pumps, and the determination of the motor size and drive type (Chapters 13–18).

4.2.2. Laminar Flow—Flow Models

Slurries can have either Newtonian behaviour (constant viscosity over a defined shear rate range) or non-Newtonian behaviour (increasing or decreasing viscosity as shear rate is varied). To use the flow curve information, a flow model is frequently fitted to the data in order that either analytical or semi-theoretical equations relating volumetric flow rate to frictional pressure gradient may be used. These flow models take the general form of

$$\tau = \tau(\dot{\gamma}, k_1 \cdots k_n)$$

where τ is the shear stress and is a function of the shear rate $\dot{\gamma}$, and model parameters k_1 to k_n. Numerous models have been developed and published but the most useful for engineering applications are the Newtonian, power law, Bingham plastic and generalised Bingham plastic (or yield-pseudoplastic) models. These models are discussed further in Chapter 7.

Estimates of the parameters appearing in these models is normally straightforward and is done through regression analysis. However, slurries tend not to give a unique flow curve which is independent either of the viscometric geometries selected or of the scale of any particular geometry. In such cases, the use of a 'mean' flow curve and/or upper bound to the shear stress/shear rate data is appropriate (Chapter 4).

Turbulent data from tube or pipe viscometers only can be used for scale-up purposes for turbulent flow pressure loss prediction in larger pipe sizes (Bowen, 1961) but cannot be used in the laminar flow models listed above. Another common reason for non-superposition of the viscometric data is the presence of the phenomenon known perhaps rather misleadingly as *wall-slip*. This needs to be identified early and quantified in terms of a *slip velocity* (Jastrzebski, 1967) because if slippage effects are occurring in a viscometer they may well occur and be important in the full-scale pipeline. Further details of wall-slip assessment in viscometry are given in Chapter 4 and its application in pipeline pressure drop prediction in Chapter 7.

4.2.3. Prediction of Laminar-Flow Breakdown

Once an appropriate flow model has been selected and its parameters estimated, it is then possible to predict the mean slurry flow velocity, for any specified pipe diameter, above which laminar flow breaks down. For Newtonian slurries the critical Reynolds number for laminar-flow breakdown in pipeflow is 2100, but for non-Newtonian slurries the breakdown point is not so well-defined.

For power law slurries, the critical Reynolds number has been predicted to be a function of the flow behaviour index only (Ryan & Johnson, 1959), and another method is available for Bingham plastic slurries (Hanks, 1963). Predictions for other flow models, including the Herschel–Bulkley (1926) model, can be made by applying the Metzner and Reed (1955) approach to the Ryan and Johnson method for predicting the critical Reynolds number.

It is important to define the limit of laminar flow for a number of reasons.

- Analytical expressions for pressure drop prediction based on flow models are available for the laminar regime only; a different set of expressions is available for the turbulent regime.
- Overall optimisation of a slurry pipeline system often indicates that it is most economical to operate the pipeline in the transitional region between the laminar and turbulent regimes.
- It may be necessary or advisable to operate some way into the turbulent regime if turbulent eddies are required to maintain the coarser size fractions of the solids in suspension.
- Inner pipe wall roughness has no effect on slurry hydrodynamics in the laminar regime, but in turbulent flow the pressure drop can be raised by a factor of

three of more if rough wall pipe is used compared with *hydrodynamically* smooth pipe.

4.2.4. Turbulent Flow

Newtonian slurries. A large number of expressions have been published which relate friction factor to Reynolds number for turbulent flow of Newtonian fluids in both smooth and rough wall pipes. These are considered in greater detail in Chapter 7.

Non-Newtonian slurries. Uncertainty in turbulent pressure drop estimation is considerably greater for non-Newtonian slurries; predictions can vary by a factor of three or more (Heywood & Cheng, 1984) even for smooth wall pipe. Many attempts have been made to correlate pressure drop data with operating variables and parameters appearing in the non-Newtonian flow models. Attempts at correlation are sometimes made on a semi-theoretical basis but generally from empiricism. Thus, Dodge and Metzner (1959) extended the von Karman (1931) type expression for friction factor to power law slurries and other equations have been published since then (Shaver & Merrill, 1959; Tomita, 1959; Kemblowski & Kolodziejski, 1973).

Rather less effort has been focussed on the turbulent flow of Bingham plastic slurries (Thomas, 1963; Hanks & Dadia, 1971). It is often argued that yield stress is unimportant in turbulent pipe flow and this suggests simplified approaches where yield stress effects are ignored. Only two approaches have been published for the Herschel–Bulkley (1926), or generalised Bingham plastic, three-parameter flow model: Torrance (1963) and Hanks (1978). Neither has been properly evaluated with experimental data.

Estimates not involving flow models. In view of this uncertainty in turbulent flow pressure drop prediction, an alternative approach is to take the prediction for water flow alone at the same mean velocity as the slurry flow and multiply this by a suitable factor. This has been done with some success with various kinds of sewage sludge and using a factor of 1·25–1·75 (Hayes *et al.*, 1973). However, some fibrous suspensions can give drag reduction when a factor of less than unity is indicated (Cheng & Heywood, 1984*a*, *b*). Unfortunately, the likelihood of drag reduction cannot usually be predicted from laminar-flow data (either from a viscometer or from a pipeline).

Another method involves the assumption that all the increase in pressure drop in turbulent flow is due to inertial effects only and viscous effects are not important. An estimate for pressure drop can then be made by multiplying the water pressure drop prediction at the same mean flow velocity by the slurry specific gravity. This will probably, in many cases, give a more reliable pressure drop estimation than using a more complicated route based on the slurry's flow curve.

Effect of pipe roughness. With the exception of Torrance (1963), the effect of pipe wall roughness on the turbulent flow of non-Newtonian slurries has yet to be studied in any depth and there exists no reliable predictive equation for pressure drop in this situation. An approximate method is to calculate the non-Newtonian Reynolds number and use this in the Moody equation to estimate the friction factor at the appropriate roughness. The smooth wall estimate for pressure drop for the non-Newtonian slurry can then be multiplied by the ratio of this rough wall friction factor to the smooth wall friction factor giving an improved pressure loss estimate.

The pipe wall roughness is seldom known with any degree of accuracy and estimates for typical materials of construction for pipe work are given in standard texts. Alternative approaches include measurement of the mean roughness (Ra value) of a sample of the pipe material directly or estimation of the roughness by conducting a turbulent water flow test on a sample of the pipe and reading off the roughness from the Moody chart.

Pressure loss from pipe fittings. Estimates for pressure loss across pipe fittings can be readily made for turbulent flow of either Newtonian or non-Newtonian slurries using well-established techniques. The approach for Newtonian laminar flow is less well-documented, and, until relatively recently, there was a dearth of information on head loss data for non-Newtonian laminar flow. Pressure losses arising from flow through pipe fittings tend to be important for relatively small pipe runs within process plants rather than long distance pipelines where they are usually insignificant compared with the loss from flow through the main pipe work and from the loss or gain arising from changes in elevation.

4.3. Start-up Behaviour

The start-up behaviour of a slurry pipeline is determined by

- the type of pumping system installed,
- the speed with which a pressure wave will be transmitted along the length of the pipeline, and
- whether the slurry has an appreciable yield stress and time-dependent (usually thixotropic) flow property.

The start-up pressure developed by the pumping system must be more than adequate to initiate a wall shear stress along the pipeline which exceeds the slurry's yield stress. Thus to predict start-up behaviour, measurement of yield stress is necessary and, because many slurries with yield stresses are also thixotropic, additional rheological measurements may be necessary.

In particular, it is important to determine if yield stress progressively increases with time as the structure in a slurry builds up at rest. This can have major consequences for a slurry pipeline which has been shutdown temporarily. If the proposed pumping system is unlikely to be able to cope with the highest yield stress levels which the slurry could develop, then calculations need to be made to establish the maximum permissible shutdown period beyond which the pumping system would be unable to restart the pipeline flow.

Yield stress is a notoriously difficult parameter to measure (Cheng, 1986). The start-up pressure gradient is directly proportional to the yield stress so it is nevertheless a very useful parameter. The *dynamic* yield stress can be estimated by extrapolation of the flow curve to zero shear rate but a more reliable measure is the *static* yield stress obtained by direct measurement using a controlled stress rheometer or a gun rheometer (see Chapter 4).

Thixotropic flow data can be obtained in a number of ways using an appropriate rheometer. These data can be used to predict both the excess pressure gradient needed to initiate flow and the time needed to clear the pipeline of the *structured* slurry using incoming sheared slurry with consequently less structure (Cheng & Whittaker, 1972; Cheng & Heywood, 1984*a*). Much of the analysis has been directed towards predicting both the start-up and steady-state behaviour of thixotropic crude oil pipelines (e.g. Ritter & Batycky, 1967), but the ideas could also be applied to thixotropic suspensions when necessary.

Predictions of the instantaneous pressure gradient along the pipeline depend on whether the pumping pressure or the flow rate is held constant. A scale-up method is available (Cheng, 1979) which does not require a detailed characterisation of the thixotropic property but rather experimental tests using a laboratory pipeline whose length-to-diameter ratio is the same as that of the full-scale pipeline. Except for short pipe runs this requirement can often unfortunately prove impractical.

5. REFERENCES

Bowen, R.L. (1961). Scale-up for non-Newtonian fluid flow. *Chem. Engng*, 24 July, **68**, 143–50.

Carleton, A.J. & Cheng, D.C-H. (1974). Design velocities for hydraulic conveying of settling suspensions. In *Proc. HT 3*, paper E5, pp. 57–74.*

Cheng, D.C-H. (1979). A study into the possibility of scale-up in thixotropic pipeflow. Report LR 317 (MH), Warren Spring Laboratory, Stevenage, UK.

Cheng, D.C-H. (1986). Yield stress; a time-dependent property and how to measure it. *Rheol. Acta*, **25**, 542–54.

Cheng, D.C-H. (1989). Rheological design for suspending solids. Report LR 711 (MP/BM), Warren Spring Laboratory, Stevenage, UK.

Cheng, D.C-H. & Heywood, N.I. (1984*a*). Flow in pipes. Part 1: flow of homogeneous fluids. *Phys. Technol.*, **15**, 244–51.

Cheng, D.C-H. & Heywood, N.I. (1984*b*). Viscometric testing and head loss correlations for pipeline design for fibrous suspensions. Report LR 502 (MH), Warren Spring Laboratory, Stevenage, UK.

Cheng, D.C-H. & Whittaker, W. (1972). A method for assessing the thixotropic property of fluids carried in pipelines. In *Proc. HT2*, paper B4, pp. 41–60.*

Dodge, D.W. & Metzner, A.B. (1959). Turbulent flow of non-Newtonian systems. *AIChE J.*, **5**, 189–204.

Durand, R. (1953*a*). Basic relationships of the transportation of solids in pipes—experimental research. In *Proc. Minnesota Int. Hydraulics Convention*, pp. 89–103.

Durand, R. (1953*b*). Hydraulic transport of coal and solid materials in pipes. In *Proceedings of a Colloquium on the Hydraulic Transport of Coal*. National Coal Board, London, UK. pp. 39–52.

Einstein, H.A. & Graf, W.H. (1966). Loop system for measuring sand–water mixtures. *Proc. ASCE* (*J. Hydraulics Div.*), **92**, 1–12.

Engelmann, H.E. (1978). Vertical hydraulic lifting of large solids—a contribution to marine mining. Paper presented at 10th Annual Offshore Technology Conference, Richardson, Texas, USA, 8–11 May, pp. 731–40.

Ercolani, D., Ferrini, F. & Arrigoni, V. (1979). Electric and thermal probes for measuring the limit deposit velocity. In *Proc. HT 6*, paper A3, pp. 27–42.*

Ferrini, F. & Pareschi, A. (1980). Experimental study of the solid bed profile in sloping pipe sections. in *Proc. HT 7*, paper F3, pp. 245–58.*

Goddard, K. & Richardson, J.F. (1969). Correlation of data for minimum fluidising velocity and bed expansion in particulately fluidising systems. *Chem. Eng. Sci.*, **24**, 363–7.

Govier, G.W. & Aziz, K. (1972). *The Flow of Complex Mixtures in Pipes*. Van Nostrand Reinhold, New York, USA.

Hanks, R.W. (1963). The laminar-turbulent transition for flow in pipes, concentric annuli and parallel plates. *AIChE J.*, **9**, 45–8.

Hanks, R.W. (1978). Low Reynolds number turbulent pipeline flow of pseudohomogeneous slurries. in *Proc. HT 5*, pp. C23–34.*

Hanks, R.W. & Dadia, B.H. (1971). Theoretical analysis of the turbulent flow of non-Newtonian slurries in pipes. *AIChE J.*, **17**, 554–7.

Hayes, J., Flaxman, E.W. & Scivier, J.B. (1973). A comprehensive scheme for sewage sludge in North West England. *Proc. Inst. Civ. Engrs* (Part 2: *Res. and Theory*), **55**, paper 7604, 1–21.

Herschel, W.H. & Bulkley, R. (1926). Konsistenzmessungen von Gummi-Benzollosungen. *Kolloid Z.*, **39**, 291.

Heywood, N.I. (1985). Selecting a viscometer. *Chem. Engr*, Jun., **415**, 16–23.

Heywood, N.I. (1986). A review of techniques for reducing energy consumption in slurry pipelining. In *Proc. HT 10*, paper K3, pp. 319–32.*

Heywood, N.I. & Cheng, D.C-H. (1984). Comparison of methods for predicting head loss in turbulent pipeflow of non-Newtonian fluids. *Trans. Inst. Meas. Contr.*, **6**, 33–45.

Iyoho, A.W., Horeth, J.M. & Veenkant, R.L. (1987). A computer model for hole cleaning analysis. In *62nd Annual Technical Conference and Exhibition of the Society of Petroleum Engineers*, Dallas, Texas, USA, 27–30 Sep. Paper SPE 16694, pp. 397–410.

Jastrzebski, Z.D. (1967). Entrance effects and wall effects in an extrusion rheometer during flow of concentrated suspensions. *Ind. Eng. Chem. Fundam.*, **4**, 445–54.

Kao, D.T. & Hwang, L.Y. (1979). Critical slope for slurry pipeline transporting coal and other solid particle. In *Proc. HT 6*, paper A5, pp. 57–76.*

Kemblowski, Z. & Kolodziejski, J. (1973). Flow resistances of non-Newtonian fluids in transitional and turbulent flow. *Int. Chem. Eng.*, **13**, 265–79.

Metzner, A.B. & Reed, J.C. (1955). Flow of non-Newtonian fluids—correlation of laminar, transition and turbulent regions. *AIChE J.*, **1**, 434–40.

Moody, L.F. (1944). Friction factors for pipeflow. *Trans. ASME*, **66**, 671–84.

Newitt, D.M., Richardson, J.F., Abbott, M. & Turtle, R.B. (1955). Hydraulic conveying of solids in horizontal pipes. *Trans. Inst. Chem. Engrs*, **33**, 93–113.

Newitt, D.M., Richardson, J.F. & Gliddon, B.J. (1961). Hydraulic conveying of solids in vertical pipes. *Trans. Inst. Chem. Engrs*, **39**, 93–100.

Noda, K., Masuyama, T. & Kawashima, T. (1980). Influence of pipe inclination on deposit velocity. In *Proc. HT 7*, paper F2, pp. 231–44.*

Okada, T., Hisamitsu, N., Ise, T. & Takeishi, Y. (1982). Experiments on restart of reservoir sediment slurry pipeline. In *Proc. HT 8*, paper H3, pp. 399–414.*

Ritter, R.A. & Batycky, J.P. (1967). Numerical prediction of the pipeline flow characteristics of thixotropic liquids. *Soc. Petr. Engrs J.*, **7**, 369–76.

Roco, M.C. & Shook, C.A. (1985). Critical deposit velocity in slurry flow. *AIChE J.*, **31**, 1401–4.

Ryan, N.W. & Johnson, M.M. (1959). Transition from laminar to turbulent flow in pipes. *AIChE J.*, **5**, 433–5.

Shaver, R.G. & Merrill, E.W. (1959). Turbulent flow of pseudoplastic polymer solutions in straight cylindrical tubes. *AIChE J.*, **5**, 181–8.
Shook, C.A. (1988). Segregation and plug formation in hydraulic hoisting of solids. In *Proc. HT 11*, paper G3, pp. 359–79.*
Shook, C.A. & McLeod, D.J. (1975). The effect of line length for inclined slurry pipelines at shutdown. *Can. J. Chem. Engng*, **53**, 594–8.
Sifferman, T.R., Myers, G.M., Haden, E.L. & Wahl, H.A. (1974). Drill-cutting transport in full-scale vertical annuli. *J. Petr. Tech.*, Nov., **XXVI**, 1295–302.
Televantos, Y. (1977). The flow mechanism of hydraulic conveying at high solids concentration. PhD thesis, University of London, UK.
Thomas, D.G. (1963). Non-Newtonian suspensions, Part 2. *Ind. Eng. Chem.*, **55**, 27–35.
Tomita, Y. (1959). On the fundamental formula of non-Newtonian flow. *Bull. JSME*, **2**, 469–74.
Torrance, B.Mck. (1963). Friction factors for turbulent fluid flow in circular pipes. *South African Mech. Engr*, **13**, 89–91.
Turian, R.M., Hsu, F-L. & Ma, T-W. (1987). Estimation of the critical velocity in pipeline flow of slurries. *Powder Technol.*, **51**, 35–47.
Von Kármán, Th. (1931). Mechanische Ahnlichkeit und Turbulenz. In *Proceedings of III International Congress of Applied Mechanics*, **85**, Stockholm, Sweden.
Wilson, K.C. (1970). Slip point of beds in solid–liquid pipeline flow. *Proc. ASCE* (J. Hydraulics Div.), **96**, 1–12.
Wilson, K.C. & Tse, J.K.P. (1984). Deposition limit for coarse-particle transport in inclined pipes. In *Proc. HT 9*, paper D1, pp. 149–62.*
Wilson, K.C., Brown, N.P. & Streat, M. (1979). Hydraulic hoisting at high concentration: a new study of friction mechanisms. In *Proc. HT 6*, pp. 269–82.*

*Full details of the Hydrotransport (HT) series of conferences can be found on pp. 11–12 of Chapter 1.

6

Pipeline Design for Settling Slurries

Kenneth C. Wilson
Queen's University, Kingston, Canada

1. INTRODUCTION

Although non-settling or homogeneous slurries behave rather like fluids of given density and viscosity, settling slurries do not. At low throughput velocities, settling solids tend to accumulate in particle-rich layers near the invert of the pipe, with a corresponding particle-lean zone near the pipe soffit. This behaviour is known as stratification, and for fully stratified flow the concentration profile can be approximated by the two-layer configuration shown in Fig. 6.1. If the net force acting on the solids-rich lower layer is not enough to keep it in motion, the lower layer becomes a stationary bed. It is generally agreed that operating a pipeline with a stationary bed of solids is undesirable; to avoid this condition, the designer must be able to predict the velocity at the limit of stationary deposition so that the operating velocity can be kept clear of the deposition zone.

Settling slurries also behave quite differently from non-settling ones at velocities

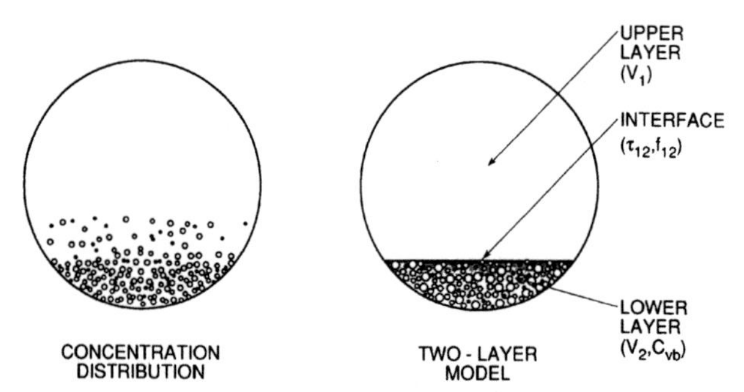

Fig. 6.1. Definition sketch for two-layer model of stratified flow.

greater than the deposition limit. As Blatch (1906) found in her settling slurry experiments, curves of pressure gradient versus mean velocity (for constant solids concentration) consistently display a minimum. Although the velocity at this minimum may be near the deposition limit for slurries of small particles, it tends to be well above this limit for coarse-particle slurries, and the two velocities must not be confused.

In the 1950s Durand (1951) and Gibert (1960) obtained, by data-correlation techniques, an equation which produced gradient–velocity curves with a minimum. This equation, like others to be discussed below, refers to friction losses in horizontal pipeline flow. The behaviour of flows other than horizontal will be referred to where appropriate, but will not be analysed in this chapter. Although Durand's equation had been widely used, it became clear by about 1970 that for many settling slurries this equation shows poor agreement with the data. As demonstrated by Babcock (1971), a major cause of the difficulty is that settling slurries are of two types: fully stratified and partially stratified or heterogeneous. For fully stratified flow the excess pressure gradient (solids effect) diminishes with velocity at a rate significantly smaller than that predicted by the Durand equation, but for partially stratified (heterogeneous) flows the observed rate can be much larger than the predicted one. In fact, both types of flow had been recognized earlier by Newitt *et al.* (1955). These workers proposed, in effect, that the flow would be fully stratified for all values of the mean velocity, V_m, less than 17 times the terminal settling velocity of the solid particles, u_∞. For this stratified flow they indicated that the excess friction gradient (beyond that for an equivalent clear-water flow) should be approximately proportional to the submerged weight of solids, giving

$$\frac{i_m - i_w}{(S_m - 1)} \simeq B \qquad (\text{for } V_m < 17u_\infty) \tag{6.1}$$

Here, i_m is the frictional hydraulic gradient of the mixture, i.e. the friction loss expressed in m of water per m of pipe. The gradient for an equal flow of water alone (expressed in the same units) is i_w. S_m is the relative density of the mixture, and $(S_m - 1)$ can also be written $(S_s - 1)C_{v,d}$ where S_s is the relative density of solids and $C_{v,d}$ is delivered volumetric concentration of solids. For their data Newitt's group found the coefficient B to be roughly 0·8. In heterogeneous (partially stratified) flow the ratio which forms the left-hand side of eqn (6.1) is less than 0·8, and diminishes with increasing V_m.

Measured or calculated values of frictional hydraulic gradient can be used to find the energy consumed in pipeline transport. For comparing different applications this is best expressed as the specific energy consumption (SEC), which denotes the energy in kilowatt-hours required to move each tonne of solids across a distance of 1 km. SEC is a simple function of frictional hydraulic gradient, relative density of solids and delivered solids concentration (Wilson & Judge, 1978), i.e.

$$\mathrm{SEC}(\mathrm{kWh\,t^{-1}\,km^{-1}}) = \frac{2{\cdot}73 i_m}{S_s C_{v,d}} \tag{6.2}$$

2. PHYSICAL MECHANISMS GOVERNING SLURRY BEHAVIOUR

The major features of settling-slurry flow have been identified above, and next it is appropriate both to mention the physical mechanisms which lie behind the observed phenomena and to indicate how these mechanisms lead to the various analytic and design techniques presented in subsequent sections of this chapter.

As described by Bagnold (1956) and detailed in section 3 below, two major physical mechanisms can act to support solid grains: fluid suspension and inter-granular contact. For fully stratified flow the velocity of the turbulent eddies is less than the fall velocity of the particles, and it can be expected that fluid suspension will be ineffective, so that virtually all of the moving solid particles are supported by sporadic granular contact. Together with the particles supported by continuous contact in stationary or moving beds, these particles are designated as *contact load.* The motion of contact-load particles can be analysed by application of force–balance techniques, as shown in section 3 below. This analytic approach was first applied in the early 1970s (Wilson, 1970; Wilson *et al.*, 1972), and was later extended to the determination of the limit of deposition and to fully stratified flow.

Applications of the force–balance model are set out in section 4 (Limit of stationary-deposit zone) and section 5 (Fully stratified flow). These sections include worked examples for the effects of particle size and density on deposition velocity, on the use of deposition velocity for initial pipe sizing, and on friction loss for fully stratified flow.

For values of V_m in excess of about $20u_\infty$ the velocity of the turbulent eddies exceeds the settling velocity of the particles, and hence the fluid suspension mechanism can also come into play. With further increases in the mean velocity, a progressively larger fraction of the solids will be suspended by the fluid forces. The contact-load fraction, or stratification ratio, diminishes accordingly. The suspended fraction of solids (but not the contact-load fraction) sets up an extra fluid pressure differential between the pipe invert and soffit (this pressure has been measured by Shook *et al.*, 1982, giving direct evidence of the suspended load). Once the suspended fraction of solids approaches unity, heterogeneous flow becomes indistinguishable, for practical purposes, from homogeneous flow. This has been

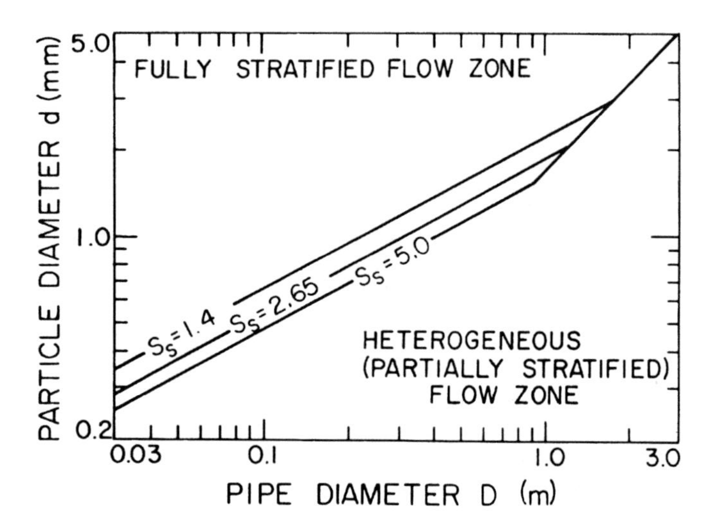

Fig. 6.2. Effect of pipe and particle diameter on flow type. (Adapted from Wilson, 1982.)

found to occur at a mean velocity of roughly $200u_\infty$ (Carstens & Addie, 1981; Gillies *et al.*, 1990), indicating that heterogeneous flow (which represents a transition zone) is limited to a single order-of-magnitude range of V_m/u_∞, from roughly 20 to 200. The analysis of heterogeneous flow is dealt with in section 6, with applications, including a worked example of friction-loss calculations, in section 7.

As indicated in a previous publication (Wilson, 1982) the boundary between fully stratified and heterogeneous behaviour can be shown graphically as a locus on a plot of pipe and particle diameter. This plot, which is based on water as the carrier fluid, appears here as Fig. 6.2. It can be seen that S_s also has an effect, but not a large one.

WORKED EXAMPLE: ESTIMATING TYPE OF SLURRY BEHAVIOUR

> For sand with $S_s = 2{\cdot}65$, Fig. 6.2 shows that slurries of grains of 2 mm size or greater will show fully stratified behaviour for all pipes up to 1 m in diameter, whereas those with grains of 0·7 mm size will behave as fully stratified only in pipes less than 0·2 m in diameter, but should be treated as heterogeneous for flows in larger pipes.

Although Fig. 6.2 gives a reasonable indication of flow type, it may not be entirely accurate for conditions near the boundary between fully stratified and heterogeneous flows. If it is necessary to make accurate calculations for such a borderline case, it is recommended that curves of frictional loss versus mean velocity can be prepared using the techniques for both fully stratified flow (section 5) and heterogeneous flow (section 7). The intercept of the curves determines the velocity for the change-over between the two flow types.

3. FORCE–BALANCE ANALYSIS

Within a stationary bed, or one sliding *en bloc*, the submerged weight of the solids, together with other forces applied to the granular mass, are transferred by continuous intergranular contacts. In the mechanics of granular media, the result of these contacts is represented by the normal and shear components of intergranular stress. The normal component is σ_p and by the Coulombic relation for cohesionless materials the shear component, τ_p, cannot exceed $\sigma_p \tan \phi$, where ϕ is the internal friction angle (Coulomb, 1776). Bagnold (1956) extended the analysis to grains above the bed which were not in continuous contact with the underlying grains. He considered the *load* or submerged weight of these grains as divided into two categories depending on the method of support. The bed load is the part whose submerged weight is carried downward by (discontinuous) intergranular contacts. These contacts set up a normal stress at the upper surface of the bed, without which the top of the bed would be constantly eroded away. For a bed of solids occupying the invert of a pipe, the total normal force acting against the pipe, F_n, can be obtained by integrating σ_p over the lower boundary of the bed. It was found that for beds occupying only a small part of the pipe, F_n is approximately equal to the submerged weight of solids, F_w. However, as the fraction of the pipe occupied by the bed increases, so does the ratio of F_n/F_w, which approaches two when the bed virtually fills the pipe (Wilson, 1970; Wilson *et al.*, 1972).

In order to set a stationary bed of solids in motion, the forces driving the bed must exceed the resisting force given by $\mu_p F_n$, where μ_p is the coefficient of mechanical friction between the solid particles and the pipe. Two driving forces must be considered, one acting within the bed and the other at its upper surface. The force within the bed is associated with the seepage flow set up by the pressure gradient. This seepage flow produces drag forces on the individual particles, creating a total force per unit length equal to the product of the pressure gradient and the cross-sectional area occupied by the bed (Wilson, 1970).

The other driving force acts at the interface (i.e. the upper surface of the bed, see Fig. 6.1) and is caused by the interfacial shear stress, $\tau_{1,2}$. This stress depends on the difference between the velocity V_1 in the zone above the bed, and the bed velocity V_2 (of course, for a stationary bed V_2 is zero). The relation is expressed by

$$\tau_{1,2} = 0.125 \rho_f f_{1,2} (V_1 - V_2)^2 \tag{6.3}$$

where ρ_f is the density of the fluid and $f_{1,2}$ is the interfacial friction factor (to be discussed below).

Fortunately, there are two configurations for which the other portions of the model can be verified without evaluating $f_{1,2}$. One involves placing a bed of solids in a short length of pipe which is tilted, and measuring the inclination at which the bed begins to move. In this case the drag forces on the individual particles in the bed are replaced by the axial component of gravitational acceleration (Wilson, 1970).

The other configuration for which $f_{1,2}$ is not important involves flows with very high solids concentration, approaching the loose-poured value $C_{v,b}$. This is coarse-particle dense-phase flow, also known as plug flow, and the force balance for this case gives the hydraulic gradient required to initiate plug-flow motion, i_p, as

$$i_p = 2\mu_p(S_s - 1)C_{v,b} \quad (6.4)$$

Very satisfactory agreement was found between the predictions of this equation and extensive experimental investigations (Wilson *et al.*, 1972). This work was subsequently extended to account for the additional velocity-dependent component of hydraulic gradient associated with the fluid friction between the moving particulate plug and the pipe wall. The friction factor, f_f, associated with this fluid resistance was found to be a function of a Reynolds number written $V_m d/\nu_f$, where V_m is mean velocity, d is particle diameter and ν_f is the kinematic viscosity of the fluid (Wilson & Brown, 1982).

Evaluation of the interfacial friction factor $f_{1,2}$, mentioned above, was initially based on considering the interface as similar to a hydraulically rough boundary. For this type of boundary the friction factor depends on the ratio of the particle size to the hydraulic radius of the area above the bed, which in turn varies with d/D, the ratio of particle size to pipe size. It seemed reasonable that the relationship should be similar in nature to that normally found for granular roughness, but with higher values of friction factor because the saltating grains of a mobile bed are more effective than fixed grains in transferring shear stress. On this basis it was proposed that the interfacial friction factor would be roughly twice that given by the Nikuradse formula for fixed-bed roughness (Wilson, 1976). As the particle size is decreased, the effect of d/D is less important in determining the interfacial friction factor. Initially, the rough-boundary law was simply extrapolated to smaller particle sizes, but recent work on this type of friction has produced a significant modification in the analysis of $f_{1,2}$ (Wilson, 1988*a*; Wilson & Pugh, 1988).

The force–balance model which has been outlined above is rather too complex for manual solution, and various computer algorithms for this model have been developed. The output from these algorithms provides information on both the limit of stationary deposition (discussed immediately below in section 4) and on the behaviour of coarse-particle flow at velocities larger than the deposition limit (discussed in section 5).

4. LIMIT OF STATIONARY-DEPOSIT ZONE

The mean velocity (i.e. discharge/total pipe area) at the limit of the stationary-deposit zone has small values at low delivered solids concentration, rises to a maximum (denoted V_{sm}) at some specific concentration and then decreases again as the delivered solids concentration rises toward the loose-poured or plug-flow value. This behaviour is shown schematically in Fig. 6.3. In some cases the designer requires the full locus of the limit of the stationary deposition zone, but for conservative design and operation it may be sufficient to know only the maximum velocity at the limit of deposition, V_{sm}. Maintaining the operating velocity above

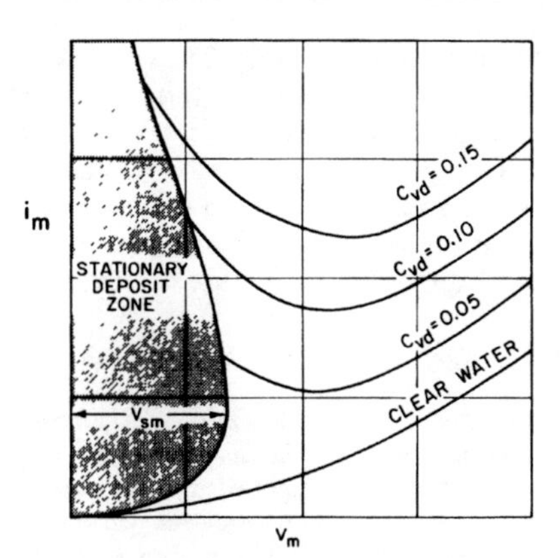

Fig. 6.3. Definition sketch for limit of stationary deposit zone.

this value ensures that deposition will not occur. The computer output shows how V_{sm} varies with internal pipe diameter, particle diameter and relative density. The effect of these variables can be expressed concisely by means of a nomographic chart (Wood, 1935; Wilson & Judge, 1978; Wilson, 1979). This chart, reproduced here as Fig. 6.4, is recommended as a practical design aid.

It should be noted, by way of explanation of the chart, that the left-hand panel

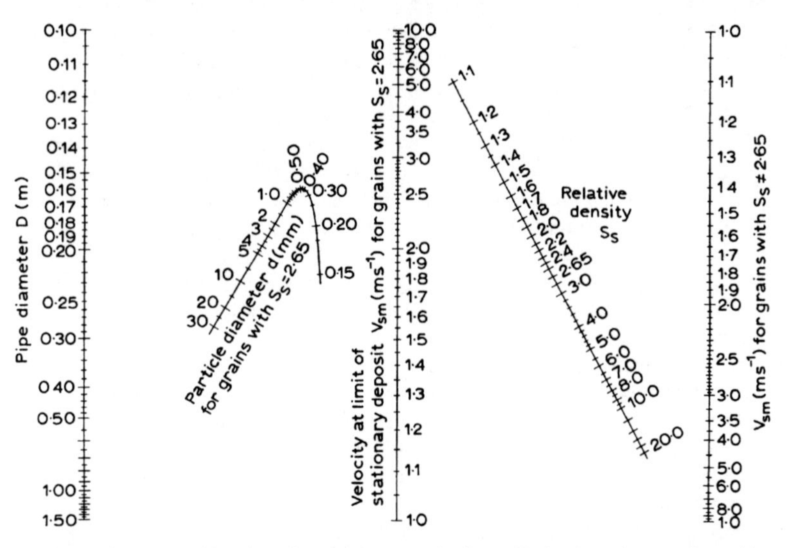

Fig. 6.4. Nomographic chart for maximum velocity at limit of stationary deposition. (Adapted from Wilson, 1979.)

deals with sand-weight materials (S_s = 2·65). The internal pipe diameter appears on the left vertical axis, with V_{sm} on the central vertical axis. The particle diameter is plotted on a curve which, on the basis of its shape, is sometimes referred to as the *Demi McDonald.* This shape is associated with the finding that for large particles, but not for small ones, the interfacial friction factor, $f_{1,2}$, increases with increasing particle size. It follows that for coarse-particle transport in a pipe of given size, V_{sm} decreases with increasing particle diameter.

WORKED EXAMPLE: EFFECT OF PARTICLE SIZE ON DEPOSITION VELOCITY

> To demonstrate the effect of particle size on V_{sm}, consider particles of S_s = 2·65 in a pipe of D = 0·4 m. This diameter is located on the left axis of Fig. 6.4 and connected by straight-edge to the particle size on the curved scale. V_{sm} is then obtained by projecting to the central vertical scale. For instance, a particle size of 0·7 mm gives V_{sm} of about 4·5 m s^{-1}, which is the largest value found for this pipe diameter and solids density. For a larger particle of, say, 5 mm the deposition-limit velocity is diminished to about 3.3 m s^{-1}.

This decrease in V_{sm} with increasing d could appear surprising at first sight, but has been fully substantiated by experimental evidence. However, when operating with centrifugal pumps it may be difficult to take advantage of the decrease of V_{sm} because of the difficulty of obtaining a stable intercept of pump and pipeline characteristics. For applications of this type, especially if control of particle size is limited, the conservative designer may wish simply to assume particles of *Murphian* size, i.e. those which give the largest value of V_{sm} for the pipe under consideration. In this case it should be noted that the values of V_{sm} obtained from Fig. 6.4 tend to be conservatively high, especially for large pipe diameters, and hence can be considered to be directly suitable for use as operating velocities.

The nomographic form of Fig. 6.4 gives an immediate indication of the sensitivity of the output to variations in input. For the sand-weight material in the 0·40 m pipe used in the example calculation it is seen that V_{sm} is scarcely affected by variations of particle diameter between 0·4 mm and 1·0 mm, whereas a change from 0·15 mm to 0·2 mm alters V_{sm} by more than 25%. The particle-diameter scale of Fig. 6.4 has not been extended below 0·15 mm since smaller particles begin to be influenced by mechanisms not included in the mathematical model on which the nomogram is based. As shown by Thomas (1979) the viscous sub-layer can have a significant effect, and all particles which are small enough to be completely embedded in this sub-layer will behave in a fashion which is no longer dependent on particle diameter. In this limiting case certain simplifications can be made in the mathematical model. Thomas (1979) found that these led directly to a formula for the shear velocity at the limit of stationary deposition, and on employing a power law approximation for the friction factor, he then obtained a corresponding expression for the deposition-limit velocity, i.e.

$$V_{sm} = 9{\cdot}0(g\nu_f(S_s - 1))^{0{\cdot}37}(D/\nu_f)^{0{\cdot}11} \tag{6.5}$$

This equation gives the minimum value of V_{sm} for small particles, assuming turbulent flow. The coefficient of eqn (6.5) applies for any consistent system of units.

Particles of relative density other than 2·65 require the use of the right-hand panel of Fig. 6.4, which, in effect re-calibrates the force–balance model on the basis that the resisting force is proportional to $(S_s - 1)$. This process is accomplished graphically by use of the inclined relative-density scale in the right-hand portion of the figure, as illustrated by the following example.

WORKED EXAMPLE: EFFECT OF PARTICLE DENSITY ON DEPOSITION VELOCITY

Consider the pipe of $D = 0{\cdot}4$ m and the particles with $d = 0{\cdot}7$ mm and $d = 5{\cdot}0$ mm, as used in the earlier example, but now with $S_s = 1{\cdot}50$ (a value that might be found for coal). If the 0·7 mm particle had been sand, the value of V_{sm}, read on the central vertical axis of Fig. 6.4, would have about 4·5 m s^{-1}, as shown previously. This point on the central vertical axis is now joined by straight-edge to $S_s = 1{\cdot}5$ on the inclined axis, and projected to the right-hand vertical axis, giving a value of $V_{sm} \simeq 2{\cdot}4$ m s^{-1} for the 0·7 mm coal. Similarly, the coal of $d = 5{\cdot}0$ mm has $V_{sm} \simeq 1{\cdot}7$ m s^{-1}.

It must be recognised that particles which differ in density from sands may also have different values of other properties, including the solids fraction in the deposit and the mechanical friction coefficient between the particles and the pipe. The computer program used in developing the nomographic chart employed values of these quantities which apply specifically to sands. Therefore the values for V_{sm} obtained from Fig. 6.4 for other materials should be treated as somewhat less accurate than values for sand. However, both the solids fraction in the bed and the particle–pipe mechanical friction coefficient occur in the force–balance model only as multiples of $(S_s - 1)$, and it is found that any change in either of these quantities merely gives rise to an apparent value of S_s. If V_{sm} has been obtained for pilot-plant tests with the material of interest, the appropriate apparent value of S_s can be found on the inclined axis and used to calibrate the nomographic chart for this material.

The preceding portion of this section has been concerned with the maximum value of the velocity at the limit of the stationary deposit zone, V_{sm}. Although this value is of primary interest in insuring deposition-free operation, there are cases where a pipeline is operated at a concentration widely different from that which produces V_{sm}. In such cases it may be possible to operate at smaller velocities and still avoid deposition. To deal with these the technique has been extended so that the complete locus of the limit of the stationary deposition zone can be plotted (Wilson, 1986), but because of space limitations this work cannot be discussed here. Another influence which can only be mentioned in this chapter is pipe inclination. It is known experimentally that moderate upward inclinations in the direction of flow tend to increase V_{sm}; this increase is most pronounced at inclination angles near 30° (Wilson & Tse, 1984). The effect of inclination on V_{sm} is included in a recently published computer algorithm (Wilson, 1988*b*). A related topic, the influence of pipe inclination on friction gradient, was studied by Wilson and Byberg (1987).

Before concluding the section on deposition limit, it should be noted that the output from the nomographic chart discussed above can provide a useful initial approach to pipe sizing. This feature can best be illustrated by an example, as follows:

WORKED EXAMPLE: USE OF DEPOSITION VELOCITY FOR INITIAL PIPE SIZING

1 200 t h^{-1} (dry weight basis) of coarse sand (S_s = 2·65) is to be transported. Head-end considerations indicate a maximum delivered volumetric concentration $C_{v,d}$ of 0·25. Assuming that pipes are available with D in increments of 50 mm, select appropriate combinations of D, $C_{v,d}$ and V_m.

To convert 1 200 t h^{-1} to m^3 s^{-1} of *solids* (Q_p) divide by the solids density of 2·65 t m^{-3}, and by 3 600 s h^{-1}, giving

$$Q_p \;=\; 1\,200/(2{\cdot}65(3\,600)) \;=\; 0{\cdot}126\,\mathrm{m^3\,s^{-1}}$$

The mixture discharge, Q, equals $Q_p/C_{v,d}$, from which the minimum Q, corresponding to the maximum $C_{v,d}$ of 0·25, is 0·503 m^3 s^{-1}.

A pipe of D = 0·35 m is considered first. Its area is 0·0962 m^2, and hence the velocity must be at least equal to 0·503/0·0962 or 5·24 m s^{-1}. For the *Murphian* particle (about 0·6 mm for this pipe diameter) Fig. 6.4 gives V_{sm} as approximately 4·2 m s^{-1}, i.e. the required velocity of 5·24 m s^{-1} is some 25% larger than the maximum V_{sm} for this pipe. This appears to be acceptable, giving one appropriate combination as D = 0·35 m, V_m = 5·24 m s^{-1} and $C_{v,d}$ = 0·25.

The next smaller pipe at D = 0·30 would require a velocity of 7·12 m s^{-1}, which is about 1·9 times the Murphian value of V_{sm}, and would be too high for efficient operation.

In the opposite direction, a pipe with D = 0·40 m would require a velocity of only 4.0 m s^{-1} from continuity considerations, but this is considerably less than V_{sm} of 4·5 m s^{-1} (found in a previous example). In this case the velocity would have to be increased to 4·5 m s^{-1} by reducing the solids concentration, giving a second appropriate combination as D = 0·40 m, V_m = 4·5 m s^{-1}, $C_{v,d}$ = 0·223.

5. FULLY STRATIFIED FLOW

As noted in section 3, for fully stratified flow the submerged weight of virtually all the particles is transferred to the invert of the pipe by intergranular contact, producing a total normal force, F_n. A driving force beyond that needed to counteract fluid resistance is required to overcome the mechanical friction force equal to $\mu_p F_n$, where μ_p is the coefficient of mechanical friction. The expression of Newitt *et al.* (1955) for the resulting excess pressure gradient (eqn. (6.1) above) is based on the assumption that F_n is proportional to the submerged weight of the solids, which in turn is assumed to be proportional to the delivered concentration $C_{v,d}$. The detailed computer modelling of stratified flows gives results which are not so simple.

Although fully stratified flows can be analysed completely by the force–balance computer algorithm, it has been found that this approach is not particularly convenient for the designer. Instead, the computer can be used to obtain generalised solutions which are then approximated graphically, or by simplifying *fit* functions (Wilson, 1988*b*). These approximations make use of three ratios or dimensionless variables: relative concentration, relative excess gradient and relative velocity. The relative concentration, C_r, is simply the ratio of the delivered volumetric concentration $C_{v,d}$ to the maximum value (without compaction) $C_{v,b}$, thus

$$C_r = C_{v,d}/C_{v,b} \tag{6.6}$$

Obviously C_r must lie in the range zero to one, with the higher value representing the plug flow of eqn (6.4). For plug flow the hydraulic gradient equals the sum of i_p (from eqn (6.4)) and the fluid gradient i_f. For values C_r less than unity the excess gradient $(i_m - i_f)$ will be less than i_p. With i_w (the gradient for an equal discharge of water) used to represent i_f, the ratio $(i_m - i_w)/i_p$ forms a suitable dimensionless measure of the excess gradient. This relative excess pressure gradient will be denoted here by ζ, i.e.

$$\zeta = \frac{(i_m - i_w)}{i_p} \tag{6.7}$$

The equation of Newitt *et al.* (1955) for fully stratified flow (eqn (6.1)) is equivalent to setting ζ equal to $BC_r/2\mu_p$, i.e. a linear proportionality in C_r, with velocity having no influence.

In reality, velocity does influence the relative excess gradient, especially near the deposition limit where there is considerable hold-up. Hold-up indicates that solid particles travel, on average, more slowly than the liquid phase, so that the in-situ or resident concentration exceeds the delivered concentration. This feature causes an increase in ζ with decreasing velocity, leading to minima in curves of i_m versus V_m at constant $C_{v,d}$. Observations of this behaviour by Fowkes and Wancheck (1969) have been discussed by Wilson (1976). The force–balance model shows that, for constant C_r, the effect of hold-up on ζ depends on velocity relative to the deposition limit V_{sm}. This relative velocity is denoted by V_r, i.e.

$$V_r = V_m/V_{sm} \tag{6.8}$$

For the fully stratified flow dealt with here, ζ is a function of both C_r and V_r. Although the force–balance model does not provide the required functional relation in closed form, it has been used to obtain simplified *fit* functions (Wilson, 1988*b*). Even at large relative velocities, where ζ is not significantly affected by changes in V_r, it was found that ζ is not a linear function of C_r (as has been indicated by eqn (1)). Instead it can be approximated by the fit function

$$\zeta \simeq 0{\cdot}5C_r(1 + C_r^{0{\cdot}66}) \qquad (\text{for } V_r \gg 1) \tag{6.9}$$

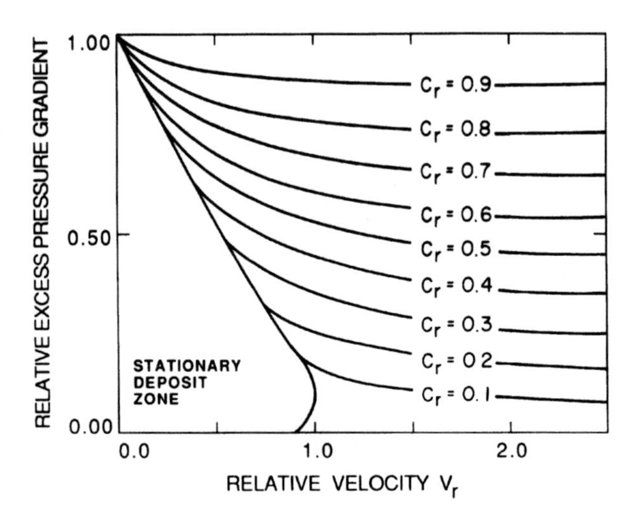

Fig. 6.5. Curves of relative excess pressure gradient. (Adapted from Wilson, 1988*b*.)

At lower values of V_r the fit functions become more complicated, and it is convenient to employ a graphical version (Wilson, 1988*b*), reproduced here as Fig. 6.5.

The fully stratified flow for which this figure applies is not an energy-efficient mode of transport, but it can eliminate problems of head-end processing and tail-end separation, and thus may be attractive for reasonably short lines. In such cases highly sophisticated computational methods are not required, and Fig. 6.5 is recommended as a design tool.

An example of the use the fit functions underlying Fig. 6.5 has been given elsewhere (Wilson, 1988*b*), based on the transport of fist-size clay balls in a pipe of 0·70 m diameter. The results were found to compare very well with field measurements. A direct application of Fig. 6.5 will be illustrated below using gravel as the material to be transported.

WORKED EXAMPLE: FRICTION LOSS FOR FULLY STRATIFIED FLOW

The gravel has diameter 5·0 mm and relative density 2·65, and the pipe through which it moves has $D = 0{\cdot}40$ m and clear-water friction factor, $f_w = 0{\cdot}013$. The mean velocity, V_m, will be 5·0 m s^{-1}, and for these conditions the head provided by the pump is equivalent to a gradient, i_m of 0·130 (m water per m pipe). It is necessary to determine the amount of solids which can be pumped ($C_{v,d}$ and Q_p) and the SEC.

In order to proceed, several quantities must be calculated. The clear-water friction gradient, i_w, is given by $f_w V_m^2/(2gD)$, i.e. 0·013(25)/(19·62(0·40)) or 0·041. The plug-flow friction gradient (eqn (6.4)) requires knowledge of μ_s and $C_{v,b}$, and typical values of 0·4 and 0·6, respectively, will be assumed, giving $i_p = 2(0{\cdot}4)(1{\cdot}65)(0{\cdot}6)$ or 0·792.

The relative pressure gradient, ζ, is then calculated from eqn (6.7), i.e.

$\zeta = (0{\cdot}130 - 0{\cdot}041)/0{\cdot}792$ or $0{\cdot}112$. The relative velocity is also required, based on V_{sm} for this combination of pipe and particles. As seen from an earlier worked example, this value of V_{sm} is approximately $3{\cdot}3\,m\,s^{-1}$. For $V_m = 5{\cdot}0\,m\,s^{-1}$, $V_r = 5{\cdot}0/3{\cdot}3$ or about $1{\cdot}5$.

Figure 6.5 can now be entered with the values of ζ and V_r found above, giving the appropriate value of C_r, i.e. $0{\cdot}10$. The delivered concentration, $C_{v,d}$, is the product of C_r and $C_{v,b}$, thus $C_{v,d} = 0{\cdot}10(0{\cdot}60)$ or $0{\cdot}06$. The tonnes per hour of solids transport in this pipe (cross-sectional area $0{\cdot}126\,m^2$) is $0{\cdot}06(5{\cdot}0)(0{\cdot}126)(2{\cdot}65)(3\,600) = 360\,t\,h^{-1}$. Using eqn (6.2), the specific energy consumption is found to be $2{\cdot}23\,kWh\,t^{-1}\,km^{-1}$.

Figure 6.5 can, of course, also be used to solve problems where $C_{v,d}$ is given and i_m is the quantity to be found. In such cases the known values of C_r and V_r are entered on Fig. 6.5, and ζ is read from the figure. This value of ζ is multiplied by i_p to obtain $i_m - i_w$, and the calculated value of i_w is added to give i_m. If a constant C_r locus is desired, various V_m are assumed and used to obtain a series of V_r values for which i_m is calculated. The locus of i_m versus V_m can then be plotted.

6. ANALYSIS OF HETEROGENEOUS FLOW

As noted previously, heterogeneous flow, which involves particle support by both intergranular contact and fluid suspension, forms a transition zone between fully stratified flow and homogeneous flow. The stratification ratio, representing the fraction of solids supported by intergranular contact, is equal to unity for fully stratified flows, is effectively zero for homogeneous flows, and lies between zero and unity for heterogeneous flows.

Stratification obviously could not occur in a weightless environment, where the concentration and velocity profiles, and hence the pressure gradients, would be the same for flows in any direction. Within the Earth's gravitational field the gradient which would exist in the absence of stratification can be approximated by averaging the frictional losses for rising and descending vertical flows (Wilson, 1972). This averaged vertical gradient is written $\bar{i}_v$ and measures the energy loss due to friction in terms of m of water per m of pipe. The hydraulic gradient for horizontal flow of a mixture, i_m, is expressed in the same units, and the excess gradient associated with stratification is $i_m - \bar{i}_v$. This quantity gives a direct measure of the additional energy required to overcome the effect of gravitational attraction at right angles to the flow.

To a first approximation the excess gradient should be proportional to the submerged weight of contact-load solids, from which it follows that for fully stratified flow (all particles travelling as contact-load) the ratio $(i_m - \bar{i}_v)/(S_m - 1)$ should be approximately constant. If the mean vertical gradient, $\bar{i}_v$, has not been measured, it is common practice to use in its place the calculated gradient for an equal discharge of water, i_w. This substitution yields the approximate relation for fully stratified flow proposed by Newitt's group (eqn (6.1)).

In the heterogeneous flow range, however, the stratification ratio is far from

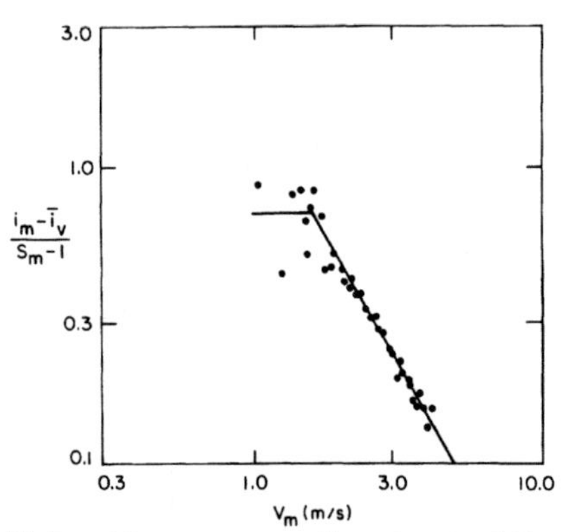

Fig. 6.6. Fully stratified and heterogeneous flow data for 0·7 mm sand in 52·4 mm pipe. (Adapted from Wilson, 1973.)

constant, decreasing markedly with increasing velocity. Newitt *et al.* (1955) correlated their heterogeneous-flow data ($V_m > 17u_\infty$) on the basis of a proportionality between $(i_m - i_w)/(S_m - 1)$ and u_∞/V_m. Similarly, algebraic rearrangement of the Durand equation gives a proportionality between $(i_m - i_w)/(S_m - 1)$ and an expression which is the product of u_∞/V_m and $C_d^{-0.25} f_w (D/d)^{0.5}$, where C_d is the particle drag coefficient, f_w is the clear-water friction factor and D and d are the pipe and particle diameters, respectively. As noted in section 1, Babcock (1971) had shown that Durand's equation had serious weaknesses. However, the point to be made here is simply that Durand's equation, like Newitt's, gives a velocity effect of V_m^{-1}.

By the early 1970s it had become obvious that reality was not so simple. Figure 6.6 shows results for a slurry of 0·7 mm sand in a pipe of internal diameter 52·4 mm. The ordinate of the figure is $(i_m - \bar{i}_v)/(S_m - 1)$, making use of the enhanced accuracy of the gradient excess based on $\bar{i}_v$. At velocities below about 1·8 m s^{-1} the points show significant scatter but no well-defined slope, as expected for fully stratified flow. (This scatter arises because the gradient excess for fully-stratified flow is not linearly related to $(S_m - 1)$, a point noted in section 5). This behaviour at velocities above 1·8 m s^{-1} is totally different, showing very little scatter and a clearly defined downward trend approximating a straight line on the logarithmic coordinates of the figure. This result may be expressed as

$$\frac{(i_m - i_f)}{(S_m - 1)} \propto V_m^{-M} \tag{6.10}$$

where i_f denotes the fluid friction gradient, i.e. $\bar{i}_v$ or, when this is not available, i_w. For the data shown on Fig. 6.6, M is about 1·7, a typical value for particles of narrow size grading (Clift *et al.*, 1982). It is important to note that Fig. 6.6 indicates clearly that M is considerably larger than the value of 1·0 required by both the Durand and Newitt correlations.

As remarked earlier, to a first approximation the excess gradient should be proportional to the submerged weight of the contact-load solids. Thus, the observed decrease in $(i_m - i_f)/(S_m - 1)$ implies a commensurate change in the stratification ratio, i.e. the fraction of solids travelling as contact load, and from a practical viewpoint $(i_m - i_f)/(S_m - 1)$ is taken as directly proportional to the stratification ratio. It follows that the remaining fraction of solids, the suspended load, increases with increasing V_m. This increase is to be expected, as the effectiveness of fluid support depends on the velocity of turbulent eddies and on the velocity gradient, and both these quantities increase with the fluid shear velocity, U_*. As U_* equals $V_m(f_f/8)^{0.5}$, and the fluid friction factor, f_f, is effectively constant within the velocity range of interest, it is seen that for given pipe size and particle properties V_m is indeed the independent variable governing stratification ratio (discussion of this point, with experimental findings, is given by Wilson, 1972).

The power-law expression of eqn (6.10) implies an abrupt change in behaviour between fully stratified flow and heterogeneous flow, occurring at a certain mean velocity which defines the threshold of fluid suspension. The formulation of Newitt *et al.* (1955) suggested that this threshold velocity is $17u_\infty$, but work done in the early 1970s (Wilson, 1972; Wilson & Watt, 1974) introduced the fluid friction factor, f_f, mentioned above, and also showed that the particle-to-pipe diameter ratio must be included (the size of typical turbulent eddies increases with pipe diameter, and small eddies cannot support large particles). Nevertheless, the identification of such a threshold velocity is not easy, especially in large pipes, where the deposition velocity may exceed the expected threshold for turbulent suspension. Indeed, it is doubtful that a clearly defined threshold exists, since, as noted previously, heterogeneous flow is simply a transition zone between fully stratified and homogeneous flow. Such transitions generally take the form of sigmoidal (ogee) curves and in the present case the logical choice for a transition curve is the integrated log-normal distribution.

The log-normal function is defined by only two parameters (the mean and standard deviation of the logarithms), and the central region of the integrated log-normal distribution does not differ greatly from an approximating power law, so that data such as shown on Fig. 6.6 can be fitted by a log-normal as well as by a power law. On the other hand, the tails of a log-normal distribution approximate straight lines on semi-logarithmic axes and it is of interest that a semi-logarithmic plot has recently been used to correlate stratification-ratio data (Gillies *et al.*, 1990).

Once the basic form of the relation between stratification ratio and velocity is seen as an ogee curve rather than a power law, it becomes clear that the major defining parameter should represent the mid-point, not some *threshold* value. This parameter can be denoted as V_{50}, indicating the value of V_m at which the stratification ratio is 0·50, i.e. half the mass of the particles is supported by intergranular contact and half by fluid suspension. From a practical standpoint it is necessary to match the 50% stratification point with a specific value of $(i_m - i_f)/(S_m - 1)$. The mechanics of the situation suggests that this value should be $0.5\mu_p$ or a little larger, and as the friction coefficient μ_p is usually found to be near 0·4, it can be expected that V_{50} will correspond to a value of $(i_m - i_f)/(S_m - 1)$ between 0·20 and 0·25.

Experiments carried out at the GIW Hydraulic Laboratory, Augusta, GA, USA (Clift *et al.*, 1982; Wilson *et al.*, 1990) indicate the value of 0·22, which will be employed below.

In evaluating V_{50}, many of the points mentioned previously in connection with the earlier concept of a threshold velocity can be used. Thus, V_{50} should vary with $(8/f_w)^{0.5}$ and depend on the diameter ratio d/D. The exponential expression suggested earlier (Wilson & Watt, 1974) is in need of some modification and has now been replaced by a relation using the hyperbolic cosine, i.e. $\cosh(60d/D)$. The most important change pertains to the influence of the particle settling velocity. Previous formulations, following Newitt and others, had considered the ratio V_m/u_∞ as a unit, implying a direct proportionality between V_{50} and the particle settling velocity. Formulations of this type are in accordance with the perception that fluid suspension is caused solely by turbulent diffusion; a mechanism which had been described earlier by Schmidt (1932) and Rouse (1937). Although diffusion is dominant in the core of the flow (Hsu *et al.*, 1980), two other mechanisms of fluid support may be important in the near-wall layers where velocity gradients are large. The first is the velocity-gradient lift which occurs in the absence of particle rotation (Einstein & El-Samni, 1949), the second is the Magnus lift which accompanies particle rotation.

Although the separation of the effects of these two lift mechanisms is not generally possible in practice, the importance of their combined operation can be seen in the experimental results of Shook (1985), who showed solids concentration increasing with height near the bottom of the pipe, behaviour which would be impossible if turbulent support by turbulent diffusion were the only cause of fluid suspension. A definitive analysis of all components of fluid suspension is not to be expected in the near future, but an approximate approach gives useful indications to the type of behaviour to be expected.

Specifically, as the particle size becomes progressively finer, the value of V_{50} does not tend toward zero, as does the particle fall velocity. Instead, for particles much smaller than the thickness of the viscous sub-layer, V_{50} approaches a value which depends on the properties of the carrier fluid but is no longer dependent on particle size or settling velocity. This small-particle behaviour can be represented by dividing u_∞, by a dimensionless measure of settling velocity, given by $N_u^{1/3}$, where N_u is defined as

$$N_u = \frac{0{\cdot}75u_\infty^3}{(S_s - 1)g\nu_f} \tag{6.11}$$

The small-particle effect on V_{50} is thus proportional to $u_\infty(N_u)^{1/3}$ or to $((S_s - 1)g\nu_f)^{1/3}$, while for larger particles the dependence on u_∞ will still be involved. Summing the two effects gives a reasonable representation of their combined influence on V_{50}.

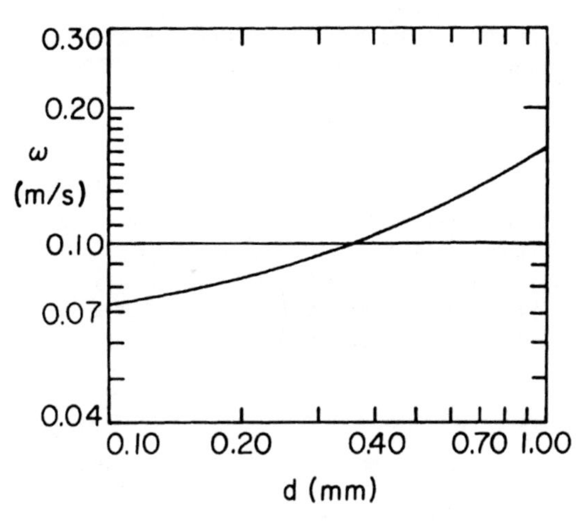

Fig. 6.7. Curve for estimating ω, based on sand–water slurries at 20°C.

7. CALCULATION OF FRICTION LOSS FOR HETEROGENEOUS FLOW

Denoted here as ω, the sum referred to above contains two numerical coefficients that require evaluation. The evaluation was carried out by comparison with experimental results obtained at the GIW Hydraulic Laboratory, Augusta, GA, USA (Clift *et al.*, 1982; Wilson *et al.*, 1990), producing the following expression for the particle-associated velocity, ω:

$$\omega = 2{\cdot}7((S_s - 1)gv_f)^{1/3} + 0{\cdot}9u_\infty \tag{6.12}$$

Although much additional experimental and analytical work is required, the resulting formulation for V_{50} appears to represent a suitable design-orientated approximation. It is

$$V_{50} = \omega\sqrt{\frac{8}{f_f}}\cosh(60d/D) \tag{6.13}$$

In applying eqn (6.13), it is convenient to have a curve giving ω versus particle diameter. Figure 6.7 shows this curve for sand particles ($S_s = 2{\cdot}65$) in water at 20°C. Equation (6.12) was used for ω and the evaluation of u_∞ followed the methods described by Clift *et al.* (1978), and employed the value 0·65 for the velocity correction factor, K_A.

The logarithm of V_{50} represents the mean of the log-normal distribution, and the second parameter of this distribution is the standard deviation. One method of applying the log-normal function to calculate the effects of particle grading is given in Wilson *et al.* (1990), but for practical use it is considered better to approximate the central portion of the integrated log-normal by a power law relation such as that employed in Fig. 6.6. Within its range of validity, which cannot extend above $(i_m - i_f)/(S_m - 1) = 0{\cdot}6$, this approximation is represented by

$$\frac{(i_m - i_f)}{(S_m - 1)} = 0{\cdot}22\left(\frac{V_m}{V_{50}}\right)^{-M} \tag{6.14}$$

For narrow particle grading $M \simeq 1{\cdot}7$ as noted previously. A useful grading parameter is the ratio d_{85}/d_{50}, where the subscripts indicate the percent of the total mass of solids which is finer than the particle size in question. Narrow gradings are represented by d_{85}/d_{50} ratios less than 1·5. Broad gradings correspond to values greater than 1·5, and have M values smaller than 1·7. For example, the broadest size grading studied by Wilson *et al.* (1990) had a d_{85}/d_{50} of about 2·8, and an M value near 0·9.

WORKED EXAMPLE: FRICTION LOSS FOR HETEROGENEOUS FLOW

As an example consider the transport of sand ($S_s = 2{\cdot}65$) at 1 500 t h^{-1} (dry weight basis) in a pipe of 0·40 m internal diameter. As with the worked example in section 5, it will be assumed that $V_m = 5{\cdot}0\,\text{m s}^{-1}$ and $f_w = 0{\cdot}013$. The frictional hydraulic gradient and specific energy consumption are to be calculated, based on sand with $d_{50} = 0{\cdot}5$ mm and $d_{85} = 0{\cdot}7$ mm.

The first step is to check that the proposed operating conditions are clear of the zone of stationary deposition. This is done by using Fig. 6.4, as illustrated in the first worked example of section 4. For either d_{50} or d_{85}, the resulting value of V_{sm} is about $4{\cdot}5\,\text{m s}^{-1}$, indicating that the proposed operating velocity of $5{\cdot}0\,\text{m s}^{-1}$ is acceptable.

The next point to note is that d_{85}/d_{50} is less than 1·5, so that the grading is narrow, and the typical M value of 1·7 can be used. From Fig. 6.7, ω is seen to be $0{\cdot}115\,\text{m s}^{-1}$ for this sand. With d_{50} employed for the particle diameter, the value of cosh $(60d/D)$ is 1·003, and with f_f set equal to 0·013, eqn (6.13) gives $V_{50} \simeq 2{\cdot}9\,\text{m s}^{-1}$.

The solids transport rate of 1 500 t h^{-1} gives Q_p of 1 500(2·65(3 600)) or $0{\cdot}157\,\text{m}^3\,\text{s}^{-1}$. With $Q = 5{\cdot}0(0{\cdot}126)$ or $0{\cdot}628\,\text{m}^3\,\text{s}^{-1}$; $C_{v,d} = 0{\cdot}25$. Thus, $(S_m - 1) = (1{\cdot}65)(0{\cdot}25) = 0{\cdot}412$. Multiplying this quantity by the right-hand side of eqn (6.14) gives $(i_m - i_f)$ as

$$(0{\cdot}412)(0{\cdot}22)\left(\frac{5{\cdot}0}{2{\cdot}9}\right)^{-1{\cdot}7} = 0{\cdot}036$$

The water gradient, i_f, has already been evaluated (in the worked example in section 5) as 0·041, giving the required hydraulic gradient of the mixture, i_m, as 0·077 (m water per m pipe).

Finally, the SEC is calculated as $2{\cdot}73 i_m/S_s C_{v,d} = 0{\cdot}317\,\text{kWh t}^{-1}\,\text{km}^{-1}$ from eqn (6.2).

It is interesting to compare the SEC for the above example of heterogeneous flow with that of the previous example of friction loss for fully stratified flow. For the fully stratified flow with gravel the specific energy consumption was $2{\cdot}23\,\text{kWh t}^{-1}\,\text{km}^{-1}$, i.e. more than seven times that for the heterogeneous flow with sand. This comparison clearly shows the economic benefit to be obtained by heterogeneous, rather than fully stratified flow.

8. NOTATION

B	Numerical coefficient in eqn (6.1) [1]
C_d	Particle drag coefficient [1]
C_r	Relative volumetric solids concentration, see eqn (6.6) [1]
$C_{v,b}$	Volumetric solids concentration in loose-poured bed (fractional) [1]
$C_{v,d}$	Delivered volumetric solids concentration (fractional) [1]
d	Particle diameter [L]
d_{50}	Mass-median particle diameter [L]
d_{85}	Diameter for which 85% (by mass) of the particles are finer [L]
D	Internal pipe diameter [L]
f_f	Darcy–Weisbach friction factor for fluid [1]
f_w	Darcy–Weisbach friction factor for equal flow of water [1]
$f_{1,2}$	Interfacial friction factor [1]
F_n	Normal intergranular force against pipe [$M\,L\,T^{-2}$]
F_w	Submerged weight of solids [$M\,L\,T^{-2}$]
i_f	Frictional hydraulic gradient of fluid (m water per m pipe) [1]
i_m	Frictional hydraulic gradient of mixture (m water per m pipe) [1]
i_p	Frictional hydraulic gradient for plug flow, see eqn (6.4) (m water per m pipe) [1]
$\bar{i}_v$	Average hydraulic gradient in ascending and descending vertical flow (m water per m pipe) [1]
i_w	Hydraulic gradient of equal discharge of water (m water per m pipe) [1]
K_A	Velocity correction factor used to determine u_∞ for non-spherical particles [1]
M	Exponent in eqns (6.10) and (6.14) [1]
N_u	Dimensionless settling velocity, see eqn. (6.11) [1]
Q	Volumetric discharge [L^3T^{-1}]
Q_p	Volumetric discharge of solids [L^3T^{-1}]
SEC	Specific energy consumption ($kWh\,t^{-1}\,km^{-1}$), see eqn (6.2)
S_m	Relative density of mixture (density of mixture/density of water) [1]
S_s	Relative density of solids (density of solids/density of water) [1]
u_∞	Terminal settling velocity of particle [$L\,T^{-1}$]
U_*	Shear velocity [$L\,T^{-1}$], equal to $V_m\,(f_f/8)^{0.5}$
V_m	Mean velocity [$L\,T^{-1}$]
V_r	Relative velocity, see eqn (6.8) [1]
V_{sm}	Mean velocity at limit of deposition, see Fig. 6.1 [$L\,T^{-1}$]
V_1	Velocity in upper zone [$L\,T^{-1}$], see eqn (6.3)
V_2	Velocity in lower zone [$L\,T^{-1}$], see eqn (6.3)
V_{50}	Value of V_m at which the stratification ratio is 0·50 [$L\,T^{-1}$], see eqn (6.13)
ζ	Relative excess pressure gradient, see eqn (6.7) [1]

μ_p Coefficient of mechanical friction between particles and pipe [1]
ν_f Kinematic viscosity of fluid $[L^2 T^{-1}]$
ρ_f Fluid density $[M L^{-3}]$
σ_p Intergranular normal stress $[M L^{-1} T^{-2}]$
τ_p Intergranular shear stress $[M L^{-1} T^{-2}]$
$\tau_{1,2}$ Interfacial shear stress $[M L^{-1} T^{-2}]$
ϕ Internal friction angle of particles [1]
ω Particle-associated velocity $[L T^{-1}]$, see eqn (6.12)

9. REFERENCES

Babcock, H.A. (1971). Heterogeneous flow of heterogeneous solids. In *Advances in Solid–Liquid Flow in Pipes and its Application*, ed. I. Zandi. Pergamon Press, Oxford, UK, pp. 125–48.

Bagnold, R.A. (1956). The flow of cohesionless grains in fluids. *Phil. Trans. Royal Soc.* (London), Ser. A, **249**, 235–97.

Blatch, N.S. (1906). Discussion of 'Works for the purification of the water supply of Washington D.C.' (Hazen, A. & Hardy, E.D.). *Trans. ASCE*, **57**, 400–9.

Carstens, M.R. & Addie, G.R. (1981). A sand–water slurry experiment. *Proc. ASCE (J. Hydraulics Div.)*, **107**, 501–7.

Clift, R., Grace, J.R. & Weber, M.E. (1978). *Bubbles, Drops and Particles*. Academic Press, London, UK.

Clift, R., Wilson, K.C., Addie, G.R. & Carstens, M.R. (1982). A mechanistically-based method for scaling pipeline tests for settling slurries. In *Proc. HT 8*, pp. 91–101.*

Coulomb, C.A. (1776). Essai sur une application des règles de maximis et minimis à l'architecture. *Memoires de mathématique et de physique presentés à l'Academie Royal des Sciences*, **VII**, 343–82. (Republished with English translation and commentary: Heyman, I. (1972). *Coulomb's Memoir on Statistics*. Cambridge University Press, Cambridge, UK.)

Durand, R. (1951). Transport hydraulique de groviers et galets en conduits. *La Houille Blanche*, **B**, 609–19.

Einstein, H.A. & El-Samni, A. (1949). Hydrodynamic forces on a rough wall. *Rev. Mod. Phys.*, **21**, 520–42.

Fowkes, R.S. & Wancheck, G.A. (1969). Materials handling research; hydraulic transportation of coarse solids. Report 7283, Bureau of Mines, Dept of the Interior, Washington, DC, USA.

Gibert, R. (1960). Transport hydraulique et réfoulement des mixtures en conduites. *Annales des Ponts et Chausées*, Mai–Juin, 307–73, Juil.-Août, 437–93.

Gillies, R., Shook, C.A. & Wilson, K.C. (1990). An improved two layer model for horizontal slurry pipeline flow. *Can. J. Chem. Engng* (accepted for publication).

Hsu, S.T., Van der Beken, A., Landweber, L. & Kennedy, J.F. (1980). Sediment suspension in turbulent pipeline flow. *Proc. ASCE (J. Hydraulics Div.)*, **106**, 1783–92.

Newitt, D.M., Richardson, J.F., Abbot, M. & Turtle, R.B. (1955). Hydraulic conveying of solids in horizontal pipes. *Trans. Inst. Chem. Engrs*, **33**, 93–110.

Rouse, H. (1937). Modern conceptions of the mechanics of fluid turbulence. *Trans. ASCE*, **102**, 436–505.

Schmidt, W. (1932). Der Massenaustausch in freier Luft und werwandte Erscheinungen. *Die Wasserwirtschaft*, **5–6**. (Also see the footnote in: Yalin, M.S. (1977). *Mechanics of Sediment Transport (2nd edn)*. Pergamon Press, Oxford, UK, Chapter 6.)

Shook, C.A. (1985). Experiments with concentrated suspensions of slurries with densities slightly greater than the carrier fluid. *Can. J. Chem. Engng*, **63**, 861–9.

Shook, C.A., Gillies, R., Haas, D.B., Husband, W.H.W. & Small, M. (1982). Flow of coarse and fine sand slurries in pipelines. *J. Pipelines*, **3**, 13–21.

Thomas, A.D. (1979). The role of laminar/turbulent transition in determining the critical deposit velocity and the operating pressure gradient for long distance slurry pipelines. In *Proc. HT 6*, pp. 13–26.*

Wilson, K.C. (1970). Slip point of beds in solid–liquid pipeline flow. *Proc. ASCE (J. Hydraulics Div.)*, **96**, 1–12.

Wilson, K.C. (1972). A formula for the velocity required to initiate particle suspension in pipeline flow. In *Proc. HT 2*, paper E2, pp. E23–E36.*

Wilson, K.C. (1976). A unified physically-based analysis of solid-liquid pipeline flow. In *Proc. HT 4*, paper A1, pp. A1–A16.*

Wilson, K.C. (1979). Deposition-limit nomograms for particles of various densities in pipeline flow. In *Proc. HT 6*, pp. 1–12.*

Wilson, K.C. (1982). The dense-phase option for coarse coal pipelining. *J. Pipelines*, **2**, 95–101.

Wilson, K.C. (1986). Effect of solids concentration on deposit velocity. *J. Pipelines*, **5**, 251–7.

Wilson, K.C. (1988*a*). Evaluation of interfacial friction for pipeline transport models. In *Proc. HT 11*, pp. 107–16.*

Wilson, K.C. (1988*b*). Algorithm for coarse-particle transport in horizontal and inclined pipes. In *Proceedings International Symposium on the Hydraulic Transport of Coal and Other Minerals (ISHT 88)*. CSIR and Indian Institute of Metals, Bhubaneswar, India, pp. 103–26.

Wilson, K.C. & Brown, N.P. (1982). Analysis of fluid friction in dense-phase pipeline flow. *Can. J. Chem. Engng*, **60**, 83–6.

Wilson, K.C. & Byberg, S.P. (1987). Stratification—ratio scaling technique for inclined slurry pipelines. In *Proc. STA 12*, pp. 59–64.*

Wilson, K.C. & Judge, D.G. (1978). Analytically-based nomographic charts for sand–water flow. In *Proc. HT 5*, pp. A1–11.*

Wilson, K.C. & Pugh, F.J. (1988). Dispersive-force modelling of turbulent suspension in heterogeneous slurry flow. *Can. J. Chem. Engng*, **66**, 721–7.

Wilson, K.C. & Tse, J.K.P. (1984). Deposition limit for coarse-particle transport in inclined pipes. In *Proc. HT 9*, pp. 149–69.*

Wilson, K.C. & Watt, W.E. (1974). Influence of particle diameter on the turbulent support of solids in pipeline flow. In *Proc. HT 3*, paper E1, pp. E1–E13.*

Wilson, K.C., Streat, M. & Bantin, R.A. (1972). Slip-model correlation for dense two-phase flow. In *Proc. HT 2*, paper B1, pp. B1–B10.*

Wilson, K.C., Clift, R., Addie, G.R. & Maffett, J. (1990). Effect of broad particle grading on slurry stratification ratio and scale-up. *Powder Technol.*, **62**, (2), 165–72.

Wood, F.M. (1935). Standard nomographic forms for equations in three variables. *Can. J. Res.*, **12**, 14–40.

*Full details of the Hydrotransport (HT) and the Slurry Transport Association (STA) series of conferences can be found on pp. 11–13 of Chapter 1.

7

Pipeline Design for Non-Settling Slurries

Nigel I. Heywood
Warren Spring Laboratory, Stevenage, UK

1. TOTAL PRESSURE LOSS ESTIMATION

Design of pipe work for transporting non-settling slurries is based on the type of slurry involved. The three basic kinds are

- non-settling slurries that behave as homogeneous non-Newtonian fluids,
- stabilised slurries, and
- slurries that show thixotropic properties (generally non-settling).

For slurries that behave as homogeneous Newtonian or non-Newtonian fluids,

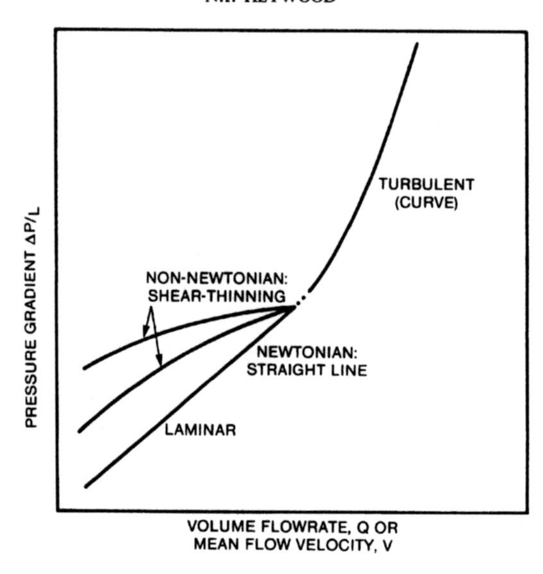

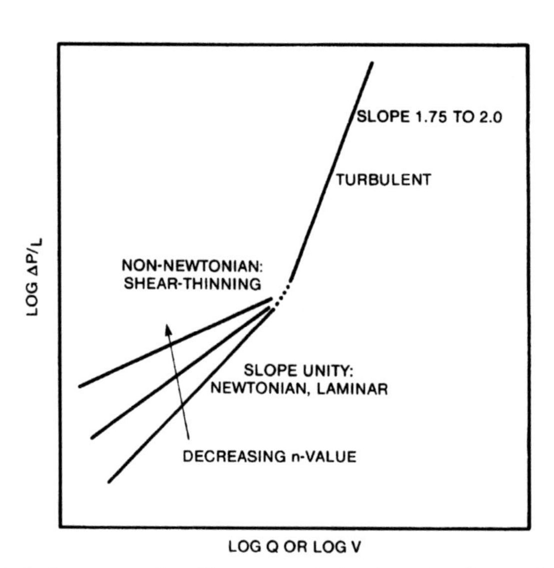

Fig. 7.1. Typical pressure-drop/flow-rate relationships for Newtonian and non-Newtonian slurries, in laminar and turbulent pipe flow.

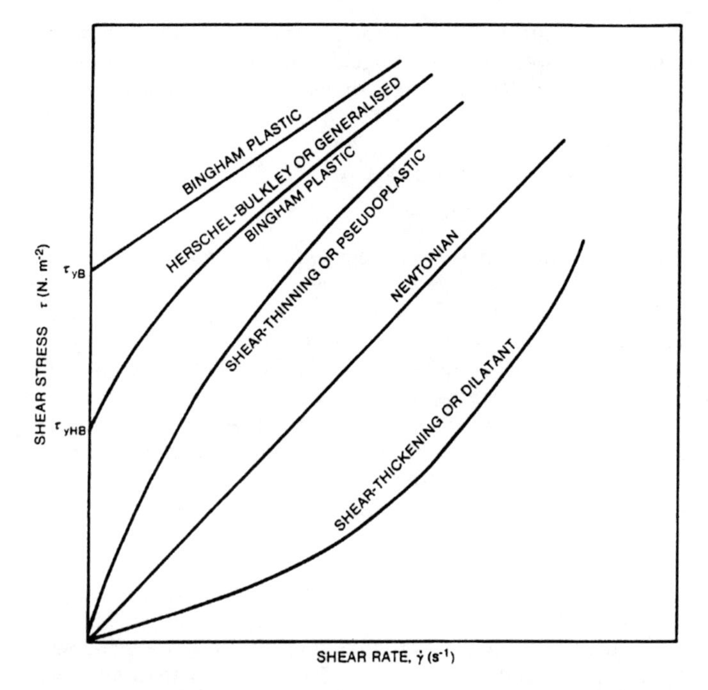

Fig. 7.2. The main classes of laminar flow behaviour of slurries. (τ_{yB} = Bingham plastic yield stress; τ_{yHB} = Herschel–Bulkley yield stress.)

design methods predict the laminar/turbulent transition and the pressure gradient in both the laminar and turbulent flow regimes. Stabilised slurries (see Chapter 1) can also often be treated as pseudohomogeneous materials for the purposes of pressure gradient prediction.

Pressure losses can be estimated using laminar viscometric data (section 2) involving flow models or by scale-up using either laminar or turbulent pipe flow test data (section 3). In addition, pressure losses arising from flow through pipe fittings are often significant and therefore require estimation (section 4).

For slurries with thixotropic properties, methods are also required for predicting the start-up pressure after shutdown and the time required to clear the pipe of gelled material so that steady flow can be resumed (section 5).

Criteria are available which determine whether a slurry type of interest can be treated as essentially *non-settling*, in which case the technology outlined in this chapter should be followed. For a *settling* slurry a quite separate approach is covered in Chapter 6. If a range of possible solids concentrations are to be considered then solids with a given particle size and size distribution, and with a given density, may give a *settling* slurry at low concentrations but an essentially *non-settling* slurry at higher concentrations. This may arise through the hindered settling effect (see Chapter 2). Suggested guidelines on the categorisation of a slurry type are outlined in Chapter 1. Further comparisons of the approaches to settling and non-settling slurry pipe flow are given by Cheng and Heywood (1984) and by Heywood and Cheng (1984*a*).

The total pressure change in a piping system, ΔP_T, is in general comprised of

$$\Delta P_T = P_f + P_{ff} + P_{st} \tag{7.1}$$

where P_f is the frictional pressure loss (see sections 2 and 3), P_{ff} is the frictional pressure loss arising from flow through fittings (section 4), and P_{st} is the pressure loss or gain resulting from elevation changes and is negative for a reduction in static head and positive for an increase in static head.

In many circumstances, for instance for the flow of low viscosity fluids (e.g. dilute slurries) through large diameter pipes, the frictional pressure loss may be insignificant compared with the static head change. On the other hand, the pipelining of viscous slurries over long distances will normally result in frictional pressure loss being the main contributor.

In the former case, estimation of P_T is very straightforward:

$$P_T \simeq P_{st} = \rho_m g \Delta z \tag{7.2}$$

where Δz is the net change in pipeline elevation, assuming that the pipeline is running full of slurry.

A typical plot of the frictional pressure gradient versus volume flow rate (or mean slurry flow velocity) is shown in Fig. 7.1 using both linear and logarithmic coordinate scales, while Fig. 7.2 shows the classical types of slurry flow behaviour commonly encountered.

2. DESIGN FROM LAMINAR FLOW VISCOMETRIC DATA

2.1. Laminar Flow

Considerable work has been undertaken to develop methods by which the frictional head losses encountered during the pipeline flow of non-Newtonian fluids may be predicted either from bench-scale viscometric data on the fluids or from data obtained using pilot-scale small-bore pipelines (Heywood, 1980 and Sutton, 1988). Laminar flow data only are normally provided using bench-scale equipment, whereas both laminar and turbulent flow data, together with information regarding the transitional flow region between these flow regimes, may be obtained using pilot-scale pipelines (see Chapter 10).

Chapter 4 discusses the different types of the laminar flow behaviour exhibited by different slurries and these are depicted in idealised form in Fig. 7.2. It should be borne in mind, however, that some slurries show a variety of flow properties over the relevant shear rate range for pipe flow.

For laminar flow, the pipeline pressure loss can be calculated from first principles for any given rheological data or model fitted to the data. The accuracy of pressure drop prediction is limited only by the accuracy of the data and the closeness with which the slurry flow properties fit any chosen model.

Many fluid models have been put forward to relate shear stress, τ, to shear rate, $\dot{\gamma}$, for bench-scale viscometric data and in the laminar flow regime. These fluid models are then used either to obtain analytical expressions for volumetric flow rate as a function of pressure gradient and other relevant variables in the laminar flow regime, or to develop methods by which turbulent flow data may be predicted from laminar flow data.

2.1.1. Flow Models

The simplest fluid model for a Newtonian slurry involves the single viscosity parameter, η_m,

$$\tau = \eta_m \dot{\gamma} \tag{7.3}$$

where τ is the shear stress $[\mathrm{M\,L^{-1}\,T^{-2}}]$, $\dot{\gamma}$ is the shear rate $[\mathrm{T^{-1}}]$, and η_m is the constant Newtonian viscosity $[\mathrm{M\,L^{-1}\,T^{-1}}]$.

This equation is valid for pipeline design provided the slurry exhibits Newtonian behaviour over the relevant shear rate range. This range is from zero, the shear rate at the centre of the pipe, to the maximum value at the wall which, for a Newtonian slurry, is given by

$$\dot{\gamma}_{max} = 8V/D \tag{7.4}$$

where V is the mean slurry flow velocity $[\mathrm{L\,T^{-1}}]$, and D is the inside pipe diameter [L].

The mean slurry flow velocity is calculated from

$$V = \frac{4Q}{\pi D^2} \tag{7.5}$$

where Q is the slurry volume flow rate $[\mathrm{L^3\,T^{-1}}]$.

To check whether a slurry exhibits Newtonian flow behaviour over the full relevant shear rate range of zero to the maximum value would require a considerable amount of rheological test work which is normally not warranted. A useful and practical approach is to calculate the wall shear rate given by eqn (7.4) from a selected combination of Q and D and perform rheological measurements over the shear rate range of the calculated wall shear rate down to one-tenth of the maximum wall shear rate.

For non-Newtonian slurries, two of the most commonly encountered flow models involve the empirical estimation of two parameters each. They are

Power Law

$$\tau = K\dot{\gamma}^n \tag{7.6}$$

where K is the consistency coefficient $[\mathrm{M\,L^{-1}\,T^{n-2}}]$, and n is the flow behaviour index [1];

and *Bingham plastic*

$$\tau = \tau_{yB} + \eta_p \dot{\gamma} \tag{7.7}$$

where τ_{yB} is the yield stress [M L^{-1} T^{-2}], and η_p is the plastic viscosity [M L^{-1} T^{-1}].

A third and very useful model for non-Newtonian flow behaviour involves three parameters (Herschel–Bulkley model, 1926), and slurries which obey this model are often referred to as yield-pseudoplastics or generalised Bingham fluids:

$$\tau = \tau_{yHB} + K\dot{\gamma}^n \tag{7.8}$$

where τ_{yHB} is the yield stress parameter in the model [M L^{-1} T^{-2}].

Again, it is important to check that the validity of any particular model for non-Newtonian flow holds over most of the relevant shear rate range. This is particularly important as many shear-thinning slurries ($n < 1$ in eqn (7.6)) can show Newtonian behaviour ($n = 1$) at low and high shear rates. The maximum shear rate occurring at the wall is, in general, for a non-Newtonian slurry given by (Govier & Aziz, 1972):

$$\dot{\gamma}_{max} = \frac{8V}{D}\left(\frac{1 + 3n'}{4n'}\right) \tag{7.9}$$

where n' is the local flow behaviour index defined as

$$n' = \frac{d(\ln \tau_w)}{d(\ln (8V/D))} \tag{7.10}$$

For a power law slurry, $n' = n$. The rule of thumb factor of 10 when defining the lowest shear rate for rheological measurements is also appropriate for non-Newtonian slurries. Further details are given in Chapter 4.

The fluid model parameters given in eqns (7.3), (7.6) and (7.7) can be readily estimated through linear regression. However, parameter estimation for the Herschel–Bulkley model (eqn (7.8)) is less straightforward. Two different methods are available. One involves non-linear least-squares regression on weighted data while, in the other method, the data are not weighted. Heywood and Cheng (1984*b*) have shown that while both methods will give best estimates of the three parameters which will then allow prediction of τ to within $\pm 2\%$ over the original viscometric shear rate range, extrapolation well outside this shear rate range leads to very different τ predictions using the two methods. Figure 7.3 shows the large differences in τ prediction obtained for an 8% digested sewage sludge.

Clearly, extrapolation outside the experimental shear rate range should be done only with caution when dealing with any type of slurry and, if shown to be unsatisfactory, should not be undertaken. Extrapolation to larger shear rates beyond $8V/D$ is not required when scaling-up from the tube viscometer laminar data to laminar flow in the full-scale pipeline but can be necessary when scaling-up to turbulent flow. However, experimental shear rate ranges employed in viscometric measurement are rarely so wide unless adequate funds are available for such work. Thus, it may be worthwhile sometimes to assess the two methods of flow model parameter estimation to determine which method gives the least uncertainty in the predictions of τ outside the shear rate range used in viscometric tests.

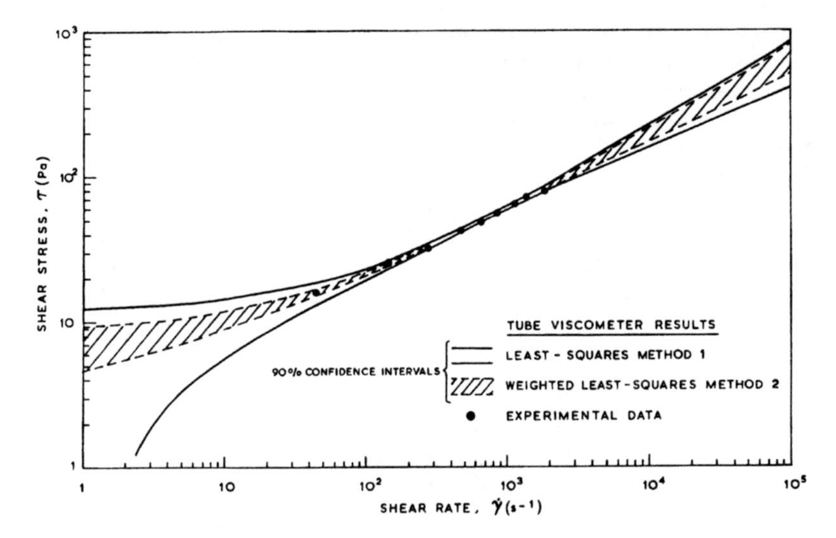

Fig. 7.3. Confidence intervals (90%) for the flow curve of an 8% by weight digested sewage sludge. (From Heywood & Cheng (1984*b*).)

2.1.2. Consideration of Yield Stress

When applying eqns (7.7) or (7.8) to experimental data a yield stress parameter, $\hat{\tau}_y$, will be obtained. Such experimental values should not necessarily be considered as an absolute property of the fluid but rather a constant which arises out of the correlation of shear stress/shear rate data (see Chapter 4). The *true* yield stress of the fluid, if it exists at all, may have a value quite removed from the value of the experimental $\hat{\tau}_y$ derived from data for which the lower limit of the viscometric shear rate range is of the order of tens or even hundreds of reciprocal seconds. A knowledge of the *true* yield stress facilitates the calculation of the start-up pressure gradient for stagnant fluid in a pipe according to the relation

$$\frac{P_f'}{L} = \frac{4\tau_y}{D} \tag{7.11}$$

where P_f' is the start-up frictional pressure drop $[\mathrm{M\,L^{-1}\,T^{-2}}]$, and L is the pipe length [L].

For fluids whose laminar flow behaviour does not conform well to any of the above four models (eqns (7.3), (7.6)–(7.8)), Metzner and Reed (1955) have suggested that pipeline data may be treated using the completely general expression

$$\tau_w = K'\left(\frac{8V}{D}\right)^{n'} \tag{7.12}$$

where n' is given by eqn (7.10) and, unlike the n parameter in the power law model, n' is allowed to vary with shear rate. Equation (7.12) is applicable to any time-dependent fluid and may be used in conjunction with correlations for pressure drop developed for the power law model provided that both n' and K' values are evaluated at the same shear stress level, corresponding to either laminar or turbulent flow conditions in full-scale pipe.

The relationship between pressure drop and volumetric flow rate in the laminar flow regime may be obtained by integration of the chosen flow equation relating shear stress to shear rate. In general, the flow rate through the pipe is given by

$$Q = \int_0^{D/2} 2\pi r v(r)\, \mathrm{d}r \tag{7.13}$$

where $v(r)$ is the local slurry velocity [L T^{-1}], and r is the radial position in the pipe, from 0 to $D/2$ [L]. The relationship between shear rate and the velocity distribution is given by

$$\dot{\gamma} = -\frac{\mathrm{d}v}{\mathrm{d}r} \tag{7.14}$$

and the shear stress at a radial position, r, is related to the overall axial pressure drop over the pipe of length, L, by equating the resultant force ($\pi r^2 P_f$) from the pressure drop over a cylinder of radius, r, with the force derived from the internal shear stress ($2\pi r\tau$)

$$\tau = \frac{r}{2}\frac{P_f}{L} \tag{7.15}$$

Partial integration of eqn (7.13) using eqns (7.14) and (7.15) leads to

$$Q = \frac{\pi}{8}\left(\frac{D}{\tau_w}\right)^3 \int_0^{\tau_w} \tau^2\dot{\gamma}\, \mathrm{d}\tau \tag{7.16}$$

where the shear stress at the wall, τ_w, is given by

$$\tau_w = \frac{DP_f}{4L} \tag{7.17}$$

Equation (7.16) is a perfectly general relationship between Q and P_f and is not dependent on the form of the fluid model adopted.

Substitution of the *Newtonian* fluid model eqn (7.3) into eqn (7.16) leads to the Hagen–Poiseuille equation:

$$Q = \frac{\pi D^4 P_f}{128\eta_m L} \tag{7.18}$$

Substitution of eqn (7.6) into eqn (7.16) gives

Power law

$$Q = \frac{\pi D^3}{32}\left(\frac{4n}{3n+1}\right)\left(\frac{\tau_w}{K}\right)^{1/n} \tag{7.19}$$

whereas eqn (7.7) gives

Bingham plastic

$$Q = \frac{\pi D^3 \tau_w}{32\eta_p}\left[1 - \frac{4}{3}\frac{\tau_{yB}}{\tau_w} + \frac{1}{3}\left(\frac{\tau_{yB}}{\tau_w}\right)^4\right] \tag{7.20}$$

and eqn (7.8) gives

generalised Bingham model

$$Q = \frac{\pi D^3}{32}\left(\frac{4n}{3n+1}\right)\left(\frac{\tau_w}{K}\right)^{1/n}\left(1 - \frac{\tau_{yHB}}{\tau_w}\right)^{1/n}$$
$$\times\left[1 - \frac{\tau_{yHB}/\tau_w}{2n+1}\left(1 + \frac{2n}{n+1}\left(\frac{\tau_{yHB}}{\tau_w}\right)\left(1 + n\frac{\tau_{yHB}}{\tau_w}\right)\right)\right] \quad (7.21)$$

Alternative forms of the above equations may be written in dimensionless form using the Fanning friction factor defined by

$$f = \frac{\tau_w}{\frac{1}{2}\rho_m V^2} \quad (7.22)$$

The relationship between friction factor and the Reynolds number for a Newtonian slurry then becomes

$$f = \frac{16}{\mathrm{Re}} \quad (7.23)$$

where the pipe flow Reynolds number is defined by

$$\mathrm{Re} = \frac{DV\rho_m}{\eta_m} \quad (7.24)$$

The pipe flow Reynolds number is a dimensionless quantity which expresses the ratio of inertial to viscous forces for slurry flow and is used to define the boundaries between the laminar, transitional and turbulent flow regimes. For Newtonian flow, the laminar regime upper limit is defined through the Reynolds number with a value of 2100. Section 2.2 provides a fuller discussion.

The friction factor/Reynolds number relationship for a power law fluid is similar to eqn (7.23), i.e.

$$f = \frac{16}{\mathrm{Re}'} \quad (7.25)$$

but here the power law Reynolds number, Re′, is given by

$$\mathrm{Re}' = \frac{\rho_m VD}{K}\left(\frac{4n}{1+3n}\right)^n\left(\frac{D}{8V}\right)^{n-1} \quad (7.26)$$

This Reynolds number is derived by using eqn (7.25) together with the definition of the Fanning friction factor given in eqn (7.22).

The friction factor equation for a *Bingham plastic model* involves the Hedström number (He) and the Bingham Reynolds number, (Re_B) defined respectively as

$$\mathrm{He} = \frac{D^2\tau_{yB}\rho_m}{\eta_p^2} \quad (7.27)$$

$$\mathrm{Re}_B = \frac{DV\rho_m}{\eta_p} \quad (7.28)$$

The relation was first published by Buckingham (1921):

$$\frac{1}{\mathrm{Re_B}} = \frac{f}{16} - \frac{\mathrm{He}}{6\mathrm{Re_B^2}} + \frac{\mathrm{He}^4}{3f^3\mathrm{Re_B^8}} \tag{7.29}$$

or alternatively

$$f = \frac{16}{\mathrm{Re_B}}\left(\frac{1}{1 - 4x/3 + x^4/3}\right) \tag{7.30}$$

where $x = \tau_{yB}/\tau_w$. Equation (7.20) is explicit in Q and is generally easier to use for calculating volume flow rate, Q, for a given pressure gradient, $\Delta P/L$, compared with eqns (7.29) and (7.30). If Q is specified an iterative technique is necessary to obtain P_f/L from either eqns (7.20), (7.29) or (7.30). Govier and Aziz (1972) and Smith (1960) have shown that an expression explicit in P_f/L may be derived if the final term in the Buckingham (1921) eqn (7.20) is assumed negligible, i.e. if $x^4 \ll 1$. It is then possible to express the friction factor in terms of a modified Reynolds number, $\mathrm{Re_{mod}}$:

$$f = \frac{16}{\mathrm{Re_{mod}}} \tag{7.31}$$

where

$$\mathrm{Re_{mod}} = \frac{1}{\dfrac{\eta_p}{DV\rho_m} + \dfrac{\tau_{yB}}{6\rho_m V^2}} \tag{7.32}$$

$$\equiv \frac{6\mathrm{Re_B^2}}{6\mathrm{Re_B} + \mathrm{He}} \tag{7.33}$$

An important point to note is that eqn (7.31) is exact if viscometric data are obtained by capillary tube and a plot of wall shear stress versus *nominal* wall shear rate, $8V/D$, is a straight line. The yield stress parameter in the flow equation

$$\tau = \frac{4}{3}\tau'_{yB} + \eta'_p\left(\frac{8V}{D}\right) \tag{7.34}$$

is then obtained from the intercept on the shear stress axis at zero nominal wall shear rate. Equation (7.31) is now in an analogous form to both the power law model case and the Newtonian case. This yield stress parameter may be far removed from the true yield stress of the slurry, which can be difficult to quantify (Cheng, 1986).

The *generalised Bingham model* equation for friction factor in laminar flow is

$$f = \frac{16}{\mathrm{Re'}}\left(1 - \frac{2\mathrm{He'}}{f(\mathrm{Re'})^2}\right)\left[1 - \frac{1}{(2n+1)}\frac{2\mathrm{He'}}{f(\mathrm{Re'})^2}\right.$$
$$\left.\times\left(1 + \frac{2n}{(n+1)}\frac{2\mathrm{He'}}{f(\mathrm{Re'})^2}\left(1 + n \times \frac{2\mathrm{He'}}{f(\mathrm{Re'})^2}\right)\right)\right] \tag{7.35}$$

where

$$\mathrm{He}' = \frac{\tau_{yHB}}{\rho_m}\left(\frac{\mathrm{Re}'}{V}\right)^2 \tag{7.36}$$

Equation (7.21) is more convenient to use if Q is required having specified P_f/L, but both eqns (7.21) and (7.35) require an iteration procedure if P_f/L is required for a given Q value. Unlike the Bingham model, no simplifying assumption can be made easily to derive an expression from either eqns (7.21) or (7.35) which is then explicit in $\Delta P/L$.

An alternative way of defining the Hedström number for a generalised Bingham slurry is given by

$$\mathrm{He}' = \frac{D^2 \rho_m}{\tau_{yHB}}\left(\frac{\tau_{yHB}}{K}\right)^{2/n} \tag{7.37}$$

The advantage of this definition is that He′ is a true material property parameter and does not include mean flow velocity. It is this definition which is used in the remainder of this chapter.

A simplified but completely rigorous approach to the estimation of the pressure drop for the laminar flow of a slurry having any rheological property is possible provided that flow curve data for the slurry have been obtained using a smaller (or larger) pipe diameter and there are no *wall-slip* effects (see Chapter 4). Figure 7.3 shows a typical flow curve obtained using a pipe or capillary tube viscometer and, in general, the curve may show a complex form and may not be amenable to the application of a simplified flow model. Although this curve has been obtained using one pipe diameter only, it is valid for any pipe diameter as long as the flow remains in the laminar regime. Thus, this master curve may be used to predict the pressure gradient for flow in any other pipe size.

WORKED EXAMPLE

An example will help to illustrate this. The pressure gradient for the flow of a given slurry is required at a flow velocity of $1\,\mathrm{m\,s^{-1}}$ in a 0·2 m diameter pipe. The so-called nominal wall shear rate is given by eqn (7.4) and is calculated to be $40\,\mathrm{s^{-1}}$. A 10 mm diameter capillary tube viscometer is available to measure the flow property of this slurry. The pressure gradient, and hence the resulting wall shear stress, needs to be estimated using the capillary viscometer at the same wall shear rate $40\,\mathrm{s^{-1}}$. As the capillary tube diameter is 10 mm this means that the pressure gradient must be measured at a flow velocity in the capillary tube of $40 \times 0{\cdot}01/8\,\mathrm{m\,s^{-1}}$ or $0{\cdot}05\,\mathrm{m\,s^{-1}}$ (i.e. at a volume flow rate of $3{\cdot}9 \times 10^{-6}\,\mathrm{m^3\,s^{-1}}$). Once the pressure gradient, P_f/L, is measured the wall shear stress is calculated from eqn (7.17). This same wall shear stress will also occur in the larger design pipe diameter and hence the pressure gradient in the large pipe size may be estimated, again using eqn (7.17).

It must be stressed that this procedure is valid only when flow in both pipe sizes is laminar. Thus, appropriately-defined Reynolds numbers should be calculated to ensure that laminar flow conditions prevail in both cases.

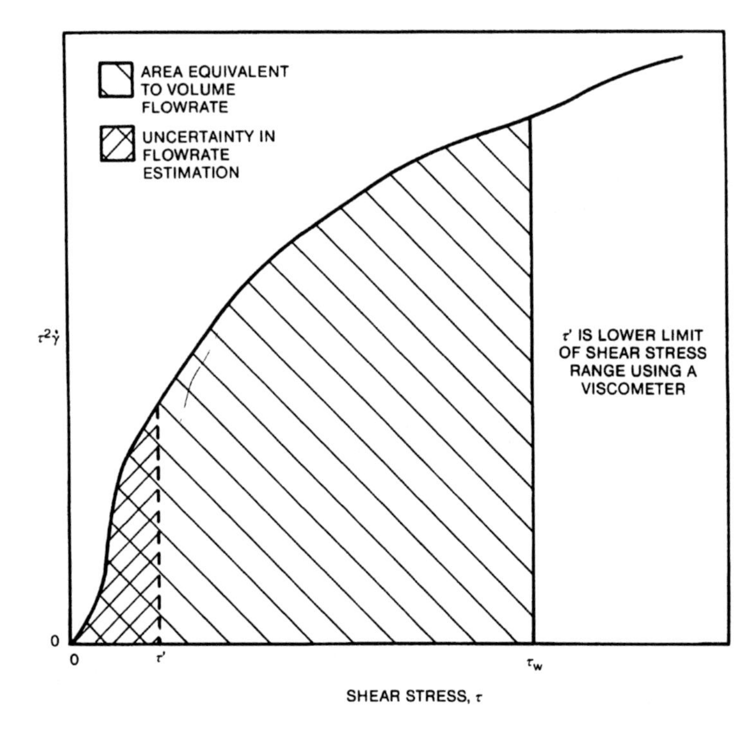

Fig. 7.4. Estimation of volume flow rate in laminar pipe flow by replotting flow curve data.

Another approach to dealing with the situation when the flow curve for the slurry has a complicated form is to replot the $(\tau, \dot{\gamma})$ flow curve data in the form of $\tau^2\dot{\gamma}$ as a function of τ, as indicated in Fig. 7.4. The flowrate, Q, for various τ_w can then be evaluated by estimating the area denoted by the hatching in Fig. 7.4 in order that the integral in eqn (7.16) may be evaluated.

In practice, of course, it is convenient to measure the flow curve down to a minimum shear stress τ', so there will always be some uncertainty in evaluating the complete area under the curve and the area shown by the double hatching can only be approximated. For accurate estimates, τ' needs to be as low as possible.

2.2. Laminar/Turbulent Transition

It is important that methods are available which can define the point at which laminar, or streamline, flow breaks down as the flow rate is raised progressively in any given pipe size. Once this point is defined for a slurry with a particular flow property, it is then possible to decide whether expressions for pressure drop prediction outlined in section 2.1 are relevant or whether turbulent flow equations need to be used, as described in section 2.3. A lower critical Reynolds number defines the breakdown of laminar flow and an upper critical Reynolds number the onset of fully turbulent flow. The flow region between these two critical Reynolds numbers is typically referred to as the transitional flow regime.

The simplest approach towards dealing with the transitional region between laminar and turbulent flow is to assume that it does not exist and that the transition, as the flow velocity is progressively increased, occurs completely at a single operating point. In this case only one critical Reynolds number need be predicted and this can be achieved by taking the intersection of the f, Re relationships for laminar and turbulent flow for the fluid model under consideration. However, this approach will not yield accurate results. It is well known that for Newtonian slurries the lower critical Reynolds number for the initial breakdown of laminar flow, Re_1, is 2100 and the upper value Re_2 for the commencement of fully turbulent flow is approximately 3000.

However, for non-Newtonian slurries, the individual values of Re_1 and Re_2 are much more difficult to predict. Various attempts have been made to develop methods for Re_1 prediction based on different flow models but Cheng (1970) has offered an approach for predicting both Re_1 and Re_2 valid for both power law slurries and generalised Bingham slurries. There are few data with which to verify Cheng's approach, because industrial pipelines tend to avoid operating in this regime. However, the most economical design often suggests that flow should be in the transitional regime and some pipelines operate just above Re_2. One of the reasons why Re_2 is difficult to define more precisely is that the fluctuations in pressure gradient prevalent in the transitional regime diminish progressively rather than abruptly as flow velocity is increased, when the fully turbulent flow regime is reached.

Ryan and Johnson (1959) suggested the use of a stability parameter to predict Re_1:

$$Z = -\frac{D/2}{\tau_w} \rho_m v(r) \frac{dv}{dr} \tag{7.38}$$

which is the ratio of the rate of energy supply to the rate of energy dissipation at any point in the pipe. The maximum value of Z is obtained if the velocity distribution for laminar flow is known. This distribution may be obtained by integration of the relevant constitutive equation. For a Newtonian fluid

$$Z_{max} = \frac{2}{3\sqrt{3}} \rho_m \frac{VD}{\eta_m} \quad \text{at} \quad \frac{r}{D/2} = \frac{1}{\sqrt{3}} \tag{7.39}$$

and thus the stability parameter is a special form of Reynolds number. Taking the critical Reynolds number as 2100, Ryan and Johnson (1959) proposed that the critical value of Z_{max} above which laminar flow is unstable is

$$Z_c = \frac{2}{3\sqrt{3}} \times 2100 = 808 \tag{7.40}$$

and it has been assumed that this value is universally valid for all fluids and

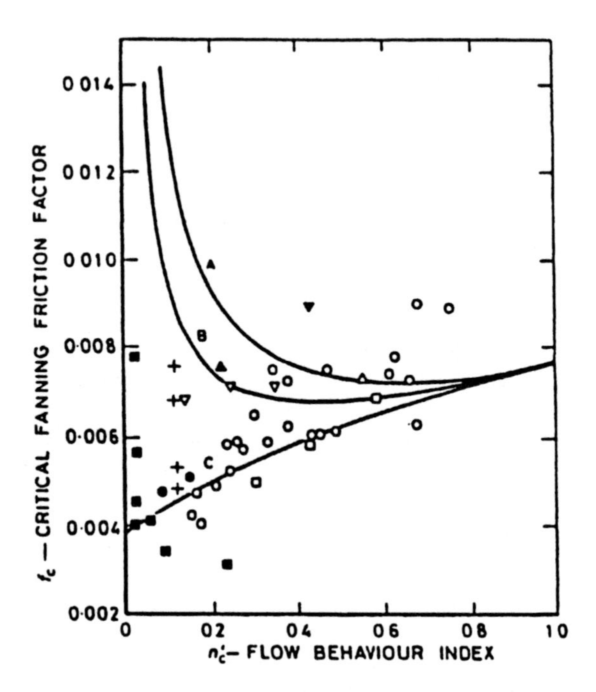

Fig. 7.5. Comparison of critical friction factor for laminar flow breakdown with experimental data. Curve A: Hanks (1963); Curve B: Ryan & Johnson (1959); Curve C: Mishra & Tripathi (1973). (From Mishra & Tripathi (1973), reproduced by permission of AIChE.)

different flow situations. When applied to the *power law model*, Ryan and Johnson obtained an expression for the lower critical Reynolds number in terms of the flow behaviour index, n, only

$$\mathrm{Re}_c' = \frac{6464n\,(n+2)^{(n+2/n+1)}}{(1+3n)^2} \tag{7.41}$$

or, alternatively, the lower critical friction factor, f_c, is given by

$$f_c = \frac{(1+3n)^2}{404n}\left(\frac{1}{n+2}\right)^{(n+2/n+1)} \tag{7.42}$$

Figure 7.5 shows how f_c varies with n_c'. An interesting conclusion from this plot is that f_c reaches a minimum at $n_c' \sim 0{\cdot}4$. Thus, for highly shear-thinning materials, f_c increases as n_c' diminishes. On the same figure is plotted a number of data taken from various sources (Mishra & Tripathi, 1973). Despite the scatter, these data suggest that f_c decreases progressively as n_c' falls from unity to zero.

Hanks (1963) and Hanks and Pratt (1967) have applied the Ryan and Johnson criterion to the Bingham model and found the lower critical Bingham Reynolds number is given by

$$(\mathrm{Re_B})_c = \frac{\mathrm{He}}{8x_c}\left(1 - \frac{4}{3}x_c + \frac{1}{3}x_c^4\right) \tag{7.43}$$

where x_c, the critical value of the ratio τ_{yB}/τ_w, is given by the relation

$$\frac{x_c}{(1 - x_c)^3} = \frac{\mathrm{He}}{16\,800} \tag{7.44}$$

To find $(\mathrm{Re_B})_c$, calculate x_c from eqn (7.44) and then substitute this value in eqn (7.43).

Mishra and Tripathi (1973) have also included a predicted curve from the Hanks analysis in Fig. 7.5 although it is not immediately obvious how eqns (7.43) and (7.44) may be represented in terms of $f_c = f(n'_c)$.

2.3. Turbulent Flow

2.3.1. Newtonian Slurries

Unlike laminar flow, in turbulent flow the relative pipe wall roughness can be important and equations for the Fanning friction factor, f, can be classified according to whether the pipe wall is essentially smooth (e.g. plastic pipe), or is sufficiently rough (Scaggs *et al.*, 1988) to affect frictional losses (e.g. galvanised steel or concrete). These equations are obtained by correlating experimental data, with or without guidance from theory. Depending on the studies, many different versions are available (Blasius, 1913; von Karman, 1931; Drew *et al.*, 1932; Nikuradse, 1932; Colebrook, 1939; Moody, 1944, 1947; Knudsen & Katz, 1958; Olujic, 1981). However, the equation by Churchill (1977) is probably the most useful, as it is a smooth function for f, valid in both the laminar and turbulent flow regimes.

$$f_{ch} = 2\left[\left(\frac{8}{\mathrm{Re}}\right)^{12} + (A + B)^{-1.5}\right]^{1/12} \tag{7.45}$$

in which

$$\left.\begin{aligned} A &= (2{\cdot}457 \ln (1/C))^{16} \\ B &= \left(\frac{37530}{\mathrm{Re}}\right)^{16} \\ C &= \left(\frac{7}{\mathrm{Re}}\right)^{0.9} \end{aligned}\right\} \tag{7.46}$$

When the pipe wall is not hydrodynamically smooth, the situation is rather different. Here, an estimate for the relative roughness of the wall, e/D, is required, as it is an additional variable in equations for the prediction of friction factor (Colebrook, 1939).

$$\frac{1}{\sqrt{f}} = -\log_{10}\left[\left(\frac{e/D}{3{\cdot}7}\right) + \frac{5{\cdot}02}{\mathrm{Re}\sqrt{f}}\right] \tag{7.47}$$

Moody (1947) suggested an approximate form of eqn (7.47),

$$f \approx 0{\cdot}001375\left[1 + \left[20\,000\, e/D + \left(\frac{10^6}{\mathrm{Re}}\right)\right]^{1/3}\right] \tag{7.48}$$

in order that the equation becomes explicit in f. Zigrang and Sylvester (1982) have compared this and other approximations to the Colebrook equation. Moody (1944) also produced a detailed chart based on the Colebrook equation giving f as a function of Re and e/D. This is reproduced in Fig. 7.6.

Other expressions include those by Wood (1966) and Churchill (1977), in which the parameter C in eqn (7.46) is modified to allow for pipe roughness:

$$C = \left(\frac{7}{\mathrm{Re}}\right)^{0{\cdot}9} + 0{\cdot}27\left(\frac{e}{D}\right) \tag{7.49}$$

Predictions of f from some of the above equations for either smooth pipe or where the ratio e/D is significant have been compared by Heywood and Cheng (1984*b*). For smooth wall pipe it was found that over a very wide range of Reynolds number (3000–2 × 10^5), f values differ by a maximum of about ±4% as shown in Fig. 7.7 and hence it is relatively unimportant which equation is selected for predicting f, although those explicit in f are obviously preferable for ease of computation. The same level of agreement was also found for equations for rough pipe.

2.3.2. Non-Newtonian Slurries

Power law and generalised Bingham plastic slurries. Just as there are many alternative equations for predicting f for turbulent Newtonian flow, so there are many available for turbulent non-Newtonian flow. As before, they are correlations of experimental data, and the functional form of the correlation may or may not be guided by theory. Most equations are based on the power law model, e.g. those developed by Dodge and Metzner (1959), Shaver and Merrill (1959), Tomita (1959), Thomas (1960), Kemblowski and Kolodziejski (1973), and Szilas *et al.* (1981). These require modification before they can be used with the Herschel–Bulkley model, while others have been derived specifically for use with the Herschel–Bulkley flow model, i.e. Torrance (1963) and Hanks (1978). All assume that flow is in a smooth-walled pipe. Those equations for f based on the power law model are

Tomita (1959)

$$\frac{1}{\sqrt{f_{\mathrm{To}}}} = 4\log_{10}\left(\mathrm{Re}_{\mathrm{To}}\sqrt{f_{\mathrm{To}}}\right) - 0{\cdot}40 \tag{7.50}$$

$$f_{\mathrm{To}} = \left(\frac{1+2n}{1+3n}\right)\frac{4}{3}f \tag{7.51}$$

and

$$\mathrm{Re}_{\mathrm{To}} = \mathrm{Re}'\, 8^{1-n}\left(\frac{1+3n}{4n}\right)^n \frac{6[(1+3n)/n]^{1-n}}{2n((1+2n)/n)} \tag{7.52}$$

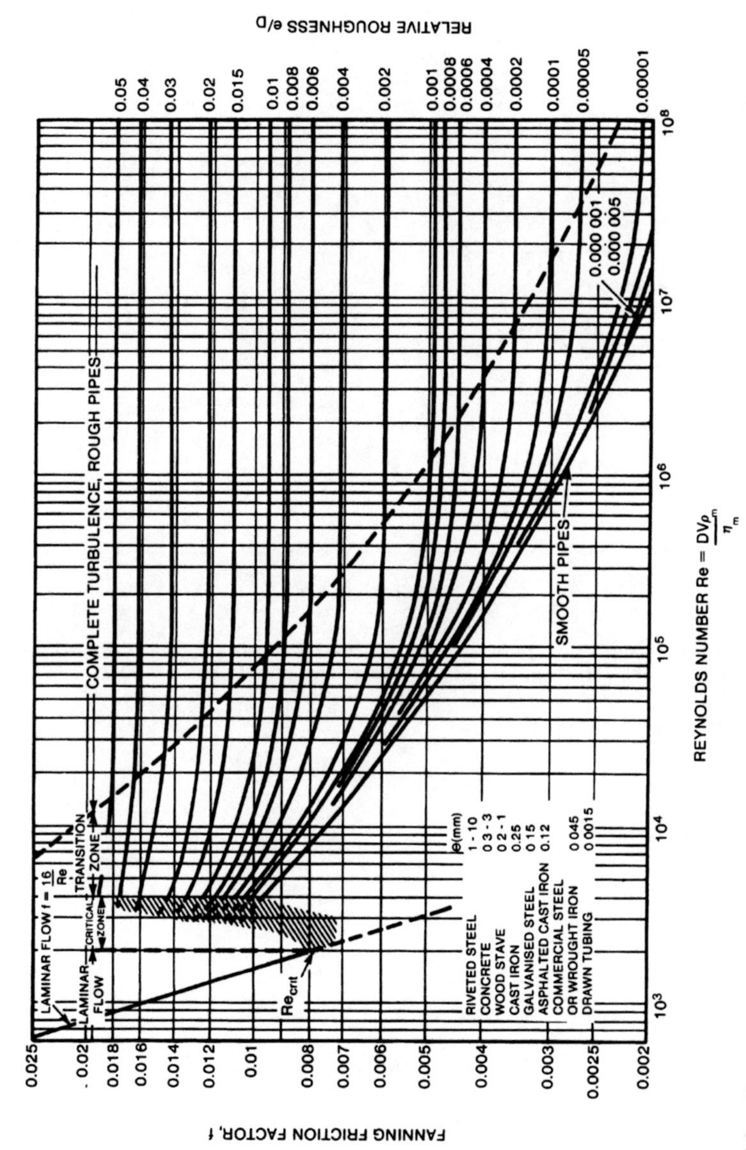

Fig. 7.6. The Moody (1944) chart for friction factor prediction for laminar and turbulent flow through smooth or rough wall pipes.

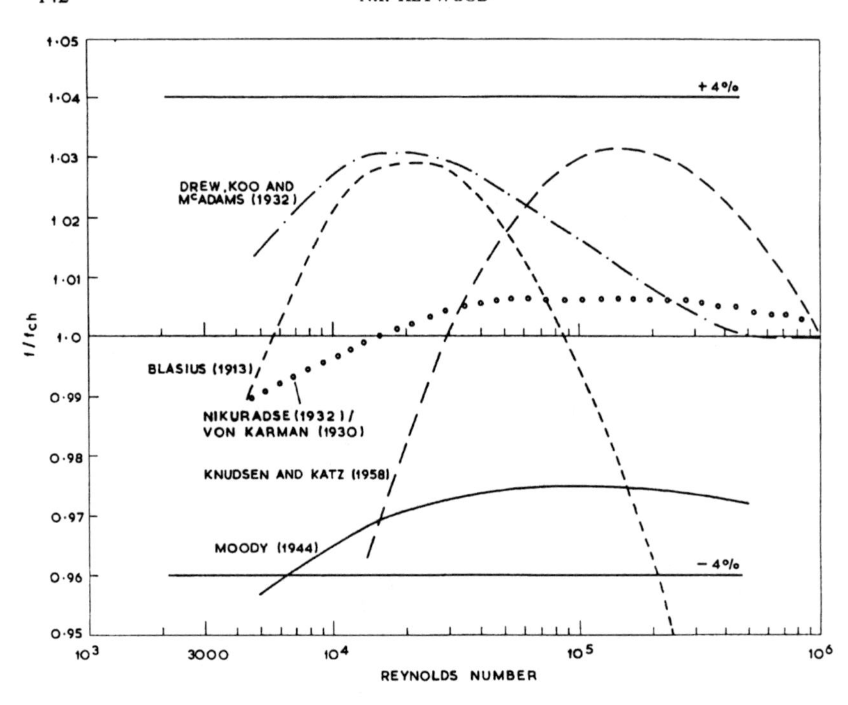

Fig. 7.7. Comparison of prediction methods for friction factor for Newtonian turbulent flow in a smooth wall pipe. (From Heywood & Cheng (1984*b*).)

Dodge and Metzner (1959)

$$\frac{1}{\sqrt{f}} = \frac{4{\cdot}0}{n^{0{\cdot}75}} \log_{10}(\mathrm{Re}' f^{1-n/2}) - \frac{0{\cdot}40}{n^{1{\cdot}2}} \tag{7.53}$$

Shaver and Merrill (1959)

$$f = \frac{0{\cdot}0791}{n^5(\mathrm{Re}')^t} \tag{7.54}$$

where

$$t = 2{\cdot}63/(10{\cdot}5)^n \tag{7.55}$$

Thomas (1960)

$$\frac{1}{\sqrt{f}} = \frac{4{\cdot}0}{n} \log_{10}(\mathrm{Re}' f^{1-n/2}) - \frac{0{\cdot}40}{n} \tag{7.56}$$

Kemblowski and Kolodziejski (1973)

$$f = \frac{E}{4} \frac{\phi^{1000/\mathrm{Re}'}}{\mathrm{Re}'^{m'}} \tag{7.57}$$

in which

$$\left.\begin{aligned} E &= 0{\cdot}0089 \exp(3{\cdot}57n^2) \\ \phi &= \exp \frac{0{\cdot}572\,(1 - n^{4\,2})}{n^{0\,435}} \\ m' &= 0{\cdot}314n^{2{\cdot}3} - 0{\cdot}064 \end{aligned}\right\} \tag{7.58}$$

Equation (7.57) holds provided that

$$\mathrm{Re}' < \mathrm{Re}'_t = 31600/n^{0\,435} \tag{7.59}$$

For $\mathrm{Re}' > \mathrm{Re}'_t$,

$$f = 0{\cdot}0791/\mathrm{Re}'^{0{\cdot}25} \tag{7.60}$$

Szilas et al. (*1981*)

$$\frac{1}{\sqrt{f}} = \frac{4}{n}\log_{10}[\mathrm{Re}'\,(4f)^{1-n/2}] + 1{\cdot}51^{1/n}\left[4{\cdot}24 + \frac{1{\cdot}414}{n}\right] - \frac{8{\cdot}03}{n} - 2{\cdot}114 \tag{7.61}$$

Irvine (*1988*)

$$f = \frac{F'(n)}{\mathrm{Re}'^{(1/(3n+1))}} \tag{7.62}$$

where

$$F'(n) = \frac{2}{8^{n-1}}\left[\frac{8n^n}{7^{7n}\,(1 + 3n)^n}\right]^{\frac{1}{3n+1}} \tag{7.63}$$

The Torrance equation based on the Herschel–Bulkley flow model is

Torrance (*1963*)

$$\frac{1}{\sqrt{f}} = 0{\cdot}45 - \frac{2{\cdot}75}{n} + \frac{1{\cdot}97}{n}\ln(1 - x) + \frac{1{\cdot}97}{n}\ln(\mathrm{Re}'\,f(n)f^{1-n/2}) \tag{7.64}$$

where $f(n) = [(1 + 3n)/(4n)]^n$.

It can be seen that many of the expressions given above are implicit in f. Mansour and Rajie (1987, 1988) have developed explicit relationships using existing equations for f as their starting point.

The equations for f developed by Hanks (1978) are not easily summarised, and the interested reader is referred to the original paper.

There are two ways of modifying the correlations based on the power law model for use with the Herschel–Bulkley model, as follows.

(1) *The Warren Spring Laboratory approach.* The importance of yield stress in turbulent flow is assumed to be negligible because of high shear rates involved and estimates for K and n in the Herschel–Bulkley model are used directly in equations for f.
(2) *The Metzner–Reed approach.* Yield stress is taken into account indirectly

using localised values of K' and n' corresponding to the relevant wall shear stress level for turbulent flow. These are calculated from a knowledge of $\hat{\tau}_y$, K and n as well as operating variables. As f is required to facilitate the calculation of K' and n' an iterative procedure is necessary.

Predictions of friction factor from eqn (7.53) using the Dodge–Metzner correlation are shown in Fig. 7.8. For materials that do not obey the power law model over the relevant shear stress range, Dodge and Metzner suggest that n' may be used in eqn (7.53) in place of n and the correlation may then be applied to any material. Provided that either n or n' values are known from laminar flow data in the wall shear stress range applicable to the turbulent flow conditions occurring in the pipe size of interest, there would appear to be no problem in using eqn (7.53) although the correlation has not been adequately tested by Dodge and Metzner for materials having n values below 0·5. In practice, however, n' or n is obtained from laminar flow data at wall shear stresses substantially below the relevant stress levels for turbulent flow. This is because either turbulent flow predictions are required for the same pipe size used for the laminar flow tests or the material is so markedly shear-thinning that even tests using small bore tubes do not result in sufficiently high wall shear stresses. The result is that the parameter n' or n used is often too low.

Heywood and Richardson (1978*a*) and Kemblowski and Kolodziejski (1973) have shown that for the pumping of flocculated clay suspensions the Dodge–Metzner correlation predicts friction factors substantially below experimental values when n obtained at low wall shear stresses is used in the correlation. Doubt on the validity of the correlation has been expressed also by Harris (1968). However, as long as the limitations of the correlation are appreciated, it may be used with confidence. Unfortunately these limitations are rarely, if ever, quoted in the numerous text books which give the prediction method.

One particular problem area concerns the pipelining of fibrous slurries. For instance, it has been observed that the turbulent flow behaviour of some types of sewage sludge exhibiting shear-thinning laminar flow property can be adequately predicted using the Dodge–Metzner approach, while other types, showing similar laminar flow property, can give rise to drag reduction in turbulent flow. Drag reduction is more commonly associated with dilute polymer solutions where the turbulent pressure drop is lower than that for solvent flow alone at the same mean flow velocity and in the same pipe size. This failure to predict neither the presence nor the magnitude of any drag reduction effect is, of course, not confined to the Dodge–Metzner correlation but applies to all methods based on knowledge of the laminar flow property of the slurry.

Heywood and Cheng (1984*b*) have compared predictions for f using a number of expressions for non-Newtonian turbulent flow. Figures 7.9 and 7.10 give comparisons of the Fanning friction factor, predicted using a number of published equations and normalised by dividing by the friction factor predicted using the Churchill (1977) approach. The predictions are compared in the plot with experimental data for the flow of digested and activated sewage sludge in a 100 mm diameter pipeline. Considerable variation in the predictions is evident and it is

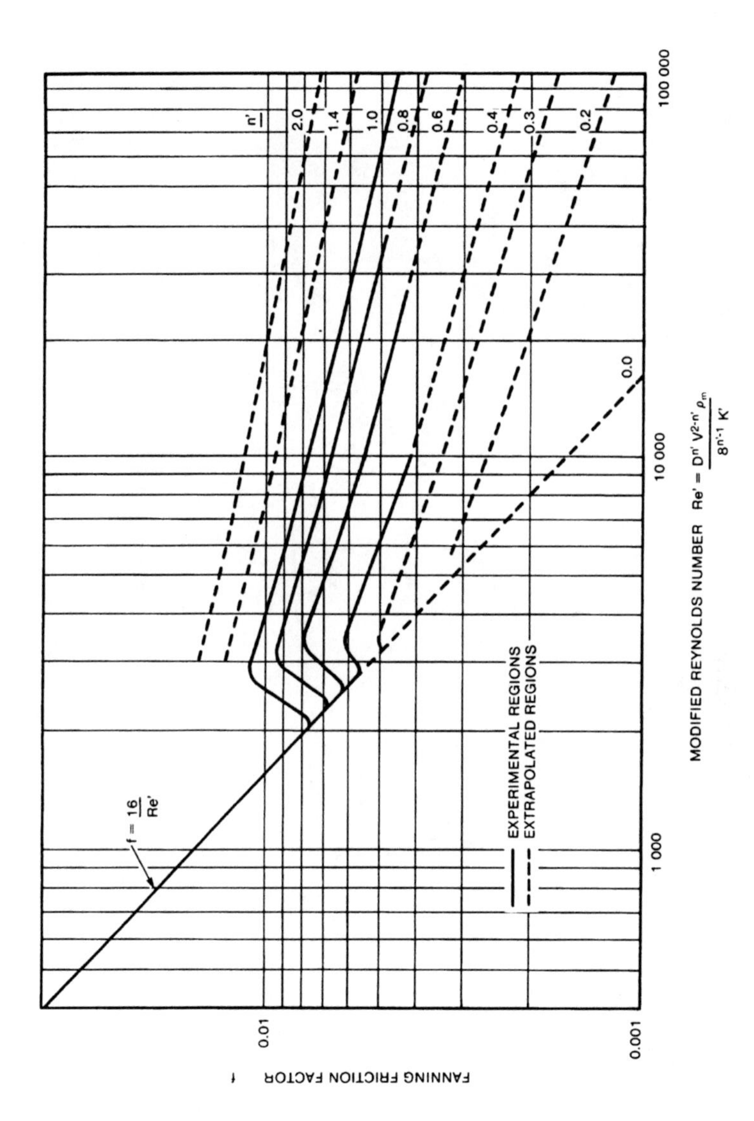

Fig. 7.8. Predictions of friction factor for turbulent flow of pseudoplastic/power law slurries in smooth wall pipe using the Dodge and Metzner correlation. (From Dodge & Metzner (1959).)

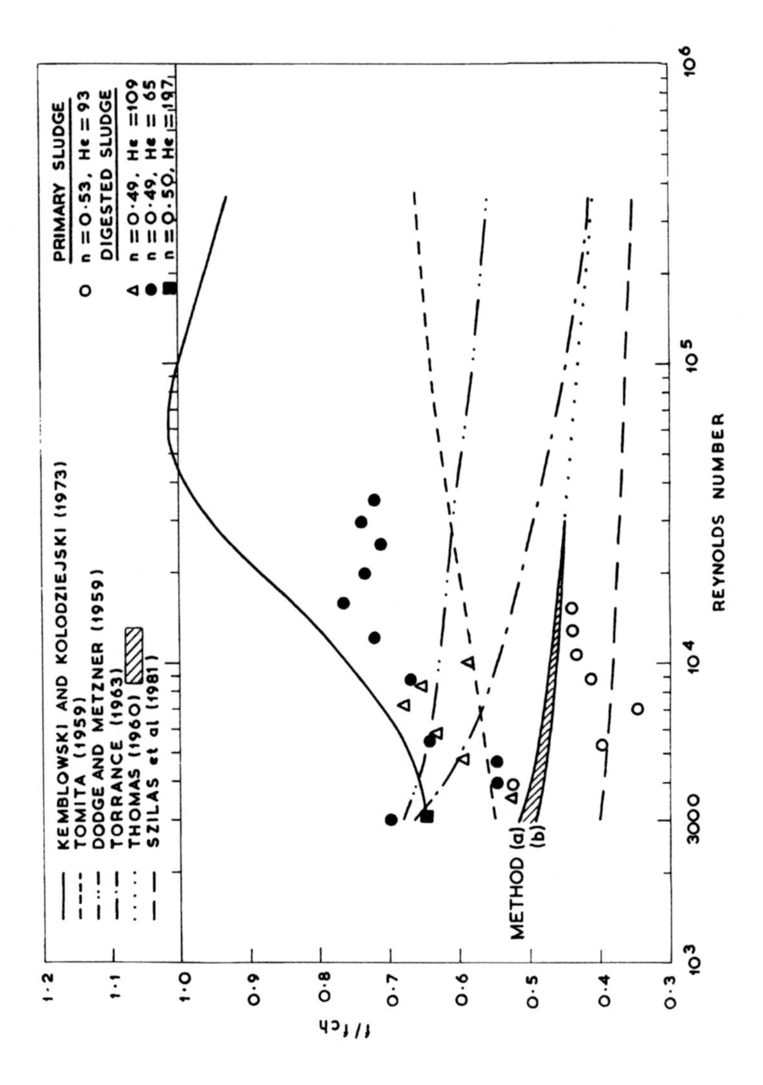

Fig. 7.9. Comparison of methods for predicting friction factor for turbulent flow of a generalised Bingham plastic slurry for $n = 0{\cdot}5$ and He = 100. (From Heywood & Cheng (1984*b*).)

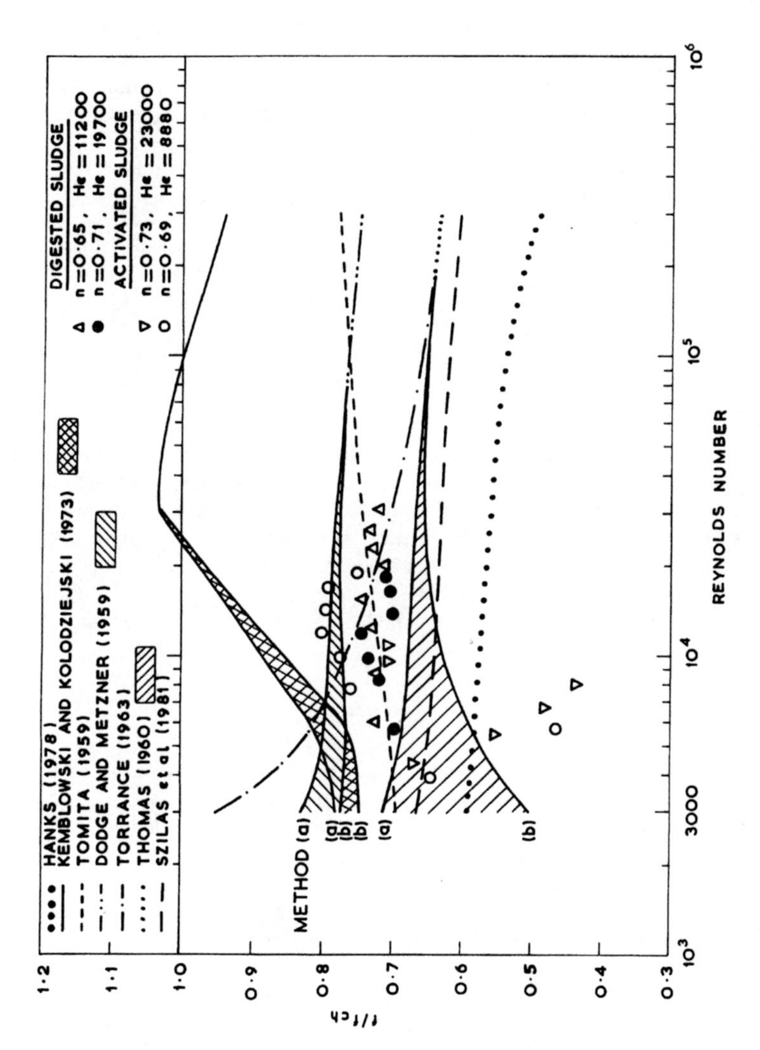

Fig. 7.10. Comparison of methods for predicting friction factor for turbulent flow of a generalised Bingham plastic slurry for $n = 0{\cdot}7$ and He = 10000. (From Heywood & Cheng (1984*b*).)

necessary to apply engineering judgement to the degree of conservatism to be adopted in head loss estimation.

Methods (a) and (b) in Figs 7.9 and 7.10 refer to the two alternative approaches which can be used when applying a predictive equation for f developed for a power law flow model to the three-parameter Herschel–Bulkley model equation as outlined above. The hatched areas in the figures indicate the difference in f values using the two methods but the same predictive equation for f. This variation diminishes as either the Reynolds number, Re′, increases or the Hedström number, He′, decreases.

Prediction of f for turbulent flow in pipe where wall roughness is important is generally based on the above relations for smooth wall pipe. Govier and Aziz (1972) categorised roughness effect into three classes:

(1) smooth wall pipe turbulence,
(2) partially rough wall pipe turbulence, where friction factors are larger than for smooth pipe and are dependent on Reynolds number, and
(3) fully rough wall pipe turbulence, where friction factors are independent of Reynolds number but are dependent upon relative roughness, e/D.

In their recommended design methods, Govier and Aziz suggest that if the pipe is rough, the pressure gradient calculated by the various methods for smooth pipes should be multiplied by the ratio of the friction factor for rough pipe to the friction factor for smooth pipe, as determined from the Moody chart for Newtonian materials at the appropriate Reynolds number consistent with the fluid model being used, i.e. $\mathrm{Re_B}$ for the Bingham plastic model or Re′ for either the power law or generalised Bingham models. If the relative pipe roughness is such that fully rough wall pipe turbulence may be expected, Govier and Aziz (1972) recommend an equation developed by Torrance (1963):

$$\frac{1}{\sqrt{f}} = \frac{1{\cdot}77}{n}\ln\left(\frac{D}{2e}\right) + 6{\cdot}0 - \left(\frac{2{\cdot}65}{n}\right) \qquad (7.65)$$

and Szilas *et al.* (1981) have incorporated a modification to their eqn (7.61). In addition, Wilson and Thomas (1985) have considered the problem.

When pipe roughness is expected to be an important factor, the relative roughness, e/D, may be estimated from standard texts (Govier & Aziz, 1972) or, alternatively, an estimate of e/D may be obtained from Fig. 7.6 by measuring friction factors for the turbulent flow of water through the pipe at known Reynolds numbers and reading off the appropriate e/D value for a given f and Re combination.

Bingham plastic slurries. For slurries whose flow curve may be adequately described by the flow model given by eqn (7.7), Thomas (1963) has suggested the use of a Blasius type equation:

$$f = B_{\mathrm{T}}\mathrm{Re_B}^{-b} \qquad (7.66)$$

in which the parameters B and b have been related empirically by Thomas to fluid properties using data for flocculated aqueous suspensions flowing in smooth tubes:

$$\left.\begin{aligned} B_T &= 0{\cdot}079\left[\left(\frac{\eta_f}{\eta_p}\right)^{0{\cdot}48} + \left(\frac{\rho_m \tau_{yB} K_{T1}^2}{\eta_f^2}\right)^2\right] \\ b &= 0{\cdot}25\left[\left(\frac{\eta_f}{\eta_p}\right)^{0{\cdot}15} + \left(\frac{\rho_m \tau_{yB} K_{T2}^2}{\eta_f^2}\right)^2\right] \end{aligned}\right\} \qquad (7.67)$$

where $K_{T1} = 1{\cdot}94\,\mu$m, $K_{T2} = 1{\cdot}73\,\mu$m, and the Newtonian viscosity of the suspending medium is given by η_f.

An alternative semi-theoretical approach is outlined by Hanks and Dadia (1971). They developed relationships for f versus Re_B at each of several values of He. Their predictions must be evaluated numerically and are summarised in Fig. 7.11.

The curves in Figure 7.11 are nearly linear in both the laminar and turbulent flow regimes. Darby and Melson (1981) have developed an expression which fits the turbulent flow regime:

$$f_T = 10^c Re_B^{-0{\cdot}193} \qquad (7.68)$$

where

$$c = -1{\cdot}378\,(1 + 0{\cdot}146 \exp(-2{\cdot}9 \times 10^{-5}\,Re_B)) \qquad (7.69)$$

Equation (7.68) does not apply when the Hedström number, defined by eqn (7.27), is less than 1000, but this is not a practical constraint for most slurries with a measurable yield stress.

Darby and Melson (1981) employed the approach taken by Churchill (1977) to combine the Buckingham equation for f_L (eqn (7.29)) for laminar flow with eqn (7.68) for f_T for turbulent flow to arrive at a single friction factor expression valid for all flow regimes:

$$f = [f_L^m + f_T^m]^{1/m} \qquad (7.70)$$

It was found that the exponent m depends on Bingham Reynolds number according to

$$m = 1{\cdot}7 + \frac{40\,000}{Re_B} \qquad (7.71)$$

Agreement between Hanks and Dadia (1971) predictions in Fig. 7.11 and those obtained from eqn (7.70) is good.

Alternatively, the turbulent flow of Bingham plastics in a smooth pipe may, in general, be represented conservatively when He $< 10^6$ by the Blasius or Nikuradse equation for Newtonian materials with the Newtonian Reynolds number replaced by the Bingham Reynolds number, Re_B. This approach is reasonable because Thomas (1963) found his experimental friction factor data fell approximately 12% below the Newtonian Blasius line, and consequently yield stress was not an important factor. For values of He $> 10^6$, the Hanks and Dadia analysis indicates values of f greater than for Newtonian fluids.

2.3.3. Rapid Assessment of Pressure Loss for Fully Turbulent Flow

The pressure loss predictive equations outlined in the previous two sections are

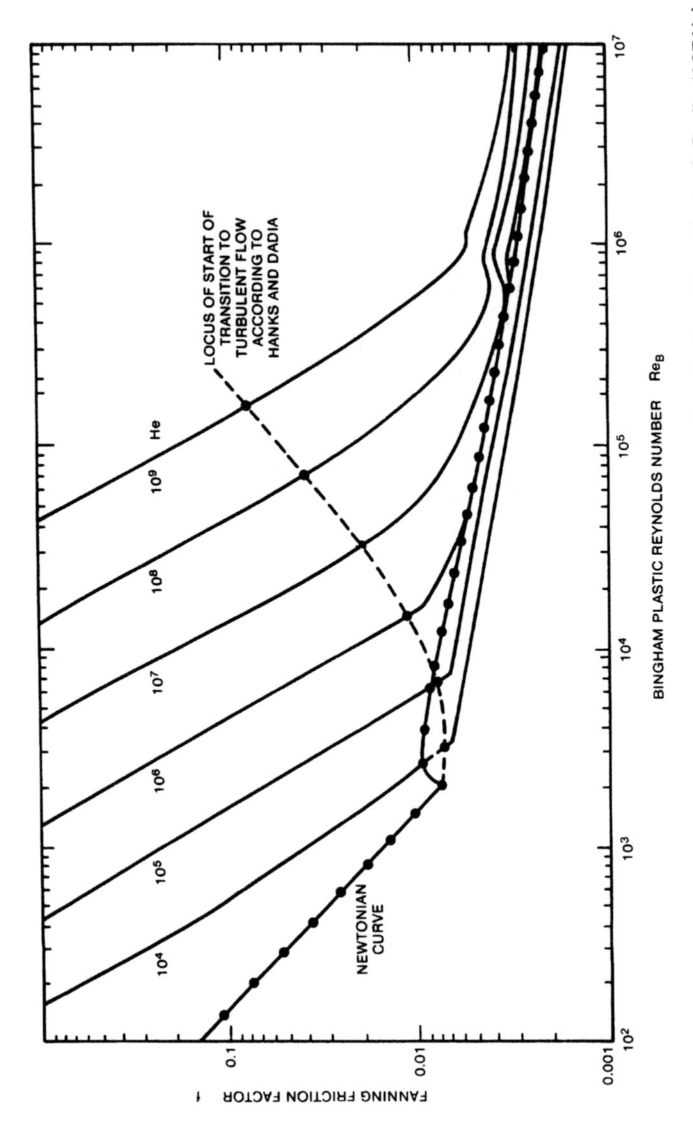

Fig. 7.11. Prediction of friction factor for Bingham plastic slurries in smooth wall pipe. (From Hanks & Dadia (1971).)

generally based on semi-empirical approaches which sometimes have a laminar flow model as the starting point. An initial estimate of pressure loss can often be made by using the head loss for water flow at the same mean velocity and adding an appropriate percentage to allow for the presence of solids. Hayes *et al.* (1973) have used this approach with some success using a percentage ranging from 25 to 75% for various types of sewage sludge, and Heywood and Cheng (1984*c*) have compared the approach with many of the expressions for f listed earlier. This is useful provided an increase in head loss when solids are present always occur. However, drag reduction, i.e. a reduction in head loss for slurry flow compared with the suspending liquid, has been observed for different types of fibrous suspensions such as raw sewage sludge and paper pulp as already mentioned.

The problem, of course, is to identify the appropriate multiple of suspending liquid pressure loss, assuming no drag reduction is occurring. One simple approach is to assume that for flow in fairly rough pipes at fairly high Reynolds numbers, in fully developed turbulent flow, the pressure loss is independent of slurry laminar flow property (see Fig. 7.6). In this case, the additional frictional pressure loss above that for flow of suspending liquid alone is a result simply of the increase in fluid density arising from the presence of denser solids in the slurry. Thus, the pressure loss for slurry

$$P_f \simeq \frac{\rho_m}{\rho_f} \times \Delta P_l \qquad (7.72)$$

2.3.4. Accuracy of Pressure Drop Prediction

The question arises: in view of the wide variation in predictions of f, which of these predictive methods should be adopted for general use? Until a comprehensive comparison is undertaken using a large number of experimental data covering a variety of fluid types, this question will remain unanswered. At the present time it seems clear that the design engineer must be aware of the many prediction methods available and that they can lead to widely differing f values. By using as many of these methods as possible, upper and lower bounds for f can be assigned and then a degree of engineering judgement must be exercised so that an appropriate f level, or reduced f range, is adopted for further detailed design.

Upper and lower bounds at selected confidence levels can be estimated for the pressure-drop/flow-velocity relationship but these are not obtained by simply substituting the upper and lower bound values of parameters evaluated experimentally in flow models into the pipeline equations for f prediction. A more complicated procedure has to be used, which in essence involves the calculation of f values for different combinations of the flow-curve/flow-model parameters within their 90% (or other) confidence intervals (Cheng, 1987). From the results it is possible to determine the upper and lower bound pressure-drop/flow-velocity curves for, say, 90% confidence intervals for finding pressure drop at any given flow velocity.

These confidence limits can also be represented on an f/f_{ch} versus Reynolds-number plot, similar to Figs 7.9 and 7.10. Figure 7.12 shows that the construction

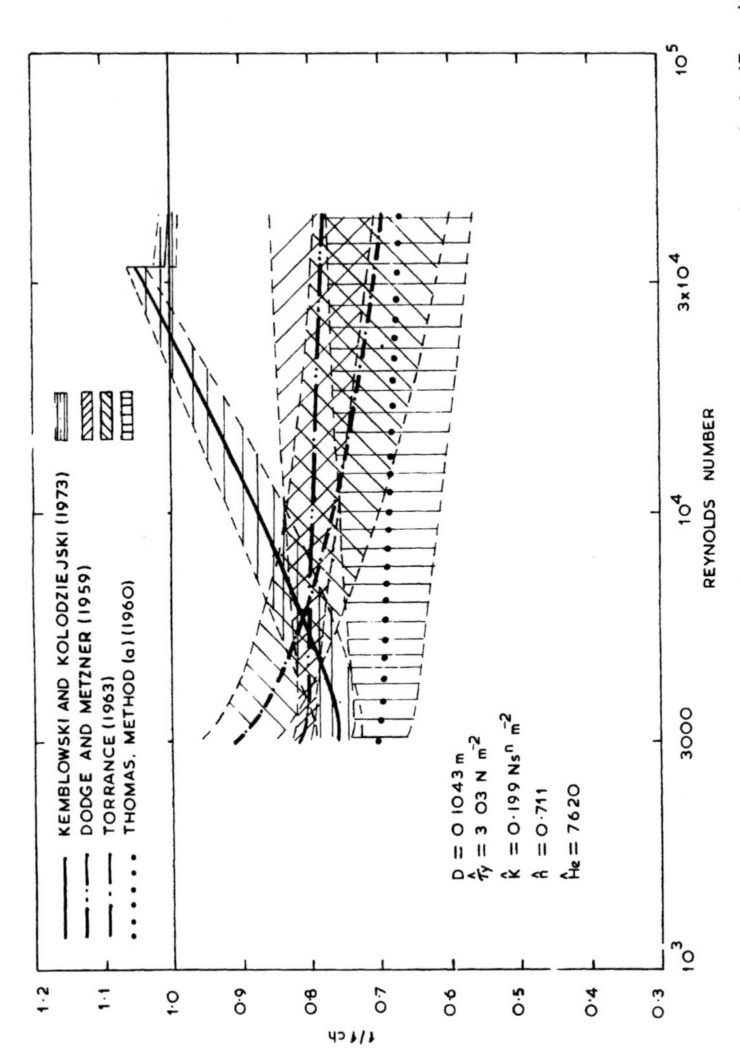

Fig. 7.12. Confidence limits (90%) for predicting friction factor for non-Newtonian slurries using various methods. (From Heywood & Cheng (1984*b*).)

of confidence intervals increases the spread of possible f values for a given Reynolds number. (The discontinuities in the Kemblowski–Kolodziejski curves are a consequence of two equations for f, eqns (7.57) and (7.60).) The uncertainty in f prediction will always be even greater than that obtained when no attention is given to confidence limits for the predictions, and once again we must conclude that for any pipeline design involving non-Newtonian turbulent flow, not only should as many prediction methods as possible be examined in order to obtain the spread of f values, but also the uncertainty in f due to uncertainty in flow model parameters must be taken into account.

2.4. Effect of Wall Slippage on Pipeline Design

When the non-Newtonian character of a fluid arises through the dispersion in a liquid of either solid particles (suspensions/slurries) or of liquid droplets (emulsions), a liquid layer can form at the pipe wall serving to lubricate the bulk flow. This effect can be treated as a discontinuity in velocity at the tube wall, and is thought to be caused by particle or droplet migration from fluid near the wall, the viscosity of which then becomes lower than that in the bulk of the fluid.

The direct consequence of *slip* is the increased flow rate through the pipe compared to that which would occur if *slip* were absent. Although true slip is not believed to occur, the phenomenon can be quantified by a *slip* velocity, V_s, such that the flow rate given for laminar flow by eqn (7.16) is enhanced by an amount proportional to this velocity.

$$Q = \frac{\pi D^2}{4} V_s + \frac{\pi}{8}\left(\frac{D}{\tau_w}\right)^3 \int_0^{\tau_w} \tau^2 \dot{\gamma}\, d\tau \tag{7.73}$$

When a *slip* coefficient β, equal to V_s/τ_w, is defined, many experimental studies in which a variety of materials have been used have shown that β is not only a function of τ_w but varies inversely with pipe diameter (Jastrzebski, 1967). It therefore seems probable that, in general, β is inversely proportional to D raised to some power p, where p may be in the range $0 < p < 1$ or even $p > 1$. By defining a corrected β value as $\beta_c = \beta D^p$, eqn (7.73) may be rearranged to give

$$\frac{32Q}{\pi D^3 \tau_w} = \frac{8\beta_c}{D^{p+1}} + \frac{4}{\tau_w^4} \int_0^{\tau_w} \tau^2 \dot{\gamma}\, d\tau \tag{7.74}$$

Thus, by using a range of pipe sizes and for a selected value of wall shear stress, τ_w, the best estimate for p may be determined by searching for the largest correlation coefficient between $32Q/\pi D^3 \tau_w$ and $1/D^{p+1}$ using various p values. Having determined β_c as the slope of the linear plot of $32Q/\pi D^3 \tau_w$ versus $1/D^{p+1}$, the total flow rate for laminar flow conditions at a specified τ_w may be calculated from eqn (7.74).

Methods by which slip velocity may be estimated using a coaxial cylinder viscometer have also been given by Cheng and Parker (1976). No guidelines are, however, currently available for estimating the extent of slip under turbulent flow conditions and the application to turbulent flow in full-scale pipe.

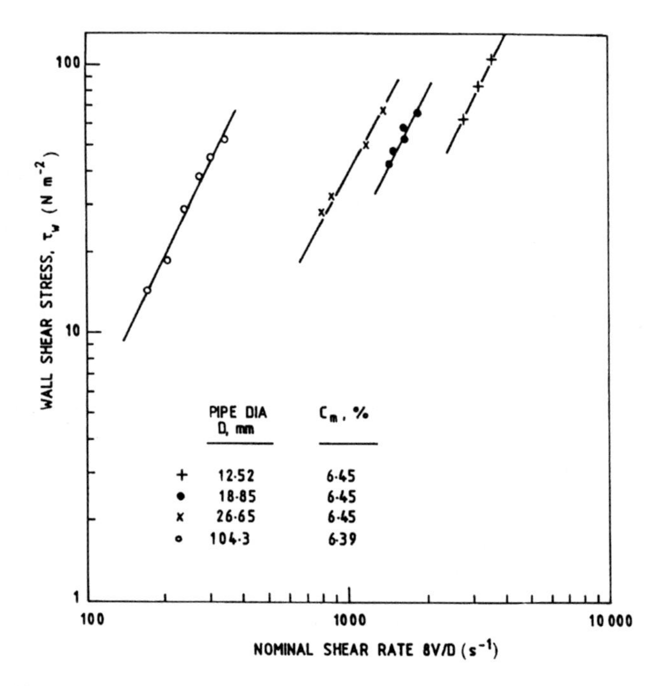

Fig. 7.13. Wall shear stresses for 6·4% by mass digested sludge obtained using various pipe sizes.

3. DESIGN BY SCALE-UP FROM PIPE FLOW DATA

3.1. Laminar Flow

An alternative method for predicting the performance of a large pipeline containing a homogeneous non-Newtonian slurry is by scale-up from small pipeline data and has been described in section 2.1.

3.2. Turbulent Flow (Bowen Method)

For turbulent flow, Bowen (1961) suggests a modification of the Blasius equation:

$$D^a \tau_w = kV^w \tag{7.75}$$

A plot of wall shear stress τ_w against flow rate allows w to be determined; then by replotting the data in the form τ_w/V^w against D, the values of k and a can be obtained. The transition between laminar and turbulent flow is given by the intercept of the curves for laminar and turbulent flow, though of course this yields a single point rather than a range. The Bowen method is independent of the laminar flow property of the slurry and the method should also be valid for elastic

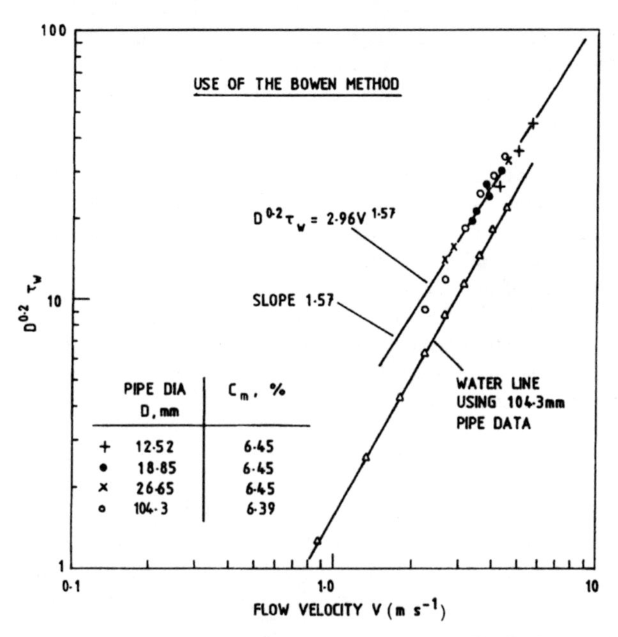

Fig. 7.14. Correlation of turbulent wall shear stress data in Fig. 7.13 using the Bowen (1961) method.

fluids. Figures 7.13 and 7.14 give an example of the use of the method for sewage sludge. Kenchington (1972) has used this method for scaling up from 27, 35 and 53 mm pipes to a 329 mm pipe.

It may be that, because industrial slurries do show variable properties, unattainably high precision is necessary at the pilot scale in order to exploit the Bowen method. So, if high accuracy is essential, large-scale test work is unavoidable.

The degree of pipe roughness will be reflected in both exponents for D and V in eqn (7.75). The problem arises, however, if scale-up from smooth-walled small bore (e.g. plastic pipe) to rough wall large-scale pipe (e.g. galvanised iron) is undertaken, or vice versa. A corrected pressure gradient could in these cases be estimated presumably by using the Govier and Aziz (1972) approach of multiplying the predicted pressure gradient by a ratio of friction factors obtained from the Moody chart (see section 2.3.2), although such a procedure has not yet been validated. The friction factor ratio would obviously depend upon the wall roughnesses of the small bore and full-scale pipe and would neglect any wall slippage effects arising from flow in the smooth-walled pipe.

4. PRESSURE LOSSES FROM PIPE FITTINGS

Estimates of pressure losses arising from flow through pipe fittings are required before total system head loss estimation can be made. This then allows the selection and sizing of an appropriate pump or pumps. Pipeline fittings may be classified as branching (tees), reducing/expanding (bushes and sudden changes in flow area) or deflecting (elbows and bends).

For the turbulent flow of Newtonian non-settling slurries much information exists with head loss often expressed in terms of the number of equivalent pipe diameters of straight pipe giving the same head loss (Perry & Chilton, 1985). Table 7.1 gives some typical loss data in terms of number of pipe diameters. These data should be used with caution as considerable variation will occur according to the exact construction of the fittings. For the turbulent flow of non-Newtonian slurries, both Cheng (1970) and Turian *et al.* (1983) have concluded that the losses do not depend significantly on the non-Newtonian character of the slurry.

Normally, frictional pressure loss from flow through a fitting is expressed through a loss coefficient, K_{ff}:

$$P_{ff} = K_{ff} \frac{V_{ff}^2}{2g} (\rho_m g) \tag{7.76}$$

where V_{ff} is the characteristic slurry velocity in the fitting. K_{ff} is approximately constant and independent of pipe Reynolds number for the turbulent flow of either Newtonian or non-Newtonian slurries and Table 7.1 also includes some typical K_{ff} values. For laminar flow in both cases K_{ff} is inversely proportional to Reynolds number.

For the laminar flow of Newtonian slurries through fittings, much less information is available than for turbulent flow. A summary (see Table 7.2) is provided by Perry and Chilton (1985) and the available information has been reviewed by Edwards *et al.* (1985) who have also provided some much needed data for the

TABLE 7.1
Pressure Losses for Turbulent Flow through Various Pipe Fittings[a]

Pipe fitting	*Number of velocity heads* (K_{ff})
45° bend	0·3
90° bend	0·6–0·8
90° elbow	1·2
'Tee' entry from leg	1·2
'Tee' entry into leg	1·8
Unions and couplings	Very small
Globe valve: fully open	1·2–6·0
Gate valve: fully open	0·15
3/4 open	1
1/2 open	4
1/4 open	16

[a]From Coulson & Richardson

TABLE 7.2
Pressure Losses for Laminar, Newtonian Flow through Various Pipe Fittings[a]

Pipe fitting	K_{ff}			
	Re = 1000	*500*	*100*	*50*
90° elbow	0·9	1·0	7·5	16
'Tee' entry to leg	0·4	0·5	2·5	
'Tee' entry from leg	1·5	1·8	4·9	9·3
Gate valve	1·2	1·7	9·9	24
Globe valve, composition disc	11	12	20	30
Plug valve	12	14	19	27
Angle valve	8	8·5	11	19
Check valve, swing	4	4·5	17	55

[a]From Perry & Chilton (1985)

laminar flow of non-Newtonian slurries. Their overall conclusion is that correlations relating K_{ff} to Re for Newtonian slurries may also be used for non-Newtonian, power law slurries modelled by the power law flow model (eqn (7.6)) provided the non-Newtonian Reynolds number Re′ is used in these correlations instead of the Newtonian Re. Steffe *et al.* (1984) have also provided data for the flow of foodstuffs through various types of fittings. Das *et al.* (1988) have supplied data for flow in bends and Ma (1987) has covered a number of types of fittings.

It is important to note that the Reynolds number ranges over which these correlations apply are generally smaller than the usual range for laminar pipe flow. Thus, the maximum Reynolds number in the case of globe valves is around 12 while for a 90° elbow, the value is 900. This compares with the usual value of 2100 for the breakdown of Newtonian laminar flow in a straight pipe.

5. START-UP BEHAVIOUR OF THIXOTROPIC SLURRIES

For slurries that possess strong interparticle attraction and form a gel when left to stand in a pipeline, the start-up pressures are very much greater than the steady-state pressure. A method for predicting both start-up pressures and the time taken for the gelled material to be expelled from the pipeline and attain steady flow has been developed by Carleton *et al.* (1974). It is based on a generalised Bingham fluid with thixotropic build-up and breakdown following the Moore (1959) model. The method is available as a computer program from Warren Spring Laboratory and can be used to predict the relationship between pressure, position in the pipe and time, both for conditions of constant applied pressure and for a sawtooth variation in flow rate (the latter represents the flow from a positive displacement pump). The program for constant applied pressure is particularly useful for estimating thixotropic parameters from measurements in a tube viscometer.

The program for constant applied pressure has been used by Cheng and Whittaker (1972) for sewage sludges in pipes up to 26 mm in diameter. The time to empty the pipe, using water as the expelling liquid, was generally -50 to $+100\%$

of the predicted value. This order of discrepancy is understandable in view of the gross inhomogeneity of the material involved.

As yet, this method has not been applied to larger diameter pipes. It can readily be anticipated that experimental verification will not be easy, not only because of the simplifying assumptions made about thixotropic behaviour and pipeline flow but also because of the intervention of other phenomena in large pipes. These include temperature or other effects resulting in non-uniform build-up, inclusion of air or liquid vapour pockets (leading to overall compressibility) and surge effects in full-scale installations.

The problems of gelling of both crude and fuel oils and of drilling mud and bentonite clay slurries are of particular concern to oil and civil engineers. Little published information is available for clay slurries. Some analyses of this problem for waxy crude oils have assumed Bingham plastic behaviour using a value for the yield stress measured after the crude oil has stood in a tube viscometer. It was found that yield stresses measured in a 6 mm tube were about 30% higher than those found in the field in a 610 mm line (Knegtel & Zeilinga, 1971). This was regarded as providing a safety margin of the order of magnitude desired. However, it is also known (Verschuur *et al.*, 1971) that the thermal history of the oil has a considerable effect on its rheological properties and that yield values measured in a small pipe do not give realistic predictions when applied to full-scale (203 mm) lines. Cawkwell and Charles (1987) have also considered this problem using a single structural parameter in relation to crude oil pipelines. This approach can be readily applied to thixotropic suspensions. Thixotropic pipe flow prediction has also been undertaken by Ritter and Batycky (1967).

Cheng (1979) has conducted a study into the possibility of using scale-up procedures to predict the pipe flow characteristics of thixotropic fluids. This study is based on the assumptions that experimental results obtained on a small, laboratory-scale pipe rig can be scaled up to a full-scale industrial pipeline without specific reference to the detailed thixotropic property of the fluid, that the fluid obeys single-structured thixotropic constitutive equations (see Cheng, 1979) and that quasi-steady-state flow pertains. Three different cases for start-up were discussed depending on whether the flow rate or the pumping pressure is prescribed, or if the pump characteristics (described by a functional relationship between flow rate and pump pressure) are prescribed. Cheng showed that for geometrically similar pipelines (i.e. D/L constant), if the initial out-going fluid and the inlet incoming fluid are similar and the flow rate is prescribed as a function of time only, then the pumping pressure is a function of time only and independent of L. The situation arising if the pumping pressure is prescribed, or if the pump characteristics are prescribed, is more complex but was also analysed. Cheng also considered how the scale-up rules might be achieved in practice.

A rather impractical situation which has to be overcome is that D/L should be constant between the full-scale pipeline and the pilot or laboratory scale rig. This means that in order to match a 0·6 m diameter, 50 km pipeline, for example, where $D/L = 12 \times 10^{-6}$, a 0·1 m diameter pilot-scale rig would have to be 8 km long, and a 0·01 m diameter laboratory rig would have to be 800 m long. Because of this, and other impracticalities, Cheng concluded his study by suggesting certain experimental work which could be conducted to check the validity of a simpler approach towards developing scale-up rules for predicting the start-up characteristics for pipelines transporting thixotropic fluids. No report of this experimental work has yet appeared in the open literature.

6. TECHNIQUES FOR REDUCING ENERGY CONSUMPTION IN SLURRY PIPELINING

Heywood (1986) has reviewed methods available to reduce head loss and hence energy consumption incurred when non-settling slurries are pumped through horizontal pipes. Some of the more obvious ways of reducing slurry viscosity levels directly include the following.

- Reducing slurry concentration.
- Increasing particle size (if Brownian motion and particle surface effects control viscosity levels).
- Broadening the particle size distribution at constant total solids concentration (Cheng *et al.*, 1990).
- Reducing the angularity of particle shape while maintaining the particle size distribution and solids concentration essentially constant; also, adding high aspect ratio fibres.
- Adding deflocculants (soluble ionic compounds) to disperse flocculated slurries (important when a significant percentage, say 5–10%, of solids are below approximately 5 μm in size) (Heywood & Richardson, 1978*a*; Horsley & Reizes, 1980).
- Injecting water (or chemical deflocculant) through the pipe wall periodically along the pipeline length to create a film lubricating layer (Horsley, 1988).
- Development of stabilised or semi-stabilised slurries by supporting coarse particles in a fine slurry medium made up of flocculated particles of low aspect ratio or consisting of a fibrous particle network (see Chapter 1).

Most of the above methods can alter a suspension or slurry formulation irreversibly in the sense that a significant cost arising from extra capital and operating requirements may be incurred to return the slurry to its original state (if it is possible at all) once the slurry has been transported through the pipeline.

A further set of techniques for non-settling slurries tends not to have this disadvantage.

- Oscillation of the slurry flow rate or pressure gradient (Round *et al.*, 1981; Round & El-Sayed 1985); vibration or oscillation of the pipeline in the direction of the pipe axis while maintaining a constant slurry flow rate (Deysarkar & Turner, 1981).
- Injecting air (or other gas, if appropriate) into the pipeline to create a three-phase mixture which would generally be readily separated on discharge from

the pipeline (Carleton *et al.*, 1973; Heywood & Richardson, 1978*b*; Farooqi *et al.*, 1980; Dzuibinski & Richardson, 1985).

The methods listed above are effective under certain conditions only. In determining whether a method may be appropriate for a particular application it is important to decide for a non-settling slurry whether turbulent or laminar flow conditions will prevail during pipe flow, and Newtonian or non-Newtonian flow property is important, if the flow is in the laminar regime. Most methods are applicable for the laminar flow of non-settling slurries.

7. CONCLUDING REMARKS

Methods have been reviewed for the calculation of pressure-drop/flow-rate characteristics in pipe flow of Newtonian and non-Newtonian slurries. The procedures in general aim to predict laminar and turbulent flow data, and the point of breakdown of laminar flow conditions heralding the onset of turbulence. In scaling-up from small-scale viscometers, the shear stress ranges of design data should approach closely those obtained in the full-scale pipeline. Thus, for flow in a large diameter pipeline where low shear stresses are encountered, it would be unwise to use data obtained from a high-speed rotational viscometer (Heywood, 1985).

The scaling-up of laminar flow data from pilot-scale pipelines presents no problem, provided there is no complication arising from a *slippage* effect at the pipe wall for dispersed systems, i.e. a plot of τ_w versus $8V/D$ is independent of pipe diameter. The uncertainty lies in the appropriate method to use when predicting turbulent head losses from laminar flow data for non-Newtonian slurries. For each of the three main fluid models, a number of alternative methods exist and it has yet to be established which of these methods gives the most reliable predictions. It seems that until the best method has been established, several of these methods should be used for the design to see by how much the predictions differ, and then to take a mean or an upper bound of the estimates, depending on the adopted design philosophy.

8. NOTATION

a	Exponent on D in Bowen eqn (7.75) [1]
A, B, C	Parameters in Churchill eqn (7.45) [1]
b, B_T	Parameters used in Thomas' eqn (7.66) [1]
c	Exponent in Darby and Melson eqn (7.67) [1]
C_m	Concentration by mass of solids, % [1]
D	Inside pipe diameter (L)
e	Absolute pipe roughness (L)
E	Parameter defined in eqns (7.58) [1]
f	Fanning friction factor defined by eqn (7.20) [1]
f_c	Lower critical friction factor for transitional flow given by eqn (7.42) [1]

f_{ch}	Fanning friction factor given by Churchill eqn (7.45) [1]
f_L	Fanning friction factor for laminar flow in eqn (7.70) and defined by eqn (7.20) [1]
f_T	Fanning friction factor for turbulent flow obtained from eqn (7.68) and defined by eqn (7.20) [1]
f_{To}	Friction factor defined by eqn (7.51) [1]
$F'(n)$	Irvine (1988) function defined by eqn (7.63) [1]
g	Gravitational acceleration [$L\,T^{-2}$]
He	Hedström number, defined by eqn (7.27) for Bingham plastic slurry [1]
$\widehat{He}$	Estimated Hedström number from regression analysis
He′	Hedström number, defined by eqns (7.36) or (7.37) for a generalised Bingham plastic slurry [1]
k	Constant in Bowen eqn (7.75)
K	Consistency coefficient in power law model, and in generalised Bingham model [$M\,L^{-1}\,T^{n-2}$]
K_{ff}	Head loss coefficient for flow through a pipe fitting, defined by eqn (7.76) [1]
K_{T_1}, K_{T_2}	Constants in Thomas' eqn (7.67) (μm) [L]
K'	Constant in eqn (7.12) related to consistency coefficient, K [$M\,L^{-1}\,T^{n'-2}$]
$\hat{K}$	Estimate of K from regression analysis [$M\,L^{-1}\,T^{n-2}$]
L	Pipe length [L]
m	Exponent in Darby and Melson eqn (7.70) [1]
m'	Exponent in eqn (7.57) defined in eqn (7.58) [1]
n	Flow behaviour index in power law model, and in generalised Bingham model [1]
n'	Gradient defined by eqn (7.10), related to flow behaviour index, n [1]
n'_c	Value of n' at commencement of transitional flow regime [1]
$\hat{n}$	Estimate of n from regression analysis [1]
p	Exponent on pipe diameter for wall-slip correlation [1]
P_f	Frictional pressure loss in straight pipe [$M\,L^{-1}\,T^{-2}$]
P'_f	Start-up frictional pressure loss in straight pipe [$M\,L^{-1}\,T^{-2}$]
P_{ff}	Frictional pressure loss from flow through fittings [$M\,L^{-1}\,T^{-2}$]
P_{st}	Pressure drop/gain arising from elevation changes [$M\,L^{-1}\,T^{-2}$]
ΔP	Pressure drop [$M\,L^{-1}\,T^{-2}$]
ΔP_l	Frictional pressure drop for flow of suspending liquid alone at the same mean velocity as the slurry flow [$M\,L^{-1}\,T^{-2}$]
ΔP_T	Total pressure drop along a pipeline [$M\,L^{-1}\,T^{-2}$]
Q	Volumetric flow rate [$L^3\,T^{-1}$]
r	Radial position in pipe [L]
Re	Newtonian Reynolds number [1]
Re_B	Bingham Reynolds number, defined by eqn (7.28) [1]
$(Re_B)_c$	Lower critical Bingham Reynolds number for laminar flow breakdown, given by eqn (7.43) [1]
Re_{mod}	Modified Bingham Reynolds number, defined by eqn (7.32) and (7.33) [1]

Re_{TO}	Reynolds number used by Tomita and defined by eqn (7.52)
Re'	Generalised Reynolds number for non-Newtonian slurry [1], defined by eqn (7.25) [1]
Re'_c	Lower critical generalised Reynolds number for laminar flow breakdown by eqn (7.41) [1]
Re'_t	Reynolds number value when Kemblowski–Kolodziejski turbulent flow prediction line meets Blasius Newtonian line [1]
t	Exponent on Reynolds number in Shaver and Merrill eqn (7.54) [1]
v	Local fluid velocity [$L\,T^{-1}$]
V	Mean slurry velocity over pipe cross-section [$L\,T^{-1}$]
V_{ff}	Characteristic slurry velocity through a pipe fitting [$L\,T^{-1}$]
V_s	Slip velocity at pipe wall defined by eqn (7.73) [$L\,T^{-1}$]
w	Velocity exponent in Bowen eqn (7.75)
x	Ratio of yield stress parameter to wall shear stress, τ_{yB}/τ_w or τ_{yHB}/τ_w [1]
x_c	Lower critical value of parameter $x = \tau_{yB}/\tau_w$ in transitional flow [1]
Z	Stability parameter; defined by eqn (7.38) [1]
Z_c	Critical value of Z [1]
Z_{max}	Maximum value of Z [1]
Δz	Net change in pipe elevation [L]
β	*Slip* parameter [$M^{-1}\,L^{-2}\,T$]
β_c	Corrected *slip* parameter, $=\beta D^p$ [$M^{-1}\,L^{p-2}\,T$]
$\dot{\gamma}$	Shear rate [T^{-1}]
$\dot{\gamma}_{max}$	Maximum shear rate in pipe flow, at pipe wall [T^{-1}]
η_f	Viscosity of suspending liquid [$M\,L^{-1}\,T^{-1}$]
η_m	Newtonian slurry viscosity [$M\,L^{-1}\,T^{-1}$]
η_p or η'_p	Plastic viscosity in Bingham plastic model [$M\,L^{-1}\,T^{-1}$]
ρ_f	Suspending liquid density [$M\,L^{-3}$]
ρ_m	Slurry density [$M\,L^{-3}$]
τ	Shear stress [$M\,L^{-1}\,T^{-2}$]
τ'	Minimum wall shear stress measured in viscometric tests [$M\,L^{-1}\,T^{-2}$]
τ_w	Wall shear stress [$M\,L^{-1}\,T^{-2}$]
τ_{yB} or τ'_{yB}	Yield stress parameter in Bingham plastic model [$M\,L^{-1}\,T^{-2}$]
τ_{yHB}	Yield stress parameter in generalised Bingham model [$M\,L^{-1}\,T^{-2}$]
$\hat{\tau}_y$	Estimate of yield stress from regression analysis [$M\,L^{-1}\,T^{-2}$]
ϕ	Parameter defined in eqn (7.58) [1]

9. REFERENCES

Blasius, H. (1913). Das Ahnlichkeitsgesetz bei Reibungsvorgangen in Flussigkeiten. *Forschg. Arb. Ing-Wes.*, **134**, Berlin, Germany.

Bowen, R.L. (1961). Scale-up for non-Newtonian fluid flow. *Chem. Engng*, 24 July, **68**, 143–50.

Buckingham, E. (1921). On plastic flow through capillary tubes. *Proc. ASTM*, **21**, 1154.

Carleton, A.J., Cheng, D.C-H. & French, R.J. (1973). Pneumatic transport of thick pastes. In *Proceedings of Pneumotransport 2*, the *2nd International Conference on the Pneumatic Transport of Solids in Pipes*. BHRA Fluid Engineering, Cranfield, UK, paper F2.

Carleton, A.J., Cheng, D.C-H. & Whittaker, W. (1974). Determination of the rheological properties and start-up pipeline flow characteristics of waxy crude and fuel oils. Paper No. IP74-009. Institute of Petroleum, London.

Cawkwell, M.G. & Charles, M.E. (1987). An improved model for start-up of pipelines containing gelled crude oil. *J. Pipelines*, **7**, 41–52.

Cheng, D.C.-H. (1970). A design procedure for pipeline flow of non-Newtonian dispersed systems. In *Proc. HT 1*, paper J5, pp. 77–96.*

Cheng, D.C.-H. (1979). A study into the possibility of scale-up in thixotropic pipeflow. Report LR 317 (MH), Warren Spring Laboratory, Stevenage, UK.

Cheng, D.C.-H. (1986). Yield stress: a time-dependent property and how to measure it. *Rheol. Acta*, **25**, 542–54.

Cheng, D.C.-H. & Heywood, N.I. (1984). Flow in pipes. Part 1: Flow of homogeneous fluids. *Phys. Technol.*, **15**, 244–51.

Cheng, D.C.-H. & Parker, B.R. (1976). The determination of wall-slip velocity in the co-axial cylinder viscometer. In *Proceedings of the 7th International Congress on Rheology*, ed. C. Klason & J. Kubat. Swedish Society of Rheology, Gothenburg, Sweden, pp. 518–19.

Cheng, D.C.-H. & Whittaker, W. (1972). A method for assessing the thixotropic property of fluids carried in pipelines. In *Proc. HT 2*, paper B4, pp.41–60.*

Cheng, D.C.-H., Kruszewski, A.P., Senior, J.R. & Roberts, T.A. (1990). The effect of particle size distribution on the rheology of an industrial suspension. *J. Mater. Sci.*, **25**, 353–73.

Cheng, R.C.-H. (1987). Confidence bands for two-stage design problems. *Technometrics*, **29**(3), 301–9.

Churchill, S.W. (1977). Friction factor equation spans all fluid-flow regimes. *Chem. Engng*, 7 Nov., **84**, 91–2.

Colebrook, C.F. (1939). Turbulent flow in pipes with particular reference to the transition region between smooth and rough pipe laws. *J. Inst. Civ. Eng.*, **11**, 133–156.

Coulson, J.M. & Richardson, J.F. *Chemical Engineering* (Vol. I), Pergamon Press, Oxford, UK.

Darby, R. & Melson, J. (1981). How to predict the friction factor for the flow of Bingham plastics. *Chem. Engng*, 28 Dec., **88**(26), 59–61.

Das, S.K., Biswas, M.N. & Mitra, A.K. (1988). Flow of non-Newtonian fluids through bends. In *Proceedings of the International Symposium on the Hydraulic Transport of Coal and Other Minerals*. Council of Scientific & Industrial Research and Indian Institute of Metals, Bhubaneswar, India.

Deysarkar, A.K. & Turner, G.A. (1981). Flow of paste in a vibrated tube. *J. Rheol.*, **25**(1), 41–54.

Drew, T.B., Koo, E.C. & McAdams, W.H. (1932). The friction factor for clean round pipes. *Trans. AIChE*, **28**, 56–72.

Dodge, D.W. & Metzner, A.B. (1959). Turbulent flow of non-Newtonian systems. *AIChEJ*, **5**, 189–204.

Dzuibinski, M. & Richardson, J.F. (1985). Two phase flow of gas and non-Newtonian liquids in horizontal pipes. *J. Pipelines*, **5**, 107–11.

Edwards, M.F., Jadallah, M.S.M. & Smith, R. (1985). Head losses in pipe fittings at low Reynolds numbers. *Chem. Eng. Res. Des.*, **63**(1), 43–50.

Farooqi, S.I., Heywood, N.I. & Richardson, J.F. (1980). Drag reduction by air injection for suspension flow in a horizontal pipeline. *Trans. Inst. Chem. Engrs*, **58**, 16–27.

Govier, G.W. & Aziz, K. (1972). *The Flow of Complex Mixtures in Pipes*. Van Nostrand Reinhold, New York, USA.

Hanks, R.W. (1963). The laminar-turbulent transition for flow in pipes, concentric annuli and parallel plates. *AIChEJ*, **9**, 45–8.

Hanks, R.W. (1978). Low Reynolds number turbulent pipeline flow of pseudohomogeneous slurries. In *Proc. HT 5*, paper C2, pp. C23–34.*
Hanks, R.W. & Dadia, B.H. (1971). Theoretical analysis of the turbulent flow of non-Newtonian slurries in pipes. *AIChEJ*, **17**, 554–7.
Hanks, R.W. & Pratt, D.R. (1967). On the flow of Bingham plastic slurries in pipes and between parallel plates. *Soc. Petr. Engrs J.*, **7**, 342–6.
Harris, J. (1968). The correction of non-Newtonian turbulent pipe-flow data. *Rheol. Acta*, **7**, 228–35.
Hayes, J., Flaxman, E.W. & Scivier, J.B. (1973). A comprehensive scheme for sewage sludge in North West England. *Proc. Inst. Civ. Engrs* (Part 2: *Res. and Theory*), **55**, paper 7604, 1–21.
Herschel, W.H. & Bulkley, R. (1926). Measurement of consistency as applied to rubber-benzene solutions. *Proc. ASTM*, **26**(II), 621–33.
Heywood, N.I. (1980). Pipeline design for non-Newtonian fluids. *Inst. Chem. Engrs Symp. Ser.*, no. 60, 33–52.
Heywood, N.I. (1985). Selecting a viscometer. *Chem. Engr*, Jun., **415**, 16–23.
Heywood, N.I. (1986). A review of reducing energy consumption in slurry pipelining. In *Proc. HT 10*, paper K3, pp. 319–32.*
Heywood, N.I. & Cheng, D.C.-H. (1984*a*). Flow in pipes. Part 2: Multiphase flow. *Phys. Technol.*, **15**(6), 291–300.
Heywood, N.I. & Cheng, D.C.-H. (1984*b*). Comparison of methods for predicting head loss in turbulent pipeflow of non-Newtonian fluids. *Trans. Inst. Meas. Contr.*, **6**, 33–45.
Heywood, N.I. & Cheng, D.C.-H. (1984*c*). Viscometric testing and head loss correlations for pipeline design for fibrous suspensions. Report LR 502 (MH), Warren Spring Laboratory, Stevenage, UK.
Heywood, N.I. & Richardson, J.F. (1978*a*). Rheological behaviour of flocculated and dispersed kaolin suspensions in pipe flow. *J. Rheol.*, **22**(6), 599–613.
Heywood, N.I. & Richardson, J.F. (1978*b*). Head loss reduction by gas injection for highly shear-thinning suspensions in horizontal pipe flow. In *Proc. HT 5*, paper C1, pp. C1–22.*
Horsley, R.R. (1988). Reduction in pipeline pressure drop due to viscosity modifiers in the transport of bauxite residue. In *Proc. HT 11*, paper J2, pp. 471–80.*
Horsley, R.R. & Reizes, J.A. (1980). The effect of zeta potential on the head loss gradient for slurry pipelines with varying concentrations. In *Proc. HT 7*, pp. 163–72.*
Irvine, T.F. (1988). A generalised Blasius equation for power fluids. *Chem. Eng. Commun.*, **65**, 39–47.
Jastrzebski, Z.D. (1967). Entrance effects and wall effects in an extrusion rheometer during flow of concentrated suspensions. *Ind. Eng. Chem. Fundam.*, **6**, 445–54.
Kemblowski, Z. & Kolodziejski, J. (1973). Flow resistances of non-Newtonian fluids in transitional and turbulent flow. *Int. Chem. Eng.*, **13**(2), 265–79.
Kenchington, J.M. (1972). The design of large pipelines for cement slurries. In *Proc. HT 2*, paper C4, pp. 41–60.*
Knegtel, J.T. & Zeilinga, E. (1971). Field tests with waxy crudes in the Rotterdam–Rhine pipeline system. *J. Inst. Petr.*, **57**(555), 165–74.
Knudsen, J.G. & Katz, D.L. (1958). *Fluid Dynamics and Heat Transfer*. McGraw Hill Book Co., New York, USA.
Ma, T-Z.W. (1987). Stability, rheology and flow in pipes, bends, fittings, valves and venturi meters of concentrated suspensions. PhD thesis, University of Illinois, Chicago, USA.
Mansour, A.R. & Rajie, O.T. (1987). A new equation for the friction factor for non-Newtonian fluids in circular ducts. *Chem. Ing. Tech.*, **59**(4), 330–2.
Mansour, A.R. & Rajie, O.T. (1988). Generalised explicit equation for the friction factor for Newtonian and non-Newtonian fluids in both circular and non-circular ducts. *Chem. Ing. Tech.*, **60**(4), 304–6.
Metzner, A.B. & Reed, J.C. (1955). Flow of non-Newtonian fluids – correlation of laminar, transition and turbulent regions. *AIChEJ*, **1**, 434–40.

Mishra, P. & Tripathi, G. (1973). Heat and momentum transfer to purely viscous non-Newtonian fluids flowing through tubes. *Trans. Inst. Chem. Eng*, **51**, 141–150.
Moody, L.F. (1944). Friction factors for pipeflow. *Trans. ASME*, **66**, 671–84.
Moody, L.F. (1947). An approximate formula for pipe friction factors. *Trans. ASME*, **69**, 1005–6.
Moore, F. (1959). The rheology of ceramic slips and bodies. *Trans. Brit. Ceram. Soc.*, **58**, 407–94.
Nikuradse, J. (1932). Gesetzmassigkeiten der turbulenten Strömung in glatten Rohren. *Forschg. Arb. Ing-Wes.*, **356**, Berlin, Germany.
Olujic, Z. (1981). Compute friction factors fast for flow in pipes. *Chem. Engng*, Dec., **88**(25), 91–3.
Perry, R.H. & Chilton, C.H. (eds) (1985). Section 5–21, *Chemical Engineers' Handbook* (6th edn). McGraw-Hill, New York, USA.
Ritter, R.A. & Batycky, J.P. (1967). Numerical prediction of the pipeline flow characteristics of thixotropic liquids. *Soc. Petr. Engrs J.*, **7**, 369–76.
Round, G.F. & El-Sayed, E. (1985). Pulsating flows of solid/liquid suspensions. I. Bentonite clay/water suspensions. *J. Pipelines*, **5**, 95.
Round, G.F., Hameed, A. & Latto, B. (1981). In *Proc. STA 6*, pp. 122–30.*
Ryan, N.W. & Johnson, M.M. (1959). Transition from laminar to turbulent flow in pipes. *AIChEJ*, **5**, 433–5.
Scaggs, W.F., Taylor, R.P. & Coleman, H.W. (1988). Measurement and prediction of rough wall effects on friction factor–uniform roughness results. *J. Fluids Engng*, **110**, 385–91.
Shaver, R.G. & Merrill, E.W. (1959). Turbulent flow of pseudoplastic polymer solutions in straight cylindrical tubes. *AIChEJ*, **5**, 181–8.
Smith, R.W. (1960). Flow of limestone and clay slurries in pipelines. *Trans. Soc. Min. Eng. AIME*, **217**, 258–65.
Steffe, J.F., Mohamed, I.O. & Ford, E.W. (1984). Pressure drops across valves and fittings for pseudoplastic fluids in laminar flow. *Trans. ASAE*, 616–19.
Sultan, A.A. (1988). Sizing pipe for non-Newtonian flow. *Chem. Engng*, 19 Dec., **95**(18), 140–6.
Szilas, A.P., Bobok, E. & Navratil, L. (1981). Determination of turbulent pressure loss of non-Newtonian oil flow in rough tubes. *Rheol. Acta*, **20**, 487–96.
Thomas, D.G. (1963). Non-Newtonian suspensions, Part 2. *Ind. Eng. Chem.*, **55**, 27–35.
Thomas, G. (1960). PhD thesis, University College of Swansea, UK.
Tomita, Y. (1959). On the fundamental formula of non-Newtonian flow. *Bull. JSME*, **2**(7), 469–74.
Torrance, B.Mck (1963). Friction factors for turbulent fluid flow in circular pipes. *South African Mech. Engr.*, **13**, 89–91.
Turian, R.M., Hsu, F.L. & Sami Selim, M. (1983). Friction losses for flow of slurries in pipeline bends, fittings and valves. *Particulate Sci. Technol.*, **1**, 365–92.
Verschuur, E., Verheul, C.M. & Den Hartog, A.P. (1971). Pilot-scale studies on re-starting pipelines containing gelled waxy crudes. *J. Inst. Petr.*, **57**(555), 139–46.
Von Karman, T. (1931). Mechanische Ahnlichkeit and Turbulenz. In *Proceedings III International Congress of Applied Mechanics*, **85**, Stockholm, Sweden.
Wilson, K.C. & Thomas, A.D. (1985). A new analysis of the turbulent flow of non-Newtonian fluids. *Can. J. Chem. Engng*, **63**, 539–47.
Wood, D.J. (1966). An explicit friction factor relationship. *Civil Eng. ASCE*, **36**(12), 60–1.
Zigrang, D.J. & Sylvester, N.D. (1982). Explicit approximations to the solution of Colebrook's friction factor equation. *AIChEJ*, **28**(3), 514–15.

*Full details of the Hydrotransport (HT) and the Slurry Transport Association (STA) series of conferences can be found on pp. 11–13 of Chapter 1.

8

Slurry Transport in Flumes

Kenneth C. Wilson
Queen's University, Kingston, Canada

1. TYPES OF FLUME TRANSPORT

As used here, the word flume comprises any artificial channel carrying flow with a free surface. Such free-surface or open-channel flows occur in pipes flowing partly full, as well as in channels of rectangular, trapezoidal or other shapes. Typical cross-sections are shown in Fig. 8.1. Sections with corners often have these rounded, either for ease of construction or to diminish the likelihood of local deposition. Materials of construction commonly include steel, concrete, wood and plastic, the last being used primarily in the form of liners.

The analysis and design of open channels carrying water or other Newtonian fluids forms a well-developed branch of hydraulic engineering. Standard texts include those of Chow (1959) and Henderson (1966). Rivers and other waterways with stationary deposits of solids comprising the bottom have also been studied extensively, as can be seen from Yalin (1972); but such studies are of limited interest for slurry-transport flumes, in which stationary deposits must be avoided. Deposits of solids in such flumes are often not self-clearing, and may grow until they cause the flow to overtop the sides of the flume. The likelihood of deposition diminishes as the slope of the flume is increased, but this increase in slope reduces

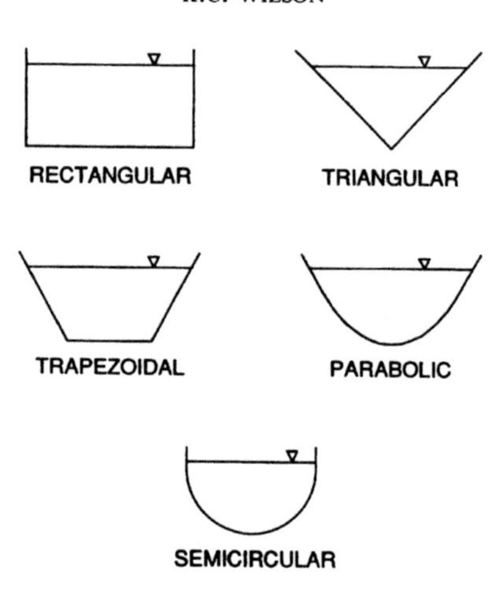

Fig. 8.1. Typical flume cross-sections.

the distance for gravity transport, limiting the conditions for which open-channel flow can be employed. Hence, slope is an extremely important design variable for flume transport of slurries.

The design process by which the slope and cross-sectional area of the flume are determined depends on the type of slurry. On the one hand, there are settling mixtures of coarse particles in water, and on the other flows of fine particles interacting with the carrier liquid to form a homogeneous mixture that is non-settling under design conditions. Slurries of the latter type can usually be considered as behaving like an equivalent fluid, often with distinctly non-Newtonian rheological properties.

In practice, the flume transport of settling slurries of coarse particles is usually associated with short-distance moving of raw materials—for example, the transport of coal or ore from the working face to storage or to a hoisting or shipping facility. Flumes for this purpose, often called launders, typically involve short distances and steep slopes. The slopes are determined by the fact that the working face will produce a wide range of particle size, and the flume must be designed to transport all sizes without deposition. The heterogeneous type of flow, intermediate between fully settling and homogeneous, is of less importance in flume transport than in pipeline flow. It has been considered previously (Wilson, 1980) and will not be dealt with here.

Flume transport with homogeneous slurry flow usually involves longer distances and flatter slopes than the coarse-particle case mentioned above. These slurries are comprised of partially processed materials or, more typically, tailings being transported to retention areas. The characteristics of such slurries are under closer control than those of the strongly settling slurries mentioned previously, but the moderate slopes that are normally used increase the risk of deposition.

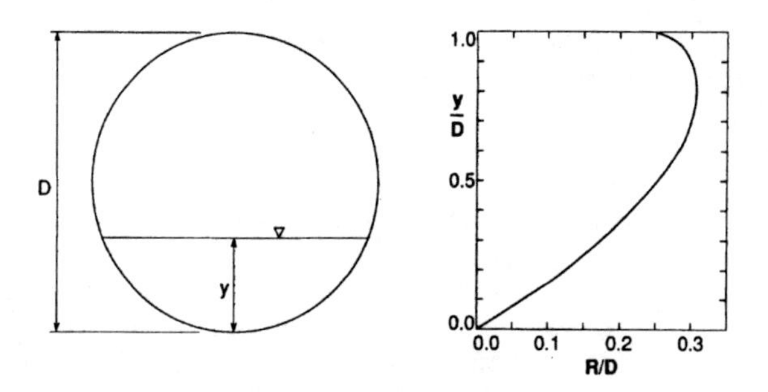

Fig. 8.2. Hydraulic radius curve for a pipe flowing partly full.

The organisation of the remainder of this chapter is based on the types of flume transport which have been mentioned above. The next section gives a review of open-channel flow of Newtonian liquids. This is followed by a section on flumes for coarse-particle settling slurries, with a final section pertaining to transport of fine-particle slurries which typically behave in a homogeneous but non-Newtonian fashion.

2. BRIEF REVIEW OF OPEN-CHANNEL HYDRAULICS

The cross-section of an open-channel flow is characterised by the area of flow A, the free surface at the top of the flow, of width B, and the wetted perimeter P, where the flow is in contact with the channel. The height or stage of flow, y, is measured from the lowest point in the cross-section. An important linear dimension characterising the flow is the hydraulic radius, R, which is defined as the ratio of area to wetted perimeter, i.e. A/P (see Chow, 1959). For example, consider a pipe of internal diameter, D, flowing partly full (Fig. 8.2). If the flow fills half the pipe, $y/D = 0{\cdot}5$, $A = \pi D^2/8$ and $P = \pi D/2$, giving $R = D/4$. If the pipe were flowing full, both the area and the wetted perimeter would have doubled, but R would again be $D/4$. In fact, R/D passes through a maximum value at y/D near 0·8, as shown on the right-hand portion of Fig. 8.2.

As seen above, for pipe-full flow $R = D/4$, or $D = 4R$. Hence, it is convenient to define an equivalent diameter, D_e, equal to $4R$. This allows standard friction relations, such as those displayed on the Stanton–Moody diagram, to be generalised to any geometry. Wherever the pipe diameter occurs, e.g. in the Reynolds number, it is replaced by the equivalent diameter of $4A/P$. An implicit assumption in this approach is that the shear stress is approximately constant over the wetted perimeter. This assumption, correct for the circular pipe, operates reasonably well for most flume sections (apart from corners, which as noted earlier are often rounded in practice).

For flows with high Reynolds number and rough walls, the friction factor depends only on the roughness ratio. This condition is normal for open channels conveying water, where the friction relation is commonly approximated by the Manning equation, which may be written

$$V = \frac{C_m}{n} R^{2/3} \times j^{1/2} \quad (8.1)$$

where n is a roughness parameter (approximately 0·012 for well-finished concrete and larger for rougher surfaces; see Chow (1959), Chapter 5) and the coefficient C_m has a value of 1·00 in SI units. The letter j is used here in place of the more usual S to denote slope. This slope refers to the inclination of the energy grade line, expressed as metres of fluid per metre run. For the usual case of uniform flow, the energy line is parallel to the channel bottom, so that the longitudinal slope of the channel (vertical drop/horizontal distance) also equals j.

The presence of a free surface introduces the question of surface waves. The wave celerity, equal to $(gy)^{0.5}$ for two-dimensional flow, is given by $(gA/B)^{0.5}$ for channel flow. The Froude number, Fr, represents the ratio of flow velocity to wave celerity, i.e. $\mathrm{Fr} = V/(gA/B)^{0.5}$. This number is closely analogous to the Mach number, Ma, used for compressible flow, which is also the ratio of velocity to a wave celerity. Supercritical flow ($\mathrm{Fr} > 1$) involves velocities greater than the propagation velocity of small waves, and hence can produce shock fronts analogous to supersonic flow ($\mathrm{Ma} > 1$). Similarly, conditions near critical ($\mathrm{Fr} \approx 1$, analogous to $\mathrm{Ma} \approx 1$) can involve troublesome instabilities, and thus are avoided wherever possible.

When channel flow is uniform (i.e. depth and area of flow do not vary with distance along the channel) then, as noted above, the slope j of the energy line equals the bottom slope of the channel. The Froude number can be used to classify slopes as steep, critical and mild. For a channel laid at the critical slope, uniform flow will have Froude number of unity, for a steep slope, uniform flow will be supercritical ($\mathrm{Fr} > 1$) and for a mild slope, it will be subcritical ($\mathrm{Fr} < 1$). It should be noted that, though a *steep* slope is greater than the critical, it may not be steep to the eye, i.e. slopes of only a few parts per thousand may be capable of maintaining uniform supercritical flow.

When the flow is not uniform, the surface profiles can be quite different for mild and steep slopes, as shown in texts on open-channel flow. The configuration of deposits are also distinct for the two cases. For a typical mild-slope configuration, such as a river flowing through a deep alluvial deposit, dunes tend to form on the top of the deposit, progressing downstream at a celerity much less than the velocity of the individual moving particles. Dunes can also be found in flumes at mild slopes, where they may be separated by lengths free of deposition. Supercritical flows in deep alluvial deposits commonly produce antidunes—sand waves which move upstream (though, of course the individual moving particles all travel in the direction of flow). Antidunes are accompanied by surface undulations of the same wavelength, and the interaction of the surface and the deposit sets up significant forces tending to move the deposit as a whole. In flumes where the deposited material is not sufficient to form a deep continuous stationary bed, the surface-interaction effect of supercritical flow tends to sweep away individual antidunes (as

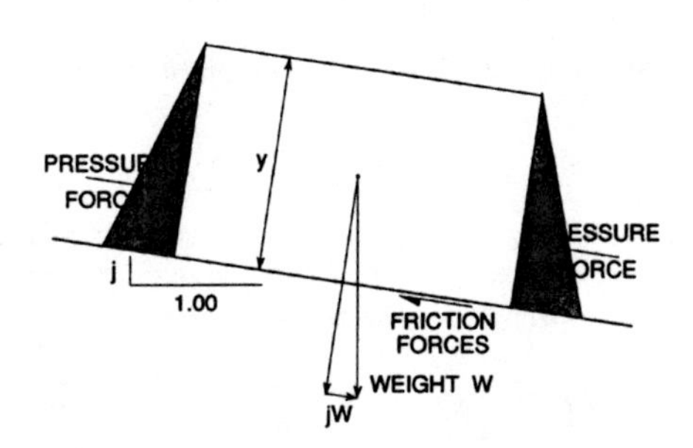

Fig. 8.3. Definition sketch for forces in flume flow.

opposed to the individual separated dunes common in subcritical flow, as mentioned above). The result is that supercritical flows in flumes are less susceptible to deposition than are subcritical flows.

3. FLUME DESIGN FOR COARSE-PARTICLE SETTLING SLURRIES

As mentioned previously, coarse-particle slurries are often produced at the working face of a mine, or in similar circumstances where there is little control of particle size. Thus, the flumes must be set at a slope great enough to ensure that large heavy particles remain in motion. This situation can be analysed using a force–balance model of the type developed for deposition in pipelines (Wilson, 1970) and described further in Chapter 6 of the present volume. For simplicity, it will be assumed here that the flumes are broad rectangular channels in which the flow can be considered as two-dimensional. A sketch of a unit length of such a flume is shown in Fig. 8.3. The depth of flow is represented by y and $\bar{C}_r$ is the average volumetric in-situ concentration of solids. The fluid density is ρ_f, and S_s is the ratio of solid/fluid density, giving a slurry weight per unit length (and breadth) of $\rho_f g y(1 + (S_s - 1)\bar{C}_r)$. As long as the flow is uniform, the pressure forces on the upstream and downstream ends of the unit length are balanced, and the net force acting downstream is given by the product of weight and slope, i.e. $\rho_f g y j(1 + (S_s - 1)\bar{C}_r)$. This driving force is resisted by the sum of two forces. One arises from fluid friction which sets up a shear stress τ_f at the boundary of the flume. For the unit breadth and length considered here, the fluid resisting force is simply equal to τ_f. The second resisting force is the mechanical sliding friction developed between the moving particles and the bottom of the channel. For the present configuration, this force is the product of the submerged weight of the solids $\rho_f g y(S_s - 1)\bar{C}_r$ and the coefficient of mechanical sliding friction, μ_p.

On algebraic rearrangement, the force balance described above yields the equation

$$j = \frac{(S_s - 1)\bar{C}_r \mu_p}{(1 + (S_s - 1)\bar{C}_r)} + \frac{\tau_f}{\rho_f g y(1 + (S_s - 1)\bar{C}_r)} \qquad (8.2)$$

The fluid shear stress, τ_f, may be written $\rho_f f_f V^2/8$, where f_f is the Darcy–

Weisbach friction factor for the fluid. The final term of eqn (8.2) thus becomes

$$\frac{f_f}{8} \times \frac{\mathrm{Fr}^2}{(1 + (S_s - 1)\bar{C}_r)} \tag{8.3}$$

When it is recalled that the Froude number should be somewhat greater than unity to promote deposition-free flow, it is possible to evaluate the term expressed by eqn (8.3) as roughly 0·01, giving the approximation

$$j \approx \frac{(S_s - 1)\bar{C}_r\mu_p}{(1 + (S_s - 1)\bar{C}_r)} + 0{\cdot}01 \tag{8.4}$$

This relation can be tested by comparison with the results of research and development work at the Hansa Hydro-Mine in Dortmund, Germany, using coarse run-of-mine coal. This work has been described by Kuhn (1980) in an article which gives a great deal of information based on practical experience. The flumes involved were constructed of various materials and had cross-sections that were triangular, trapezoidal and semicircular. Kuhn recommended a volumetric solids concentration of 0·25 and flume slope of 6°, i.e. $j \approx 0{\cdot}10$. The latter may be taken as a conservative recommendation, as the test results cited generally appeared to give satisfactory performance for inclinations of 4° ($j = 0{\cdot}07$). For the plastic-lined trapezoidal flumes performance remained satisfactory down to 2.5° ($j = 0{\cdot}044$).

Comparison with eqn (8.4) can be made using $\bar{C}_r = 0{\cdot}25$ and typical relative density of about 1·65, based on coal at $S_s = 1{\cdot}40$ with 20% rock (Kuhn's figure) at $S_s = 2{\cdot}65$. The friction coefficient, μ_p, can be evaluated from experimental results; for example, the Saskatchewan Research Council found that a value of μ_p near 0·32 is reasonable for coarse coal (Shook, 1981). Substituting these figures into eqn (8.4) gives a value of j of 0·055 or 3.1°, which appears to be in good accord with the Hansa experience with run-of-mine coal. The larger value of 6° proposed by Kuhn was based on controlled tests with rock alone, and, as can be seen from eqn (8.4), larger values of relative solids density require greater slopes. In fact, if the coal–rock mixture used above is replaced by rock and all other quantities left unchanged, the slope obtained by eqn (8.4) becomes 5.9°.

In view of the inherent lack of precision in the engineering of flumes for coarse-particle slurries, it is considered that eqn (8.4) constitutes an adequate approximation for this purpose. It should be noted that the required slope is not influenced significantly by scaling effects. However, the slope increases with solids density ratio and depends directly on the mechanical friction factor, μ_p. The latter quantity depends on the type of solids and on the material of the flume (or liner), and can be determined by simple tests (see Chapter 11).

4. FLUME DESIGN FOR HOMOGENEOUS SLURRY FLOW

As mentioned in section 1, homogeneous slurry flow usually involves longer distances and flatter slopes than those associated with coarse-particle settling slurries. Typical applications of homogeneous slurry flow are found in transporting tailings from production facilities to retention areas.

Homogeneous slurries are also called *non-settling*, indicating a lack of deposition within the range of design conditions. However, settling can occur, and design parameters must be selected to avoid such an occurrence. For example, many slurries settle if they are left at rest, and the slow shearing motions typical of laminar flow may exacerbate this tendency (Highgate & Whorlow, 1967; Thomas, 1979). For this reason, laminar flow is not recommended for flume transport of slurries. However, in most cases slurry properties are such that laminar flow would not occur within the range of velocities and channel sizes which are of commercial interest. If there is any doubt in this regard, the laminar–turbulent transition point should be calculated using the non-Newtonian rheological techniques to be considered later in this section.

Settling can also occur in turbulent flow, but here it is counteracted by the turbulent eddies, which have vertical velocities measurable in terms of the shear velocity U_*, equal to $(gRj)^{0{\cdot}5}$. As the shear velocity itself varies with the mean velocity V, practical design rules can be set up to ensure that V or U_*, or both, exceed specified minimum values.

The minimum value of V is expressed in terms of the Froude number of the flow. As mentioned above in reviewing open-channel hydraulics, supercritical flows in flumes are less susceptible to deposition than are subcritical flows. Although there is no difficulty with supercritical flow in straight channels, care must be taken at changes of alignment, which produce surface waves in supercritical flows. Thus, bends should be provided with sections of higher freeboard, or even be completely enclosed. To ensure that supercritical flow is maintained it is appropriate to require a lower-limit value of Fr somewhat greater than unity. In this regard Faddick (1986) proposed that Froude numbers less than 1·2 should be avoided, and his Froude-number criterion will be adopted here. In addition to setting the lower limit on velocity V, it also provides a convenient method for estimating flume size.

For this purpose it will be assumed that the design discharge, Q, is known, and that the geometry of the flume section has been selected. For example, the section chosen for the flow might be a semicircle, a trapezoid in the form of a half-hexagon or a rectangle of given breadth-to-depth ratio. With the geometry specified in this way, all features of the section are determined by a single dimension, say the equivalent diameter D_e. The flow area equals a known coefficient times D_e^2, and surface breadth B is proportional to D_e, so that, with fixed Froude number, the velocity is proportional to $D_e^{0{\cdot}5}$. Thus Q is proportional to $D_e^{2{\cdot}5}$, or

$$D_e = k_1 Q^{0{\cdot}4} \tag{8.5}$$

where the coefficient k_1 can readily be calculated for any specific channel geometry. Of course, this coefficient depends on the value of Fr. Since the proposed value of 1·2 is a lower limit of acceptable Fr, the values of D_e calculated from eqn (8.5) on this basis will represent the upper size limit for a flume to carry the design flow. The points mentioned above can best be illustrated by an example.

WORKED EXAMPLE

The design discharge, Q, is specified as $0{\cdot}77\,\mathrm{m^3\,s^{-1}}$, and the flow section is to be a semicircle, giving $B = D$, $A = 0{\cdot}393D^2$, $R = 0{\cdot}25D$ and $D_c = D$. The condition that Fr = 1·2 then gives (in SI units) $Q = 0{\cdot}925D^{2{\cdot}5}$ or

$$D = 1{\cdot}032Q^{0{\cdot}4} \tag{8.6}$$

Substitution of the design discharge of $0{\cdot}77\,\mathrm{m^3\,s^{-1}}$ indicates a diameter of 0·930 m. If this diameter is not commercially available, the next smallest available size should be used, so that Fr will exceed 1·2. In this example it is assumed that concrete pipe of internal diameter 0·915 m (36 in.) is available. Using half-full pipe of this size the velocity will be $2{\cdot}34\,\mathrm{m\,s^{-1}}$, for a Froude number of 1·25.

Although modification may be required later in the design process, the flume size found above can now be used to obtain a reasonable initial estimate of the required slope. This slope is calculated from the Manning equation (eqn (8.1)). For the present example with $n = 0{\cdot}012$ for well-finished concrete, $V = 2{\cdot}34\,\mathrm{m\,s^{-1}}$, $R = 0{\cdot}25D$, i.e. 0·229 m, and the slope j is found to be 0·0056.

The value of the slope calculated by the method shown above can be compared with the available elevation difference and the required length of line, giving an immediate check on the viability of the flume transport system. If the available elevation difference is too small, it will probably be necessary to shift to a pressurised system using pumps and if the elevation difference is greater than required, it may be appropriate to use a smaller flume set on a steeper slope. Of course, the method that has just been outlined can only be an approximate one, as it does not take into account either the properties of the carrier fluid or those of particles prone to settling. Thus, if the fluid has a large resistance to deformation, the estimate of slope may have to be revised upward, using a method of calculation more sophisticated than eqn (8.1). Likewise, if the elimination of particle settling requires an increase in the value of U_*, then design changes will be required, including a larger slope. This question of particle settling will be considered further below, after dealing with the effect of fluid rheology on friction calculations.

The rheological properties of non-Newtonian fluids, including fine-particle slurries, cannot be calculated from first principles, but must be determined by experimental tests. As described in Chapter 4, these tests are usually carried out either in rotational viscometers or in pipes of modest diameter. Rotational viscometers involve laminar flow, and scaling the results to turbulent flows in pipes or flumes may produce considerable uncertainties. Testing in pipes is strongly recommended, especially if the pipe diameter is large enough to place at least part of the test range in the turbulent regime. Experimental results of this type are often presented as graphs of τ_w versus $8V/D$. For laminar flow of any given fluid, the Rabinowitsch–Mooney scaling law shows that data from pipes of all sizes must plot as a single line on this axis system provided there is no *wall-slip*. Representative non-Newtonian behaviour is illustrated in Fig. 8.4 which is a plot of data for a

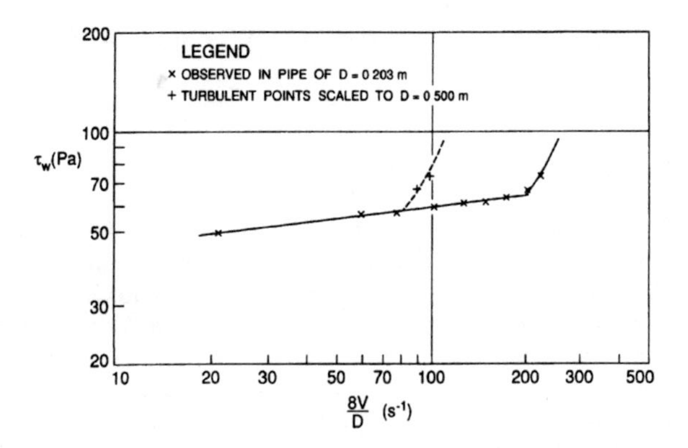

Fig. 8.4. Plot of τ_w versus $8V/D$ for clay slurry (data from GIW Hydraulic Laboratory, Augusta, GA, USA).

clay–water slurry tested in a 203 mm pipe at the GIW Hydraulic Laboratory, Augusta, GA, USA. On the logarithmic axes of this figure the locus of laminar flow approximates a straight line of slope less than unity (a straight line of unit slope represents laminar flow of a Newtonian fluid). Turbulent flows, on the other hand, plot as a series of essentially parallel lines (one for each pipe diameter) with a common slope of approximately two. Figure 8.4 shows the laminar–turbulent intercept for the experimental pipe size. For each value of τ_w, the turbulent flow velocity V_1, in the experimental pipe of size D_1 can be scaled to velocity V_2 in a similar pipe of size D_2 by use of the equation (Wilson, 1986)

$$V_2 = V_1 + 2{\cdot}5U_* \ln (D_2/D_1) \qquad (8.7)$$

where U_* is the shear velocity. The values of $8V_2/D_2$ plotted in Fig. 8.4 apply to the scale-up of the data to a larger pipe size (500 mm for this example). Downward extrapolation of this scaled turbulent-flow line gives the laminar–turbulent intercept for the larger pipe size (for the example $8V_2/D_2 = 82$ at transition, equivalent to $V_2 = 5{\cdot}12\,\mathrm{m\,s^{-1}}$).

Use of this method with the equivalent diameter of the flume, D_e, in place of D_2 can provide a reasonable estimate of the laminar–turbulent transition velocity for the flume. As mentioned previously, this velocity is required to ensure that the flume flow is turbulent. Should there be no experimental pipe-flow data in the turbulent range, methods are available to approximate the required turbulent-flow locus from laminar viscometric data (Wilson & Thomas, 1985; Thomas & Wilson, 1987).

Pipes used for viscometric testing usually have hydraulically smooth walls, but this is not generally the case for flumes. For example, the value of Manning's *n* of 0·012 which was used above implies a rough wall with roughness size about 0·6 mm. For fluids considerably more viscous than water, frictional behaviour is likely to lie in the transition range, requiring the use of more complicated friction formulations, such as the Colebrook–White equation, which can be written

$$\frac{V}{U_*} = -2{\cdot}43 \ln\left(\frac{e}{3{\cdot}7D_e} + \frac{0{\cdot}89\mu_f}{\rho_f U_* D_e}\right) \tag{8.8}$$

Here, e is roughness size and μ_f is fluid viscosity. For a non-Newtonian fluid, μ_f must be replaced by the equivalent viscosity, μ_e. This is the viscosity which would be possessed by a Newtonian fluid (of the same density) which gives a smooth-wall friction factor the same as that obtained by testing the non-Newtonian. For a Newtonian fluid, it is obvious that μ_e is simply equal to the fluid viscosity, and does not depend on specific flow conditions, but for a non-Newtonian μ_e is a variable. As shown elsewhere (Wilson, 1988), μ_e depends on the shear velocity, U_*.

Curves of μ_e versus U_* vary widely for different slurries, and must be obtained empirically from test results. As the accuracy of pipe-flow data is generally better than that of data from open channels, testing in a pilot-plant pipe loop is the best method for establishing these curves. For any specified U_*, the value of μ_e obtained from the appropriate curve can be substituted in eqn (8.8), together with the known, or assumed, values of e and D_e.

CONTINUATION OF WORKED EXAMPLE

The technique will be illustrated by continuing the previous example design. The equivalent diameter of 0·915 m and the slope of 0·0056 as estimated from the Manning equation produce a shear velocity of 0·112 m s^{-1}. It will be assumed that the slurry to be transported has density ρ_f of 1 600 kg m^{-3} and behaves like the slurry of this density plotted in Fig. 8.5 (to be discussed below). At $U_* = 0{\cdot}112$ m s^{-1} the plot shows $\mu_e = 0{\cdot}0041$ Ns m^{-2}. With these quantities, and $e = 0{\cdot}6$ mm, (appropriate for concrete), eqn (8.8) gives $V/U_* = 20{\cdot}7$, or $V = 2{\cdot}32$ m s^{-1}. This is only slightly below the velocity of 2·34 m s^{-1} required to obtain the design discharge, and the required values can be obtained by increasing the slope of the flume from 0·0056 to 0·0057.

For the conditions of the example problem, it is clear that the previous calculation using the Manning equation gave a very good friction approximation. However, the validity of the approximation would decrease for less rough walls or larger values of μ_e. As a general rule, the Manning equation is suitable for initial approximations, to be refined later by use of eqn (8.8).

The remaining point which should be considered in design is to ensure that U_* exceeds the lower limit needed to prevent particle settling. For Newtonian flows with low solids concentrations, this lower limit can be taken as some coefficient (2·5 is suggested as an appropriate value) times the terminal fall velocity of a particle. For non-Newtonian fluids, however, the tendency for particle settling is much harder to quantify, and the concept of terminal fall velocity may be of dubious value. In such cases it is better to rely on pipe tests, which indicate particle settling by the associated increase in friction loss, manifesting itself as a sharp increase in μ_e as U_* is diminished. Thus, the same curves used to obtain μ_e for the friction

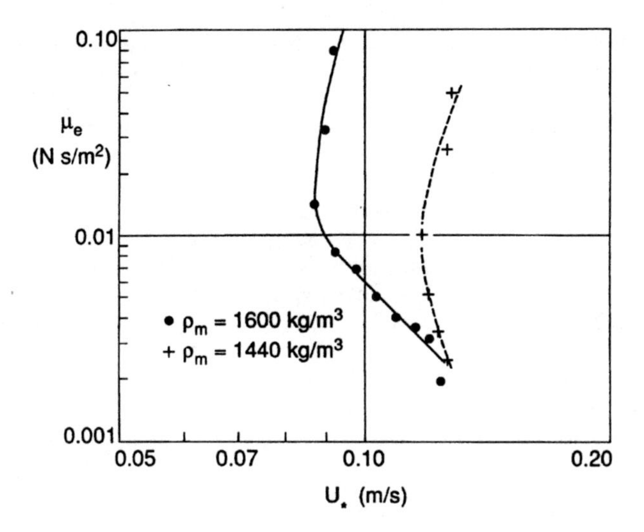

Fig. 8.5. Plots of equivalent viscosity versus shear velocity for limestone slurries (data from Schriek *et al.*, 1973). (●) $\rho_m = 1\,600\,\mathrm{kg\,m^{-3}}$, (+) $\rho_m = 1\,440\,\mathrm{kg\,m^{-3}}$.

calculations can also be used to indicate solids settling, and hence to set the allowable lower limit of U_*.

The examples of μ_e curves shown in Fig. 8.5 are based on data obtained by the Saskatchewan Research Council for fine limestone slurries in a pipe of 0·138 m inside diameter (Schriek *et al.*, 1973). The points shown on the left-hand curve of this figure are for a volumetric solids concentration of 33·1%, corresponding to a mixture density of $1\,600\,\mathrm{kg\,m^{-3}}$. It can be seen that, as U_* is decreased from an initially high value, μ_e rises, moderately at first, but increasing to a vertical rise as the solid particles settle out. There is even an upper limb of the curve with the opposite slope, corresponding to the rise in hydraulic gradient with decreasing velocity found for pipes with a deposit of solids. As a result, the shear velocity exhibits a minimum value, equal to $0{\cdot}087\,\mathrm{m\,s^{-1}}$ for the data plotted as the left-hand curve of Fig. 8.5. At this point deposition has already taken place, and thus the allowable lower limit for U_* must be significantly greater. If a factor of, say, 1·2 is adopted, the allowable minimum U_* would be $0{\cdot}104\,\mathrm{m\,s^{-1}}$. On this basis the value of $U_* = 0{\cdot}112\,\mathrm{m\,s^{-1}}$ found in the example calculations above would be satisfactory and no further design modifications would be required.

It is interesting to note the effect of slurry consistency. The Saskatchewan data also includes runs with a more concentrated slurry of the same material ($\rho_m = 1\,720\,\mathrm{kg\,m^{-3}}$) and a more dilute slurry ($\rho_m = 1\,440\,\mathrm{kg\,m^{-3}}$). The curve of μ_e versus U_* for the more concentrated slurry has been omitted from Fig. 8.5, because it coincides in places with the curve for $\rho_m = 1\,600\,\mathrm{kg\,m^{-3}}$. In fact the minimum U_* for the concentrated slurry is only $0{\cdot}078\,\mathrm{m\,s^{-1}}$, i.e. the tendency for deposition is less for $\rho_m = 1\,720\,\mathrm{kg\,m^{-3}}$ than for $1\,600\,\mathrm{kg\,m^{-3}}$ (for which the minimum U_* was $0{\cdot}087\,\mathrm{m\,s^{-1}}$). For the less concentrated slurry ($\rho_m = 1\,440\,\mathrm{kg\,m^{-3}}$), the values of μ_e and U_* have been plotted in Fig. 8.5, forming the curve to the right on that figure. In this case the minimum observed value of U_* is significantly greater, i.e. $0{\cdot}117\,\mathrm{m\,s^{-1}}$. On applying the factor of 1·2 mentioned previously, the suggested minimum allowable value of U_* now becomes $0{\cdot}142\,\mathrm{m\,s^{-1}}$. Thus, for the less-concentrated slurry the design of the example above would not provide sufficient U_* to avoid particle settling. Redesign will be required; the slope must be steeper, the flow velocity will be higher, and there may be a reduction in flume size. For the design discharge of the example problem, calculations show that a diameter of 0·838 m (33 in.), is appropriate, laid at a slope of 0·0097 or, say, 0·0100.

CALCULATION SCHEME

The design process for homogeneous slurry flow, described above, may be recapitulated in point form as follows.

(1) Knowing the design discharge, and the type of flume geometry to be used, the Froude number criterion ($\mathrm{Fr} \geqslant 1{\cdot}2$) is employed to estimate the flume size. If necessary, the next smallest commercially available size is selected.
(2) An initial estimate of the required flume slope is made using the Manning equation (eqn 8.1) and the previously estimated flume size. These estimates of size and slope can be used to check whether the case under study is suitable for flume transport.
(3) A sample of the slurry is tested, preferably in a pipe loop capable of obtaining turbulent flow. From the results, the laminar–turbulent intercept for the flume is calculated in order to verify that the prototype flow will be turbulent.
(4) Using the test data the equivalent viscosity for turbulent flow, μ_e, is plotted against U_*. Using the dimensions and shear velocity found from steps (1) and (2) above, μ_e is obtained and the required slope is recalculated using eqn (8.8), iterating if necessary.
(5) The allowable lower limit of U_* is estimated; either as a multiple of terminal fall velocity or, if possible, from the μ_e curves. If this value is exceeded, the design is satisfactory. Otherwise, further calculations are carried out to obtain the required increase in slope, and possible reduction in flume size.

5. NOTATION

A Flow area of flume [L^2]
B Width of free surface [L]
C_m Coefficient in Manning equation (1·000 in SI units)
$\bar{C}_r$ Average in-situ volumetric concentration of solids
D Pipe diameter [L]

D_e Equivalent diameter $(4A/P)$ [L]
D_1 Diameter of experimental pipe [L]
D_2 Diameter of prototype pipe [L]
e Roughness size [L]
f_f Darcy–Weisbach friction factor for fluid [1]
Fr Froude number [1]
g Gravitational acceleration $[\mathrm{L\,T^{-2}}]$
j Channel slope (vertical drop/horizontal distance) [1]
k_1 Coefficient in eqn (8.5) $[\mathrm{L^{-0.2}T^{0.4}}]$
Ma Mach number [1]
n Roughness parameter in Manning equation
P Wetted perimeter [L]
Q Volumetric discharge $[\mathrm{L^3\,T^{-1}}]$
R Hydraulic radius (A/P) [L]
S_s Relative density of solids (solids density/fluid density) [1]
U_* Shear velocity $[\mathrm{L\,T^{-1}}]$
V Mean velocity $[\mathrm{L\,T^{-1}}]$
V_1 Mean velocity in pipe of diameter D_1 $[\mathrm{L\,T^{-1}}]$
V_2 Mean velocity in pipe of diameter D_2 $[\mathrm{L\,T^{-1}}]$
W Weight $[\mathrm{M\,L\,T^{-2}}]$
y Height measured from bottom of channel [L]

μ_e Equivalent viscosity of non-Newtonian $[\mathrm{M\,L^{-1}T^{-1}}]$
μ_f Viscosity of Newtonian fluid $[\mathrm{M\,L^{-1}T^{-1}}]$
μ_p Mechanical friction factor [1]
ρ_f Fluid density $[\mathrm{M\,L^{-3}}]$
ρ_m Mixture density $[\mathrm{M\,L^{-3}}]$
τ_f Fluid shear stress $[\mathrm{M\,L^{-1}T^{-2}}]$
τ_w Shear stress at wall $[\mathrm{M\,L^{-1}T^{-2}}]$

6. REFERENCES

Chow, Ven te (1959). *Open-Channel Hydraulics*. McGraw Hill, New York, USA.

Faddick, R.R. (1986). Slurry flume design. In *Proc. HT 10*, pp. 143–7.*

Henderson, F.M. (1966). *Open Channel Flow*. Macmillan, New York, USA.

Highgate, D.J. & Whorlow, R.W. (1967). The viscous resistance to motion of a sphere falling through a sheared non-Newtonian liquid. *Br. J. Appl. Phys.*, **18**, 1019–22.

Kuhn, M. (1980). Hydraulic transport of solids in flumes in the mining industry. In *Proc. HT 7*, pp. 111–22.*

Schriek, W., Smith, L.G., Haas, D.B. & Husband, W.H.W. (1973). *Experimental Studies on Solids Pipelining of Canadian Commodities* (Report II, *Experimental Studies on the Hydraulic Transport of Limestone*). Report no. E 73-10, Saskatchewan Research Council, Saskatoon, Canada.

Shook, C.A. (1981). *Lead Agency Report II for Coarse Coal Transport MTCH Hydrotransport Cooperative Programme*. Saskatchewan Research Council, Saskatoon, Canada.

Thomas, A.D. (1979). Settling of particles in a horizontally sheared Bingham plastic. Paper presented at 1st National Conference on Rheology, Melbourne, Australia.

Thomas, A.D. & Wilson, K.C. (1987). New analysis of non-Newtonian turbulent flow—yield-power-law fluids. *Can. J. Chem. Engng*, **65**, 335–8.
Wilson, K.C. (1970). Slip point of beds in solid–liquid pipeline flow. *Proc. ASCE (J. Hydraulics Div.)*, **96**, 1–12.
Wilson, K.C. (1980). Analysis of slurry flows with a free surface. In *Proc. HT 7*, pp. 123–32.*
Wilson, K.C. (1986). Modelling the effects of non-Newtonian and time-dependent slurry behaviour. In *Proc. HT 10*, pp. 283–9.*
Wilson, K.C. (1988). Effect of non-Newtonian slurry properties on drag reduction and coarse particle suspension. In *Proceedings of the 10th International Congress on Rheology* (Sydney, Australia) **1**, 110–15.
Wilson, K.C. & Thomas, A.D. (1985). A new analysis of the turbulent flow of non-Newtonian fluids. *Can. J. Chem. Engng*, **63**, 539–46.
Yalin, M.S. (1972). *Mechanics of Sediment Transport*. Pergamon, Oxford, UK.

*Full details of the Hydrotransport (HT) series of conferences can be found on pp. 11–12 of Chapter 1.

9

Transient Pipe Flow Behaviour

Alan G. Huggett
BP Engineering, London, UK

1. INTRODUCTION

When a velocity change occurs in a flowing fluid some of the kinetic energy is converted into pressure energy giving an effect called *hydraulic surge* or *water hammer*. In most pipelines this can be caused by delivery valve closure, pump start-up or shutdown. There is normally the potential to create pressures significantly above or below those occurring during steady conditions. With this potential to cause mechanical failure of pressure containing equipment it is prudent for the designer to include hydraulic surge considerations from the earliest stage of the design. With such early consideration it is possible to modify either the design or operating conditions to remove the risk of failure.

Hydraulic surge in fluids not containing solids is a wide subject and is adequately covered by books such as Streeter (1975) and Wylie and Streeter (1983) and conferences contained in the bibliography. Therefore, this chapter will explain only the principles, give some general advice on the approach for multi-phase systems and point to references for more specific guidance. Should a detailed study be needed there are numerous specialist organisations with the computer programs needed for the analysis and staff skilled in the modelling techniques used for single phase fluids. The slurry-system designer may need to guide such a specialist in the unusual characteristics of slurry systems, making the understanding of the principles of hydraulic surge of value in the interface with these specialists. At the end of this chapter is a section showing what hand calculations can usefully be carried out to ensure that a system is sensible before asking a specialist to check a near-final design for safety.

2. HYDRAULIC SURGE PRINCIPLES

When the velocity of a fluid is reduced by an amount, ΔV, the accompanying pressure rise, ΔP, is given by the equation

$$\Delta P = -\rho_f a_0 \Delta V \tag{9.1}$$

where a_0 is the pressure wave speed [L T^{-1}], ΔP is the pressure rise [M L^{-1} T^{-2}], ΔV is the velocity change [L T^{-1}], and ρ_f is the density of the fluid [M L^{-3}].

If the fluid velocity increases then there is a pressure decrease. In the extreme case of a fluid being stopped completely the head rise is known as the *Jacowski* pressure rise. In a simple system of a pipeline discharging through a terminal valve, which is considered to close fully, the full Jacowski pressure rise will be seen if the valve closes within the time that a pressure wave takes to travel to the other end and return. This time is referred to as the pipeline period and is calculated from

$$\tau = 2L/a_0 \tag{9.2}$$

where L is the pipeline length [L], and τ is the pipeline period [T]. From this it can be seen that the pressure wave speed is the most important unknown parameter for the designer since it affects both the magnitude and the speed of the propagation of pressure changes.

3. PRESSURE WAVE SPEED

3.1. Liquids

In circular pipes running full of liquid not containing solids or gases the pressure wave speed, a_0, is given by

TABLE 9.1
Reported Values of Bulk Moduli

Material	*Density* ($kg\,m^{-3}$)	*Bulk modulus* (*GPa*)	*Reference*
Coal	1 400	13·3	Hunt & Liou (1984)
Iron Ore	5 000	17·2	Hunt & Liou (1984)
Sand	2 650	16	Bechteler & Vogel (1981)
Water	1 000	2·06	Bechteler & Vogel (1981)

$$a_0^2 = \frac{\dfrac{1}{\rho_f}}{\left(\dfrac{1}{K_f} + \dfrac{D\psi}{Et}\right)} \tag{9.3}$$

where D is the inside pipe diameter [L], E is the Young's modulus for the pipe wall [$M\,L^{-1}T^{-2}$], K_f is the bulk modulus of the liquid [$M\,L^{-1}T^{-2}$], t is the pipe wall thickness [L], and ψ is the pipe restraint factor [1]. ($\psi = (1 - \nu^2)$ for completely restrained pipes or $\psi = (1 - \nu/2)$ for pipes restrained at one end, where ν is Poisson's ratio [1] for the pipe material.)

In practice there is little error in assuming that ψ is unity.

3.2. Solid–Liquid Slurries

Numerous equations have been proposed for incorporating the effect of solids into eqn (9.3): they include Wood and Kao (1966), Thorley and Hwang (1979), and Thorley (1980). Bechteler and Vogel (1981) have assessed these three equations, both theoretically and experimentally using water-based sand and coal slurries. The equation given by Thorley and Hwang (1979) is used here as the numerator, is well behaved and is equal to the reciprocal of slurry density.

$$a_0^2 = \frac{\left(\dfrac{1}{C_f\rho_f + C_p\rho_p}\right)}{\left(\dfrac{C_f}{K_f} + \dfrac{C_p}{K_p} + \dfrac{D}{Et}\right)} \tag{9.4}$$

where C_f is the volume fraction of liquid [1], C_p is the volume fraction of solid particles [1], K_p is the bulk modulus of the solid particles (see Table 9.1) [$M\,L^{-1}T^{-2}$], and ρ_p is the density of the solids [$M\,L^{-3}$].

These equations generally lead to a wave speed greater than that experienced with water for coal slurries and less than water for iron-ore slurries. These are borne out by measurements on real systems. Also, these equations assume that the particles in the slurry are uniformly distributed; if they are not, slightly different wave speeds will result. The settled condition has been investigated by (among others) Bechteler and Vogel (1981) and Shook and Hubbard (1973).

3.3. Solid–Liquid–Gas Slurries

Bechteler and Vogel (1981) have included the effects of entrained air:

$$a_0^2 = \frac{\left(\dfrac{1}{C_f\rho_f + C_p\rho_p + C_g\rho_g}\right)}{\left(\dfrac{C_f}{K_f} + \dfrac{C_p}{K_p} + \dfrac{C_g}{K_g} + \dfrac{D}{Et}\right)} \tag{9.5}$$

where C_g is the volume fraction of gas [1], K_g is the bulk modulus of the gas [M L^{-1} T^{-2}], and ρ_g is the density of gas [M L^{-3}].

4. TRANSIENT SCENARIOS

There are several operating changes that can cause flow changes and hence pressure transients. The most common of these are valve opening/closing and pump start/stop.

When a valve closes in a flowing pipeline the fluid upstream experiences a pressure rise and the fluid downstream a pressure fall. In the simple, classic case of instantaneous valve closure these pressure changes will be step changes with time and will be transmitted at the pressure wave speed along the line in both directions. In practice, the closure will occur over a finite time and the shape of the pressure/time variation will depend on the flow/time change which caused it and thus on the characteristics of the valve. Typical flow characteristics of ball, globe and gate valves are shown in Fig. 9.1. It is important to note that the last 10% of the closure has the most significant effect.

If the pipeline period (evaluated from eqn (9.2)) is greater than the valve movement time then the type of valve and its characteristic will not influence the total pressure rise although the rate of pressure rise will be important to any control devices and the mechanical load on supports. If the pipeline period is less than the valve movement time then the shape of the characteristic will influence the peak pressure. Special care should be exercised in specifying valves whose effective closure time is only a fraction of the total movement time, e.g. gate valves.

Pressure falls can be as damaging as pressure rises. Often the negative head changes lead to theoretical pressures less than the fluid vapour pressure. In such cases a vapour cavity forms. Large diameter pipelines are more susceptible to damage by reduced pressure conditions. More commonly vapour cavities create damage by their collapse. This occurs if there is a situation which can re-pressure the line. This could be caused by refilling from an elevated section of line or tanks or by restarting a pump to continue operation. Often this refilling occurs at a higher rate because the pump discharge pressure is lower than normal. When the cavity volume reaches zero the effect is of a truly instantaneous valve closure. These positive pressures can be higher than with normal valve closure and have step pressure changes. Therefore, they should be avoided if at all possible.

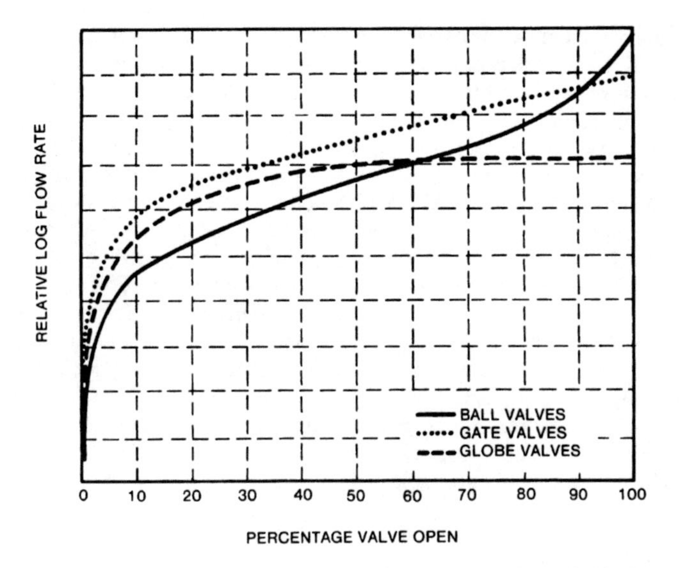

Fig. 9.1. Relative flow rate against percentage valve opening for typical ball, gate, and globe valves.

Pump start/stop has the capability of producing similar positive and negative pressure waves with the attendant damage potential.

The use of reciprocating pumps in slurry systems can result in special problems associated with repeated surge effects.

5. POINTS TO WATCH

From the above, it becomes apparent that the significant factors in creating high surge pressures are wave speed and rate of change of flow.

There is little that can be done to change the fluid wave speed except to introduce gas into the system. This may not be an acceptable solution since it may have deleterious effects on the process or equipment. It may also be difficult to guarantee the quantity injected or entrained.

Sometimes it is possible to modify the rate of change of velocity in a transient. This is particularly true of the effective valve closure time. Even valves with the most suitable closure characteristics create most of the flow change over the last quarter of their movement. Thus, two part closures where the first part is fast and the last part slow can be beneficial.

With reciprocating pumps, surge is inherent in the operation of the pump. The action presupposes acceleration and deceleration of the fluid leaving the cylinder and the action of a valve arrangement to allow the cylinder to be refilled. All of these will produce velocity changes which should be kept as small as possible. One part of the cycle that can be particularly damaging is the timing of the closure (and opening) of the valve between the cylinder and the line. If it is open at a time when there is low pressure in the cylinder, fluid will rush into the cylinder from the high-pressure line and be stopped when the valve closes. Since the velocity could be several times the normal velocity, so is the surge pressure generated.

Where surge is a problem, it is particularly important to avoid smaller bore dead ends. When a surge pressure wave reaches a pipe junction it propagates along both lines. It reduces in ratio to the new area divided by the old area. When the wave propagates it induces a flow by pressure difference. At the instant of the wave arriving at the dead end the forward velocity is stopped creating another surge effect on top of the pressure wave. *This leads to a potential doubling of the pressure.*

Particularly with pump generated surges, one solution is to add surge accumulators. These are vessels which contain a gas pocket connected as closely as possible upstream of the surge generation device. When a surge occurs, the excess fluid flows into the vessel compressing the gas. By removing excess fluid (which is nearly incompressible) the line pressure is reduced. The volumes and costs can be quite small when the transients are rapid.

6. SCOPE OF CALCULATIONS

The techniques discussed in this chapter can be used to obtain estimates of the surge behaviour. Computer simulation is required for a rigorous dynamic analysis. The figures discussed in this section were obtained by such a means.

WORKED EXAMPLE

For a simple continuous pipeline from a pump to a delivery point as shown in Fig. 9.2, most of the points discussed above can be illustrated.

The pipeline is carrying a coal–water mixture through a 305 mm (12 in.) schedule 40 line at a mass flow rate of 600 t h^{-1}. It is assumed that a valve closes at the discharge in 12 s and that the length of the pipeline is 34 km. Additional data are

C_f = 0·5 (by volume)
C_p = 0·5 (by volume)
D = 0·303 m (11·938 in.)
E = 207 GPa
K_f = 2·15 GPa
K_p = 13·3 GPa (from Table 9.1)

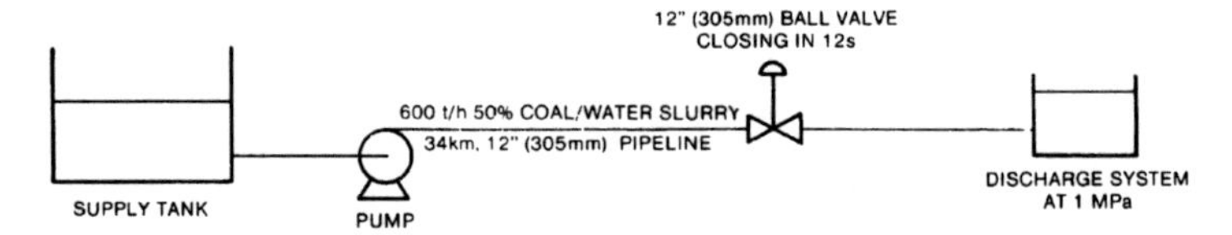

Fig. 9.2. Schematic for illustrative example.

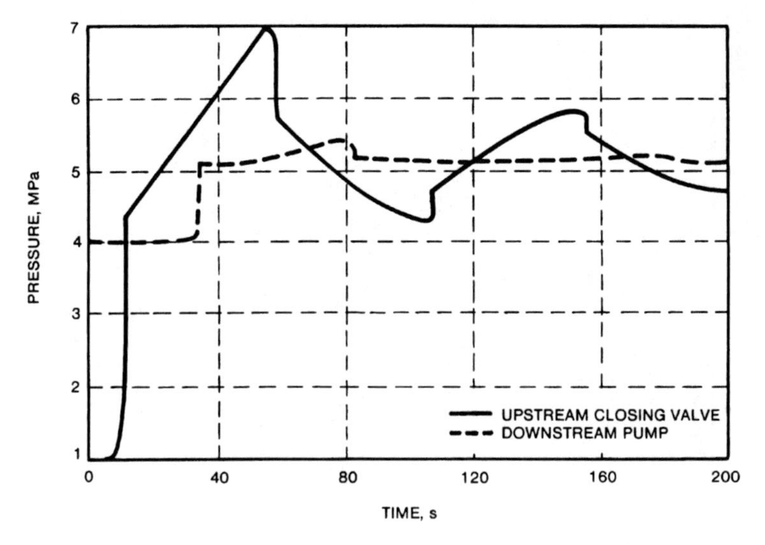

Fig. 9.3. Pressure in slurry pipeline—no air entrainment.

t = 0·0 103 m (0·406 in.)
ρ_f = 1 000 kg m^{-3}
ρ_p = 1 400 kg m^{-3}

The volumetric flow rate is 600 000/((0·5 × 1 000) + (0·5 × 1 400)) = 500 m^3 h^{-1}, which equates to a superficial velocity of 1·93 m s^{-1}.

From eqn (9.4) the wave speed, a_0, for the mixture is 1 420 m s^{-1} and from eqn (9.2) the pipeline period, τ, is 47.8 s.

Thus, if the slurry were stopped by a valve which effectively closed faster than the pipeline period, we would have a Jacowski pressure rise (calculated from eqn (9.1)) of 3·29 MPa. If the pressure immediately upstream of the closing valve (say at the end of the line) were 1·0 MPa g, the pressure would rise to 4·29 MPa g.

In practice, this immediate rise in pressure is not the complete picture. The results of a computer simulation of the system is shown in Fig. 9.3. After the initial rise to 4·29 MPa g, the pressure at the upstream valve, shown by the solid line, continues to rise as the higher pressure fluid in the line continues to flow towards the valve and is stopped, resulting in further increase in pressure. This effect is known as *packing*.

The pressure at the pump is shown by the broken line in Fig. 9.3. There is a lower pressure rise $\tau/2$ s later than at the valve resulting from the reflection of the pressure wave. The pressure reaches 5·2 MPa g, which is the pump shut-in head. After a further $\tau/2$ s, there is a slight further rise as the higher pressure at the valve induces reverse flow which in turn creates pressure rise as the flow is stopped. Once the pressure wave from the valve has reached the pump, it is reflected as a negative wave which reaches the valve about $\tau/2$ s later ($\tau/4 + \tau/2 + \tau/2$ = 59·8 s). There-

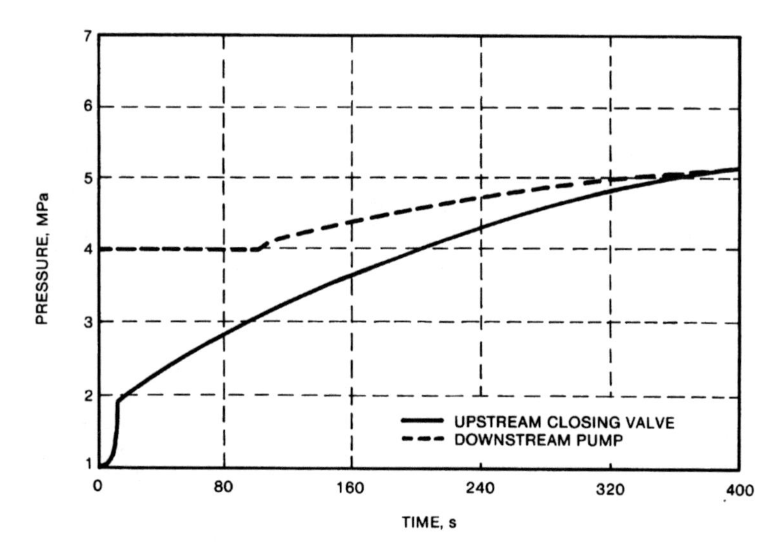

Fig. 9.4. Pressure in slurry pipeline—0·1% (by volume) air entrained.

after, at the valve, there is an inverse of the earlier rises, but of reduced magnitude due to friction effects.

Equation (9.5) can be used to evaluate the reduction in wave speed with the addition of gas. If it is assumed that 0.1% by volume of air is entrained in the fluid (with the air having a bulk modulus of 0·1 804 MPa) the wave speed drops to 374 m s^{-1} and the pipeline period increases to 181·8 s. This reduces the Jacowski pressure rise to 0·86 MPa g, giving a pressure immediately upstream of the valve of 1·86 MPa g immediately after the valve has closed. However, this reduction in pressure may not be as good as it looks because there is a period after the valve is closed when the fluid packs due to the original fluid friction effects in the line. The results of a computer simulation of this situation is shown in Fig. 9.4. In this figure it can be seen that the much longer pipeline period results in a smooth ramp up to the pump shut-in head at all points in the line after the initial surge rise.

Further examples and more detailed calculations are to be found in Ricks (1982).

7. NOTATION

a_0 Pressure wave speed $[\mathrm{L\,T^{-1}}]$
C_f Volume fraction of liquid [1]
C_g Volume fraction of gas [1]
C_p Volume fraction of solid particles [1]

D Inside pipe diameter $[\mathrm{M\,L^{-1}T^{-2}}]$
E Young's modulus for the pipe wall $[\mathrm{M\,L^{-1}T^{-2}}]$
K_f Bulk modulus of the liquid $[\mathrm{M\,L^{-1}T^{-2}}]$
K_g Bulk modulus of the gas $[\mathrm{M\,L^{-1}T^{-2}}]$
K_p Bulk modulus of the solid particles $[\mathrm{M\,L^{-1}T^{-2}}]$
L Pipeline length [L]
ΔP Pressure rise $[\mathrm{M\,L^{-1}T^{-2}}]$
t Pipe wall thickness [L]
ΔV Velocity change $[\mathrm{L\,T^{-1}}]$

ν Poisson's ratio for the pipe material [1]
ρ_f Density of the liquid $[\mathrm{M\,L^{-3}}]$
ρ_g Density of gas $[\mathrm{M\,L^{-3}}]$
ρ_p Density of the solid particles $[\mathrm{M\,L^{-3}}]$
τ Pipeline period [T]
ψ Pipe restraint factor [1]

8. REFERENCES

Bechteler, W. & Vogel, G. (1981). Theoretical considerations and laboratory investigations about water hammer propagation in slurry pipelines. In *Proc. STA 6*, pp. 131–7*.

Hunt, W.A. & Liou, C.P. (1984). Transient analysis: a design tool for slurry pipelines. In *Proc. STA 9*, pp. 63–9.*

Ricks, B.L. (1982). Application of pressure transient analysis in a long distance slurry pipeline. *J. Pipelines*, **3**, 73–86.

Shook, C.A. & Hubbard, L.T. (1973). An experimental study of transient slurry flow. *Can. J. Chem. Engng.*, **51**, 607–12.

Streeter, V.L. (1975). Fluid Mechanics. 6th Edition, McGraw Hill, New York, USA.

Thorley, A.R.D. (1980). Transient propagation in slurries with hold-up. *Proc. ASCE, J. Hydr. Div.*, **106**, HY8.

Thorley, A.R.D. & Hwang, L.Y. (1979). Effects of rapid change in flowrate of solid–liquid mixtures. In *Proc. HT 6*, pp. 229–42.*

Wood, D.J. & Kao, T.Y. (1966). Unsteady flow of solid–liquid suspensions. *Proc. ASCE* (*J. Eng. Mech. Div.*), **92**, 117–34.

Wylie, E.B. & Streeter, V.L. (1983). *Fluid Transients.* Feb Press, Ann Arbor, MI, USA.

*Full details of the Hydrotransport (HT) and the Slurry Transport Association (STA) series of conferences can be found on pp. 11–13 of Chapter 1.

9. BIBLIOGRAPHY

International Conference on Pressure Surges, BHRA Fluid Engineering, Cranfield, UK.
1st, Canterbury, UK, 6–8 Sept. 1972.
2nd, London, UK, 22–24 Sept. 1976.
3rd, Canterbury, UK, 25–27 Mar. 1980.
4th, Bath, UK, 21–23 Sept. 1983.
5th, Hannover, Germany, 22–24 Sept. 1986.
6th, Cambridge, UK, 4–6 Oct. 1989.

10

Flow Loop Studies

Randall G. Gillies
Saskatchewan Research Council, Saskatoon, Canada

1. INTRODUCTION

Experimental studies are usually considered before any substantial investment is committed to a slurry transport facility. This chapter deals with the design and operation of pilot-plant flow loops and indicates what types of information may be obtained from flow loop experiments.

The basic flow loop arrangement and experimental techniques described in this chapter have been used and found to produce good quality data at the Saskatchewan Research Council. During 25 years of operation, the Saskatchewan Research

Council's flow loops have been used to test a wide variety of slurries from highly heterogeneous mixtures such as coarse mineral ore in water slurries to non-settling mixtures such as coal–water fuels and powdered coal in crude oil.

2. THE NEED FOR PILOT-PLANT STUDIES

A small laboratory device such as a concentric cylinder viscometer or a capillary tube viscometer is often the only experimental equipment required to estimate energy requirements for a pipeline system designed to carry a single-phase fluid. Continuum models (i.e. Newtonian, Bingham, power-law fluid) are usually adequate for extending viscometric data for single-phase fluids and some non-settling slurries to pipeline flows.

However, for settling slurries, more complex multi-phase flow models are almost always required. In their present state of development, multi-phase flow models are filled with empirical terms and actual pipeline flow data are usually required to calibrate or verify these terms.

Pipe flow data are obtained either in a *once-through* flow system where the feed materials (solids and carrier fluid) are supplied continuously during the course of the experiment or in a *recirculating* flow loop where the contents of the pipe are returned directly to the inlet of the pump. Due to the substantial volume of feed materials required, once-through systems are usually located at an operating mine or mill. A recirculating flow loop will require a relatively small amount of slurry and may be located at a small pilot-plant facility.

Pilot-plant testing is advisable if a new slurry pipeline installation is planned. The installation may be a short transfer line to transport ore from mine to mill or it may be a cross-country pipeline moving a coal–water fuel to market. Also, pilot-plant data are often sought if major modifications are to be made to equipment or operating procedures in an existing facility. A properly designed pilot-plant flow loop will have much greater flexibility and will be capable of producing data over a wider range of conditions than an operating pipeline. This wide range of test conditions is required to develop a good understanding of the flow behaviour of the slurry.

Pilot-plant flow loop studies are commonly used to

- determine the frictional pressure gradient for the flowing slurry,
- determine the upper concentration limit for practical handling of a slurry,
- determine the slurry's deposition velocity (the velocity at which a stationary bed of solids forms along the bottom of the pipe),
- determine pipeline velocity and concentration distributions for the flowing slurry,
- test and calibrate flow metering and concentration sensing devices,
- assess pump performance,
- measure particle degradation rates,
- measure pump and pipe wear rates, and
- identify any special problems such as pipeline restart difficulties.

3. FLOW LOOP DESIGN

3.1. General Design Considerations

An example of a flow loop piping layout is shown in Fig. 10.1 and a typical flow loop pumping arrangement is shown in Fig. 10.2. This basic design is used for all of the Saskatchewan Research Council's flow loops which range in size from a 50 mm diameter pipe with a pipe loop volume of $0{\cdot}06\,m^3$ litres to a 500 mm diameter pipe with a volume of $30\,m^3$ litres.

Figures 10.1 and 10.2 show a *closed loop* arrangement with the mixture returning directly from the flow loop to the pump inlet. The other commonly used flow loop

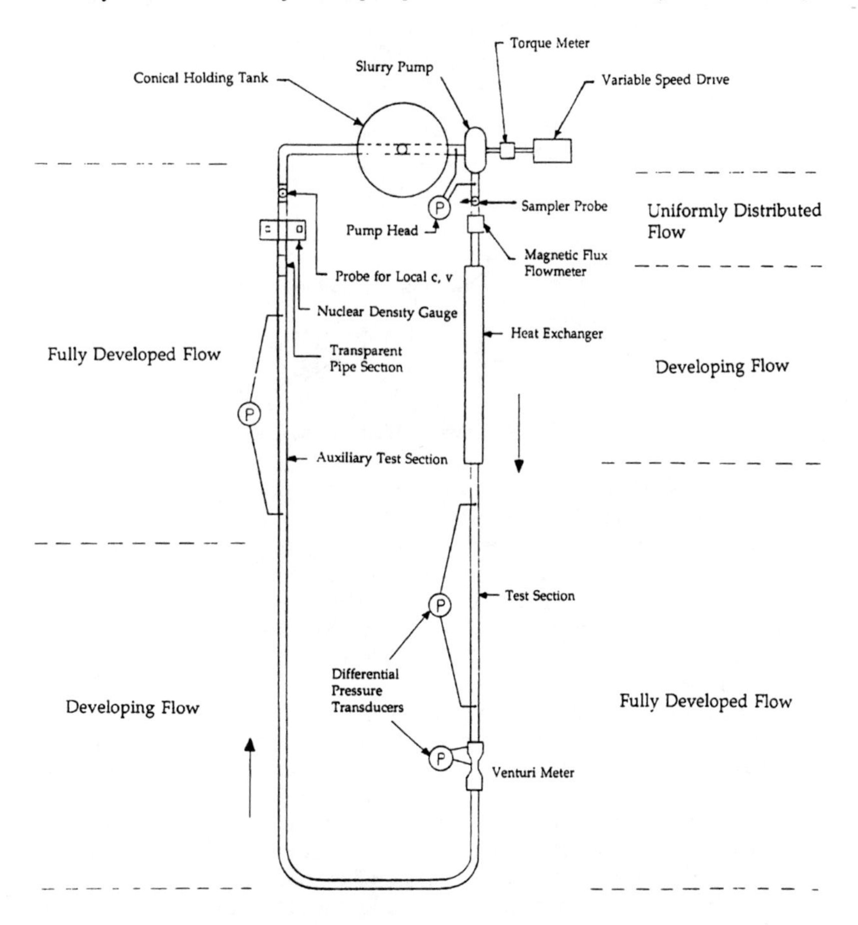

Fig. 10.1. Plan of the typical layout of equipment on a flow loop.

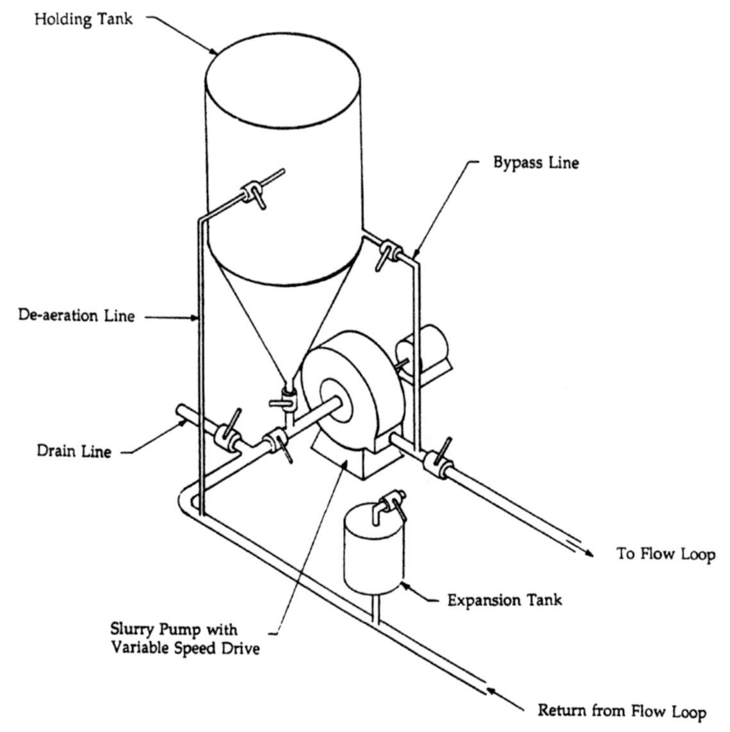

Fig. 10.2. Pumping circuit for flow loop.

arrangement is referred to as *open loop*. In an open loop system the mixture is discharged from the flow loop into the top of a holding tank. The mixture flows through the holding tank and enters the pump through a connection at the bottom of the holding tank. The slurry level in the holding tank must be high enough to ensure that air is not entrained in the mixture entering the pump. The slurry level must also be high enough to provide the suction head required for proper pump operation. The main advantage of an open loop configuration is the ease with which samples may be collected.

The Saskatchewan Research Council has found that closed loop systems perform better than open loop systems when testing settling slurries. With an open loop system there will be some differential settling in the holding tank. (The coarse solids will move through the tank at a higher rate than the fine solids and the carrier fluid.) As a result, the slurry in the pipeline will contain solids which, on average, are coarser than the sample used to prepare the slurry. Also, the solids concentration will be higher in the pipeline than in the flow loop system as a whole. These complications are avoided by using a closed loop arrangement.

While the flow loop design discussed in this chapter is primarily for horizontal flow testing, inclined and vertical pipe flows may also be tested by adding non-horizontal sections of pipe to the flow loop. Gravitational effects contribute significantly to the total pressure gradient for inclined and vertical pipes and, therefore, the mixture density needs to be known before the frictional component of the pressure gradient can be determined. The flow loop designer will have to install instrumentation which provides accurate measurements of the slurry density in the non-horizontal test sections. The designer will also have to include provisions for flushing the solids from steeply inclined sections of pipe to prevent plugging upon shutdown of the flow loop.

The flow loop of Fig. 10.1 contains a double pipe heat exchanger, two sections of pipe for measuring frictional head loss, a transparent pipe section and space for a number of slurry instruments. These instruments may include magnetic flux (electromagnetic) flowmeters, venturi meters, probes for measuring velocity and concentration distributions, radiometric gauges for determining slurry density and probes for detecting solids deposition. Instruments which are used routinely at the Saskatchewan Research Council and have proven to be effective for flow loop applications are discussed briefly in this chapter. The reader should refer to Chapter 11 for a more detailed discussion on the subject of slurry flow instrumentation.

Selecting the piping length is a critical element in designing a versatile test loop. If the flow loop is too long, sample requirements will be large and sample acquisition and handling will be costly. If the loop is too short, the experimenter will not be able to obtain accurate measurements for fully developed flow conditions. A test section, over which the frictional head loss is measured, must be preceded by a disturbance-free approach section. Experience has shown that the test section should be 60–100 pipe diameters long. The approach section should be at least 50 pipe diameters in length.

The location of slurry instruments in a flow loop is also an important design consideration. A slurry tends to be well mixed as it leaves a centrifugal pump and instruments that require uniformly distributed flow should be installed near the pump discharge. A sub-sampling device might be located at the pump discharge if the samples are to be used to estimate the solids concentration or average particle size distribution for the mixture delivered by the pipeline. Devices which require fully developed flow should be installed near the end of a long disturbance-free pipe section. Devices for determining velocity and concentration profiles and devices for detecting solids deposition are examples of instruments requiring fully developed flow conditions.

The slurry density and the slurry flow rate must be known accurately for flow loop data to be of any practical use. It is beneficial to have at least two methods available for determining each of these very important parameters. The frictional head loss for the flowing slurry is obviously also an important measurement. It is recommended that pressure differentials be measured over two sections of pipe to provide some verification of the quality of the head loss data.

3.2. Flow Loop Pumping Circuit

The arrangement shown in Fig. 10.2 may be used for testing both slowly and rapidly settling slurries. The essential features of the system are a slurry pump, a conical slurry holding tank, a de-aeration line, a flow loop bypass line, a drain line and an expansion tank.

A centrifugal slurry pump is normally used to circulate the mixture through the flow loop although, in some cases, a positive displacement pump (plunger or progressing cavity) might be used. The pump should have a variable speed drive to allow slurry testing over a wide range of flow rates.

When selecting a centrifugal pump, consideration has to be given to the type of stuffing box used to seal the pump shaft. A stuffing box with a liquid flush will quickly dilute the slurry in a recirculating flow loop. This dilution is usually unacceptable and a dry box arrangement is often the preferred option.

A large conical tank is connected to the pump inlet section of the flow loop. The tank volume should exceed the volume of the flow loop to allow the piping to be completely filled or flushed without having to stop to refill the tank.

A de-aeration line is used to facilitate the removal of entrained air from the slurry prior to a flow loop test. The de-aeration pipe should be connected to the top of the return section of the flow loop. At this location, the flow is fully developed and the air is likely to be travelling in the form of large bubbles near the top of the pipe. A small pump (not shown) may be required to transfer the aerated slurry from the top of the flow loop into the holding tank.

A bypass line is installed to allow slurry to be pumped directly back into the holding tank. The bypass line is used to provide mixing in the tank and also to facilitate the loading of coarse solids by washing the material from the tank into the flow loop. The flow should enter the tank tangentially to provide a good mixing action in the tank.

The drain line is used to allow air or carrier liquid to escape when pumping prepared slurry from the holding tank into the flow loop. It is also used to flush spent slurry from the flow loop at the end of a series of tests.

The expansion tank is required to control the flow loop system pressure when operating with a completely closed pipe loop arrangement. Pressurised air is supplied at the top of the expansion tank to provide the required suction head for the pump.

3.3. Plugging Considerations

Unless the loop is to be used only for non-settling slurries, consideration has to be given to the problems of plugging. Because a flow loop will be used to determine limiting operating conditions, plugging is likely to occur on numerous occasions over the lifetime of the test rig. Provisions for unplugging the flow loop should be considered in the initial design. The liberal use of removable pipe coupling devices is recommended for easy disassembly and unplugging of the flow loop.

Vertical pipe sections should be avoided to reduce plugging problems. If a vertical section is required, then provisions should be made for flushing the solids before stopping the flow.

Plugs which occur in the pump suction section are particularly troublesome because they effectively starve the pump and prevent it from developing any head. For this reason, piping between the holding tank and the pump inlet should be kept as short as possible. Alternatively, an auxiliary line should be provided to supply carrier fluid directly to the inlet of the slurry pump to help initiate flow in the main loop.

3.4. Slurry Valves

A flow loop requires a number of valves to provide the flexibility needed to test a variety of slurries over a wide range of operating conditions. Knife gate, butterfly and ball valves have been found to perform satisfactorily as on/off valves in slurry flow loops. Iris valves which close when a hydraulic fluid is pumped into a flexible bladder are useful as slurry throttling valves. Pinch valves, which operate by mechanically pinching a short length of flexible hose, are also useful as throttling devices but they tend to plug and for this reason, iris valves are preferred when the slurry contains coarse particles.

3.5. Temperature Control

The operating temperature is an important variable because changes in the viscosity of the carrier fluid may have dramatic effects on the flow behaviour of a slurry. The energy input to the pump will ultimately be dissipated as heat which has to be removed from the flowing slurry to maintain a constant operating temperature. Therefore, temperature control is an essential element of any definitive slurry flow test. The operating temperature may be controlled by pumping a chilled heat transfer fluid through the annulus of a double pipe heat exchanger.

4. SLURRY TESTING PROCEDURES

4.1. Preparing the Flow Loop

All pressure transducers should be calibrated against standard devices such as U-tube manometers or dead-weight testers. The flow metering devices and density sensing devices should be calibrated prior to installation in the flow loop.

Rust and other deposits should be removed from the flow loop prior to a series of tests to ensure that the pipe wall roughness does not change during the slurry tests. This is easily accomplished by circulating a sand in water mixture through the flow loop at a high velocity. This flow loop scouring procedure should be followed by a test with clear water. The clear water frictional head loss versus flow rate data may be used to establish the effective wall roughness for subsequent slurry tests.

4.2. Filling the Flow Loop

For a slowly settling slurry, the flow loop tank may be used as a preparation vessel or as a holding tank for slurry prepared elsewhere. The main slurry pump may be used to maintain the slurry in suspension in the tank. This is done by closing the valves on the flow loop discharge and return sections and opening the valves at the bottom of the tank and in the bypass line. Alternatively, a mechanical mixer may be used to disperse the slurry in the tank.

A slowly settling slurry may be transferred from the holding tank to a dry flow loop or to a flow loop filled with carrier fluid. With the pump running, the valves in the flow loop discharge section and the drain line are opened and the bypass valve is closed. This starts the flow of slurry into the flow loop. As soon as the flow loop is filled with slurry, the valve in the pump inlet section is opened and the drain line valve is closed. The flow loop is then pressurised by shutting the valve in the bottom of the holding tank and by applying air pressure to the expansion tank.

A different procedure is used to fill the flow loop with a rapidly settling slurry. At the start of the loading operation the flow loop and the conical section of the holding tank are filled with carrier fluid. With the valve at the bottom of the tank, the valve in the bypass line and the valves in the pump inlet and discharge sections open, the pump is operated to establish flow through the flow loop and also through the bypass line. Dry solids are dropped into the tank from above and are washed into the flow loop by the fluid entering through the bypass line. When the required amount of solids have been loaded, the flow rate is reduced and any solids remaining in the tank are allowed to settle into the flow loop. The excess carrier liquid is then drained from the tank.

Prior to the collection of slurry data, the flow loop should be operated at a low flow rate to remove entrained air and to establish the desired operating temperature. If the flow loop contains a transparent section of pipe, then it is quite simple to determine the best flow rate for removing entrained air. When the proper flow rate is established, the entrained air will rise to the top of the pipe and form large bubbles which will travel slowly around the loop. These bubbles are readily removed through the de-aeration line.

4.3. Solids Concentration Determination

There are two distinct definitions of slurry concentration:

(1) the in-situ concentration which is the ratio of the volume of solids in a section of pipe to the total volume of the pipe section; and
(2) the delivered concentration which is the volume fraction of solids in a mixture discharged from the flow loop.

The in-situ concentration reflects the local conditions of solids loading in a pipe and is often used in models for estimating the frictional head loss for a slurry. The delivered concentration is important because it ultimately determines the capacity of a pipeline system. In a definitive slurry flow loop test, both in-situ and delivered solids concentration will have to be known.

The in-situ concentration may be determined by using a radiometric density gauge (described in Chapter 11). ^{137}Caesium radioisotopes are used in the Saskatchewan Research Council's density gauges.

If the internal volume of a horizontal flow loop is known, then the in-situ concentration can be determined by measuring the amount of solids added to the loop.

The delivered concentration may be determined by measuring the density or moisture content of samples of slurry discharged from the flow loop. For small diameter flow loops (100 mm, or less) it is usually best to divert temporarily the full pipe flow into a sampling vessel. A rotating plate sample (illustrated in Chapter 11) may be used to divert the pipeline flow. A sub-sampler is normally used for pipelines larger than 100 mm. If a sub-sampling device is used, then an isokinetic sampler probe of the type described in Chapter 11 should be used. The sampler probe should be located near the discharge of the pump where the slurry is well mixed and the effects of solids segregation are likely to be minimal.

A probe which measures local concentrations and velocities (also described in Chapter 11) may be used to determine both in-situ and delivered concentrations. The in-situ concentration is obtained by integrating local values of concentration over the pipe cross-section. The delivered concentration is obtained by integrating the product of local concentration and local velocity over the pipe cross-section.

4.4. Flow Rate Determination

The most direct method for determining slurry flow rate is to divert the flow loop discharge into a weighed vessel for a short but accurately measured period of time. The displaced slurry has to be replaced by mixture from the holding tank or from the expansion tank. This creates a problem if a centrifugal pump is used because the performance of the pump is affected. The pump's suction head is reduced as the mixture level drops in the tank. Also, in the case of settling slurries, the mixture in the tank will have a lower density than the mixture in the flow loop and when this low density mixture enters the pump, the pump discharge pressure will drop. As a result, the slurry flow rate will decrease during the flow determination exercise. Because of this complication, other methods are usually used to monitor the slurry flow.

Magnetic flux flowmeters are suitable for most aqueous slurries. Their accuracy is generally very good and their use is recommended for flow loop testing.

If the density of the flowing mixture is known, then kinetic energy meters (i.e. venturi meters, flow nozzles, wedge meters) may be used to determine the slurry flow rate.

If a positive displacement pump is used on the flow loop then the slurry flow rate may be determined from the speed of the pump. The use of a positive displacement pump is sometimes the best alternative for determining flow rate if the carrier fluid is a petroleum crude oil or petroleum product.

4.5. Pressure Sensing

The determination of the pressure differential between two points is one of the most important measurements to be made in a flow loop test. These points may be two locations on a section of straight pipe for pipeline frictional head loss determination, the tappings on a venturi meter or pressure tappings used to measure the head rise produced by a centrifugal pump.

For accurate test results the use of carefully calibrated differential pressure transducers is recommended. The pressure-sensing lines connecting the pressure transducer to the pipe tappings should be filled with carrier fluid. Frequent flushing will be required to keep solids and entrained gases from entering the sensing lines and adversely affecting the accuracy of the differential pressure determinations.

Diaphragms are sometimes used to isolate the slurry from the fluid in the pressure sensing lines. Unfortunately, there is some hysteresis associated with the pressure transmission characteristics of diaphragms. Because flow loops are usually short, the pressure differentials are small and the errors introduced by using diaphragms can be unacceptably large. Therefore, the use of pressure sensor diaphragms is not recommended for pilot-plant flow loops.

4.6. Deposition Velocity Determination

A properly designed slurry pipeline will operate at a velocity which exceeds the solids deposition velocity. A poor estimate of deposition velocity will seriously affect the performance of a slurry pipeline. A flow loop test will provide a direct measurement of the deposition velocity for a particular slurry pipeline design situation.

The deposition velocity is best determined visually by observing the slurry flow through a transparent pipe section. Either glass or transparent plastic piping sections may be installed in a flow loop.

Electronic deposit detection probes of the type described in Chapter 11 are also useful for monitoring solids deposition conditions in flow loops.

4.7. Wear Rate Determination

The expected life of a slurry pipeline installation will be highly dependent on the abrasiveness of the slurry. Flow loop tests can provide useful information about pipe, pump and valve wear rates.

Wear rate determinations are usually performed by weighing a pipe section or piece of equipment (i.e. pump casing or impeller) before and after a period of continuous operation of the flow loop. The flow loop may have to be operated for several weeks to obtain sufficient weight loss to determine the wear rate accurately. Rounding of particles will occur when solids are circulated for a prolonged period and the wear rate will decrease. It is usually necessary to replace the solids at frequent intervals to obtain realistic wear rate information.

4.8. Particle Degradation Rate Measurement

Particle degradation rates may be measured by removing samples of slurry from the flow loop (see Chapter 12). The degradation rate is measured by determining the particle size distribution of samples removed at various time intervals.

The particle degradation rate measurements obtained in flow loop studies must be regarded as qualitative results. This is because a flow loop will normally have a much larger ratio of pump passes to distance travelled than an operating pipeline. Since the pump contributes considerably to particle degradation, flow loop degradation rates may be considerably higher than those in actual operating pipelines.

4.9. Pump Performance Tests

Pump performance can be monitored if a torque meter is connected to the pump shaft, a pressure sensor is installed to measure the head rise across the pump and the pump speed is monitored. The slurry flow rate will also have to be measured to complete the pump efficiency determination.

The density of the delivered mixture will have to be known if the pressure rise across the pump is to be expressed as a slurry head rise.

11

Some Experimental Techniques Applicable to Slurries

Nigel P. Brown
The British Petroleum Company plc, London, UK

1. INTRODUCTION

This chapter discusses some experimental techniques that have been developed for use with slurries. While in general they often contain commercially available equipment the system is often not available as a *package*. Guidance on the selection of commercially available measuring equipment is provided in Chapter 23.

2. SETTLING TENDENCY

2.1. Introduction

Measurement of the settling tendency of particles in a slurry under conditions of no-flow is a technique that is a useful means for classifying slurries. For slurries in which the particles are coarse and their relative density large, the tendency to settle can readily be ascertained from observation. However, for slurries that are *stabilised*, or where the particles are comparatively small and the relative density close to unity, the tendency for the particles to settle is less readily ascertained. For these slurries experimental techniques have been developed to enable comparisons to be made.

2.2. Stabilised Slurries

Stabilised slurry technology is discussed in Chapter 1.

Qualitative assessments of the stability of samples of stabilised coal–water slurries have been reported by Lawler *et al.* (1978) and Duckworth *et al.* (1983). Lawler *et al.* froze samples of slurry in cylindrical paper containers. When the slurry was solid the container was peeled away and the coarse coal exposed by spraying the column with warm water. Duckworth *et al.* devised an elegant technique whereby the coal below 20 μm was removed and replaced by an equal volume of cement with a similar size. The coal–cement mixture was blended to form a slurry of the required concentration, poured into a cylindrical mould and allowed to set. The solidified sample was then sectioned in the axial plane and at several cross-sections. In both techniques the location of the coal is ascertained by visual inspection.

The distribution of coal in pipelines containing both flowing and static Stabflow slurries (see Brookes & Snoek, 1986) has been measured using a scanning densitometer (described by Brown, 1988). The densitometer used the attenuation of gamma-photons to measure the concentration profile of the coal in the slurry at 15 vertical locations across a (nominal) 300 mm pipe. Although the technique measured the average coal concentration across a chord of the pipe the resolution was sufficient to identify the higher coal concentration that would be presented by lumps of coarse coal that may have settled through the fine-particle slurry.

2.3. Fine-Particle Slurries

The settling tendency of fine-particle slurries can be qualitatively assessed using a Brookfield Helipath (Stoughton, MA, USA) viscometer. This commercially available instrument is essentially a conventional rotary viscometer in which the cylindrical bob is replaced with a T-shaped spindle. As the T-bar is rotated it is lowered at constant speed through the sample. These speeds are selected such that as the T-bar is lowered a helical path is cut through the undisturbed sample. Comparison of the torque-depth data enables comparative assessment of settling to be made.

3. FLOW PATTERN IN SETTLING SLURRIES

3.1. Introduction

Flow patterns are phenomena exhibited by slurries in pipe flow in which the particles settle under conditions of no-flow; they are discussed in detail in Chapter 3.

3.2. Electrical Probes

The use of sensors reduces the amount of subjective interpretation of flow patterns. Kazanskij (1979) describes a technique that measures fluid conductivity that reportedly enables the difference between a stationary and moving deposit to be distinguished. Ercolani *et al.* (1979) describe two types of probe that can be flush mounted in the bottom of a pipe using a single hole. The probes, shown in Fig. 11.1, measure changes in fluid conductivity and heat transfer caused by the presence of particles in the slurry. The output signal from the two types of probe are presented for a sand with mean particle diameter (d_{50}) of 3 mm. Output from the conductivity probe enabled a range of flow patterns to be distinguished and in particular the condition where a stationary deposit started to move. The flow features could more readily be distinguished from the probe based on measuring heat transfer to the slurry. Although the characteristics of the output signal were dependent upon flow pattern, calibration against visual observation is, however, advisable. The use of probes enabled the effect of average velocity and concentration of solids on the conditions at the limit of a stationary deposit to be

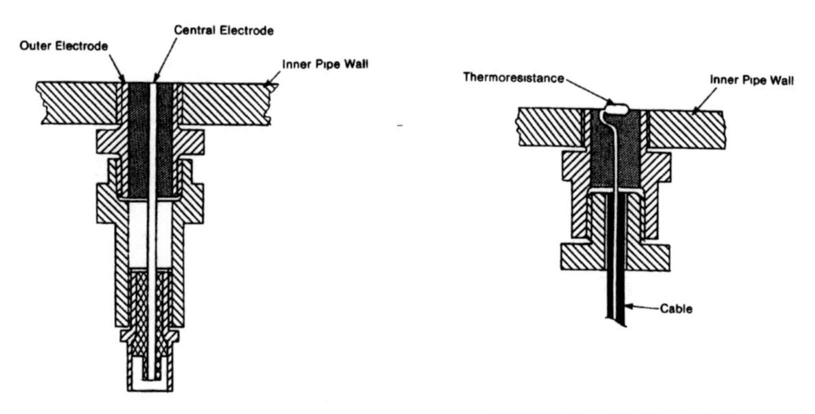

Fig. 11.1. Electric and thermic probes used by Ercolani *et al.* (1979) to detect flow patterns. (Reproduced by permission from *Proceedings of the 6th International Conference on the Hydraulic Transport of Solids in Pipes (1979)*, published by BHRA (Information Services), The Fluid Engineering Centre, Cranfield, MK43 OAJ, UK.)

accurately determined. Gillies *et al.* (1984) reported the use of both types of probe on fine sand and coal slurries.

Ultrasonic transducers mounted in the top of a pipe have been used by Ferrini and Pareschi (1980) to measure the depth of a deposit after shut-down. Ultrasonic transducers mounted on the bottom of the pipe have been reported by many researchers including Faddick *et al.* (1979) and Lazarus and Lazarus (1988). In this deployment the Doppler technique is used to detect the motion of a sliding deposit offering the potential for optimising transport conditions and preventing pipeline blockage.

3.3. Pressure Measurements

Measurements of differential pressure between the top and bottom of a horizontal pipe have been used by Shook *et al.* (1982*a*, *b*) to measure the homogeneity of flowing fine-sand slurries in pipes with an inside diameter of 159 mm and 495 mm. The arrangement of the pressure tappings is shown in Fig. 11.2. With the connecting pipe work to the transducer filled with water the measured differential pressure is given by

$$\Delta P = D(\rho_m - \rho_f)g \tag{11.1}$$

where D is the inside diameter of the pipe [L], g is the gravitational acceleration [$L T^{-2}$], ΔP is the measured differential pressure [$M L^{-1} T^{-2}$], ρ_f is the density of fluid [$M L^{-3}$], and ρ_m is the density of slurry [$M L^{-3}$].

The slurry density will be the same as that computed from the in-situ concentration if the particles are suspended, a reduction in pressure indicating the

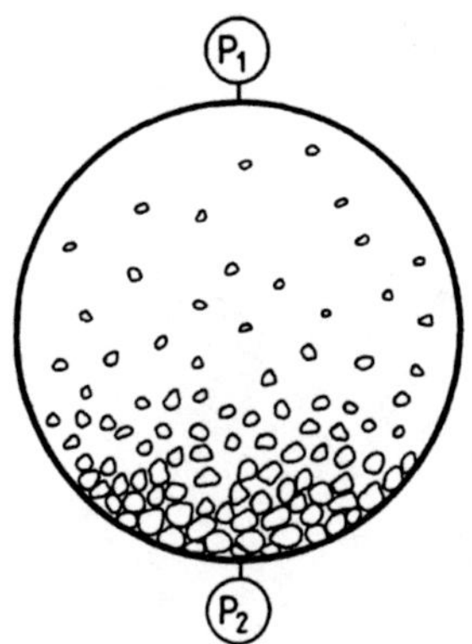

Fig. 11.2. Vertical pressure difference measurements used by Shook *et al.* (1982*a*) to detect bed-load transport.

presence of bed-load. In practice, the pressure difference is small and careful experimental technique is required to obtain meaningful results.

4. PARTICLE–WALL FRICTION

4.1. Introduction

Prediction of the pipeline frictional pressure drop for segregated coarse-particle slurries (such as those considered in Chapter 5) requires an estimate of the kinetic friction between the particles and the pipe wall. The coefficient of friction is defined as

$$\mu_p = \tan(\tau/b) \tag{11.2}$$

where μ_p is the coefficient of friction [1], τ is the shear stress [M L T^{-2}] and b is the normal force [M L T^{-2}]. As defined, the friction coefficient is independent of the velocity at which the shear takes place. In practice, two coefficients are commonly observed; the static value, measured at the onset of motion, and the kinetic value obtained during subsequent slow motion. It is possible to measure the friction coefficient using a number of techniques that include rotating tube apparatus (Briscoe *et al.*, 1983) and the Shearometer (Jacobs & Tatsis, 1986). However, shear cells and tilting-tube apparatus have received considerable attention.

4.2. Shear Cells

Commercial apparatus used to measure the internal shear resistance of soils, such as the Jenike shear cell or conventional shear box (described by Liu and Evett, 1984) can be adapted to measure the friction coefficient of a slurry against a plane surface. The lower part of the cell in the conventional arrangement is replaced with a flat specimen of the material against which the particles are to be sheared. The standard 100 mm diameter Jenike shear cell is suited to testing granular materials

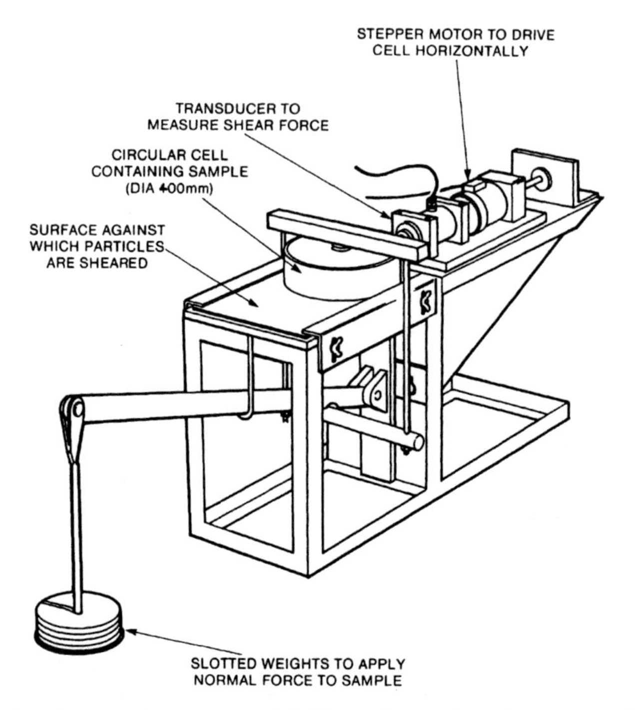

Fig. 11.3. Modified Jenike shear cell (400 mm diameter) used to measure the coefficient of particle–wall friction (shown in the set-up position). (Courtesy of Warren Spring Laboratory, Stevenage, UK.)

and stabilised slurries containing particles below approximately 10 mm under dry and flooded conditions. Larger diameter cells, such as those described by Heywood and Kirby (1988) and shown in Fig. 11.3, can accommodate larger particles. Experiments with shear cells have to be carefully executed, especially with fine materials, to avoid erroneous results caused by particles becoming trapped under the frame of the cell. Results with sands are reported by many workers including Rowe (1962), Butterfield and Andrawes (1972), and Cumolióglu (1975).

4.3. Tilting-Tube Apparatus

Pipe geometries not only present a complex distribution of stresses but also lower levels of normal stress than can be comfortably accommodated by most designs of shear cell. Wilson (1970, 1972) reports the use of a tilting-tube apparatus to measure the friction coefficient. More recently, Eyler (1981) and Gillies *et al.* (1984) have constructed equipment based on the same principle. The apparatus reported by Gillies *et al.* is shown in Fig. 11.4. The end of the tube is loaded with a bed of particles, restrained by the retractable stopper. The tube is carefully flooded and

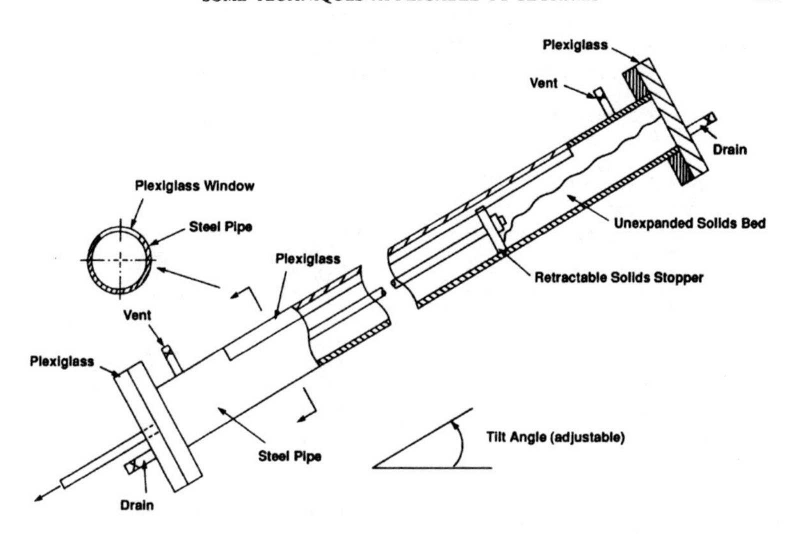

Fig. 11.4. Tilting-tube apparatus used by Gillies *et al.* (1984) to measure the coefficients of friction. (Courtesy of Saskatchewan Research Council, Saskatoon, Canada.)

any air vented. The apparatus is then inclined to a predetermined angle and the stopper retracted slightly to ascertain if there is any tendency for the bed to slide. The procedure is repeated at increasing tilt angles until the onset of motion is detected. Although conditions within tilting-tube friction testers can closely resemble pipeline conditions at the point of incipient bed motion, experimenters report that the measurements are often difficult to perform.

5. SLURRY SAMPLING

5.1. Introduction

Whenever a sample is removed for analysis measures must be taken to ensure that it is representative of the bulk from which it was removed. A comprehensive review of theoretical considerations, practical techniques and commercial sampling equipment, covering a wide range of applications is given by Cornish *et al.* (1981). Where practicable, preference is usually directed towards removing a total discharge (or full-flow) sample rather than a sub-sample as problems with representativeness are considered less worrisome. In large-scale operations, full-flow sampling is often impracticable as high flow rates result in large volume samples and resort has to be made to withdrawing a small sub-sample.

TABLE 11.1
Minimum Mass of Sample to be Taken for Particle Size Analysis[a]

Maximum size of material present in substantial proportion (more than 10%, by mass) retained on test sieve (mm)	*Minimum mass of sample to be taken for sieving (kg)*
63	50
50	35
37·5	15
28	5
20	2
14	1
10	0·5
6·3	0·2
Passing 2, 0·6, 0·3, or 0·063	0·1

[a] Recommended by BSI (1975).

5.2. Guidance on Sample Size

Statistical considerations imply that the size of the sample increases with increasing top particle size. Standards such as BS 1017 (BSI, 1960) and ASTM D2234-86 (ASTM, 1986) provide useful guidance on sample size for coals; it is reasonable to assume that they can be adapted for use with other crushed minerals. Guidance on the minimum sample size of soils to be taken for particle size analysis for civil engineering purposes is provided in BS 1377 (BSI, 1975) and reproduced here in Table 11.1. In practice, the sample size must be verified and frequently established by experiment.

5.3. Full-Flow Sampling

Shook *et al.* (1981, Fig. 2) showed a pneumatically actuated apparatus for removing a full-flow sample from an open pipe loop. Especially at high flow rates, obtaining a sample of reasonable volume can be difficult. Sliding plate samplers provide a convenient means of controlling the sample size. Haas *et al.* (1977) describe a two-port *rotating-plate* sampler, shown in Fig. 11.5, constructed around 100 mm (nominal bore) pipe. The plate on which the inlet port is attached is rotated from the undiverted position, above the outlet port, to the sampling port using a pneumatic cylinder. When the sample has been collected the inlet port is rotated back to align with the outlet port. Flexible pipe is used to connect the main pipeline to the rotating plate. This type of sampler can be adapted to higher pressure applications by suitably restraining the sliding plate.

5.4. Sub-sampling

Samples removed by a probe at a velocity greater than the undisturbed velocity upstream of the probe are likely to be lower than the true local concentration (see,

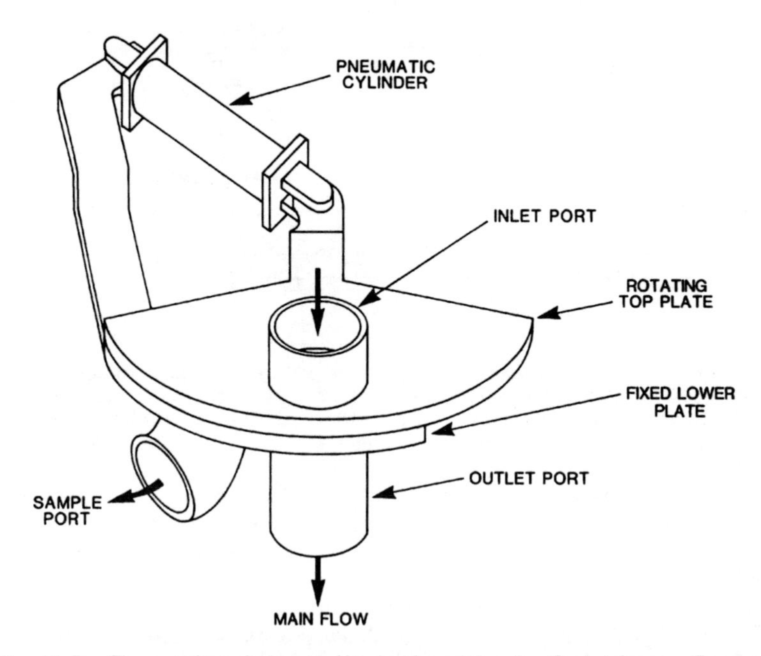

Fig. 11.5. The rotating-plate sampler developed by the Saskatchewan Research Council. (Courtesy of Saskatchewan Research Council, Saskatoon, Canada.)

for example, Nasr-El-Din *et al.*, 1986) due to a tendency to sample more fluid than particles. Removing the sample at a lower velocity can result in higher concentration of solids and in extreme circumstances a blocked probe. The ideal condition is to remove samples isokinetically. Experiments performed by Nasr-El-Din *et al.* (1986) show that the design of the probe entrance and the rate of sample removal is less critical for slurries in which particle inertia is not dominant. Earlier work reported by Nasr-El-Din *et al.* (1984) showed that reliable samples could be taken by a thin-walled L-shaped probe directed into the flow with a tip angle of less than 20°. Using a probe with a 5° tapered entrance constructed from 8 mm bore tube (with a wall thickness of 0·2 mm), sand with a mean diameter of 0·72 mm was sampled at concentrations up to 25% (by volume) from a 52 mm pipe. The same probe has been used (Nasr-El-Din *et al.*, 1986) to sample irregular polystyrene chips (of relative density 1·06) with a mean diameter of 1·4 mm at concentrations up to 34% (by volume) from bulk flows at 3·4 m s^{-1}. The local velocity in these flows was measured using a separate velocity probe (Brown *et al.*, 1983). Thin-walled probes suffer from severe abrasive wear; Nasr-El-Din and Shook (1985) provide guidance on the design of two straight probes, using a 45° entrance and a circular port, that can be readily withdrawn from the flow.

Cooke and Lazarus (1988) report the use of a 24·8 mm diameter sample probe

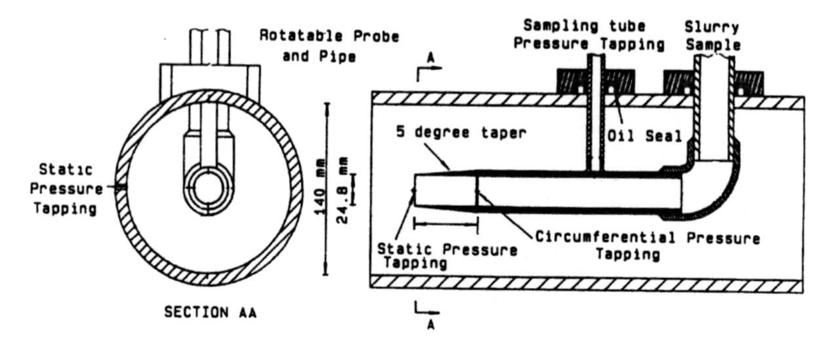

Fig. 11.6. The isokinetic sampling probe developed by Cooke and Lazarus (1988). (Reproduced by permission from *Proceedings of the 11th International Conference on the Hydraulic Transport of Solids in Pipes (1988)*, published by BHRA (Information Services), The Fluid Engineering Centre, Cranfield, MK43 OAJ, UK.)

within a 140 mm pipe. The stagnation pressure in the sampling tube is used as a reference for measuring the free-stream approach velocity, thereby removing the need for separate velocity measurements. The sampling probe is shown in Fig. 11.6.

Gillies *et al.* (1981) (also, Shook *et al.*, 1982*c*) used the centre-stream sampling device shown in Fig. 11.7 to remove samples of slurry isokinetically from the centre of a 250 mm pipe during particle degradation experiments carried out with a centrifugal pump. Slurry is removed with a 100 mm diameter sample tube from the pump discharge and returned to the suction side of the pump. To ensure that the sample is removed isokinetically the sample flow rate is measured by a magnetic flowmeter and controlled by a hydraulically operated pinch valve. Samples collected

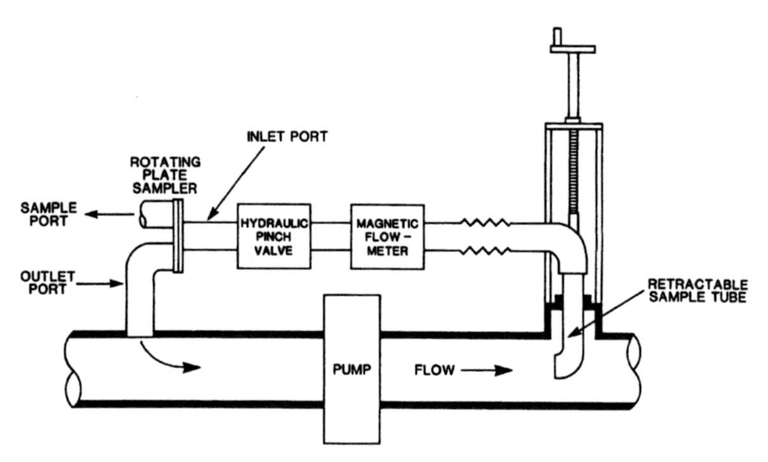

Fig. 11.7. Centre-stream sampler developed at the Saskatchewan Research Council. (Courtesy of Saskatchewan Research Council, Saskatoon, Canada.)

for particle size early in the tests were found to be representative of the coal loaded into the pipe loop. The top size of coal used with the sampler was 50 mm. The sampler is designed to be withdrawn from the flow when the sample has been collected.

6. PARTICLE, LIQUID AND SLURRY VELOCITY

6.1. Introduction

The electromagnetic flowmeter is commonly used to measure slurry flow rate, or superficial velocity (flow rate divided by flow area), and can be used with confidence without calibration with a wide range of slurries. Brook (1962) described pressure measurement devices including whistle and bend meters that have been used for measuring total flow. These devices are of restricted practical use as they require in-service calibration. The venturi meter offers considerable promise for the measurement of flow rate providing that limitations are appreciated and the in-situ solids concentration can be measured by an independent technique. The comparatively recent development of probes enables local particle velocity to be measured.

6.2. Venturi Meters

Flowmeters that use the venturi principle are discussed in Chapter 23. They can be used to measure flow rate or concentration provided that the complementary measurement is also available. The geometry of the flow tube in a venturi meter is shown schematically in Fig. 11.8; design details can be found elsewhere (BSI, 1981). The increase in pressure as the flow accelerates into the throat is readily measured with a differential transducer. Reduction of wear by weld deposition of a hardened metal lining has proved highly satisfactory, although it was found that the discharge coefficient was reduced by 10% from its usual value.

Shook and Masliyah (1974) reported results from a theoretical and experimental study conducted with both horizontal and vertical venturi meters using slurries that show distinct two-phase behaviour. Theoretical modelling carried out on

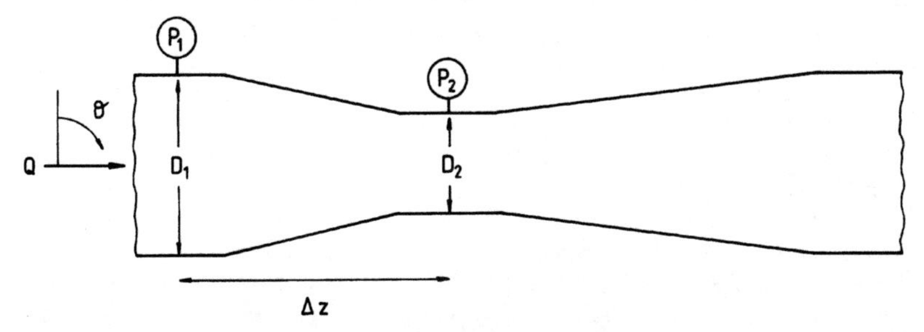

Fig. 11.8. Geometry of the flow-tube in a venturi flowmeter.

slurries that pose extreme cases, where inter-phase slip and relative solids density is large, indicated that discharge coefficients greater than those for water could be expected. Experimental measurements with lead-shot–water slurries at concentrations up to 6% (by volume) gave values of discharge coefficients at or below 1·1. However, the discharge coefficient for sand–water slurries, at volumetric concentrations up to 25%, were indistinguishable from those measured with water. Experimental studies reported by Shook (1982) carried out with 13 mm gravel in 250 mm (nominal bore) pipeline confirmed earlier studies conducted in small diameter pipes that the discharge coefficient approached that for water flowing alone as the character of the slurry flow became more homogeneous. Further confirmation of the use of clear fluid discharge coefficients for slurry flows consisting of sands (of median particle diameter 0·29, 0·33, 0·55 and 2·4 mm) at concentrations up to 35% (by volume) is reported by Gillies *et al.* (1988) in pipes of (nominal) diameter 50 and 250 mm.

The usual equation, given below, for clear fluids can be used to predict slurry flow rate with the density term being that of the mixture (obtained from in-situ measurements).

$$Q = C\frac{\pi D_2^2}{4}\left(\frac{2(P_1 - P_2) - 2g\Delta z\rho_m\cos(\theta)}{\rho_m(1 - (D_2/D_1)^2)}\right)^{0.5} \qquad (11.3)$$

where C is the discharge coefficient [1], D_1 is the diameter of the inlet pipe [L], D_2 is the diameter of throat [L], g is the gravitational acceleration [L T^{-2}], P_1 is the pressure at inlet [M L^{-1}T^{-2}], P_2 is the pressure at throat [M L^{-1}T^{-2}], Q is the volumetric flow rate [L^{3}T^{-1}], Δz is the axial distance between pressure stations [L], ρ_m is the density of slurry [M L^{-3}], and θ is the angle of the axis from the vertical [Θ].

6.3. Tracer Techniques

Techniques for using liquid tracers to measure the bulk flow of water in conduits are discussed in considerable detail elsewhere (BSI, 1980). Measurement of the time from the injection of a pulse of tracer into the flow to its detection downstream is a particularly useful technique. However, for segregated slurry flows the response measured downstream can be highly asymmetric. Considerable difficulty can be experienced in interpreting the response to obtain the transit time. Kao and Kazanskij (1979) present results (originally reported elsewhere in Russian) where a liquid tracer was used with a coal slurry. As presented, the results show that, provided the centroid of the response signal is used for the arrival time of the tracer at the measurement station tracer, determinations appear to be in good agreement with timed discharge samples.

The use of a radioactive liquid tracer (^{56}manganese) to measure local fluid velocity in high concentration glass ballotini (spheres)–water slurries is described by Bantin and Streat (1971). Local particle velocities were also measured (Bantin & Streat, 1972) by introducing radioactive labelled ballotini into the flow. The velocity was calculated by measuring the transit time of the particles between two detectors mounted on the outside of the pipe. The sodium in the glass was activated (to ^{24}sodium) by neutron bombardment in a research nuclear reactor.

6.4. Electrical Probes

Cross-correlation techniques extract the time difference between two continuous electrical signals. Measurement of the separation of the sensors enables velocity to be calculated. Such techniques offer considerable advantages for slurry systems as they require no calibration and have the potential to reject spurious signals. Beck *et al.* (1974) and Beck (1981) describe sensors applicable to velocity measurements in slurry systems.

Cross-correlation has been used to measure local particle velocity using an electrical probe (Brown *et al.*, 1983) that physically resembles the pitot tube used for clear fluids. The local concentration of particles in the fluid is reflected by the

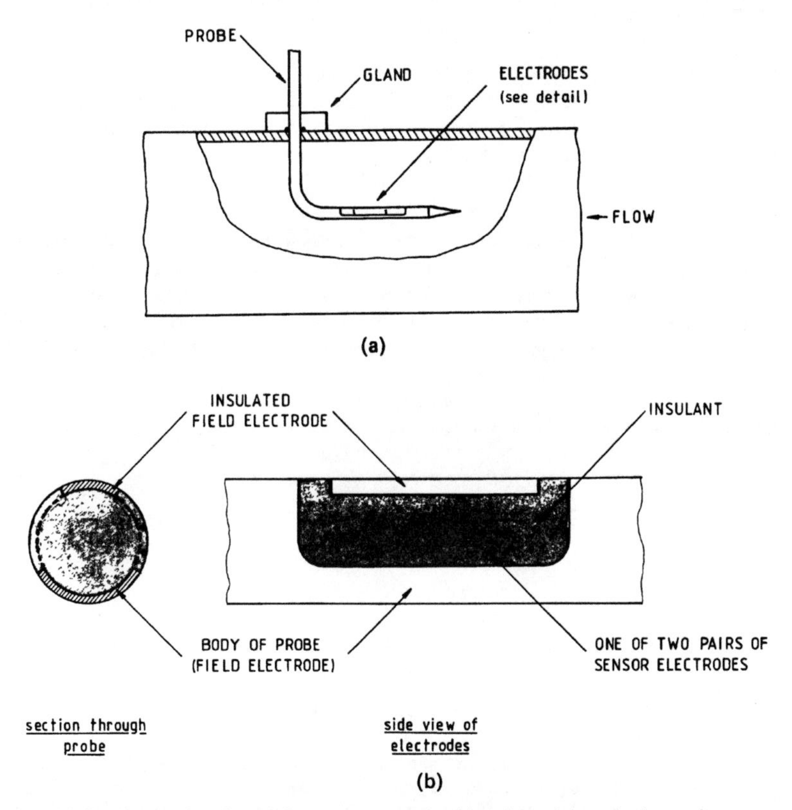

Fig. 11.9. Probe developed for measurement of local particle velocity and concentration. (After Nasr-El-Din *et al.*, 1987)

local resistivity of the slurry. The local resistivity of the fluid is measured by two pairs of electrodes contained along the body of the probe. The electrical signals are cross-correlated to obtain the characteristic transit time of the particles past the electrodes. During early development work in a 50 mm (nominal inside diameter) pipe the electrical field was supplied by two electrodes flush-mounted into the wall of the plastic pipe. For application in large diameter pipes a self-contained probe, shown schematically in Fig. 11.9, has been developed (Gillies *et al.*, 1984; Nasr-El-Din *et al.*, 1987) in which the potential gradient is supplied by electrodes in the body of the probe. The technique has been used on a wide range of slurries with particle sizes greater than approximately 200 μm in (nominal) pipe diameters up to 495 mm. This probe has also been used to measure local particle concentration.

7. CONCENTRATION OF THE SOLIDS

7.1. Introduction

In many slurry applications the concentration of solids is commonly used in preference to slurry density. Primarily for economic reasons, concentrations by mass (or weight) are used in operational installations whereas volumetric concentrations are most frequently reported in research papers. Concentration of solids by volume is defined as the volume of solids in the slurry sample divided by the total volume occupied by the sample. An analogous definition is used for mass concentration. Table 11.2 provides a ready means of converting between concentrations of solids by volume and mass and density of the two-phase mixture.

The in-situ and delivered concentrations are two values of concentration commonly encountered in slurry pipeline operations. The in-situ concentration, occasionally referred to as the *transport* concentrations is that of solids flowing in

TABLE 11.2

Interrelation between Slurry Density (ρ_m) and Concentration by Volume (C_v) and Mass (C_m)

	C_v	C_m	ρ_m
C_v =	1	$\dfrac{100C_m}{C_m + s(100 - C_m)}$	$\dfrac{100(\rho_m - \rho_f)}{(\rho_p - \rho_f)}$
C_m =	$\dfrac{100C_v s}{(C_v s) + (100 - C_v)}$	1	$\dfrac{\rho_s(\rho_m - \rho_f)}{\rho_m(\rho_p - \rho_f)}$
ρ_m =	$\dfrac{C_v(\rho_p - \rho_f)}{100} + \rho_f$	$\dfrac{100\rho_p}{(100s) - C_m(s - 1)}$	1

C_m is the concentration of solids by mass (expressed as a percentage) [1]. C_v is the concentration of solids by volume (expressed as a percentage) [1]. s is the relative density of the particles, (ρ_p/ρ_f) [1]. ρ_f is the density of fluid [M L^{-3}]. ρ_m is the density of slurry [M L^{-3}]. ρ_p is the density of particles [M L^{-3}].

the pipeline. With the exception of pseudohomogeneous flows (see Brown *et al.* 1983, p. 601) a slip velocity is likely to exist between the phases due to the solids and liquid travelling at different velocities. In contrast, the delivered, or *bucket* concentration is the concentration as measured at the exit of the pipeline, where the velocity of the two phases is identical. The numerical difference between the two concentrations is therefore dependent upon the relative velocity of the fluid and particles. Where the fluid velocity is greater than that of the particles, such as in horizontal heterogeneous flows and counter-gravity flow of coarse particle slurries, the in-situ concentration will be greater than the delivered. The two concentrations are defined mathematically by integrals over the flow area of the pipe:

$$C_{\mathrm{v,d}} = \int_A \frac{c_{\mathrm{v}} v_{\mathrm{p}}}{A V_{\mathrm{m}}} \, \mathrm{d}A \tag{11.4}$$

and

$$C_{\mathrm{v,i}} = \int_A \frac{c_{\mathrm{v}}}{A} \, \mathrm{d}A \tag{11.5}$$

where A is the flow area of pipe [L^2], c_{v} is the local concentration of solids (by volume) [1], $C_{\mathrm{v,d}}$ is the delivered solids concentration (by volume) [1], $C_{\mathrm{v,i}}$ is the in-situ solids concentration (by volume) [1], V_{m} is the superficial slurry velocity [$\mathrm{L\,T^{-1}}$], and v_{p} is the local particle velocity [$\mathrm{L\,T^{-1}}$].

Three spatial resolutions of concentration are reported: averaged over the entire cross-section of the pipe, averaged along a chord through the pipe and local measurements restricted to a small area of the cross-section, or volume. Measurements performed with a counter-flow or venturi meter provide concentration measurements that are usually taken to be averaged over the flow area (but more correctly are volume-averaged over the length of pipe between the pressure-measuring stations). Measurements made with a narrowly collimated beam of gamma-photons provide chord-averaged measurements. Local measurements can be made using techniques that include electrical probes, isokinetic sampling, radiometric attenuation employing tomographic techniques, or visual observation as reported by Scarlett and Grimley (1974). Nasr-El-Din *et al.* (1986) compare measurements obtained from radiometric attenuation along a chord, and local measurements using probes employing sampling and electrical techniques.

7.2. Sampling

Sampling offers the prospect of the most direct means of measuring concentration and is discussed earlier in this chapter. The concentration of the solids is necessarily the delivered value and can be evaluated directly from heating the sample to constant mass or alternatively, obtained from measurements of the volume and mass of the sample, and the density of the components.

When large samples are removed the measurement of sample volume can be a source of significant error. This can be reduced by using a volumetric tank similar to that used for metering water (see, for example, Hayward, 1979, p. 143). Since it is not always practicable to take a sample to fill the tank to the calibrated capacity a useful practice is to weigh the sample as collected and add liquid of known density to the calibrated volume.

7.3. Gravimetric

Gravimetric or *pipe weighing* techniques involve weighing a section of pipe that has been calibrated with a fluid of known density. The pipe is usually horizontal. Two designs are commonly used; based either on a straight section of pipe, or a U-shaped measuring element. In most applications (see, for example, Okude & Yagi, 1974; Debreczeni *et al.*, 1978; Kao & Hwang, 1979) the weighed section of pipe is isolated using flexible couplings. Instruments that use a U-shaped measuring tube can be sensitive to changes in pipeline pressure and under abrasive conditions the bend is susceptible to erosive wear. Poor alignment of the flexible couplings and anchorage of the fixed pipe work can result in velocity dependence of the indicated mass. These details have been carefully addressed in an instrument described by Kachel (1986) (see also Chapter 23). This instrument uses hydrostatic bearings to isolate the weighed section of pipe. The results of tests carried out in a closed pipe loop using coarse coal and sand–water slurries indicate that measurements were independent of flow rate and well within the limits of resolution of error of the techniques used for comparison ($\pm 0{\cdot}5\%$).

7.4. Counter-Flowmeter

The counter-flowmeter consists of an inverted, vertically mounted, U-shaped pipe, shown schematically in Fig. 11.10. Pressure measurements in the two limbs provide

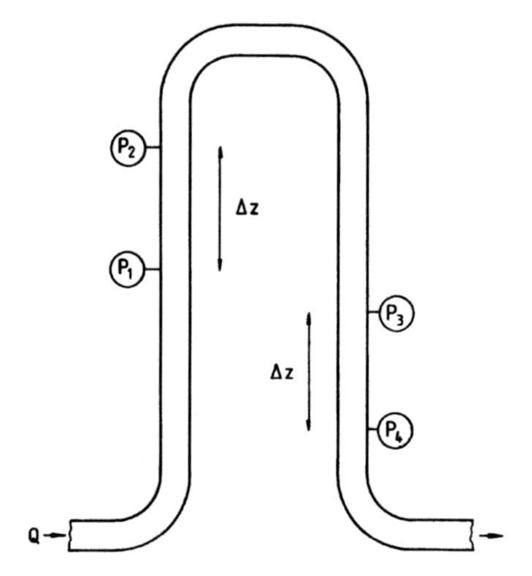

Fig. 11.10. Schematic diagram of a counter-flowmeter.

the potential to use the meter to determine the concentration and flow rate of the flowing slurry. Construction of the device must ensure that the pressure gradient is fully developed at the downstream pressure station in each limb. Little guidance appears to be available to enable the length of approach pipe to be estimated. Simplicity of construction and operation offers considerable attraction for large operations (see, for example, Van der Veen, 1972). However, vertical pipe work is generally to be avoided because of the possibility of blockage and the accumulation of air at the high point could pose problems.

As a result of the settling tendency of the particles the concentration of solids in the two limbs is not equal. The analysis given by Clift and Clift (1981) shows clearly that the average in-situ concentration is obtained from the differential pressure measurements.

$$\begin{aligned} C_{v,i} &= ((C_{v,i})_u + (C_{v,i})_d)/2 \\ &= \left(\frac{(P_1 - P_2) + (P_4 - P_3)}{2\Delta z g(\rho_p - \rho_f)} - \frac{\rho_f}{(\rho_p - \rho_f)}\right) \end{aligned} \tag{11.6}$$

where $C_{v,i}$ is the average fractional in-situ volume concentration of solids in the meter [1]; $(C_{v,i})_d$ is the fractional in-situ volume concentration of solids in the down-flowing limb [1]; $(C_{v,i})_u$ is the fractional in-situ volume concentration of solids in the up-flowing limb [1]; g is the gravitational acceleration [L T^{-2}]; P_n is the pressure measurement at station n (see Fig. 11.10) [M L^{-1}T^{-2}]; Δz is the vertical separation of pressure measurement stations on a limb [L]; ρ_f is the density of fluid [M L^{-3}]; and ρ_p is the density of particles [M L^{-3}].

The authors show that the average in-situ concentration measurement can be used to approximate the delivered concentration. The error in this approximation is shown to increase with particle size and density, but decreases with increasing solids concentration. The difference being of the order 1% for 100 μm particles of sand (density 2 650 kg m^{-3}) at a volumetric concentration of 20%.

The frictional pressure losses in the two limbs is considered to be equal and independent of solids concentration. The usual practice is to assume that the shear stress can be approximated by the value for liquid alone. Clift and Clift (1981) showed that the error in the wall shear stress measured by a counter-flowmeter increases with particle size and density and increases weakly with concentration. For sand particles the error reaches 1% for particles a little larger than 100 μm, and exceeds 10% for 1 mm particles. Although the counter-flowmeter appears to be satisfactory for measurement of concentration, flow measurements appear less satisfactory.

7.5. Radiometric Attenuation

Optical techniques have been used to measure in-situ concentration for dilute slurries where transmission of light is feasible. Outside the laboratory, most *real* slurries are opaque to light and resort has to be made to a more penetrating form of radiation. Gamma-photons emitted from radioactive isotopes such as ^{241}americium, ^{60}cobalt, and ^{137}caesium are commonly used. Commercial devices utilise

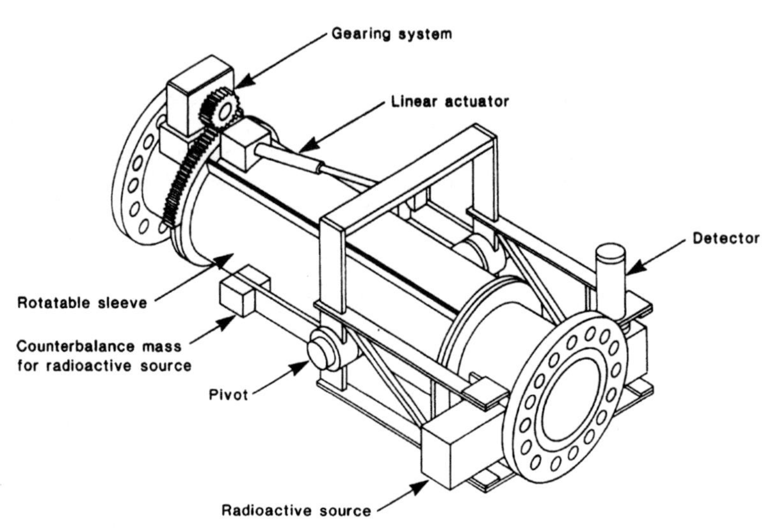

Fig. 11.11. Scanning densitometer. (Reproduced from Brown, 1988, Fig. 3.)

fan-shaped beams to obtain an average in-situ concentration (see Chapter 23 and Wiedenroth, 1979). The equipment shown in Fig. 11.11 uses a highly collimated beam of gamma-photons to measure the solids concentration along a chord through the cross-section. Using the equipment a concentration profile can be obtained either vertically or transversely across the pipe. The device and techniques required to obtain the chord-averaged concentration are fully described elsewhere (Brown, 1988). Chord-averaged measurements can be refined to extract values of local concentration from the spatially averaged values using the technique of tomography. Measurements of local solids concentration made in this manner have been reported by Korbel *et al.* (1976). The scanning equipment used to perform the measurements is described by Przewlocki *et al.* (1979).

The application of two different gamma-emitting isotopes (^{241}americium and ^{137}caesium) in a densitometer is described by Fanger *et al.* (1978). The authors claim that the use of two radioactive sources emitting gamma-photons at two energies enables the densitometer to be less sensitive to changes in the chemical composition of the slurry. Measurements of the absorption at two different energies extends the capability of the attenuation technique to determining the composition of three-phase systems such as manganese-nodules–sea-water–air. The capability is further extended by a third source to enable measurements of velocity to be made by cross-correlating two attenuation signals.

A more complex three-component sensor is described by Verbinski *et al.* (1979). The device uses three separate measurement techniques: gamma-photon attenuation, neutron moderation, and fluid conductivity, to determine the composition of a coal–rock–water slurry. In tests conducted in a pipe loop under conditions where the chemical composition of the solids is known, the composition is reported to be measured to an accuracy of 2%.

7.6. Electromagnetic Flowmeter

Peters and Shook (1981) report that fluctuations in the output voltage in the frequency range between 6 and 30 Hz from a conventional electromagnetic flowmeter are not only dependent upon the velocity of the slurry but also the concentration of particles. The relation between the output voltage signal and concentration was found to be non-linear and dependent on particle and fluid properties.

7.7. Electrical Probes

The most recent development of the electrical probe described for velocity measurements (shown in Fig. 11.9) has been used to measure local in-situ concentration of particles in flowing slurries. Nasr-El-Din *et al.* (1986, 1987) report local concentration measurements in a wide range of slurries. The measurements are independent of local velocity and can be made in conducting pipe work. Fluid temperature and chemical composition do not affect measurements. Calibration is not required for low concentrations of particles that are no larger than the spacing of the electrodes. For larger particles calibration can be readily carried out using a small fluidised bed.

8. DIFFERENTIAL PRESSURE

Differential pressure measurements are usually obtained with fluid-filled transmission lines. The line pressure is either sensed by direct contact with the transmission fluid or isolated by an elastomer or deformable steel diaphragm (isolated measurement techniques are considered in Chapter 23). Direct contact between the transmission fluid and the slurry requires a tapping point to be made through the wall of the pipe. Pressure tappings are usually made at 45° from the vertical (on horizontal pipes) to prevent air from disturbing the measurements and to discourage the ingress of particles. The tapping hole must be burr free, which can often be achieved by careful drilling using a drill re-ground with a larger than normal tip angle. A useful technique that can be used to make the connection to a steel pipe is to weld a pipe nipple to the pipe, make up a ball valve on to the nipple and drill the tapping hole through the open valve. With care this technique can be used to drill a tapping into a *live* pipe.

Traditional measurement of differential pressure has relied on manometers of various configurations connected to the pipe by clear-fluid filled transmission lines. Manometers have offered many advantages including cost, simplicity of construction, and ease of calibration. However, they are suited only to low-pressure application and can become difficult to read with time as the indicating fluid becomes hydrolysed at the interface with the transmission fluid. Differential pressure transducers have become a natural replacement for the manometer. Although there is significantly less movement of fluid in the transmission lines than with a manometer, solids and air can tend to migrate into the lines. Various techniques have been proposed to mitigate the problems including a continuous purge of clear fluid into the pipeline to discourage the ingress of solids and air, and the use of catchpots and filters. A continuous purge of fluid tends to be less satisfactory as it is difficult to maintain a balanced flow to each side of the sensing device over a long period. The usual procedure for making pressure measurements where the transmission fluid is in direct contact with the slurry involves isolation of the transducer and purging the transmission lines with fluid. The inclusion of a flow sensor in the purge supply line is a useful means of detecting blocked or partially blocked tappings.

9. TECHNIQUES APPLICABLE TO COAL

9.1. Introduction

Whenever determinations are made on a mass of coal, complications arise due to the moisture content of the coal and the representativeness of the sample. This is due to the diversity of minerals broadly referred to as coals and both their porous nature and chemical complexity.

Complications also arise with other materials such as sewage sludge, wood fibres, and some food products. Moisture can reside both within the internal structure of the solid and on the exterior surface. All moisture can be removed by application of heat under controlled conditions; however, excessive heating can cause volatiles to be liberated and also promote oxidation. The coal industry has long recognised the complexity of performing measurements on coal and test procedures are detailed by standards organisations including the British Standards Institution (BSI) and the American Society for Testing and Materials (ASTM).

9.2. Moisture Content

The total moisture associated with a sample of coal is frequently measured in two steps. The sample is first air dried to remove the free moisture, making the sample more manageable; the residual moisture can then be driven-off using an oven. The procedure for measuring the free moisture is detailed in BSI (1960). The residual moisture in the air-dried sample can be determined using one of the procedures detailed in BSI (1973). The total moisture is then obtained from application of the following expression (in which the moisture contents are expressed as percentages):

$$\text{TOTAL moisture} = \text{FREE moisture} + \text{RESIDUAL moisture} \times \left(1 - \frac{\text{FREE moisture}}{100}\right) \tag{11.7}$$

ASTM (1982) also provides procedures for determining the total moisture in a coal sample.

In practice, these standardised tests are highly time consuming. For use with stabilised coal slurries (composed of medium-rank coals) two alternative tests were developed for moisture determinations that could be more rapidly executed without compromising results (Brown, 1989). A moisture tester using calcium carbide was used to determine the surface moisture of coal samples that were crushed to pass a 10 mm screen (for a description of the tester and suitable procedure, see, for example, Liu and Evett, 1984). A domestic microwave oven was found to be satisfactory for determining the total moisture content of fine slurries with an average particle size of 50 μm. The technique worked well for coal passing a 10 mm screen. Sample volumes required for larger particle sizes were inappropriate for the capacity of the oven. Larger particles also tended to explode on heating due to rapid expansion of internal water. The total moisture content (or coal concentration) of slurries containing larger sized coal was determined by carefully heating a minimum sample volume of 12 litres to constant mass in a convection oven.

9.3. Density

The density can be readily measured using a pycnometer (see, for example, Liu and Evett, 1984). Accurate and reproducible results with coal have been obtained from determinations performed with a 250 ml graduated flask using 150 g samples with a top size of 200 μm. The density of coal is usually reported on a dry basis, from a sample that has been used for a total moisture determination. Immediately the dry sample starts to cool to ambient temperature the moisture content increases until equilibrium is reached with the environment. In practice, it has been found more convenient to determine the density on samples at equilibrium conditions and correct the density to dry conditions using eqn (11.8). The equilibrium moisture content was expediently measured by carefully drying the finely divided sample in a microwave oven.

$$\rho_d = \frac{\left(1 - \frac{m}{100}\right)}{\left(\frac{1}{\rho_e} - \frac{m}{100\rho_f}\right)} \tag{11.8}$$

where ρ_d is the density of coal on a dry basis [M L^{-3}], ρ_e is the density of coal determined at equilibrium moisture content [M L^{-3}], ρ_f is the density of water [M L^{-3}], and m is the equilibrium moisture content (expressed as a percentage) [1].

$$m = \left(1 - \frac{\text{mass of coal after drying to constant mass}}{\text{mass of coal at equilibrium conditions}}\right) 100$$

10. REFERENCES

ASTM (1982). D3302-82: Test method for total moisture of coal. Vol. 05.05., American Society for Testing and Materials, Philadelphia, PA, USA.

ASTM (1986). D2234-86: Methods for collecting gross sample of coal. Vol. 05.05., American Society for Testing and Materials, Philadelphia, PA, USA.

Bantin, R.A. & Streat, M.S. (1971). Mechanism of hydraulic conveying at high concentration. Paper 15e, presented at Symposium on Solids Transport in Slurries, Atlantic City, NJ, USA, 29 Aug.–1 Sept.

Bantin, R.A. & Streat, M.S. (1972). Mechanism of hydraulic conveying at high concentrations in vertical and horizontal pipes. In *Proc. HT 2*, paper B2, pp. 11–24.*

Beck, M.S. (1981). Correlation in instruments: cross correlation flowmeters. *J. Phys. E., Sci. Instrum.*, **14**.

Beck, M.S., Mendies, P.J., Walecki, T. & Gatland, H.B. (1974). Measurement and control in hydraulic transport systems using cross-correlation measurement systems and fluidic diverters. In *Proc. HT 3*, paper F5, pp. 69–80.*

Briscoe, B.J., Radwan, H. & Streat, M. (1983). Model experiments on sliding friction for application in hydraulic conveying of solids. *Can. J. Chem. Engng*, **61**, 769–75.

Brook, N. (1962). Flow measurement of solid–liquid mixtures using venturi and other flow meters. *Proc. Inst. Mech. Engrs*, **176**(6), 127–40.

Brookes, D.A. & Snoek, P.E. (1986). Stabflow slurry development. In *Proc. HT 10*, paper C3, pp. 89–100.*

Brown, N.P. (1988). An instrument to measure chord-averaged concentration profiles in two-phase pipeline flows. *J. Pipelines*, **7**, 177–89.

Brown, N.P. (1989). Two ways to check moisture content: when speedy results are the goal, these tests replace elaborate standards. *Chem. Engng*, **96**(8), 173–4.

Brown, N.P., Shook, C.A., Peters, J. & Eyre, D. (1983). A probe for point velocities in slurry flows. *Can. J. Chem. Engng*, **61**, 597–602.

BSI (1960). BS 1017: Methods for sampling of coal and coke. Part 2: Sampling of coke. British Standards Institution, London, UK.

BSI (1973). BS 1016: Methods for analysis and testing of coal and coke. Part 1: Total moisture of coal. British Standards Institution, London, UK.

BSI (1975). BS 1377: Methods of test for soils for civil engineering purposes. British Standards Institution, London, UK.

BSI (1980). BS 5857: Methods for measurement of fluid flow in closed conduits, using tracers. Parts 1 to 7. British Standards Institution, London, UK. (Equivalent International Standard—ISO 2975/1-1974: Measurement of water flow in closed conduits—tracer methods.)

BSI (1981). BS 1042: Differential pressure devices. Section 1.1: Specification for square-edged orifice plates, nozzles and venturi tubes inserted in circular cross-section conduits running full. British Standards Institution, London, UK.

Butterfield, R. & Andrawes, K.Z. (1972). On the angles of friction between sand and plane surfaces. *J. Terramechanics*, **8**(4), 15–23.

Clift, R. & Clift, D.H.M. (1981). Continuous measurement of the density of flowing slurries. Int. J. Multiphase Flow, 7(5), 555–61.

Cooke, R. & Lazarus, J.H. (1988). Isokinetic sampling probe for slurry flows. In *Proc. HT 11*, paper C1, pp. 117–30.*

Cornish, D.C., Jepson, G. & Smurthwaite, M.J. (1981). *Sampling Systems for Process Analysers*. Butterworths, London, UK.

Cumalióglu, I.M. (1975). An experimental study of the coefficient of friction between flooded granular particles and solid surfaces. MSc thesis, University of London, UK.

Debreczeni, E., Meggyes, T. & Tarjan, I. (1978). Measurement methods in an experimental rig for hydraulic transport. In *Proc. HT 5*, paper G1.*

Duckworth, R.A., Pullum, L. & Lockyear, C.F. (1983). The hydraulic transport of coarse coal at high concentration. *J. Pipelines*, **3**, 251–65.

Ercolani, D., Ferrini, F. & Arrigoni, V. (1979). Electric and thermic probes for measuring the limit deposit velocity. In *Proc. HT 6*, paper A3, pp. 27–42.*

Eyler, L.L. (1981). Sliding coefficient of friction, experimental apparatus and measurement procedure. Report ESE-81-100, Pacific Northwest Laboratory, USA.

Faddick, R.R., Pouska, G., Connery, J., Di Napoli, L. & Punis, G. (1979). Ultrasonic velocity meter. In *Proc. HT 6*, paper B4, pp. 113–24.*

Fanger, H.U., Michaelis, W., Pepelnik, R. & The, H.L. (1978). Application of gamma-ray absorptiometry in the hydraulic transport of solids. In *Proc. HT 5*, paper G5, pp. 43–58.*

Ferrini, F. & Pareschi, A. (1980). Experimental study of the solid bed profile in sloping pipe sections. In *Proc. HT 7*, paper F3, pp. 245–58.*

Gillies, R.G., Haas, D.B., Small, M.H. & Husband, W.H.W. (1981). Coarse coal in water slurries—full scale pipeline tests. Report no. E 725-7-C-81, Saskatchewan Research Council, Saskatoon, Sask., Canada.

Gillies, R., Husband, W.H.W., Small, M. & Shook, C.A. (1984). Some experimental methods for coarse coal slurries. In *Proc. HT 9*, paper A2, pp. 33–46.*

Gillies, R.G., Small, M.H., Eyre, D. & Shook, C.A. (1988). Slurry concentration measurements using venturi meters. In *Proc. HT 11*, paper C3, pp. 157–72.*

Haas, D.B., Husband, W.H.W. & Shook, C.A. (1977). Hydraulic transport of coarse coal at high concentrations in a 4-inch pipeline loop. *CIM Bull.*, **70**(781), 109–11.

Hayward, A.T.J. (1979). *Flowmeters, a Basic Guide and Source-Book for Users*. The Macmillan Press Ltd, Basingstoke, UK.

Heywood, N.I. & Kirby, J.M. (1988). Laboratory measurement techniques for the bulk properties of wet granular solids. Report LR 543 (MH), Warren Spring Laboratory, Stevenage, UK.

Jacobs, B.E.A. & Tatsis, A. (1986). Measurement of wall shear stresses for high concentration slurries. In *Proc. HT 10*, paper H1, pp. 267–73.*

Kachel, G.C. (1986). The slurry monitor, a new development in the precision determination of the density of slurries flowing in pipelines. Paper presented at Symposium on Slurry Flows and Measurement—Instrumentation, Anaheim, CA, USA, 9 Dec.

Kao, D.T. & Hwang, L.Y. (1979). Critical slope for slurry pipeline transporting coal and other solid particles. In *Proc. HT 6*, paper A5, pp. 57–74.*

Kao, D.T. & Kazanskij, I. (1979). On slurry flow velocity and solid concentration measuring techniques. In *Proc. STA 4*, pp. 102–20.*

Kazanskij, I. (1979). Critical velocity of depositions for fine slurries—new results. In *Proc. HT 6*, paper A4, pp. 43–56.*

Korbel, W., Michalik, A., Przewlocki, K., Parzonka, W. & Wolanski, Z. (1976). Determination of the polyfractional solids distribution in a pipe. In *Proc. HT 4*, paper A4, pp. 51–61.*

Lawler, H.L., Pertuit, P., Tennant, J.D. & Cowper, N.T. (1978). Application of stabilised slurry concepts of pipeline transportation of large particle coal. In *Proc. STA 3*, pp. 164–78.*

Lazarus, M.D. & Lazarus, J.H. (1988). Development of an ultrasonic Doppler bed load velocimeter. In *Proc. HT 11*, paper C4, pp. 173–85.*

Liu, C. & Evett, J.B. (1984). *Soil Properties, Testing Measurement and Evaluation*. Prentice-Hall, Englewood Cliffs, NJ, USA.

Nasr-El-Din, H. & Shook, C.A. (1985). Sampling from slurry pipelines: thick-walled and straight probes. *J. Pipelines*, **5**, 113–24.

Nasr-El-Din, H., Shook, C.A. & Esmail, M.N. (1984). Isokinetic probe sampling from slurry pipelines. *Can. J. Chem. Engng*, **62**, 179–85.

Nasr-El-Din, H., Shook, C.A. & Colwell, J. (1986). Determination of solids concentration in slurries. In *Proc. HT 10*, paper F1, pp. 191–8.*

Nasr-El-Din, H., Shook, C.A. & Colwell, J. (1987). A conductivity probe for measuring local concentrations in slurry systems. *Int. J. Multiphase Flow*, **13**(3), 365–78.

Okude, T. & Yagi, T. (1974). The spatial solid concentration and the critical velocity. In *Proc. HT 3*, paper E2, pp. 15–27.*

Peters, J. & Shook, C.A. (1981). Electromagnetic sensing of slurry concentration. *Can. J. Chem. Engng*, **59**, 430–7.

Przewlocki, K., Michalik, A., Korbel, K., Wolski, K., Parzonka, W., Sobota, J. & Pac-Pomarnacka, A. (1979). A radiometric device for the determination of solids concentration distribution in a pipeline. In *Proc. HT 6*, paper B3, pp. 105–12.*

Rowe, P.W. (1962). The stress–dilatancy relation for static equilibrium of an assembly of particles in contact. *Proc. Royal Soc.*, **A269**, 500–27.

Scarlett, B. & Grimley, A. (1974). Particle velocity and concentration profiles during hydraulic transport in a circular pipe. In *Proc. HT 3*, paper D3, pp. 24–37.*

Shook, C.A. (1982). Flow of stratified slurries through horizontal venturi meters. *Can. J. Chem. Engng*, **60**, 342–5.

Shook, C.A. & Masliyah, J.H. (1974). Flows of a slurry through a venturi meter. *Can. J. Chem. Engng*, **52**, 228–33.

Shook, C.A., Haas, D.B., Husband, W.H.W., Small, M. & Gillies, R. (1981). Pipeline flow of coarse coal slurries. *J. Pipelines*, **1**, 83–92.

Shook, C.A., Gillies, R., Husband, W.H.W., Small, M. (1982*a*). Pipe slope and roughness effects for fine sand slurries. *Proc. Amer. Soc. Mech. Eng.* (ASME, Materials Handling Div.), 1–8.

Shook, C.A., Gillies, R., Haas, D., Husband, W.H.W. & Small, M. (1982*b*). Flow of coarse and fine sand slurries in pipelines. *J. Pipelines*, **3**, 13–21.

Shook, C.A., Gillies, R., Haas, D., Husband, W.H. & Small, M. (1982*c*). Flow of coarse coal in 150, 250 and 500 mm pipelines. In *Coal: Phoenix of the 80s* (Proc. 64th CIC Coal Symp.), ed. A.M. Al Taweel. CSChE, Ottawa, Canada, pp. 262–7.

Van der Veen, R. (1972). The development of an accurate concentration meter for two-phase flow. In *Proc. HT 2*, paper J2, pp. 13–20.*

Verbinski, V.V., Cassapakis, C.G., de Lesdernier, D.L. & Wang, R. (1979). Three-component coal slurry sensor for coal–rock–water concentrations in underground mining operations. In *Proc. HT 6*, paper C4, pp. 161–8.*

Wiedenroth, W. (1979). Methods for the determination of the transport concentration and some problems associated with the use of radiometric density meters. In *Proc. HT 6*, paper B2, pp. 89–104.

Wilson, K.C. (1970). Slip point of beds in solid–liquid pipeline flow. *Proc. ASCE* (*J. Hydraulics Div.*), **96**, HY1–12.

Wilson, K.C. (1972). A formula for the velocity to initiate particle suspension in pipeline flow. In *Proc. HT 2*, paper E2, pp. 23–36.*

*Full details of the Hydrotransport (HT) and the Slurry Transport Association (STA) series of conferences can be found on pp. 11–13 of Chapter 1.

12

Particle Size Degradation in Slurries

Randall G. Gillies
Saskatchewan Research Council, Saskatoon, Canada

1. INTRODUCTION

Particle size is perhaps the most important variable to be considered in the operation of slurry pipelines. Frictional pressure gradients, solids deposition velocities and slurry dewatering characteristics are known to be strongly dependent on particle size. Particle size reduction occurs as a result of collisions with other particles and collisions with pump and piping surfaces. Pipeline operating costs and the quality of the material delivered by the pipeline may be affected by particle degradation processes that occur during hydraulic transport.

Particle degradation is of great importance during hydraulic transport of coal. When transporting raw coal slurries, degradation of the coal particles and pulverisation of the associated rock material will have adverse effects on washery plant performance. For long distance transport of conventional coal slurries, the separation costs are increased by excessive particle size degradation. In the case of highly concentrated coal–water fuels, particle–particle shear induced degradation alters the rheological properties of the mixture with resulting increases in the pipeline pressure gradient and the reader should refer to the work of Ercolani *et al.* (1988) for a detailed discussion of this particular type of coal slurry degradation. This chapter focuses attention on particle size reduction resulting from particle collisions with equipment surfaces.

While the majority of degradation studies have been concerned with coal hydrotransport, particle degradation is also of concern in other operations such as the following:

- pipeline transportation of salt crystals,
- hydrotransport of overburden from open pit mining operations,
- pipeline transportation of tar sand to synthetic crude oil extraction plants, and
- the release of clays during pipeline transportation of mineral ore and tailings materials.

At present, the principles governing particle size degradation are only beginning to be understood. It is known that two forms of degradation, particle breakage and surface abrasion, are important for slurry pipeline applications. Hard, brittle solids tend to break while softer materials such as lignitic coals tend to degrade through abrasion. For some mixtures, such as clay lumps in water, surface degradation is accelerated by the penetration of the carrier fluid into the particles. The particle degradation rate for a slurry can be considered to be the change in particle size distribution (PSD) with time. This is expressed mathematically later in this chapter (eqn (12.1)). It is often impossible to predict particle degradation rates with confidence. If degradation is considered to be important for a particular design situation, then it is very likely that tests will have to be conducted to obtain reliable estimates. The purpose of this chapter is to provide some insight into the factors which affect degradation, and hence, the factors which must be considered when planning a degradation study.

2. DEGRADATION TESTS

Particle degradation data are obtained by monitoring changes in the PSD of the solids flowing in a pipeline. The test equipment may be a once-through pipeline system or a recirculating flow loop. In the case of a flow loop, samples are obtained at various times as the material is pumped around the test rig. In a once-through system, time effects are studied by removing samples at various points along the

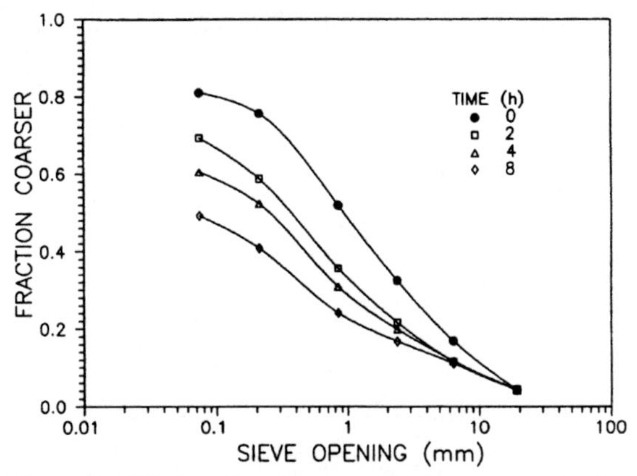

Fig. 12.1. Change in PSD for a bituminous coal during recirculation in a 150 mm flow loop. (Adapted from Gillies *et al.*, 1981.)

pipeline. (Slurry sampling techniques are discussed in Chapter 11.) The PSDs of the samples are usually determined by using a series of sieves. The size distribution of the finer particles may be determined by using an electronic particle size analyzer or a sedimentation pipette.

Recirculating flow loops (such as those discussed in Chapter 10) are often used instead of once-through systems to generate degradation data because flow loops require relatively small amounts of slurry sample. Furthermore, flow loops usually provide greater ranges of test conditions than once-through systems. The main disadvantage is that there is a large number of pump passes per unit length of pipeline travel in a typical flow loop.

Short once-through pipeline systems are sometimes used in attempts to overcome the disadvantages of recirculating flow loops. Unfortunately, the amount of degradation may be overestimated because only the initial rate of degradation is obtained from a test in a short once-through pipeline. The initial rate may be high due to breakage along existing fractures and the loss of rough edges from the particles.

Figures 12.1 and 12.2 show changes in the PSD of a coal sample during prolonged recirculation in flow loops. In the study by Gillies *et al.* (1981), 150 and 500 mm diameter flow loops were used to study the degradation of a western Canadian bituminous coal. In each case, slurry velocity was $3\,m\,s^{-1}$. Centrifugal pumps were used to circulate the slurries. The particles passed through the pump on the smaller loop approximately 120 times per hour. In the larger loop, the particles experienced 80 pump passes per hour. From a comparison of the two figures, it is apparent that particle degradation is occurring at a considerably higher rate in the smaller flow loop. However, it is not readily apparent whether the degradation process is due to particle breakage or surface attrition. The model which is described in the following section is useful for examining the degradation process and for quantifying the rates.

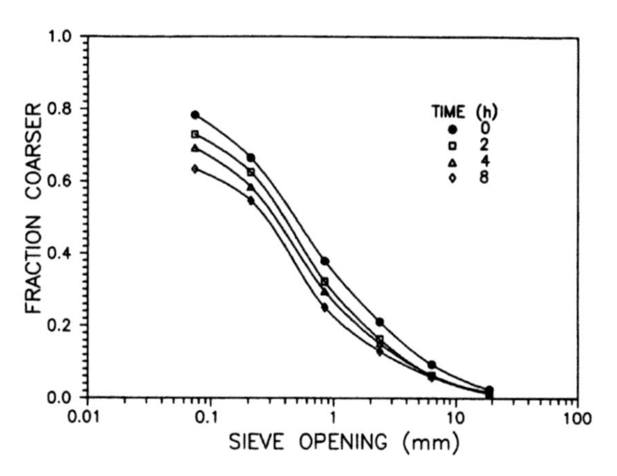

Fig. 12.2. Change in PSD for a bituminous coal during recirculation in a 500 mm flow loop. (Adapted from Gillies *et al.*, 1981.)

3. MODELLING OF DEGRADATION TEST RESULTS

3.1. Milling Model

Karabelas (1976) studied degradation by shearing coal particles in a concentric cylinder viscometer; no settling occurred as the fluid used had the same density as coal. Karabelas determined that the degradation rates increased with increasing particle size and that the degradation process could be described by the equations which are commonly used to model rod and ball-mill grinding. These equations have been summarised by Austin and Luckie (1971/72), (see also Chapter 25 of this handbook). Shook *et al.* (1978, 1979) and Gillies *et al.* (1981, 1982) showed that Karabelas' approach could be used to model particle degradation which takes place in slurry pipelines.

The mass fraction of solids, w_i retained on a screen i (with opening d_i) and passing screen $i - 1$ (with opening $d_i\ 2^{0.5}$) varies with time as follows:

$$\frac{\mathrm{d}w_i}{\mathrm{d}t} = \sum_{j=1}^{i-1} S_j b_{i,j} w_j - S_i w_i \tag{12.1}$$

where b is the particle breakage distribution function and S is the selection rate parameter. This equation implies that the rate at which particles from sieve interval j enter the smaller sieve interval i depends on the amount of solids in sieve interval j, the rate at which particles in interval j degrade and the fraction of particles from interval j that fall into interval i upon breakage ($b_{i,j}$). The rate at which solids are lost from interval i depends on the fraction of the solids which are in interval i and the rate at which particles in interval i degrade.

The breakage distribution function is usually assumed to be independent of

TABLE 12.1
Summary of Breakage Distribution Coefficients for Pipeline Degradation Tests

Source of degradation data	*Breakage distribution coefficient, $b_{i,1}$*					
	i = 2	*i = 3*	*i = 4*	*i = 5*	*i = 6*	*i = 7*
A[a]	0·5	0·1	0·0	0·0	0·0	0·0
B[b]	0·6	0·0	0·0	0·0	0·0	0·0
C[c]	0·45	0·15	0·05	0·0	0·0	0·0
D[d]	0·5	0·2	0·1	0·05	0·05	0·0
E[e]	0·5	0·2	0·1	0·1	0·05	0·0

[a] A—Canadian lignite, recirculation in a 50 mm diameter pipeline flow loop, Shook *et al.* (1978).
[b] B—Canadian lignite and sub-bituminous coals, recirculation in a 100 mm diameter pipeline flow loop, Shook *et al.* (1979).
[c] C—Western Canadian bituminous coal, recirculation in 150 and 500 mm diameter pipeline flow loops, Gillies *et al.* (1981).
[d] D—Western Canadian bituminous coal, single pass through a 400 mm centrifugal pump, Gillies *et al.* (1982).
[e] E—Pittsburgh bituminous coal, recirculation and one-pass tests in a 150 mm pipeline, Pipilen *et al.* (1966).

particle size such that

$$b_{i,j} = b_{i-j+1,1} \qquad (12.2)$$

thus reducing the number of parameters that must be specified to complete the modelling exercise.

An iterative approach is used to obtain values for S and b from experimental data of the type shown in Figs 12.1 and 12.2 where the PSD is measured as a function of time (or as a function of number of pump passes). According to Shook *et al.* (1979), the most satisfactory values for S and b are obtained from tests which use narrowly sized solids as the starting material.

3.2. Breakage Distribution Function

Shook *et al.* (1979) showed that if surface abrasion is the primary degradation process, then $b_{2,1}$ should have a value of approximately 0·6 and all other values of $b_{i,1}$ should be near zero. As material is removed from the surface, some particles fall into the next smaller sieve interval and the fines fall through all the sieves. Therefore, only $b_{2,1}$ has a non-zero value. However, if the particles are brittle and produce a range of small fragments on breakage then the breakage distribution matrix would be expected to have a broad range of elements with non-zero values.

Table 12.1 gives the breakage distribution coefficients, $b_{i,1}$ which have been found to fit degradation data for various types of coals. For the bituminous coal samples of Figs 12.1 and 12.2, both degradation mechanisms appear to be important and the breakage distribution coefficients given in row C of Table 12.1 may be used to correlate the data.

TABLE 12.2
Selection Rate Correlating Parameters for Degradation of Western Canadian Bituminous Coal Slurries in 150 and 500 mm Flow Loops

	Pipe diameter	
	150 mm	*500 mm*
k_1 (min^{-1})	0·0054	0·0027
k_2 (min^{-1})	0·003	0·0015
d_m (mm)	0·21	0·84

3.3. Selection Rate Parameter

The selection rate parameter depends on the nature of the particles and also on the equipment in which the degradation occurs. For the bituminous coals of Figs 12.1 and 12.2, the selection rate parameters are time dependent and particle-size dependent. The parameters decrease approximately exponentially with time and also decrease with decreasing particle size.

The following empirical expression may be used to correlate the particle degradation data shown in Figs 12.1 and 12.2:

$$S_i = k_1 f(d_i) \exp(k_2 t) \tag{12.3}$$

with t measured in minutes and d_i in mm, and where

$$f(d_i) = \frac{d_i}{d_m} \quad \text{for } d_i < d_m \tag{12.4}$$

and

$$f(d_i) = 1 \quad \text{for } d_i > d_m \tag{12.5}$$

The values for the correlating parameters k_1, k_2 and d_m are given in Table 12.2. The differences in the two sets of parameters show that the degradation process occurs at a higher rate in the smaller flow loop.

By using the milling model, it is possible to select correlating parameters that describe the effects of time and PSD for one degradation test. However, it is much more difficult to determine how factors such as variation in solids properties, changes in pump design and operation and changes in piping configuration affect the degradation rate. These factors are considered in a qualitative manner in the following section.

4. FACTORS AFFECTING DEGRADATION

4.1. Nature of Solids

Degradation rates are highly dependent on the nature of the solids which are being transported. In a German coal degradation study using a 200 mm flow loop, Klose

(1984) reported that the degradation rates are strongly influenced by coal particle composition.

In a study of the feasibility of using hydraulic transport systems for underground stowage of German coal washery tailings, Geller and Gies (1982) showed that the degradation rates were very sensitive to the composition of the tailings. Degradation rates were very high for tailings with high shale and low sandstone contents.

Using a 100 mm diameter flow loop, Shook *et al.* (1979) showed that there was considerable variation in both the nature and rates of degradation for different samples of western Canadian coals. Table 12.1 shows the variation in the breakage distribution coefficients which have been found to fit degradation data for various coals. The nature of the degradation process for a particular solids material can be inferred by examining these coefficients. For lignitic and sub-bituminous coals (rows A and B of Table 12.1), surface abrasion is the dominant degradation mechanism and the coefficients are zero or near zero for $i > 2$. Bituminous coals (rows C, D and E of Table 12.1) gave a wider range of non-zero breakage distribution coefficients. This indicates that particle breakage is important for these more brittle coals.

Mikhail *et al.* (1982) were successful in correlating the amount of coal micro-fissuration (as measured by methane desorption rate tests) with pipeline flow loop degradation rates. The results indicated a linear relationship between the coal micro-fissuration index and the amount of fines generated during pipeline attrition tests.

4.2. Particle Size Effects

Because of inertial effects, large particles undergo larger impact forces than smaller particles during contact with pump parts and pipe bends. As a result, particle degradation rates are higher for coarse particles than for finer particles. The experiments by Shook *et al.* (1978, 1979) and Gillies *et al.* (1981) showed clearly that the particle breakage selection rates increase with increasing particle size. This particle size dependence is verified in the flow loop experiments described by Klose (1984) and Sold (1982).

4.3. Solids Concentration

Shook *et al.* (1979) found that there was no significant correlation between particle degradation rate and solids concentration for coal slurries. This conclusion was reached after performing tests in a 100 mm diameter by 100 m long flow loop. A centrifugal pump was used to circulate the slurries. A lignite–water slurry was tested at concentrations of 25, 30, 35 and 40% solids (by mass) and a sub-bituminous coal slurry was tested at 30, 40 and 50% solids (by mass). The maximum particle size was 19 mm.

Other workers have reported increased *cushioning* of the coarse particles by fine particles as the solids concentration increases. However, the cushioning effect does not appear to be significant at normal slurry pipeline concentrations.

TABLE 12.3
Particle Degradation Selection Rate Constants for Western Canadian Coal Slurries during Recirculation in Pipeline Flow Loops with Centrifugal Pumps. The Degradation Rate Constants are for Particles Coarser than 1 mm

Pipeline diameter (mm)	*Coal type*	*Pump impeller tip speed* ($m\,s^{-1}$)	*Initial selection rate,* S_i (min^{-1})	*Time per pump pass,* t (*min*)	$S_i t$
50	Lignite	13	0·015	0·11	0·0016
100	Lignite	18	0·0058	0·47	0·0027
500	Bituminous	10	0·0027	0·77	0·0021
150	Bituminous	12	0·0054	0·51	0·0027
100	Bituminous	18	0·0077 to 0·014	0·47	0·0036 to 0·007

4.4. Effects of Time

Particle breakage rates are usually highly time dependent with rapid initial breakage along flaws in the particle structure. Abrasion rates are also usually high initially as sharp edges and corners are removed. Lower abrasion rates occur after the particles have been worn to a *rounded* shape.

The time dependence of coal degradation are shown in several studies including those by Shook *et al.* (1978, 1979), Gillies *et al.* (1981) and Geller and Gies (1982).

4.5. Centrifugal Pump Effects

The work by Shook *et al.* (1978) using a flow loop with two different lengths of pipe showed that most of the degradation occurred in the centrifugal pump with much less degradation occurring in the piping.

Particle degradation rates appear to be highly dependent on the pump impeller tip speed (tangential velocity of the outer edge of the impeller). Pipilen *et al.* (1966) found that the degradation rate varied with the pump speed to the power 2·5. Table 12.3 shows that the degradation rates for western Canadian coals as determined by Shook *et al.* (1978, 1979) and Gillies *et al.* (1981) were also very sensitive to the pump impeller tip speed. There is a strong correlation between the tip speed and the amount of degradation per pump pass (as indicated by the initial selection rate parameter multiplied by the time interval between pump passes). The fact that the pump impeller tip speed was higher for the 150 mm flow loop than for the 500 mm flow loop explains, at least partially, the differences in the degradation rates of Figs 12.1 and 12.2.

4.6. Pipeline Velocity

The tests by Shook *et al.* (1978) showed that the degradation rates in piping and bends are small in comparison to those in centrifugal pumps. It seems reasonable to assume that degradation rates in bends will depend on particle inertia.

Therefore, the rate should be strongly dependent on velocity. In straight pipe sections, the effects of velocity are not well understood. Geller and Gies (1982) found that the degradation rate actually decreased as the velocity increased in straight sections of horizontal pipe carrying coal washery tailings slurries.

While the direct effects of pipeline velocity are small the indirect effects are very important, particularly if centrifugal pumps are used. A high pump speed will be required if a pipeline is designed to operate at a high velocity. The overall degradation in the pipeline system can be minimized by operating the pipeline near the minimum velocity required to prevent deposition of solids.

5. DESIGN CONSIDERATIONS

5.1. Pump Selection

Centrifugal pumps have been identified as major sources of degradation in slurry pipeline systems. If particle degradation is to be minimised, then the pipeline designer may want to consider alternative means for producing the pressure required to maintain flow in the slurry pipeline system. Brookes and Snoek (1986) showed that reciprocating piston pumps could be used to pump slurries which contained coarse coal particles. Pumps originally designed for concrete were adapted for coarse coal and used to propel stabilised (Stabflow) slurries.

In any pumping system, whether centrifugal pumps or other pressure generating devices are used, steps which reduce overall energy consumption will also reduce degradation. Therefore, the maximum pipe diameter should be selected which still allows operation at a velocity exceeding the solids deposition velocity.

5.2. Carrier Fluid Modification

The work on stabilised slurries, reported by Brookes and Snoek (1986), showed that solids deposition velocities could be reduced to zero by using a fine coal slurry as a *carrier fluid* for coarse coal particles. This allows the use of larger diameter pipes thus reducing pipeline velocities and energy consumption rates. Presumably, this would also reduce particle degradation rates.

Further work is required to assess whether fine particles, which are present in high concentrations in stabilised slurries, provide a *cushioning* effect to reduce degradation of coarse particles. Shook *et al.* (1979) found that increasing the fines concentration did not help to *cushion* the coarser particles. However, the concentrations of fines were relatively low in the experiments of Shook *et al.* and it is possible that degradation would be reduced if the fines concentration was high enough to alter the slurry's rheological properties.

5.3. Using Degradation Test Results

The data required to determine the breakage distribution and selection rate parameters for a particular slurry application may be obtained from flow loop tests.

The selection rate parameters are highly dependent on the characteristics of the pipeline equipment, and the flow loop tests must be designed carefully to ensure that the results are useful. The tests should provide quantitative information about the effects of pump size and speed, number of pumps, pipe length and mixture flow rate on the selection rate parameters for the particular slurry application.

The effects of equipment size are poorly understood at present. For this reason, the pipeline designer should seek degradation data from full-scale equipment. It is especially important that centrifugal pump degradation data be obtained using equipment similar to the actual pipeline system. The pump impeller tip speed is an especially important parameter for degradation studies.

In practice, it is difficult to obtain independent pump and pipe selection rate parameters. Historically, centrifugal pumps have been used for flow loop studies of particle degradation. It is very difficult to design experiments which isolate the effects of the centrifugal pump from that of the piping. For example, if the flow loop piping is lengthened, then the pump speed will have to increase to maintain a constant flow rate. The degradation rate is highly dependent on pump speed, and therefore the effects of the additional piping will be difficult to separate from the effects of the increased pump speed. The use of reciprocating piston pumps might be considered in future degradation studies. Alternatively, rotating toroidal wheels can be used to simulate pipeline flow and to measure the degradation which is attributable to the piping. Worster and Denny (1954) used this type of apparatus to investigate the particle size degradation of coal.

After suitable degradation rate data are obtained, eqn (12.1) may be used to estimate the PSD of the mixture delivered by the pipeline for various design alternatives. The selection rate parameters are likely to be time dependent and therefore a numerical method of integration (i.e. fourth order Runge–Kutta) will be required to obtain estimates of the PSD from eqn (12.1).

6. NOTATION

b Breakage distribution parameter [1]
d_i Diameter of particles in screen interval i [L]
d_m Correlating parameter [L]
k_1 Correlating parameter [T^{-1}]
k_2 Correlating parameter [T^{-1}]
S Selection rate parameter [T^{-1}]
t Time [T]
w_i Mass fraction of solids [1]
w_j Mass fraction of particles retained between screens j and $j-1$ [1].

7. REFERENCES

Austin, L.G. & Luckie, P.T. (1971/72). Methods for determining breakage distribution parameters. *Powder Technol.*, **5**, 215–22.

Brookes, D.A. & Snoek, P.E. (1986). Stabflow slurry development. In *Proc. HT 10*, paper C3, pp. 89–100.*

Ercolani, D., Carniani, E., Meli, S., Pellegrini, L. & Primicerio, M. (1988). Shear degradation of concentrated coal-water slurries in pipeline flow. In *Proc. STA 13*, pp. 159–79.*

Geller, F.J. & Gies, R. (1982). The influence of degradation in the hydraulic conveyance of coarse washed dirt on the dimensioning of dewatering installations. *J. Pipelines*, **2**, 157–67.

Gillies, R.G., Haas, D.B., Small, M.H. & Husband, W.H.W. (1981). Coarse coal in water slurries—full scale pipeline tests. Report no. E 725-7-C-81, Saskatchewan Research Council, Saskatoon, Canada.

Gillies, R., Haas, D.B., Husband, W.H.W., Small, M. & Shook, C.A. (1982). A system to determine single pass particle degradation by pumps. In *Proc. HT 8*, paper J1, pp. 415–32.*

Karabelas, A.J. (1976). Particle attrition in shearflow of concentrated slurries. *AIChEJ* **22**, 765–71.

Klose, R.B. (1984). Changes in the particle size distribution (PSD) of pipelined coal suspensions. In *Proc. HT9*, paper C3, pp. 111–28.*

Mikhail, M.W., Mikula, R.J. & Husband, W.H.W. (1982). Prediction of size degradation of coarse coal during slurry pipeline transportation. In *Proc. STA 7*, pp. 399–404.*

Pipilen, A.P., Weintraub, M. & Orning, A.A. (1966). Hydraulic transport of coal. Report 6743, Bureau of Mines, Dept. of the Interior, USA.

Shook, C.A., Haas, D.B., Husband, W.H.W. & Small, M. (1978). Breakage rates of lignite particles during hydraulic transport. *Can. J. Chem. Engng*, **56**, 448–54.

Shook, C.A., Haas, D.B., Husband, W.H.W. & Small, M. (1979). Degradation of coarse coal particles during hydraulic transport. In *Proc. HT 6*, paper C1, pp. 125–36.*

Sold, W. (1982). Particle degradation of ROM coal and washery tailings up to 63 mm size with horizontal hydrotransport. *Bulk Solids Handling*, **2**, 261–5.

Worster, R.C. & Denny, D.F. (1954). Transport of solids in pipes. Report SP496, BHRA, Cranfield, UK.

*Full details of the Hydrotransport (HT) and the Slurry Transport Association (STA) series of conferences can be found on pp. 11–13 of Chapter 1.

13

Classification of Slurry Pumps

N. I. Heywood
Warren Spring Laboratory, Stevenage, UK

1. INTRODUCTION

A very wide range of slurry pumps is available but they can be broadly classified into two main categories: centrifugal and displacement (Fig. 13.1). The main difference between the two categories is that in the latter, discrete volumes of slurry are isolated between the moving and stationary parts in cavities in the rotor, screw, cylinder or diaphragm, with a seal or fine clearance between the moving

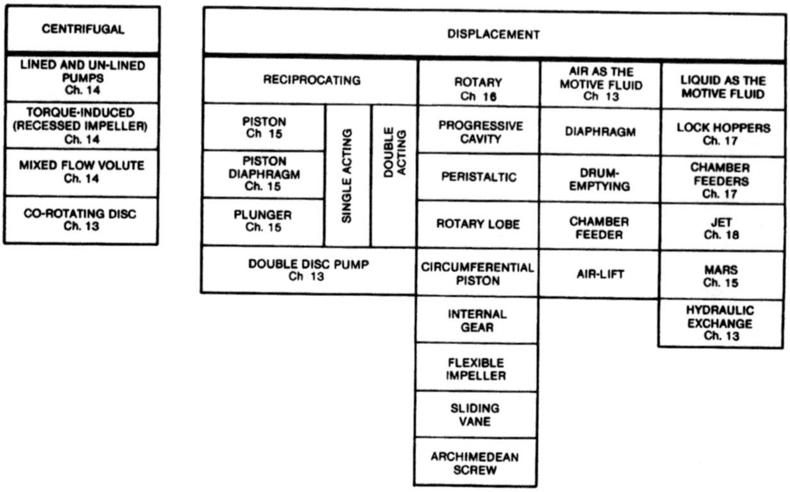

CENTRIFUGAL

LINED AND UN-LINED PUMPS Ch. 14

TORQUE-INDUCED (RECESSED IMPELLER) Ch. 14

MIXED FLOW VOLUTE Ch. 14

CO-ROTATING DISC Ch. 13

DISPLACEMENT

RECIPROCATING

PISTON Ch 15

PISTON DIAPHRAGM Ch. 15

PLUNGER Ch. 15

SINGLE ACTING

DOUBLE ACTING

DOUBLE DISC PUMP Ch 13

ROTARY Ch 16

PROGRESSIVE CAVITY

PERISTALTIC

ROTARY LOBE

CIRCUMFERENTIAL PISTON

INTERNAL GEAR

FLEXIBLE IMPELLER

SLIDING VANE

ARCHIMEDEAN SCREW

AIR AS THE MOTIVE FLUID Ch 13

DIAPHRAGM

DRUM-EMPTYING

CHAMBER FEEDER

AIR-LIFT

LIQUID AS THE MOTIVE FLUID

LOCK HOPPERS Ch. 17

CHAMBER FEEDERS Ch. 17

JET Ch. 18

MARS Ch. 15

HYDRAULIC EXCHANGE Ch. 13

Fig. 13.1. A general classification of slurry pumps.

component and the pump casing. These cavities are moved from the suction to the discharge branch by direct mechanical action. Thus, the flow path is not continuous as it is with the centrifugal pump and this results in a pulsating flow.

The next five chapters provide detailed information on most of the important designs of slurry pump. Chapter 14 covers centrifugal pumps while Chapters 15–18 focus on important categories of displacement pump. The purpose of this chapter is to give an overall introduction to slurry pumps and to provide some information on some pump types not covered in the following chapters.

Several useful reviews are available. Odrowaz-Pieniazek (1979, 1982) and Odrowaz-Pieniazek and Steele (1986) have considered most pump types for in-plant transfer, long distance applications or for dredging and hoisting, Willis and Truscott (1978) undertook a survey of commercially available designs, while Thompson *et al.* (1973) considered both short and long distance transport in the mining industry. More general descriptions of available designs are provided by Karassik *et al.* (1976), Pollack (1980), and Warring (1984*a*, *b*). Other publications have addressed pump selection (Neerken, 1978; Walker & Goulas, 1981; Vanderpan, 1982; Davidson, 1986). Jones (1980) has compared four different pump designs for slurry filtration applications.

2. PUMP PERFORMANCE AND SUCTION LIMITATIONS

2.1. Pump Performance

The (power) efficiency of a pump, E, is the ratio of the power transferred to the liquid, W_h, by the pump (and often referred to as hydraulic power) to the power supplied to the pump. The power supplied is that which would be measured by a brake dynamometer (hence, brake horsepower) or from measurements of pump-shaft speed and torque measured in the shaft between the motor (or other prime mover) and the pump.

$$E = W_h/(2\pi NT) \quad (13.1)$$

where T is the torque in the pump shaft [M L^{2}T^{-2}], N is the speed of rotation of pump shaft (rev s^{-1}) [T^{-1}], and W_h is the power transferred to the liquid by the pump [M L^{2}T^{-3}].

$$W_h = \Delta h \rho_m g Q \quad (13.2)$$

where ρ_m is the slurry density [M L^{-3}], g is the gravitational acceleration [L T^{-2}], Q is the volumetric slurry flow rate [L^{3}T^{-1}], and Δh is the total dynamic head of the pump (m water) [L].

$$\Delta h = (h_d)_t - (h_s)_t \quad (13.3)$$

where $(h_d)_t$ is the total head measured at the discharge flange (m water) [L], and $(h_s)_t$ is the total head measured at the suction flange (m water) [L].

The total head on the suction side, $(h_s)_t$, of an *installed* pump is obtained from measurements on the operating pump:

$$(h_s)_t = (h_s)_g + h_a + (h_s)_v \quad (13.4)$$

where $(h_s)_t$ is the total head on the suction side (m water) [L], $(h_s)_g$ is the reading of a gauge on the suction side (m water) [L], h_a is the atmospheric pressure (m water) [L], and $(h_s)_v$ is the velocity head at the location of the gauge (m water) [L].

$$(h_s)_v = (Q/(\pi D^2/4))^2/(2g) \quad (13.5)$$

where D is the inside diameter of the pipe at the location of the gauge [L].

At the *design stage*, before installation, the total head on the suction side can be estimated from

$$(h_s)_t = (h_s)_s - (h_s)_f \quad (13.6)$$

where $(h_s)_s$ is the static head on the suction side (m water) [L], and $(h_s)_f$ is the frictional head loss in the pipework on the suction side (m water) [L].

All heads must be measured from, or corrected to, a datum at the same elevation. It is convenient to use the centre of the shaft for a horizontal pump or the eye of a vertical pump. The above expressions are equally applicable to estimating the head on the discharge side. Techniques to estimate static head are considered in Chapter 5. Techniques for frictional head are introduced in Chapter 5 and covered in detail in Chapter 6 for settling slurries and Chapter 7 for non-settling slurries.

2.2. Suction Limitations

Every pump requires a minimum suction head to operate satisfactorily, referred to as the required net positive suction head $(NPSH)_R$. Manufacturers usually publish this information in graphical form as it depends on pump and impeller geometry, volumetric throughput and speed. For correct operation of the pump it is essential that the net positive suction head available $(NPSH)_A$ from the system is equal to, or greater than, the $(NPSH)_R$. If the $(NPSH)_A$ is inadequate the absolute pressure in the pump will fall below the vapour pressure of the liquid in the slurry at the operating temperature and vapour bubbles will appear. Damage to the pump results from the energy released as the vapour bubbles collapse rather than (as commonly thought) during cavitation when they form.

The consequences of cavitation tend to be most important for centrifugal pumps and are readily seen when non-elastic metals are used (e.g. grey cast iron). General pitting and often severe erosion of the impeller vanes can result. Unfortunately, cavitation is quite common in slurry pumps but because materials of construction can have sufficient ductility, the characteristic traces of damage do not always appear on the wetted parts of the pump. Instead, there are signs of secondary effects such as damage to keys, loosening of rotors, shaft breakage resulting from fatigue and problems with stuffing boxes and mechanical packings. An additional early indication of cavitation is the presence of air in the discharge slurry when the inlet stream is bubble-free.

Sometimes cavitation occurs because the pump speed is too high in an attempt to maintain a high flow rate and prevent particles settling out in the pipe work. It is quite possible (Crisswell, 1983) that the suction pressure in the eye of the impeller is below the $(NPSH)_R$. When such low pressure conditions are transferred to the volute in rubber-lined pumps, loose liners collapse and come into contact with the impeller causing severe damage.

In an existing installation, the $(NPSH)_A$ can be estimated from

$$(NPSH)_A = h_a + (h_s)_g + (h_s)_v - h_{svp} \tag{13.7}$$

where $(NPSH)_A$ is the available net positive suction head (m water) [L], h_a is the atmospheric pressure (m water) [L], $(h_s)_g$ is the reading of a gauge on the suction side (corrected to the centre of the shaft for a horizontal pump or the eye of a vertical pump), (m water) [L], $(h_s)_v$ is the velocity head at the location of the gauge, given by eqn (13.5), (m water) [L], and h_{svp} is the saturated vapour pressure of the liquid at the operating temperature (m water) [L].

For water the saturated vapour pressure (svp) may be obtained from a thermodynamic equation (Reid *et al.*, 1987):

$$h_{svp} = (2\,256{\cdot}24)\exp\left(\frac{(-7{\cdot}76451b + 1{\cdot}45838b^{1{\cdot}5} - 2{\cdot}77580b^3 - 1{\cdot}23303b^6)}{(1 - b)}\right) \tag{13.8}$$

where

$$b = 1 - ((T + 273{\cdot}16)/647{\cdot}3) \tag{13.9}$$

where T is the temperature (°C); $2 \leqslant T \leqslant 374$.

At the *design stage*, the $(NPSH)_A$ can be calculated from

$$(NPSH)_A = (h_s)_s - (h_s)_f - h_{svp} \tag{13.10}$$

where $(h_s)_s$ is the static head on the suction side (m water) [L], and $(h_s)_f$ is the frictional head in the pipe work on the suction side (m water) [L].

Estimation of static and frictional heads is considered after eqn (13.6).

3. CENTRIFUGAL PUMPS

3.1. General Features

Centrifugal slurry pumps are frequently used, partly because they tend to be cheaper than a displacement pump which can develop the same head and throughput. They are designed with larger passages than *clean* liquid centrifugal pumps to reduce blockage risk. In addition, wall thicknesses of wet-end parts (casing, impeller, etc.) are greater. Other distinctive features include extra large stuffing boxes, replaceable shaft sleeves and impeller back vanes to keep solids away from the pump's stuffing box (Dalstad, 1977).

Delivered head: is limited to the range 25 m of water per stage (Gandhi *et al.*, 1980) to 30 m (Crisswell, 1983). This corresponds to a maximum impeller tip speed in the range 20–28 $m\,s^{-1}$. Parts life on pumps running faster than these limits drops in proportion to the square of the impeller speed. Maximum heads to 100 m water are possible but wear rates are normally unacceptable. To increase overall head developed, several pumps can be operated in series, subject to a casing pressure limitation of around 7 MPa.

Flow rate: depends on pipe size. The limit for one pump is around 3 000 $m^3\,h^{-1}$, but several pumps can be operated in parallel to increase throughput.

Efficiencies: typically 35–65%. Significantly lower than centrifugal pumps designed for *clean* liquid or chemical duties.

Particle size: the maximum particle size that can be passed depends on the pump size, although for rubber-lined pumps the limit is around 3–6 mm. Large unlined pumps can handle up to 200 mm (Odrowaz-Pieniazek, 1982).

Solids concentration: limits on solids concentration depend on the slurry's rheological properties. For fine-particle flocculated slurries, the limit is around 25–30% solids by volume, but if the solids are deflocculated the limit is much greater. The limit for fibrous or stringy material is much lower: 6–8% for paper stock and 4–6% for sewage sludge (Willis & Truscott, 1978). The mixed-flow volute design (see Chapter 14) can handle up to 10% sewage sludge.

Special designs: there are several special designs of centrifugal slurry pump which attempt to cope with some of the problems of pumping large solids and abrasive slurries. The torque induced, or recessed impeller, pump and the mixed-flow volute (helical screw) are two examples described in Chapter 14.

3.2. Co-rotating Disc Pump

A further design which can pump high viscosity, fine-particle slurries is the co-rotating disc pump (Fig. 13.2) whose impeller consists of several parallel and

Fig. 13.2. The co-rotating disc pump (Reproduced by permission of Discflo Corporation, Santee, CA, USA.)

equally spaced rotating circular discs which create pressure and flow through boundary layer and viscous drag effects. This results in a pump with characteristics which lie between centrifugal and rotary displacement pumps. However, whereas more conventional centrifugal pumps can pump slurries having viscosities up to about 1 Pa s, it is claimed (Rendall, 1988) that the upper viscosity limit for the disc pump is around 65 Pa s. Typical slurries reported to have been pumped include coal slurry, paper fibre, kaolin slurry, ceramic glaze slurry and drilling mud. The particle size in the slurry affects the choice of disc spacing in the pump; this is usually selected to be greater than the maximum particle size. The pump can also handle slurries containing much higher volumes of gas than other types of rotodynamic pump. Tadmor *et al.* (1985) give design configurations, design equations and experimental performance data for the pump.

4. DISPLACEMENT PUMPS

4.1. General Features

Displacement pumps cover a very wide range of designs. They can be classified into four main categories (see Fig. 13.1). Reciprocating pumps draw slurry into the pump chamber on the inlet stroke through a non-return valve. On the discharge stroke, the slurry is pressurised and forced out of the chamber through another non-return valve. Several chambers or cylinders can be used to increase throughput and to smooth-out pulsations. The intensity of the pulsations varies greatly depending on the details of the pump design. Sometimes a pulsation damper is required at the pump discharge (Ekstrum, 1981). Rotary pumps generally develop pulsations of much lower intensity, have no valves, and operate through rotary rather than reciprocating action. Pumps using air as the motive fluid operate either with the air in direct contact with the slurry or separated from it by a flexible diaphragm. Pumps using liquid as the motive fluid generally use either water or hydraulic oil and are often used for highly abrasive slurries. Again, the liquid may be in direct contact with the slurry or separated by a flexible diaphragm.

The various types of reciprocating pump are designed to pump medium to high flow rates of slurries (up to 900 $m^3 h^{-1}$) having a wide range of abrasivities, whereas the various classes of rotary pumps handle lower flow rates of slurries over a wide viscosity range mainly in the process industries for in-plant slurry transfer duties. Unlike centrifugal slurry pumps, the output from either reciprocating or rotary displacement pumps is normally relatively insensitive to the discharge pressure requirement and so both pump types can usually be readily used as a volumetric metering device.

4.2. Reciprocating Pumps

Reciprocating displacement pumps can be either single-acting or double-acting. With single-acting pumps, only one stroke of the pump conveys slurry through an open valve while the return stroke draws new slurry into the cylinder, whereas double-acting pumps draw in at one end and discharge at the other end of the cylinder during each stroke of the piston. Plunger pumps are single-acting only where the piston, or *plunger*, is normally washed to prevent ingress of abrasive particles into the packing. They are used for abrasive slurries where high pressures are required and are typically of the *triplex* design, i.e. three cylinders each containing a plunger are mounted in parallel either vertically or horizontally. Double-acting piston pumps, on the other hand, are usually of the *duplex*, or twin-cylinder design. Piston diaphragm pumps are either single- or double-acting, and employ flexible diaphragms to transmit pressure from hydraulic oil to the slurry, so minimising wearing parts. Henshaw (1981) provides a useful review of most aspects of reciprocating pumps.

The independence of flow rate with respect to discharge head and the high overall pump efficiency (including drive train—of the order of 85%) are very important advantages of reciprocating pumps. The maximum flow rate per pump is sometimes limited to relatively low discharge pressures which are insufficient for long distance slurry pipeline applications and therefore several pumps operating in parallel may be needed to achieve the high flow rates and working pressures required. Pumps with discharge pressure up to 35 MPa are available.

The maximum particle size that a reciprocating pump can handle is determined by the non-return valve seal requirements. According to Gandhi *et al.* (1980) piston or plunger pumps are appropriate for particle sizes up to 3 mm in conjunction with discharge pressures up to 15 MPa. For particle sizes in the range 3–6 mm special valve designs are used. These larger particle sizes are likely to reduce the maximum discharge pressure substantially to around 1 MPa (Willis & Truscott, 1978).

4.2.1. Mud Pumps

The first piston pumps were those developed for the upstream oil industry. Known as *mud pumps*, they are often of the double-acting, duplex type and are used for high-pressure pumping of lightly abrasive drilling muds at oil wells. For long distance slurry transport, the stroke rate must be kept low and these pumps can be used only for slurries with a Miller index (Miller, 1974) below about 30 (see section 5.2). Single-acting, triplex pumps are taking over drilling mud pumping applications. Because of the single-acting function, high stroke rates of 160 min^{-1} are required to achieve similar flow rates to the double-acting, duplex pump. Parts of the triplex pump are consequently subjected to high wear rates and have to be replaced regularly. For long distance slurry transport applications, the triplex pump stroke rate needs to be lower at around 70 min^{-1}. Bhambry and Wallrafen (1987), Miller (1985, 1989) and Wallrafen (1983) describe commercial designs and application areas. Brandis *et al.* (1986) performed trials on a triplex piston pump using fine coal slurry and Giacomelli *et al.* (1987) considered the detailed design of a prototype piston pump for coal slurry.

4.2.2. Hydraulically Driven Piston Pumps

Several designs of mobile, hydraulically driven pumps are available. These were originally developed for the hoisting of concrete through considerable vertical distances in the construction industry but they are now marketed in a wide range of industries where the requirement is to pump highly viscous pastes and sludges, some of which may be unsaturated, i.e. they contain substantial amounts of occluded air. They can be used only in a feeding capacity and cannot be operated in series.

A particular design feature of many of these pumps is the type of valving arrangement. Conventional ball valves are offered but the transfer tube system is also frequently employed. Stiff sludges generally have to be forced into the pump body using a single or twin auger combined with a hopper (Fig. 13.3).

A typical design of hydraulically driven piston pump consists of two delivery cylinders and two hydraulic drive cylinders which are flanged to either side of a flushing water box. The delivery cylinders are fitted with rubber pistons. The outlet end of the delivery cylinders terminate in the material receiving hopper in which is mounted the hydraulically powered transfer tube. On the delivery stroke, one piston is pushing material through the transfer tube and into the delivery line whilst the other piston is sucking in material from the hopper. At the end of their respective strokes, the pistons reverse their directions and, at the same time, the transfer tube swings over to cover the other cylinder operating and the cycle is repeated. Carroll (1986), Fehn (1984, 1986) and Kronenbourg (1988) describe variations in pump design and applications for this pump class.

4.2.3. Piston Diaphragm Pumps

Piston diaphragm pumps develop the required pressure in an oil phase and this

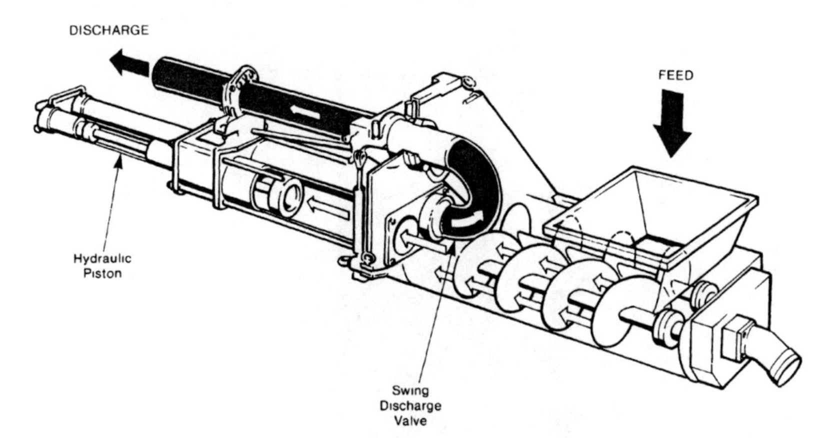

Fig. 13.3. A hydraulically operated piston pump showing the transfer tube valve design. (Reproduced by permission of Putzmeister-Werk Maschinenfabrik GmbH, Filderstadt, Germany.)

pressure is transmitted to the slurry via the action of a reciprocating piston and one or two diaphragms. They are typically used for the transport of very abrasive media, such as iron ore, copper ore, bauxite, red mud and sand. They have been compared with other types of reciprocating slurry pump by Bhambry and Wallrafen (1987), Holthuis and Simons (1980, 1981), and Schaaf (1981). For further information, refer to Chapter 15.

Wear of the diaphragms can be reduced by using a flexible hose placed in line between the two non-return valves as shown in Fig. 13.4. Here, the piston displaces hydraulic oil onto the diaphragm, and the diaphragm in turn acts on the pump hose via a second actuation liquid (e.g. water). Flow rates up to $100\,m^3\,h^{-1}$ and discharge pressures up to 16 MPa are possible.

4.2.4. Plunger Pumps

Plunger pumps (Duggan *et al.*, 1984; Hughes, 1986; Stepin & Korotkov, 1988) are used for very abrasive slurries with Miller numbers (see section 5.2) above 60 when the wear rate of pistons and cylinders in conventional piston pumps would be unacceptably high. For further information, refer to Chapter 15.

4.2.5. Double Disc Pump

The readily transportable double disc pump (Fig. 13.5) is used for modest, on-site slurry transfer. The pump is a type of piston diaphragm pump, but with this design slurry is allowed to flow onto either side of the diaphragms. It can be considered as valveless and glandless, can be run dry and is self-priming. It is used for low flow rate and low-pressure applications and can pass large solids (Westwood, 1986) such as pieces of metal, stone, wood and rags embedded in slurry which may be fibrous or stringy. A 50 mm port size pump can handle solids up to 12 mm, while a 75 mm port size can pump up to 19 mm solids. The pump operates through the

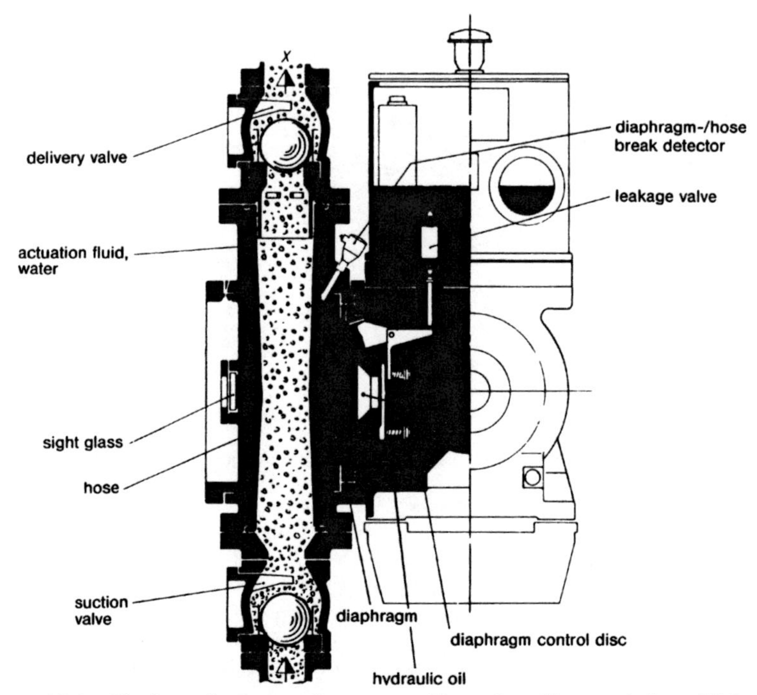

Fig. 13.4. The hose diaphragm piston pump (Reproduced by permission of Feluwa Pumpenfabriken, Cologne, Germany.)

rotation of a shaft which reciprocates two rubber discs in opposition to each other. A large cavity is created between them at one end of the stroke, and closed at the other end of the stroke, producing a positive suction and discharge sequence. The pumps are designed to pump anything from water up to high viscosity pastes.

4.3. Rotary Pumps

The range of rotary pumps used for slurry transfer in process plant applications is described in detail in Chapter 16. Rotary pumps consist of chamber elements actuated by relative rotation of the drive shaft and casing, and which has no separate inlet and outlet valves. Such a description is satisfactory for rotary pumps such as progressive cavity, rotary lobe, circumferential piston, internal gear and flexible vane pumps but does not cover the increasingly important development of peristaltic pumps for slurry applications.

4.4. Pumps Using Air as the Motive Fluid

4.4.1. Diaphragm Pumps

This category of pump makes use of the energy in compressed air to act as the

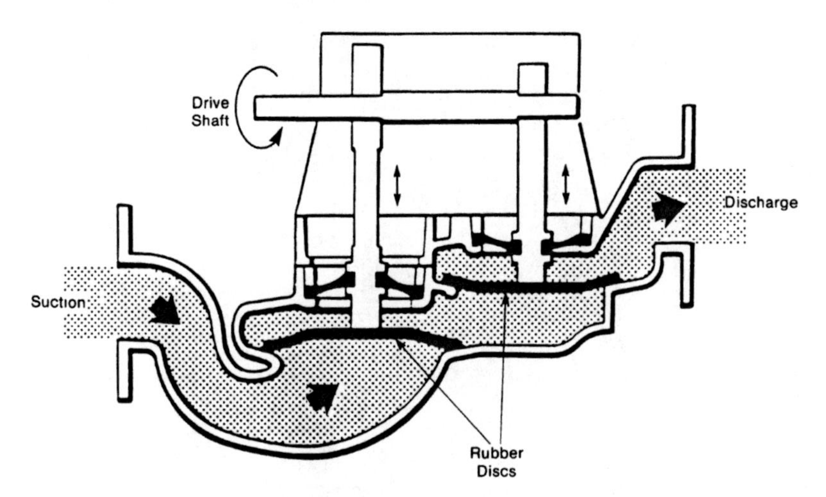

Fig. 13.5. The double disc pump. (Reproduced by permission of SSP Pumps Ltd, Eastbourne, UK.)

motive power for slurry transfer. Probably the most important example of this pump type is the air-operated diaphragm pump (Fig. 13.6) which finds widespread use in relatively inexpensive, low-maintenance applications either in-plant or on-site, where either permanent installation or portability is necessary and a compressed air source is available up to 0·85 MPa. Some 20–25 manufacturers offer various designs of this pump. The pump works on a double-acting, reciprocating principle and consists of two chambers each fitted with a diaphragm which separates the slurry chamber from the air chamber, suction and discharge valves. Air pressure in one chamber pushes out the slurry through the outlet valve, while the diaphragm in the other chamber is pulled by a connecting rod, thus sucking fresh slurry through the inlet valves. When the supply pressure is shifted to the second chamber, it displaces the slurry previously sucked in and fresh slurry is induced into the first chamber.

Dynamic head: up to 75 m of water.

Suction head: self-priming with suction heads up to 8 m of water.

Flow rate: maximum flow rates of 55 $m^3 h^{-1}$ are possible; the pump speed is varied by throttling the air supply.

Efficiency: relatively low so tends to be used in intermittent flow applications.

Particle size: solids up to 5–10 mm. Ball valves are most commonly used for viscous and/or gritty slurries but flap valves are supplied by some manufacturers for large solids up to a similar size to the pipe diameter.

Solids concentration: up to 90% by weight, but very variable depending on slurry viscosity.

Application areas: broad range, e.g. paint, resins, adhesives, concrete, plaster, yeast slurries and dog food; useful where self-priming is required and for various

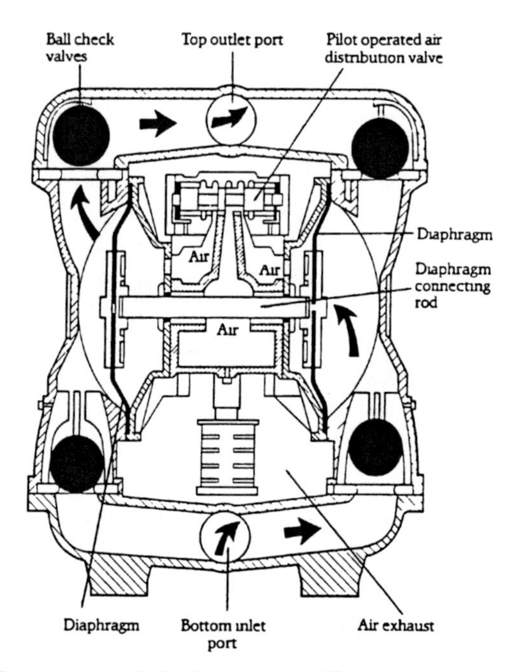

Fig. 13.6. The air-operated diaphragm pump. (Reproduced by permission of Kecol Pumps Ltd, Bridgnorth, Shropshire, UK.)

submerged duties where a light and portable pump is required (Huddle & Burton, 1986).

4.4.2. Drum-Emptying Pumps

Air-operated, drum-emptying pumps (Fig. 13.7) are designed to remove viscous fluids including non-settling slurries and pastes from standard 200-litre capacity drums. The pump consists of a pair of air-driven rams which are connected to a disc assembly with a central opening connected to a delivery pipe. The disc diameter is only slightly smaller than the inner diameter of the drum and, as the disc assembly is forced downwards into the drum by the compressed air rams, the material within it is forced up the delivery pipe. The pump is particularly useful for pumping high-viscosity pastes such as mastics and sealants. Flow control is good and is achieved by varying the compressed air flow rate to the pump.

4.4.3. Air Displacement Chamber Feeder

This pump (Fig. 13.8) can handle considerably larger solids than the air-operated diaphragm pump. It is mostly used for dredging but is also useful for picking up solids such as slag from basins, pits and dumps. The system consists of three main components. The submerged pump body comprises three pressure chambers, each with an inlet valve and a spherical delivery valve of anti-abrasive rubber. The other

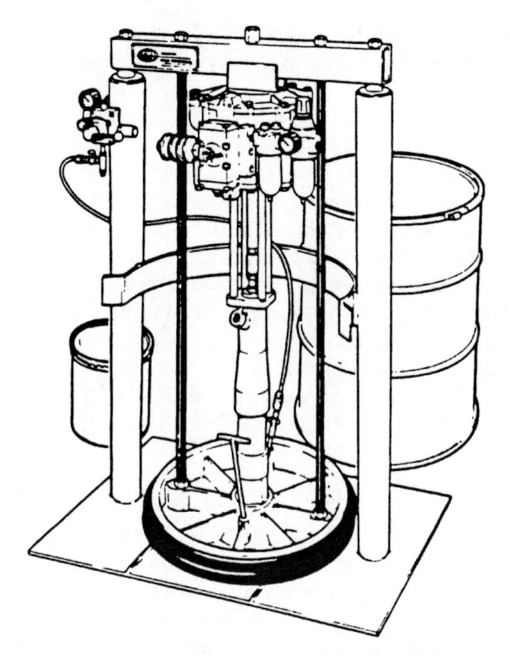

Fig. 13.7. The air-operated, drum-emptying piston pump. (Reproduced by permission of Pyles Business Unit, Graco Inc., MI, USA.)

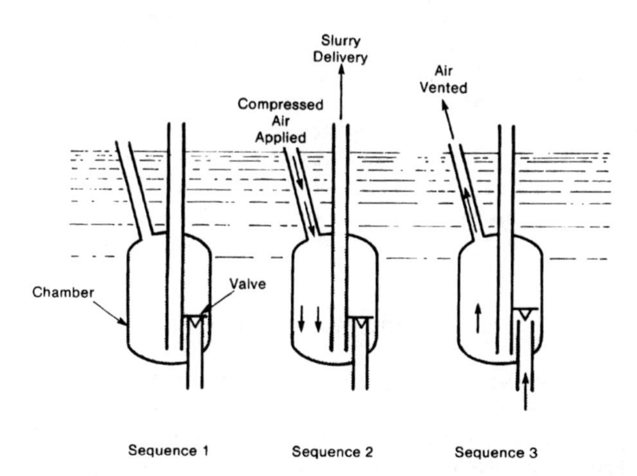

Fig. 13.8. An air-displacement chamber feeder. (Adapted from Lumbard (1972).)

two components are a distributor, which controls the inlet and the discharge to the atmosphere of the compressed air from each cylinder of the pump body, and an air compressor.

Each chamber in the pump is filled in turn with slurry; as soon as one cylinder is filled, the inlet valve automatically closes by its own weight (Fig. 13.8). The pump cycle consists of three sequences. In the first sequence, the chamber is filled with slurry. During the second sequence, compressed air is supplied to the top of the chamber, forcing the slurry out through the delivery pipe which is common to all three chambers. The third sequence starts when the chamber is almost empty, at which point the air is discharged to the atmosphere through the distributor and the cycle restarts. Operating speeds can be varied by controls at the distributor but are usually slow in order to maintain the slurry/air interface. Normal cycle duration is 6 s, i.e., 9 cycles min^{-1} or three cycles per min per chamber. The system is available in several sizes and can pump up to 3 000 $m^3 h^{-1}$ at pressures up to 0·4 MPa (Lumbard, 1972). Solids up to 200 mm can be handled by the largest systems.

4.4.4. Air-Lift Pump

The air-lift pump (Fig. 13.9) is well-established for liquid raising but only relatively recently has the pump principle been applied to hoisting both settling and non-settling slurries. The pump is potentially useful for the dredging of river estuaries and harbours, the mining of minerals from the ocean floor (Kaufman & Rothstein, 1970; La Motte, 1970) and the recovery of coal in mine shafts (Mills, 1978). In addition, sludges have been raised (Storch, 1975) and mud has been dredged from estuaries (Berleur & Giot, 1986). Some fundamental studies have been carried out on glass particles and coal (Kato *et al.*, 1975; Weber & Dedegil, 1976).

In attempting to predict the performance of air-lift pumps, little is known about the effect of the slurry's flow properties on lifting ability and pump efficiency. Pump performance can be predicted with reasonable confidence for low-viscosity Newtonian slurries in turbulent flow based on data for water (Stenning & Martin, 1968), but this is not so for Newtonian or non-Newtonian slurries in laminar flow in the riser. Theory has been developed for non-Newtonian fine particle slurries (Heywood & Charles, 1978) and small-scale trials have been undertaken using flocculated kaolin slurries (Heywood *et al.*, 1981). For in-plant applications, the pump is particularly useful when corrosive or abrasive slurries are to be hoisted over modest distances. The pump is extremely inexpensive and requires little maintenance.

4.5. Pumps Using Liquid as the Motive Fluid

Chapter 17 covers in detail lockhoppers and chamber feeders (and variations such as hydrohoists, hydrofeeders and hydrolifts) which all rely on filling a chamber, vessel or pipe with slurry and displacing it with water using a high pressure centrifugal pump. Chapter 18 describes the principle and main application areas of jet pumps.

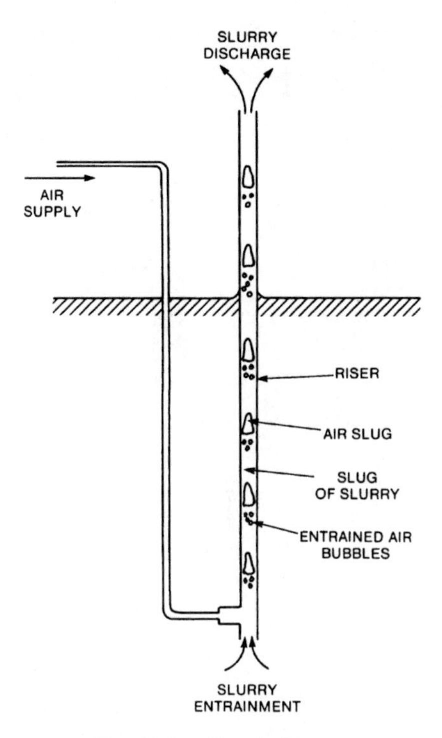

Fig. 13.9. The air-lift pump.

Other types of liquid displacement pump include the Mars oil barrier piston pump (Rouse, 1982; and described in Chapter 15) and the hydraulic exchange pump.

4.5.1. Hydraulic Exchange Pump

This pump uses cylindrical diaphragms as the barrier between slurry and the pressurised hydraulic oil (Zoborowski & Radloff, 1978; Zoborowski & Hastings, 1986). Its principle of operation is shown in Fig. 13.10. The slurry side of the pump consists of the inlet piping connection and inlet non-return valves, the outlet non-return valves and outlet piping connection, and the pressure vessels containing the diaphragm bags and bag holders. The pumping power is transmitted from an electric motor driving one or two hydraulic pumps (depending on the required capacity) through the hydraulic system to the diaphragm bags. The bags are filled and evacuated alternately with the hydraulic oil so causing the slurry surrounding the bags to be brought into the pressure vessels and then expelled.

The relatively slow cycle rate of 10 cycles min^{-1} helps to minimise the wear of the slurry non-return valves. This cycle rate is fixed. To vary the slurry flow rate, volume settings on the hydraulic pumps are adjusted. The pump can deliver up to

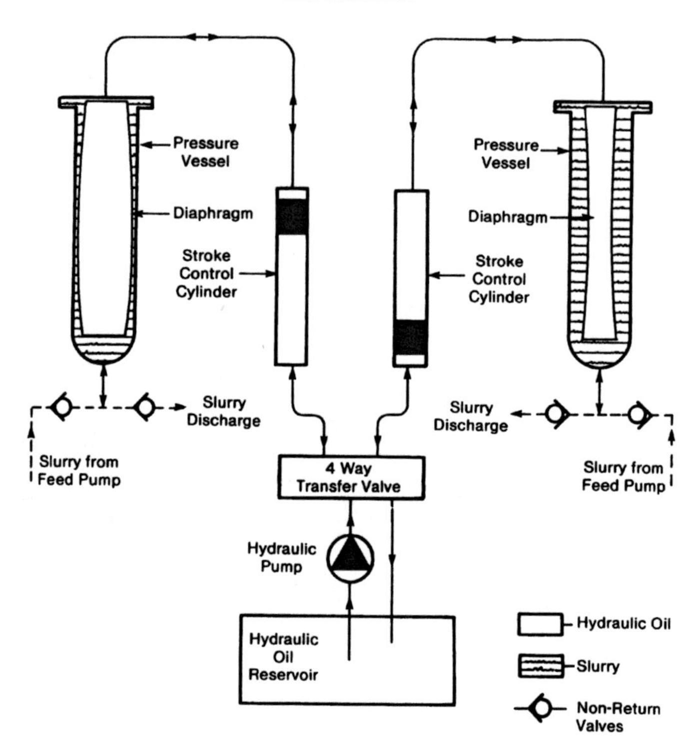

Fig. 13.10. The operating principle of the hydraulic exchange, cylindrical diaphragm pump. (Reproduced by permission of Zimpro Inc., Rothschild, WI, USA.)

$180\,m^3\,h^{-1}$ at pressures up to 23 MPa and has been used in a variety of mining installations, including the pumping of red mud waste from bauxite processing (Want *et al.*, 1984).

5. SOME CONSIDERATIONS IN SLURRY PUMPING

5.1. Pump Priming

Some pump designs are inherently self-priming while others are not irrespective of the flow properties of the slurries to be pumped and hence the ease, or otherwise, with which slurry can enter the pump casing. It is generally true that provided slurry can be introduced into a pump, then it can be pumped out, albeit at often a rather low hydraulic efficiency depending on the detail of the pump design, pump speed, etc.

A centrifugal slurry pump is unable to draw slurry into its casing because the pump cannot develop a suction head below ambient. Therefore, the pump must be primed by gravity feeding of the slurry or as a result of pressure generated by an upstream pump. On the other hand, many designs of displacement pump have the capability of developing a suction head as well. In order to develop this suction head, the pump should generally not be run dry; most should be started wet except for the peristaltic pump. Dry running is generally discouraged because much wear damage can take place in a very short time period. However, because peristaltic pumps make use of a lubricating liquid such as glycerol or oil where the rollers or shoes impinge onto the collapsible hose, little if any wear will result inside the hose if the pump is run dry. Hence the pump can run dry for long periods of time, if necessary, to draw slurry into the hose.

5.2. Pump Wear

The potential for a slurry to cause progressive wear is one of the most important considerations when selecting a pump. The word *wear* obviously has many connotations, but in terms of slurry pumping Miller (1986) defines it as 'the gradual deterioration of any part in the system to the point of danger or uselessness'. Obviously wear of important components can cause significant operational problems such as loss of system performance long before danger or uselessness are reached. An alternative definition is 'the progressive volume loss of material from a surface arising from all causes'. The two main causes of wear are erosion involving mechanical action by either solid particles (abrasive wear) or cavitation, or corrosion, involving chemical or electrochemical action. Both erosion and corrosion can take place simultaneously, the relative proportions depending on the particular application. Hence, the choice of construction materials is usually a compromise between abrasion resistance, mechanical strength and cost. Figure 13.11 gives the approximate relationship between the hardness values of common ores and pump materials of construction (Wilson, 1980).

Miller number. Miller (1974) developed an abrasivity tester to measure the relative abrasivities of slurries, but the tester does not appear to be commercially available. The test gives a Miller number which has been found to be useful in estimating the service life of pumps and valves, and also for determining the rate at which abrasivity changes with time owing to particle attrition. The Miller number consists of two values: the first represents abrasivity and is obtained by reciprocating a wear block on a V-shaped tray confining the slurry particles to the path of the block. The second value is a measure of the particle attrition, i.e. the loss or gain in abrasivity as particles break down during the test period. For further details of the test, see Chapter 20. Typical Miller numbers are given in Table 13.1 (Miller, 1974). The second value is negative if there is a reduction in abrasivity as a result of particle attrition and positive if there is an increase in abrasivity.

When purchasing a pump, a guarantee will often be sought from the vendor regarding pump wear rate. Usually it is best to try to approach vendors with some experience in specifying a pump for the specific application. Information on likely

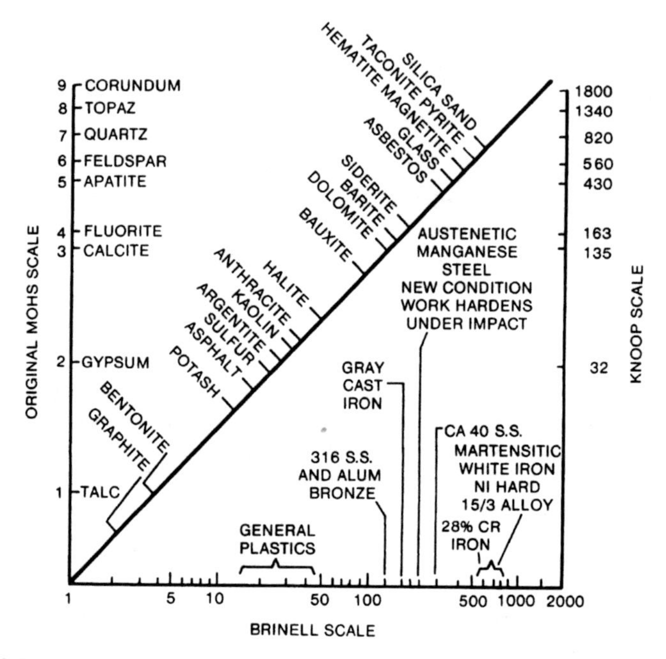

Fig. 13.11. The hardness values of various particulate solids and pump materials of construction. (Adapted from Wilson, 1980.)

wear rates may then be available and the vendor may be reasonably confident about providing a guarantee. If the application is novel, it is useful to consult any relevant literature on typical wear rates. Vendors who show some understanding and technical appreciation of the potential wear problems can then be approached with the appropriate questions.

Most of the published experimental studies of wear rates appear to concentrate on centrifugal pumps (Want, 1980; Faddick & Studebaker, 1981; Wiedenroth, 1983). Wear data from large centrifugal slurry pumps have been compared with model predictions (Roco & Addie, 1984, 1987). Such published information can be useful to identify the reasons for pump wear and remedial action can sometimes be taken by altering the operating conditions or the slurry characteristics.

6. PUMP SELECTION

Selecting a suitable slurry pump depends largely on the application, in terms of flow rate, head, etc. and on the physical properties of the slurry, such as maximum particle size, abrasiveness and viscosity. The main parameters affecting pump choice are essentially the discharge pressure requirement and the slurry abrasiveness. Odrowaz-Pieniazek (1979) has provided a selection table based on three

TABLE 13.1
Typical Miller Numbers for Slurry Abrasivity[a]

Material	*Solids concentration (% by mass)*	*Miller number*
Detergent	65	0 − 0
Potash	50	0 − 0
Sulphur–water	25	1 − 37
Gebo coal − 30 Mesh	25	6 − 26
− 16 Mesh	50	10 − 27
Potash–brine	25	10 + 1
	50	11 + 2
Oilwell standard mud	0·25 Sand	10 − 4
Lignite	25	13 + 4
BITU coal slurry	(as received)	14 − 6
Calcium carbonate	25	14 − 12
Black Mesa coal − 30 Mesh	25	26 − 14
− 16 Mesh	25	21 − 16
Limestone	(as received)	27 − 3
Magnetite	25	67 − 4
Sand–shell	50	67 − 7
Microshporite	(as received)	76 − 13
Urn Sand − 70 Mesh	12·5	93 − 13
	25	104 − 14
	50	112 − 10
− 100 Mesh	25	137 − 15
Mill tailings	25	159 − 11
	50	155 − 9
Blackfill − 100 Mesh	25	180 − 11
	50	217 − 15
Pyrite	50	194 − 4
Corundum − 400 Mesh	25	241 + 21
− 200 Mesh	25	1040 − 12
Copper tailings	(as received)	314 − 15
Copper concentrate	50	322 − 16
Chalcopyrite	50	436 − 22
Wabush tailings	50	848 − 21

[a]From Miller (1974).

parameters and also on particle size and throughput. Three levels of slurry abrasiveness are covered according to the slurry Miller number. In addition to these selection parameters, pump efficiency may also be an important consideration; for long distance pipeline applications it is, of course, of prime importance, but for in-plant usage, pump efficiency is often less critical than ensuring that a pump is reliable and does not block.

An important consideration is that pumping equipment is almost always available in discrete unit sizes defined by maximum throughput, inlet and discharge pipe sizes, maximum delivered head and power rating. Centrifugal pumps are usually considered for higher flow and lower head applications than positive displacement

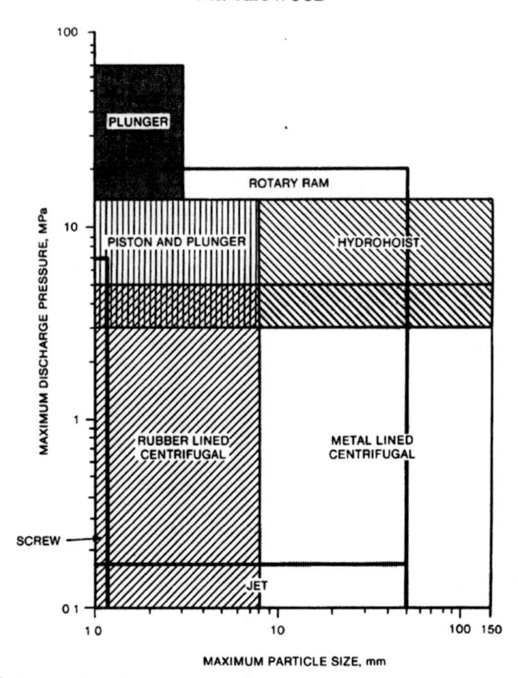

Fig. 13.12. Slurry pump selection based on maximum discharge pressure and maximum particle size. (Adapted from Gandhi *et al.*, 1980.)

pumps. Consequently, the upper limit on slurry viscosity for centrifugal pumps is generally considerably lower than that for most displacement pumps. Centrifugal pumps usually have a lower capital cost than displacement pumps for a similar duty and also have lower maintenance costs. However, they tend to have higher operating costs because of their lower mechanical–hydraulic efficiencies. At the lower flow rate end, the range of rotary pumps (Chapter 16) such as progressive cavity, peristaltic and lobe pumps, or the air-operated diaphragm pump will be applicable.

Gandhi *et al.* (1980) have also developed an initial selection chart for hydraulic conveying applications (Fig. 13.12). This chart indicates that there are certain combinations of maximum particle size and discharge pressure for which only one type of pump is probably appropriate, while with other combinations more than one pump type could be used. Selection, particularly in the latter case, then requires more detailed considerations of the volume flow rate required, slurry abrasivity (and also its tendency to degrade) and capital and operating costs of alternative pumping systems. With regard to slurry abrasivity, a useful rule of thumb is that, for Miller numbers below about 30, piston pumps can be specified, but for abrasivities above 60 plunger pumps are the preferred choice for long distance slurry pipe lining at relatively high flow rates.

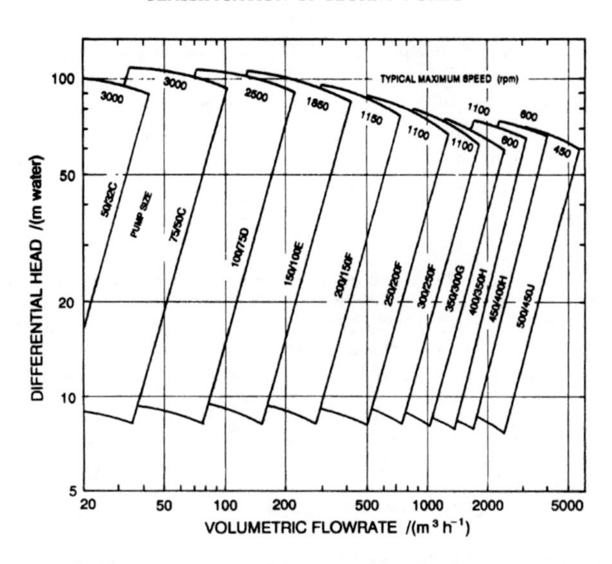

Fig. 13.13. Coverage chart for typical range of centrifugal slurry pumps. (Pump size is indicated as suction diameter (mm)/discharge diameter (mm).) (Reproduced by permission of Denver Process Equipment, Leatherhead, UK.)

Several papers give practical guidance on selecting pumping systems for the hydraulic conveying of solids or slurry transfer over significant distances rather than in-plant. Paterson and Watson (1979) give a discussion of the choice between slurry pumps and feeders for NCB's Hordern colliery mine waste pipeline while Graham and Odrowaz-Pieniazek (1978) give an account of pump selection considering most aspects, including availability, for an emergency pumping system following a Zambian copper mine disaster. Raisbeck (1988) discusses two main alternatives for silt removal from an open-cast copper mine: a high lift reciprocating pumping installation or a series of lower lift, centrifugal pump stations.

7. PUMP SIZING

Pump sizing is important to ensure that the required throughput is developed and starts with reference to *coverage charts* (Fig. 13.13). Reference is then made to individual pump characteristics. When considering these curves, the main requirement is to obtain the best efficiency at the design head and throughput, bearing in mind that there will be a trade-off between efficiency and capital cost. Other considerations include $(NPSH)_R$ for each unit. Both coverage charts and pump characteristics are speed dependent. Allowances must also be made for likely slurry variability, such as solids concentration, particle size, size distribution and shape, all of which can influence greatly frictional pressure losses in the downstream pipe work.

Frequently, a range of pump sizes will be able to provide the required discharge pressure and flow rate combinations, and also be capable of achieving an acceptable efficiency. However, if the slurry is abrasive and/or if the solids in the slurry are liable to degrade readily or are shear-sensitive, then a larger pump operating at a slower speed may be preferable to a smaller pump running at a higher speed.

8. REFERENCES

Berleur, E. & Giot, M. (1986). Application of the air-lift principle to solve maintenance problems. In *Proc. HT 10*, paper D2, pp. 109–21.*

Bhambry, I.K. & Wallrafen, G. (1987). Positive displacement pumps for long distance transport of solids. *Bulk Solids Handling*, 7(5), 687–94.

Brandis, U., Klose, R. & Simons, P.W.H. (1986). Investigation into the transport behaviour of highly concentrated fine-grained coal–water slurries (dense coal) using a 470 m^3/hr triplex piston pump. In *Proc. HT 10*, paper A4, pp. 31–8.*

Carroll, E. (1986). Pumps for stiff and aggressive wet solids. In *Proceedings of the International Conference on Positive Displacement Pumps*. BHRA Fluid Engineering, Cranfield, UK, paper B1.

Crisswell, J.W. (1983). Selection and operation of slurry pumps. *World Coal*, Apr., **9**(2), 47–53.

Dalstad, J.I. (1977). Slurry pump selection and application. *Chem. Engng*, 25 Apr., 101–6.

Davidson, J. (ed.) (1986). *Process Pump Selection: A Systems Approach*. Institution of Mechanical Engineers, London, UK.

Duggan, P.A., Duha, J.L. & Laskey, H.K. (1984). Ingersoll Rand test facility for mainline slurry pumps and their components. In *Proc. STA 9*, pp. 85–90.*

Ekstrum, J.D. (1981). Sizing pulsation dampeners for reciprocating pumps. *Chem. Engng*, 12 Jan., **88**(1), 111–18.

Faddick, R.R. & Studebaker, A.R. (1981). Abrasive wear of centrifugal pumps. In *Proc. STA 6*, 145–55.*

Fehn, B. (1984). Two cylinder piston pumps with hydraulic drive: technical characteristics and application options. In *Proc. HT 9*, paper H3, pp. 377–96.*

Fehn, B. (1986). Transport of stiff sludge and filter cake with twin-cylinder piston pumps. In *Proceedings of the International Conference on Positive Displacement Pumps*. BHRA Fluid Engineering, Cranfield, UK, paper B3.

Gandhi, R.L., Snoek, P.E. & Carney, J.C. (1980). An evaluation of slurry pumps. In *Proc. STA 5*, pp. 267–75.*

Giacommelli, E., Graziani, F., Mezzedimi, V., Pignone, N. & Verdelli, A. (1987). Designing a line of reciprocating pumps for coal slurry, constructing and testing a prototype unit. In *Proc. STA 12*, pp. 191–9.*

Graham, J.D. & Odrowaz-Pieniazek, S. (1978). The design and operation of a slurry pumping system from 1 000 m underground to the surface. In *Proc. HT 5*, paper J2.*

Henshaw, T.L. (1981). Reciprocating pumps. *Chem. Engng*, 21 Sept., **88**(19), 105–23.

Heywood, N.I. & Charles, M.E. (1978). The pumping of pseudo-homogeneous, shear-thinning suspensions using the air-lift principle. In *Proc. HT 5*, paper F5.*

Heywood, N.I., Michalowicz, R.A. & Charles, M.E. (1981). A preliminary experimental investigation into the air-lift pumping of shear-thinning suspensions. *Can. J. Chem. Engng*, **59**, 42–51.

Holthuis, C.H. & Simons, P.W.H. (1980). The Geho diaphragm pump—a new generation of high pressure slurry pumps. In *Proc. HT 7*, paper A3, pp. 17–31.*

Holthuis, C.H. & Simons, P.W.H. (1981). The economics of positive displacement pumps. In *Proc. STA 6*, pp. 151–60.*

Huddle, F.G. & Burton, J.D. (1986). Air driven, double diaphragm pumps; history and

performance review and recent experimental findings. In *Proceedings of the International Conference on Positive Displacement Pumps*. BHRA Fluid Engineering, Cranfield, UK, paper A2, pp. 5–14.
Hughes, C.V. (1986). Coal slurry pump development—mainline pumps for the Belovo–Novosibirsk pipeline. In *Proc. STA 11*, pp. 273–6.*
Jones, J.R. (1980). The selection of pumps for filtration duties. *Filtration and Separation*, Mar./Apr., **17**(2), 120–8.
Karassik, I.J., Krutsch, W.C., Fraser, W.H. & Messina, J.P. (eds) (1976). *Pump Handbook*. McGraw-Hill Book Co., New York, USA.
Kato, H., Mujazawa, T., Timaya, S. & Iwasaki, T. (1975). A study of an air lift pump for solid particles. *Bull. JSME*, **18**, 286–94.
Kaufman, R. & Rothstein, A.J. (1970). *6th Annual Preprints* (Vol. 2). Marine Technology Society, Washington, DC, USA, 29 Jun.–1 July.
Kronenbourg, J. (1988). New Putzmeister developments. *Bulk Solids Handling*, **8**(3), 359–62.
La Motte, C. (1970). *Ocean Industry*, **5**(7).
Lumbard, D. (1972). A new patented dredging system by compressed air. In *Proc. HT 2*, paper K5, pp. 53–60.*
Miller, J.E. (1974). Miller number. *Chem. Engng*, 22 July, 103–6.
Miller, J.E. (1985). Reciprocating pumps for slurry service. *Lubr. Eng.*, **41**(6), 45–50.
Miller, J.E. (1986). The mechanics of wear in slurry pumping. Paper presented at the 3rd International Symposium on Solid–Liquid Flows (Annual Winter Meeting, ASME), Chicago, IL, USA, 28 Nov.–2 Dec.
Miller, J.E. (1989). Bucking up reciprocating pump performance. *Chem. Engng*, Sept., **96**(8), 149–57.
Mills, L.J. (1978). Hydraulic mining in the USSR. *Mining Engr*, Jun.
Neerken, R.F. (1978). Selecting the right pump. *Chem. Engng*, 3 Apr., **85**.
Odrowaz-Pieniazek, S. (1979). Solids handling pumps—a guide to selection. *Chem. Engr*, Feb., **341**, 94–101.
Odrowaz-Pieniazek, S. (1982). Pumps for abrasive media. *Chartered Mechanical Engr*, Apr., 21–7.
Odrowaz-Pieniazek, S. & Steele, K. (1986). Advances in slurry pumps, Parts 1 and 2. *Chem. Engr*, Feb., **422** 34–7; Mar., **422**, 30–3.
Paterson, A.C. & Watson, N. (1979). The NCB pilot plant for solids pumping at the Hordern colliery. In *Proc. HT 6*, paper H1, pp. 353–66.*
Pollack, F. (1980). *Pump Users' Handbook* (2nd edn). Trade and Technical Press Ltd, Morden, Surrey, UK.
Raisbeck, J.K. (1988). De-silting Nchanga open pit. *Mining Magazine*, July, **159**(1), 26–9.
Reid, R.C., Praufnitz, J.M. & Poling, B.E. (1987). *The Properties of Gases and Liquids* (4th edn). McGraw-Hill, New York, USA.
Rendall, J. (1988). Pump those problem fluids. *Proc. Engng (London)*, Apr., 49–50.
Roco, M.C. & Addie, G.R. (1984). Modelling erosion wear in centrifugal slurry pumps. In *Proc. HT 9*, paper G1, pp. 291–316.*
Roco, M.R. & Addie, G.R. (1987). Erosion wear in slurry pumps and pipes. *Powder Technol.*, **50**, 35–46.
Rouse, W.R. (1982). Operating experiences with a residue disposal system. In *Proc. HT 8*, paper A6, pp. 77–90.*
Schaaf, M. (1981). Piston and piston diaphragm pumps for long distance hydraulic conveying. *Bulk Solids Handling*, **1**(2), 301–5.
Stenning, A.H. & Martin, C.B. (1968). The analytical and experimental study of air-lift pump performance. *J. Engng for Power, Trans. ASME*, **90**, 106–10.
Stepin, B.S. & Korotkov, Y.F. (1988). Testing on Ingersoll Rand Canada pump on high concentration coal slurry. In *Proc. STA 13*, pp. 75–89.*
Storch, B. (1975). Extraction of sludges by pneumatic pumping. In *Proceedings of the 2nd*

Symposium on Jet Pumps, Ejectors and Gas Lift Techniques. BHRA Fluid Engineering, Cranfield, UK, paper G4.
Tadmor, Z., Mehta, P.S., Valsamis, L.N. & Yang, J.C. (1985). Corotating disc pumps for viscous liquids. *Ind. Eng. Chem. PDD*, **24**, 311–20.
Thompson, T.L., Frey, R.J., Cowper, N.T. & Wasp, E.J. (1973). Slurry pumps—a survey. *CIM Bull.*, Jan., **66**(729), 102–8.
Vanderpan, R.I. (1982). Proper pump selection for coal preparation plants. *World Coal*, **8**(3), 54–6.
Walker, C.I. & Goulas, A. (1981). Computer-aided slurry pump selection. In *Proceedings of the 7th Technical Conference of the British Pump Manufacturers' Association*. BHRA Fluid Engineering, Cranfield, UK, paper 15.
Wallrafen, G. (1983). Piston pumps for the hydraulic transport of solids. *Bulk Solids Handling*, **3**(1), 99–104.
Want, F.M. (1980). Centrifugal slurry pump wear—plant experience. In *Proc. HT 7*, paper H1, pp. 310–14.*
Want, F.M., Zoborowski, M.E. & Lauters, C.L. (1984). Alcoa of Australia Ltd, Kwinana Refinery Pilot Plant for red mud disposal. In *Proc. HT 9*, paper F1, 237–50.*
Warring, R.H. (1984*a*). *Pumping Manual* (7th edn). Trade and Technical Press Ltd, Morden, Surrey, UK.
Warring, R.H. (1984*b*). *Pumps: Selection, Systems and Applications* (2nd edn). Trade and Technical Press Ltd, Morden, Surrey, UK.
Weber, M. & Dedegil, Y. (1976). Transport of solids according to the air-lift principle. In *Proc. HT 4*, paper H1, pp. 1–24.*
Westwood, J.B. (1986). The double disc pump and its application. In *Proceedings of the International Conference on Positive Displacement Pumps*. BHRA Fluid Engineering, Cranfield, UK, paper A3.
Wiedenroth, W. (1983). The evaluation of wear of pipes, elbows and centrifugal pump components. In *Proc. STA 8*, pp. 193–200.*
Willis, D.J. & Truscott, G.F. (1978). Solids handling pumps—a survey of current practice, problems and developments. In *Proc. HT 5*, paper F6.*
Wilson, G. (1980). Selecting slurry pumps: how to specify the best. *World Mining*, Feb., **33**(2), 43–5.
Zoborowski, M.E. & Hastings, J. (1986). Vertical hoisting of settled sludge solids from 2 500 level via a single pump lift. In *Proc. HT 10*, paper D1, pp. 101–7.*
Zoborowski, M.E. & Radloff, D.E. (1978). Transportation of slurried minerals by a high pressure cylindrical diaphragm pump. In *Proc. STA 3*, pp. 155–63.*

*Full details of the Hydrotransport (HT) and the Slurry Transport Association (STA) series of conferences can be found on pp. 11–13 of Chapter 1.

14

Centrifugal Slurry Pumps

Denis Delaroute
Denver Process Equipment Ltd, Leatherhead, UK

1. THE VERSATILE CENTRIFUGAL PUMP

Centrifugal pumps are mechanically compact and relatively cheap machines. As such, they are the first choice of pump design for many applications, their use varying from emptying washing machines to pumping liquid rocket propellants, with other designs considered only when operating conditions exclude their use.

In general, centrifugal slurry pumps are a development of centrifugal pumps used for single-phase liquids and are characterised by robust design, use of wear resistant materials, large passageways for the free passage of solids and ease of maintenance.

With centrifugal slurry pumps there are two fundamental points that users should bear in mind. Firstly, centrifugal pumps are velocity machines in which energy is imparted to the liquid by acceleration—they primarily generate a head of pumped liquid and not a pressure. Pump handling liquids that are identical except for density will generate the same head of liquid, but the pressure, and hence the power absorbed, varies with liquid density. Therefore, when the slurry density varies, it is not good practice to specify centrifugal slurry pump duties in terms of pressure. Secondly, centrifugal pumps do not pump at a fixed flow rate. The flow rate is determined by the interaction of the flow rate–head characteristic of the pump and that of the pipe work system. If the characteristic of either varies from that predicted, then the flow rate will vary. When specifying a pump it is the responsibility of the user to supply the system characteristic.

In this chapter reference is made only to the effect of slurries on the operation of a centrifugal pump. For a general description of the hydraulics of clean liquid pumps, refer to a standard text (for example, Stepanoff, 1957).

2. OPERATIONAL LIMITS

2.1. Flow Range

Centrifugal slurry pumps of hard metal or elastomer lining materials are generally used for flow rates ranging from 1 to 10 000 $m^3 h^{-1}$. There are no theoretical limits to this range, but in practice efficiencies drop rapidly for small pumps (peak efficiencies of less than 40%) and the cost of building very large pumps rises dramatically, making multiple pumps in parallel an economic proposition.

2.2. Head Range

2.2.1. Metal Pumps

Increasing speed and hence head generated increases the hydraulic forces on the impeller, reducing the life of bearings and shaft sealing through shaft deflection.

Centrifugal forces and the mechanical strength of the impeller also limit the head that can be generated.

Bearing and shaft seal life generally limit peripheral velocities to 40 m s^{-1} but, with proper application, the impeller mechanical limit of 50 m s^{-1} can be achieved. These velocities equate to differential heads of approximately 100 m and 150 m of pumped liquid.

2.2.2. Elastomer Linings

Friction between the liquid and internal surfaces of the pump causes heat. In metal pumps this energy is readily dissipated by conduction through the casing. In pumps lined with elastomers, which are good insulators, heat conduction is reduced and if the flow velocity is too high the elastomer will thermally degrade and fail. Furthermore, although elastomer impellers are bonded to metal skeletons the impeller is subject to stretching and will tear if the centrifugal forces become too high.

Peripheral velocities are therefore generally limited to 17·5 m s^{-1} with a head limitation of approximately 45 m, though the use of metal impellers can slightly increase this.

2.2.3. Multi-staged Pumps

Higher heads can be obtained by running pumps in series. For metal pumps there are no limits beyond the design casing pressure. Elastomer-lined pumps are limited by the loss of wear resistance due to decreasing resilience of the linings, as they are compressed under higher pressures.

Practically, metal pumps are limited to a head of 700 m and elastomers to 250 m. This corresponds to seven units in series.

2.2.4. Viscosity

Because of the relatively large passageways through slurry pumps, viscosity has less effect on slurry pumps compared with clean-liquid pumps. However, there are no published data available and the safest course of action is to derate in line with the standard methods such as those of the Hydraulic Institute (1983).

2.2.5. Temperature

In general, metal slurry pumps can operate at temperatures up to 80°C as standard, but with high temperature seals temperatures up to 135°C are possible. Elastomers are limited to temperatures shown in Table 14.1.

3. PUMP DESIGN

3.1. Hydraulic Requirements

The principle constraint on the hydraulic design of centrifugal slurry pumps is the mechanical requirement to pass the maximum particle size in the slurry through the

TABLE 14.1
Maximum Operating Temperatures for Elastomers

Material	*Maximum operating temperature* (°C)
Polyurethane	60
Natural rubber	80
Butyl rubber	110
Neoprene (synthetic rubber)	135
Nitrile rubber	140

impeller and casing. Secondly, the impeller and casing need generous wear allowances. Relatively large passageways have to be allowed through the impeller and a large impeller-tip to casing clearance is necessary. The number of vanes is typically limited to five such that the thick cross-section does not choke the eye of the impeller. The vane width between shrouds is held constant, so that a particle clearing the eye will pass through the pump without blocking.

These requirements take the hydraulic design away from the ideal, but within these restrictions the hydraulic design procedure is the same as for clean liquid pumps and the aim is to reach the highest possible efficiency. High efficiency is a requirement not only for low power consumption but also for minimising wear, since low efficiency is indicative of high turbulence and hence wear within the pump (Wiedenroth, 1988).

The single entrance, end suction pump (Fig. 14.1) provides the simplest design of centrifugal pump, with only one shaft seal. As such it is the standard design for slurry pumps.

One of the main causes of loss of efficiency in a centrifugal pump is the movement of liquid from the region of high pressure at the periphery of the impeller to the impeller entrance. In clean liquid pumps this recirculatory flow is reduced by the use of neck rings of various designs. These are essentially flow-restricting devices relying on tight running clearances and are not suitable for slurry applications as they quickly wear. An alternative is to add expelling vanes externally to the front of the impeller shroud (see Fig. 14.1). These vanes generate a head opposing the main vanes and thus reduce the flow around the impeller. This device is less prone to wear than neck rings, and can be adjusted to take up wear by axial movement of the impeller. Expelling vanes are also added to the rear shroud where they reduce the pressure on the gland and help extend the shaft seal life.

3.2. Mechanical Requirements

In general, slurry pumps are subject to higher loads than clean-liquid pumps. Consequently they require a much stronger frame and bearing assembly and a heavier shaft with minimum impeller overhang. The causes of the higher loads in slurry pumps are as follows.

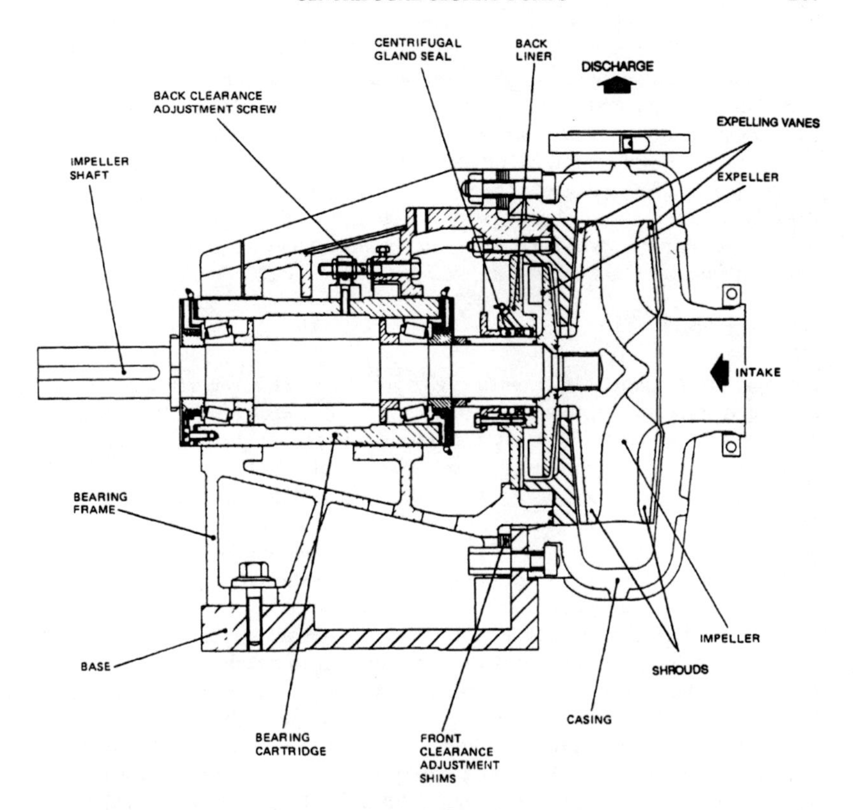

Fig. 14.1. Unlined slurry pump. (Courtesy of Denver Process Equipment Ltd, Leatherhead, UK.)

- *Hydraulic loads*: impeller radial loads are higher due to the higher liquid density.
- *Shock loads*: particles hitting the impeller and particles that are sheared when trapped between the impeller and casing causing high shock loads.
- *Vee belt drives*: as impeller materials are difficult to machine, slurry pump duties are generally gained by speed change rather than reducing impeller diameters; belt drives are therefore generally used. Modern wedge belt designs are extremely efficient but impose high lateral shaft loads.

3.3. Bearing Shaft Assembly

As the impellers of slurry pumps are subject to higher loads than clean-water pumps it is essential that the shaft is of robust design. The shaft flexibility factor (SFF) relates the shaft diameter, D (mm), to the cantilevered length, L (mm), and is defined as L^3/D^4(mm^{-1}). This is a measure of the susceptibility of the shaft to deflection which is critical to shaft sealing and bearing life. Typically, for a clean-liquid pump the SFF will be in the range 1–5 mm^{-1}, depending on pump size. Equivalent figures for slurry pumps are 0·2–0·75 mm^{-1}. The SFF can be used to evaluate the robustness of different shaft designs (Bloch & Johnson, 1985).

The bearing life (expressed in hours) for a 10% bearing failure rate is referred to as the B_{10} life. Typically, bearings are sized to give a minimum B_{10} life of 40 000 h over the standard range of the pump, with bearing application factors 30% higher than standard for clean-liquid pumps to cover shock loads.

3.4. Wet Ends

Although slurry pumps are designed to give good service life, wetted parts are wear items, and will consequently require replacement on a regular basis, the frequency of which will depend on the slurry abrasiveness. Therefore, due consideration has to be given to the cost of those parts and the ease with which they can be replaced.

In metal pumps, a lined casing design allows the volute wear parts constructed in hard metal to be of lighter construction since the wear parts do not form the pressure containment vessel as they do with an uncased design (Fig. 14.1). However, the uncased design is much simpler and should show significant capital cost savings as well as being easier to maintain. The uncased designs are also advantageous if a *back pull-out* design is used. In this design, the bearing frame and rotating element can be removed from the casing which remains in the pipe work, thus greatly simplifying inspection procedures.

The life of the wear parts is further increased by the ability to take up wear readily that occurs between the pump casing and the front and back shrouds. Recirculation around the front of the impeller decreases pump efficiency and promotes wear and therefore this running clearance needs to be maintained. Wear at the back of the impeller causes excessive pressure at the shaft seal which should be kept to a minimum. With most designs the rotating elements can be moved axially to take up one wear clearance, and with some designs, particularly those which can be serviced by back pull-out, both clearances can be adjusted.

In the design shown in Fig. 14.1 these clearances are readily adjusted upon assembly and maintenance. The *shaft side* clearance is closed by the adjusting screw on the bearing carriage assembly and the *suction side* clearance by removing casing shims and re-tightening casing bolts. Pumps that allow both clearances to be readily adjusted therefore offer considerable advantages.

3.5. Shaft Seals

3.5.1. General

The principle cause of operational difficulties with centrifugal slurry pumps is the shaft seal. Most slurry pumps use packed gland shaft seals with packing running on replaceable hard shaft sleeves.

Although manufacturers of mechanical seals have put considerable effort into developing mechanical seals with hard faces, they are very expensive and unreliable in most slurry applications.

For pumps with expelling vanes on the back of the impeller (such as those shown in Fig. 14.1) the internal pressure on the shaft seal should be the same as the inlet pressure to the pump, because the expelling vanes generate the same head as the main vane of the impeller. In practice, because the expelling vanes are much narrower they are not as efficient as the main vanes. Therefore, the shaft seal pressure is higher than the pump inlet pressure. The shaft seal pressure can be reduced by the use of a differential impeller whose main vanes are of smaller diameter than the expelling vanes. The latter generate a higher head relative to the main vanes. Alternatively, or in conjunction with a differential impeller, an expeller can be mounted on the shaft behind the main impeller, effectively pumping back into the casing.

For pumps without impeller expelling vanes, the pressure at the shaft seal should be treated as that at the pump discharge although in practice the pressure will be less owing to the swirling of the liquid between the casing and the impeller.

3.5.2. Packed Glands

The traditional packed gland is not generally suitable for slurry duties because it requires a low flow of liquid to weep to atmosphere to cool and lubricate the packing. This flow introduces solids into the packing and causes rapid wear and failure. Where the pump inlet pressure is very low, and either the slurry solids have lubricating properties or the solids concentration is very low (less than 1% by mass), the use of standard packed glands may be possible.

3.5.3. Water-Flushed Packed Glands

To keep solids from entering the packing and to ensure a source of liquid to cool and lubricate the packing, a stream of flushing liquid (usually water) can be introduced into the bottom of the stuffing box as shown in Fig. 14.2. Here a lantern restrictor creates a back pressure and ensures a positive high-velocity flow into the pump casing, as well as a flow to the packing.

It is essential that the flow to the gland is constant as even short periods without flushing water, such as on start-up, will introduce solids into the gland. The gland must also be adjusted regularly to ensure that the flushing water is passing into the pump, rather than escaping through the gland packing having mixed with slurry. It is stressed that the important factor is the flow of flushing liquid through the gland, and not flushing water pressure. A blocked pipe to the gland will give a high pressure, but no flow!

Where the flow of water to a standard flushed gland (see lower inset in Fig. 14.2) is unacceptably high, a low-flow flushed gland may be acceptable. One or more rings of packing are then introduced between the injection point and the pump casing thus reducing the water requirement (see upper inset in Fig. 14.2). However, this gland is more difficult to pack and adjust, and because the internal packing ring is behind the lantern ring, it can be forgotten and eventually extrude into the pump casing. This then makes it ineffective.

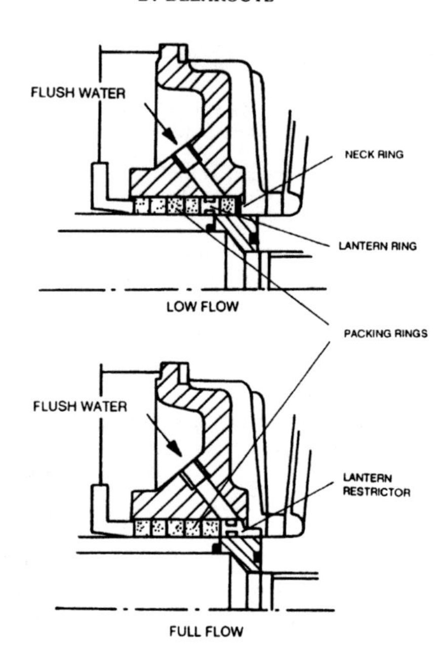

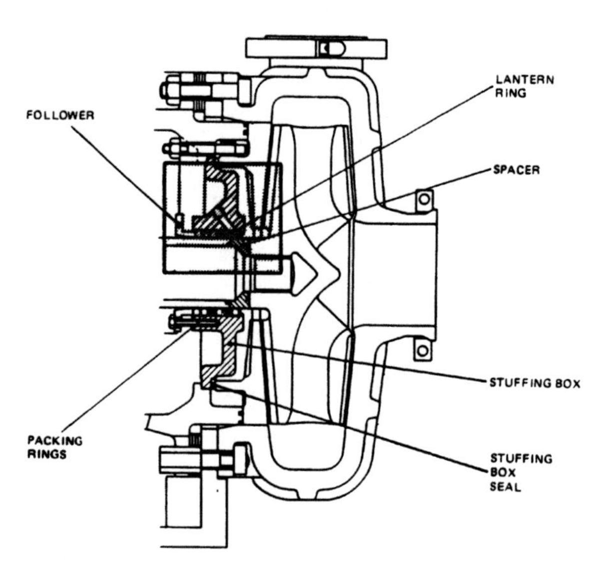

Fig. 14.2. Water-flushed packed gland. (Courtesy of Denver Process Equipment Ltd, Leatherhead, UK.)

3.5.4. Expeller-Aided Packed Gland

The provision of flushing water is expensive, both in capital and running cost, and difficulties can be experienced with ensuring a constant supply. In certain circumstances, the addition of flushing water to the slurry may be unacceptable. To eliminate the need for flushing water, expeller-aided packed glands have been developed as previously outlined (shown in Fig. 14.1). Here, the aim is to reduce the internal shaft seal pressure to atmospheric pressure, such that slurry does not pressurise the gland. The gland is then lubricated by grease, which has to be regularly applied to the gland through a grease nipple.

For a given pump speed there will be a limit to the pump inlet head that is acceptable and higher inlet heads will cause the shaft seal to leak. To ensure correct pump selection it is essential that the inlet head for any application is stated if an expeller seal is to be used. Where the inlet head is too high for an expeller seal alone, an expeller and differential impeller combination may be used as previously described. The inlet head that an expeller-aided packed gland can handle is dependent on design. As a general guide, the inlet head, expressed as a percentage of the differential head, varies between 7% at shut valve to 14% at best efficiency point.

The main disadvantage of an expeller seal is that it is a dynamic device and ineffective when the pump is stopped. Ideally, suction and discharge valves should be closed upon shutdown, such that the gland is not pressurised by slurry. The expeller also absorbs power in generating the sealing head. This is not as high as might be expected, because fewer rings of packing are used and the gland friction is lower. Nevertheless, they are slightly less efficient than pumps with standard packed glands. Because the limitation for an expeller seal is the pump inlet head, they are not suitable for multi-staged pump applications where one pump pressurises the inlet of a second pump.

3.5.5. Mechanical Seals

Mechanical seals would seem to offer the perfect solution to shaft sealing but this has not proved to be the case, except for slurries of low abrasion or low concentration where wear of the seal faces and cooling are less problematic.

Mechanical seals wear because they rely on a thin film of liquid to lubricate and cool the seal faces. Although the flow through the seal is extremely low, it introduces solids between the faces. To combat this, hard seal faces such as silicon carbide are used. However, wear is still a problem and the cost of using these materials, bearing in mind that the seals are large relative to the pump size, are prohibitive. Typically, a mechanical seal can add 50% to the price of a pump.

Cooling is a problem with slurry mechanical seals because it is not desirable to have a stream of slurry flowing over the seal. Therefore the whole of the area surrounding the mechanical seal has to be opened up to allow circulation of slurry within the housing area. Even with these modifications, cooling the seal faces is still problematic, and often seal faces crack through overheating. Obviously an external quench of clean-liquid flush can be applied to the seal, but there is then little advantage over water-flushed packed glands which are far cheaper.

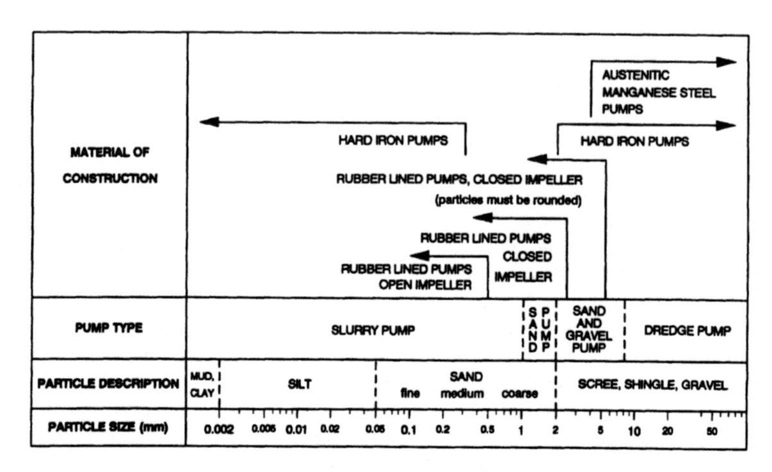

Fig. 14.3. Effect of particle size (of sand-hardness particles) on pump construction.

It should also be mentioned that packed glands form a hydrodynamic shaft bearing which helps to carry impeller loads, but this is not true with mechanical seals. Consequently far greater care has to be taken in selecting pumps for use with mechanical seals to ensure that shaft deflection is acceptable.

4. CONSTRUCTION MATERIALS

4.1. General Guidelines

The choice of wear parts for a slurry pump is made on the basis of balancing the resistance to wear and cost of manufacture. There are two strategies that can be followed. Either the material has to be very hard to resist the cutting action of impinging solids, or it can be elastic, such that it absorbs the shock and rebounds.

The effect of particle size (of sand-hardness particles) on the material of construction is shown schematically in Fig. 14.3. For slurries containing large particles (approx. > 4 mm) the momentum of the particle is such that elastomers fail rapidly owing to the material cutting. For small particles (approx. $< 75\,\mu$m) there is generally no difference in wear between hard metal and elastomer wear parts. Between these two particle sizes elastomer-lined pumps can show a considerable wear-life advantage over metal pumps. However, these are general guidelines and other factors such as angularity and density of the particles will affect the range over which elastomers show advantage.

4.2. Metals

The hardness of some commonly pumped minerals, expressed in terms of comparative hardness (Mohs' scale), Knoops' scale and Brinell hardness number (Bhn) are given in Chapter 13. Traditionally, the material of choice for hard metal slurry pumps has been Nihard II, or Nihard IV where greater ductility is required to handle the shock loads created by large particles. These materials have a minimum Bhn of 500. More recently, these alloys have been superseded by the high-chrome irons, which whilst being just as wear resistant, offer a much greater corrosion resistance owing to the chrome content. The change to the chrome iron has been caused mainly by the rising price of nickel making chrome irons more competitively priced against Nihards. Where greater shock resistance is required, such as on large dredge pumps, softer materials with higher ductility such as manganese steels are used.

Chrome irons to BS 4844 grade 3D (22–28% Cr) (BSI, 1986) have been successfully used on slurries with pH values in the range 2–14, though below a pH of 3 the rate of corrosion accelerates. For acidic duties (pH less than 7) and particularly phosphoric acid applications in fertiliser plants the duplex stainless steel CD4 MCU is widely used. However, it is considerably softer than high-chrome iron having a hardness of 275 Bhn and therefore is more prone to wear.

4.3. Elastomers

For good wear resistance, an elastomer requires resilience, tear resistance and, as previously discussed, good thermal stability. These requirements are generally best met by compounds of natural and synthetic rubber. Rubbers also exhibit good chemical resistance and can be used on acids and alkalies. However, rubber is more expensive to produce than cold-poured materials such as polyurethane, because it has to be pressure moulded and vulcanised. In practice it has been found that the ideal elastomer hardness is approximately 45° Shore. For a comparison between the Shore and other hardness scales see Shah (1984, p. 439). Softer materials are more wear resistant, but are limited in mechanical strength, reducing maximum permissible impeller speeds and slurry temperature.

Polyurethanes are generally harder (60° Shore), have good tear resistance and resilience but lack the thermal stability of rubbers. They are therefore excellent on low-velocity applications such as pipe linings and agitator propellers. However, they are much less acceptable for higher velocity application in pumps, where failure through heat build-up in the elastomer is more likely, particularly with high ambient temperature.

Where slurry temperatures exceed 80°C, natural rubbers are not suitable and synthetic rubbers such as butyl have to be considered (see Table 14.1). Although these have good thermal stability, they lack the tear resistance and resilience of natural rubber, and in general do not wear as well. They are also more expensive.

When selecting elastomer-linked pumps, any oversize tramp material entering the pump will cause almost immediate failure of the lining.

4.4. Ceramics

Ceramics have a high wear resistance but are extremely brittle and expensive to

Fig. 14.4. Torque-induced recessed impeller pump, with a cup-shaped impeller. (Courtesy of Wemco Pumps, Northampton, UK.)

produce. Their use has therefore been very limited on slurry applications, and they are not considered for general slurry applications.

5. SPECIAL DESIGNS

5.1. Vortex Impeller Pumps

Vortex impeller pumps (Fig. 14.4) are also known as *recessed impeller* pumps. They are characterised by open impellers with large clearances between the front of the impeller and the casing.

They are widely used for applications such as raw sewage, where stringy materials might wrap around a standard impeller vane causing the pump to block. They are also widely specified for handling delicate particles where degradation is to be avoided. A typical application is carbon particles in carbon-in-pulp gold-recovery plants.

In applications where the slurry contains gas the open impeller design is useful in preventing the formation of an air or gas lock in the pump. In this situation, gas in the liquid moves to the centre of the inlet pipe, owing to the centrifugal forces induced by flow pre-rotation as it approaches the pump, and is trapped in the eye of the impeller. A stable cone of gas gradually builds up at the low-pressure area in the eye of the impeller, throttling the flow until the liquid velocity through the annulus is sufficient to stop the cone growing any larger. This situation is less likely

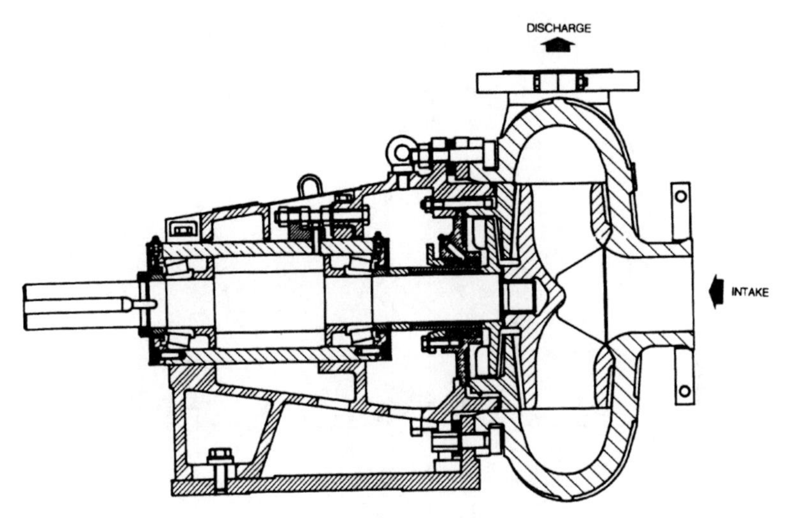

Fig. 14.5. Gravel pump. (Courtesy of Denver Process Equipment Ltd, Leatherhead, UK.)

to arise with a vortex impeller, presumably because of the recirculating flows within the casing that cause the gas cone to break up.

It is not clear where the benefits of this type of impeller come from, because even though particle contact on the impeller vane may be less, this type of impeller is less efficient (typical peak efficiency 55%) indicating increased shear within the casing and presumably greater particle–particle interaction.

5.2. Gravel and Dredge Pumps

Gravel and dredge pumps are similar to slurry pumps, but have wider impellers with fewer vanes to enable a larger particle size to be passed (an example is shown in Fig. 14.5). Generally they have better suction characteristics but show lower peak efficiencies (typically around 60%). Because of the wider impeller and the consequent higher bearing loads, these pumps usually cannot generate heads as high as the equivalent slurry pumps. Standard slurry pumps should be used unless they will not pass the particle size required.

5.3. Vertical Sump Pumps

Vertical sump pumps are normally adaptations of horizontal slurry pumps, as shown in Fig. 14.6. It is essential for these applications where the pump head is immersed in the slurry that no bearings are below the slurry level in the sump. The impeller is therefore carried on a cantilevered shaft with the bearings above the support plate. No shaft seal is required as any leakage is simply recirculated to the

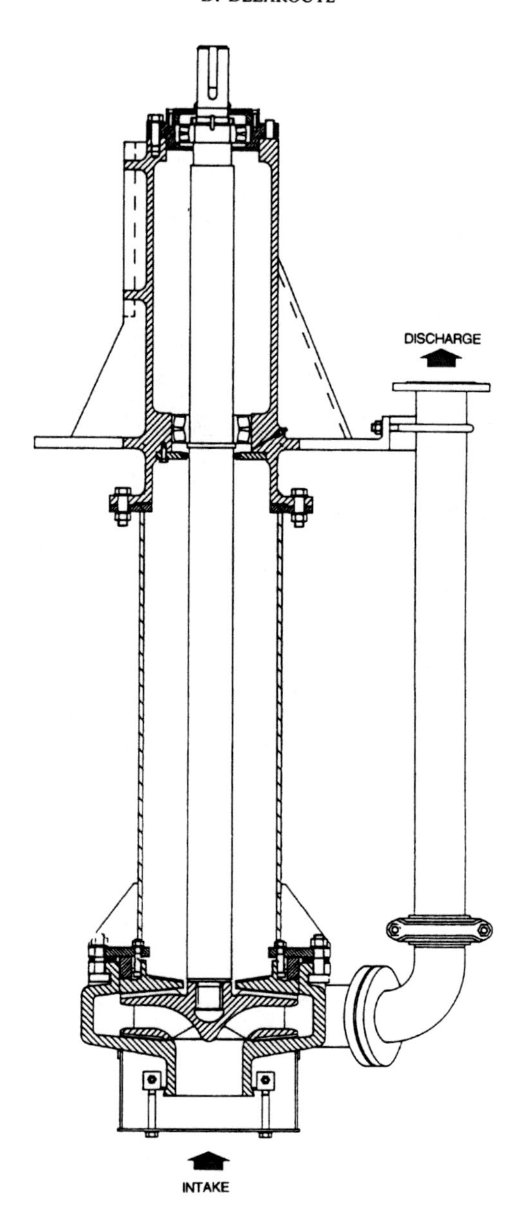

Fig. 14.6. Vertical sump pump. (Courtesy of Denver Process Equipment Ltd, Leatherhead, UK.)

sump. Differential, or top and bottom entry impellers are often employed to reduce this leakage to a minimum.

The depth to which these pumps can be used is dependent on the diameter of the shaft and the spread between the bearings. Sump depths of 2 m can be accommodated without problem and depths to 3 m are possible if the pump does not take air (when it is often said to *snore*). Greater depths are possible using longer tail pipes, within the suction capabilities of the pump, but in order to prime, the sump level must be above the impeller.

A problem encountered frequently with vertical sump pumps is that the fundamental frequency of the unit often falls within the pump operational speed range, particularly with longer shafts. If the pump operates within 20% of the fundamental frequency then the pump can be subjected to severe vibrations. Therefore, it may be necessary to lengthen or shorten the pump depth to ensure an adequate margin of safety. Users should bear in mind that the installed fundamental frequency of the set depends upon the pump foundations, and it is the customer's responsibility to ensure that the installed fundamental frequency does not pose a problem for pump operation.

In addition to partially submersible sump pumps, there are several designs of fully submersible pump where the entire pump, consisting of casing, drive shaft and motor, are immersed often in several metres of slurry or supernatant liquid. These are used, for example, to empty sewage sludge lagoons, or to fluidise and remove compacted sediments contained under supernatant liquid. An example of a fully submersible pump is shown in Fig. 14.7.

5.4. Froth Pumps

Froth pumps are vertical shaft slurry pumps with integral sumps (see Fig. 14.8) generally used on the output from froth flotation cells. Centrifugal pumps cannot pump froths, and therefore require sufficient volume in the sump to hold the froth until broken down to a liquid. The froth is broken down by being sheared in the sump either by reverse swirl spirals at the entrance to the pump, or by mechanical agitators on the pump shaft, or by use of water sprays. To provide enough volume within the sump for froth retention, a froth application factor is applied for pump sizing. This factor is higher the more stable the froth and is multiplied by the flow rate to size the unit. However, power is calculated from the liquid flow rate. Froth factors can generally be estimated only from experience with the slurry and must be supplied by the user.

Froth pumps have been used as general slurry pumps that are self-regulating, in that when the sump empties, the pump partially air locks until the level in the sump rises again. The cantilevered shaft design limits the head that can be generated.

5.5. Mixed Flow Volute Pumps

Mixed flow volute pumps are a cross between a screw pump and a centrifugal pump. The entrance to the impeller is the screw and the exit the centrifugal part

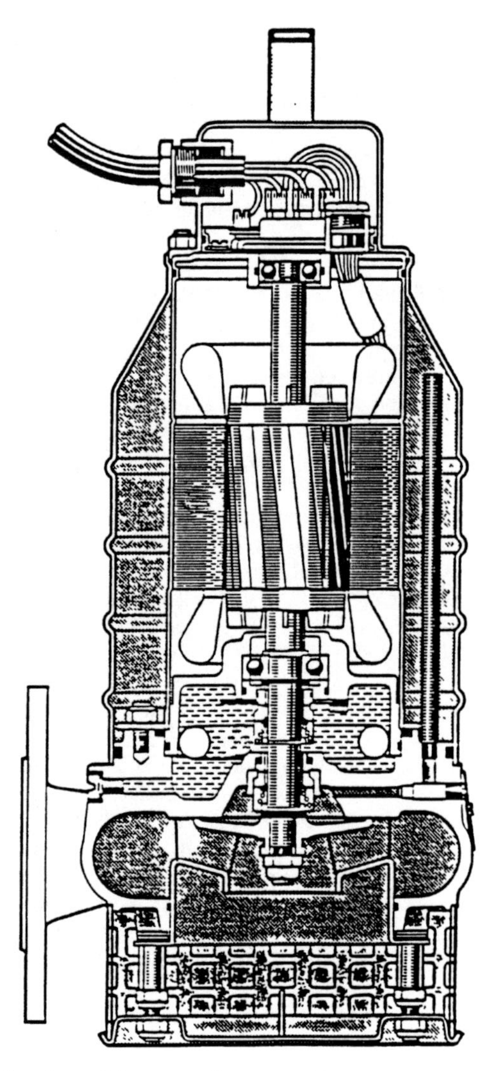

Fig. 14.7. Fully submersible pump. (Courtesy of Flygt Pumps Ltd, Nottingham, UK.)

of the pump as shown in Fig. 14.9. This arrangement allows the pump impeller to be cut to suit different duties at a given speed. The design has found application in sewage and handling soft solids, with minimal damage to solids. Peak efficiencies are comparable to standard centrifugal pumps, but rely upon the seal between the impeller and casing. The rubbing action in this area quickly opens the gap in the

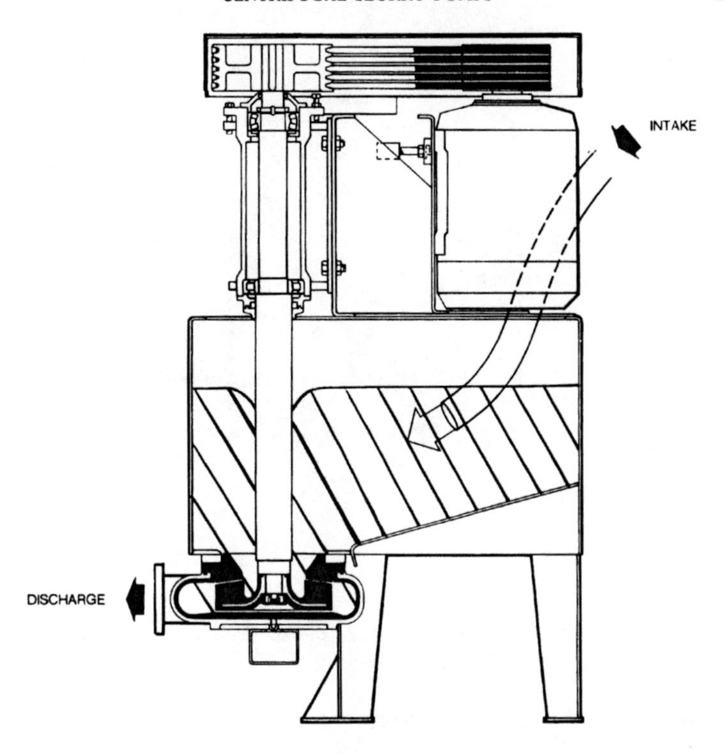

Fig. 14.8. Froth pump.

presence of abrasive solids and causes large drops in efficiency. It has not therefore found widespread use in slurry applications, but, because it is aimed largely at sewage sludge applications in the water industry, a fully submersible design of the pump is therefore also available.

6. PUMP CHARACTERISTICS

6.1. Typical Characteristics

A typical pump characteristic curve for a slurry pump is shown in Fig. 14.10. The pump characteristic provides the relationship between differential head, efficiency and net positive suction head required, $(NPSH)_R$, against volumetric flow rate and pump speed. These curves are produced from tests carried out on water and have to be derated for the effects of solids. In general, solids cause the pump to generate less head and to be less efficient. Peak efficiencies vary with pump size; a typical value for a pump with an entrance diameter of 150 mm and above is about 75%.

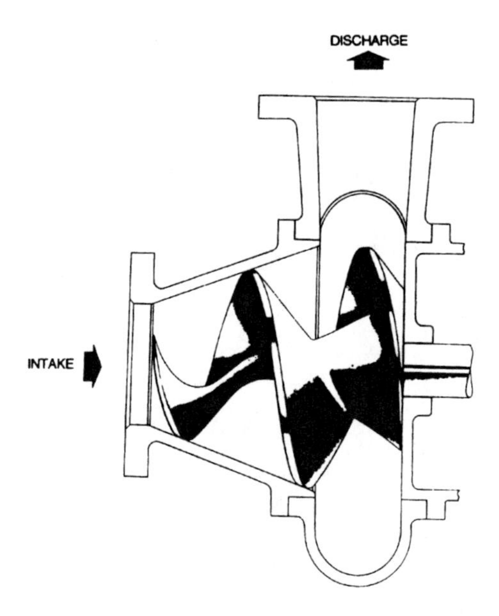

Fig. 14.9. Wet end of a mixed flow volute pump. (Courtesy of Hidrostal Process Engineering Ltd, Newbury, UK.)

6.2. Derating Head and Efficiency

There are two methods of derating pumps. Non-settling slurries can be treated as a pseudohomogeneous liquid. For Newtonian slurries that are fully described by a shear independent viscosity, the pump can be derated using charts published by the Hydraulic Institute (1983). Non-Newtonian liquids present a greater challenge for which there is no general consensus as to how they should be treated. Consequently, applications using such liquids should be referred to an experienced pump supplier who is familiar with the application.

Alternatively, and in particular for settling slurries, where it is not possible to define a shear-independent viscosity, deration factors can be calculated from formulae given by Cave (1976) or Sellgren (1979). These require knowledge of the liquid carrier density and average particle size, density and concentration. Cave's formula is purely empirical, and Sellgren's is semi-empirical. Both have been found to give similar results over the normal range of operation, using the average weight particle size and in the case of Sellgren, using an empirical formula to link particle size to drag coefficient. Both formulae and experience with pump applications show that the head and efficiency deration factors are similar over most of the normal range of operations.

Figure 14.11 can be used to obtain approximate deration figures. The broken line indicates use. The figure is entered with the average particle size

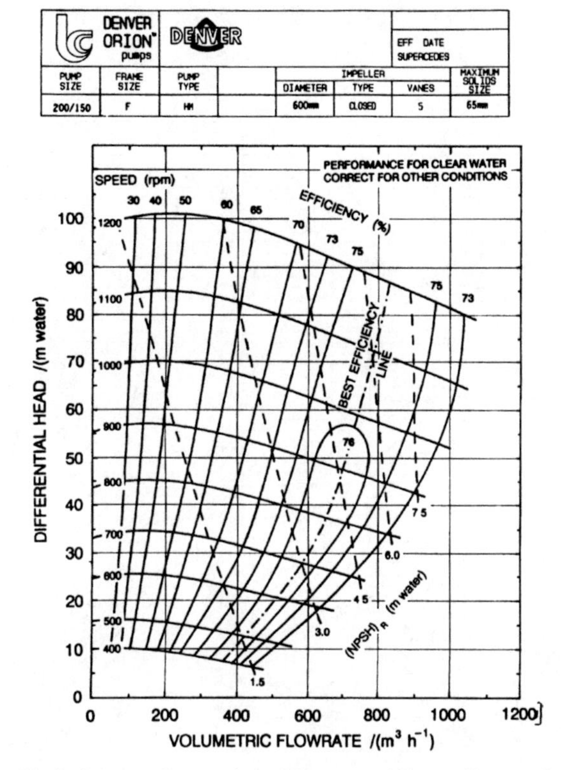

Fig. 14.10. Typical pump characteristic. (Courtesy of Denver Process Equipment Ltd, Leatherhead, UK.)

in the slurry on the lower horizontal axis. On encountering the line representing the density of the solids a horizontal path is taken to the line representing the mass concentration of solids. The deration factor is read from the upper horizontal axis.

6.3. Net Positive Suction Head

The $(NPSH)_R$ characteristics, shown in Fig. 14.10, are again based on tests conducted on water. The effect of solids is, in general, to lower the $(NPSH)_R$ as the pressure differential across a solid particle tends to push the particle into the vapour cavities formed, suppressing cavitation. However, there is no way to quantify this improvement and as such the safest approach is to treat the water $(NPSH)_R$ as the slurry requirement.

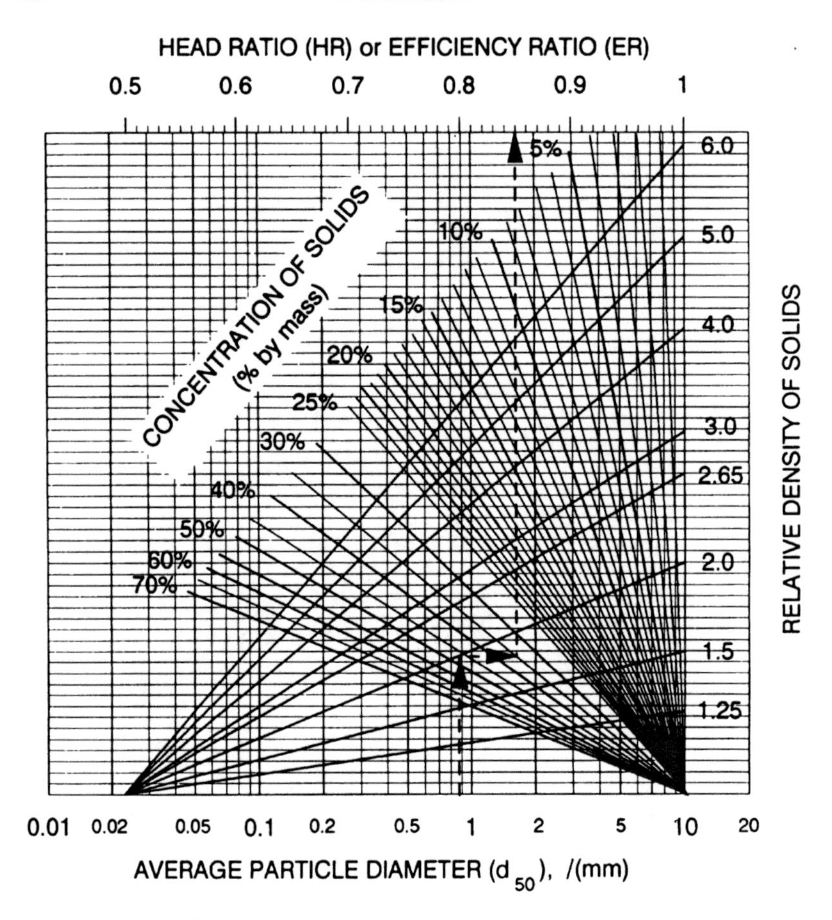

Fig. 14.11. Graphical representation of the deration of water performance for slurry duties in terms of differential head and efficiency. (Adapted from Cave (1976).)

$$\left(HR = \frac{\text{Differential head on slurry}}{\text{Differential head on water}} \qquad ER = \frac{\text{Efficiency on slurry}}{\text{Efficiency on water}} \right)$$

7. PUMP SIZING

7.1. The Best Efficiency Point

The best efficiency line for the pump running at different speeds is shown by the chained line in Fig. 14.10. The best efficiency point is located at the intersection of the best efficiency line and the line relating differential head to volumetric flow rate at a particular pump speed. It is the optimum operating point for the pump. This is particularly true for higher speeds, as at this point for any given speed the radial loads imposed on the impeller are at a minimum. Consequently bearing and shaft seal life are at a maximum. Furthermore, if expeller-aided glands are used, the inlet head capabilities are at a maximum. For lower flows, bearing and seal life is reduced as is the expeller seal performance. For centrifugal slurry pumps generally, wear is least at the best efficiency point because here the flow is least turbulent and the flow over the impeller vanes most efficient.

7.2. Entrance Velocity

Although the best efficiency point is hydraulically the optimum operating point, consideration has to be given to the velocity with which the solids enter the impeller as the flow has to be turned through 90°. Particles that enter with a high momentum will strike the exposed shroud with a high impulse force, and cause excessive wear.

For coarse abrasive particles, it is normal to limit the entrance velocity to $4{\cdot}5\,\mathrm{m\,s^{-1}}$, without regard to the best efficiency point. As a guide, very abrasive duties are defined as slurries with sharp particles of average particle size greater than 0·5 mm (which includes mill discharge and crushed bottom ash duties). For the majority of normal slurry duties, the limit on entrance velocity is raised to $6{\cdot}5\,\mathrm{m\,s^{-1}}$ or the best efficiency line, whichever is higher. Using these limitations the appropriate pump size can be selected.

7.3. Peripheral Velocity

Pump users generally specify impeller speed limits to minimise pump wear rates. However, to generate a given head a certain peripheral velocity will be required which will be approximately the same for any pump design. Thus, the rotational speed of the pump is immaterial, unless the particles are coarse (typically >4 mm) in which case the velocity of impact at the vane entrance (which is proportional to rotational speed) becomes the limiting factor.

Imposing rotational speed limits on manufacturers can be counter-productive, in that they may force the manufacturer to select oversized pumps, running well away from best efficiency. Therefore, in general, pump speed limits should be stated as peripheral impeller velocities and not rotational speed. Where peripheral velocity limits are given to manufacturers, and the head cannot be generated within these limits, then series pumps must be considered. Whilst this may limit wear rate, it may not be decreased sufficiently to compensate for the wear on additional pumps and the cost of flush water required at the pump gland of multi-staged units.

7.4. Multiple Pumps

Where head or flow requirements are greater than can be achieved from a single unit, pumps can be used in parallel to increase flow or in series to increase head.

There are no specific problems with operating pumps in parallel, but for flows over 10 000 $m^3 h^{-1}$, partial flow, parallel pumps may offer lower capital and operating costs, particularly if stand-by pumps are required.

There may be a number of disadvantages in splitting a single-stage duty into two pumps in series, in an attempt to reduce wear. Capital costs are increased significantly, and gland sealing in the second or subsequent stages is more problematic owing to the high inlet heads. High pressure, water-flushed packed glands or mechanical seals would be required with control systems that ensure flushing water is applied before start-up.

8. ANCILLARY EQUIPMENT

8.1. Drives

By far the most common drive is the squirrel cage induction motor, which is relatively cheap and reliable. It is common practice to have a minimum power margin over the calculated absorbed power of 15%. This margin allows for uncertainties in the duty calculations and duty modifications at a later date.

For centrifugal pumps, there is little, if any, advantage in the use of star/delta starters, as the current step from star to delta is similar to that starting direct on line.

With belt drives, it is normal to select four pole motors, as this provides the cheapest drive arrangement.

Much less common are diesel and petrol engine drives, but where the pump is remote from electrical power lines, their expense and high maintenance costs may be a secondary consideration. For engine drives it is common to specify a minimum 25% margin on absorbed power. This reflects the relative difficulty and expense of changing an engine drive, should the power available prove inadequate.

8.2. Drive Arrangements

There are several drive arrangements available for electric motors with belt drives. The most common are the side mounted and overhead mounted motors. Overhead mounting is generally the cheapest, and lifts the motor off the floor away from spillage. The size of motor however, is limited by the size of the pump frame, and overhead motors may make servicing of the pump, particularly if it is a back pull-out design, more difficult. For this reason, side-mounted motors, with slide rails for belt tensioning, are the preferred arrangement.

8.3. Transmission

8.3.1. Fixed-Speed Drives

Because it is difficult to machine impellers manufactured from hard metals, and impossible for elastomers, duty points are gained by speed change through the use of Vee belt drives. Given the variability of slurries, this arrangement has the added advantage that within the power available from the drive, the duty can be *fine tuned* following installation or to allow for process changes at a later date.

Provided the belts are tensioned correctly, modern Vee belt drives are extremely reliable, with a life expectancy of 40 000 h, and very efficient, with power losses typically less than 0·5%. Their disadvantages are that they require re-tensioning after an initial running-in period and give only discrete speeds, depending on the pulley ratios available. For very low speeds, particularly on large dredge pumps, the speed ratio and power may be too large for Vee belt drives, and a gear box transmission may be required. The speed ratio limit is generally about 5:1 for belt drives.

8.3.2. Variable Speed Drives

There are significant advantages in the use of variable speed drives for centrifugal slurry pumps (Odrowaz-Pieniazek & Steele, 1986), particularly on longer pipe lines (over 200 m). Only their cost, which is considerable, prevents wider use. With variable speed drives a centrifugal pump can be given the characteristics of a positive displacement pump by tying the speed to a flowmeter signal. Changes in concentration or particle size then have minimal effect on flow rate. Should the line start to block, the speed will increase to keep the flow velocity constant and help prevent blockage.

In the past, fluid couplings were widely used for speed control, but more recently electronic drives, particularly variable frequency, have found widespread use. Variable frequency drives have many advantages. They can be used with standard pump and motor sets without modification and can be retrospectively fitted. They provide some built-in control facilities such as speed or power limits and can be readily tied into control systems. Their cost can be offset against the starter equipment they replace.

9. REFERENCES

Bloch, H.P. & Johnson, D.A. (1985). Downtime promotes upgrading of centrifugal pumps. *Chem. Engng*, 25 Nov., **92**(24), 35–41.

BSI (1986). BS 4844: Specification for abrasion-resisting white cast iron. British Standards Institution, London, UK.

Cave, I. (1976). Effects of suspended solids on the performance of centrifugal slurry pumps. In *Proc. HT 4*, paper H3, pp. H35–52.*

Hydraulic Institute (1983). Standards for centrifugal, rotary and reciprocating pumps. Hydraulic Institute, Cleveland, Ohio, USA.

Odrowaz-Pieniazek, S. & Steele, K. (1986). Advances in slurry pumps, Part 2. *Chem. Engr*, Mar., **423**, 30–3.

Sellgren, A. (1979). Performance of a centrifugal pump when pumping ores and industrial minerals. In *Proc. HT 6*, paper G1, pp. 291–304.*

Shah, V. (1984). *Handbook of Plastics Testing Technology*. J. Wiley & Sons, Chichester, UK.

Stepanoff, A.J. (1957). *Centrifugal and Axial Flow Pumps*. John Wiley & Sons Inc., New York, USA.

Wiedenroth, W. (1988). Wear of solids handling centrifugal pump impeller. In *Proc. HT 11*, paper K2, pp. 507–22.*

*Full details of the Hydrotransport (HT) series of conferences can be found on pp. 11–12 of Chapter 1.

15

Reciprocating Slurry Pumps

Peter E. Snoek & Ramesh L. Gandhi
Bechtel Inc., San Francisco, USA

1. INTRODUCTION

1.1. General

A reciprocating pump is a positive-displacement machine. It traps a fixed volume of liquid at or near suction conditions, compresses it to discharge pressure and pushes it out through the discharge connection. This is accomplished by the reciprocating motion of a piston, plunger or diaphragm.

A reciprocating pump can be driven by a rotating drive such as an electric motor, engine or turbine. Alternatively, it can be driven directly by a motive fluid by means of differential pressure. The motive fluid may be steam, air or other gases and hydraulic oil.

1.2. Mud Pumps

Long-distance transportation of slurries is generally accomplished by using reciprocating slurry pumps. However, the initial application of these pumps in slurry service occurred nearly a century ago in circulating fluids for oil-well drilling.

Normal drilling muds consist of colloidal suspensions of clays in water with chemical additives to control viscosity and other properties. For obtaining the high pressures required for pumping this slurry, reciprocating, gear driven, triplex or duplex piston pumps were used. Since a breakdown during operations could have serious consequences including the loss of the hole, the mud pumps were sturdily constructed so they could sustain service under heavy loads.

1.3. Use in Long-Distance Slurry Pipeline Systems

Slurry pipelines have been used for the transport of dredgings and storm and sanitary sewage for a number of years. However, it was only in the 1950s that the technology was sufficiently developed to allow this means of transport to be used for long-distance transport in the mineral industry. Notable examples during the decade from 1950 to 1960 include the following:

- 9 km limestone pipeline in Trinidad,
- 10 km coal pipeline in France,
- 116 km gilsonite line in the state of Utah, USA, and
- 174 km coal pipeline in the state of Ohio, USA.

All four of these systems used positive displacement pumps to achieve the high discharge pressures required. However, it was the last system that firmly established the technical and economic viability of long-distance slurry transport. The system used three pump stations to transport over 1 Mty^{-1} through a 273 mm (10·75 in.) o.d. pipeline to an existing power plant near Cleveland, Ohio for a period of about six years.

In the following three decades, many additional systems have been constructed for transporting other materials such as limestone, copper concentrate, iron concentrate, phosphate and various tailings. As larger and more remote mining projects were developed, slurry pipelines also became larger as well as longer.

2. DESIGN AND CONSTRUCTION

2.1. General

As indicated in the previous section, reciprocating pumps have been used for most of the long-distance slurry pipelines constructed to date. Although the capital cost is higher for reciprocating pumps than for centrifugal units, reciprocating pumps have several desirable features.

- Reciprocating pumps can meet any reasonably desired system discharge pressure requirement. Pumps are in operation today that produce pressures up to 17·9 MPa. Units are available which are capable of producing discharge pressures over 34·5 MPa.
- The overall efficiency of the pump, including the drive train, is relatively high —in the order of 85%.
- The flow rate of the pump is almost independent of system pressure.
- Pipeline flow rate can be determined without the use of a flowmeter.

On the other hand, there are a number of disadvantages that are also associated with these units.

- The maximum installed power per pump is limited to about 1·5 MW. Therefore a high flow, high discharge pressure pump station will require a large number of reciprocating pumps in parallel to obtain the desired flow conditions.
- Pump maintenance costs are relatively high. Skilled labour is required for operation and maintenance.
- Variable speed drives are required to vary flow rates and to start the pumps.
- Pulsating flow is produced which requires greater attention to station piping design to avoid vibration and fatigue problems.
- The maximum size of particles that can be pumped is restricted by the check valve seal requirements. (In practice, this is not usually a disadvantage as the particle size for mining operations is often set by a concentration step which results in a relatively fine product.)

2.2. Pump Classification

Reciprocating pumps are usually classified by their features.

- Power end—power driven (using a crank and throw mechanism) or direct acting (connected to a steam engine).
- Fluid end orientation—horizontal or vertical.
- Single-acting or double-acting.

- Configuration of fluid end—piston, plunger or diaphragm.
- Number of cylinders—simplex (one), duplex (two) or multiplex (more than two).

The size of a power driven pump is normally designated by listing first the diameter of the plunger (or piston) and then the stroke length. For a direct driven pump, the same convention is followed, except that the diameter of the drive piston precedes the fluid-end-element diameter.

2.3. Piston Pumps

Of the different types of reciprocating pumps that are available for slurry service, piston pumps are generally selected for pumping low to moderately abrasive slurries (i.e. with a Miller number below 50, see Chapter 13) for pressures under 14 MPa. Various configurations are available depending on the number of cylinders and whether the pump is a single-acting or double-acting type.

Fig. 15.1 shows a cross-section of a typical double-acting slurry piston pump. On the left is the power end where rotative motion of a driver (electric motor, diesel engine, etc.) is converted into a reciprocating motion. This is accomplished by means of a crankshaft which transmits energy through a connecting rod to the crosshead which moves back and forth along the crosshead guide.

On the right is the fluid end of the pump. The crosshead extension is directly connected to a piston rod which moves the piston within a replaceable cylinder liner. The piston is comprised of three basic elements: (1) the sealing elastomer, (2) the fabric section, and (3) a back-up metal plate section. Its speed is variable throughout the stroke, depending on the position of the connecting rod (see Fig. 15.2 for details of piston construction). Since the unit shown is a double-acting pump, slurry is discharged on both the forward and backward movements of the piston.

Figure 15.3 shows the fluid end construction. As pictured, the piston rod passes through a stuffing box which contains rings of packing to prevent slurry from leaking from the cylinder. Figure 15.4 shows details of the stuffing box construction. Henshaw (1981) reviews many other designs.

An oily water flushing system is normally provided to remove any solid material which collects on the packing to minimise wear of the packing as well as the piston rod. A suction and discharge valve is found at each cylinder end. These valves open by the liquid differential pressure and allow flow in only one direction. Each valve is comprised of a plate and a tapered seat which fits into the cylinder. As shown in Fig. 15.5, they may have a variety of shapes. The plate movement is controlled by a spring or retainer. Valve covers allow access to the valves without disturbing the cylinder.

Figure 15.6 shows a performance curve for a typical piston pump. The maximum pump discharge pressure and capacity are shown as a function of the piston diameter. For a given pump power-end rating, constant connecting rod loading and a constant crankshaft displacement, as the diameter increases, the flow rate also increases but the allowable pressure is reduced.

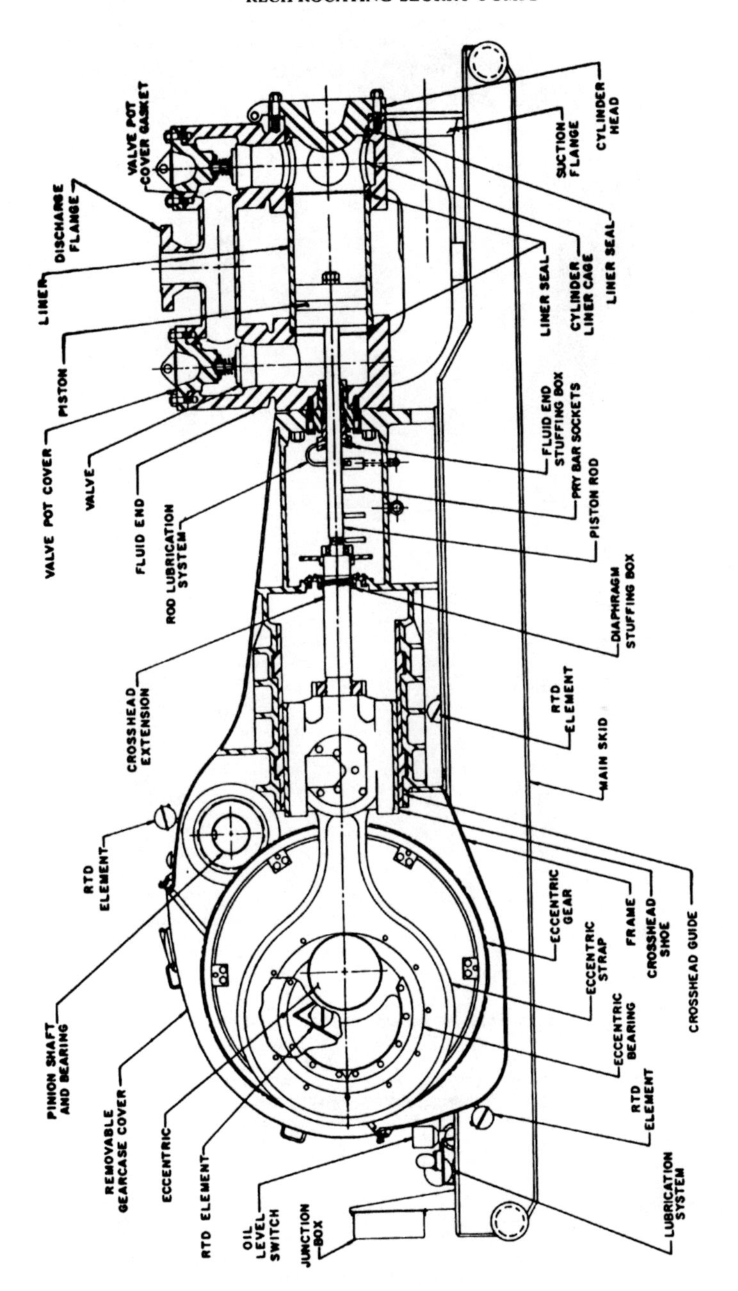

Fig. 15.1. Cross-section of double-acting slurry piston pump.

1. RUBBER COMPOUND FOR SLURRY APPLICATION
2. SUPPORT-AND-WEAR RING
3. REINFORCING FABRIC
4. RETAINER RING
5. SNAP RINGS
6. PISTON BODY

Fig. 15.2. Slurry pump piston.

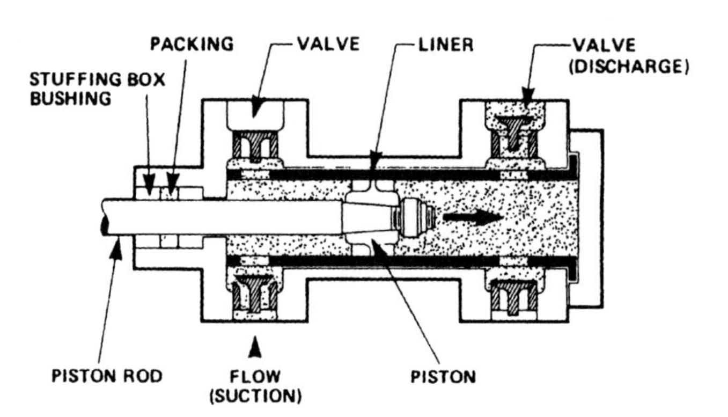

Fig. 15.3. Piston pump—fluid end.

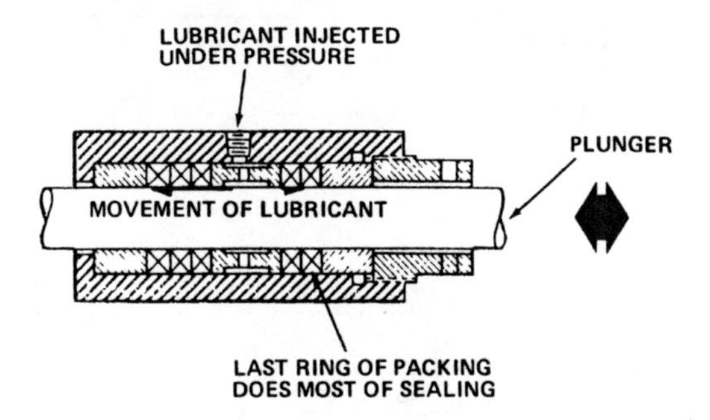

Fig. 15.4. Stuffing box.

As the piston strokes, the discharge pressure fluctuates with the changing flow rate. Figure 15.7 shows an oscilloscope trace of the varying discharge pressure over one revolution of a duplex piston pump. It should be noted that the sudden opening of a pump discharge valve caused a water hammer *shockwave*.

Figure 15.8 traces the typical sawtooth suction pressure transient for the same

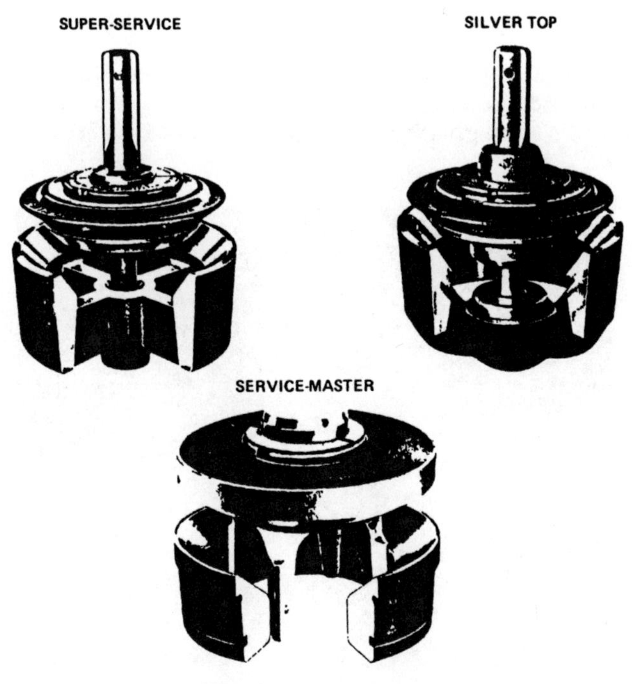

Fig. 15.5. Pump valves.

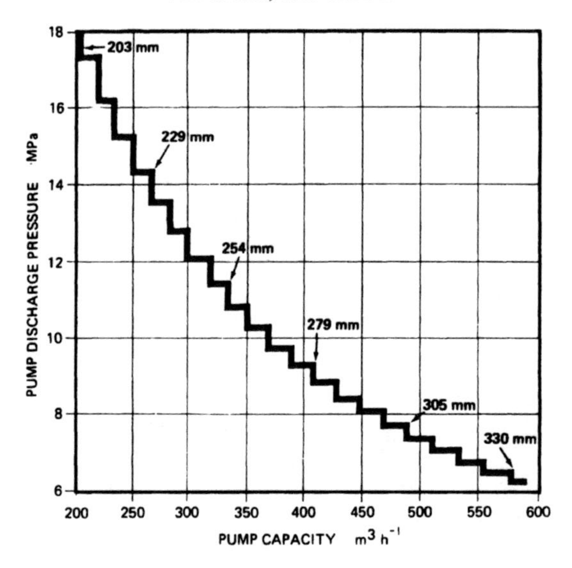

Fig. 15.6. Typical piston pump performance curve as a function of piston diameter.

pump. As a piston starts on a suction stroke, the suction valve opens and the suction pressure drops. The pressure then builds up until the next suction valve opens. A sudden opening of the pump suction valve caused another water hammer *shockwave*.

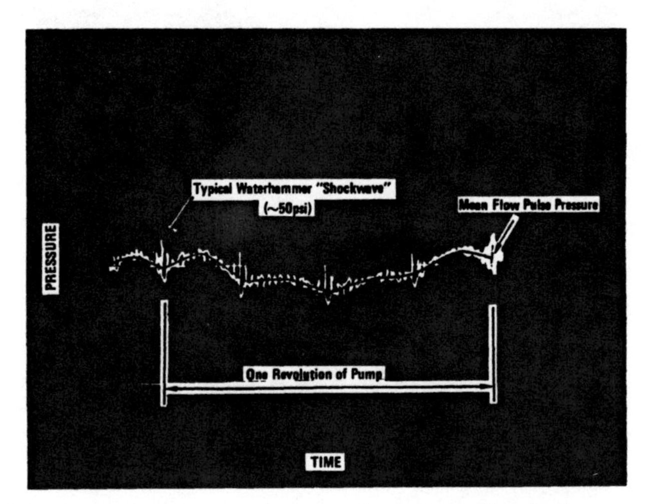

Fig. 15.7. Trace of discharge pressure transient of a positive displacement pump.

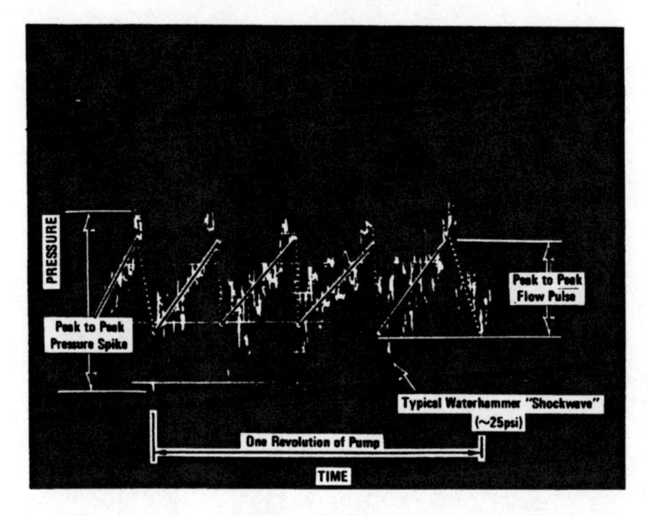

Fig. 15.8. Trace of suction pressure transient positive displacement pump. (Note—centrifugal booster pump provides positive suction pressure to positive displacement pump.)

2.4. Plunger Pumps

With abrasive materials, high wear of the pistons, cylinder liners, piston rods and packing may occur. For this service, plunger pumps are often used. A plunger is a smooth rod and can be only single-acting. While the sealing elements move with a piston pump, they are stationary in a plunger unit. This type of pump maintains a clear liquid barrier between the plunger and packing by means of an internal flushing system. Plunger pumps are also used for high discharge pressures (over 14 MPa) when the piston rod diameter becomes so large that a double-acting piston pump will pump very little on its return stroke.

Figure 15.9 shows a cross section of a typical horizontal slurry plunger pump. The power end is similar to that discussed earlier for a piston pump but instead of a piston, a plunger is moved by the crosshead extension rod. Depending on the construction, the pump may be of simplex, duplex or triplex design.

Figure 15.10 gives details of the plunger pump fluid end. A packed stuffing box is again provided but water is injected into a flush bushing placed ahead of the packing. The bushing is designed so that the injected liquid encircles the plunger and flows through the clearance space between the bushing and plunger, thereby cleansing the plunger surface and providing a barrier to the slurry. Lantern rings are used to provide an annular space between the sections of packing so that the injected fluid can freely flow to the rod surface.

In earlier installations, the flush system provided water only during the suction stroke when the pressure in the fluid end was below the supply pressure. A check valve was installed in the flush line to prevent backflow during the discharge stroke.

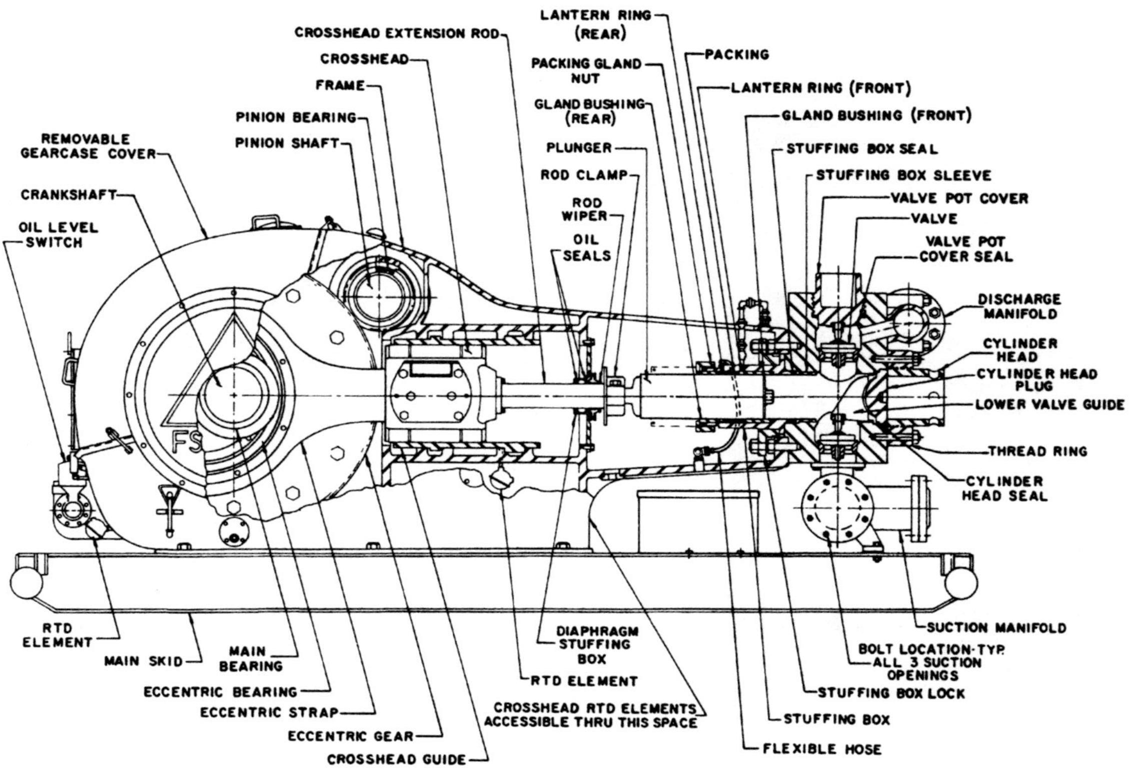

Fig. 15.9. Cross-section of a typical slurry plunger pump.

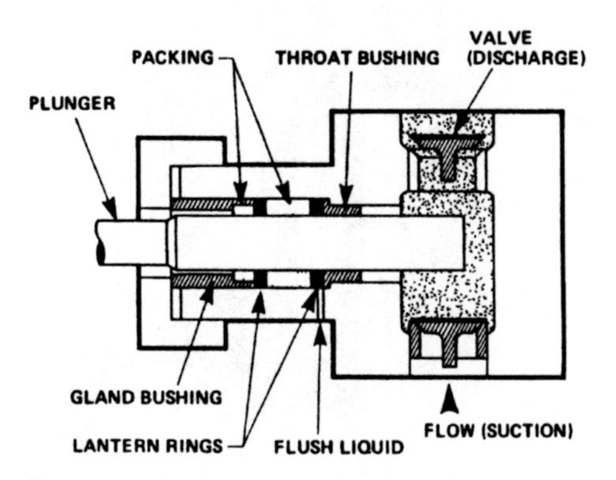

Fig. 15.10. Plunger pump—fluid end.

However, because the check valves were subject to constant cycling, high wear rates were often experienced.

An improved system is to use a synchronised flush in which a separate flush pump provides high-pressure water to each slurry pump during either the suction or the discharge stroke. Alternatively, one large flush pump capable of satisfying the requirements of all the operating slurry pumps can be installed. The flushing rate may range up to 5% of pump displacement with 3% considered a typical figure. A major factor is the cleanliness of the flushing fluid.

Several long distance slurry pipelines have used vertical plunger pumps instead of the more common horizontal units. The vertical pumps possess the advantage of minimising eccentric wear of stuffing box components caused by the weight of the plunger. Although they require less floor area, vertical units result in a loss of suction head and are less accessible for servicing. Figure 15.11 gives details of the fluid end for a vertical plunger pump. Since the solids in the slurry usually settle, a vertical unit should require less flushing water to keep the packing clear of solids.

2.5. Piston Diaphragm Pumps

In order to achieve higher pump availability along with reduced maintenance costs, several pump designs have been developed which use an intermediate transfer fluid to avoid contact between the abrasive slurry and the pump moving parts. Of these, the piston diaphragm pump and the Mars pump are the best known examples.

Figure 15.12 shows the fluid end of a duplex double-acting piston diaphragm unit. The power end is similar in construction to that used for a conventional slurry piston pump. In this case, however, the driving piston transmits energy to hydraulic fluid (oil) which is in contact with a rubber or metal diaphragm. The diaphragm and the pump suction and discharge valves are the only parts in contact with the slurry.

By using large piston diameters and long stroke lengths, diaphragm pumps can be operated at relatively low speeds (40–50 strokes min^{-1}). This in turn allows an

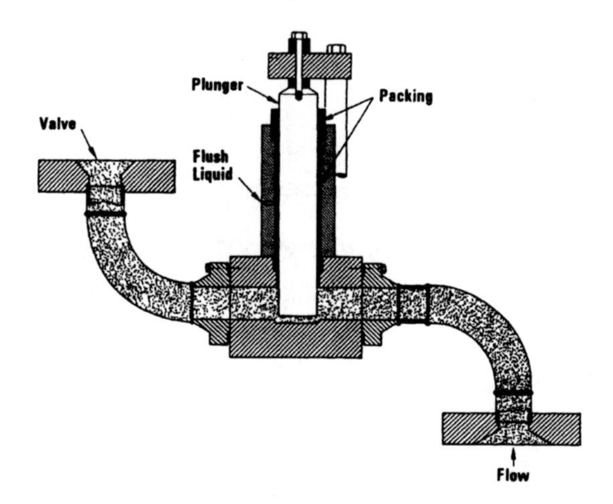

Fig. 15.11. Vertical plunger pump—fluid end.

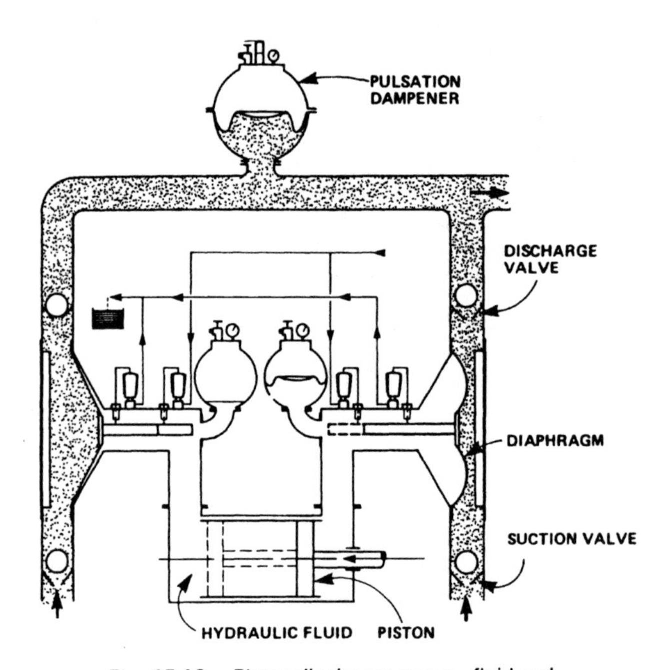

Fig. 15.12. Piston diaphragm pump—fluid end.

increased valve size to be used which results in low valve velocities. The low pump speed also produces a low valve opening and closing rate.

2.6. Mars Pump

Figure 15.13 shows a cross-sectional outline of a Mars pump. The power end is on the right driving a piston in a cylinder similar to a conventional piston pump. The difference is that a measured volume of oil is interposed between the slurry and the piston. The oil chamber is constructed to prevent slurry from penetrating into the working cylinder during pumping or rest. This is achieved by the difference in specific gravity between oil and slurry and the relatively slow movement of the interface due to the large cross-sectional area. The pump typically runs at 45–50 rev min^{-1}, although there is a high-speed/low-pressure model designed to run at up to 100 rev min^{-1}.

Since the only working parts exposed to slurry wear are the slurry valves, the Mars pump offers relatively low operating costs and is of particular interest for abrasive solids; slurries with Miller indices up to 500 can be pumped. However, the special feature (protective oil barrier) leads to a higher capital cost than for typical piston or plunger pumps. Also there have been problems with solids build-up in the isolation section.

2.7. Design

2.7.1. Power Requirement

The power required by the pump, W, (customarily referred to as brake horsepower) is given by the following formula:

$$W = QP/E_m \tag{15.1}$$

where W is the power $[ML^2T^{-3}]$, Q is the flow rate $[L^3T^{-1}]$, P is the developed pressure $[ML^{-1}T^{-2}]$ and E_m is the mechanical efficiency [1].

2.7.2. Displacement

Displacement Q_0 is the calculated capacity of the pump with no slip losses. For a single acting plunger or piston pump it is

$$Q_0 = AnNL \tag{15.2}$$

where Q_0 is the volumetric displacement $[L^3T^{-1}]$, A is the cross-sectional area of plunger or piston $[L^2]$, n is the number of pistons or plungers [1], N is the rotary speed of the pump $[T^{-1}]$ and L is the stroke of pump [L].

For a double-acting plunger or piston pump it is

$$Q_0 = (2A - a)nNL \tag{15.3}$$

where a is the cross-sectional area of the piston rod $[L^2]$.

2.7.3. Volumetric Efficiency

The volumetric efficiency is the ratio of the fluid pumped per stroke to the pump displacement per stroke.

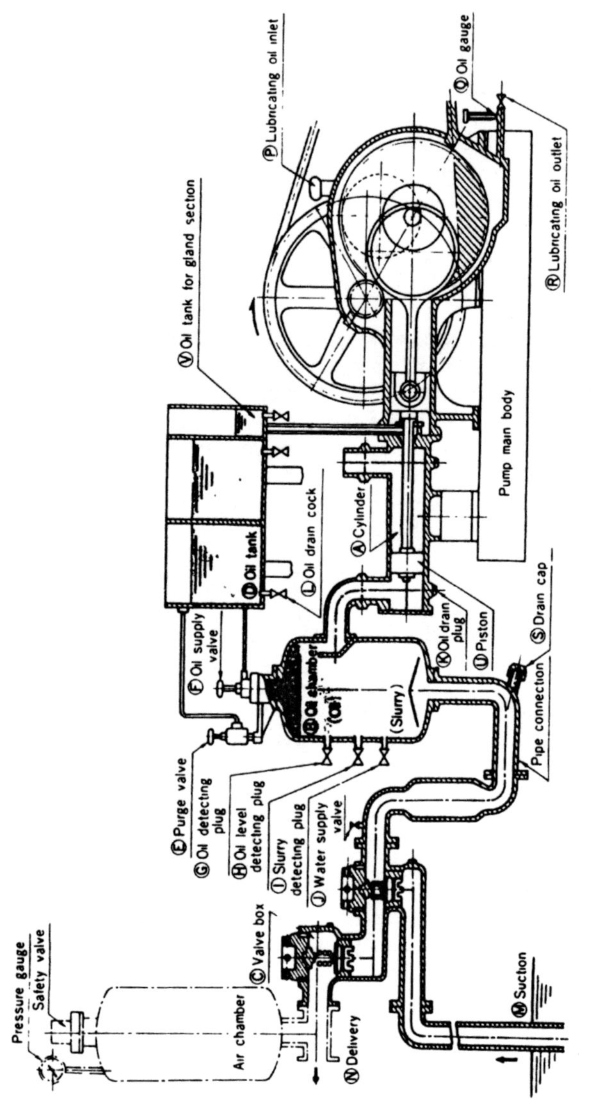

Fig. 15.13. Cross-section—Mars pump.

At each stroke, the cylinder fills with fluid under suction pressure. On the delivery stroke, the fluid is compressed to the discharge pressure and then delivery is made through the discharge valve. The remaining fluid in the fluid chamber at the end of the delivery stroke must re-expand to suction pressure before the suction valve will open. Therefore, some effective plunger or piston travel is lost on the suction stroke resulting in a reduction in the amount of fluid being taken into the cylinder.

The volumetric efficiency, E_v, of a reciprocating pump is based on volume pumped at discharge pressure, relative to piston displacement. It is calculated as follows:

$$E_v = 1 - PB(1 + C/\delta) - S \tag{15.4}$$

The volumetric efficiency based on volume expanded to suction pressure after being pumped to discharge pressure, relative to piston displacement is calculated as follows:

$$E_v = \frac{1 - PB(1 + C/\delta)}{1 - PB} - S \tag{15.5}$$

where B is the compressibility of the slurry; change in volume per unit volume per unit change in pressure $[M^{-1} L T^2]$, C is the fluid chamber volume in the passages between valves when piston/plunger is at the end of discharge stroke $[L^3]$, δ is the volume displacement per piston $[L^3]$, P is the discharge pressure minus suction pressure, in bar $[M L^{-1} T^{-2}]$, S is the valve slip, expressed as a decimal [1], and E_v is the volumetric efficiency expressed as a decimal [1].

EXAMPLE: CALCULATION OF VOLUMETRIC EFFICIENCY

Find the volumetric efficiency, based on suction volume, of a plunger pump under the following conditions.

Pump type: 228·6 mm × 304·8 mm (9 in. × 12 in.) triplex plunger
Fluid pumped: 30% (by vol.) iron concentrate slurry
Suction pressure: 3·45 bar
Discharge pressure: 124·1 bar
Slurry temperature: 20°C
$C = 2452\,mm^3$
$\delta = 12\,510\,mm^3$
$S = 0{\cdot}02$
B water $= 4{\cdot}34 \times 10^{-5}\,mm^3\,mm^{-3}\,bar^{-1}$
B iron concentrate solids $= 6 \times 10^{-7}\,mm^3\,mm^{-3}\,bar^{-1}$

Compute compressibility of the slurry:

$$B \text{ slurry} = (\text{volume fraction water})(B \text{ water}) + (\text{volume fraction solids})(B \text{ solids}) \tag{15.6}$$

$$B \text{ slurry} = 0{\cdot}7\,(4{\cdot}34 \times 10^{-5}) + 0{\cdot}3\,(6 \times 10^{-7})$$
$$= 3{\cdot}056 \times 10^{-5}$$

From eqn (15.5), volumetric efficiency:

$$E_v = \frac{1 - (124{\cdot}1 - 3{\cdot}45)(3{\cdot}056 \times 10^{-5})(1 + 2452/12510)}{1 - (124{\cdot}1 - 3{\cdot}45) \times 3{\cdot}056 \times 10^{-5}} - 0{\cdot}02$$

$$= 0{\cdot}979$$

2.7.4. Valve Slip

Valve slip is the fraction of fluid going back through the valve while closing and/or seated. This loss may vary between 2 and 10% depending upon the condition of the valve and its design.

2.7.5. Rod Loading

A power pump is designed for a rod loading limit. This is the force that is applied to the plunger or piston and the bearing system. The rod load is

$$RL = PA \tag{15.7}$$

where RL is the rod load [MLT^{-2}], P is the discharge pressure–suction pressure [$ML^{-1}T^{-2}$], and A is the cross-sectional area of piston or plunger [L^2].

With double-acting pistons, the piston rod area, a, is subtracted on the forward stroke.

The bearing system is rated for a specific load at design speed. Higher plunger loads will result in shorter bearing life.

2.8. Construction

2.8.1. Cylinder

The cylinder is the part of the fluid end where pressure is developed. Cylinders on many horizontal pumps have the suction and discharge manifolds made integral with the cylinder. Vertical pumps usually have separate manifolds. Instantaneous pressure within the cylinder may be two to three times the design pressure. When the slurry has entrained gas that can be released because of inadequate suction pressure, the cylinder pressure may be four to five times the design pressure. Cylinders may be made from castings or forgings. Because of the abrasive nature of some slurries, cylinder liners are used in the case of piston pumps. The liners should be easily replaceable. Their length is slightly longer than the pump stroke.

2.8.2. Plunger

The plunger may be made out of Colmonoy No. 6 or stainless steel with chrome plating. Ceramic plungers may also be considered. The plungers may be solid up to 127 mm in diameter. Above this size they are generally hollow.

2.8.3. Piston

Pistons are made of cast steel with reinforced elastomer faces.

2.8.4. Stuffing Box

The stuffing box bore is machined smooth to ensure packing sealing. The stuffing box bushings are normally made out of bronze and have a very smooth finish with approximately 1·0–2·0 μm diametrical clearance per mm of plunger diameter.

The plunger packing is V or chevron-shaped. A packing unit consists of top and bottom adapter with a seal ring. A lower injection ring is used for flushing solids away from the packing.

2.8.5. Valve

A slurry valve must be strong enough to resist the load imposed by the fluid pressure and must be able to seal against this pressure in the presence of solid particles. Sufficient support must be given in the metal-to-metal valve seal bearing area without too much reduction of the flow-through area to avoid high fluid velocities which may produce erosion.

The valves are provided with elastomeric (e.g. polyurethane) inserts that extend the life of the valve disc and the valve seat. A valve guide and spring are also provided. The valve guides keep the valve in place and the spring provides a positive return and reduces knocking and excessive surges. The valve assembly as well as the valve seats are replaceable wear parts.

2.8.6. Relief Valves

The relief valves should be set at 25% above the operating pressure for a double-acting duplex pump and 10% above the operating pressure for triplex and multiplex pumps.

3. OPERATION

3.1. Flow Characteristics

A reciprocating pump produces pulsating flow. Table 15.1 compares the variation in flow rate over one pump revolution for duplex, triplex and quintuplex single-acting pumps. It illustrates the advantage of having a large number of cylinders per pump.

The fluid in the suction and discharge pipes will have to be accelerated to match the flow through the pump. Sufficient pressure must be available to accelerate the slurry on the suction side of the pump to prevent cavitation in the suction pipe and/or the pumping chamber.

TABLE 15.1
Flow Variation versus Number of Pump Cylinders for a Single-Acting Pump

Pump	*Number of cylinders*	*Flow variation (%)*
Duplex	2	160
Triplex	3	25
Quintuplex	5	7
Septuplex	7	4
Nonuplex	9	2

TABLE 15.2
Values of Constant k_1

Pump type	*Constant,* k_1
Single-acting duplex	0·200
Triplex	0·066
Quintuplex	0·040
Septuplex	0·028
Nonuplex	0·022

The acceleration head, h_a, required by a reciprocating pump can be estimated using the following:

$$h_{acc} = LVN\,k_1/(k_2\,g) \tag{15.8}$$

where h_{acc} is the acceleration head (in water) [L], L is the actual length of suction pipe [L], V is the average velocity through suction pipe [LT^{-1}], N is the rotary speed of the pump [T^{-1}], k_1 and k_2 are constants (see Tables 15.2 and 15.3) [1], and g is the gravitational acceleration [LT^{-2}].

The equation for acceleration head does not take into account the system elasticity and velocity of pressure wave. It is recommended only for pipe lengths of less than 15 m.

If two or more pumps operate in parallel, the acceleration for the common suction line is calculated assuming that all pumps act as one large unit.

The use of a suction dampener reduces the effective length of suction pipe to the distance between the pump cylinder and the dampener. Thus the acceleration head is virtually eliminated by the use of a suction stabiliser close to the pump suction manifold. The effective pipe length may be taken as 10 pipe diameters.

3.2. Net Positive Suction Head

The net positive suction head (NPSH) is the difference between suction pressure and vapour pressure measured at the pump suction while the pump is in operation. The NPSH is required to lift the suction valve and to overcome the friction losses in the suction pipe and the acceleration head. It is convenient to express NPSH for reciprocating pumps in pressure units. The pump losses for valve lift may be taken as 0·014 MPa.

TABLE 15.3
Values of Constant k_2

Fluid	*Constant,* k_2
Non-compressible slurries	1·4
Most liquids	1·5
Compressible slurries (CO_2 etc.)	2·5

The available NPSH is calculated using the following:

$$(\text{NPSH})_A = h_a + (\rho_m/\rho_f)h_z - (h_{svp} + h_e + (h_s)_f + h_{acc}) \quad (15.9)$$

where $(\text{NPSH})_A$ is the available net positive suction head (m water) [L], h_a is the atmospheric pressure (m water) [L], ρ_m is the density of the slurry [M L^{-3}], ρ_f is the density of water [M L^{-3}], h_z is the height of liquid in the slurry feed tank above the pump cylinder (negative if below pump level) (m slurry), h_{svp} is the saturated vapour pressure of the fluid (for water given by eqn (13.8)) (m water) [L], h_e is the entrance loss (m water) [L], $(h_s)_f$ is the frictional head loss in the pipework on the suction side (m water) [L] and h_{acc} is the acceleration head of the slurry (from eqn (15.8)) (m water) [L].

The NPSH required is equal to the $(\text{NPSH})_A$ at the speed where the pump just begins to knock.

The problem of knocking due to low suction head is eliminated in the case of long-distance slurry pipelines by the use of a charge pump. A centrifugal slurry pump is used to maintain a pressure of 0·17 MPa or more at the reciprocating pump inlet.

3.3. Torque Characteristics

For fixed suction and discharge pressures, a reciprocating pump requires an input torque that is independent of pump speed. The torque at very low speeds may be about 150% of the average full-load torque due to improper lubrication at low speeds. An external forced lubrication system, started prior to the start of the pump, will reduce the torque at very low speeds to a value equal to the average full-load torque.

A constant torque requirement is an important consideration in the design of slurry pump stations. The pump flow rate can be varied only by changing the pump speed. If a constant speed drive unit such as an electric motor is used, then a variable speed drive will be needed to change pump speed. The variable speed drive system must be capable of providing full-load torque at all speeds.

3.4. Prime Mover Selection

Electric motors, diesel engines and turbines can be used for driving reciprocating pumps. Electric motors are generally used where electric power is readily available. The electric motors usually operate at 1000–1800 rev min^{-1}. Slurry pumps operate at speeds in the range of 60 to 120 rev min^{-1}. A speed reduction gear is normally provided between an electric motor and the pump.

A diesel drive may be considered in remote locations. The diesel engine can be directly connected to a pump, thereby eliminating the need for a variable speed drive. The disadvantages of diesel engines are

- higher maintenance costs,
- larger space requirements,
- need for storage of diesel fuel,

- increased noise,
- environmentally less attractive, and
- higher capital cost (the cost of a diesel engine is higher than that of an electric motor of the same power rating).

Alternating current (AC) motors are generally used because of their economy, simplicity, ruggedness and low maintenance requirements. The speed of an AC motor remains constant. The pump speed variation with an AC motor can be achieved by a mechanical drive, electromechanical drive, fluid couplings or solid-state controllers.

3.5. Flow Control Options

3.5.1. Mechanical Variable Speed Drives

Mechanical variable speed drives are the simplest, least expensive and oldest devices for flow control. They tend to be lightweight, efficient and easy to maintain. These devices use belts or chains with adjustable sheaves.

Belt drives are available over a power range up to 75 kW. Reduction ratios of up to 10 : 1 may be achieved. The belt drive efficiency is usually 95%. The accuracy of speed control with belt drives will be about 5%.

Chain drives are more durable and can transmit much higher torques and provide better speed accuracy compared with belt drives. However, chain drives provide no shock-load protection and are suited only to low-speed operation. They are also more expensive.

3.5.2. Electromechanical Variable Speed Drive

An electromechanical clutch or wound rotor motor can be used to vary the speed of the pumps. These are easier to control and their controls can be remotely located. The drawback of electric clutches is that when the speed is reduced, there is a proportionate amount of energy loss in the form of heat. These devices are available over a large range of pump power. Similar heat loss occurs with wound rotor motors. The speed of a wound rotor motor is reduced by directing a portion of the energy through external resistors via slip rings. The energy is dissipated as heat by the resistors. The energy loss due to slip is equal to the ratio of operating speed to design speed times the energy requirement at full speed at a given discharge pressure. Electric clutches and wound rotor motors can be used to achieve speed reductions of up to 50% of design speed. Power recovery methods are available for recovering power from the wound rotor motors which can be fed back to the power grid instead of being wasted as heat in the external resistors. The power recovery involves a rectifier and inverter.

3.5.3. Adjustable Speed Fluid Drives

These devices produce a slip between the driving and driven member using hydro-viscous or hydrokinetic means. The speed reduction is proportional to the slip. The extra energy is lost as heat. The fluid drives have high power capabilities, smooth torque transmission and are inherently safe to operate in hazardous environments.

They are large, need to dissipate heat and are more expensive compared with mechanical and electromechanical drives. At full speed there is a 2–3% loss due to slip. Use of clutches can eliminate the slip loss.

A hydrokinetic system contains two rotating elements, an impeller and a runner. The impeller transmits energy supplied by the motor into kinetic energy in a moving mass of fluid. The runner converts the kinetic energy of the fluid into rotational energy which is used to drive the load. The impeller and runner are mounted facing each other in a common housing such that the input and the output shafts are coaxial. A change in the amount of fluid going from the impeller to the runner changes the power imparted to the runner. This results in a reduction in the speed of the runner. A scoop-tube is used to control the volume of fluid between the impeller and the runner. A mineral oil is generally used as the driving fluid.

In a hydroviscous fluid coupling, fluid shear is used to transmit torque. Friction discs are mounted on the driver and the drive shafts. The oil film thickness between the shearing friction discs is the primary variable controlling speed.

The hydraulic fluid couplings have the following advantages:

- soft, virtually no load start-up of driver,
- gradual acceleration of even the highest inertias without oversizing standard motors,
- stepless speed regulation,
- elimination of motor overloads and burn-outs,
- reduced stress on connected motor and the pump during start-up and shutdown,
- simple maintenance,
- wear-free power transmission, and
- high availability.

It should be noted that hydroviscous drives are much more susceptible to problems arising from oil viscosity changes and are therefore less reliable than hydrokinetic drives.

3.5.4. Solid-State Controllers

Solid-state controllers are used with AC or direct current (DC) motors to vary the speed of the motors. Since motor speed is varied, there is no mechanical slip.

DC motors have been successfully used as slurry pump drives. The motor speed is regulated by a heavy duty silicon controlled rectifier which converts AC power supply into DC power. Electronic controllers for DC motors are relatively simple, and often cost only a little more than other drive packages such as mechanical, electromechanical or fluid transmissions. The speed of the motor is varied by varying the voltage applied to the DC motor armature. DC motors are more expensive than AC motors and require more maintenance. The pump can be operated at virtually any speed using a variable speed DC drive. The efficiency of the motor does not decrease significantly at lower speeds.

With an AC motor, both the voltage and frequency of the input current are varied to control the motor speed. The basic components of a variable frequency

AC controller are a power converter, power inverter and a control regulator. The power converter changes AC current to DC. The power inverter alters the DC to variable voltage–frequency AC power. The regulator controls the actions and response of the converter and inverter. The variable speed AC drives are complex and are correspondingly the most expensive type of drives. These devices can, however, be used to convert a constant speed AC motor into a variable speed AC drive.

The selection of prime mover should be based on the initial installed cost as well as operating cost of the pumps. One should also consider the availability and technical skill of the available labour. Availability and cost of spare parts for maintenance should also be included in the analysis.

3.6. Pulsation Control

A reciprocating pump delivers a pulsating flow. One way of smoothing the pulsations is to use a pulsation dampener, which is nothing more than an air chamber that *cushions* the flow by alternate compression and expansion of the trapped air. The chamber is often equipped with a flexible membrane to isolate the air from the process fluid, so that the cushioning air will not dissolve in the fluid.

The dampener stores the slurry when the pump output exceeds the average flow rate and releases the slurry stored in it when the pump output drops below the average flow rate. By pre-charging (pressurising) the flexible membrane, the total fluctuation in dampener pressure can be reduced. A pre-charge pressure equal to 50–70% of the pump discharge pressure is normally used to reduce pressure fluctuations due to pulsating flow.

Dampener manufacturers should be contacted for sizing of the dampeners for a given pumping installation (Miller & Miller, 1980; Ekstrum, 1981).

3.7. Expendable Parts Life

The fluid end parts exposed to slurry are subject to wear and require periodic replacement. Reciprocating pumps are built such that the wear parts can be easily replaced. The expendable wear parts for three major types of reciprocating slurry pumps are listed in Table 15.4.

The parts life depends upon the discharge pressure, slurry abrasivity and the pump speed. The actual relationship between these variables and wear parts life is not well established. Past experience and vendor information serve as the best guides in projecting parts life. Parts life reported for pumps used in selected pipeline installations are reported in section 4.

4. EXAMPLES OF RECIPROCATING PUMP SYSTEMS

4.1. Black Mesa Coal (Double-Acting Piston Pump)

4.1.1. System Description

The Black Mesa coal pipeline system originates in the Navajo–Hopi Indian reservation in north-east Arizona and traverses some 439 km to a 1580 MW power

TABLE 15.4
Expendable Wear Parts versus Pump Type

Part	*Piston pump*	*Plunger pump*	*Diaphragm pump*
Valves and seats	Yes	Yes	Yes
Piston	Yes	No	No
Piston rod[a]	Yes	No	No
Stuffing box	Yes[b]	Yes	No
Packing	Yes[b]	Yes	No
Plunger	No	Yes	No
Bushings	No	Yes	No
Diaphragm	No	No	Yes

[a] Piston rod is exposed to slurry in double-acting piston pumps only.
[b] Stuffing box and packing are required only in double-acting piston pumps.

plant in southern Nevada. It is constructed of mostly 0·457 m diameter pipe and has a rated capacity in excess of 5·0 Mt year^{-1}. It is currently the longest slurry pipeline system in the world. Initial coal deliveries began in November 1970.

At the mine site preparation plant, the coal is ground in a two-stage dry-impact-crushing/wet-rod-mill grinding process to a typical size consisting essentially of < 1·4 mm with 18–20% < 44 μm. Water is added to obtain a solids concentration of approximately 48% (by mass). The pipeline product is stored in four slurry holding tanks each holding about 2 h of pipeline capacity.

4.1.2. System Operation

The pipeline system is designed to pump 600 t h^{-1} of coal in *tightline* operation without surge tankage. (In other words, the pumps take suction directly from the incoming pipeline.) Four pump stations are used, each containing three pumping units in parallel except for station No. 2 where a fourth pump is required to overcome a large static lift. Each unit is a double-acting duplex piston pump which is driven by an electric motor through a gear reducer. The pumps have special fluid ends for handling coal slurry with liners composed of alloys to resist corrosion and erosion.

At the initial pump station, coal slurry is transferred from the storage tanks by centrifugal booster pumps, which provide about 480 kPa suction pressure to the mainline units. Two 305 mm (12 in.) diameter piston pumps operate with the third in standby service. Figure 15.14 shows a view of the first pump station. The discharge pressure ranges from about 3·45 to 5·52 MPa; driver is 1·3 MW. At station 2, three 250 mm ($9\frac{7}{8}$ in.) diameter piston pumps operate. Discharge pressures are in the 8·6–11·4 MPa range; driver is 1·2 MW. The mainline pumps at stations 3 and 4 are identical in size and number to those installed at station 1.

Variable speed fluid couplings are used for changing pump speed. Station 1 operates on flow control and the booster stations on suction pressure control—normally 690 kPa.

Expendable parts on the pump fluid end are normally changed out at prescribed intervals to minimise system down-time. Table 15.5 (from Montfort, 1981) lists

Fig. 15.14. Initial pump station—Black Mesa system.

parts life for these items and shows that there is a significant difference in life between the high pressure station 2 and the other three stations, particularly for valve parts.

4.2. Valep Phosphate (Single-Acting Piston Pump)

4.2.1. System Description

The Valep pipeline system was the first long-distance pipeline installed for phosphate concentrate. Initiating commercial operations in March 1979, this 120 km long, 229 mm (9 in.) diameter pipeline in the state of Minas Gerais, Brazil, is

TABLE 15.5
Expendable Pump Parts Life—Black Mesa Coal Pipeline System

Fluid end item	*Parts life (h)*	
	High-pressure station	*Lower-pressure stations*
Valve insert, soft	600	1 400–2 700
Valve body	1 200	4 000–6 000
Valve seat	2 400	4 000–6 000
Packing, new piston rod	500	600–800
Packing, second change	100	300–600
Packing, third change	None	100–200
Piston rod	600	1 000–1 400
Piston rubbers	600	1 000–4 000
Cylinder liners	5 000	8 000–10 000

designed to transport 2 Mt year^{-1} dry phosphate from the mine site facilities at Tapira to a fertilizer plant at Uberaba (Fister *et al.*, 1979).

At the mine site, slurry is prepared as a concentrate of about 61% solids by mass. The particle size distribution is $<0{\cdot}25$ mm with approximately 35% $<44\,\mu$m.

4.2.2. System Operation

The Valep system uses only a single pump station which contains three mainline pumps—two operating and one spare. The pumps are driven by 680 kW (915 hp) electric motors through a variable speed fluid coupling and a gear reducer and are rated at 132 m^3 h^{-1} at 15 MPa.

Originally, all units were plunger pumps. Subsequently, they were converted to piston pumps due to high maintenance costs caused by difficulties in the plunger flushing system.

Reported parts life for the piston units are as follows.

- piston rubber: 700 h
- piston assembly: 2200 h
- piston liners: 2900 h

4.3. Samarco Iron Concentrate (Horizontal Triplex Plunger Pump)

4.3.1. System Description

The Samarco pipeline system transports iron concentrate from the Germano mine near Belo Horizonte in the state of Minas Gerais, Brazil, to a pellet plant and filtering facility on the Atlantic Ocean at Ponta Ubu (Hill *et al.*, 1978; Jennings & Rizzone, 1980; Oliveiro, 1986). The pipeline is mostly 508 mm (20 in.) in diameter, 396 km in length and is rated for an ultimate production of 12 Mty^{-1} dry concentrate. Since initiation of operations in 1977, the system has operated in batch mode because the concentrator facilities were not sized for this maximum throughput.

The pipeline facilities include two pump stations, one at the Germano mine site and a booster station approximately midway on the pipeline (Fig. 15.15). At Germano, slurry is transferred from four agitated storage tanks to the mainline pumps at a concentration of 60–68% by weight and a size consisting of 0–6% $>74\,\mu$m and 70–96% $<44\,\mu$m. The transported material is a hematite slurry with a relatively high Miller number abrasivity index of about 150.

4.3.2. System Operation

Each pump station contains seven triplex plunger pumps in parallel with 230 mm (9 in.) diameter plungers and a 305 mm (12 in.) stroke. Normally five operate at station one and six at station two. Each pump is capable of pumping 193 m^3 h^{-1} at a discharge pressure of 13·8 MPa and is driven by a 930 kW electric motor through a variable speed fluid coupling and a gear reducer.

A small, separately driven auxiliary triplex plunger pump is provided with each mainline unit to supply high-pressure flush water to the plungers. The flush water is injected at about 690 kPa above the pipeline pressure with the flow rate to each

Fig. 15.15. Samarco iron concentrate pump station.

plunger controlled by orifices. The complete pump package was instrumented so that units could be started, stopped and speed controlled from the system control centre at Germano.

The Samarco profile is such that the first section of line generally assumes the concave shape of a large valley while the shape of the second section is converse. With a batch operation, the flow remains constant but pump pressure varies significantly with the location of the slurry–water interface. Bladder type pulsation dampeners are provided at each pump to overcome this variability. On the suction side, 0·23 m^3 suction stabilisers are used to accommodate the varying flow to the pumps. On the discharge side, two 0·15 m^3 dampeners are provided for each pump. They are set to operate over different pressure ranges to accommodate the varying discharge pressure which is dependent upon the location of a slurry batch in the pipeline.

Some difficulty was experienced with the initial fluid end valves. Improved life was obtained with the installation of a different design which provided for contact between the beveled edge of the valve plate and the seat, as well as on the seat ribs. Reported pump parts life with the modified valves are as follows.

- valve seats: 1200 h
- valves: 1200 h
- plungers: 2000 h
- packing: 1500 h
- pulsation dampeners: 6000 h

4.4. Bougainville Copper Concentrate (Vertical Triplex Plunger Pump)

4.4.1. System Description

The Bougainville slurry pipeline system is located on the island of Bougainville, which is part of Papua New Guinea (Deason, 1975). Extending from the Bougainville Copper Ltd mine to port facilities at Anewa Bay, the 152 mm (6 in.) o.d., 27·5 km pipeline transports about 0·9 Mt y^{-1} of copper concentrate in a batch operation. The system began commercial operations in January 1972 and was the first major slurry pipeline designed solely on the basis of bench-scale tests without confirmatory loop tests prior to start-up.

Slurry is received from the two concentrator storage tanks at a concentration of about 65% solids (by mass). The particle size consists of 100% $< 250\,\mu m$ with 70–84% $< 44\,\mu m$.

4.4.2. System Operation

The Bougainville pipeline system uses only a single pump station at the mine site (Fig. 15.16). Slurry laboratory tests indicated the concentrate to be relatively corrosive and abrasive so plunger-type units were selected for this application. Two vertical triplex plunger pumps, one operating and one standby, were installed. The pumps have a 165 mm (6·5 in.) bore and 229 mm (9 in.) stroke and are powered by 520 kW variable speed wound rotor motors with speed control provided by liquid rheostats. Small high-pressure pumps mounted on the same base as the mainline pumps deliver flushing water to the main plungers during the suction stroke.

Pump discharge pressures vary from about 9·65 MPa when pumping slurry to 4·6 MPa when pumping water. Bladder-type pulsation dampeners were originally installed on both the suction and discharge sides of the mainline pumps but the discharge dampeners periodically failed due to their inability to regulate bladder pressure compatible with the varying discharge pressure. Hydril accumulators were subsequently installed.

4.5. New Zealand Iron Sands (Diaphragm Pump)

4.5.1. System Description

The New Zealand Iron Sands pipeline system began operations in 1986 and is located approximately 40 km south of Auckland (McLush & Pope, 1987). With a

Fig. 15.16. Bougainville copper concentrate pump station.

rated capacity of 1·5 Mt y^{-1}, the 219 mm diameter, 18 km long system transports an iron sand concentrate from the Maioro open cut mine to a steel mill at Glenbrook. The concentrate has an average relative density of 4·5 and a size consist of 1% > 300 μm and 1% < 53 μm. The slurry concentration is about 48·5% solids (by mass).

4.5.2. System Operation

Although the pipeline length is fairly short, the system was designed for an operating velocity of nearly 4 ms^{-1} which results in a high pressure drop. Accordingly, two pump stations were installed, each containing two piston diaphragm pumps which are driven through a common gearbox, fluid drive variable speed coupling and a 1·4 MW electric motor (Fig. 15.17). Normally the pumps in station one operate at full speed and those in station two at a controlled speed which is dictated by suction pressure requirements. The pump discharge pressure is 9·0 MPa.

Fig. 15.17. New Zealand iron sands pump station.

4.6. Odate–Noshiro Tailings (Mars Pump)

4.6.1. System Description

This pipeline system delivers up to 0·5 Mt y^{-1} of dry base metal tailings from a mine site near Odate approximately 71 km to a tailings storage basin at the Noshiro beach on the Sea of Japan. The 300 mm pipe diameter delivers solids of relative density 2·8 and a size consist of $< 210\,\mu m$ at a concentration of about 19% solids (by mass). The system went into operation in 1968 (Couratin, 1969).

4.6.2. System Operation

The single pump station at Odate contains three Mars pumps, two variable speed units and one constant speed with each driven by a 370 kW electric motor. The pumps are double-acting, duplex units with a 225 mm piston diameter and 450 mm stroke and run at 50 rev min^{-1}. Two pumps operate with the third used as a standby unit. The discharge rate is 164 $m^3 h^{-1}$ at the maximum discharge pressure of 4·9 MPa.

5. REFERENCES

Couratin, P. (1969). If you can't store it, pump it. *World Mining*, May, 22–7.

Deason, D. (1975). Bougainville line transports slurried copper concentrate. *Pipe Line Industry*, Feb., 49–50.

Ekstrum, J.D. (1981). Sizing pulsation dampeners for reciprocating pumps. *Chem. Engng*, 12 Jan., **88**(1), 111–18.

Fister, L.C., Finerty, B.C. & Hill, R.A. (1979). The World's first long distance phosphate concentrate slurry pipeline. In *Proc. STA 4*, pp. 60–3.*

Henshaw, T.L. (1981). Reciprocating pumps. *Chem. Engng*, 21 Sept., **88**(19), 105–23.
Hill, R.A., Jennings, M.E. & Derammelaere, R.H. (1978). Samarco iron ore slurry pipeline. In *Proc. STA 3*, pp. 134–41.*
Jennings, M.E. & Rizzone, M.L. (1980). The Samarco pumping system. In *Proc. STA 5*, pp. 276–86.*
McLush, S. & Pope, P.B. (1987). The New Zealand steel development iron sand slurry pipeline. In *Proc. STA 12*, pp. 411–18.*
Miller, J.E. & Miller, J. (1980). Pulsation control for slurry pumps. In *Proc. STA 5*, pp. 287–95.*
Montfort, J.G. (1981). Operating experience is described for Black Mesa coal–slurry pipeline. *Oil and Gas J.*, 27 July, **79**(30), 192–200.
Oliveiro Jr., M.C. (1986). Samarco's iron concentrate pipeline. In *Proc. STA 11*, pp. 231–41.*

*Full details of the Slurry Transport Association (STA) series of conferences can be found on p. 13 of Chapter 1.

16

Rotary Slurry Pumps

Nigel I. Heywood
Warren Spring Laboratory, Stevenage, UK

1. INTRODUCTION

The purpose of this chapter is to review the main classes of rotary positive displacement (PD) pump which are used for various types of solid–liquid mixture. Because the clearances in some designs are very small, many of these pumps are suitable for fine particle, relatively non-abrasive slurries only. However, the progressive cavity and peristaltic designs are examples where relatively large particles can be handled with ease. This chapter will also consider whether the pump is readily primed, if it can run dry (and if so, for how long) and the types of slurry which are commonly handled.

Rotary PD pumps are used mainly for in-plant, low to medium flow rate applications and are rarely specified for pipe runs in excess of 100–200 m. Like other types of PD pump, most rotary pumps can also be used to meter the flow fairly accurately, although the details of how the metering calibration is affected by discharge pressure and slurry rheological properties have yet to be investigated and catalogued for the majority of commercial designs. Hence, a pump used for metering duties must be calibrated under appropriate conditions.

One of the main advantages of rotary pumps over reciprocating pumps is the absence of valves. This eliminates an important potential wear problem and consequently reduces maintenance costs, and makes them highly suited to hygienic applications. However, the absence of valves also means that the maximum discharge pressure is limited to a relatively low value compared with reciprocating pumps. As with all PD pumps, it is normally advisable to fit a pressure relief valve downstream from the pump to allow for partial constriction or complete blockage of the discharge pipe.

With most rotary pump designs it is better to select a larger pump running more slowly than a smaller pump to achieve the same throughput for any slurries that are considered abrasive. Just how large a pump should be selected obviously depends on overall economic considerations in addition to the degree of slurry abrasivity. Pump sizing could be more straightforward if pump manufacturers supplied pump performance data for non-Newtonian materials. The use of effective viscosity (Steffe & Morgan, 1985) is useful up to a point, but a related parameter is also required to assist in the determination of slip corrections and viscous power requirements. Selection considerations have also been outlined by Neerken (1980) and Davidson (1986).

2. CLASSIFICATION OF ROTARY SLURRY PUMPS

2.1. Progressive Cavity Pump

This pump type is also known as an eccentric screw or helical rotor and is typically referred to by the trade name *Mono* in the UK and *Moyno* in the USA. The original design was developed by Moineau in France in the 1930s, but there are now some 40 manufacturers worldwide.

Principle of operation. The progressive cavity pump is a variety of the screw pump and consists of a resilient stator or sleeve in the form of a double internal helix and a single start helical screw rotor. The rotor turns within the stator with a slightly eccentric motion as shown in Fig. 16.1, where the diagrams numbered 1–4 show the progress of a slurry through the pump. The index mark on the left face of the rotor indicates the rotor's changing position.

The rotor, of constant cross-section and having half the pitch of the stator, makes an interference fit inside the stator and creates a continuously forming cavity as it rotates which passes from the suction to the delivery side of the pump. The rotor maintains a positive seal along the length of the stator and this seal progresses continuously immediately behind the cavity through the pump so giving uniform positive displacement.

Advantages. The advantages of the pump are that it is self-priming (although

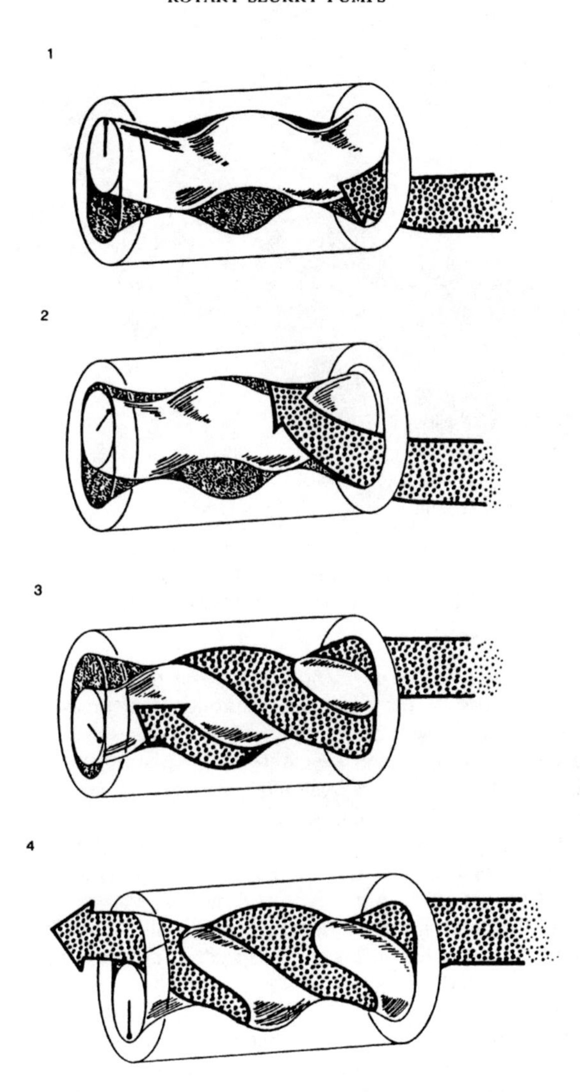

Fig. 16.1. Principle of operation of progressive cavity pump.

should not be run dry because damage to the stator can occur rapidly), it gives a uniform discharge with very little pulsation to the flow, it can operate in reverse so pumping material against the seal if air ingress into the pumped material is a problem, and it can handle shear-sensitive material with relatively little damage. The design does not require valves and the end discharge port allows straight through flow (Fig. 16.2) or a 90° bend can be accommodated.

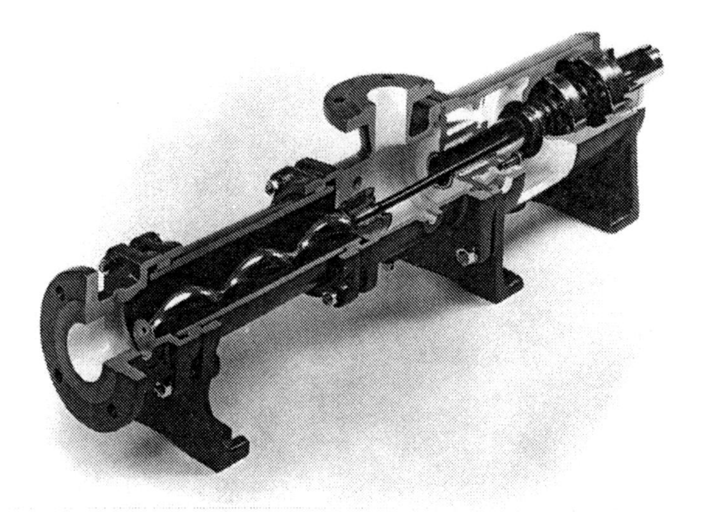

Fig. 16.2. Typical construction of a progressive cavity pump. (Reproduced by permission of Mono Pumps Ltd, Audenshaw, Manchester, UK.)

Pump construction and performance. The pump drive is normally via flexible couplings at each end of the drive shaft although sometimes rigidly mounted flexible drive shafts are used. Either a packed gland or a mechanical seal can be used in the stuffing box located in a separate gland section. Various combinations of rotor and stator materials are generally provided by most manufacturers to resist corrosion and abrasion. Pump speed is generally governed by the abrasivity of the material handled; the greater the abrasivity the lower the speed. Materials used for the rotor include cast iron, stainless steel for food and chemical use and specially lined pumps for highly corrosive duties. The stator is made from an elastomer including natural rubber, nitrile, hypalon and viton. The larger pumps can handle up to $250\,m^3\,h^{-1}$ at maximum speed, at pressures up to 4 MPa and temperatures up to 180°C.

Application areas. The pump use is widespread. Typical materials pumped include whole, friable food such as strawberries and baked beans, cosmetic creams, sewage, beach sand, chalk slurry, gypsum plaster, vitreous china slips and coal slurry tailings from washery plants.

Some manufacturers now offer pumps designs which incorporate a hopper section on the suction side carrying an auger which is rotated with the rotor, as shown in Fig. 16.3. The hopper and auger are designed to force the high-viscosity material into the pump body at a constant rate. Without this design many stiff pastes would not be able to flow into the pump body.

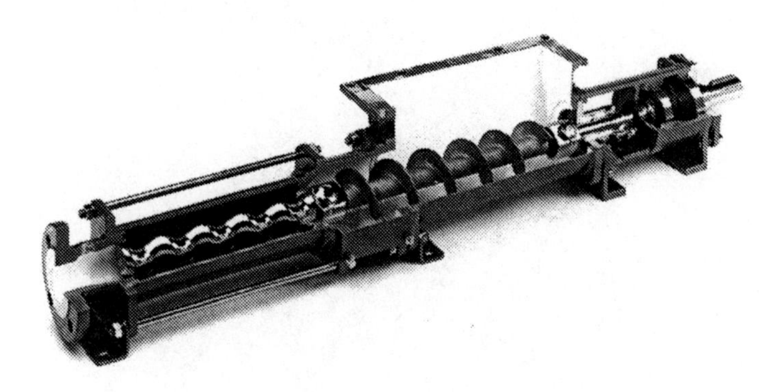

Fig. 16.3. The Mono pump showing the hopper and auger facility for feeding highly viscous and stiff slurries and pastes. (Reproduced by permission of Mono Pumps Ltd, Audenshaw, Manchester, UK.)

2.2. Peristaltic Pump

2.2.1. The Peristaltic Hose Pump

The peristaltic hose slurry pump was developed in the mid-1970s from the smaller pump designs typically used in laboratories for the transfer of liquids at relatively low rates. A commercial design is shown in Fig. 16.4.

Principle of operation. The heart of the pump consists of a compressible hose normally made from elastomer and reinforced with nylon or steel braiding. The tube is compressed continuously along its length by two or three sliding shoes or bearing-mounted rollers. The shoes or rollers are mounted on a rotating assembly as shown in Fig. 16.5. As the rotor turns, the advancing shoe or roller progressively squeezes the hose so driving the slurry before it. The hose returns to its normal shape after the roller has passed thus creating a reduced pressure which draws in more slurry which in turn will be driven forward by the next roller. Sometimes the hose does not expand quickly enough so limiting the pump throughput. This can often be overcome by applying a reduced pressure to the outside of the hose, compared with the line pressure. The action of the shoes or rollers on the hose is usually lubricated by oil or glycerol. Some designs mount the shoe or roller on a spring mechanism which provides some *give* in the stroke when large, incompressible particles are carried into the pump, as shown in Fig. 16.5. Without this design

Fig. 16.4. An example of a peristaltic pump. (Reproduced by permission of Rosewater Engineering Ltd, Stonebroom, UK.)

feature there have been cases where the pump shaft has fractured when oversize particles have entered the pump.

Advantages. One of its main advantages is that it is self-priming and can develop suction pressures approaching the slurry vapour pressure. Thus, in principle, water can be drawn up vertical distances of 8–9 m. The pump can also be primed with slurries over similar distances if the density of the slurry is low and a low flow rate is induced on the suction pipe, so minimising friction losses. Slurries can be metered relatively accurately using a peristaltic pump although there is little information available on how the metering performance is affected by slurry viscosity and pump discharge pressure.

The pump can also be run dry almost indefinitely and is reversible so allowing pump blockages to be freed on occasion. There are no seals or valves and therefore the main wear component is the hose. Unfortunately, hose life can be very variable, typically from 1000 to 8000 h, and is difficult to predict, so unexpected leakage of slurry can occur into the pump body in the event of hose fracture. Thus, hose design is very important.

Disadvantages/limitations. Because of its pumping action, quite severe pulsations in the flow can occur, particularly at low flow rates. This can be partially overcome by employing a damper downstream from the pump.

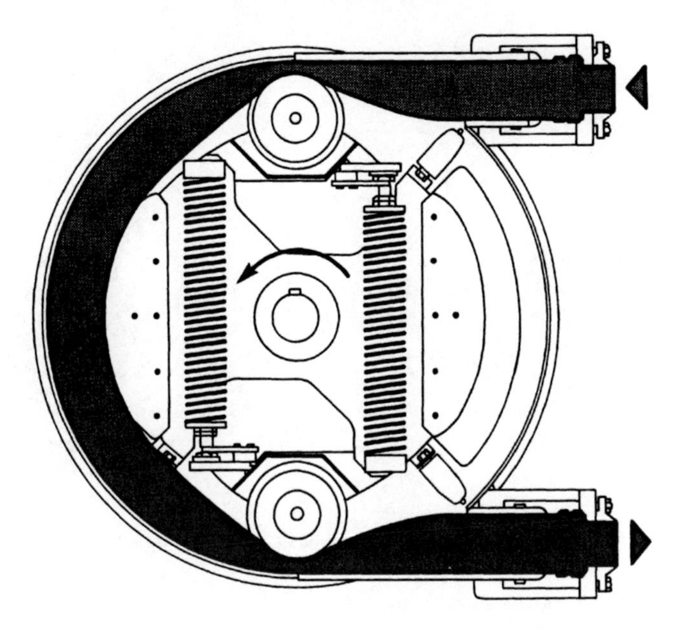

Fig. 16.5. The peristaltic pump design with spring action shoes. (Reproduced by permission of Abel Pumps Ltd, Essen, Germany.)

Pump construction and performance. One commercial example of the hose material is the patented *Tri-Laminar* hose used in the Dutch Bredel peristaltic pump. This hose consists of a thick wall outer covering of hard natural rubber and an embedded braided nylon reinforcement together with a softer elastic inner case or a perbunan inner core for food and similar duties. Pumps are available commercially with port sizes up to 100 mm, developing pressures up to 1·6 MPa and achieving flow rates approaching 40 $m^3 h^{-1}$ for low-viscosity slurry.

Application areas. Of all the rotary pump designs available, peristaltic pumps are suitable for pumping a wide range of both abrasive and non-abrasive, viscous slurries over short distances. Examples include paint, chocolate, meat products, various types of mortar, sewage, and mine tailings. Very little has been published on user experience of this pump type but Sweeney and Gaines (1983) of Pfizer Inc., IL, USA, have described how Bredel pumps of various sizes are used to transfer 75% (by weight) barium sulphate slurry. The service life of the hose was found to range from 1000 to 5000 h depending on the application. They report that only 1 h is needed to replace the hose, including the time to drain and clean the housing and rotor. However, they do not describe whether advance warning of hose failure is possible before contamination of the process fluid occurs.

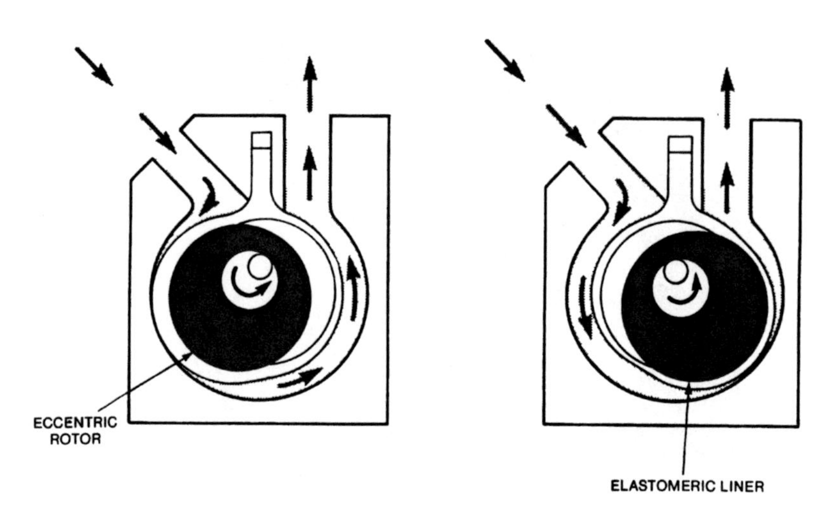

Fig. 16.6. The operating principle of the orbital lobe pump.

2.2.2. The Orbital Lobe Pump

Principle of operation. This pump design employs the peristaltic principle but using a single rotor mounted on the shaft rather than two or three separate rollers or shoes. The rotor is mounted on an eccentric shaft which rotates within an elastomeric liner creating a peristaltic action on the slurry trapped between the liner and the pump body, as illustrated in Fig. 16.6. Shaft seals are again eliminated by isolating the slurry from the rotating parts and valves are unnecessary as the momentum of the slurry flow through the pump prevents backflow as the rotor reaches the top of its stroke and crosses from discharge to suction.

Advantages. The pump is self-priming and, in addition, it can, the manufacturers claim, run dry for relatively long times without damage to the liner. The pump requires no shaft sealing and so there is no leakage risk unless the liner is breached. Pump maintenance costs tend to be low because the only wear item is the liner which is usually inexpensive to replace. The low shear action of the pump can reduce the possibility of the breakdown of shear-sensitive slurries.

Pump construction and performance. Flow rates from around 0·6 up to $8\,m^3\,h^{-1}$ are achievable and the pump can develop continuous maximum discharge pressures up to 0·3 MPa. Most of the pump body is normally fabricated using various plastics (including polyethylene, polypropylene, PTFE and PVC) or stainless steel and the liner is made from a range of elastomers for duties up to 120°C. These elastomers include natural rubber, neoprene, chloroprene, Buna N, Hypalon, Nordel and Viton.

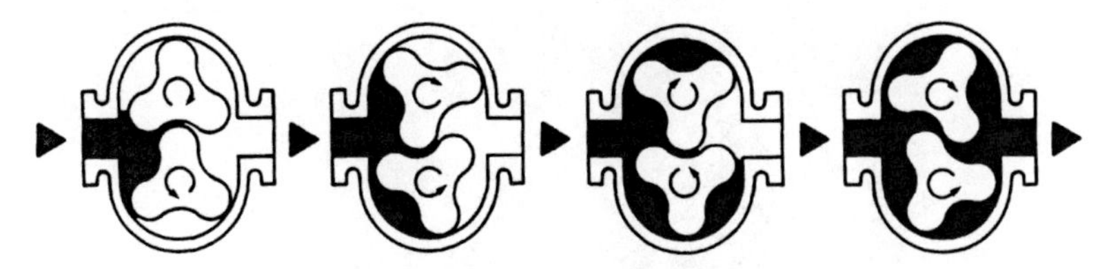

Fig. 16.7. The operating principle of the tri-lobe pump.

Application areas. Manufacturers claim the pump is suitable for not only corrosive liquids but also abrasive slurries. Fine-particle slurries such as lime or tile glaze slurries as well as some foodstuffs can be pumped.

2.3. Lobe Pump

Principle of operation. It is possible to design a single tri-lobe rotor pump but most commercial designs employ two rotors with either two or three lobes. The principle of operation of the pump is illustrated in Fig. 16.7. Slurry enters the inlet port and is gently drawn into the pockets formed between the rotors and the contoured rotor casing. The pockets move with the rotors in both clockwise and anti-clockwise directions. Unlike either external or internal gear pumps, there is no contact between the rotating lobes. A typical commercial design is shown in Fig. 16.8.

Advantages. As the rotors do not touch, lobe pumps can be run dry for long periods if necessary provided that the seals are flushed. However, slurry should normally fill the pumping chamber before the pump is started. Their self-priming capability can also be improved significantly in this way as the presence of slurry will partially seal the clearances and thereby reduce gas leakage across the rotors.

Pump construction and performance. The lack of contact between the rotating lobes is achieved by synchronising their movement with an exterior gearbox which is an integral part of the pump. The pump is frequently manufactured from stainless steel for hygienic duties in the food, biotechnology and pharmaceutical industries. Where hygiene is particularly important, a design is available which eliminates from the pumping chamber rotor fixing nuts, splines and front cover recesses which are normally used. For abrasive duties, the rotors may be made from polyurethane or rubber and the pump casing from ductile iron which is easier to machine and less brittle than cast iron.

Differential pressures up to 2 MPa can be developed and maximum flow rates of $200\,m^3\,h^{-1}$ are possible with available commercial designs.

Application areas. These pumps can handle fine-particle suspensions, but because of the fine tolerances between the rotors and the casing, the upper rigid particle size which can be allowed to pass through the pump is quickly reached. Two lobe rotors tend to be used for pumping fragile solids.

Fig. 16.8. The tri-lobe pump. (Reproduced by permission of SSP Pumps Ltd, Eastbourne, UK.)

Lobe pumps are also used in the water industry. Westwood (1986) claims that this pump type has replaced centrifugal pumps on many activated sludge transfer duties and suggests that the reason for this is that the speed of lobe pumps is appreciably lower than centrifugal and recessed impeller pumps and so noise and vibration levels are lower and the pumping action creates low shear.

Other materials pumped include muds, tile glazes, paints, chocolate, diced vegetables, meat, tomato purée and paste, and preserves. Jones (1980) has compared lobe pumps with centrifugal, progressive cavity and diaphragm pumps for slurry filtration duties.

2.4. Circumferential Piston Pump

Principle of operation. The principle and action of the circumferential piston pump (see Fig. 16.9) are similar to those of the lobe rotor pump but the design of the rotors imparts a very smooth, low shear action on the pumped material resulting in low material degradation. Figure 16.10 shows a commercial design.

Advantages. High volumetric efficiency is achieved by minimising the amount of slip in the pump head. This has the consequence of reduced wear between the

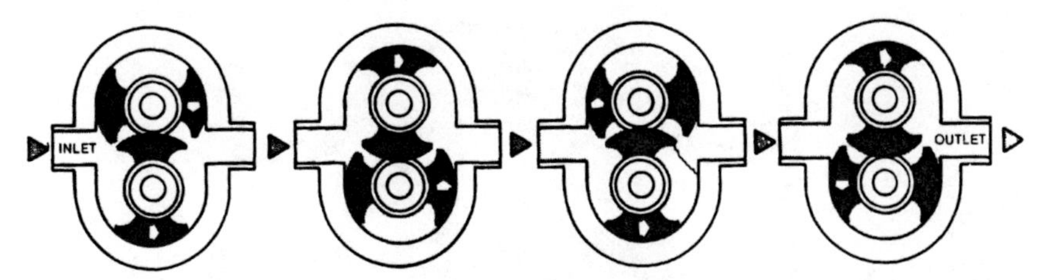

Fig. 16.9. The operating principle of a circumferential piston pump.

rotors and the pump casing. In addition, high levels of accuracy are possible when volumetrically metering materials with the pump.

Pump construction and performance. Stainless steel is the main construction material for widespread use in hygienic applications. Maximum differential pressures are of the order of 1·5 MPa. Maximum flow rates are around 180 $m^3 h^{-1}$. Some designs incorporate a modification which reduces the pulsation level of the pump discharge.

Fig. 16.10. The circumferential piston pump showing one type of rotor design. (Reproduced by permission of SSP Pumps Ltd, Eastbourne, UK.)

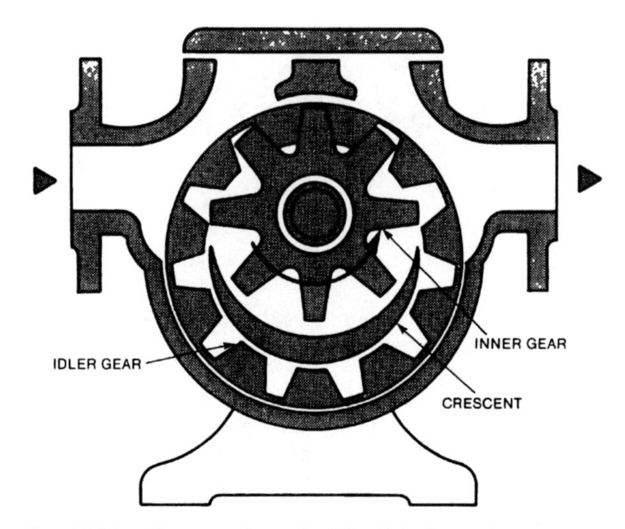

Fig. 16.11. The operating principle of the internal gear pump.

Application areas. The pump is used frequently in the food industry and can pump materials with a wide viscosity range from about 0·001 Pa s to 100 Pa s. Steffe and Morgan (1985) discuss the sizing of a circumferential piston pump for fruit paste which exhibits non-Newtonian flow properties.

2.5. Internal Gear Pump

Principle of operation. There are two basic types of gear pump. The internal gear pump tends to be more widely used for slurries than the external gear pump. Figure 16.11 shows the general operating principle. As the rotor rotates with its internal gear, an idler gear is driven inside a totally enclosed casing. The idler pin on which the idler rotates is located eccentrically with respect to the rotor. As the rotor rotates so the rotor and idler gears unmesh and slurry fills the space between the gear teeth. Further on the rotor and idler teeth mesh once again, providing the pressure required to force the slurry through the discharge port. To assist in the minimisation of backflow from the outlet to the inlet side, the pump is often provided with a crescent-shaped partition fitted to the casing between the gears.

Advantages. The pump is self-priming and can pump slurries with entrained air. The pump is bi-rotational and can be operated in either direction, assisting in the clearance of any blockages. Unlike an external gear pump, the generation of large local pressures and shear stresses does not occur in an internal gear pump and hence its action is more gentle. Thus, high-viscosity and shear-sensitive slurries are handled better with an internal gear pump.

Pump construction and performance. The pump's bearings are generally

Fig. 16.12. The operating principle of the flexible vane pump.

lubricated by the pumped fluid so for slurry pumping applications the bearings for the two gears are sometimes mounted externally and separately lubricated. For abrasive slurry duty, both the main bearing and the idler bearing are sometimes made from ceramic material. The crescent-shaped partition allows typical commercial designs of this pump to develop pressures up to 1·6 MPa at flow rates up to 160 $m^3 h^{-1}$ using speeds up to 1800 rev min^{-1}. Slurries with viscosities up to 80 Pa s can be pumped. The maximum allowable pump speed depends on the slurry viscosity and suction conditions. The pump is run relatively slowly for slurries that are at all abrasive. Some designs also allow axial adjustment of the rotor to compensate for wear, or to minimise friction losses in the pump when pumping high-viscosity slurries.

Application areas. The pump type is generally suitable for fine-particle slurries such as chocolate, paint and printing inks.

2.6. Flexible Impeller Pump

Principle of operation. Flexible impeller pumps have a single rotor mounted on an axis eccentric to the pump casing as shown in Figs 16.12 and 16.13. The blades of the rotor are flexible and made from a variety of synthetic elastomers. During pump operation these blades always remain in contact with the pump casing, transporting the pumped material in the pockets between the blades.

The flexing action of the blades is controlled partly by a cam inside the pump-housing between the intake and discharge ports, and partly by blade thickness, material and geometry. Each blade starts to deflect as it mounts the cam ramp at the centre of the discharge port and extends to full length again as it leaves the cam after passing the intake port. The increase in volume between two adjacent blades at the intake port creates a reduced pressure that causes material to flow into the expanded volume. The volumetric contraction at the discharge port, in turn, forces material into the discharge line.

Advantages. Rapid self-priming is possible over a wide speed range. This is because there is usually sufficient residual liquid to wet the blade tips. However, the pump should be installed with flooded suction so that the blades never run dry. It is claimed by Watkins (1989) that fluids containing very large amounts of entrained air (up to 90%) can be pumped. The pump is capable of emptying remaining

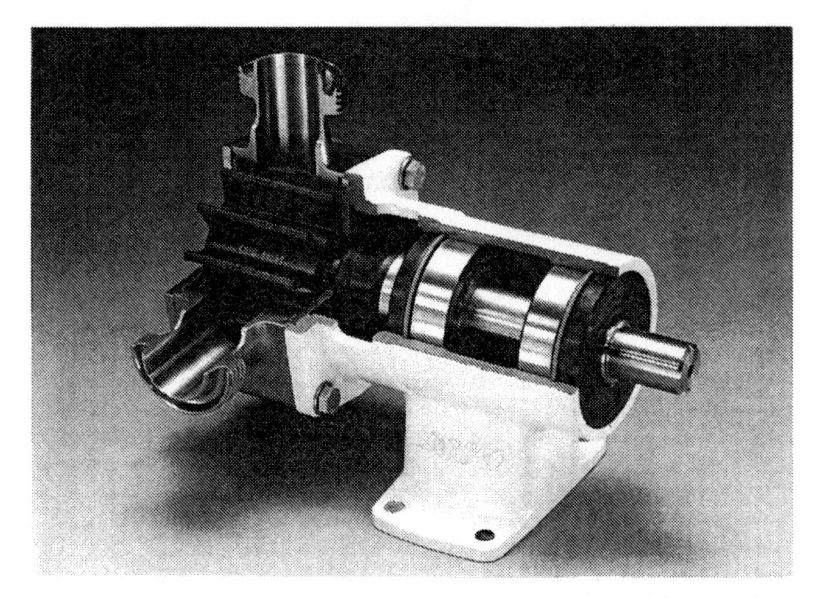

Fig. 16.13. The flexible vane pump. (Reproduced by permission of ITT Jabsco, Costa Mesa, CA, USA).

amounts of material from a container because it handles an air and material mixture without air-locking (McLean, 1982).

Unlike most PD pumps, a pressure relief valve is not normally required and, because the pump can be operated over a wide speed range up to 5000 rev min^{-1}, it can handle slurries with a wide viscosity range.

Disadvantages/limitations. The pump should never be run dry as this will cause rapid wear of the blade tips. The pump cannot meter as accurately as other PD pumps because the impellers flex. Because the discharge pressure is limited to around 0·2 MPa the pump cannot be used for high-pressure duties such as the transfer of slurry through a long length of pipe.

Pump construction and performance. Pump body materials include bronze, glass-filled epoxy resin and stainless steel. The cam that deflects the impeller blades often includes grid bars that cover part of the inlet port, so limiting the size of solids entering the pump. These bars also support the impeller blades as the blades pass over the inlet and outlet ports. This support helps to control both the amount of blade distortion at the ports, so increasing impeller life (Watkins, 1989) and the wear on the sharp edges of the inlet and outlet ports.

The pump is capable of handling very low viscosity slurries without loss of suction but developing pressures up to 0·2 MPa. Flow rates achieved are typically in the range 0·06–30 $m^3\ h^{-1}$.

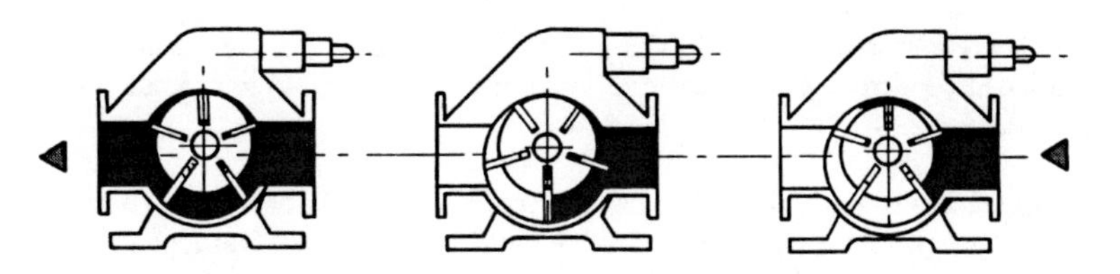

Fig. 16.14. The operating principle of the sliding vane pump.

Application areas. The pumps are very useful for small-duty dirty-water applications where suspended solids are particularly abrasive. The impeller blades deform elastically and so can absorb the cutting action of both large and fine abrasive particles, including metal powder, silt and sand (Watkins, 1989). However, the cam can sometimes be particularly prone to wear. The pump is also useful for emptying drums and other short-distance transfer applications.

Other application areas include the food and pharmaceutical industries. The fluid temperature which can be handled by the pump is limited by the maximum service temperature of the elastomer used for the impeller. This is of the order of 80°C for neoprene and nitrile impellers, 90°C for polyurethane and 95°C for fluorocarbon impellers (Warring, 1984). Watkins (1989) has compared the advantages and disadvantages of flexible impeller pumps with other PD pumps and centrifugal pumps.

2.7. Sliding Vane Pump

Principle of operation. This pump consists of a circular rotor with radial slots which are mounted eccentrically in what is essentially a circular casing (Fig. 16.14). Each rotor slot contains a rigid vane which is free to slide in a radial direction. The pump rotation throws the vanes outwards so that the tips always rub against the inner surface of the casing, so providing a seal. Pockets are created between adjacent vanes which alternately expand and contract in volume.

Advantages. The flow is generally quite smooth and non-pulsating for a PD pump. Cleaning by flushing through is straightforward; this can be assisted by reversal of the shaft rotation. Worn vanes can be rapidly replaced. The vanes are self-adjusting as they progressively wear. The pump is self-priming and some but not all designs can be run dry. Variable delivery characteristics are possible by adjustment of the rotor spindle and the casing, thus altering the capacity of the pockets between the vanes.

Disadvantages/limitations. The simple single-cell vane pump is unbalanced hydraulically and has a low volumetric efficiency because internal leakage is difficult to control. This can be particularly critical at low pump speeds when the radial pressure exerted by the vane tips to create the seal can be relatively low. This can be overcome by spring-loading the vanes.

Pump construction and performance. Flow rates up to $150\,m^3\,h^{-1}$ are typical with slurry viscosities up to 100 Pa s. Particles of maximum size 0·5 mm can be handled at concentrations up to 40%. Differential pressures up to around 1·4 MPa are possible.

Application areas. The pump is not commonly used for slurries but some manufacturers offer designs which can handle abrasive slurries. Hardened, replaceable liners are supplied against which the tips of the vanes press. Typical slurries include paint and other coatings, pulp and paper, and various foodstuffs.

2.8. Archimedean Screw Pump

Principle of operation. Single Archimedean screw pumps consist of an inclined spiral helix which turns slowly in a semi-circular close-fitting trough. The screw and trough are inclined at an angle of about 30° to the horizontal. The pumps lift only to the top of the shaft and the lift obtainable is normally about half the screw length.

Pump construction and performance. Flow rates of up to 20 000 $m^3 h^{-1}$ are possible with heads up to 12 m. Practical considerations will tend to limit the screw length to around 15 m. High heads can be achieved by arranging a number of screws to pump in series. Because of their very simple construction, maintenance costs are often very low.

Application areas. Archimedean screws are frequently used in the water industry for raising various slurry types such as activated sludge and raw sewage. In these applications the trough will probably be constructed in concrete.

3. REFERENCES

Davidson, J. (ed.) (1986). Positive displacement pumps: rotary. In *Process Pump Selection: A Systems Approach.* Institution of Mechanical Engineers, London, UK, pp. 39–62.
Jones, J.R. (1980). The selection of pumps for filtration duties. *Filtration and Separation,* March/April, **17**(2), 120–8.
McLean, M.G. (1982). How to select and apply flexible-impeller pumps. *Chem. Engng,* 20 Sept., **89**(19), 101–6.
Neerken, R.F. (1980). How to select and apply positive displacement rotary pumps. *Chem. Engng,* 7 Apr., **87**(7) 76–87.
Steffe, J.F. & Morgan, R.G. (1985). Pipeline design and pump selection for non-Newtonian fluid foods. *Food Technol.,* 78–85.
Sweeney, T. & Gaines, A. (1983). Peristaltic hose pumps solve transfer problems with high solids slurries. *Chem. Proc.,* Dec., 28–9.
Warring, R.H. (1984). *Pumping Manual* (7th edn). Trade and Technical Press, Morden, Surrey, UK, pp. 87–105, 119–31.
Watkins, D.L. (1989). Plagued by pumping problems? Consider the flexible-impeller pump. *Chem. Engng,* Jan., **97**(1), 94–8.
Westwood, J.B. (1986). Rotary lobe pumps—a new concept in sewage sludge pumping. In *Proceedings of the International Conference on Positive Displacement Pumps.* BHRA Fluid Engineering, Cranfield, UK, paper B2.

17

Lockhoppers and Pipe Feeders

Michael Streat
Loughborough University of Technology, UK

1. INDIRECT FEEDING TECHNIQUES

Methods of indirect pumping of particulate solids have found widespread application in the handling of coarse, abrasive or corrosive materials in the mining industry. For example, the hydraulic transportation of bauxite mineral has been achieved by indirect pumping using lockhopper feeders, thereby ensuring that neither the mineral nor the corrosive transfer water is put into direct contact with the moving parts of a conventional centrifugal or positive displacement pump (Yamaguchi *et al.*, 1970). The major advantage of indirect feeding techniques for most industrial applications is that wear, attrition and erosion can be minimised in the transfer pumps and in the transport pipeline.

The method of indirect feeding particulate solids into a pipeline depends largely on particle size. Finely divided materials with a top size of about 2 mm can be transported using either vertical or horizontal lockhoppers whereas coarse materials of particle size greater than about 2 mm and up to about 50 mm are usually fed into a pipeline from horizontal pipe or tube feeders. Vertical lockhoppers are considerably larger in diameter than the transfer pipeline and require careful design of the bottom cone outlet to ensure particulate mass flow. Vertical lockhoppers are usually designed as pressure vessels and may require some means of isolating the transfer water from the process slurry if fine particles are to be transported. This provision is unnecessary for the handling of coarse solids and it is more common to use horizontal tube feeders in this case. Horizontal tube feeders are usually long lengths of pipe, of precisely the same diameter as the transport pipeline itself. Long lengths of pipe can be configured as hairpins to conserve space and then suitably valved in order to allow proper sequencing of the flow.

Individual lockhoppers and tube feeders are inherently discontinuous in operation and therefore require valving and associated control systems. The batchwise nature of the process necessitates the sequencing of three independent chambers to facilitate semi-continuous delivery of solids. The discharge vessel, either a lockhopper or tube feeder, is normally filled at low pressure and the slurry is subsequently transmitted to the transport pipeline using a high-pressure flow of clean water from a conventional pump. The high-pressure valves are specially designed to avoid blockage and serious erosion. The efficient design of high-pressure valves is one of the major mechanical engineering advances in recent years.

2. LOCKHOPPER FEEDERS

2.1. Introduction

Lockhopper feeders are essentially cylindrical pressure vessels with a conical base designed to promote mass flow of the slurry. They are usually installed vertically to assist loading with either slurry or dry solids. Loading usually takes place under conditions of low pressure; either the slurry is pumped in or the solids gravity fed. The solids are discharged as a slurry by pressurising the lockhopper with a suitable motive fluid. Depending upon the application, the discharge into the pipeline may or may not be assisted by a mechanical feeding device such as a screw or *pocket* feeder. Mining applications tend to transport particulate solids using water as the transfer fluid and the most advanced lockhopper equipment is supplied by Hitachi of Japan. Their lockhoppers do not tend to use mechanical assistance at the discharge. Valves for the motive fluid tend to be of the same design as for clean fluids; specially designed valves may be necessary for slurry duty or where solids may be loaded.

2.2. The Hitachi Hydrohoist

Description. Sakamoto (1967, 1974, 1976) and Sakamoto *et al.* (1976, 1978, 1979, 1982) have extensively described the Hitachi *hydrohoist* and the principle of the technique is shown schematically in Fig. 17.1. The vertical type lockhopper design was developed to transport finely divided mineral ore suspensions. Three vertical lockhopper chambers are interconnected to provide virtually continuous flow of slurry in the pipeline. Floats are placed at the boundary between the slurry and the transfer water in order to avoid mixing. The lockhoppers can be filled with slurry at low pressure and sequentially discharged at high pressure with a centrifugal pump, thus ensuring that this pump is never in contact with abrasive solids. The application of a vertical lockhopper system for the transportation of a bauxite slurry will be discussed in section 4.2.

Operation. The flow sheet is shown schematically in Fig. 17.1. Valves A and D close only on clear water. Valves B and C are one-way slurry valves sealing under high pressure and preventing flow in much the same way as the valves on a positive displacement pump. The operation of the lockhoppers is cyclic; the next operation to occur in lockhopper 3 will be the one described below for lockhopper 1.

Lockhopper 1. Filling with low pressure slurry. Valve A_1 is closed. Valve C_1 is closed by the high pressure in the slurry discharge line. Valves B_1 and D_1 are open. Low-pressure slurry is pumped into the lockhopper forcing the float up and displacing the water (also at low pressure).

Lockhopper 2. The lockhopper is about to commence discharge of the slurry. Valve A_2 is opened and valve D_2 is closed when the arrival of the float is detected by the upper sensor. High-pressure water is pumped into the lockhopper. The pressurised slurry closes valve B_2. The float is forced down and high-pressure slurry leaves the lockhopper via valve C_2.

Lockhopper 3. The lockhopper has just ceased delivery of slurry. The arrival of the float at the bottom of the chamber is detected by the lower sensor; valve A_3 is closed and valve D_3 is opened. The lockhopper is pressurised by the high-pressure water which closes valve C_3. Low-pressure slurry flows into the lockhopper through valve B_3 forcing the float up the chamber and displacing the contained water.

2.3. The Hydrolift

The vertical hoisting of mineral ore bodies has been the subject of research in both Australia and the Republic of South Africa. The hydrolift principle was first suggested by Oedjeo (1964) in Australia and a slightly adapted version of the technique was later developed by Laubscher and Sauermann (1972) in the Republic of South Africa. A schematic diagram of this lockhopper technique is shown

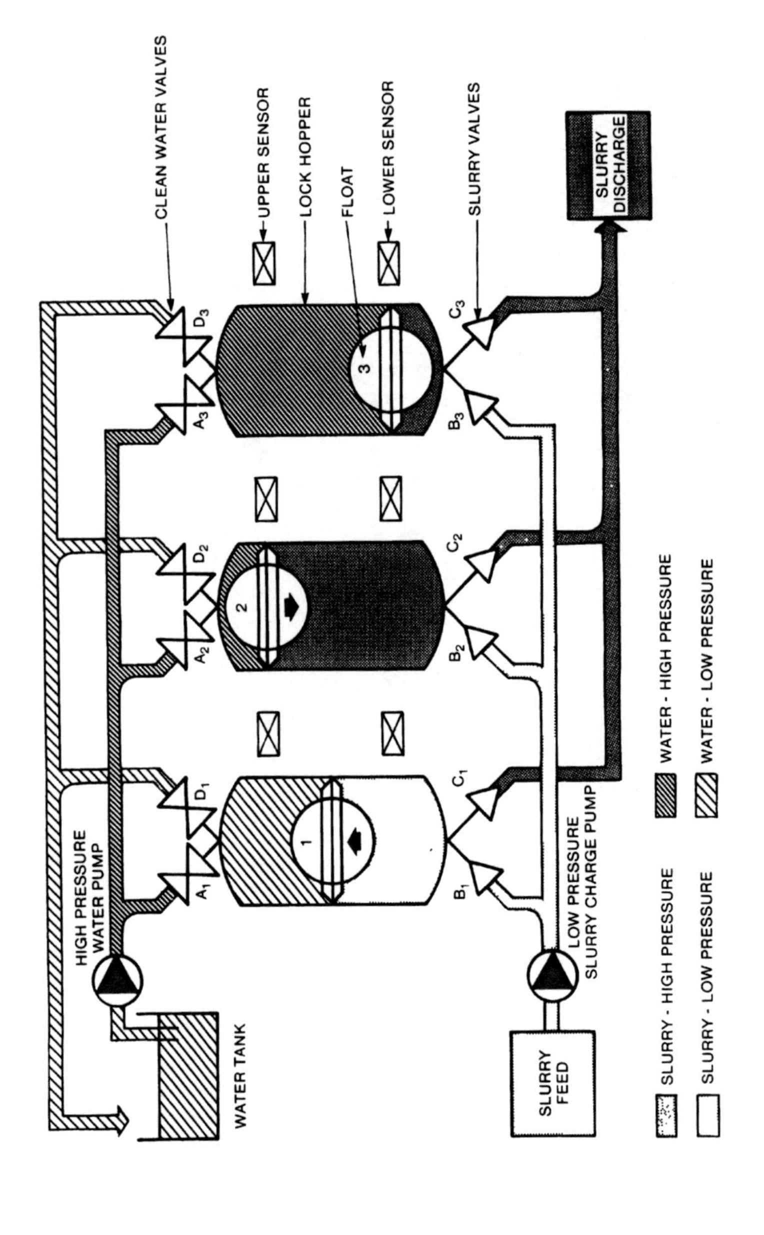

Fig. 17.1. Schematic flow sheet of the vertical type lockhopper Hitachi hydrohoist showing sequence of operation. (Adapted from Sakamoto (1974).)

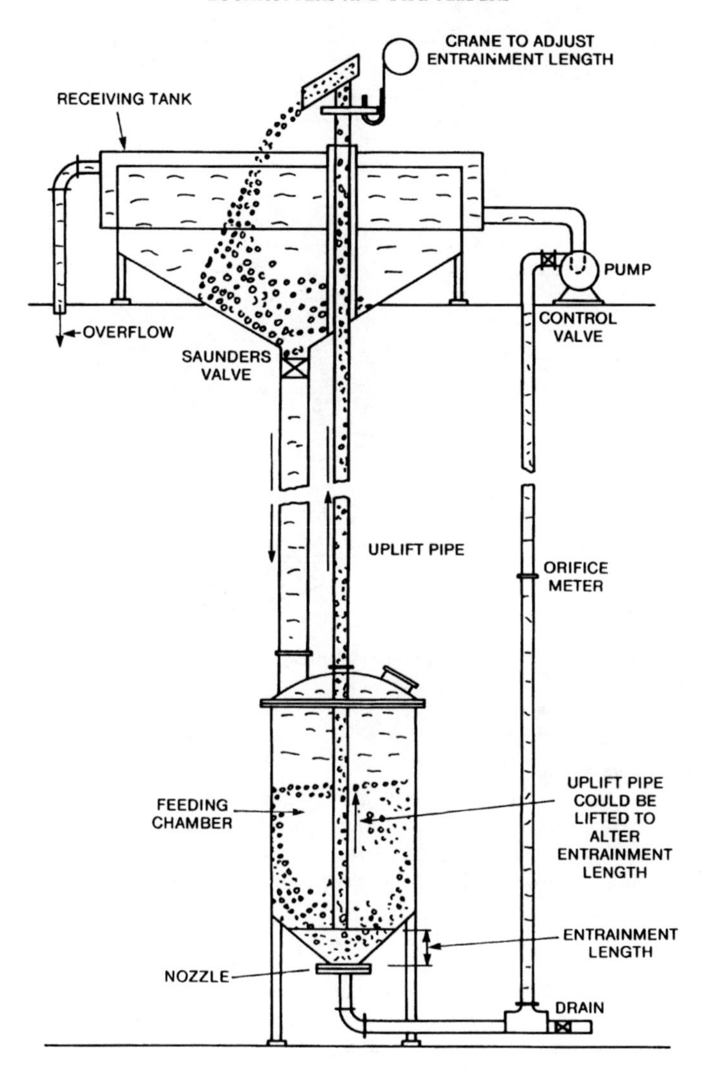

Fig. 17.2. Hydrolift. (Adapted from Oedjoe (1964).)

in Fig. 17.2. The so-called *Hydrolift* uses a pressure vessel and stand-pipe principle to discharge particulate solids in vertical flow and the device is capable of handling coarse particles in the size range up to about 25 mm over relatively short distances.

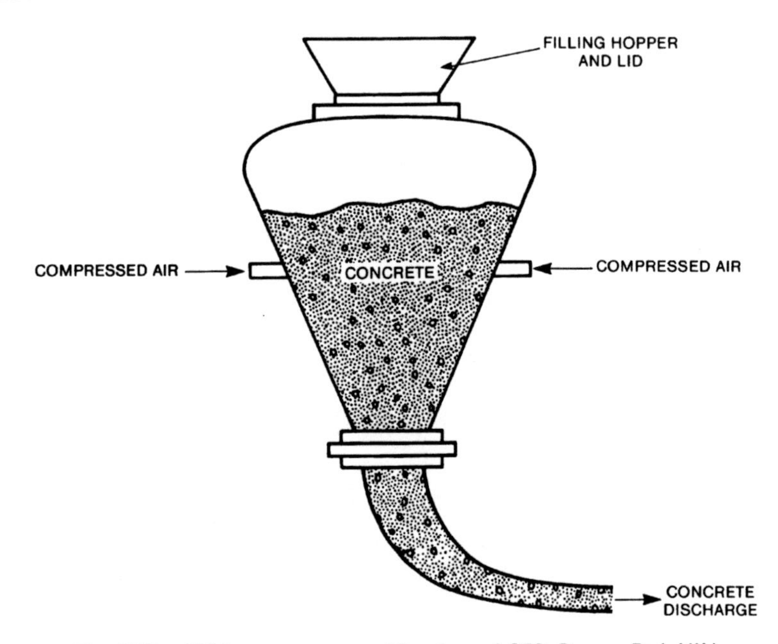

Fig. 17.3. SEM concrete pump. (Courtesy of CES, Bourne End, UK.)

2.4. Other Systems

SEM Concrete Pump
A simple example of lockhopper feeding is the SEM concrete pump as illustrated in Fig. 17.3. Here, concrete is pumped from a conical pressure vessel by the action of compressed air. The SEM pump can deliver concrete at the rate of 15–25 $m^3 h^{-1}$ over horizontal distances of 150 m and vertical distances of 15 m.

Kamyr Chamber Feeder

Description. A novel high-pressure chamber feeding system has been developed by Kamyr Inc. in the USA to transfer solids from a low-pressure stream of water to a high-pressure transport stream of water. This system is in operation on a large scale in the USA for feeding wood chips to pressure digesters and its use has been extended to coal transportation (Funk *et al.*, 1978). The feeder as shown in Fig. 17.4 can be described as a four port rotary valve. Opposing horizontal ports are at high pressure and pressure balance is achieved because of the opposing forces.

Operation. The moving part, which is the rotor, contains four in-line holes which act as chambers as well as flow passages. The feeder principle is illustrated in Fig. 17.5. There is always a downwards continuous flow of water and a hori-

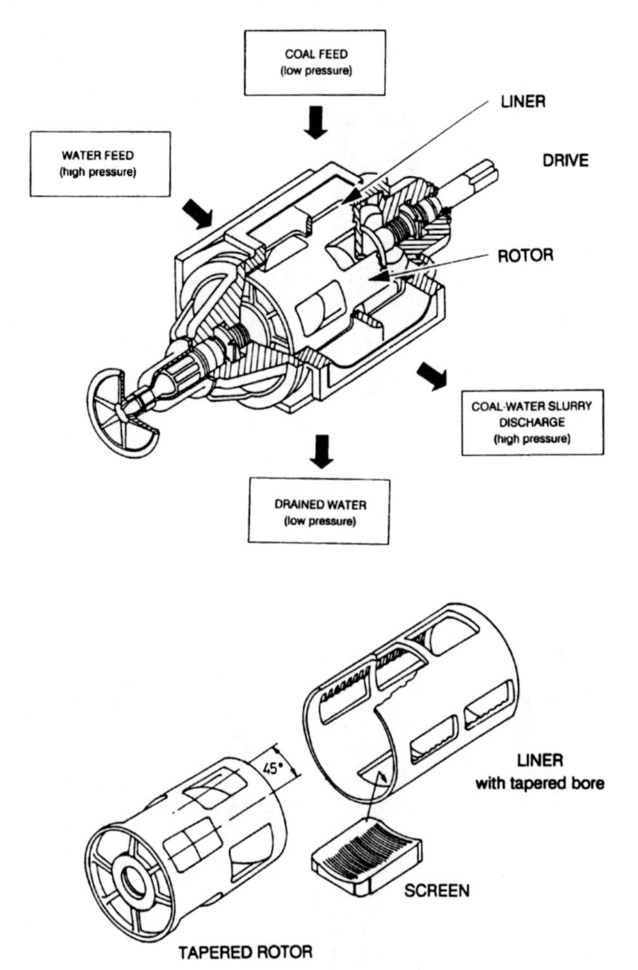

Fig. 17.4. Sectional diagram of the Kamyr feeder. (Adapted from Funk *et al.* (1978).)

zontal continuous flow of water. Solids are fed by gravity into the downward water flow and are stopped in the vertical chamber by a screen in the bottom housing port as shown in Fig. 17.6. As the rotor turns, the screen is wiped clean and the filled hole moves into line with the high-pressure horizontal flow. The solids are carried into the transfer pipeline by the high-pressure water thereby diluting the solids. Since there are four holes in the rotor located at every 45°, the filling and emptying of chambers is virtually continuous. The principle is similar to a multiple pipe feeder except that the high- and low-pressure paths remain fixed and the pipes, i.e. chambers, are switched in and out of the flow paths by the rotor. The Kamyr feeder

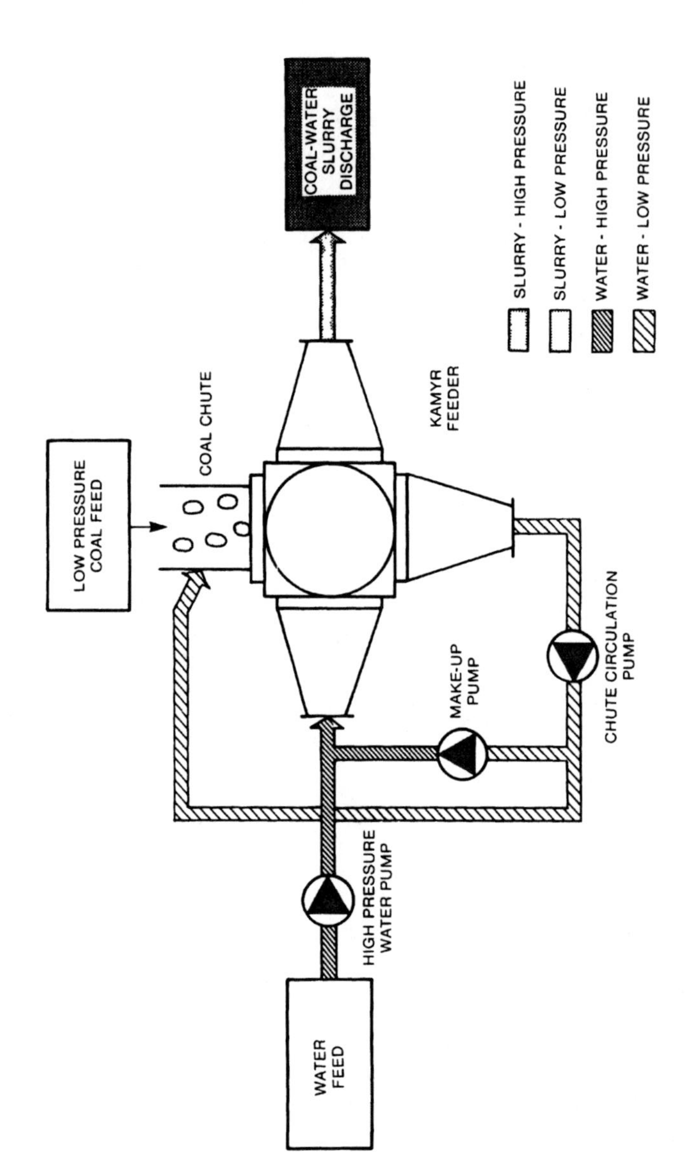

Fig. 17.5. Schematic flow sheet for the Kamyr feeding system. (Adapted from Funk *et al.* (1978).)

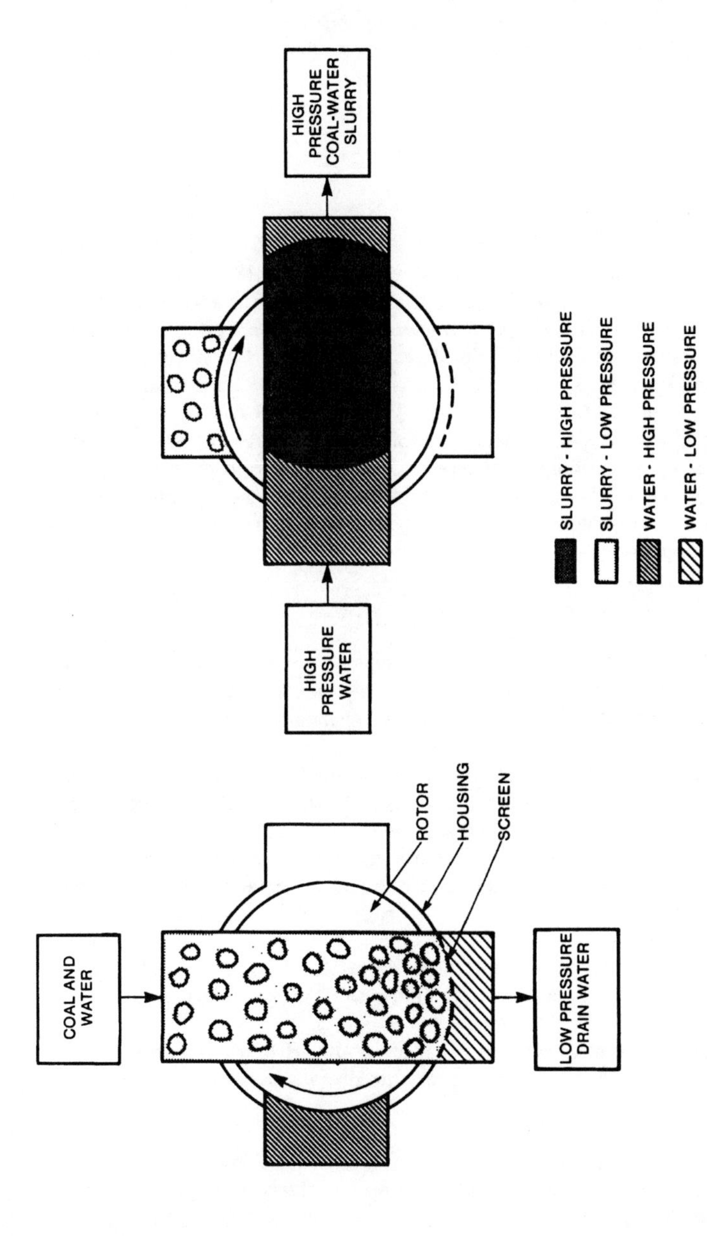

Fig. 17.6. Operating sequence of the Kamyr feeder; left, chamber filling with drained solids at low pressure; right, solids discharged with high-pressure water. (Adapted from Funk *et al.* (1978).)

eliminates the use of independent high- and low-pressure valves and also affords considerable savings of space over conventional lockhopper and pipe feeders. The feeder is capable of handling coal at flow rates of $500\,t\,h^{-1}$ at operating pressures up to about 2·5 MPa.

3. TUBE AND PIPE FEEDERS

3.1. Introduction

Tube and pipe feeders were extensively developed in the 1950s and some of the earliest applications are to be found in the USA and in England, e.g. Woodend colliery, though more extensive development took place in Eastern Europe, particularly in the coalfields of Czechoslovakia, Hungary, Poland and Germany (Hrbek & Kupka, 1970; Kocsanyi, 1972; Szivak *et al.*, 1978). Similar developments occurred at about the same time in Japan. The concept is basically similar to that of the lockhopper, except that horizontal pipes are employed in place of vertical pressure vessels. Cyclic filling of the pipe sections under low pressure and discharging at high pressure provides almost continuous flow of slurry. Horizontal pipe feeders can transport coarse-particle suspensions with rather more ease than vertical lockhoppers.

3.2. Hitachi Horizontal Pipe Feeder

Description. The Hitachi pipefeeder consists of three pipes, each having four plate valves, a low-pressure slurry pump, a high-pressure water pump and control system. A schematic diagram of a typical arrangement is shown in Fig. 17.7.

Operation. The three pipes shown diagramatically in Fig. 17.7 indicate the three basic sequences of operation. Operation is cyclic with the next sequence of pipe 3 being that currently occurring in pipe 1. All valves close only on clear water. In the case of discharge valves C and D they close only when the pipe full of slurry has passed. The sequencing of the valves is initiated by detection of the interface between slurry and water.

Pipe 1. Filling with low-pressure slurry. Valves A_1 and C_1 are closed, B_1 and D_1 open. Low-pressure slurry is pumped to fill the pipe thereby discharging the water.

Pipe 2. Commencement of slurry discharge. Valves B_2 and D_2 are closed, A_2 and C_2 are opened. High-pressure water forces the slurry (now at high pressure) into the discharge pipe.

Pipe 3. End of slurry discharge. Valves A_3 and C_3 are closed and valves B_3 and D_3 are opened when the pipe is completely empty of slurry. Low-pressure slurry is pumped into the pipe and the contents (water at low pressure) are displaced.

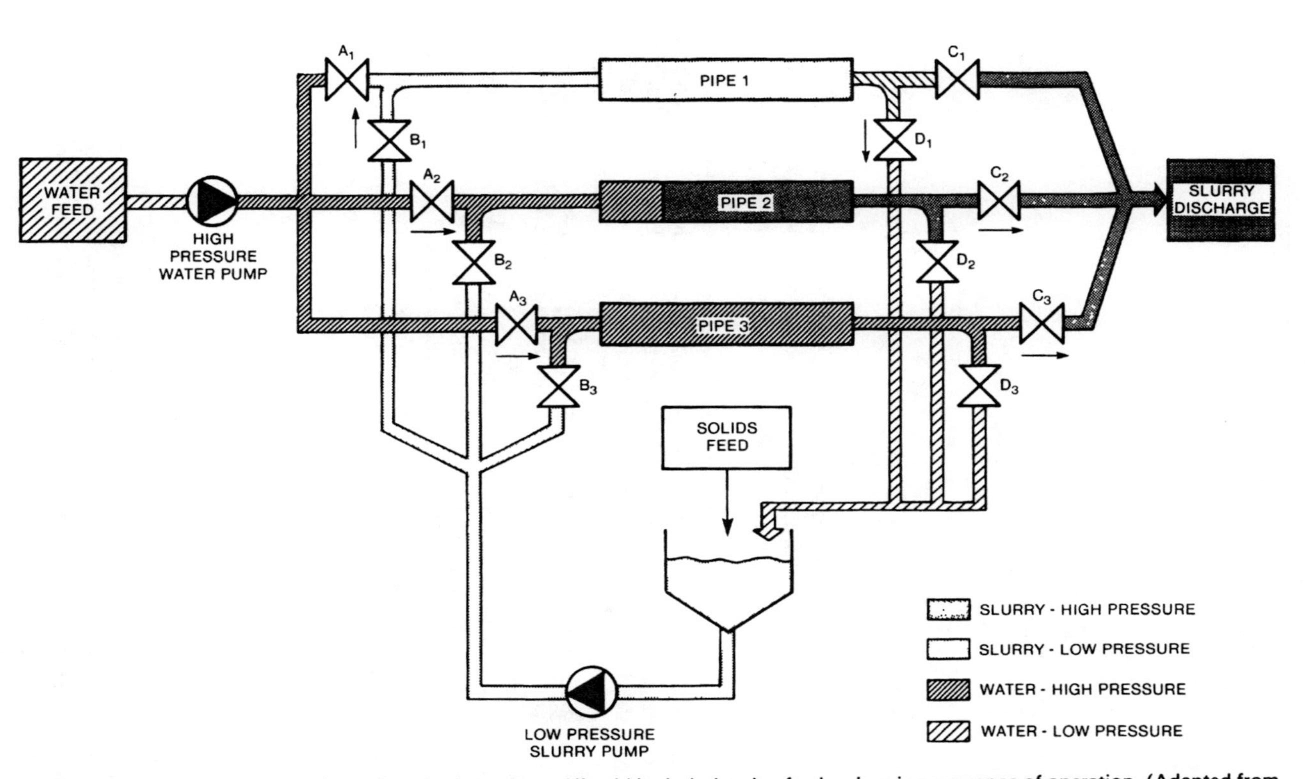

Fig. 17.7. Schematic flow sheet of the horizontal type Hitachi hydrohoist pipe feeder showing sequence of operation. (Adapted from Sakamoto (1976).)

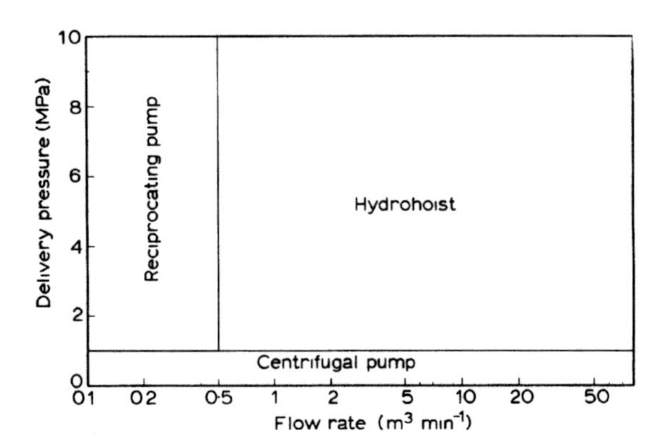

Fig. 17.8. Scope of applications for the Hitachi hydrohoist. (Adapted from Sakamoto (1976).)

By repeating the above operations for each of the three feed pipes, semi-continuous transport is achieved. The operations are sequenced by a system controller. The plate valves are specially designed for the purpose and are opened and closed by hydraulic-oil pressure.

The plate valve is a special bi-directional sliding plate block valve. The sealing surfaces are overlaid with stellite which prevents abrasion due to operation and assures long life. The plate valves also have a bypass to equalise the pressure. The inside pressure of the feed chamber is equalised by increase and decrease of the pressure before opening of plate valves A or D (Fig. 17.7). Since the diameter of the plate valve is equal to that of the feed pipe, coarse particles can easily be passed. Coarse particles that settle to the bottom of the horizontal feed pipe can easily be removed when high-pressure water starts to flow.

All plate valves are designed to close only in clear water and therefore the opening and closing frequency of the valve is not normally more than twice per minute and thus erosion is negligible. Large flow rates of solids are possible at high pressure when compared with conventional slurry pumps (see Fig. 17.8). Degradation of coarse coal is reported to be small. Operating experience with horizontal Hitachi pipe feeders is given in Table 17.1

4. APPLICATIONS

4.1. Coal

Lockhoppers and pipe feeders have found extensive application in the hydraulic transportation of coal–water suspensions in horizontal and vertical pipes. Although there are many direct pumping options involving the use of either centrifugal or positive displacement pumps for the conveying of finely divided coal–water suspensions, indirect methods of feeding are widely used for coarse

TABLE 17.1
Hitachi Hydrohoist Installations as of April 1987

Site	Specifications		Material (size range)	Number of hydrohoists	Total operation (h)	Year of supply
	Flow rate ($m^3 h^{-1}$)	Pressure (MPa)				
Vetical type						
Yokohama Plants of Showa Denko K.K.	80	3·5	Bauxite slurry (<2 mm)	1	120 000	1967
Yokohama Plants of Showa Denko K.K.	136	4·5		1	61 000	1970
Queensland Alumina Ltd (Australia)	170	5·6		3	107 000	1971
Hitachi Smelter of Nippon Mining Co.	30	5·5	Sulphide ore	1	69 000	1976
Alumina Espanola, S.A. (Spain)	326	5·5	Bauxite slurry (<2 mm)	2	51 000	1978
Alumina Contractors Ltd (Republic of Ireland)	326	5·5		2	27 000	1980
Queensland Alumina Ltd (Australia)	170	5·1		2	28 000	1981
Kaiser Aluminium & Chemical Corp. (USA)	680	6·0		1	—	1981
Horizontal type						
Yoshima Colliery of Furukawa Co., Ltd (Japan)	($80 t h^{-1}$) 318	5·2	Raw coal (<50 mm)	1	273	1962
Sunagawa Coal Mine of Mitsui Coal Mining Co., Ltd (Japan)	($100 t h^{-1}$) 300	8·3	Raw coal (<35 mm)	1	29 000	1965
Rinkai Construction Co., Ltd	1 300	1·0	Fine sand (<2 mm)	1	—	1979
Electric Development Power Co., Ltd (Sakuma) (Japan)	83·4	5·0	Sand (<10 mm)	1	7 000	1980

coal–water slurries. There has been considerable development of hydraulic transportation of run-of-the-mine coal with particle size up to about 50–60 mm. Indirect methods of feeding tend to overcome problems of coal size reduction by erosion, abrasion and attrition. Lockhoppers and pipe feeders have found application for vertical hoisting and horizontal conveying from the coal face to the surface, for the transportation of coal slurries in the washery and for the disposal of tailings. Horizontal pipe feeders have been preferred in most coal mines, though the technique adopted will depend to some extent on the method of coal winning. Several coal mines have adopted hydraulic mining procedures, notably in Germany and Eastern Europe (Harzer, 1974; Kuhn, 1976, 1978; Harzer & Kuhn, 1982). Jet cutting at the coal face produces a dilute coal–water suspension which is amenable to hydraulic transportation to the surface.

An early application of pipe feeders in coal mining was evaluated at the Egercsehi coal mine in Hungary. Test work on the hydraulic transportation of waste rock and tailings started in 1961 and an operational plant was commissioned in 1966 to transport waste over a vertical distance of 200 m to the spoil bank.

The success of this project led to the evaluation of the Hungarian design of pipe feeder in the Gneisenau coal mine in Dortmund, Germany. The Gneisenau hydromine used high-pressure jet cutting at the coal face. The coal is crushed to pass a 60 mm screen, partially dewatered and directed to an intermediate, flooded storage hopper. Slurry is discharged from the hopper into a two-chamber pipe feeder by a centrifugal pump. The solids are separated from surplus water in a clarifier and fed to the temporary storage hopper. Clarified water is used for cutting jets and to discharge the pipe feeders. The slurry is transported over a horizontal distance of 1200 m to the shaft bottom followed by a vertical lift of 700 m to the surface. The system transported 1 500 t day^{-1} solids at an average superficial velocity of 3·7 m s^{-1} in the hoisting pipe of internal diameter 196 mm. A two-chamber pipe feeder system was installed in this prototype project and to prevent pressure surges during changeover from filling (0·3 MPa in the chamber) to transport (9 MPa in the chamber), a pressure compensation system had to be developed and used (Prettin & Gaessler, 1976).

The Hansa hydromine in Germany went into production in May 1977 and is the largest hydraulic mining project to date. Slurries containing waste rock and run-of-the-mine coal with a top size of 60 mm were transported over a horizontal distance of 600 m followed by a vertical lift of 840 m to the surface. The system transported approximately 4 500 t day^{-1} solids as a slurry with a solids content of 27% (by volume). The superficial velocity of the slurry was 4·8 m s^{-1} in the transport pipe of internal diameter 250 mm. Because slurry flow was not constant, intermediate storage was provided underground. A three-chamber pipe feeder was installed at the Hansa mine and this allowed semi-continuous hoisting of coal slurry to the surface. The coal slurry with nearly constant solids to water content was pumped alternatively into one of the three pipe loops by a low-pressure centrifugal pump. High-pressure water discharged the slurry in the opposite direction out of the loop into the vertical shaft. The Hansa hydromine remained in operation for three years but was finally closed down in late 1980 because unexpectedly difficult geological conditions underground precluded the development of further economical seams.

4.2. Minerals

Hydraulic transportation of mineral ore bodies has been used for bauxite suspensions, sulphide ores, limestone and lead/zinc minerals. Many mineral ore bodies are abrasive and are therefore not amenable to direct pumping with conventional centrifugal or positive displacement pumps. For example, bauxite slurry is strongly alkaline and is usually handled at high temperatures (about 130°C) and is therefore severely abrasive. The vertical version of the Hitachi hydrohoist has been successfully applied to this problem and several operating plants are in existence in Japan (see Table 17.1). The relative density of the bauxite slurry is 1·5. The slurry contains particles with a relative density of 2·4 and an average size of 300 μm (95% are < 1 mm) at a concentration of 50% (by mass). The vertical hydrohoist relies on a spherical float to separate the aggressive bauxite slurry from the motive driver fluid which in this particular case is a sodium aluminate solution of concentration similar to the mother liquor of the suspension. This concept overcomes problems of mixing even at the minimum level encountered in this system. A photograph of the bauxite slurry plant at the Showa Denko plant in Yokohama is shown in Fig. 17.9.

The hydraulic conveying of lead/zinc ores has been reported by Hrbek and Gibian (1974) at the Kank mine near Kutna Hora in Czechoslovakia. The system consists of a single lockhopper feeder (shown schematically in Fig. 17.10) located underground. The lockhopper is operated in a batchwise manner. When the solids have been discharged, surplus water is discharged by opening the drain valve. This valve is closed and a controlled volume of solids are loaded into the lockhopper by isolating the high-pressure water supplies and the slurry discharge valve and then opening the hopper valve. The lockhopper is then pressurised with high-pressure water, supplied from the surface, and the solids are fed into the transportation pipeline with a screw feeder. Automatic operation is achieved with a logic controller. Lead/zinc ore of relative density in the range 2·9–3·1 and particle size up to about 50 mm has been transported a horizontal distance of 370 m underground to the bottom of a shaft and then hoisted a vertical distance of 441 m using high-pressure clean water.

Hitachi have reported studies on the hydraulic transportation of limestone suspensions containing coarse particles of about 2 mm in diameter in a matrix of finely divided material of about 50 μm. By adjusting the rheological properties of the finely divided limestone slurry of approximate concentration 50–60% by mass, it is possible to support coarse particles under 16 mm diameter, i.e. their hindered settling velocity is effectively zero. Tests showed that it was possible to convey this mixture using horizontal pipe feeders at relatively low head loss values. There is, however, no recorded large-scale industrial application of this technique for limestone or indeed other non-Newtonian suspensions.

Other mineral applications include the hydraulic conveying of finely divided and

Fig. 17.9. Vertical hydrohoist for bauxite slurry using four lockhoppers at Showa Denko, Yokohama, Japan.

coarse sand. The Hitachi hydrohoist has been used for this application and details are to be found in Table 17.1.

4.3. Waste Disposal

An important application of hydraulic conveying is the disposal of mine tailings, the disposal of fly ash from furnaces and the removal of wastes from chemical plants. This topic has been reviewed by Szivak *et al.* (1978). The transportation of solid wastes often involves short distances between the point of origin and the point

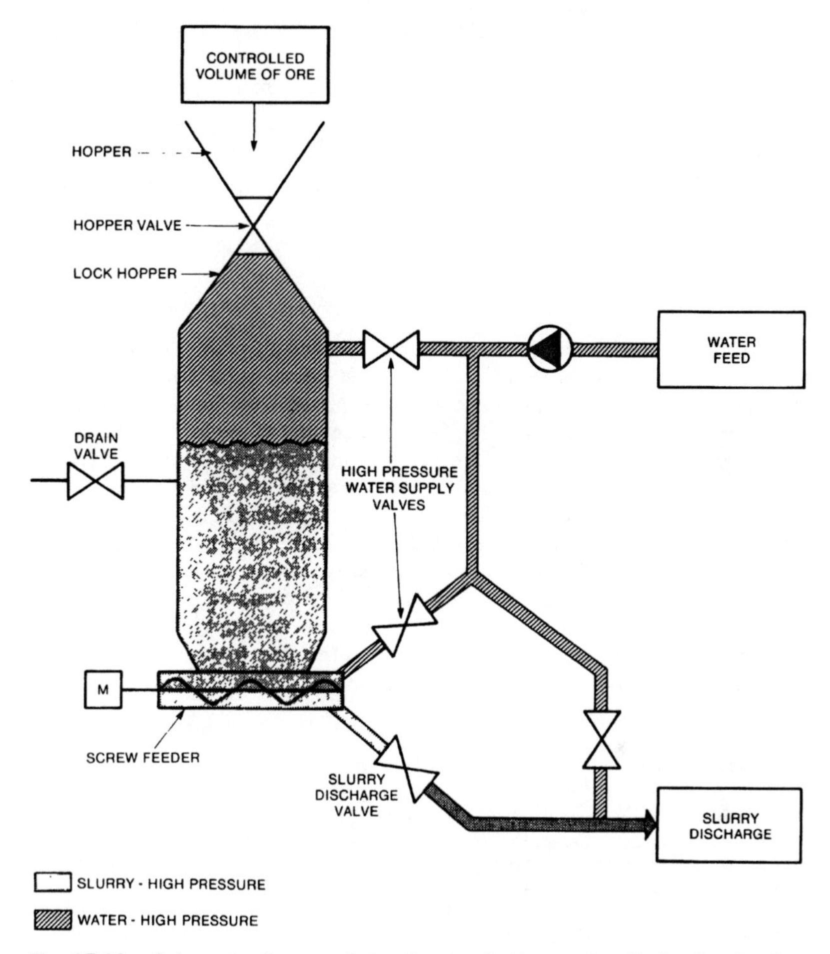

Fig. 17.10. Schematic diagram of the chamber feeder used at Kank mine Czechoslovakia. (Adapted from Hrbek & Gibian (1974).)

of deposition. An integrated materials handling system is suggested with the following features:

- the energy needed to transport the particles of solid waste is solely transmitted by water or other inert liquid stream during collection, transportation and deposition;
- power consumption is as low as possible;
- operational facilities should be automated;

- wear-resistant materials should be used to minimise maintenance costs; and
- all plants and facilities should satisfy the latest environmental considerations.

The practical exploitation of lockhoppers and pipe feeders to transport waste materials is not yet widespread but it remains a feasible alternative to belt conveyors, trucks, etc. in urban areas and also in inaccessible mining regions throughout the world.

5. CONCLUSIONS

This chapter has reviewed the status of lockhopper and pipe feeder systems for the hydraulic transport of both finely divided and coarse solid suspensions. The literature on this subject was extensive in the period from about 1960–80 and thereafter there has been a noticeable fall in the amount of published information. This suggests that this aspect of hydraulic conveying is mature and very few further advances have occurred in the last 10 years. Part of the explanation may be attributed to the closing down of the hydraulic coal mining ventures in Germany thereby reducing the demand for this technology and lessening the need for new developments. Other fields, such as minerals transportation, are likely to arise and it is possible that the heightened awareness of the environmental impact of hazardous waste disposal could bring renewed interest in the application of indirect feeding systems of the type discussed above and this may precipitate a revival of industrial activity and the development of improved design concepts.

6. REFERENCES

Funk, E.D., Barrett, M.D. & Hunter, D.W. (1978). Pilot plant experiences with run of mine coal injection and pipelining. In *Proc. HT 5*, pp. F39–46.*

Harzer, H. (1974). Hydraulic hoisting in the mining industry of the Federal Republic of Germany. In *Proc. HT 3*, paper A2, pp. 17–22.*

Harzer, J. & Kuhn, M. (1982). Hydraulic transportation of coarse solids in a continuous system from underground production face to the end product in the preparation plant. In *Proc. HT 8*, pp. 461–70.*

Hrbek, J. & Gibian, E. (1974). Hydraulic transport of Pb Zn Ore from 450 m depth to surface. In *Proc. HT 3*, paper A4, pp. 39–54.*

Hrbek, J. & Kupka, F. (1970). Special problems of hydraulic transport of heterogeneous materials made possible by means of feeders. In *Proc. HT 1*, paper B5, pp. 41–52.*

Kocsanyi, L. (1972). High pressure transport of coal and other mining products by means of pipe feeders. In *Proc. HT 2*, paper H4, pp. 69–80.*

Kuhn, M. (1976). New possibilities and current developments of hydraulic transport in the mining industry of the Federal Republic of Germany. In *Proc. HT 4*, paper E4, pp. 49–61.*

Kuhn, M. (1978). Feeding of solid matter into hydraulic conveying systems. In *Proc. HT 5*, pp. F1–16.*

Laubscher, B. & Sauermann, H.B. (1972). Performance of the hydrolift feeder. In *Proc. HT 2*, paper H3, pp. 53–68.*

Oedjoe, D. (1964). Hydraulic lifting of discrete solids with 'The Hydro-Lift'. PhD thesis, University of New South Wales, Wollongong, Australia.

Prettin, W. & Gaessler, H. (1976). Bases of calculation and planning for the hydraulic transport of run-of-mine coal in pipelines according to the results of the hydraulic plants of Ruhrkohle AG. In *Proc. HT 4*, paper E2, pp. 13–36.*

Sakamoto, M. (1967). Hydraulic transport of granular solids by Hitachi Hydro-Hoist. *Hitachi Rev.*, **16**, 439–46.

Sakamoto, M. (1974). Development of hydrohoists: slurry feeders without contamination. In *Proc. HT 3*, paper A1, pp. 1–16.*

Sakamoto, M. (1976). A pilot plant for capsule and slurry transport. In *Proc. HT 4*, paper E6, pp. 79–100.*

Sakamoto, M., Uchida, K. & Honma, I. (1976). Hydraulic transportation systems using hydrohoists. *Hitachi Rev.*, **25**, 237–42.

Sakamoto, M., Nagawa, Y., Uchida, K. & Kamino, Y. (1978). A hydraulic study of coarse particles including fine particles with hydrohoist. In *Proc. HT 5*, pp. D79–90.*

Sakamoto, M., Uchida, K. & Kamino, Y. (1979). Vertical type hydrohoist for hydraulic transportation of fine slurry. In *Proc. HT 6*, pp. 257–68.*

Sakamoto, M., Uchida, K. & Kamino, Y. (1982). Transportation of coarse coal with fine particle–water slurry. In *Proc. HT 8*, pp. 433–43.*

Szivak, A., Illes, K. & Varga, L. (1978). Up-to-date hydraulic transport systems for the delivery of industrial wastes. In *Proc. HT 5*, pp. F17–38.*

Yamaguchi, T., Yamada, Y., Sakamoto, M. & Uchida, K. (1970). Investigation of hydrohoist for pumping bauxite slurry. *Hitachi Rev.*, **19**, 185–90.

*Full details of the Hydrotransport (HT) series of conferences can be found on pp. 11–12 of Chapter 1.

18

Jet Pumps

Anthony W. Wakefield
AW Wakefield, Stamford, UK

1. INTRODUCTION

The jet pump is one of the most misunderstood devices in hydraulics and its relevance to the transport of slurries widely underestimated. Applications can vary widely and two examples are shown in Figs 18.1 and 18.2, one for a high-throughput dredge pump handling sand and stone, the other transferring carrots at a much lower flow rate.

In this chapter, terms peculiar to jet pumps are defined by the context and printed in italic at their first occurrence.

Jet pumps are known, often according to industry or principal application, also as eductors inductors, or ejectors. If all the *motive* energy is input in kinetic form as a high-velocity stream of fluid, then the device is a *jet pump*. If not, it may be a venturi pump or some sort of device dependent on the Coanda effect. So-called parietal jet pumps may fall into this category. Such related devices are not treated here (but refer to Wakefield, 1987).

For some reason, many who would hesitate to manufacture their own centrifugal pump are not deterred from attempting a DIY jet pump. Such endeavours are usually only partially successful. This chapter will help in the understanding of the jet pump, but more importantly it will allow the more positive consideration at the early stages of a project of the possible contribution a jet pump might make.

Fig. 18.1. Pump for jet pump dredge, for production rates in the range 300–1000 t h^{-1} of sand and stone.

Fig. 18.2. Annular (ring) jet pump handling carrots at typical production rates between 3 and $10\,t\,h^{-1}$.

2. BRIEF DESCRIPTION

The parts of a solids-handling jet pump are shown in Fig. 18.3(a). Figures 18.3(b–d) illustrate adaptations of the same pump to varying functions. Note that the motive jet, nearly cylindrical when discharging in air but conical when submerged, lies on the centre-line of the jet pump. Alternative configurations of jet pump considered elsewhere (Wakefield, 1987) use multiple jet arrays or an annular jet.

A stream of fluid possessing only kinetic energy is called a *jet*. It may be generated by a rotodynamic pump or other means but is usually produced conveniently and efficiently in a *nozzle* (contraction) from a high-pressure source. The nature of the contraction governs the structure of the jet. The *motive flow* of (usually) clean liquid crosses an *entrainment zone* within which it grows by incorporation of an *induced flow* of a second fluid, in the present case, slurry. The *combined flow* is collected by a tube, called the *mixing chamber* or *mixer*, in which energy sharing between the two flows proceeds to completion.

If advantage is taken of the shape of the jet pump characteristic to control the flow in the pipeline, the mixer exit velocity will be above that for optimum pipeline flow. A well-designed enlargement, or *diffuser*, is necessary to recover the maxi-

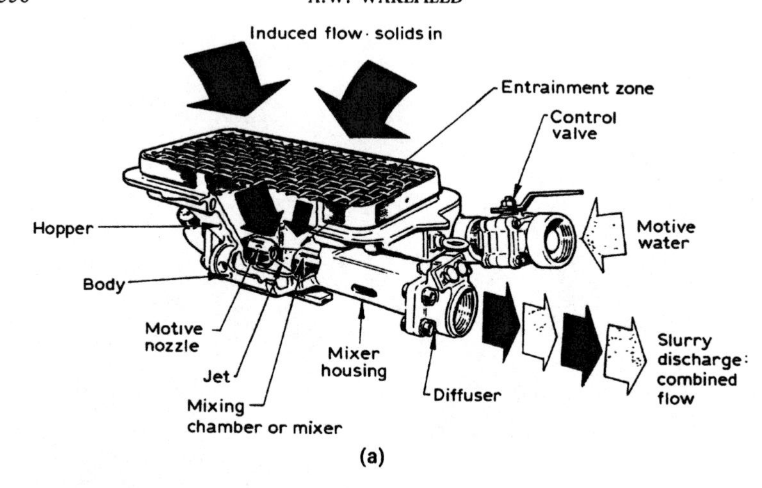

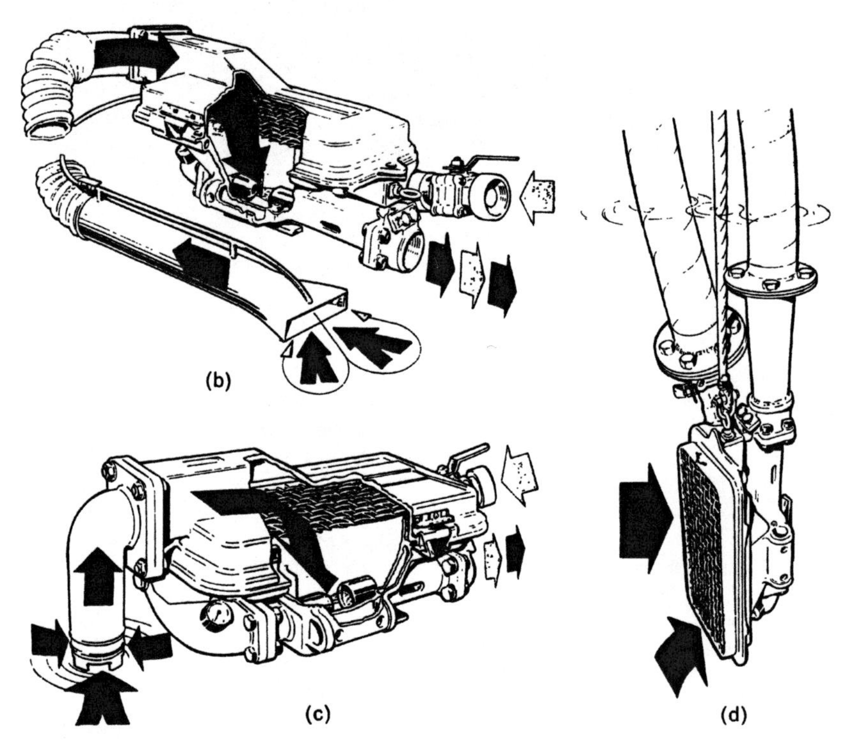

Fig. 18.3. The solids-handling jet pump. (a) Showing parts, configured as a hopper unit; (b) connected to a suction pipe; (c) used as a sump pump; (d) emptying a tank.

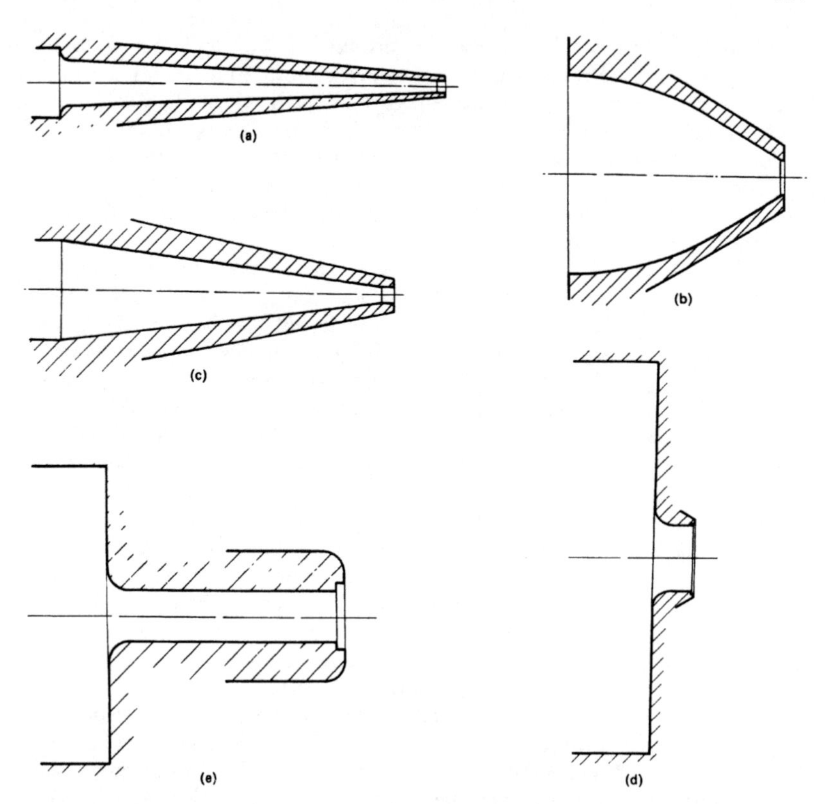

Fig. 18.4. Nozzle designs. (a) Long, fine-tapered nozzle; (b) Rouse nozzle; (c) medium-tapered nozzle; (d) radiused nozzle; (e) radius plus parallel.

mum proportion of the fragile kinetic energy of the combined flow. The velocity of flow in the mixer, although higher than that in the pipeline, is much lower than that at the exit of the nozzle. The entrainment zone is frequently contained in a *body* or *suction chamber* from which a *suction pipe* or *duct* leads to the slurry source of the induced flow.

3. ATTENTION TO DETAIL IS REPAID IN PERFORMANCE

3.1. Motive Nozzle

A very long, gradual taper, say 5° included angle as in Fig. 18.4(a), allows the growth of a turbulent boundary layer and the velocity profile will tend towards a parabolic form. Shear stresses will be present within the jet which will in any case be fully turbulent but the shear stress between the jet and surrounding fluid will be relatively low. The interface between a turbulent jet and the ambient fluid is corrugated (Thorne & Theobald, 1978; Hoyt & Taylor, 1985). Energy loss in the nozzle tends to be high because of the length and high velocity. If energy loss is excluded a discharge coefficient close to unity is secured by a length of less than half a diameter of parallel at the exit end. Because of the energy loss, however, and the changing friction characteristics of practical engineering materials in a working environment, the flow rate cannot be accurately determined from knowledge of the pressure difference between entry and exit.

Rapid contraction has been shown (Hoyt & Taylor, 1985) to linearise the jet which will then be in a sort of super-critical laminar flow; the velocity profile is then essentially rectangular. Internal shear stress is nil but that between the jet and surrounding fluid high. Any irregularity in the approach flow, however, caused by a bend perhaps, will be accelerated in the contraction, spoil the one-dimensional velocity characteristic, and any benefit dependent on jet perfection will be lost. If the exit is sharp-edged as in the Rouse nozzle of Fig. 18.4(b), then the discharge coefficient will be less than unity and will vary according to the state of wear of the orifice.

The nozzle of moderate taper of Fig. 18.4(c) lies between the previous two. The discharge coefficient is less affected by wear and is therefore more predictable though still not the best achievable. The radiused contraction of Fig. 18.4(d) gives a very small energy loss of contraction and the geometrical discharge coefficient (excluding energy loss) may be unity. The overall discharge coefficient may be of the order of 0·985. Thus, the flow rate through the nozzle can be determined accurately by measurement of the pressure difference.

The assumption is made that a parallel section of nozzle may be legitimately considered as a section of parallel pipe, as shown in Fig. 18.4(e), and does not affect the performance of the contraction other than to reduce the pressure drop across it. It also ignores the laminarising effect of the contraction and is likely to give a discharge coefficient slightly low, though not by enough to be readily detectable. In any case, this super-critical phase cannot persist and a turbulent boundary layer will thicken as the parallel section is lengthened.

Since it is likely that the corrugated surface of a turbulent jet assists entrainment, allowing the jet time to re-turbulate may be beneficial. It has been observed (Wakefield, 1972) that the hydraulic efficiency of a jet pump could be optimised by varying not only the axial gap between nozzle exit and mixer inlet but also the length of the parallel section of the nozzle. The latter might also be expected from the observed dependence of the quality of the jet on the length of the nozzle (Obot *et al.*, 1984; Trabold *et al.*, 1987).

The shear stresses between the jet and the surrounding fluid produce tiny vortices which themselves act on both the surrounding fluid and the fluid of the jet. Thus the jet expands, absorbing surrounding fluid and slowing down. It has been amply demonstrated (Rajaratnam, 1976; Massey, 1983) that momentum is conserved during this process. Not predictable is the angular rate of expansion of the jet. For a single fluid, the included angle is likely to lie in the range 20–30° (Massey, 1983; Landau & Lifschitz, 1987) according to the velocity profile and turbulence charac-

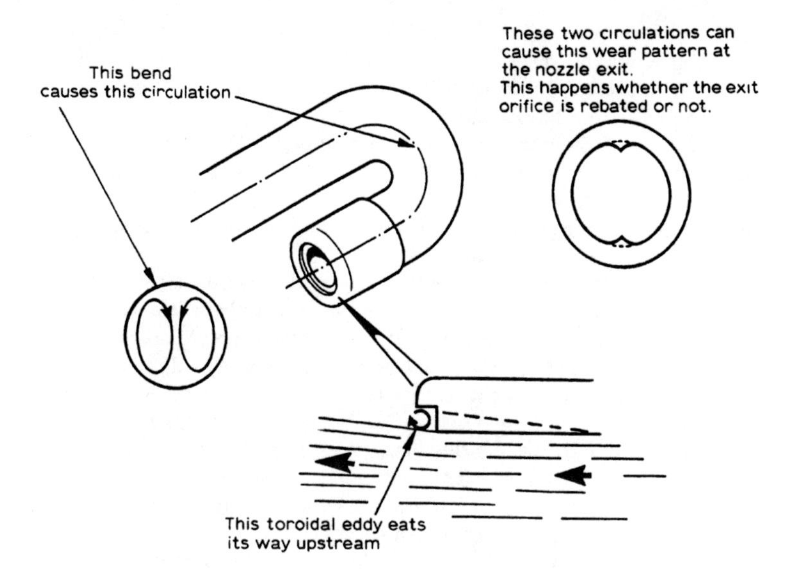

Fig. 18.5. Circulation and toroidal tip vortex.

teristics of the jet. (This angle is taken as that at which velocity has fallen to 1% of that at the axis.) It is probable that the slight corrugation of the turbulent jet, aiding entrainment, will result in a greater angle. A tendency to very high shear stresses may not be associated with greater entrainment because the micro-scale pressure differences required for the acceleration of the entrainment vortices cannot depress local pressure to an absolute pressure lower than the vapour pressure of the fluid. Micro-boiling occurs, with attendant slip because the shear stress sustainable by the vapour is negligible.

Just off the tip of the nozzle a toroidal vortex forms, (Fig. 18.5). Loaded with grit, this eats the bore of the nozzle back from the exit. The illustration shows how the toroid, and the wear pattern, may be distorted by circulations induced by upstream features. Nozzle replacement is necessary as soon as this erosion becomes excessive. Clearly, a short nozzle, fine tip or tapered bore are all unsuitable.

3.2. Entrainment Zone and Body

As the nozzle wears, the expansion of the jet starts earlier, the expanding jet eventually meeting the inlet orifice of the mixer. It can be shown that the divergence angle of the jet is smaller if the density of the induced flow is greater than that of the motive flow. Therefore, the first evidence of excessive wear might be the refusal of the jet pump to start, although once started it could continue to run. A severely worn nozzle can give rise to vortex shedding, inducing vibrations severe enough to dismantle the jet pump and adjacent pipe work.

Clearly some distance will be required before the induced fluid reaches the axis of the pump. Over this length the central core of the jet remains undisturbed, continuing with unaltered velocity (Culbreth & Legoff, 1985), but becoming progressively narrower until it disappears.

The apex angle of this central cone of motive liquid has been shown to vary with nozzle profile (Obot *et al.*, 1984; Trabold *et al.*, 1987). With the preferred long nozzle the cone will project into the mixer, so that the jet does not become fully developed before entering, though it is likely to do so before the point at which the expanding jet meets the wall of the mixer.

The process of entrainment described starts immediately the jet leaves the nozzle but there is an initial transition zone before the jet is fully developed. Thus the virtual origin (source) of the jet is a little forward of the rearward projection of the cone to the axis. This helps to keep the diameter of the jet slightly smaller than it would otherwise be by the time it reaches the mixer inlet. Impact with the pump body round the mixer inlet, or missing the mixer altogether, wastes energy, so a compact jet is helpful in spacing the nozzle back from the mixer. Some slurries contain coarse particles for which the mixer may have been sized. To prevent obstruction the nozzle–mixer gap must then not be smaller than the mixer diameter. This distance is just feasible in the design described.

Experience has shown wear in the mixer to be dependent on internal circulations. An eccentric nozzle, uneven hardness from one side to the other, nozzle deterioration or uneven feed into the mixer shortens its life. This last effect must be avoided by dimensioning the body as a plenum: kinetic energy must be negligible in the induced flow. If the slurry is conveyed to the suction chamber by pipe work, diffusion must be included to drop the velocity to something of the order of $1 \mathrm{~m~s^{-1}}$, but the lower the better.

To prevent internal arching where large solids are present, the mixer must be of a re-entrant pattern, such as those shown in Figs 18.6(a–c).

3.3 Mixing Chamber

It has been shown (Kentfield & Barnes, 1972), that a better hydraulic efficiency occurs with energy sharing at constant area rather than at constant pressure. That is, the mixer has parallel walls. This is fortunate since the mixer can be a simple tube, both costing less and being easier to make. The mixer must be long enough for energy sharing to be completed within it (Wakefield & Ruckley, 1975). Too long, and energy losses (due to friction) will reduce performance. More energy is lost than is predicted by simple application of the usual pipe-friction-factor equations because of the enhanced turbulence of mixing; this must be allowed for in design (Silvester & Vongvisesomjai, 1972; Wakefield & Ruckley, 1975; Wakefield, 1986).

Particularly lethal in the presence of coarse material is a tapered mixer inlet. This provides the perfect springing for an internal arch, stabilised by the nozzle (Fig. 18.6(a)).

The hydraulic efficiency is dependent on Reynolds number in the mixer (Wake-

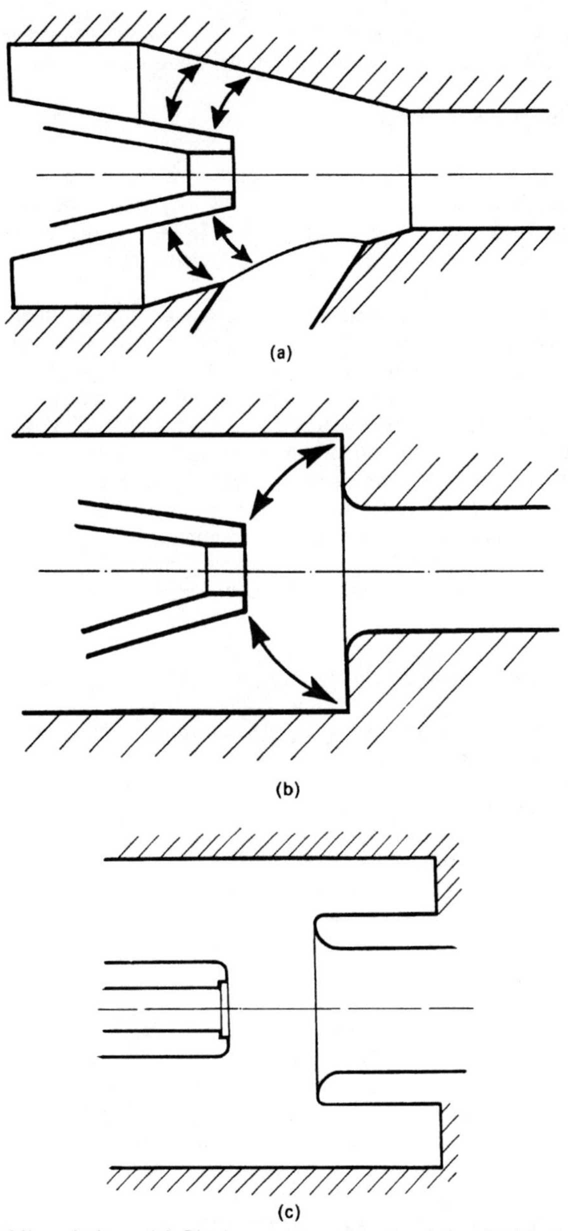

Fig. 18.6. Mixer designs. (a) Blockage at tapered mixer inlet; (b) blockage at non-re-entrant mixer inlet; (c) re-entrant mixer inlet.

field & Ruckley, 1975; Wakefield, 1987) but a jet pump can continue to function at Reynolds numbers of below 2000, albeit with poor predictability.

3.4 Suction Duct

The suction duct or pipe will be passing solids at about twice the concentration which exists in the discharge pipeline. This flow may need to be considered using soil mechanics rather than pipe flow principles. In such a case, the duct may have to diverge continuously to avoid providing a springing for an internal arch of solids.

It may be necessary for the concentration of solids within the suction duct to be limited by the use of dilution jets. In a dredging application similar jets may be required for disintegration or mobilisation of the deposit and the same jets may frequently combine these functions. A jet pump immersed in solids tends to dewater the particles surrounding its suction inlet and fluidisation jets may be required to counteract this. As depth of immersion in solids increases so will the tendency of surrounding solids to lock together in an arch. This is worse for gravel than sand, for example, and for irregular rather than rounded material. Special arch-breaker jets may on occasion have to be fitted.

Bear in mind that the jet pump will pump only material which can get into it. It is possible to calculate accurately the performance of the system only once material is in.

3.5. The Poorly Designed Jet Pump

The solids-handling jet pump of Fig. 18.7 shows all the faults described. Specifically, note the following:

- narrow, tapered nozzle with fine tip;
- confined, non-plenum body;
- tapered, non re-entrant mixer inlet; and
- short parallel section to mixer, with diffusion starting before conclusion of energy sharing.

3.6. Cavitation

Cavitation occurs in a jet pump for much the same reason as in a centrifugal pump: local accelerations reduce absolute pressure in the fluid to its vapour pressure. In a jet pump this starts in the entrainment zone and continues for as long as the turbulence of energy sharing is sufficiently violent. Since cavitation bubbles form and collapse within the body of liquid they cause no damage to the pump. However, they may be seeded by solid particles, in which case damage may be caused to the solid surface. This would be a disadvantage in coal transport where attrition is to be avoided but an advantage in handling sand and gravel where the extra surface-cleaning afforded can be beneficial: deliberate operation in cavitation

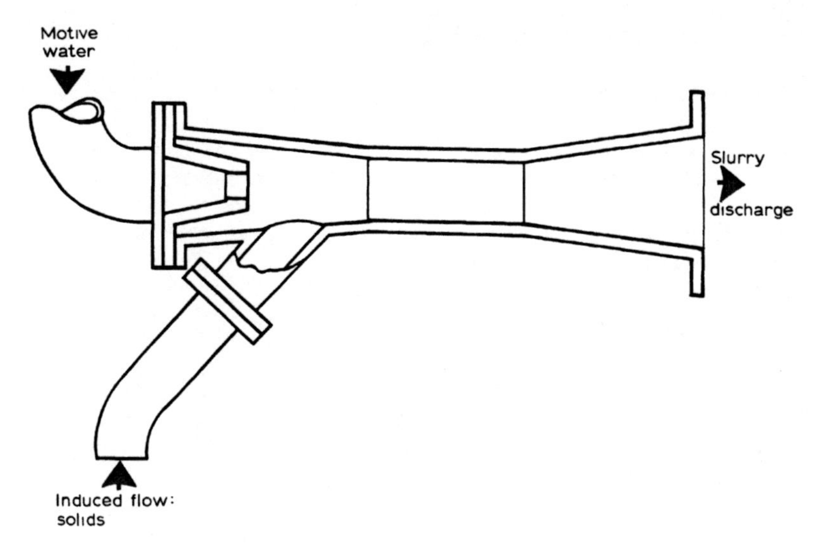

Fig. 18.7. The poorly designed jet pump.

mode allows an economical combination of pump and scrubber. The crackling or rattling sound of cavitation is distinctive.

The higher the motive pressure the more likely cavitation is to occur, and the higher the pressure in the suction chamber or around the pump required to stop it. In the nozzle of a jet pump the high velocity results in turbulence sufficient to cause cavitation and damage to the bore if suitable, erosion-resistant materials are not specified.

If the motive pressure (and flow, and power) are increased above the start of significant cavitation no increase occurs in flow ratio (flow rate of motive fluid/flow rate of induced slurry). Indeed, it may decrease. The increment of energy, which is absorbed in vapourising some of the fluid, is lost as heat and vibration in the ensuing impact of collapse. No part of the increment appears as increased flow ratio; since the flow scarcely alters, the developed pressure remains nearly the same and head ratio (head developed at the discharge/head available at the nozzle) declines. Efficiency, the product of head ratio and flow ratio, therefore also falls.

Fortunately, it is easy to calculate and design for cavitation. Sometimes a two-stage jet pump will be used, the first stage feeding and pressurising the body of the second, suppressing cavitation and allowing it to run at a high motive pressure. When pumping out of a lake or the sea, it may be more efficient to immerse the pump deeply, even though it then has further to lift; elsewhere, the pump may be placed at the bottom of a deep hopper.

4. HYDRAULIC PERFORMANCE

4.1. Disadvantages

Detractors are justified in citing low hydraulic efficiency. The sudden deceleration from nozzle to mixer is similar to that at a sudden enlargement, and a further deceleration occurs in the diffuser. Energy losses are therefore unavoidable and these must be of the order of 50% of the energy supplied. The only comparable loss in a centrifugal pump, for example, is that in the diffuser and volute. A jet pump starts with a 50% disadvantage compared with a centrifugal pump.

4.2. Advantages

Since the jet pump must usually be driven by a centrifugal pump, little advantage seems to lie in simply adding another cost item in respect of capital, maintenance and energy. Jet pumps are confined, therefore, to applications in which other advantages supervene:

- flexibility of location of the pump;
- utilisation of free source of input energy from waste fluid, or at least avoidance of the need for a prime mover;
- low cost, especially in special materials against corrosion;
- application to awkward fluids such as those which might undergo a change of state, e.g. steam;
- high shear stress breaking down time-dependent thixotropic properties—this can result in a substantial net energy reduction, sometimes over 90%;
- seam-free construction, for example, for radioactive applications; and
- intrinsic safety in hazardous locations.

In the handling of slurries, however, further major advantages become evident. A jet pump can be designed so that solids travel at low speed when changing direction and in a straight line when at high speed. The rate of internal wear can thereby be very low. Wearing parts can be designed to be simple and therefore cheap. Just as important, this simplicity allows them to be made very hard: machining may be unnecessary.

Most significant, however, for the successful pumping of settling slurries, is the shape of the pump characteristic. This acts to ensure stability at line velocities which would otherwise risk instability and blockage. In fact it has proved possible to run settling slurry pipelines at any practical concentration of likely particle size and density, at any velocity. Quarry dredges can dispense with operators and even work better un-manned. This is explained as follows.

4.3. The Hydraulic Characteristic of the Pure Jet Pump

Figure 18.8(a) is the typical near-parabolic pipeline friction curve for clean water in turbulent flow. The abscissa, V, is the mean velocity of flow in the pipeline. The ordinate i is (non-dimensional) frictional hydraulic gradient.

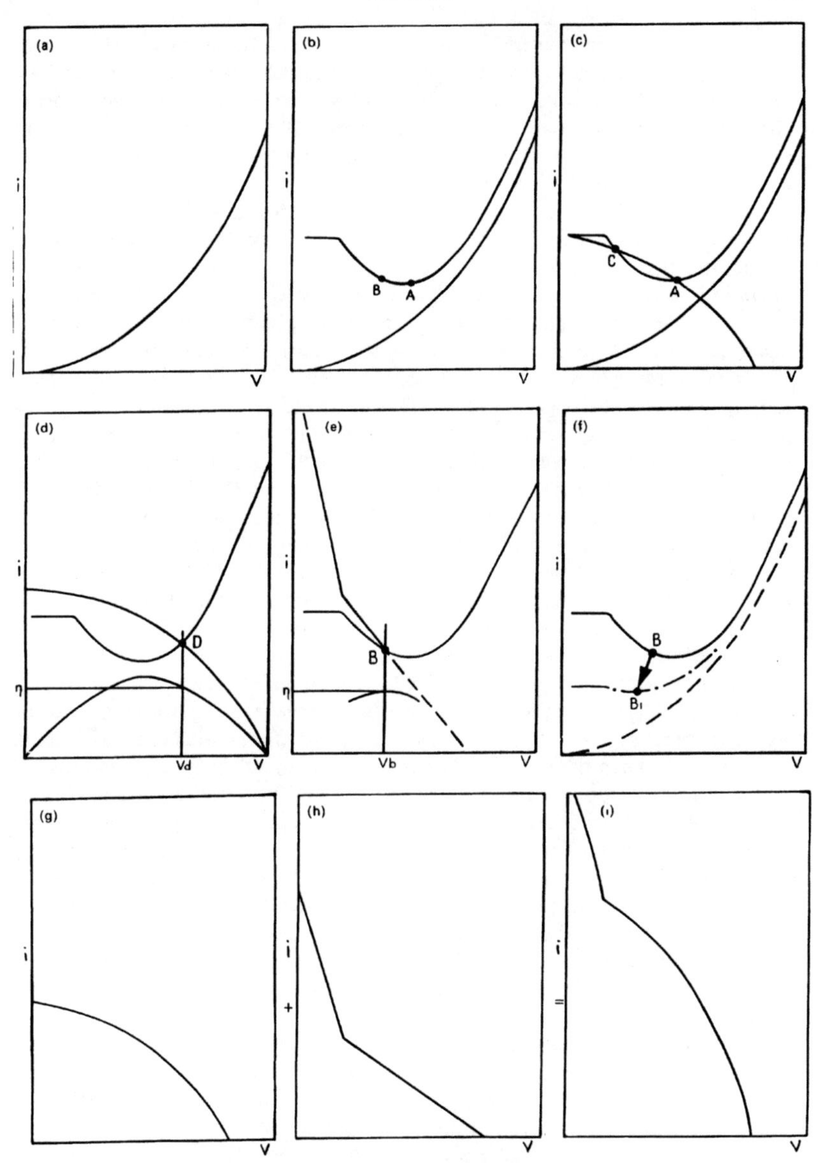

Fig. 18.8. Hydraulic characteristics. (a) Typical water-only friction curve; (b) slurry curve superimposed; (c) centrifugal pump characteristic curve added; (d) centrifugal pump curves adjusted; (e) jet pump curve and friction curve; (f) dynamic friction curve; (g) typical centrifugal pump head-flow curve; (h) typical head-flow curve for the type of jet pump described; (i) combined, hybrid pump head-flow curve.

Figure 18.8(b) adds the friction loss for a typical settling slurry. At the right hand, high velocity, end the solids curve is close to the water curve, elevated by a factor more or less equal to the specific gravity of the mixture. If we were able to choose our operating point freely it would seem to be advantageous to pick the minimum head, point A. But point B will consume less power, the product of pressure and volumetric flow rate.

Figure 18.8(c) shows how a typical curve for a centrifugal slurry pump intersects the pipeline curve at point A and again at a point C. The very existence of a point C indicates that a pipeline blockage will result. It is no more than a conceptual operating point because the smallest excursion to the left brings the developed head lower than the head applied by the pipeline. Inertia keeps things moving but the flow decelerates as progressively more material settles out and forms the dunes which will eventually reach the top of the pipe and prevent further flow. We have to resign ourselves to working at point D (Fig. 18.8(d)), far enough to the right of point A for fluctuations never to carry the operating point to the left of A, even for an instant. The pump curve will then be as shown in Fig. 18.8(d), with only one intersection, at D. The corresponding pipeline velocity is V_d. Also on this graph is drawn the efficiency curve for the centrifugal pump. Note that the operating point is at an efficiency, η, appreciably below the optimum.

Now let us forget for the moment centrifugal pumps and think about conceiving the ideal pump characteristic. The new characteristic would intersect the pipeline curve at B in Fig. 18(e). Then, it would have a gradient which carried it clear of the friction curve at all points to the left of B, the only intersection. The curve to the right of B is of little significance, but let us think a bit more about the part to the left. It will be remembered that the slurry curve turns upwards as velocity decreases because the solids settle out and build dunes. Blockage will occur approximately when the top of a dune reaches the top of the pipe. Therefore, we should, if we can, stop it positively by bending the left-hand end of the pump curve upwards. As the gap between dune crest and pipe soffit narrows, the pressure available will increase. The more the gap narrows the greater the increase, and the gap cannot close. The top of the dune will be planed off. In a final and decisive touch, we shall make the pipeline friction curve dynamic, as shown in Fig. 18.8(f). As a fluctuation moves the operating point fractionally to the left, we shall specify that the friction curve shall redraw itself a little lower. If the fluctuation moves the operating point a lot, we shall have the curve move a lot lower. Then there is no possibility of a second intersection occurring.

Is this all wishful thinking? It so happens that, within certain, fairly broad limitations, for a given jet pump geometry we are able to construct a mathematical model satisfactory for engineering purposes. We are able to tell the model what shaped curve we want out of it, including the turn-up at the left and the dynamic pipeline characteristic. The latter arises because a developed head rise causes an induced flow fall and it is the induced flow which is carrying the solids in. So the concentration in the pipeline immediately starts to decrease. Hence the new, lower curve. At the same time we can normally fix our operating point at the best efficiency point for the jet pump.

In Fig. 18.8(e), note the new, low velocity V_b corresponding to the operating point B. Abrasive wear is proportional to the velocity raised to a power between 3 and 5; 4 is a typical value. If we operate at half the velocity the pipeline can be expected to last about 16 times as long—and it does. In fact, the increase in capital cost of the larger pipeline required to give this velocity is such that the best compromise is a velocity about two-thirds of that generally accepted as necessary when using an uncontrolled centrifugal slurry pump.

Finally, it is possible to eliminate the disadvantage of low hydraulic efficiency and frequently obtain a greatly superior system efficiency (defined, for example, as tonnes of solid material delivered per unit total cost). This may, in some cases, be twice as high as for a centrifugal slurry pump acting alone and is achieved by hybridisation.

4.4. The Jet-Centrifugal Pump Hybrid

4.4.1. Introduction

The correct way to view the marriage of a jet pump and slurry handling centrifugal pump is

- not to see the jet pump as a feeder for the centrifugal pump, though it does serve this purpose;
- not to see the jet pump as a suction-side booster to relieve the centrifugal pump of suction loading and keep it out of cavitation, though it does fulfil this function;
- not to see the centrifugal pump as a distance booster for the jet pump, though it certainly is this;
- but to consider the two together as a single pump.

The centrifugal pump is frequently necessary because the required developed head is too great for the efficient application of a jet pump alone. The motive pressure would have to be such that the pump would be running in cavitation, with an associated low induced flow and low capacity. It is clearly most desirable, however, that the pipeline should be free from the risk of blockage: shutdown should be followed by reliable start-up without the need to flush the line.

4.4.2. The Hydraulic Characteristic of the Hybrid

The hybrid is the combination of jet pump and centrifugal slurry pump. The centrifugal pump curve of Fig. 18.8(g) is combined with the jet pump curve of Fig. 18.8(h) to give the hybrid curve of Fig. 18.8(i). It has been amply proved in practice that the full safety properties are retained provided at least 25–50% of the total power input goes into the jet pump and not more than 50–75% into the centrifugal slurry pump. This ratio will depend on relative efficiencies of jet pump and centrifugal slurry pump and on details of pipeline and composition of slurry. A bonus is that the centrifugal pump is now being fed at a constant rate and consequently lasts longer.

4.4.3. Advantages and Disadvantages

The stability of the whole system means that the operators can normally be dispensed with without the need for costly control systems. And low variation, high concentration and operation at the best efficiency point mean high dredging or system efficiency. Conventional hydraulic efficiency is not the primary criterion.

The typical hybrid mineral line, or sand transfer, or bypassing installation would be either of the following.

- A single jet pump recovering and feeding directly into a centrifugal slurry or gravel pump.
- A multiple jet pump recovering and delivering into a feed regulating sump; then, a single jet pump acting as a hybrid with a centrifugal slurry pump for the distance function.

The argument is encountered that

- the object of a solids transfer system is to transfer solids and not water;
- a jet pump adds water (by the motive jet);
- the efficiency of a solids pumping system is proportional to the concentration of solids; and
- the hydraulic efficiency of a jet pump is low.

Ergo, a jet pump is fundamentally unsuitable for pumping solids, and the point of this chapter evaporates.

However, it can be shown without much difficulty that the efficiency of a transport system handling settling solids peaks at a definite concentration of solids and that for most installations this peak occurs at a concentration lower than that at which the solids can be induced by a jet pump to flow into a pipe. However, it can be shown that reducing the flow rate of water into a jet pump by using a small nozzle, low flow and high pressure sacrifices the upward refraction of the curve at zero induced flow (Figs 18.8(e) and (h)) which we want to retain. It may therefore

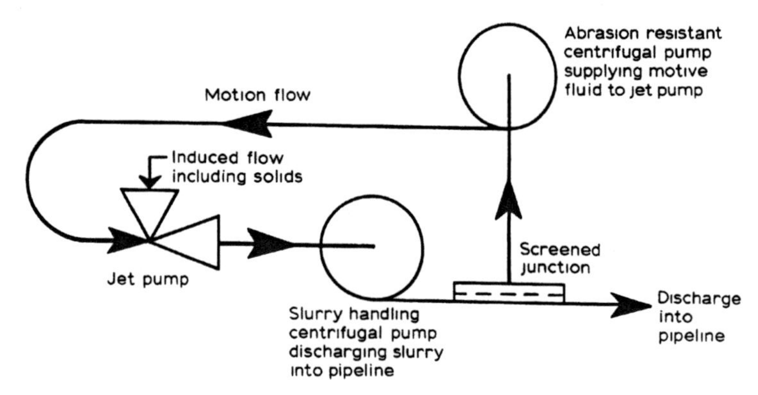

Fig. 18.9. Diagram of recirculating hybrid system.

appear, in certain cases, to be advantageous to run the line at a higher concentration than would appear possible with a jet pump. In such cases recirculation could be the answer. The jet pump feeds, as before, into a gravel or slurry distance pump, but a slurry pump deriving its flow from the inlet to the distance pump drives the jet pump, as shown in Fig. 18.9. The junction will be screened at low velocity to exclude material from the motive pump which would not pass through fluidisation nozzles which might be used in the slurrying process, or in deposit disintegration in the case of dredging. The penalties are wear in the jet pump motive system and main nozzle, an enclosed screen to monitor and maintain, and a motive pump of inferior hydraulic efficiency.

5. PRACTICAL APPLICATIONS

Table 18.1 contains a list of applications for which jet pumps have been used for two-phase flow.

6. SPECIFYING A JET PUMP

6.1. Primary Criteria

The initial criterion in pump selection may be the

- particle size to be handled, or
- required mass flow rate, or
- head to be developed.

Usually one of these will be all-important and the others of secondary significance. In cases where all are important it may not be possible precisely to satisfy all.

The motive requirements have to be determined. Sometimes a limited motive supply may be available and it may be acceptable to modify the initial stipulations to avoid the cost of a new motive installation.

A jet pump is conveniently discussed in terms of ratios. The fundamental geometric ratio is that of the nozzle exit cross-sectional area to the mixer cross-sectional area, commonly (and here) referred to as the R value. For the consideration of the performance of solids-handling, in common with other pumps, performance ratios are often used; for jet pumps the mass flow ratio is most relevant (Fish, 1970; Silvester & Vongvisesomjai, 1972).

6.2. Mixing Chamber

6.2.1. Starting with Required Particle Size

The passage size must accommodate the biggest particle. If this is only occasional the mixer need only be a little larger, say 10%. If the solids being handled are all

TABLE 18.1
Applications of Jet Pumps

Agricultural

Borehole cleaning and pumping
Sanding cranberries

Cable laying

Excavation or exposure of pipes and cables
Trenching

Coastal management

Beach cleaning
Sand and silt-traps
Sand bypassing

Dredging

Boosters
Dredge pumps
Hopper vacuum dewatering

(*Petro-*) *Chemical*

Coating of filters
Lifting of catalyst
Mixing of powder in gas
Mixing of slurries
Mixing of suspensants
Pumping of waste
Pond aeration
Pond neutralisation
Transfer of abrasive and/or corrosive materials

Docks and harbours

Clearance of gates
Clearance under jetties
Dredging of channels
Removal of sand bars

Emergency services

Recovery of oil spillage
Smothering fire with sand

Engineering

Degaussing magnets
Machine sumps
Product transfer
Spillages of grit, fuel, scale, product
Swarf

Fish farming

Aeration
Creation of current to carry effluent away
Damage-free fish sampling or transfer
Removal of sludge

Food processors

Damage-free product transfer in processing
Effluent pumping
Feed from reception hopper into plant

Foundry

Pneumatic grit and scale recovery
Pneumatic sand transport
Sand surface abrasion

General contracting

Bentonite removal
Driving well-points
Filter media
Grit and shot blast abrasive
Gritting and gravelling roofs
Pumping-out foundations
Sand bar removal in rivers
Tank desludging

Marine

Cleaning out spillage from under conveyors
Dry cargo transfer in air
Emergency cargo transfer
Hold and tank ventilation
Slops transfer

Marine maintenance

Removal of blasting abrasive and scale
Removal of external fouling in dry dock
Salvage

Mineral dressing

Aeration
Clearance of thickeners
Scrubbing
Slurry preparation

Mining

Aeration of tailings ponds
Backfilling of shafts and tunnels
Gulley and stope cleaning
Pumping explosive slurry
Removal of mine waste

Nuclear

Magnox chips
Radioactive fuel pellets in viscous medium
Radioactive sand

TABLE 18.1. *contd.*

Removal of sand from cooling water intakes and filters	Product scrubbing Re-handling of product
Oil exploration, production	*Steel mills*
Clearance of tie-ins, valve farms and templates Clearance of tubulars, legs and cells Emergency pipeline repair systems Excavation for initial installation Excavation for inspection of mud-line members General-purpose diver-held excavation tool General-purpose rov-manipulated tool	Blasting abrasive collection and return Scale removal in slurry Use of waste steam *Third World food processors* Product handling Reticulated vacuum systems for spillage Sump and gulley clearance *Water and sewage*
Power stations	Aeration
Bottom ash removal Fly ash pumping Fly ash spillage	Desilting irrigation channels Digester cleaning Grit handling from grit channels Removal of filter media from pots Removal of sand from slow sand filters
Quarrying	Replacement of sand in slow sand filters
Clearing of stockpile conveyors Dredging of aggregates Long-distance pipelining between quarry and plant	Reservoir recovery Scrubbing of sand from slow sand filters Silt traps

of one size the mixer must be at least three-times the particle size or flow may effectively cease and at least six-times the particle size if efficiency is to be unimpaired. A particle size close to, or exceeding, the nozzle size will reduce output by interference with the jet but not disasterously—say by 10%.

Potatoes act like pebbles but carrot-like particles wedge and the diameter of the mixer needs to be nearly twice that of a tapered particle. Stringy organic material, such as sewage and green beans, align themselves with the flow and readily pass through the mixer. Hanging-up precludes the use of a re-entrant mixer inlet when handling fibrous matter. Flaky materials are more difficult to get into the jet pump, but once in the pipeline cause less resistance than rounded materials. The mixer must be bigger in relation to the particle size in the case of flaky material. Crushed rock or sea shell may fall into this category.

If abrasive solids are being handled the mixer bore will wear. It is usually assumed for such applications that the mixer bore is 1·025-times the nominal size. The mixer may be regarded as notionally worn-out when the maximum bore exceeds the nominal or new diameter locally by 10% or more. Maximum wear occurs about half-way along the mixer.

6.2.2. Starting with Required Mass Flow Rate of Solids

Table 18.2 gives an idea of the mass flow of solids which a jet pump of the type described will typically handle. Under good conditions, greater flow rates are

TABLE 18.2
Target Solids Mass Flow Rates

Mixer diameter				*Flow rate of induced solids*	
Pure jet pump		*Hybrid system*		*Typical* (th^{-1})	*Maximum* (th^{-1})
(*mm*)	(*in.*)	(*mm*)	(*in.*)		
27		32	1·25	5	18
32	1·25	38	1·5	10	25
38	1·5	44	1·75	15	35
44	1·75	51	2	25	47
51	2	63·5	2·5	40	60
63·5	2·5	76	3	55	100
76	3	102	4	90	140
89	3·5	114	4·5	120	190
102	4	127	5	160	250
114	4·5	139	5·5	200	315
127	5	152	6	240	390
152	6	203	8	350	560
178	7	229	9	460	800
203	8	254	10	600	1000
229	9	279	11	760	1300
254	10	305	12	925	1600

possible. A jet pump operating against a high head may manage less than these figures owing to cavitation.

A high head requires a high velocity in the mixing chamber and therefore a small mixer diameter for a given output, so the above figures will vary somewhat according to the type of application. As a guide, if the pipeline length is more than 500 diameters, take the next diameter down, and if the jet pump is being used as the first stage of a jet-pump–centrifugal-pump hybrid or for a similar low-head duty, take the next larger diameter. The golden rule is, if in doubt go larger.

The pipeline can become blocked by oversize material. The jet pump is resistant to the increasing head due to low velocity and settling solids, but large solids can still pack to form a mechanical plug. The high head reserve of the jet pump will then simply pack the plug even tighter. If a single size is to be pumped at high concentration the pipeline must be five- or six-times the particle size. Occasional, very large particles will not cause trouble provided that the length of the major axis is less than the pipe diameter.

Irregular material, which is conveyed more easily once taken in, is more difficult to feed. A mixing chamber under-sized by one-third can reduce output by five-sixths, even if the maximum particle is substantially smaller than the mixer diameter.

6.2.3. Starting with Required Developed Head

The mixer size cannot be estimated from required head alone. However, the mixer diameter, as estimated from the target mass flow rate, will be modified if the required head is unusually high or low. If the head is very high the mixer will be smaller, provided that cavitation can be suppressed. If it cannot, the output of a given size of pump will be reduced and the pump size must therefore be increased. So the mixer may end up not being reduced at all. If the head is low, as in a centrifugal pump hybrid or booster installation, the mixer will be larger by about one size.

The internal finish of the mixer influences the flow rate developed. It is normally assumed that a solids-handling mixer will be cast and therefore hydraulically rough. However, liquid-only applications, or those in which the solid content is essentially non-abrasive, like food or organic sludge, permit the use of a machined or extruded mixer. This allows a higher maximum induced flow to be achieved, should that be an advantage.

6.3. Pipeline Size

As a first guide to pipeline size, take twice the mixer diameter, then the next standard or preferred size up. For example, a 102 mm mixer will usually feed a 250 mm pipeline. In the case of a low-head pump, as in a hybrid, the pipeline will be twice the mixer diameter, or perhaps one size less. This could apply also if the required pipeline velocity were exceptionally high. Examples of this include brick-sized chunks of vitrified mill-scale and broken uranium fuel rods, where the pipeline and mixer had a common diameter. In an unusually high-head installation or where the solids will not settle, thus allowing a low pipeline velocity, the pipeline diameter could be as much as three-times the mixing chamber diameter. The diffuser expansion ratio is also likely to be greater at small sizes, say around 50 mm pipeline.

6.4. Nozzle Diameter

A small R value is associated with low developed head, high induced mass flow, high cavitation resistance, high motive pressure and low motive flow (and therefore pipeline velocity). The converse, in all respects, applies to a large R value. The R value is varied to get the desired pipeline velocity or to match a particular pump-set or keep within the range of low-cost end-suction motive pumps.

A convenient source of information on a nozzle profile at the same time appropriate to the jet pump and permitting accurate flow prediction may be obtained from BS 1042 (BSI, 1987).

6.5. Diffuser Energy Loss (or Recovery) Coefficient

Because of the range of outlet diameters possible for a given jet pump, it is recommended that the diffuser should be considered as part of the pipework (ESDU, 1973; Miller, 1978).

TABLE 18.3
Immersion Required to Suppress Cavitation

Motive head (*m*)	*Depth of immersion* (*m*)
65	None
100	5
150	13
225	24

6.6. The Jet Pump Equation or Characteristic

The equation is most conveniently derived primarily from consideration of momentum (Wakefield & Ruckley, 1975; Wakefield, 1986). Other, extremely comprehensive approaches have been published (ESDU, 1985).

6.7. Cavitation

A rough guide to prevention by immersion is given in Table 18.3, though a more comprehensive guide is available from the literature (Bonnington, 1958; Wakefield, 1972). If the immersion is less then the maximum induced flow will also be less.

The figures in Table 18.3 apply for an R value of 0·16. If the R value is higher the jet pump is more tolerant and if lower it is less tolerant. However, note that this does not in itself imply that a pump with a high R value is less efficient; it will develop a greater discharge head from a given motive head and a lower induced flow may therefore be acceptable.

At elevated temperatures, say about 70°C in the case of water, vapour pressure becomes important and must be allowed for to avoid cavitation. Vapour pressure of water can be simply calculated (Kaye & Laby, 1986), but its effect may be variably masked by that of dissolved gases coming out of solution. For normally aerated water, a value of 3 kPa is usually sufficient.

7. CONCLUDING REMARKS

Systematic development for the economic transmission of abrasive solids is taking place and the possibilities, such as hybridisation with other types of pump, or the combination of one form of jet pump with another, are beginning to be perceived. Such investigation into a theoretical basis adequate for modern needs for both complete pumps and individual components continues around the world.

The ability of the jet pump to operate embedded in a mineral deposit offers the possibility of totally concealed sand transfer, beach replenishment and sand by-passing installations. Coastlines have traditionally been managed by a process of opposition to the forces of nature. Information not available 200 years ago is now at hand and the embedded jet pump will play a major role in the more understanding phase of coastal management just starting worldwide.

The solids-handling jet pump has never achieved the acceptance of the centrifugal pump; for this to occur, application design must be simplified and stereotyped to an extent which has not been possible in an atmosphere of inadequate understanding. It is hoped and expected that current developments will provide the remedy.

Ranges of solids-handling jet pumps are available from a few manufacturers, but quality both of product and application design varies exceedingly. Sizes of heavy-duty pump available at short notice cater for pipe sizes from 25 to 600 mm, for solids flow rates from 1 to well over 100 th^{-1}. Jet pumps for less demanding duties are available for lesser and greater outputs.

8. REFERENCES

Bonnington, S.T. (1958). The cavitation limits of a liquid/liquid jet pump. Publication RR605, BHRA Fluid Engineering, Cranfield, UK.

BSI (1987). BS 1042: Measurement of fluid flow in closed conduits. Part 1: Pressure differential devices. Section 1.1: Specification for square-edged orifice plates, nozzles and venturi tubes inserted in circular cross-section conduits running full. British Standards Institution, London, UK.

Culbreth, W.G. & Legoff, J.P. (1985). Determination of the zone of flow establishment in a submerged, axisymmetric jet. Paper presented at the *Winter Annual Meeting*, ASME, Miami Beach, FL, USA, 11–17 Nov.

ESDU (1973). Performance of conical diffusers in incompressible flow. Data Item 73024, ESDU International, London, UK.

ESDU (1985). Ejectors and jet pumps—design and performance for incompressible liquid flow. Data Item 85032, ESDU International, London, UK.

Fish, G. (1970). The solids-handling jet pump. In *Proc. HT 1*, paper L1, pp. 1–16.*

Hoyt, J.W. & Taylor, J.J. (1985). Effect of nozzle boundary layer on water jets discharging in air. Paper presented at the *Winter Annual Meeting*, ASME, Miami Beach, FL, USA, 11–17 Nov.

Kaye, G.W.C. & Laby, T.H. (1986). *Tables of Physical and Chemical Constants* (15th edn). Longman, Harlow, UK.

Kentfield, J.A.C. & Barnes, R.W. (1972). The prediction of the optimum performance of ejectors. *Proc. UK Inst. Mech. Engrs*, **186**, 54–72.

Landau, L.D. & Lifschitz, E.M. (1987). *Fluid Mechanics: Course of Theoretical Physics* (Vol. 6, 2nd edn). Pergamon Press, Oxford, UK.

Massey, B.S. (1983). *Mechanics of Fluids, Free Turbulence* (5th edn). Van Nostrand Reinhold, UK.

Miller, D.S. (1978). *Internal Flow Systems*. BHRA Fluid Engineering, Cranfield, UK.

Obot, N.T., Mujumdar, A.S. & Douglas, W.J.M. (1984). The near field behaviour of round jets at moderate Reynolds numbers. *Can. J. Chem. Engng*, **62**, 587–93.

Rajaratnam, N. (1976). *Developments in Water Science, The Integral Momentum Equation*. Elsevier, Amsterdam, The Netherlands.

Silvester, R. & Vongvisesomjai, S. (1972). The jet pump using liquids of different density. In *1st Symposium on Jet Pumps and Ejectors*, Institution of Chemical Engineers/BHRA Fluid Engineering, Cranfield, UK, paper 11.

Thorne, P.F. & Theobald, C.R. (1978). The effect of nozzle geometry on the turbulent structure of water jets—a photographic study. In *4th Symposium on Jet Cutting Technology*, BHRA Fluid Engineering, Cranfield, UK, paper A4.

Trabold, T.A., Essen, E.B. & Obot, N.T. (1987). Entrainment by turbulent jets issuing from sharp-edged inlet round nozzles. *Trans. ASME, J. of Fluid Engng*, **3**, 248–54.

Wakefield, A.W. (1972). Practical solids-handling jet pumps. In *1st Symposium on Jet Pumps and Ejectors*, Institution of Chemical Engineers/BHRA Fluid Engineering, Cranfield, UK, paper 12, pp. 183–202.

Wakefield, A.W. (1986). The handling of sand deposits and littoral drift by submerged jet pump, with reference to the Nerang River Project (1st part). In *Proc. HT 10*, paper D3, pp. 123–42.*

Wakefield, A.W. (1987). *An Introduction to the Jet Pump* (3rd edn). Available on request from A.W. Wakefield, Wakefield House, Little Casterton Road, Stamford, PE9 1BE, UK, or from J.B. Green, Glenflo Jet Pumps, Box 53, Route 7, Louisiana 70438, USA.

Wakefield, A.W. & Ruckley, A.L. (1975). Performance of solids-handling jet pumps at low Reynolds numbers. In *2nd Symposium on Jet Pumps and Ejectors*. BHRA Fluid Engineering, Cranfield, UK, paper A3, pp. 23–33, X18 to X25.

*Full details of the Hydrotransport (HT) series of conferences can be found on pp. 11–12 of Chapter 1.

19

Pipelines and Fittings

Philip B. Venton
Williams Brothers–CMPS Engineers, Chatswood, NSW, Australia

1. INTRODUCTION

Basic decisions relating to the pipeline design must be made early in the design process, and these to a large extent dictate the direction of pipeline design. For example, if the pipeline length or system hydraulic gradient requires a high operating pressure, the pipeline will be constructed from steel; if the slurry is abrasive, a wear-resistant lining must be provided. Once the basic process decisions and hydraulic parameters have been determined, pipeline design can proceed.

2. PIPELINE MECHANICAL DESIGN

2.1. Overview of Design Considerations

Pipeline design is related to designing a pipe system which has sufficient mechanical strength, and where required, sufficient corrosion and erosion resistance for the service required of it by the designer, modified by any additional requirements of the design code or the regulatory authorities, and which can economically satisfy other performance requirements including corrosion and erosion.

In addition to establishing the design life of the project, hydraulic design factors and design corrosion–erosion rates, basic factors which must be established prior to pipeline design include the following.

2.1.1. Pipeline Material Selection

The designer must make a decision early in the project on the material to be used for the pipeline, since this can dictate many aspects of the pipeline design. For example, if the pipeline is subject to significant erosion, it will probably be constructed above ground to permit access for replacement. Installed above ground, the pipeline will be unrestrained, and so will require special support, consideration of expansion, jointing and environmental matters.

2.1.2. Pipeline Route

The route length and profile are basic parameters for pipeline design. It is normal to select the minimum practical length route to minimise the project capital and operating cost. However, this may be modified by process limitations including allowable slope, and by negotiations with landowners along the route.

Long-distance pipeline routes are selected where possible across country, avoiding roads, rail and other infrastructure which impose additional design limitations (increasing the cost) and because these corridors carry many municipal services, which increases the risk of third-party damage to the slurry pipeline.

2.1.3. Pipeline Profile

The pipeline must provide the mechanical strength to satisfy the actual pressure within the pipe. This is a function of both the pipeline hydraulic gradient and the route profile. The internal pressure in metres of slurry is the distance between the elevation of the design hydraulic grade line and the pipeline elevation at any distance along the pipeline. In rugged terrain, the pipeline-wall thickness may be greater in the middle of the pipeline than at the start to satisfy the internal pressure condition, rather than the initial pump station discharge pressure. Figure 19.1 shows how the pipe-wall thickness is varied along a portion of the Ok Tedi copper-concentrate pipeline to satisfy the pipeline internal pressure.

Pipeline shutdown conditions also need consideration. If the pipeline commences at a high elevation, the shutdown pressure (when the terminal valve is closed) may impose a higher pressure for design than the pipeline operating pressure condition. The last 20 km of the Ok Tedi pipeline in Fig. 19.1 is thickened to provide the

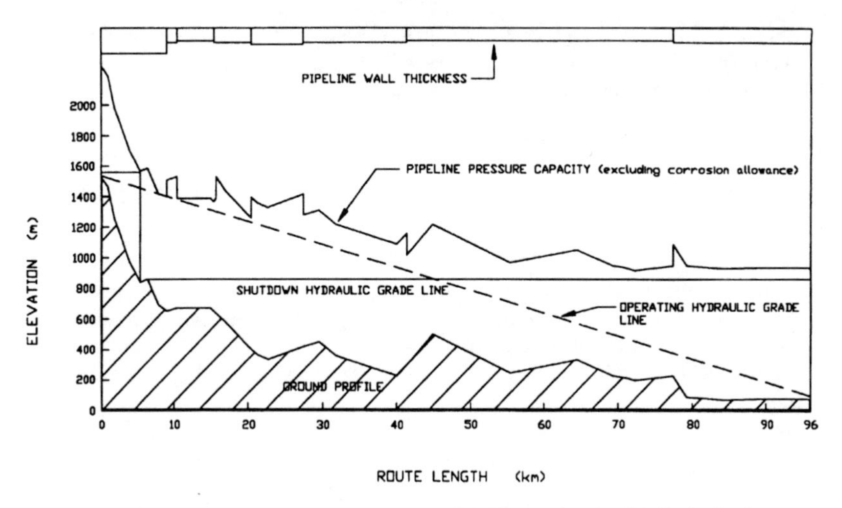

Fig. 19.1. Pressure profiles and pipe-wall thickness for the Ok Tedi pipeline.

strength necessary to allow the pipeline to be shut down against 800 m of static head (approx. 14 MPa pressure).

2.1.4. Pipeline-Wall Thickness Telescoping

Because the hydraulic grade line in a slurry pipeline is constant for a given pipeline diameter and flow rate, the pressure along the pipeline reduces at a predictable rate with increasing distance from the pump station. The designer may choose to reduce the pipeline-wall thickness in steps as the operating pressure decreases, to minimise the capital cost of pipe required for the project.

2.1.5. Environmental Considerations

Most countries now require an environmental impact statement to be prepared and accepted prior to approval being granted for a project. Buried pipelines are known for their environmental friendliness because of their safety record, and because after construction and restoration, the land through which they pass can be returned to its original use. Special consideration needs to be given to interference with land use during construction, construction of road, rail, river and foreshore crossings, and interference which may occur during future pipeline maintenance.

2.1.6. Design Code

Each country has a design code which defines the basic rules which govern mechanical design, construction, operation and maintenance of the long-distance pipeline systems. The ANSI/ASME Code B31.11, the Slurry Transportation Pipeline Code, is written specifically for the slurry pipeline industry (ANSI/ASME, 1986). It is based on the well-developed gas and liquid petroleum pipeline codes (ANSI/ASME, 1989*a*, *b*). In addition to providing specific rules intended to set safe design standards for high-pressure pipelines, the Code provides a useful design check list for short or low-pressure pipelines where the operating pressures are insufficient to require design code compliance.

2.2. Design Wall Thickness

The nominal wall thickness for straight steel pipe under internal pressure is calculated using

$$t_n = t + A \tag{19.1}$$

where t_n is the nominal wall thickness (mm), t is the design wall thickness (mm), and A is the sum of allowances for threading, corrosion and any additional allowance for protection of the pipeline (mm).

ANSI/ASME (1986) nominates that the design pipe-wall thickness be calculated using Barlow's formula:

$$t = \frac{PD_o}{2S} \tag{19.2}$$

where P is the design internal pressure (MPa), D_o is the outside diameter of pipe (mm), and S is the applicable allowable stress (MPa).

The allowable stress is defined by the design code. ANSI/ASME (1986) defines the allowable stress as

$$S = 0{\cdot}80\, E\, \mathrm{SMYS} \tag{19.3}$$

where E is the weld joint efficiency (1·0 for butt welds), and SMYS is the specified minimum yield stress of pipe (MPa).

The constant 0·80 represents the ratio of design hoop stress to SMYS. ANSI/ASME (1986) has relaxed this factor compared with the maximum value normally applied in the oil and gas industry of 0·72, in recognition of the generally non-hazardous nature of the water-based slurry being transported. The designer should consult the design code applicable to the location of the project being undertaken to determine the specific requirements permitted in that location.

The required thickness for pipelines constructed from other materials can be calculated using the above formulae and the structural properties for the material. An alternative design formula used for lower pressure piping systems is

$$t_{min} = \frac{PD_o}{2S + P} \tag{19.4}$$

where t_{min} is the minimum wall thickness (mm).

Because of other design factors applying to plastic pipes, the manufacturers' design literature should be consulted for specific design data when pipelines using these materials are being designed.

2.3. Structural Design

Several other factors impose loads on an operating pipeline and should be considered prior to finalising the pipeline-wall thickness selection. These factors may require additional wall thickness or alternative design to ensure design limits are not exceeded.

2.3.1. Stress Analysis

Design codes require compliance with design rules limiting the operating stresses within a pipeline which result from operating pressures and additional loadings caused by thermal expansion, soil stresses and backfill loading, bending stresses, and any other condition which can increase the operating stress level in the pipeline.

As operating stress levels in pipeline systems increase, the margins provided by conservative design are reduced. The designer should identify those locations subject to additional loadings, and evaluate the stress levels either manually or by one of the commercially available computer programs to ensure compliance.

2.3.2. Surge Analysis

Surge, or water hammer, occurs when the steady-state flowing conditions are suddenly changed as a result of sudden pipeline start, pump failure, or rapid valve closure. It is mandatory under B31.11 (ANSI/ASME, 1986) to undertake a surge analysis to demonstrate that surge transient pressures do not exceed the maximum allowable pressure by more than 10%. In short-distance pipelines the surge analysis can be undertaken manually, but in longer distance systems it is generally more efficient and more accurate to use one of the commercially available computer models. Development of an accurate pipeline system model is essential if the surge analysis is to produce accurate results, and this requires the designer to have a thorough understanding of the surge phenomena before attempting the analysis.

Surge pressures may be alleviated by heavier wall pipe, pressure relief valves, or by control system operation.

3. MATERIALS

3.1. Scope of Construction Materials

Common slurry transportation pipeline materials include

- steel,
- HDPE (high-density polyethylene),
- ABS (acrylonitrile butadiene styrene),
- uPVC (unplasticised polyvinyl chloride), and
- FRP (fibre-reinforced plastic).

Pipe from each of these materials may be supplied or installed either by itself or in combination with another material to provide the pipeline with additional characteristics. For example, an elastomeric lining in steel pipe can provide erosion or corrosion resistance, or a steel sleeve installed on a polyethylene pipe can provide mechanical strength.

TABLE 19.1
Design SMYS for Steel to API 5L Specification[a]

API Grade	*SMYS (psi)*	*SMYS (MPa)*
5L Grade A	30 000	206·85
5L Grade B	35 000	241·33
5L X42	42 000	289·59
5L X46	46 000	317·17
5L X52	52 000	358·54
5L X56	56 000	386·12
5L X60	60 000	413·70
5L X65	65 000	448·18

[a]API (1988).

Fibre-reinforced cement pipe was in the past used in fine tailings slurry pipelines, but its use is now limited since most Western countries have prohibited the use of asbestos fibres which were used to reinforce the cement.

3.2. Steel Pipe

Pipe is made to standard dimensions, using standard material grades in accordance with a national or defacto international standard. Steel pipe for pressure applications is generally specified to conform with the American Petroleum Institute's specification 5L (API, 1988). This specification governs the dimensions, materials and manufacture of high-pressure steel pipe for use in the oil and gas industry. API (1988) permits the use of a wide range of steel grades allowing the designer to minimise project cost by selecting the highest grade for his project to minimise the wall thickness, steel mass and hence pipe cost. Table 19.1 shows standard steel strengths for pipe supplied to API specifications.

Steel pipe is supplied in lengths of 6, 12 and 18 m (random, double random and triple random lengths). The standard length for pipelines is 12 m, with 18 m lengths being supplied to order for long-distance pipelines, where the longer lengths increase construction productivity, reducing construction cost.

By convention, steel pipe diameters are reported as *nominal* diameter but are measured by outside diameter (o.d.) as shown in Table 19.2. The diameter standards were developed using English units, and metric equivalents are a hard conversion of the dimensions in inches. The o.d.'s greater than 12 in. are the same as the nominal diameter, while those 12 in. and smaller have o.d.'s greater than the nominal diameter.

3.2.1. Joints

The technique selected for joining steel pipes depends on the pipeline performance requirements.

If the pipeline is subject to wear and must be turned periodically to equalise the wear, then the jointing system must allow the line to be broken and rejoined with

TABLE 19.2
Nominal and Actual Steel Pipe Diameters

Nominal Diameter		*Outside Diameter*	
(*mm*)	(*in.*)	(*mm*)	(*in.*)
50	2	60·3	2·375
80	3	88·9	3·5
100	4	114·3	4·5
150	6	168·3	6·625
200	8	219·1	8·625
250	10	273·1	10·75
300	12	323·1	12·75
350	14	355·6	14·0
400	16	406·4	16·0
450	18	457·2	18·0

a minimum of effort. Two or four bolted Victaulic coupling systems illustrated in Fig. 19.2 provide a flexible mechanical joint which allows quick assembly and disassembly and which allows for angular misalignment, a factor which is important when the pipeline must be rotated.

High-pressure pipelines require a high quality joint. Design codes require the joint quality be measured by non-destructive testing as part of the construction quality control. Butt welds provide a joint which has the same mechanical properties as the parent pipe, and which can be non-destructively inspected using X- or

Fig. 19.2. Cut-away photograph of a Victaulic joint. (Courtesy of Victaulic Company of America, USA.)

Fig. 19.3. Zaplock® joint on a 6 in. (150 mm) pipe. (Courtesy of A.J. Lucas Constructions, Australia.)

gamma-ray radiography to provide a permanent record of the joint quality. In addition, the butt weld dimensions are similar to those of the parent pipe simplifying the protective coating of the joint after acceptance.

Welding is generally not a practical alternative for internally lined pipelines because the internal pipe wall temperature reached during the welding process causes the lining to burn or to disbond. Two jointing techniques have been developed for pipes of this type.

The mechanical joint (Zaplok®, Crimplock® and Positive Seal®). The mechanical joint relies on an interference fit between the pin end of the pipe and the coupling (which may be formed by belling one end of each pipe (Zaplok® and Crimplock®), or by using a manufactured coupling in the case of the Positive Seal® process). Zaplok® joints (Fig. 19.3) have been tested for suitability in slurry service.

Tests have demonstrated that these joints provide adequate strength for high-pressure pipelines. They are adequate for moderate velocity homogeneous slurry pipelines but should be used with caution for slurry transportation systems subject to erosion, because the discontinuity at the internal joint can exaggerate the erosion rate. The pipeline should be constructed with the pin end of the joint in the direction of flow.

Mechanical joints are assembled by machine using an epoxy compound as a lubricant/sealant to aid assembly, and to provide a bubble-tight joint.

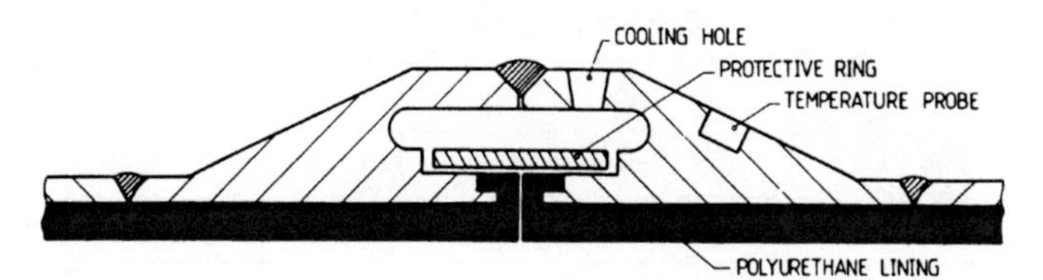

Fig. 19.4. Cross-section through the McConnell–Dowell coupling showing the location of a coolant hole and temperature probe used during the welding process. (Courtesy of McConnell–Dowell Constructors, Melbourne, Australia.)

The McConnell-Dowell coupling (Fig. 19.4). This coupling was developed for the 200 mm New Zealand Steel iron–sand pipeline, which is lined with poly urethane and which operates at 9–10 MPa. It was considered necessary to bury the pipeline to provide an environmentally acceptable transportation system, as well as to provide the owner with a low-maintenance pipeline system. Flanges and mechanical joints were considered but each had limitations.

The welding coupling was proven to offer a joint which could be inspected by radiography, and which could limit the internal pipe wall temperature to 100°C, below the temperature at which disbondment occurs. Since its use on the New Zealand Steel pipeline, the coupling has been developed for the 400 mm nominal pipesize (ns) Nerang River sand bypassing scheme, and the 850 mm ns Bougainville tailings pipeline.

3.3. Thermoplastic Pipe

Thermoplastic (or plastic) and FRP pipe have many advantageous properties for slurry transportation, including

- corrosion resistance,
- wear (erosion) resistance,
- relatively lightweight,
- flexibility,
- ease of jointing, and
- low absolute roughness.

The pipes are manufactured with a wide range of compatible fittings, enabling the designer to plan the piping system using standard techniques.

Plastic properties impose limits on their use which must be considered, including the following.

3.3.1. Creep Rupture

Thermoplastic materials, particularly HDPE, exposed to continuous stress suffer from creep. Creep effectively reduces the hoop stress at rupture in proportion to the duration of the stress. HDPE pipe classified as Type 50 is typically specified.

TABLE 19.3
Permissible Working Pressure of Type 50 HDPE Related to Design Life and Working Pressure[a]

Temp (°C)	*Design life (years)*	*Design pressure for pipe class (m of water)*					
		Class 3	*Class 4·5*	*Class 6*	*Class 9*	*Class 12*	*Class 15*
20	1	41·1	62·1	82·8	124·2	165·6	207·0
	5	40·2	60·3	80·4	120·6	160·8	201·0
	10	33·4	57·6	76·8	115·2	153·6	192·0
	25	32·4	48·6	64·8	97·2	129·6	162·0
	50	30·0	45·0	60·0	90·0	120·0	150·0
30	1	31·0	46·5	62·1	93·1	124·2	155·3
	5	30·1	45·2	60·3	90·5	120·6	150·8
	10	28·8	43·2	57·6	86·4	115·2	144·0
	25	24·3	36·5	48·6	72·9	97·2	121·5
	50	22·5	33·7	45·0	67·5	90·0	112·5
40	1	22·8	34·2	45·5	68·3	91·1	113·9
	5	22·1	33·2	44·2	66·3	88·4	110·6
	10	21·1	31·7	42·2	63·4	84·5	105·6
50	1	14·1	21·1	28·2	42·2	56·3	70·4
	5	13·7	20·5	27·4	41·0	54·7	68·4

[a] Courtesy of Hardie iplex, Sydney, Australia. Harditube Polyethylene Pipes and Fittings Product Catalogue.

Type 50 pipe is designed on the basis of extrapolation of the creep-stress–time curves to provide a design life of 50 years. Once the creep rupture limit for this material is reached the rate of deterioration accelerates. Where the design life is less than 50 years, pipe having a lower pressure rating may be safely used for design economy.

A new polymer, Type 63 medium-density polyethylene which offers significantly improved creep rupture performance, is being introduced by manufacturers around the world. It is understood that the derating factors applying to the new material are significantly less conservative than applying to the Type 50 polymers.

3.3.2. Temperature Down-rating

As the polymer is thermoplastic, the design pressure rating must be reduced with increasing operating temperature. The down-rating will vary with the material used. The creep rupture design life is also reduced with increasing operating temperature. Table 19.3 shows a typical design-life–pressure relationship with design temperature, based on Type 50 HDPE material.

The pressure capacity of ABS pipe is down-rated with increasing temperature in a similar manner to HDPE. Manufacturers should be consulted to determine the appropriate pipe selection for particular operating conditions.

TABLE 19.4
Coefficient of Linear Expansion

Material	*Coefficient of linear expansion* ($mm\,m^{-1}\,°C^{-1}$)
Steel	0·012
FRP	0·02
ABS	0·101
HDPE	0·18

3.3.3. Thermal Expansion

The coefficient of thermal expansion of thermoplastic pipe is much higher than that of steel (Table 19.4), and plastic pipe designs must make provision to relieve stresses imposed by thermal expansion resulting from diurnal or operating temperature changes. This is generally achieved by continuously supporting the pipe (e.g. on the ground), and allowing the pipe to *snake* between anchors to relieve expansion stresses. When the pipe must be supported above ground, conventional anchors, guides and expansion loops can be incorporated.

Both HDPE and ABS pipes include colour and chemical additives to protect them from embrittlement by continuous exposure to UV light. HDPE pipe is generally supplied coloured black, and ABS is generally light coloured. In hot climates the lighter coloured ABS material can minimise the temperature rise, and hence the need to down-rate the pipe in pipelines installed above ground.

The limited strength of thermoplastics restrict their use to pipelines with relatively low operating pressures (1·5 MPa). This restricts the pump spacing and effectively limits the length of pipeline constructed from this material. An alternative approach when the useful properties of the thermoplastic (corrosion and erosion resistance) are required in a long-distance pipeline is to install the thermoplastic pipe in a steel sleeve. This offers some economy by allowing the thermoplastic pipe-wall thickness to be reduced to the minimum required for wear and allowing the sleeve to provide the necessary mechanical strength.

A small tolerance between the plastic pipe outside diameter and the sleeve pipe inside diameter allows free installation without requiring the carrier pipe to expand excessively under operating pressures before being supported by the sleeve.

Thermoplastic pipe dimensions comply with national standards set by the manufacturing country, and, because of the relative ease with which dies for an extrusion process can be manufactured, generally offer a wider range of standard diameters than is offered for steel pipe. For example, pipe manufactured to AS 1159 (AS, 1979) is offered in outside diameters of 50, 63, 75, 90, 110, 125, 140, 160, 200, 225, 250, 280 and 315 mm. Larger diameters are available in increments of 50 mm. Pipe-wall thicknesses are offered in pressure classes 3, 4·5, 6, 9, 12 and 15, where the pressure class is equal to the permissible working pressure in bars for a design life of 50 years. The designer is referred to technical information from the proposed pipe vendor for detailed design of the pipeline and associated fittings.

TABLE 19.5
Comparison of Flow Velocity in Two HDPE Pipe Classes

	250 Class 4·5	*250 Class 15*
Inside diameter (mm)	230·2	186·2
Velocity ($m\,s^{-1}$)	1·67	2·55
Pressure gradient ($Pa\,m^{-1}$)	89	255
Power ($kW\,km^{-1}$)	6·2	17·7

The relatively low tensile strength of thermoplastic materials requires a relatively large wall thickness for moderate internal pressures. Because the pipe outside diameter remains constant for each pressure range the inside diameter is reduced significantly at higher pressure ratings. This can impact on the pipeline hydraulic design by reducing the flow area. For example, Table 19.5 compares the inside diameters, flow velocities and the pressure gradient for water flowing at 250 $m^3\,h^{-1}$.

3.4. Fibre-Reinforced Plastic Pipe

FRP pipe is not commonly used for mineral slurry transportation, because better wear resistance is offered by thermoplastic materials. However, it is used in less aggressive environments, and in environments which are subject to chemical attack. Because the strength of this pipe is provided by the glass reinforcing filaments embedded in a plastic resin, its strength is relatively unaffected by operating temperature.

This pipe can be supplied with special resin-rich linings to improve the chemical or wear resistance of the pipe. FRP pipe has been manufactured with a polyurethane lining for enhanced wear resistance, although this product is not a stock line.

4. PIPELINE CONSTRUCTION

4.1. Special Considerations

Slurry pipelines have special requirements which require consideration in the design stage because of the impact that these factors can have on the pipeline construction. These include

- pipeline slope limitation,
- slurry settlement from the crest of hills to the low point, and
- the need for clean-out points in some short distance tailings systems.

4.1.1. Pipeline Slope Limitation

Most slurry pipeline designs will include a maximum slope limitation on the pipeline to ensure that the solids will not migrate down a sloping pipeline on shutdown possibly forming a plug which could cause restart difficulties. To comply

Fig. 19.5. Drop box used to dissipate excess energy in a tailings flume constructed for BP Minerals. (Courtesy of Victaulic Company of America, USA.)

with the slope limitation, the route must be selected and graded to the specified slope prior to construction. In mountainous terrain, this specification can cause a massive task to select a graded route at 10–20% slope, since, to achieve the slope, the route must cut across the contours at a low angle and major excavation may be required to construct a safe platform for both the pipeline and for the construction machinery.

For example, the Savage River iron ore pipeline (McDermott *et al.*, 1968) constructed in Tasmania, Australia, in 1967 was built to a 10% slope. The route traverses rough mountainous country, and even with careful route selection and major earthworks, some 3% of the 85 km pipeline was installed above ground on trestles or spans through regions where the grade could not feasibly be maintained by route selection and earthworks.

This problem may become more severe for flumes. These must be constructed at a specified slope to maintain hydraulic stability and consequently require accurate route survey to select a route having the design slope. When the natural gradient is greater than the hydraulic gradient, the flume must be elevated to achieve the design slope and drop boxes used to dissipate the excess energy (Fig. 19.5).

4.1.2. Settlement of Solids from High Points

Even when the pipeline is constructed within the slope specification there can be some migration from the high points toward the low points when the pipeline is shut down. This can occur as a result of convection currents set up between the high and low points as the fine solids settle, promoting downward flow along the bottom of the pipe, with a corresponding upward flow of clear liquid along the upper surface of the pipe. The effect is exaggerated if dribble flow is permitted in the pipeline during shutdown.

This effect may cause the pipeline performance to become unpredictable and, it will more importantly, cause the formation of concentrated slugs of slurry which may transport through the pipeline in laminar flow. In this regime the lack of turbulence will cause coarse-solids fractions to settle and gradually lag in the flow until they leave the slurry and are picked up and concentrated by the following slug of lean fluid. The solids in this slug can behave unpredictably and could cause the line to become plugged. When settlement of this type occurs, the slurry is transported in discrete slugs of high and low concentration slurry which will not remix until they reach the terminal facilities.

4.1.3. Buried or Above Ground

Whenever a pipeline is to be constructed there develops a debate as to whether the pipeline should be constructed above ground or buried. In general, a short distance, short-life pipeline can be economically constructed above ground, because there are no special performance or pipeline life requirements which must be satisfied by design and construction techniques necessary for longer life systems. If the pipeline is to be constructed for a long life, the cost to construct the substantial supports, anchors, expansion loops etc. is typically the same order as the cost to construct a fully buried pipeline.

4.2. Above-Ground Construction

If a pipeline is to be constructed above ground, the following must be considered.

4.2.1. Pipe Supports

An above-ground pipeline must be properly supported if its mechanical performance is to be satisfactory. Support design must consider thrust loads at changes in direction, span length, pipe anchors, expansion loops, guides, and access for pipe-coating maintenance.

4.2.2. Pipeline Expansion

The temperature changes which will occur in an above-ground pipeline as a result of diurnal temperature changes and changes in pipeline operation will cause longitudinal stresses to develop as a result of pipeline expansion and contraction. These stresses must be accommodated by properly anchoring and guiding the pipeline, and by providing expansion loops or joints at appropriate intervals.

4.2.3. Pipeline Corrosion Protection

Above ground pipelines are subject to external corrosion. Coating maintenance can become a major operating expense unless the design makes provision for a properly applied high-quality protective coating system. In addition, the pipeline must be installed a sufficient height above the ground so that when coating maintenance is required it can be undertaken without major construction work.

Short distance, mechanically jointed pipelines may be protected by hot-dip galvanising.

4.2.4. Access

An above-ground pipeline makes an effective barrier across the land, preventing livestock, wildlife and vehicular access. If the pipeline is to be constructed above ground, a detailed study is required of the need for, and the frequency of bridges or underpasses constructed for access.

4.2.5. Security

An above-ground pipeline is visible and is liable to damage from external sources including impact by vehicles. The risk to the pipeline from all sources of damage, and the risk to the community in the event of pipeline failure should be considered in the design, and steps taken to minimise the risk where necessary. Designers of the Savage River iron ore pipeline considered there was a risk of the pipeline being accidentally or wilfully damaged by rifle fire, and steel jackets were provided on certain above-ground sections to provide additional protection.

4.3. Buried Construction

Buried construction is generally economic only for those pipelines designed with a long life, where the pipeline, once buried, requires no further attention until it reaches the end of its life. Buried construction is universally used for oil, gas and water distribution pipelines, and for most oil, gas, water and slurry transmission pipelines.

Buried construction offers many advantages. There are no restrictions to normal land use over a buried pipeline other than a restriction to building construction. For pipelines constructed cross country, this is of great significance to the pipeline owner and the landholder alike. The pipeline is effectively restrained throughout its length. This simplifies the structural design of the pipeline. By burying the pipeline at a depth greater than typically used by other service operators (750–1 200 mm cover), the pipeline is effectively protected from damage by external interference.

4.4. River and Stream Crossings

The route of a long-distance pipeline invariably involves construction across streams, ranging in size from a minor gully or creek to a major river. Crossings may be buried or supported on new or existing bridge structures.

4.4.1. Considerations for Buried Construction

Buried river crossings require special consideration because they are costly to construct and maintenance is costly if the construction is found to be inadequate.

The crossing location should be carefully selected to allow the pipeline slope limitation to be achieved with minimum cost and a hydrographic survey should be undertaken to determine the crossing profile. A geotechnical investigation may be necessary to determine the river bed conditions as they affect the construction cost and provide support for the pipeline. Abnormal conditions to which the river may be subjected should be investigated and the need for pipeline anchoring determined. Future river uses should be considered; for example, if the river is used for shipping the management authority may have plans for future dredging which require the pipeline to be buried at additional depth. The river bank restoration requirements must be investigated and specified, and it may be necessary to provide a layer of stone armour over the excavation to protect the backfill from erosion.

River crossing construction by its nature means the crossing pipe will be installed in a location where it is not possible to inspect the pipe prior to backfilling to ensure the pipe and coating are not damaged by the installation technique. For short crossings the pipe may be protected by additional wrapped mechanical coatings (rockshield), together with complete encirclement by wood slats. Longer crossings, and any crossings which are to be installed by dragging, are normally protected by continuous concrete coating. The concrete may be applied pneumatically (gunite), or cast in forms, with a minimum thickness of 50 mm. Reinforcement should be provided to maintain the concrete integrity during installation, particularly when the pipeline is required to flex. Because the risk of damage to the protective coating is higher than conventional land construction, designers will normally specify that all river-crossing pipe be provided with a double application of the protective coating beneath any concrete or rockshield.

4.4.2. Lift and Carry Construction Technique

This is the simplest approach to a river crossing illustrated in Fig. 19.6, applicable to narrow or shallow crossings where construction machinery can walk alongside the trench for the width of the crossing, or where land-based cranes can reach across the crossing. Prior to installation the pipeline trench is excavated to the design profile, and the crossing pipeline fabricated with sag and over bends at each end to the same profile. The crossing pipe is lifted and carried into position, and then lowered into the trench.

4.4.3. Bottom or Top Pull Construction Technique

The bottom or top pull technique is borrowed from offshore pipeline construction and is used for longer crossings. The pipe is assembled in long strings and hauled into position in the prepared trench by a winch installed on the opposite shore. As the name implies, in bottom pull installation the pipe has negative buoyancy and is fully supported by the trench bottom throughout its length. In top pull installation, the pipe is provided with positive buoyancy by floats attached to the pipeline allowing the line to be towed from the fabrication site to the crossing. When the pipeline is in position the floats are released, allowing it to settle into place. Figure 19.7 shows pipeline installation by the top pull method.

Depending on the design trench profile the transition between river bed and

Fig. 19.6. Stream crossing using lift and carry construction technique. (Courtesy of Williams Brothers–CMPS Engineers.)

normal pipeline levels may be accomplished by allowing the pipe to sag at a natural curvature within the stress limits of the pipe, or it may be necessary to construct a temporary sheet pile wall to allow the riser pipe to be welded to the crossing pipe at the edge of the river.

4.4.4. Directional Drilling Construction Technique

Directional drilling is a recent development built upon tools and experience developed in the upstream oil industry. This technique allows the crossing to be installed without any river bed excavation and without interference with other river users. Where suitable construction equipment is available, directional drilling offers cost-competitive construction, and in some special circumstances where the design makes the cost of conventional construction techniques excessive, it may still be economic to pay to mobilise directional drilling equipment internationally.

Figure 19.8 illustrates the principle of directional drilling. The drilling rig is installed on one bank of the river, and the pipe assembled in strings on the other. The directional drilled crossing is installed in two stages.

The first stage requires drilling a pilot hole at the design profile for the entire crossing. The drill enters the ground at an angle of 15–20° to the horizontal, and at a predetermined point, the drill bit characteristics are altered to develop a horizontal bend below the river bed. Drilling continues horizontally until the far bank where another deviation changes the pipeline direction to bring the drill bit

Fig. 19.7. River crossing using top pull technique. (Courtesy of Williams Brothers–CMPS Engineers.)

to the surface. Special tools have been developed which allow accurate control of the deviation and measurement of the position of the drill bit to confirm the design profile is being followed.

In the second stage, the drill bit is replaced by a reaming bit to enlarge the hole sufficiently for the pipeline to be installed. The pipeline is attached to the drill string, and the drilling operation proceeds in reverse, using the drilling machine to winch the pipeline through the drilled hole.

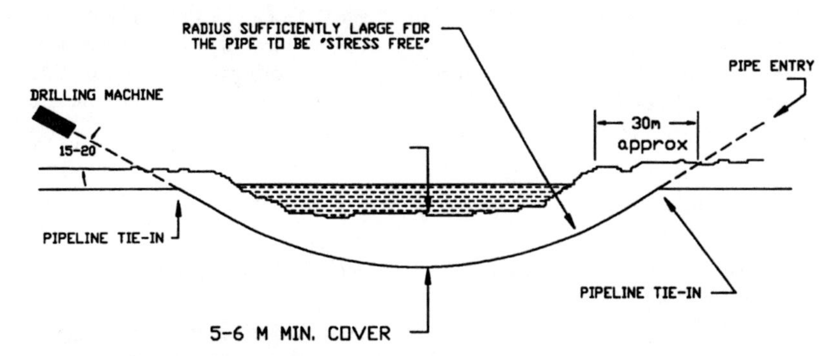

Fig. 19.8. Principle of directional drilling to achieve a river or stream crossing.

The drilled bend radius is designed to be achieved by the pipeline flexibility, within the allowable wall stress limits specified by the design code. Once the crossing pipe is in position, the pipe ends are cut and the crossing tied into transition bends and into the pipeline.

Directional drilling can be used in other difficult crossing situations such as beach and swamp crossings, and, by allowing controlled installation at depth, construction beneath congested areas.

4.5. Pipeline Bends and Fittings

In a slurry transportation system, bends should always be designed at the maximum practical radius to minimise the risk of introducing regions of high wear which can shorten the economic life of the pipeline. Approaches to changing the pipeline direction depend on the pressure rating of the pipeline.

4.5.1. Low-Pressure Pipelines

A range of standard pipe fittings are manufactured for all types of pipe available for slurry handling systems. Bends are supplied in various radii, with end connections which are compatible with the pipeline jointing system. Bends with special radii can be manufactured to order, and when wear is a problem, so can bends with special linings. Other techniques are used to reduce the installed cost, including angular misalignment, mitre bends and pipeline flexibility.

Certain mechanical joints allow angular misalignment, principally to simplify construction. Shouldered Victaulic couplings typically allow a change of direction of up to 3° per joint. Angular misalignment should not be used when handling abrasive slurries because the internal discontinuity can cause increased local wear.

Mitre bends, fabricated from pipe, are permitted in low-pressure pipeline systems. They are low cost and can be fabricated to the bend angle required. The end has limited strength and is prohibited in pipeline systems operating at hoop stress levels higher than 20% of SMYS, and, between 10 and 20% of SMYS, the maximum change in direction using a mitre bend is limited to 12·5°. The discontinuities inherent in a mitre bend can cause areas of local high rates of erosion.

Pipelines are quite flexible, and when long radius bends can be tolerated, pipe flexibility can be utilised to change the pipeline direction. This method is frequently used with unrestrained HDPE pipelines laid on the ground.

4.5.2. High-Pressure Pipelines

High-pressure slurry pipelines have performance limitations which restrict pipeline bends to

- forged elbows (short (1 diameter) and long (1·5 diameter) radius), and
- pipeline flexibility.

Pipeline flexibility is used to provide long radius direction changes, particularly for vertical bends which are required to follow undulating ground. The bend radius must be sufficiently large to prevent stresses in the pipe wall from exceeding the allowable limits under the design code.

Stresses in a slurry pipeline consist primarily of circumferential, longitudinal, thermal, bending, torsional and shear stresses. Allowable stress combinations vary between design codes, and are also dependant on the pipeline restraint and sustained loading conditions. The designer should consult the appropriate code for stress limits and load combinations to determine the minimum bend radius and hence the maximum bending stress permissible.

In cross-country pipeline construction, bends made from pipe offer an inexpensive and flexible method of constructing bends of varying deflection angles to meet the requirements of the construction site, and this method is permitted by the design codes, provided the finished bend meets specified quality standards including reduction in pipe diameter (5–6% maximum), absence of buckling or mechanical damage. Bending machines include tractor-mounted shoes around which the pipe is bent and dedicated bending machines. Larger diameter and thin-walled pipes may require an internal mandrel to prevent the pipe from collapsing during the bend. The codes permit minimum bend radii of 18 pipe diameters (ANSI/ASME, 1986) for pipe diameters smaller than 300 mm, although considerable skill is required to achieve this radius without deforming the pipe beyond the limits allowed by the code. The minimum bend radius allowable increases with increasing pipe diameter.

Standard long radius forged bends are supplied with 45 and 90° changes in direction. These may be used except for pipelines which are intended to be cleaned by pipeline scrapers (or *pigs*), or inspected by internal inspection tools.

When longer radius or special deflection angles are required, hot factory bends, manufactured using an induction heating and bending machine, may be specified.

4.5.3. Plastic Pipe Fittings

Manufacturers offer a wide range of fittings in all diameters which allow the designer freedom to engineer an adequate piping system. Standard mechanical joints, fusion butt welded joints and solvent cemented joints have been developed for thermoplastic and FRP pipes.

Mechanical joints include Victaulic couplings and flanges. These require attachment of a stub fitting to each end of the pipe using fusion welding or a proprietary glued joining technique. Flanged connections are provided with a steel back-up ring behind the flange stub to compress the flange faces uniformly. These jointing techniques require no special tools or experience and are especially suitable for short-distance pipelines and for pipelines which require frequent relocation or rotation to equalise wear.

Solvent cement joints can be used for ABS and uPVC pipes. Glued joints using bell and socket connections or moulded fittings are commonly used for FRP pipes.

Butt fusion jointing is used for HDPE pipe. This technique requires the use of a portable machine which squares pipe ends, heats and softens the ends, and then brings the pipes together under a controlled pressure to fuse the joint. This technique produces a joint having the same strength as the parent pipe, and because it does not require any extra fittings or materials, generally offers the most economic joint for long pipelines.

5. HYDROSTATIC TESTING

Part of any pipeline construction is a hydrostatic test to prove the structural integrity and the leak-tightness of the pipeline. The approach to the test, the test pressure, and the test acceptance criteria adopted for a particular pipeline design depend on

- the pipeline design and its operating pressure,
- the impact of a pipeline failure on the surrounding environment,
- the cost of failure (resulting from loss of product, and from the cost of locating and rectifying the failure), and
- the implications of failure on the construction contract.

The pipeline profile must be considered when designing a hydrostatic test programme; if the test pressure was measured at the high point, the additional pressure developed by static head could cause pipe at the low point to fail. For a high-level test, the test sections are chosen to limit the pressure difference between high and low points to 10%.

The minimum test, applicable to short distance above ground pipelines, involves filling the pipeline with water at the operating pressure, with visual inspection to identify leakage.

A buried high-pressure pipeline is typically subjected to a two part test. The strength test is conducted first for a period of approximately 4 h to prove the mechanical strength of the pipeline. This test will typically pressurise to either 1·25-times the maximum allowable operating pressure (MAOP), or to a level which develops a hoop stress equal to 90–100% of SMYS at the low point in the pipeline. Provided there is no evidence of leakage, the pressure is lowered to 1·1-times MAOP and held for a period of 24 h to prove the pipeline is free of leaks.

6. REFERENCES

ANSI/ASME (1986). Code B31.11: *Slurry transportation piping systems*. Distributed by American National Standards Institute, New York, USA.

ANSI/ASME (1989*a*). Code B31.8: *Gas distribution and piping systems*. Distributed by American National Standards Institute, New York, USA.

ANSI/ASME (1989*b*). Code B31.4: *Liquid transportation systems for hydrocarbons, liquid petroleum gas, anhydrous ammonia and alcohol*. Distributed by American National Standards Institute, New York, USA.

API (1988). Specification 5L: *Specification for line pipe*. American Petroleum Institute, Washington, DC, USA.

AS (1979). AS 1159: *Polyethylene pipe for pressure applications*. Standards Association of Australia, Sydney, Australia.

McDermott, W.F., Cowper, N.T., Davis, R.A. & Wasp, E.J. (1968). Savage River Mines. The world's first long distance iron ore slurry pipeline. Paper presented at AIME Fall Meeting, Minneapolis, MN, USA, 18–20 Sept.

20

Erosive Wear of Pipeline Systems

Melissa J. McKibben & Clifton A. Shook
University of Saskatchewan, Saskatoon, Canada

1. INTRODUCTION

Erosion is inevitable in all slurry pipelines. Despite its importance, the mechanisms by which it occurs are not well understood and compared to, say, pressure drop measurements, the quantity of experimental-wear data is still rather meagre. A primary obstacle to systematic study is that erosion is usually a slow process. This means that long-term experiments are required, making research very expensive. This has led to a dearth of data collected in actual pipelines (Shook *et al.*, 1981). Furthermore, differences in experimental conditions have led to disagreement concerning the role of some of the important variables.

Wear rates are usually calculated as the loss in pipe thickness, volume, or weight per unit time. These absolute rates, which are commonly used in pipeline testing (Tarjan & Debreczeni, 1972), can be easily converted to specific wear rates such as wall thickness lost per unit solids throughput. Various rates have been used in the literature and care must be taken, when comparing these wear rates, that the units are the same. Relative wear rates, or the wear compared with that of a reference material, may be determined by simultaneous testing of different materials but these are not common in the literature (Truscott, 1975).

The ultimate engineering goal is to predict wear rates and to minimize erosion costs by selecting the proper pipe material and by specifying suitable process variables. Before these goals can be achieved, the mechanisms by which wear occurs must be identified and somehow correlated with actual wear rates.

2. TYPES OF WEAR IN SLURRY FLOW

There are many types of particle–wall interactions which could cause a surface to erode. The main forms include wear by adhesion, abrasion, plastic deformation and fatigue. Simple models exist for each type of wear (Bitter, 1963; Rabinowicz, 1965; Sarkar, 1980). Other wear mechanisms, such as delamination or micro-flaking, have also been proposed. In general, however, more than one type of wear occurs at a given time and the situation can be complicated further by particle–particle interactions and particle–fluid interactions. For slurry erosion, we can simplify things somewhat by considering dilute and dense slurry flows separately.

2.1. Dilute Slurry Flows

In dilute slurries (concentration by volume $< 5\%$), particle–particle interactions are minimal and can be neglected. However, particles approaching the wall interact with the fluid so that fluid drag and lubrication forces will affect particle trajectories and velocities. In a dilute slurry, a single particle can either impact the wall or slide along the surface. With impact erosion, wear occurs when the impingement energy exceeds the plastic limit of the surface, producing plastic deformation (Bitter, 1963). The deformed surface is then subject to further impacts which lead to chipping-off of material (Altaweel, 1982). Impact erosion can also occur if the surface is fatigued. Here, repeated loadings eventually cause fatigue cracks which break up forming large fragments and leaving pits (Rabinowicz, 1965). Impact erosion is most common in elbows and pumps where the impact angle is high.

If a particle slides along the wall or strikes it at a small angle, wear is due mainly to adhesion and abrasion. Adhesion occurs when the particle is in contact with the wall and under a normal load. Strong adhesive forces are set up between the atoms of the two surfaces and when the contact is broken the break is not at the original interface so that fragment transfer occurs. Abrasion or cutting wear occurs when a particle cuts or ploughs a groove into a softer surface, thereby deforming it and eventually removing wear particles (Sarkar, 1980). This will occur if the local stress exceeds the limiting shear strength (Bitter, 1963). Abrasion is the most common form of erosion in slurry pipelines.

The rates of the various types of erosion depend on the type of material being eroded. Hard, brittle materials are very resistant to abrasive erosion so that wear occurs predominantly by deformation. Soft, ductile materials, however, are highly resistant to plastic deformation and abrasive wear is predominant (Bitter, 1963).

2.2. Dense Slurry Flows

As the concentration of the slurry increases, particle–particle interactions become more important. As with single particles, when the particle velocity is directed toward the pipe wall, wear occurs mainly by plastic deformation and fatigue. However, wear can also occur by random impact in which fluctuating particle velocities near the wall are generated by turbulence in the flow and by interparticle collisions (Roco *et al.*, 1987). Any one of the types of wear mentioned previously can be responsible for erosion due to random impact. Additionally, wear resulting from a concentrated slurry occurs if particles in a sliding bed are in contact with the wall. In this case, the normal load is fairly high and erosion occurs mainly by adhesion and abrasion.

One approach used to model wear in dense slurry flows relates the energy dissipated by the various particle–wall interactions to the wear rate (Roco *et al.*, 1987). To do so, a reliable model of the flow pattern is required so that one can determine the type and number of particle–wall interactions.

A microscopic study of the wear patterns on a given surface can help to determine the hydrodynamic characteristics of the flow and to identify the type of wear which has occurred. Pitting is characteristic of impact erosion at high angles whereas ripples or dune-like patterns are sometimes seen when the impact angle is relatively small (Altaweel, 1982). Particles flowing nearly parallel to the wall produce scratches in the direction of flow and will even smooth out surface discontinuities (Wright & Garrett, 1986).

2.3. Erosion–Corrosion

It is important that any simultaneous corrosion process be considered when studying erosion. The products of corrosion are easily removed by any of the wear processes and this results in the exposure of fresh metal. This in turn often causes an increase in the corrosion rate (Truscott, 1975).

Techniques aimed at reducing corrosion include the use of inhibitors such as chromates, which promote formation of a passive film. Sometimes one can use an oxygen-scavenging species, such as sodium sulphite, to de-aerate the slurry and slow the corrosion process (Hoey & Bednar, 1981). For example, Samarco Mineraco in Brazil operates a 390 km iron concentrate pipeline which utilises sodium sulphite to remove free oxygen lime to control the pH (Oliveiro, 1986). Methods do exist for determining the corrosion tendency of a given wall–fluid combination (ASTM, 1989, tests G5–87 and G59–78). Also, methods for monitoring corrosion in slurry lines, such as the electrochemical method, can be used (Ferrini *et al.*, 1982). By using such techniques, it should be possible to determine loss of material due to corrosion in a slurry (Postlethwaite *et al.*, 1972).

In many instances, corrosion can be virtually eliminated by using plastics as the wearing material. An increasing number of pipelines for abrasive slurry flows now use metal pipes for strength with plastic liners to resist corrosion and erosion. The resulting pipeline is often far superior to any steel pipe but it is also more expensive (Johns, 1988). Paint coatings are much more economical but they usually provide only corrosion protection and are quickly worn-off (Truscott, 1975).

3. METHODS OF MEASURING WEAR

There are several wear measurement techniques, which differ widely in cost and convenience. Unfortunately, few of the methods are suitable for use with large industrial pipelines.

3.1. Weight Loss Method

With an accurate balance, the weight loss of test samples can be measured. This technique gives an average weight loss for the sample as opposed to a local one. Although the method is reliable, difficulties with samples which absorb water (i.e. plastics) do arise. Care must be taken to ensure that samples are completely cleaned and dried before and after erosion. The reproducibility of the whole experimental procedure must be investigated carefully to maintain the highest possible precision. Because the drying process can be very time consuming, a weighing technique based on the theory of drying and fluid absorption has been developed (Jacobs, 1982). This is not yet in common use, however.

3.2. Measurement of Wall Thickness Change

The most common device for these measurements is the ultrasonic gauge. By calibrating the instrument with specimens of the particular material of known thickness, measurements with a precision of 0·025 mm can be made. The outer surface of the pipe may have to be modified slightly, depending on the shape of the gauge probe. Unfortunately, the method cannot be used on lined or coated pipes (Wiedenroth, 1984). Ultrasonic thickness measurements will provide local wear rates which are particularly useful in segregated flows. In these slurries, wear varies significantly with position. It is the maximum wear rate which is of greatest importance to a pipeline operator because it determines shutdown and repair costs.

A second method for determining local wear uses radio-tracers. Radioactivity is transferred to the surface of a representative sample of pipe and a counter is used to detect changes in radiation (Rabinowicz, 1965). The advantages of this method are that extreme resolution is attained, it can be used in complex geometries, and accurate rates can be found in minutes instead of weeks. The obvious disadvantage is the complexity and expense associated with the technique.

If the inconvenience of shutdown and disconnecting the pipe is acceptable, special micrometers are available to measure pipe wall thickness with fairly high accuracy (Henday, 1988*a*). In some instances, only the locations of wear attack are desired. By coating the area to be worn with a paint, it may be possible to detect problem areas without providing actual wear rates (Wiedenroth, 1984).

4. SIMULATING PIPELINE EROSION

Because wear is such a slow process, tests in actual pipelines are very costly and time consuming. Consequently, many attempts have been made to produce a small scale, inexpensive and rapid experiment to simulate pipeline erosion. Because these simulations are numerous, only a few will be mentioned here.

While some of these methods have failed to simulate pipeline wear, especially the various rotary types (Truscott, 1975), they have been successful in other aspects such as ranking various materials for wear resistance or investigating parameters such as impact angle. Wright and Garrett (1986) found it useful to examine the wear surface in a simulation microscopically and compare it with one found in an actual pipeline to help determine whether or not the wear mechanisms were the same. It should be noted, however, that no simulation to date is capable of replacing wear data collected in actual slurry pipelines. The relationship between a simulation and pipeline wear must be provided by the mechanics of the slurry flow. Thus, any study contributing to the understanding of slurry hydrodynamics is certainly useful when interpreting erosion measurements.

4.1. The Miller Test

This test was originally designed to simulate wear in slurry handling equipment, particularly pumps (Huggett & Walker, 1988; see also Chapter 13). It has since become an ASTM procedure (ASTM, 1989, test G75–89) which gives an indication of the relative abrasion tendency of a given slurry on a standard wall. In the test, a normal force is applied to the test slurry as 27% chromium iron wear blocks are reciprocated beneath it. These samples are weighed at intervals and the data are used to calculate the *Miller Index of Abrasivity*, a dimensionless number used for comparing various slurries. Alternatively, the standard-shaped wear blocks can be constructed from any material and the slurry abrasion response (SAR) number can be determined for a slurry of interest. The SAR number is related to the rate of loss of mass of the tested material. Because the total test time is low (about 24 h) and only a small amount of slurry is required (2–3 kg), the results cannot be used to predict pipe wall erosion quantitatively.

4.2. Impacting Jets

In this simulation, particles in a gas or liquid stream are directed towards a flat plate at various velocities and angles (Rao & Buckley, 1984). Unfortunately, most of the research to date has used gaseous streams and the effects of fluid density and viscosity have yet to be established. The advantage of the technique is that certain variables, such as velocity and impact angle, can be determined precisely and varied widely. This test can also be used to compare materials but it should be noted that most materials wear differently under directional impact than they do under sliding impact as previously mentioned.

4.3. Slurry Pot Tester

In this simulation, samples of material are placed around the circumference of a small chamber or pot. Slurry is pumped through the chamber and particles are flung outwards against the samples by means of a small impeller within the chamber (Huggett & Walker, 1988). A continuous, once-through slurry flow can be used or the slurry can be recirculated to study the effects of particle degradation. Test time is 30 min–12 h. Again, the device is useful for screening materials but not for predicting actual erosion rates unless the particle trajectories can be defined.

4.4. Other Simulations

Numerous other simulations have been designed and have been found to be useful in certain situations. De Bree *et al.* (1982) constructed a rotating disc apparatus consisting of two discs with a slurry sample between them. The discs carried material specimens and were rotated in opposite directions. The slurry itself did not rotate but was agitated enough to create a homogeneous mixture. Velocities and impact angle could be determined accurately and the results agreed with those in the classical study of Bitter (1963).

In another simulation, a pipe containing a slurry was rotated about its longitudinal axis for an extended period of time. Here it is possible to compare wear rates of different pipe materials but it is not possible to predict actual wear rates because of the complicated nature of the flow (Henday, 1988*a*).

Shook and Pilling (1987) used a Couette flow simulation device. By suspending the particles in either a fluid of equal density or a fluid with a yield stress, the effects of gravity and the centrifugal force were eliminated. The slurry mixture in the annular space was sheared by rotating one cylinder while keeping the other stationary. Although this method may be useful for simulating the particle–wall interaction in a liquid, it does not simulate the normal force which arises from the immersed weight of the particles.

Roco *et al.* (1987) have used a device rather similar to the Miller test to simulate and evaluate erosion by friction. Samples of material are placed on a horizontal plate which can be reciprocated. As the plate moves, a settled slurry on top of the plate slides parallel to the samples, just as it would in pipe flow. The wear mechanism should be similar to that occurring in the so-called *sliding bed* flow.

5. FACTORS AFFECTING EROSION

Many variables are known to affect erosion. While it is possible to control or measure many of these, others such as particle and material surface features are more difficult to quantify. This makes comparison of various investigations almost impossible. Controlled experiments still may be required to define the effect of the more important factors. Once these effects have been determined, it is then possible at least, to reduce the cost of the inevitable wear.

5.1. Velocity

According to a simple theory of wear for a single particle, the volume of material removed is proportional to the distance the particle travels on the surface multiplied by the normal force. For a horizontal pipe, the particle–wall force will be produced by gravity and reduced by fluid lift and lubrication effects (Shook *et al.*, 1990). Since velocity is the distance travelled per unit time, the volume removed per unit time (absolute erosion rate) should be proportional to the velocity if the normal force does not vary with velocity (Shook *et al.*, 1981). Previous studies have found that the absolute erosion rate is proportional to velocity to the power n, where n can lie anywhere between 1 and 3·5 (Truscott, 1975). Accordingly, the volume removed per unit mass of solids throughput (specific wear rate) varies with velocity to the power $(n - 1)$. From the simple theory of wear, the normal force should thus vary as $(n - 1)$.

Tarjan and Debreczeni (1972) suggested n is 1 but gave no test results to support the statement. Shook *et al.* (1981) found n to be 1 for abrasive wall wear in pipeline flows. However, another pipeline test, performed at the British Hydromechanic Research Association on mild steel pipes (Truscott, 1975), found n to be 3·3 in heterogeneous flow while the experiments of Karabelas (1978) using sand slurries on stainless steel pipes gave values of n ranging between 2 and 3. Spencer and Sagues (1986) found n to have a value near 2 for coal and sand jets impacting on stainless steel while the equations of Bitter (1963) for cutting and deformation erosion show that wear is not a simple function of velocity.

This scatter of n values can be attributed to the variety of wear mechanisms occurring in the different experimental situations as well as to other factors such as the type of slurry itself. A velocity exponent around 2 supports the idea that absolute wear rates are directly related to the kinetic energy of the impacting particles (Spencer & Sagues, 1986).

In any case, the velocity profile will affect the distribution of wear in pipes. Jvarsheishvili (1982) found that reducing the velocity caused an increase in the unevenness of wear around the circumference of a horizontal pipe. In his system, wear on the bottom zones was approximately twice that of the upper part. In

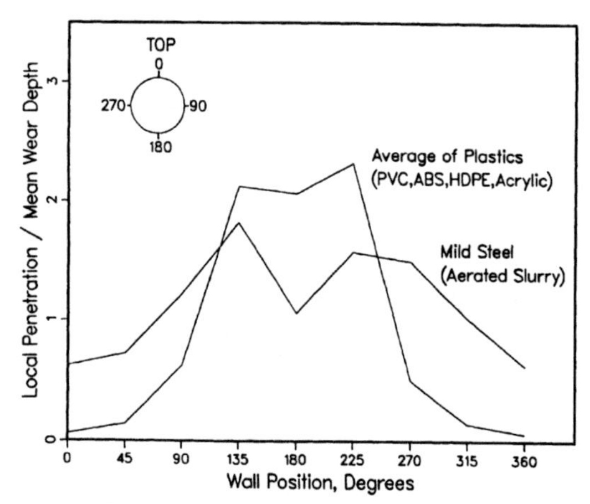

Fig. 20.1. Wear as a function of wall position.

industry, pipelines are normally operated just above deposition velocity and operators have noticed higher wear on pipe bottoms, as expected. Typically, it has been found useful to rotate a pipe by about 120° to prolong its life (Wright & Garrett, 1986). Figure 20.1 shows wear distributions for mild steel and various plastics (ABS, PVC, acrylic and high-density polyethylene) for a 20% slurry of 0·5 mm sand. The slurry in the steel experiments was aerated and therefore the wear near the top of the pipe can be attributed to corrosion.

5.2. Particle Concentration

As with velocity, one must be careful when discussing concentration effects to take note of the units used to define wear rates. Concentration effects inferred from specific wear rates will certainly have a different meaning from those based on annual erosion rates. If we consider a rate based only on the mass or volume of wall material lost, from the simple wear theory for a single particle, it seems reasonable to suggest that the wear rate should be directly proportional to slurry concentration if each impact is equally effective (Altaweel, 1982). This assumes the concentration at the wall is proportional to the bulk concentration. This is the type of relationship which many investigations have found. On a specific erosion rate basis, this means that the effect of concentration on wear is very weak if existent at all (Spencer & Sagues, 1986).

This theory neglects particle–particle interactions which become very important at high concentrations. High concentrations may reduce the frequency of pipe wall impacts and, in fact, some studies have noticed that after a certain concentration is reached, the wear rate no longer increases with increasing concentration (Truscott, 1975).

Changes in concentration can also affect the slurry flow regime and thus alter the wear profile. A slurry with a uniform concentration distribution should give an even wear pattern while sliding bed flow would cause severe wear on the bottom (Henday, 1988*a*).

5.3. Impact Angle

The effect of impact angle can be studied readily using impacting jets. In horizontal slurry pipelines, the impact angle is low (< 20°—Jvarsheishvili, 1982) and therefore is not a very important variable unless pipe joints are poor. However, impact angle becomes important when dealing with flow through bends, tees, valves and pumps where substantial direction changes occur (Henday, 1988*a*). These situations generally lead to localised erosion.

As mentioned previously, the effect of impact angle is highly dependent on the type of material being eroded. For hard, brittle materials, such as ceramics, wear due to deformation is much larger than wear due to cutting (Bitter, 1963) and studies have found that erosion is highest at angles of about 90°. For softer, ductile materials, such as rubber, cutting wear is predominant and the highest wear rates are found to occur at angles around 30° (Altaweel, 1982).

5.4. Particle Characteristics

Particle shape will affect wear rates greatly. This is especially true in the case of abrasion in which a sharp, angular particle sliding along the surface will do more damage than a rounded one, especially if a normal load is applied to the particle. Unfortunately, particle shape is difficult to quantify. For hydrodynamic purposes, particle volumes and surface dimensions can be used to calculate various shape coefficients and shape factors (Allen, 1981 and Chapter 2). This analysis of shape typically involves comparing the shape of a particle to that of a sphere of equivalent volume. Obtaining the required measurements can be difficult, however. A technique which can be used is particle imaging in which a two-dimensional image is projected from the surface of a particle. The analysis consists of measuring the perimeter, area and external angles such that it is possible to calculate the sharpness of the particle protuberances (Steward & Ruther, 1989). Enough particles must be analysed to provide a statistically representative sample. The usefulness of this technique in erosion situations is still under investigation.

Particle size is also important. If we consider the kinetic energy of a particle, we would expect larger particles to produce more wear while smaller particles would be more likely to follow stream lines (Altaweel, 1982). According to Rabinowicz (1965), for abrasive wear there is a strong dependence of wear rate on particle size until a critical size is reached. Above this size, the rate is independent of particle size. In general, however, the consensus is that wear increases with particle size but there is little agreement on the actual relationship (Truscott, 1975). This could very well be due to the fact that the shapes of the various sizes of particles used in a given study were not identical so that the effect of particle size was masked by shape differences.

In any case, particle size will affect the velocity and concentration flow profiles and therefore the type and number of interactions occurring at the wall. Shook *et al.* (1990) observed very different velocity distributions at the pipe wall for different sizes of particles. For a given particle size, however, the product of local velocity and concentration (particle flux) was relatively insensitive to mean flow velocity at velocities above deposition. However, the particle flux varied significantly with position for coarse sand and was considered to be the cause of the slightly reduced wear rates at 180° in Fig. 20.1.

Particle hardness is easy enough to measure but its effect on erosion rates is difficult to investigate independently because, as with particle size, one cannot vary the hardness without changing other parameters such as surface features or even particle density (Henday, 1988*a*). For this reason, there is little information on the effect of grain hardness on pipe wear. It appears that the erosion rate increases with particle hardness although once the particles are harder than the surface to be worn, the dependence on particle hardness is very weak (Rabinowicz, 1965; Karabelas, 1978). In any case, for erosion to be significant, the hardness of the particle must be greater than the hardness of the wall.

Finally, particle degradation during flow or testing must be considered because the investigation of particle features is further complicated if particle shape and size are changing with time. Degradation increases with flow velocity and it may be affected by mixture concentration (Truscott, 1975). The effect of the fines generated by degradation on wear rates is not well understood although for some slurries, such as coal, the presence of fines tends to reduce the impact frequency of large particles with the wall. This would reduce wear rates (Henday, 1988*a*). Because of degradation, the particles used in long-term, recirculating experiments must be replaced continually in order to give representative rates. For non-recirculating flows, wear rates will tend to be higher at the upstream end of the pipeline.

The way in which degradation was taken into account has not been documented in many investigations. In recirculating flows, the highest velocities and most abrupt changes often occur in the pumps so that degradation is probably related to the number of pump passes. For long-term wear tests, studies have replaced the particles every 50–100 running hours (Shook *et al.*, 1981). Of course, the appropriate interval depends on the shape and friability of the particles being used. Rounded particles would degrade much more slowly than sharp, angular ones.

Particle degradation can be monitored by performing size distribution and shape analyses before and after an experiment. In simulations, the changes in wear rate with time resulting from particle attrition can be measured. For example, the Miller test can be used to indicate whether particles become more or less abrasive as attrition occurs (Johns, 1988).

5.5. Pipe Characteristics

The material used to fabricate the pipe is the principal factor under the control of the designer to obtain wear resistance. Pipes can be made from unlined metals or they can be constructed from, or lined with, hard metals, plastics, rubber, or ceramics.

Many plastics are not as wear resistant as steel. Shook *et al.* (1981) found that high-density polyethylene was better than steel for fine sand (0·15 mm) but inferior for coarse sand (0·5 mm). Adams (1986) reported that extra-high molecular weight, high-density polyethylene outlasted mild steel by 4 to 1 for a slurry of 0·5 mm sand at a high velocity. He recommended that, in general, particles in polyethylene pipelines be small enough or move fast enough to remain suspended in the slurry.

As mentioned previously, one of the most popular modern approaches uses steel pipes with plastic liners which are more resistant to wear and corrosion. The most widely used liner appears to be polyurethane which is highly abrasion resistant and light in weight. It is not brittle or easily damaged during handling. Its main limitation is cost, although recent developments in the manufacturing of polyurethane liners have shown that cost can be substantially reduced by curing it at 20°C instead of using the conventional thermosetting technique which requires curing at 100°C. This process actually increases the wear resistance and allows the pipe to be joint welded instead of flanged (Owada *et al.*, 1986).

Johns (1988) found polyurethane to have a lifetime 10- to 20-times that of various steels for coarse and fine sand mixtures. Henday (1988*b*) found polyurethane to have only about one-quarter the wear of most steels when transporting coarse sand (d = 1 mm) and granite chips (d = 12 mm). Other plastics, such as polyethylene and nylon, performed much more poorly than steel. Still another study found polyurethane to be the best material for fine ore particles in comparison with rubber, alumina and high chromium white cast iron (Wright & Garrett, 1986). Most recent studies agree that polyurethane is more wear resistant than the rubber which had suffered the least wear in numerous previous experiments. In addition to the instances quoted above, polyurethane is being used in various field locations with good results. These include Du pont's *Adiprene* to transport china clay in the UK and Goodyear's *Neothane* to transport mine tailings in New Mexico (Truscott, 1975).

Ceramics are often very wear resistant, probably because of their extreme hardness. However, they are generally more costly and more awkward to use because of their weight and brittleness. Two of the most common wear-resistant ceramics are alumina and basalt. Henday (1988*b*) found alumina ceramics to be the most wear-resistant material in his tests with granite chips and only slightly inferior to polyurethane in his coarse sand experiments. Basalt was found to be superior to steel and cast iron in tests with very fine ash particles (Wright & Garrett, 1986).

It is generally accepted that the harder a pipe is, the more wear resistant it will be (Steward & Heckroodt, 1988). The hardnesses of many common pipe and abrasive materials have been listed by Altaweel (1982), Wright and Garrett (1986), and Henday (1988*b*) and are given in Table 20.1. The Vickers hardness is determined by indenting a material with a diamond pyramid of specified shape using a load that will produce an indentation with a conveniently measurable size and area (Rabinowicz, 1965). It should be noted that the applicability of hardness tests carried out on plastics is questionable as their wear-resistant properties are mainly due to their resilience (Henday, 1988*b*).

Several material properties have been measured in search for a relationship

TABLE 20.1
Common Material Hardnesses

Material	*Vickers hardness* ($kg\,mm^{-2}$)
Mild steel	145
Stainless steel	185
Cast iron	240
Heat treated steel	700
High chromium white cast iron	892
Basalt	2500
Alumina	3000
Coal	10–70
Silica sand	800
Pyrites	1100–1300
Quartz	1200–1300
Silicon carbide	2100

with pipe wear. Steward and Heckroodt (1988) considered elongation, rebound resilience, tensile strength, tear strength and molecular weight. Others have examined crystal structure, melting temperature, yield stress and impact strength (Altaweel, 1982; Rao & Buckley, 1984). According to Rao and Buckley (1984), an analysis of the energy absorption characteristics of a material, and experimental verification using material properties, are necessary to predict erosion. Very little work has dealt with the effects of pipe roughness and data in this area are definitely lacking.

Because of the complexity of particle–wall interactions described earlier, and in view of our incomplete understanding of these, it is desirable to have a simple technique available to rank various pipe materials for abrasion resistance. Many of the simulations described earlier are capable of doing this. However, one must bear in mind that workmanship in pipe fabrication can play an important role in wear resistance. Therefore, a bench-scale operation should test a representative section of the pipe in question to provide useful results.

Finally, because the pipe diameter plays an important role in determining the velocity and concentration flow profiles and the forces at the wall, it can be expected to have some effect on erosion rates. Karabelas (1978) suggested that if the particles are distributed less uniformly in larger diameter pipes, somewhat greater erosion would occur at the bottom of the pipe. However, recent research has shown that particles are distributed somewhat more evenly in larger pipes (Gillies, 1989).

More important than concentration distribution is the ability of a flow to suspend particles through fluid turbulence. The contact load concept described in Chapter 6 shows how the non-suspended fraction of the immersed particle weight transmits a force to the pipe wall. With fine particles this fraction is small and a lubricating force appears to repel the particles from the wall to some extent (Shook *et al.*, 1990). With coarse particles, the particle–wall force is higher. For a given slurry, the fraction of the particle weight not supported by turbulence depends upon the ratio (U_∞ / V) where U_∞ is the particle settling velocity and V is the flow velocity (Gillies *et al.*, 1990). Since V increases with pipe diameter, the effect of pipe diameter varies with particle settling velocity.

5.6. Fluid Viscosity

Very little work has been devoted to examining the effect of fluid viscosity on pipe wear. It has been reported that as the viscosity decreases, wear increases drastically and that changing the viscosity of the fluid will alter particle–fluid interactions, possibly in the fluid boundary layer near the wall (Altaweel, 1982). More investigation of these effects is definitely required.

6. EROSION IN PUMPS AND FITTINGS

Wear in pumps and pipeline components is even more complicated than wear in straight pipes. By coating the inside of various components with paint, the points of high wear can be determined (Wiedenroth, 1984). However, the actual erosion rate cannot be measured by this method. The methods for measuring wear mentioned previously can be used on various components but the techniques which give a local wear rate are preferable because of the localised nature of erosion on, say, a pump impeller.

Wear rates in pumps have been found to be highly dependent on the particle size, concentration and pump rotational speed. In general, impellers of rotary pumps wear much faster than casings while for various types of positive displacement pumps, valve wear tends to be the limiting factor (Willis & Truscott, 1978).

Attempts have been made to predict the flow pattern to determine the location of maximum wear for various flow conditions (Roco *et al.*, 1987). Once these locations are known, it should be possible to modify some of the pump dimensions (casing width, tongue dimension) to reduce the areas and the maximum intensity of wear (Roco *et al.*, 1987). In this manner, the optimum pump design for given flow conditions can be determined.

When dealing with wear in elbows, the impact angle becomes very important as do velocity, particle diameter and material hardness (Shimoda & Yukawa, 1983). Wear in bends is usually very localised and failure occurs primarily due to particle impact in one or two spots, their location being largely dependent on particle size and velocity (i.e. the momentum of the particles). Again, the velocity exponent n has been found to vary but appears to lie between 2 and 3 (Mills *et al.*, 1982). Fluid viscosity also plays an important role with elbows. Increasing the viscosity has been found to reduce the impact angle (Altaweel, 1982). The optimum choice of elbow material will often be different from that of the pipeline because of the higher impact angles. In other words, impact wear will be much more dominant than abrasive wear and so a softer, more ductile material may be more durable (Agarwal *et al.*, 1983).

In addition to the erosion in the pipe bend or fitting itself, fittings cause flow disturbances which influence wear rates downstream. Again, once velocities and impact angles can be predicted theoretically, these locations can be determined. However, because of vortices and swirls in these flows, the problem is difficult (Nasr-El-Din & Shook, 1987). In general, fittings that experience large flow fluctuations will have a much shorter life than the pipeline itself.

7. REFERENCES

Adams, W.I. (1986). Polyetheylene pipelines for slurry transportation. In *Proc. STA 11*, pp. 287–92.*

Agarwal, V.K., Mills, D. & Mason, J.S. (1983). A comparison of the erosive wear of steel and rubber bends in pneumatic conveying system pipelines. Paper presented at the 6th International Conference on Erosion by Liquid and Solid Impact, University of Cambridge, Cambridge, UK, 5–8 Sept. paper 60.

Allen, T. (1981). *Particle Size Measurement* (3rd edn). Chapman and Hall, London, UK.

Altaweel, A. (1982). Erosion of materials using recirculated coal–liquid mixtures. Report for the Atlantic Research Laboratory and the National Research Council of Canada, University of Nova Scotia, Halifax, NS, Canada.

ASTM (1989). *Annual Book of ASTM Standards*, Section 3, Vol. 03.03. American Society for Testing and Materials, Philadelphia, PA, USA, pp. 286–93.

Bitter, J.G.A. (1963). A study of erosion phenomena, Parts 1 and 2. *Wear*, **6**, 5–21, 169–90.

De Bree, S.E.M., Rosenbrand, W.F. & de Gee, A.W.J. (1982). On the erosion resistance in water–sand mixtures of steels for application in slurry pipelines. In *Proc. HT8*, pp. 161–80.*

Ferrini, F., Giommi, C. & Ercolani, D. (1982). Corrosion and wear measurements in slurry pipelines. In *Proc. HT8*, pp. 133–44.*

Gillies, R.C. (1989). PhD thesis (in progress). University of Saskatchewan, Saskatoon, Canada.

Gillies, R.C., Shook, C.A. & Wilson, K.C. (1991). An improved two layer model for coarse particle slurry flow. *Can. J. Chem. Engng*, **69**, 173–8.

Henday, G. (1988*a*). Predicting wear in slurry transport systems. Paper 38, presented at Antiwear 1988, London, UK, 20–22 Sept.

Henday, G. (1988*b*). A comparison of commercial pipe materials intended for the hydraulic transport of solids. Report No. 3526, BHRA, The Fluid Engineering Centre, Cranfield, UK.

Hoey, G.R. & Bednar, J.S. (1981). Erosion–corrosion of AISI C1020, Type 304 and Type 440C Alloys and Ni-hard cast iron in coal–water slurries. Report for Energy, Mines and Resources Canada, Ottawa, Ontario, Canada.

Huggett, P.G. & Walker, C.I. (1988). Development of a wear test to simulate slurry erosion. In *Proc. HT 11*, pp. 495–506.*

Jacobs, B.E.A. (1982). The measurement of wear rates in pipelines carrying abrasive slurries. In *Proc. HT 8*, pp. 145–60.*

Johns, F.E. (1988). Investigation of alternative wear materials for handling extremely abrasive slurry. Paper presented at Pipe Coatings and Linings Seminar, Cranfield, UK, 11 July.

Jvarsheishvili, A.G. (1982). Unevenness of pipeline wear in slurry systems. *J. Pipelines*, **3**, 35–41.

Karabelas, A.J. (1978). An experimental study of pipe erosion by turbulent slurry flow. In *Proc. HT 5*, paper E2, pp. E15–24.*

Mills, D., Mason, J.S. & Tong, K.N. (1982). The influence of product concentration on the

erosion of bends in pneumatic conveying lines. In *Proceedings of the American Society of Mechanical Engineers*, ASME, Phoenix, AZ, USA, pp. 89–99.
Nasr-El-Din, H. & Shook, C.A. (1987). Effect of a 90° bend on slurry velocity and concentration distributions. *J. Pipelines*, **6**, 239–52.
Oliveiro, M.C. (1986). Samarco's iron concentrate pipeline. In *Proc. STA 11*, pp. 231–41.*
Owada, S., Nakai, Y. & Ikeda, Y. (1986). Manufacture and property of room temperature curing type of polyurethane lined steel pipe for abrasive slurry transportation. In *Proc. STA 11*, pp. 293–302.*
Postlethwaite, J., Tinkler, E.B. & Hawrylak, M. (1972). Corrosion studies in slurry pipelines. In *Proc. HT 2*, paper G2, pp. G15–24.*
Rabinowicz, E. (1965). *Friction and Wear of Materials*. John Wiley & Sons Inc., New York, USA.
Rao, P.V. & Buckley, D.H. (1984). Solid impingement erosion mechanisms and characterization of erosion resistance of ductile metals. *J. Pipelines*, **4**, 193–205.
Roco, M.C., Nair, P. & Addie, G.R. (1987). Test approach for dense slurry erosion. Special Technical Publication 946, American Society for Testing and Materials, Philadelphia, PA, USA.
Sarkar, A.D. (1980). *Friction and Wear*. Academic Press Inc., New York, USA.
Shimoda, K. & Yukawa, T. (1983). Erosion of pipe bend in pneumatic conveyor. Paper presented at Proc. 6th International Conference on Erosion by Liquid and Solid Impact, University of Cambridge, Cambridge, UK, 5–8 Sept. Paper 59.
Shook, C.A. & Pilling, F.E. (1987). Simulating boundary erosion with slurry Couette flow. *J. Pipelines*, **6**, 83–8.
Shook, C.A., Haas, D.B., Husband, W.H.W. & Small, M. (1981). Relative wear rate determinations for slurry pipelines. *J. Pipelines*, **1**, 273–80.
Shook, C.A., McKibben, M.J. & Small, M. (1990). Experimental investigation of some hydrodynamic factors affecting slurry pipeline wall erosion. *Can. J. Chem. Engng*, **68**, 17–23.
Spencer, D.K. & Sagues, A.A. (1986). Erosion by sand and coal slurries. Paper presented at 3rd Berkeley Conference 'Corrosion–Erosion–Wear of Materials at Elevated Temperatures', Berkeley, California, USA, 19–31 Jan.
Steward, N.R. & Heckroodt, R.O. (1988). Polyurethane as a wear resistant material in slurry handling applications. In *Proc. HT 11*, pp. 523–30.*
Steward, N.R. & Ruther, H. (1989). A direct method to analyze particle degradation. A report for Hydroabrasive Research Department of Surveying, University of Cape Town, Republic of South Africa.
Tarjan, I. & Debreczeni, E. (1972). Theoretical and experimental investigation on the wear of pipeline caused by hydraulic transport. In *Proc. HT 2*, paper G1, pp. G1–14.*
Truscott, G.F. (1975). A literature survey on wear on pipelines. Publication TN 1295, BHRA Fluid Engineering, Cranfield, UK.
Wiedenroth, W. (1984). An experimental study of wear of centrifugal pumps and pipeline components. Paper presented at the Energy Sources Technology Conference, New Orleans, LA, USA, 12–16 Feb.
Willis, D.J. & Truscott, G.F. (1978). Solids handling pumps — a survey of current practice, problems and developments. In *Proc. HT 5*, paper F6, pp. F81–108.*
Wright, G.J. & Garrett, G.G. (1986). Materials evaluation in wear performance trials in pipelines for hydraulic conveying of fused ash. MINTEK 1986, Report M242, Council for Mineral Technology, Roudberg, Republic of South Africa.

*Full details of the Hydrotransport (HT) and the Slurry Transport Association (STA) series of conferences can be found on pp. 11–13 of Chapter 1.

21

Corrosion of Slurry Pipelines

Bryan Poulson
NEI International Research and Development Ltd, Newcastle-upon-Tyne, UK

1. INTRODUCTION

Pipelines must be protected from corrosion both internally and externally. Corrosion can occur when the pipe is being used or when it is idle. The technology of protecting the external surface of a pipe is well developed and usually relies on a coating system plus cathodic protection. During idle periods it is important to prevent localised pitting caused by oxygen concentration cells; various methods have been proposed essentially keeping the inside of the pipe dry or free from oxygen.

Where slurry pipelines are present the most difficult problem is preventing corrosion caused by the slurry. Coating and resistant materials are sometimes used, especially for short pipelines. Protection on long pipelines is usually by inhibition although de-aeration can be very important.

This chapter begins with a brief description of corrosion caused by slurries. This is followed by a discussion on laboratory techniques to measure corrosion rates. Corrosion preventative measures for both inside and outside pipelines are detailed. Finally techniques to monitor corrosion of the pipeline are discussed.

2. CORROSION OF SLURRY PIPELINES

A slurry pipeline is no different from any other pipeline in that its outside surface will corrode if not protected. Similarly, during periods of non-use, corrosion of the inside of the pipe can occur if both air and water get in. Methods of dealing with external corrosion (coatings and cathodic protection) and corrosion during idle periods (limit ingress of air or water) are well developed and are dealt with here in the appropriate section. The emphasis in this section is the corrosion of carbon steel pipelines which is caused by the slurry.

If the slurry is acidic (low pH) and cannot be treated to raise its pH then the corrosion rate will be high (Gandhi *et al.*, 1975) (Fig. 21.1); coatings or alternative materials would need to be used.

For near neutral (pH approx. 7), uninhibited slurries the corrosion rate is critically dependent on the oxygen content of the liquid (Bomberger, 1965; Gandhi *et al.*, 1975) (Fig. 21.2) and the flow velocity (Postlethwaite *et al.*, 1974, 1986) (Fig. 21.3). Various solids produce different corrosion rates (Postlethwaite *et al.*, 1974) but the cause of this is not fully understood. The details of the corrosion mechan-

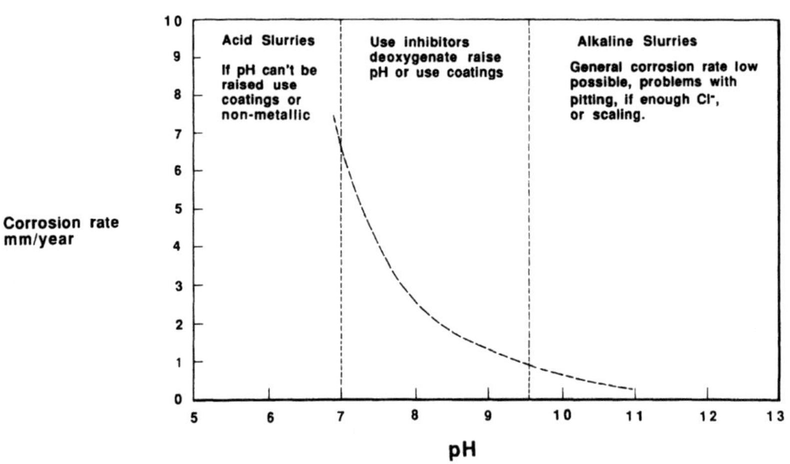

Fig. 21.1. Effect of pH on corrosion rate in an iron concentrate slurry, obtained from laboratory stirred flask experiment. (Adapted from Gandhi *et al.* (1975).)

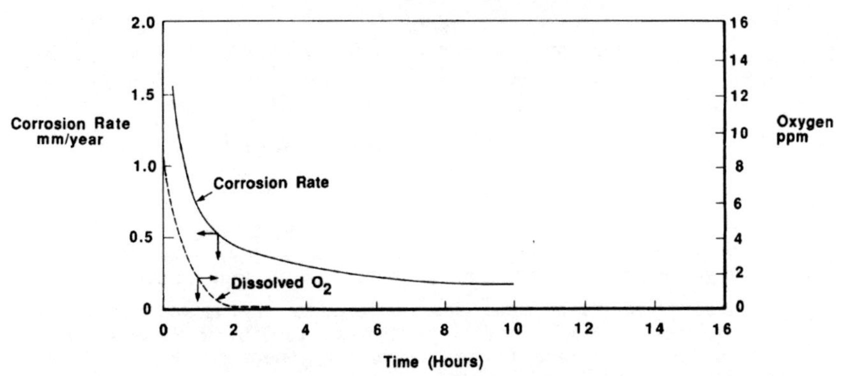

Fig. 21.2. Effect of oxygen content on corrosion rate for an iron concentrate slurry. (Adapted from Gandhi *et al.* (1975).)

ism (Gandhi *et al.*, 1975; Postlethwaite *et al.*, 1986) and the calculation of a corrosion rate (Postlethwaite *et al.*, 1974; Poulson & Chexal, 1991) are detailed in Table 21.1. It is clear that high corrosion rates can occur which can be estimated from a simplified equation:

$$\text{corrosion rate (mm year}^{-1}) = 0{\cdot}625 \times O_2 \text{ content (ppm)} \times \text{velocity (m s}^{-1})$$

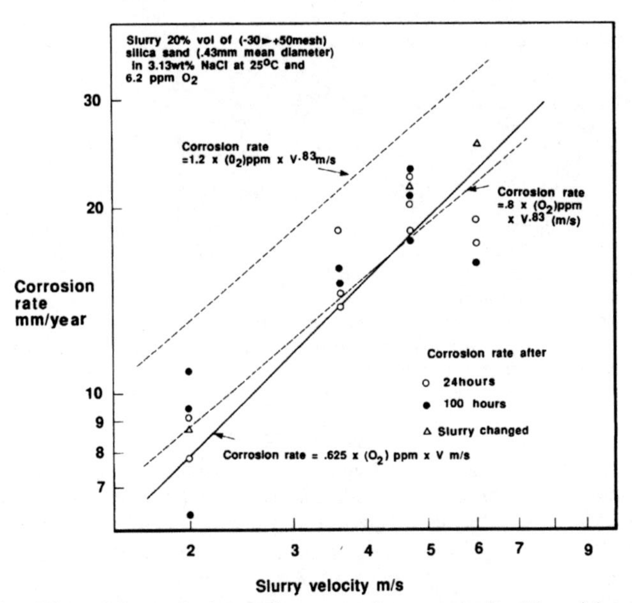

Fig. 21.3. Effect of flow velocity on slurry corrosion rate obtained from laboratory rig test. (Adapted from Postlethwaite *et al.* (1986).)

TABLE 21.1
Mechanism and Prediction of Corrosion Rate Due to Slurries

metal dissolution
$2Fe \rightarrow 2Fe^{++} + 4e^-$

$4e^-$

$2Fe(OH)_2 + O_2 + H_2O$

oxygen reduction

$2Fe(OH)_3$

$O_2 + 2H_2O + 4e^- \rightarrow 4OH^-$

$FeOOH + H_2O$

(1) Rate of corrosion is controlled by the transport of oxygen to the surface.

(2) Without inhibitors slurries can remove any corrosion product which might restrict transport of oxygen to surface; this can then be equated to the product of the local mass transfer coefficient (K) and the oxygen concentration in the liquid (ΔC).

(3) Mass transfer coefficients can be obtained from non-dimensional correlations between the Reynolds number, Sherwood number, and Schmidt numbers. For slurries in pipes an approximate solution is

$$K = 0{\cdot}01 D^{0\,66} \nu^{-0{\cdot}66} V$$

Where D is the diffusion coefficient of oxygen; ν the kinematic viscosity, taken to be that of water, and V the velocity of the slurry.

(4) The corrosion rate can then be obtained by factoring the rate of oxygen transport by appropriate constants to allow for the following.
(a) Anodic and cathodic currents being equal, rates depend on the number of electrons involved in each reaction (n) and the molecular weight of the reactants (M) as well as the density of iron (ρ_{Fe}).
(b) Getting corrosion rate into sensible units.

Then:

$$\text{Corrosion rate (mm/y)} = 0{\cdot}01 D^{0\,66} \times \nu^{0\,66} \times \frac{V}{100} \times \Delta C \times \frac{n_c}{n_a} \times \frac{M_a}{M_c} \times \frac{10}{\rho_{Fe}} \times 3600 \times 24 \times 365$$

$$= 0{\cdot}01 \times (2{\cdot}06 \times 10^{-5})^{0\,66} \times (0{\cdot}01)^{-0{\cdot}66} \times \frac{V}{100} \times \Delta C \times \frac{4}{2} \times \frac{32}{58} \times \frac{10}{7{\cdot}9} \times 3600 \times 24 \times 365$$

$$= 0{\cdot}625 \times O_2 \text{ content (ppm)} \times V\ (\text{m s}^{-1})$$

(5) This will be the maximum possible rate at the start of the pipe. It will be reduced if $Fe(OH)_2 \rightarrow Fe(OH)_3$ occurs at the surface and uses up some of the oxygen. Depletion of the oxygen will also occur due to the following.
(a) Any reactions with the slurry.
(b) The corrosion process, which will cause the rate to decay as $\exp(-4Kl/Vd(1 - C_v))$ where l is the length down the pipe, d the pipe diameter and C_v is the volume fraction of solids in the slurry.

This will decay as the oxygen is used up (Gandhi *et al.*, 1975; Poulson & Chexal, 1991) or reacts with the slurry (Bomberger, 1965). This has applications in that one of the measures to reduce corrosion, discussed later, is de-aeration. In the present context it suggests that the aggressivity of a non-acid slurry is dependent on its oxygen content. Thus, it has been reported (Wallis, 1975) that saturated brine, which has an oxygen solubility of 0·1 ppm (compared with approximately 8 ppm for water) can be used to transport slurries without significant problems.

There are slurries which are sufficiently alkaline, or made so, e.g. by the addition of lime, so that significant protection is afforded by a protective oxide (see Fig. 21.9 later in this chapter) or the formation of a scale. The former situation can lead to pitting if sufficient chlorides are present, while the latter can lead to a scaling problem. Thus the Bougainville copper concentrate line was treated with lime additions to raise the pH to above 9·5 and experienced pitting problems (Coale *et al.*, 1976), while lime overdosing led to calcium carbonate deposits at the start of the Samarco iron concentrate line at pH $>$ 10 (Jennings, 1981). Severe pitting, leading to a continuous channel on the crown of the pipe and penetration, has been described (Hassan *et al.*, 1978) for a limestone slurry pipeline. Corrosion was attributed to differential aeration but exactly how this occurred was not specified.

3. EXPERIMENTAL METHODS OF MEASURING SLURRY CORROSION RATES

There are basically three approaches to carrying out laboratory tests to obtain data relevant to slurry pipelines.

Simple non-representative test geometry. It has not yet proved possible to identify hydrodynamic parameters which would allow the practical simulation of slurry pipes. Equivalence of mass transfer rates would seem the best possibility (Poulson, 1987; Postlethwaite & Lotz, 1988). Various test geometries have been used (Fig. 21.4). Very few of these have been compared with actual pipe data. An exception is the work of Bomberger (1965) who correlated the results of laboratory tests with weight loss measurements on 10–13 kg sections of pipe (Fig. 21.5). Tests such as these can give useful data on inhibitors and other preventative measures; they cannot give quantitative design data unless some empirical correlation has been performed.

Recirculating rig studies. These have been used extensively by Postlethwaite *et al.* (1974, 1986) and others (Heyashi *et al.*, 1984; Corradetti *et al.*, 1990). Their main disadvantage is that the slurry character can become less aggressive because the solids get smaller and more rounded (Wasp *et al.*, 1977; Jacobs & James, 1984). Slurry attrition can be minimised by the choice of pump, the test duration and refreshment of the slurry. The advantage of such tests is the potential to generate quantitative design data.

Once through loop. Rarely used, because of cost, but eliminates the problem of slurry degradation.

In slurries the total metal loss will be through a combination of wear and corrosion. Most conventional ways of measuring corrosion (Table 21.2) would actually measure both wear and corrosion. There are two ways to obtain the corrosion component:

- use electrochemical techniques (see Table 21.2), or

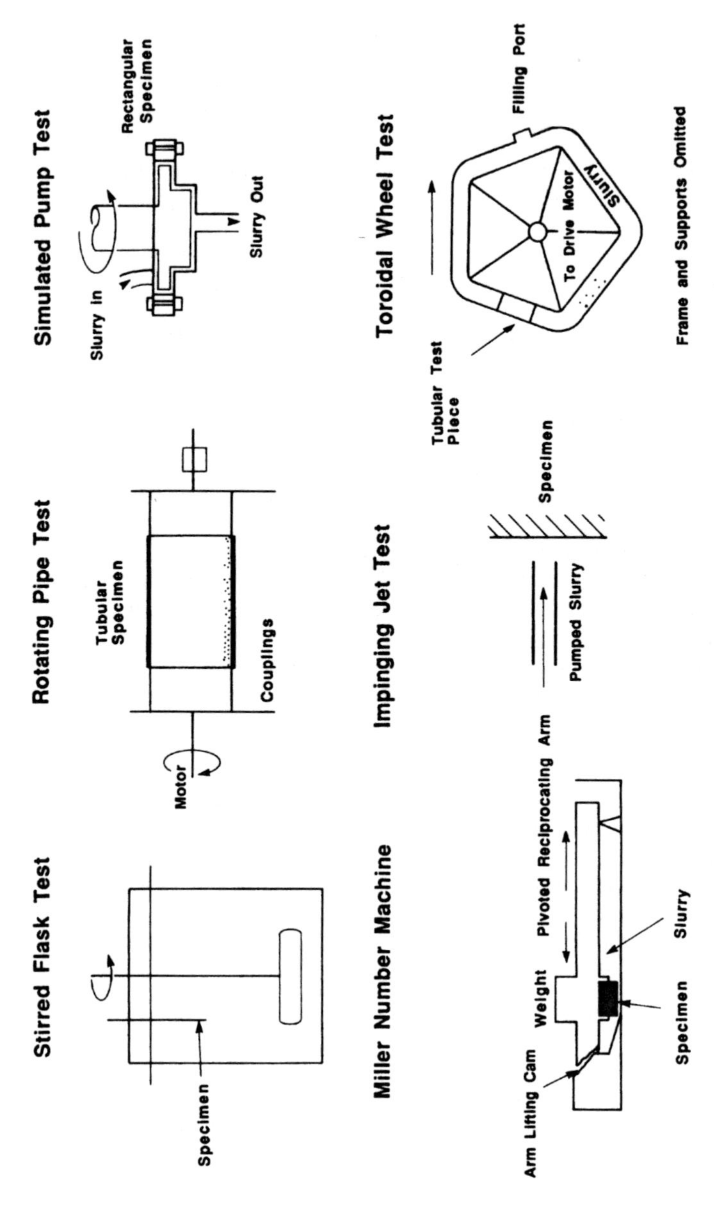

Fig. 21.4. Schematic illustration of some test geometries used to perform corrosion tests with slurries.

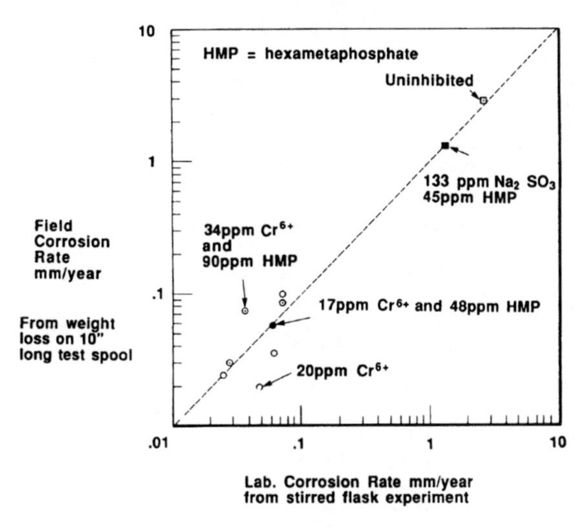

Fig. 21.5. Comparison of laboratory and field corrosion rates in coal slurries. (Data from Bomberger (1965).)

- measure total loss rate, and wear rate when corrosion suppressed (by cathodic protection, deoxygenation or inhibitors) and obtain the corrosion component by subtraction.

4. PREVENTING INTERNAL CORROSION OF THE PIPE

Any corrosion can be prevented at a cost. What is needed is the most economic method of achieving the design life of the pipeline. These economic considerations have been treated by Gandhi *et al.* (1975) and Wasp *et al.* (1977) whose treatment is summarised in Table 21.3. Inhibitors and de-aeration plants are more favourable for long pipelines and coatings for short pipelines.

4.1. Choice of Materials

In general, long pipelines are constructed out of carbon steels. However, it has been shown (Heyashi *et al.*, 1984) that increases in chromium levels reducing corrosion, or hardnesses of the solids, reducing wear, can produce substantial benefits in certain cases (Fig. 21.6).

With short pipelines, the reduced pressure requirements and the limited amount of pipe material required give the possibility of using materials other than carbon steels for very corrosive or erosive slurries. Various materials have been used for short slurry pipelines depending on the conditions. For example, asbestos cement pipes have been used for fly ash slurry discharges (Baker & Jacobs, 1979). Glass-fibre-reinforced pipes have been used with linings of

TABLE 21.2
Techniques for Measuring Wear and Corrosion (Poulson, 1983, 1987)

Technique	*Advantages and disadvantages*
Surface activation	Continuous monitoring of rate with high resolution (0·1 μm). Expensive; position of highest rate must be known if this is to be measured.
Ultrasonics	Potential for continuous monitoring with high precision; 1 μm in ideal cases, possibly 10 μm in other situations. Accurate and reproducible positioning of probe needed and protection of outside surface.
Metrology	High resolution (approx. 1 μm) on suitable geometries but limited to interrupted exposures.
Resistance inserts	Potential for continuous-monitor high-precision but untried in this application and practical problems; probably limited to certain geometries.
Weight loss	Needs suitable test section; simple measurement giving average rate; no indication of pitting unless metallographic examined.

Electrochemical techniques

The basic equation from which the corrosion rate, I_{corr}, can be obtained from the current, I, measured at potential, E, is

$$I = I_{corr} \exp\left[\frac{E - E_{corr}}{0{\cdot}434 b_a}\right] - I_{corr} \exp\left[\frac{E_{corr} - E}{0{\cdot}434 b_c}\right]$$

where b_a and b_c are slopes of anodic and cathodic curves. From this has developed a number of different approaches:

(1) *Tafel extrapolation*. With this method the linear portion of curve is extrapolated from high potentials to E_{corr} to obtain I_{corr}.

(2) *Polarisation resistance*. This method utilises the slope of the E/I curve (R_p) close to E_{corr} (or impedance techniques) in relation of the form

$$I_{corr} = \frac{b_a b_c}{2{\cdot}303(b_a + b_c)} \frac{I}{R_p}$$

Unknowns are obtained empirically, by other techniques, e.g. (1) or by educated guessing.

(3) Curve fitting or Barnartt's three-point technique. This method uses data at both high and low values of E with graphical, computer or unique situations.

(4) De-aeration technique. This was developed by Postlethwaite *et al.* (1974) for situations where corrosion is controlled by transport of oxygen. It measures the current needed in de-aerated solution to keep a potential reached in aerated solution; this is equivalent to corrosion current.

rubber (Eriksson & Sellgren, 1978) or high-purity aluminium oxide ceramic tiles (Olsen, 1975; Dempster, 1980). Cast and spun irons have good abrasion resistance but are not widely used because of brittleness problems. Stainless steel is used in specialised applications such as food processing; it has recently

TABLE 21.3
Money Available for Corrosion Protection (after Wasp *et al.*, 1977)

a is the running cost of a de-aeration plant ($£\,y^{-1}$)
C_v is the volume fraction of solids
CR is the uninhibited corrosion rate ($mm\,y^{-1}$)
CR_i is the corrosion rate after inhibition ($mm\,y^{-1}$)
d is the inside diameter of the pipeline (m)
f is the fraction of time the pipeline is operating
i is the fractional interest rate
L is the length of the pipeline (km)
P_{steel} is the price of steel used in the pipeline ($£\,t^{-1}$)
V is the operating velocity of the pipeline ($m\,s^{-1}$)
Y is the life of the pipeline (y)
ρ_m is the density of the slurry ($kg\,m^{-3}$)
ρ_{steel} is the density of the steel in the pipeline ($kg\,m^{-3}$)

Annual throughput of solids = $V(\pi d^2/4)C_v\rho_m f(60 \times 60 \times 24 \times 365) = V\pi d^2 C_v \rho_m f(7884)$ ($t\,solids\,y^{-1}$)

Cost of steel needed for corrosion allowance = $(\pi d)YP_{steel}(\rho_{steel}/1000)(CR/1000)(1000) = (\pi d)YP_{steel}\rho_{steel}(CR/1000)$ ($£\,km^{-1}$)

Annuity of the savings in capital cost of corrosion allowance for steel = $(\pi d)YP_{steel}\rho_{steel}(CR - CR_i)/1000\,i(1-(1+i)^{-Y})^{-1}$ ($£\,km^{-1}\,y^{-1}$)

Inhibitors
The money available for inhibition, M_i, is therefore the annuity of the savings on steel costs divided by the throughput of solids

$$M_i = \frac{YP_{steel}\rho_{steel}(CR - CR_i)/1000\,i(1-(1+i)^{-Y})^{-1}}{VdC_v\rho_m f(7884)} \quad (£\,t\,solids^{-1}\,km^{-1})$$

Coatings
The money available for coatings, M_c, is given by

$$M_c = \frac{(\pi d)YP_{steel}\rho_{steel}(CR/1000)L}{\pi dL} = YP_{steel}\rho_{steel}(CR/1000) \quad (£\,m^{-2})$$

De-aeration
The money to build a de-aeration plant, M_d, is given by

$$M_d = (\pi d)YP_{steel}\rho_{steel}((CR - CR_i)/1000)L - a(1-(1+i)^{Y})/i \quad (£)$$

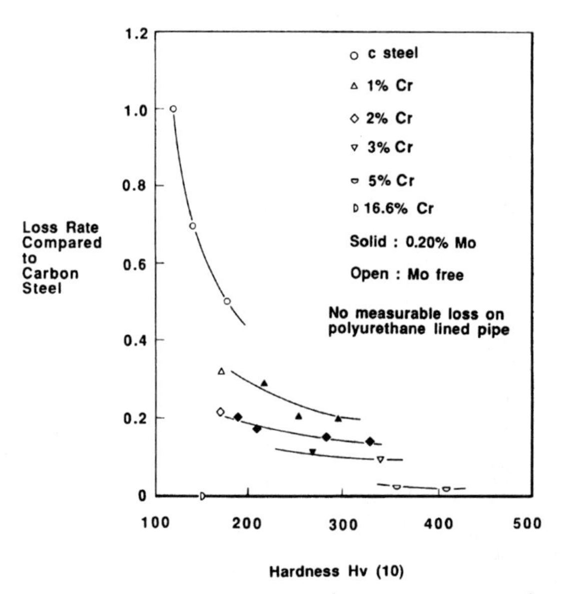

Fig. 21.6. Results of rig tests: 30% Fe_3O_4 (mean size 50 μm) at 2 ms^{-1}. (Adapted from Heyashi *et al.* (1980).)

replaced polyurethane coated pipe for a micaceous residue slurry (Coombs & Palmer, 1984), for pressures above 1 MPa. Various plastic materials have also been used depending on temperature and required mechanical properties. For example, where the pressure requirement was below 1 MPa, high-density polyethylene pipe was used to replace polyurethane coated steel (Coombs & Palmer, 1984). The situation is not one where specific recommendations can be made.

4.2. Internal Coatings

Internal coatings have the advantage that the mechanical strength is supplied by the steel pipe. A variety of materials has been used depending on whether erosion or corrosion was the principal problem; these are detailed in Table 21.4. Cast basalt is widely and successfully used under erosive conditions (Ellis, 1975; Lehrke & Nonnen, 1975). There is some recent evidence (Jacobs & James, 1984; Henday, 1988) (Table 21.4) that basalt is prone to erode in harder slurries and that pure alumina or alumina-based coatings (e.g. 50% Al_2O_3/33% ZrO/16% SiO_2/1% Na_2O) have the best resistance to abrasive wear; but even this is not always resistant enough (Table 21.4). Care must be taken to ensure cast sections are joined satisfactorily and that if tiles are used, joints are staggered to prevent preferential erosion along the joints (Olsen, 1975). Rubber is used as a coating material (Davis, 1981) but appears to have been superseded by polyurethane coatings. There appears to be some variation in the resistance of polyurethanes to erosion (Jacobs & James, 1984;

TABLE 21.4
Comparison of Pipeline Materials (Jacobs & James, 1984)

Material	*Wear rate* ($mm\,y^{-1}$)
Results of laboratory rig tests: 10% emery (mean size 0·15 mm) at 4 m s⁻¹	
Rubber 'a'	0·13
Zirconia/alumina ceramic	0·15
Ni-hard iron	0·19
Polyurethane 'a'	0·20
Polyurethane 'b'	0·22
Rubber 'b'	0·35
Sintered alumina	0·40
Rubber 'c'	0·61
HDPE (high-density polyethylene) 'b'	0·67
HDPE (high-density polyethylene) 'a'	0·87
uPVC (unplasticised polyvinyl chloride)	1·27
316 stainless steel	1·29
Mild steel 'a'	1·57
Polypropylene	1·59
Mild steel 'b'	1·69
ABS (acrylonitrile–butadiene–styrene)	2.52
Results of Horden Colliery pipeline trial: flint aggregate (mean size 22 mm) at 4 m s⁻¹	
Cast alumina ceramic	8.2
Polyurethane 10 mm	12
Polyurethane 25 mm	27
Flame hardened steel	40
Basalt	61
Carbon steel	94
ABS	410

Corradetti *et al.*, 1990) (Table 21.4) but at present it is one of the most attractive options.

4.3. Inhibitors

There have been no inhibitors tested for acid slurries. The position with respect to near-neutral slurries appears to be similar to that for near-neutral waters, with two, possibly important, exceptions. In some slurries, particularly coal, there is a reaction between the inhibitor and the solid which is transported. For example, it has been reported (Bomberger, 1965) that 90% of hexametaphosphate added to a coal slurry disappeared immediately. Hexavalent chromium reacts with coal (Bomberger, 1965) but much more slowly. There is a similar consideration—does the inhibitor affect the solids. Sastri *et al.* (1979) have shown that inhibitors can influence the properties of coal. There is also the potential problem that the slurry could mechanically remove any inhibiting films and interfere with the inhibitor action.

Usually inhibitors for near-neutral solutions are complex mixtures of chemicals, each of which produces a different effect but which synergistically interact. For an inhibitor to influence the corrosion rate in slurry systems it must either

- reduce the transport of oxygen to the surface—this is normally the rate-controlling reaction, or
- change the rate-controlling step from one of oxygen transport to one of metal dissolution.

The first long-distance slurry pipeline was the 250 mm (10 in.) Cadig to Eastlake coal pipeline. Excellent papers by Bomberger (1965) and Swan *et al.* (1963) describe the development and substantiation of an inhibitor regime based on hexavalent chromium (added as sodium dichromate). Initially this was at a 20 ppm Cr^{6+} level which was reduced to 14 ppm Cr^{6+} by the initial reaction with the coal, then slowly along the length of the line until none was detected at about 80 miles (128 km) (Fig. 21.7). Since oxygen was being depleted along the pipe this was found to be satisfactory economically, although higher levels of Cr^{6+} would have reduced the rate further. Significantly, it was found that sodium sulphite (Na_2SO_3) did not appear to work in practice (Fig. 21.5) and that additions of hexametaphosphate were not effective enough (Bomberger, 1965).

In general, chromates can produce environmental problems and are becoming unacceptable; limits for discharge as low as 0·05 ppm have been quoted (Sastri & Malaiyandi, 1983). Techniques to remove chromates have been suggested but the tendency is towards alternative inhibitors. This situation has been reviewed by Jacques and Neil (1977) (see Table 21.5) and more recently by Sastri *et al.* (1979, 1983). Recent experimental work has shown that molybdate (added as Na_2MoO_4) shows good promise for coal slurries if part of a low level multi-component system (Table 21.6) (Sastri *et al.*, 1986).

4.4. De-aeration

In near-neutral slurries, when corrosion is controlled by the transport of oxygen to the surface, it has long been known that removal of oxygen reduces the corrosion rate. This is standard technology in other industries, particularly boilers. However, the solids in the slurry make the operation more difficult to perform. There are basically two techniques to remove oxygen: mechanical methods (steam, vacuum, or inert gas) and chemical methods (chemical additives and resin beds).

The use of additions of sodium sulphite ($Na_2SO_3 + 0{\cdot}5O_2 \rightarrow Na_2SO_4$) did not reduce the corrosion rate of pipes exposed to coal slurries to an acceptable degree (Bomberger, 1965). It was suggested that specific reactions with the coal were the cause of the problem. However, the reaction at ambient temperatures is very slow and it is normal practice to add trace amounts of cobalt as a catalyst; it was not clear if this had been done. The use of sodium sulphite was reported (Wallis, 1975) as successfully reducing the corrosion rate of steel in slurries being transported by brine.

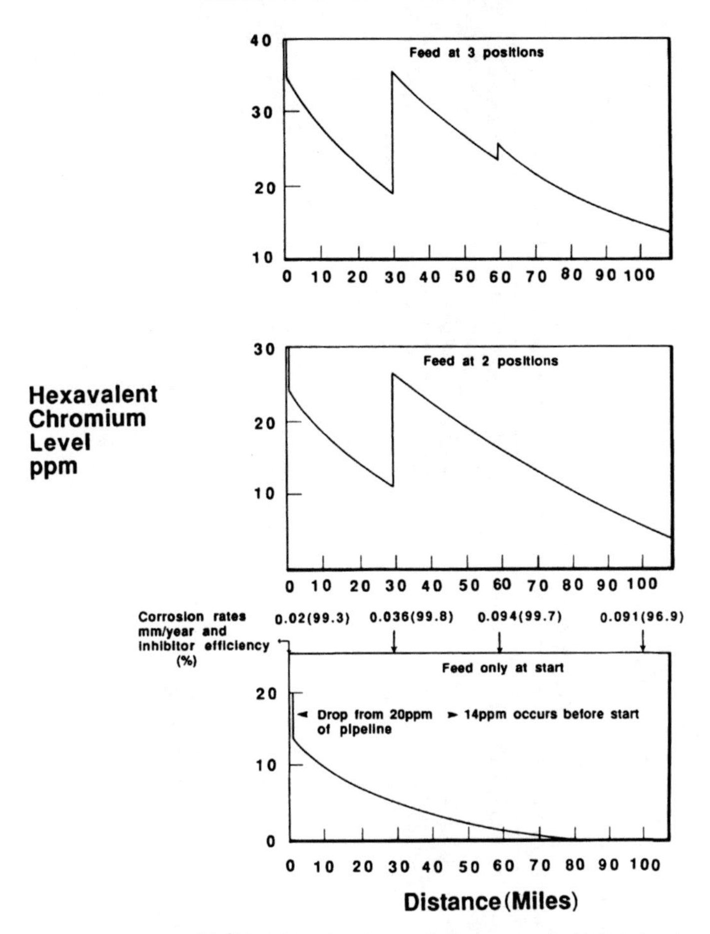

Fig. 21.7. Reduction of Cr^{6+} inhibitor levels at various stages of inhibitor development programme for coal slurry pipeline. (Adapted from Swan *et al.* (1963).)

Mechanical de-aeration has apparently been used on a 64 km pipeline in Turkey for the transport of copper and pyrite concentrate (Baker & Jacobs, 1979) but details were not given of how or its success.

Postlethwaite (1978, 1979, 1987) has long argued the case for using de-aeration to control corrosion, and, with increasing concern about the possible toxicity of inhibitors, de-aeration should receive more attention.

TABLE 21.5
Summary of Corrosion Inhibitor Characteristics in Slurry Pipeline System Application (Jacques & Neil, 1977)

Inhibitor composition	*Corrosion inhibition effectiveness*	*Cost/performance*	*Environmental acceptability*	*Ease of handling and feeding*	*Typical active feed concentrations* ($mg\ litre^{-1}$)	*Typical corrosion rates to mild steel* ($mm\ y^{-1}$)
High chromate	A	D	D	A	500–1 000	0–0.03
Low chromate	C	B	D	A	20–80	0.08–0.13[a]
Chromate zinc	A	A	C	A	2–30	0.03–0.05
Polyphosphate	C	C	B	C	10–75	0.13–0.38
Polyphosphate zinc	B	B	B	C	5–50	0.08–0.25
Organic phosphate	D	F	B	A	5–20	0.13–0.51
Organic phosphate zinc	C	C	B	A	2–15	0.13–0.38
Catalysed sodium sulphate	C	C	B	D	10–18 Variable with O_2	Few data

A—excellent, B—good, C—fair, D—poor, and F—unacceptable.
[a]Potential for local attack.

TABLE 21.6
Effectiveness of Various Non-chromate Inhibitors (Sastri *et al.*, 1986). Steel Samples Rotated on 600 rev min^{-1} Stirrer in 3·45 g litre^{-1} NaCl, 0·45 g litre^{-1} $NaSO_4$ with 15% Coal Slurry 75% of which was between 0·42–1·17 mm

System	*Inhibitor (ppm)*					*Corrosion rate*	*Inhibition (%)*
	HEDP[a]	*MBT*[b]	Na_2MoO_4	$ZnSO_4$	Na_3PO_4	($mm\,y^{-1}$)	
A	—	—	—	—	—	6·0	0
B	—	—	20	—	—	1·9	68
C	—	5	20	—	—	1·0	83
D	25	—	20	—	—	0·6	90
E	—	—	20	—	25	0·4	93
F	—	5	20	5	25	0·3	95

[a]HEDP 1—hydroxyethylidene 1·1 diphosphonic acid.
[b]MBT 2—mercaptobenzothiazole.

5. PROTECTING THE EXTERNAL PIPE SURFACE

5.1. External Pipe Coating

Until the late 1970s/early 1980s, the most common pipe coating was based on a type of coal tar. For example, prior to 1976 this was the only type of coating approved by the American Water Works Association (AWWA) (Harris, 1983) and was the only coating which would pass the British Gas cathodic disbondment test (Wood, 1975).

The position has changed rapidly. By 1983, the AWWA also approved fusion-bonded epoxy (FBE) coatings and polyethylene-tape-based systems (Harris, 1983). The change in British Gas specification for transmission pipes has been even more dramatic. In 1977 it was 100% coal tar, but by 1983 all new distribution pipes were being protected by fusion-bonded epoxy (Gray *et al.*, 1983). Table 21.7 is a summary of some of the coatings commonly used, and an assessment (Coulson & Temple, 1983) based on a number of different tests.

There are, of course, many alternatives and the coating should be selected on the basis of the specific application. Economics should only be used to choose between systems which can effectively fulfil the requirements. Figure 21.8 shows the type of coating properties which can be required from the mill coating operation to being operational.

It must be emphasised that variations in coating composition, method of application, and care during transport and construction can produce variations in quality at least as great as between coating systems. The effect of surface finish on the subsequent behaviour of a fusion-bonded epoxy coating in cathodic disbonding tests demonstrates this well; a grit-blasted surface passing, and a shot-blasted surface failing (Gray *et al.*, 1983). The various stages in shop-coating pipes and field-coating welds, with fusion-bonded epoxy have been well detailed (Gray *et al.*, 1983).

TABLE 21.7
Some Common Coating Systems and Assessment (Coulson and Temple, 1983)

Coating system	*Number of tests performed*	*Number of tests passed*	*Pass rate (%)*
Coal tar enamel	26	10	38·5
Coal tar polyester	19	2	10·5
Extruded polyethylene	26	18	69·2
Fusion bond epoxy	26	23	88·5
Polyethylene tape (PS)	26	5	19·2
Polyethylene tape (BB)	23	4	17·4
Polyethylene tape (HH)	20	5	25
PVC tape	20	3	15
Polyurethane	26	21	80·8
Polyurethane tar	26	19	73·1

Tests included
Chemical resistance in diesel fuel, distilled water, gasoline, methanol, methylethylketone, 10% NaCl, 0·1% NaCl
Cathodic disbondment at 21, 35 and 50°C
Flexibility
Hardness
Impact at −50, −30, −10, 0, 20, 50 and 75°C
Penetration
Soil stressing
Water adsorption
Weathering

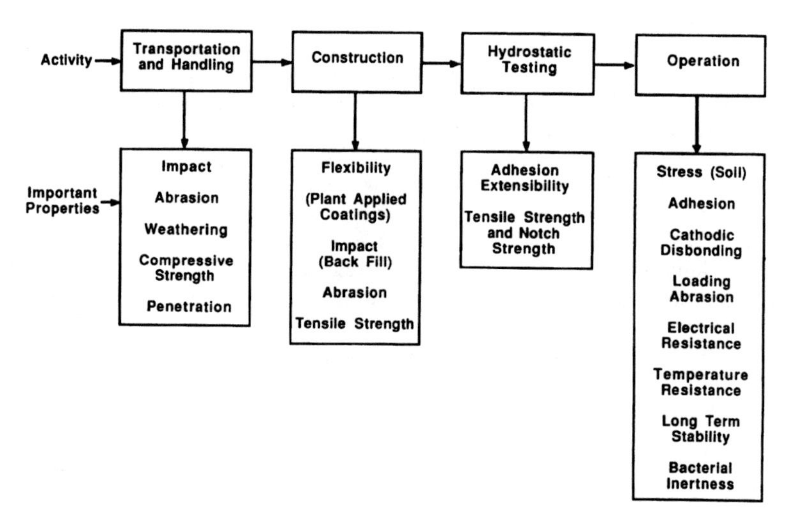

Fig. 21.8. Coating property requirements. (Adapted from Coulson and Temple (1983).)

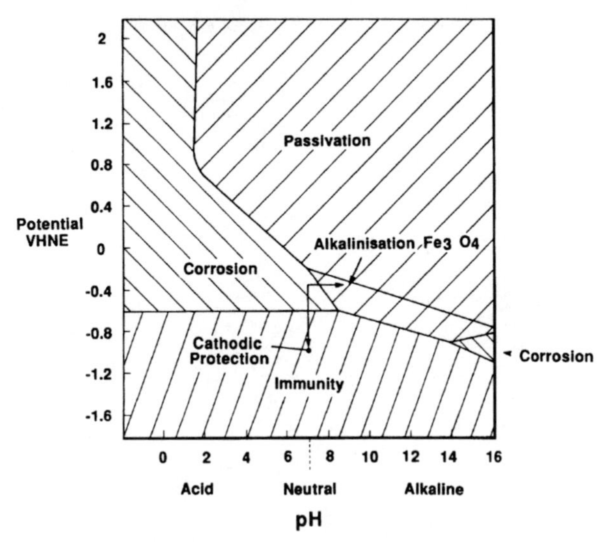

Fig. 21.9. pH-potential diagram for iron role of cathodic protection and alkalinisation.

5.2. Cathodic Protection

Cathodic protection renders a metal thermodynamically stable and immune to corrosion by depressing its potential to a suitable value (Fig. 21.9). To achieve this protection requires a certain cathodic protection current from either a sacrificial anode or from an applied current source (Fig. 21.10). Bare pipelines would require too much current and in nearly all cases cathodic protection is used on coated pipelines if the degree of security required is greater than that obtained by a coating alone.

The usual criterion for satisfactory cathodic protection is when the pipe to soil (or sea) potential is more negative than some agreed value; usually $-850\,\text{mV}$ versus a copper/copper-sulphate reference electrode, which is equivalent to $-800\,\text{mV}$ versus silver/silver chloride or $-900\,\text{mV}$ versus a calomel mercury/mercury chloride reference electrode. In anaerobic conditions, these values are 100 mV more negative and empirical factors such as $-1\,\text{mV}\,(°\text{C})^{-1}$ increase in temperature above 25°C has been suggested for pipes at higher than ambient temperatures.

It is traditional that pipe-to-soil potential measurements are made at some set position above the pipe (Fig. 21.10). Such a reading does not allow for the potential drop which occurs between the measuring position and the pipe (Poulson *et al.*, 1974; Gray *et al.*, 1983). This gives an impression that the cathodic protection is more effective than it really is. Various methods of improving the method of measurement are indicated in Fig. 21.10. Cathodic protection is not without possible problems, including the following.

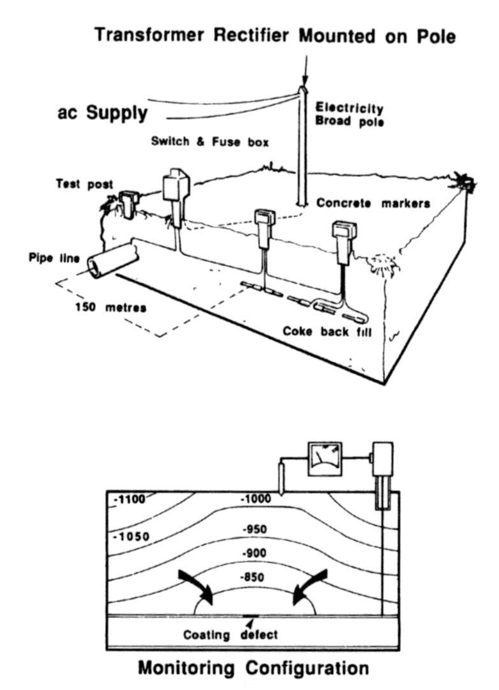

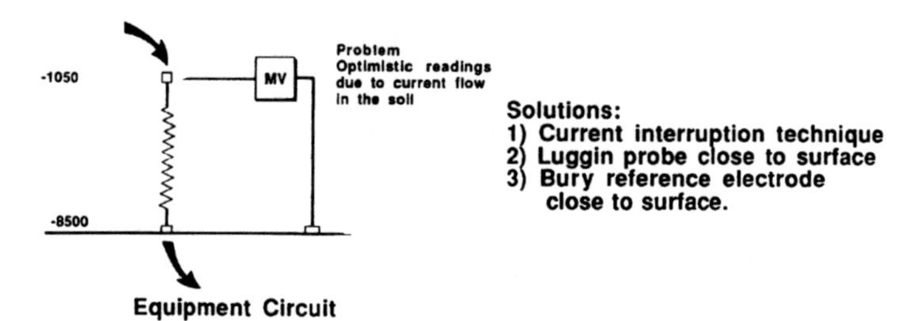

Fig. 21.10. Conventional method for cathodic protection and monitoring. (Adapted from Gray *et al.* (1983).)

- Cathodic disbonding of coating. This is controlled by careful surface preparation of pipe, choice of resistant coating and elimination of over protection.
- Cathodic protection interference, i.e. stray currents from the pipeline causing corrosion of nearby unprotected structure. The incorporation of adjustable resistive bonds, or sacrificial metal shields which are coupled to the unprotected structures can prevent this problem (Morgen, 1959).
- Stress corrosion cracking has been a problem with warm cathodically protected

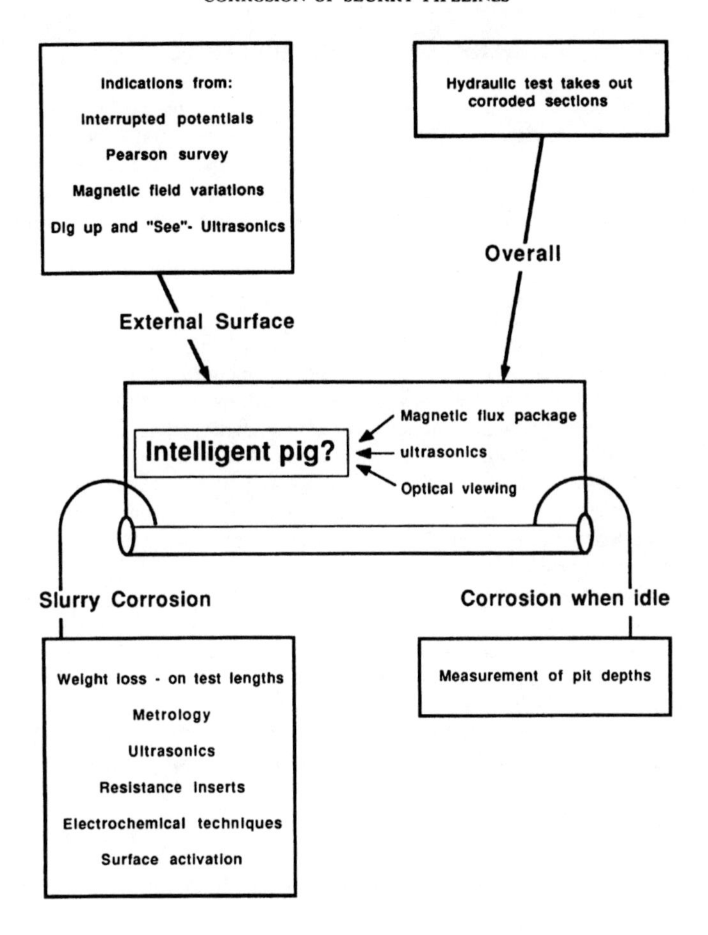

Fig. 21.11. Summary of corrosion monitoring of slurry pipelines.

carbon steel pipelines (Poulson & Arup, 1975). Specifically, failures of natural gas distribution pipes downstream of compressors have been attributed to the formation of carbonate–bicarbonate solutions. Failures of cathodically protected district heating pipes, which were insulated with foam concrete, were caused by a caustic environment. Careful control of the potential, shot-peening, and the use of an improved coating system, possibly containing a leachable inhibitor in the primer, have been suggested remedies.

6. MONITORING CORROSION OF SLURRY PIPELINES

The various methods used to prevent corrosion need to be decided upon and substantiated at the design stage. Nevertheless, it is essential that during operation the effectiveness of these various measures is checked. Figure 21.11 is a summary of the various possibilities available to assess external and internal corrosion rates. It is most important to obtain base line (before operation) data.

7. SUMMARY

Any corrosion problem can be solved at a cost. Slurry pipelines are no exception. External surfaces are protected by a combination of coatings and cathodic protection. Fusion-bonded epoxy coatings appear to offer significant advantages but do need greater care in application, coating field welds and repairs.

The corrosivity of slurries can differ widely. Acid slurries need alternative materials or coated pipes. Polyurethane appears to be an effective internal coating for many applications tending to replace rubber. Cast basalt and ceramic alumina based coatings are other alternatives.

For neutral slurries, inhibitors have been widely used but increasing environmental constraints have caused a revaluation of the situation. Complex molybdate-based inhibitor mixes are a possibility if chromates cannot be removed at the end point. Alternatively, de-aeration, although underdeveloped for slurry pipelines, appears to offer substantial promise.

Any preventative methods need to be assessed under realistic conditions and their effectiveness checked by a well-planned corrosion-monitoring scheme.

8. REFERENCES

Baker, P.J. & Jacobs, B.E.A. (1979). *Slurry Pipeline Systems*, BHRA Fluid Engineering, Cranfield, UK.

Bomberger, D.R. (1965). Hexavalent chromium reduces corrosion in a coal–water slurry pipeline. *Mater. Protect.*, **4**, 43.

Coale, R.P., Thompson, T.L. & Ehrlich, R.P. (1976). Bougainville copper concentrate slurry pumping system. *Trans. Soc. Min. Eng., AIME*, **260**, 289–97.

Coombs, V.T. & Palmer, N.A.H. (1984). A case history of a complex multiroute slurry transport system for kaolin products. In *Proc. HT 9*, paper F3, pp. 265–80.*

Corradetti, C., Ercolani, D., Alo, M. & Maceratesi, S. (1990). Internal corrosion control in coal–water slurry pipelines. In *Proceedings of the 8th International Conference on Internal and External Protection of Pipes*. BHRA, Cranfield, UK, pp. 91–105.

Coulson, K.E.W. & Temple, D.G. (1983). An independent laboratory evaluation of external pipeline coating. In *Proceedings of the 5th International Conference on Internal and External Protection of Pipes*. BHRA, Cranfield, UK, paper A3.

Davis, B. (1981). Natural rubber outlasts steel. *Rubber Dev.*, **34**, 83.

Dempster, D. (1980). The pipe with an armour plate. *Plastic and Rubber Weekly*, **819**, 8.

Ellis, T.J. (1975). Cast basalt—fit and forget. In *Proceedings of the 1st International Conference on Internal and External Protection of Pipes*. BHRA, Cranfield, UK, p. X20–X22.

Eriksson, B. & Sellgren, A. (1978). Development of slurry transportation technology in Sweden. In *Proc. HT 5*, paper J5.*

Gandhi, R.L., Ricks, B.L. & Audi, T.C. (1975). Control of corrosion–erosion in slurry pipelines. In *Proceedings of the 1st International Conference on Internal and External Protection of Pipes*. BHRA, Cranfield, UK, paper G4, pp. 39–52.

Gray, D., Brown, A. & Argent, C.J. (1983). Corrosion control on high pressure gas pipelines. In *Proceedings of the 5th International Conference on Internal and External Protection of Pipes*. BHRA, Cranfield, UK, paper G1.

Harris, G.M. (1983). External protection coatings for steel water pipelines: a review of AWWA Standards. In *Proceedings of the 5th International Conference on Internal and External Protection of Pipes*. BHRA, Cranfield, UK, paper G3.

Hassan, U., Jewsbury, C.E. & Yates, A.P.J. (1978). *Pipe Protection—A Review of Current Practice*. BHRA Fluid Engineering, Cranfield, UK.

Henday, G. (1988). A comparison of commercial pipe materials intended for the hydraulic transport of solids. Research Report 2988, BHRA, The Fluid Engineering Centre, Cranfield, UK.

Heyashi, H., Sampei, T., Oda, S. & Ohtomo, S. (1984). Some experimental studies on iron concentrate slurry transport. In *Proc. HT 9*, paper D2, pp. 149–62.*

Jacobs, B.E.A. & James, J.G. (1984). The wear rates of some abrasion resistant materials. In *Proc. HT 9*, paper G3, pp. 331–44.*

Jacques, R.P. & Neil, W.R. (1977). Corrosion in coal slurry pipelines: causes, cures and costs. In *Proc. STA 2*, pp. 124–35.*

Jennings, M.E. (1981). Samarco's 396 km pipeline—a major step in iron ore transportations. Mining Engng, **33**, 178–82.

Lehrke, W.D. & Nonnen, F.A. (1975). Internal protection of pipes against abrasion and corrosion by Abresist fused cast basalt. In *Proceedings of the 1st International Conference on the Internal and External Protection of Pipes*. BHRA, Cranfield, UK, paper G2, pp. 21–28.

Morgen, J.H. (1959). *Cathodic Protection*. Leonard Hill, London, UK.

Olsen, E. (1975). Wear resistant ceramic lining for piped materials handling. In *Proceedings of the 1st International Conference on the Internal and External Protection of Pipes*. BHRA, Cranfield, UK, paper G5, pp. 53–64.

Postlethwaite, J. (1978). The evaluation of slurry corrosion–corrosion inhibitors in pilot plant test loops. In *Proc. STA 3*, pp. 189–94.*

Postlethwaite, J. (1979). Electrochemical studies of inhibitors in aqueous slurries of sand, iron ore and coal. *Corrosion*, **35**(10), 475–80.

Postlethwaite, J. (1987). The control of erosion–corrosion in slurry pipelines. *Mater. Perf.*, Dec., 41.

Postlethwaite, J. & Lotz, U. (1988). Mass transfer at erosion–corrosion roughened surfaces. Can. J. Chem. Engng, **66**, 75.

Postlethwaite, J., Tinker, E.P. & Hawrylak, M.W. (1974). Erosion–corrosion in slurry pipelines. *Corrosion*, **35**(8), 235–90.

Postlethwaite, J., Dobbin, M.H. & Bergeuin, K. (1986). The role of oxygen mass transfer in the erosion–corrosion of slurry pipelines. *Corrosion*, **42**, 514.

Poulson, B. (1983). Electrochemical measurements in flowing solutions. *Corrosion Sci.*, **23**(4), 391–430.

Poulson, B. (1987). Predicting the occurrence of erosion–corrosion. In *Plant Corrosion: Prediction of Materials Performance*, ed. J.E. Strutt & J.R. Nichols. Ellis Horwood, Chichester, UK, pp. 101–32.

Poulson, B. & Arup, H. (1975). The stress corrosion cracking of cathodically protected hot pipelines. In *Proceedings of the 1st International Conference on Internal and External Protection of Pipes*. BHRA, Cranfield, UK, p. X37.

Poulson, B. & Chexal, B. (1991). Depletion of concentration driving force and the decay of corrosion down a pipe. *Brit. Corrosion J.*, (in press).

Poulson, B., Henriksen, L.C. & Arup, H. (1974). Cathodic protection and stress corrosion cracking of mild steel district heating distribution systems. *Brit. Corrosion J.*, **9**(2), 91–5.

Sastri, V.S. & Malaiyandi, M. (1983). Spectral studies of some novel oxygen scavengers and their use in corrosion control of coal slurry pipelines. *Can. Met. Quart.*, **22**, 241–5.
Sastri, V.S., Hoey, G.R. & Whalley, B.J.P. (1979). Review of the effects of corrosion inhibitors in coal–water slurry pipelines. *Can. Met. Quart.*, **18**, 435–40.
Sastri, V.S., Beaupric, R. & Desgagne, M. (1986). Molybdate as a pipeline corrosion inhibitor for coal-water slurry systems. *Mater. Perf.*, June, 45.
Swan, D., Bomberger, D.R. & Barthauer, G.L. (1963). Corrosion control achieved on coal slurry pipeline. *Mater. Protect.*, **2**, 26.
Wallis, R.M. (1975). *Proceedings of the 1st International Conference on Internal and External Protection of Pipes.* BHRA, Cranfield, UK. (Discussion on Paper A2, p. X51.)
Wasp, E.J., Kenny, J.P. & Gandhi, R.L. (1977). *Solid–Liquid Flow Slurry Pipe Line Transportation.* Trans Tech Publications, Clausthal-Zellerfeld, Germany.
Wood, L.J. (1975). *Proceedings of the 1st International Conference on Internal and External Protection of Pipes.* BHRA, Cranfield, UK. (Discussion on paper D3, p. X72.)

*Full details of the Hydrotransport (HT) and the Slurry Transport Association (STA) series of conferences can be found on pp. 11–13 of Chapter 1.

22

Valves for Slurry Pipeline Service

David A. Brookes
BP Engineering, London, UK

1. GENERAL PRINCIPLES

1.1. Avoid Using Valves

The first rule on the application of valves in slurry service is to avoid using them wherever possible. Any slurry valve has the potential for leakage, blockage and

TABLE 22.1
Published Operational Experience with Valves on Slurry Duty

Valve type	Duty	Reference
Knife gate	FGD systems[a]	O'Keefe (1984) O'Keefe (1986)
Diaphragm	FGD systems, manganese dioxide slurries	Sigrist & Gehriger (1987)
Pinch	Gold slimes	Sabbagha (1982)
	Gold slimes, fly ash, copper slurry	Sauermann & Webber (1980)
	FGD systems	Schneider (1982) O'Keefe (1984)
Plug	FGD systems	O'Keefe (1984)
	Coal slurries (fine)	O'Keefe (1986)
	Coal slurries—letdown	O'Keefe (1981)
	Iron ore slurry	Jennings (1981)
Ball	Coal slurries—letdown	Patton (1981)
Butterfly	FGD systems	O'Keefe (1984) O'Keefe (1986) Subramanian (1982)

[a]FGD—flue gas desulphurisation.

excessive maintenance. In comparison to valves in clean service, which are used with confidence, for a slurry service the need for each valve should be rigorously reviewed at the design stage when flowsheets and basic layouts are available. In some cases it may be preferable to have temporary shutdowns to isolate equipment rather than have installations with multiple valves used only for very occasional isolation. For example, siting block and bypass valves around a control valve where it is better to use a single isolation valve upstream of a control valve even if this means temporary shutdowns for maintenance. This is particularly true on low pressure, non-toxic, non-corrosive services where spillage on dismantling can be tolerated.

1.2. Fundamentals of Valve Selection

Valve selection is a function of the service and the price the user is willing to pay. Many in-plant slurry systems with economical, low-pressure piping, conveying non-toxic and non-corrosive slurries do not merit expensive valves. The particular service has a significant effect on the types of valves selected; for less arduous duties (low temperature/pressure) diaphragm and pinch valves are typical for block and throttling. The more arduous the duty the more specialised the required valves become. Although a particular vendor may service only a limited industrial sector, a considerable wealth of application and operational experience resides with valve manufacturers and suppliers. Table 22.1 lists published operational experience with valves on slurry service. This table is not intended as a primary source for selection, rather as a source of published information.

In general, valves which require the flow to make changes of direction are unsuitable for slurry service. Ease of maintenance must be considered in applications where wear is of concern. Broadly speaking, the market is split between valves originally designed and developed for in-plant slurry systems with relatively low pressures (below 1 MPa), and high-pressure long-distance pipeline applications. Since the mid-1970s, designs for high-pressure, long-distance product pipeline applications have been required. These are usually developments from the petrochemical industry adapted from clean service to handle the solids content. Frequently, these later types of valves are full-bore (straight through) with no restrictions enabling the pipeline to be pigged.

With regard to slurry particle size and valve selection, no definite rules can be formulated other than the obvious one that large particles in slurries will become jammed if the valve has any narrow restriction which is not cleared in the valve fully open position. Throttling valves will generally not work on coarse (+ 10 mm) particle slurries. Fine particles can also jam narrow clearances increasing operating torque.

It is important to recognise the fundamental difference between isolation (block) and throttling (control) valves in slurry service. Most slurry valves are strictly two position isolation valves—either fully open or fully closed. Any attempt to effect a partial closure throttling action will result in premature failure through increased wear.

1.3. Slurry Characteristics and Material Selection

The mechanisms of erosive wear and corrosion have been detailed in Chapters 20 and 21. This section will give only brief additional notes on material selection specific for valves.

In general, the design of the valve and the method of effecting isolation and sealing governs whether metal, elastomers, or combinations of these materials can be used. Tight tolerances for precision sealing systems cannot usually be achieved with elastomers and machined metal components or ceramics must be used. Hence, close tolerance valves, such as globe valves, cannot use elastomers except for general body internal lining. As a general rule, elastomers and polymers with good wear characteristics have a service temperature limitation of 120°C. Hence, many hot-slurry petrochemical plant duties, e.g. fluid catalytic cracking, must use all metal valves. Some particularly sharp, hard particle slurries can severely *cut* softer elastomers although modern polyurethanes are more resilient. Other than these limits the best wear characteristics are obtained with elastomer linings such as soft natural rubber, synthetic rubbers formulated for greater corrosion resistance and polyurethanes. Polyurethane in particular can be formulated with a wide variety of characteristics and is available under various trade names; it is more expensive than rubber but has excellent wear characteristics.

Where steel and alloy components are necessary, special measures can be taken to improve wear characteristics of critical wear areas with harder alloys (e.g. tungsten carbide, Stellite and Colomnoy), deposition of hard alloys (by weld or spray techniques) or ceramic inserts. Deposition techniques are becoming more sophisticated with very robust but expensive materials.

Ceramic components have been used with some success in particularly abrasive conditions, e.g. ceramic balls and seats in ball valves. Their application should be carefully reviewed as some slurries, e.g. large particle impacts, can damage ceramics and they are sensitive to thermal shock caused by rapid temperature change.

One of the most interesting developments in wear resistance is *metal composites*. These are the subject of much current worldwide research aimed at exotic alloy applications, e.g. aerospace. Metal composite compounds with excellent wear characteristics will soon be available for critical components in valves.

2. ISOLATION VALVES

2.1. Preliminary Considerations

Isolation valves can often be eliminated if the consequences of no valves can be tolerated.

- Can the system be shutdown to effect temporary isolation (swinging a blank flange)?
- Is it essential to drain and flush before shutdown?
- Can the carry over of slurry be tolerated when a system pump is shutdown to stop flow?
- Is a rapid stoppage to flow essential or can the system pump be shutdown and allowed to run down?
- Can a tank always be emptied completely to avoid pump suction valves?

In slurry service, an isolation valve that permits seepage flow of fluid can result in a potential pipeline blockage caused by the build-up of a high concentration plug of solids upstream of the closed valve.

2.2. Knife Gate Valves

2.2.1. Design and Duty Considerations

Along with diaphragm valves the knife gate valve is the simplest and cheapest design of isolation valve available. The designs are based on the fundamental principle of a gate valve but avoid any unnecessary areas where solids can become trapped. This has led to the development of a slim valve using a thin blade, or gate, passing through a large stuffing box. The thin blade *cuts through* the slurry and lands on a seat which is flush or protrudes slightly into the valve bore. Seats can either be metal–metal or incorporate a resilient elastomer seal. To reduce abrasive and/or corrosive wear a wide variety of alloys and elastomer protected designs are now available. These may use solid stainless steel, stainless steel or polyurethane

Fig. 22.1. Pneumatically actuated knife gate valve. (Courtesy of DeZurik, UK.)

lined components in appropriate combinations. The latter design is particularly good for abrasive applications. Deflector cones can also be fitted to direct solids away from the seats. A valve suitable for slurry service is illustrated in Fig. 22.1.

Designs are essentially low pressure, up to 1 MPa, in sizes to 610 mm (24 in.), although simpler constructions up to 1·83 m (72 in.) are available. Larger sizes use a fabricated body construction, smaller sizes are usually cast. A wide range of bolted connections and fittings is available to exploit the slim design similar to a butterfly valve.

Special designs are available with the gate operating in a membrane to eliminate stuffing box leakage. These designs are limited to low-pressure duties.

2.2.2. Advantages

The slim face-to-face dimensions make knife valves ideal for fitting very close to tanks or pump sumps on suction lines to keep dead legs to an absolute minimum.

Fig. 22.2. Hydraulically actuated parallel/conduit gate valve. (Courtesy of Putzmeister, UK.)

The designs are usually best in uni-directional flow applications where the line pressure forces the blade against the seat. However, guide designs have been developed for omni-directional applications.

2.2.3. Limitations

The principle disadvantage of knife gates is the very large stuffing box which can be prone to leakage. It is also easy to damage or distort the blade if pressure is applied in the wrong direction.

2.3. Parallel/Conduit Gate Valves

2.3.1. Design and Duty Considerations

In slurry service this type of valve has traditionally been associated with relatively low pressure, non-arduous service such as powder or fine slurry tank isolating valves. The aperture in the gate could become jammed or damaged by any large, hard particles. However, developments in concrete pumping have required much more rugged valves for longer distance multiple pipeline applications in the construction industry—an example of such a valve is shown in Fig. 22.2. The valve operates by cutting the slurry as shown in Fig. 22.3.

Valves now available are capable of handling line pressures over 15 MPa in special applications in sizes up to 250 mm (10 in.). The blade slides are specially hardened alloys up to 100 mm thick operating in a massive fabricated body. The seats on the body are usually hard faced using weld deposit techniques; some designs have incorporated pressure energised seat rings.

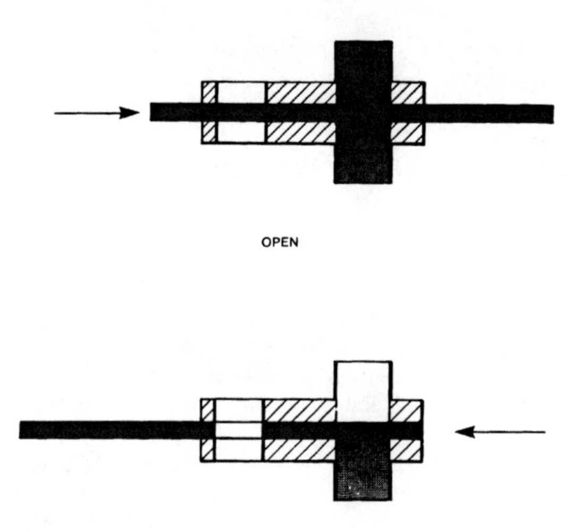

Fig. 22.3. Flow isolation is achieved using a parallel/conduit gate valve. (Courtesy of Putzmeister, UK.)

2.3.2. Advantages

All designs feature very high capacity and fast actuation by hydraulic rams which enable hard particles to be sheared on closure. The valves can shear 10-mm diameter steel bolts with ease. In pumped concrete service with a wide particle size distribution the valves have a very slight but acceptable leakage rate as the leakage paths become blocked by the slurry matrix. Fabrication techniques enable a wide range of customised designs to be available.

2.3.3. Limitations

On narrowly sized, fine particle slurries the leakage rate could be unacceptable, particularly on change over. Similarly, on clear-water duties at the full line pressure, leakage experienced with concrete type valves is unacceptable and prohibits use for isolation when pipelines are hydraulically tested.

2.4. Swinging Transfer Tube Valves

2.4.1. Design and Duty Considerations

This type of valve is a special version of a concrete pump valve known as a *Delta* or *Rock* valve. The valve literally swings a short cranked tube between two closely adjacent pipelines. Like the concrete parallel slide valves, these valves are of very rugged construction and powered by fast-acting hydraulic rams. The seats and sealing systems are similarly hard faced and pressure energised to take up any wear.

The valves are useful for special high-pressure (15 MPa) applications where the advantages of a single, quick actuated, moving tube replacing several conventional isolating valves can justify the relatively high cost.

2.4.2. Advantages

There are also no *dead legs* to block with slurry or harden—very important for concrete pumping.

2.5. Rotating Disc Valves

2.5.1. Design and Duty Considerations

This valve concept originates from highly erosive steam boiler blow-down duties. The valves share some of the features of valves developed for the concrete pumping industry. They utilise a metal–metal seated disc which slides across the face of the pipeline aperture. An example of this type of valve is shown in Fig. 22.4. The disc is both spring and pressure energised to force the metal–metal lapped seats together in a very close tolerance seal with potentially very low leakage rates. Each actuation causes the valve disc to rotate on its seat. This has the dual effect of shearing and wiping away surplus slurry while grinding in or lapping the seal face.

The valve is suitable for the most abrasive, high temperature (900°C) and high pressure (ANSI Class 2500; approx. 40 MPa) applications up to 610 mm (24 in.).

2.5.2. Limitations

High cost, but this could be justified for difficult applications. The cavities in the body probably make it unsuitable for very large particle slurries or for those which harden or set.

2.6. Diaphragm Valves

These valves are the workhorses of low pressure, low temperature abrasive slurry systems. A simple diaphragm membrane is pushed across the bore to effect a seal. There are two basic patterns.

2.6.1. Weir Pattern

The weir pattern, illustrated in Fig. 22.5, effects a tight shut-off with a short diaphragm movement (as shown in Fig. 22.6) resulting in long diaphragm life and low maintenance; good for throttling.

2.6.2. Straight-Through

The straight-through design is shown in Fig. 22.7. This pattern is better for use with high solids content, coarser particles, high viscosity and more abrasive slurries. The method of sealing, shown in Fig. 22.8, requires longer diaphragm movement which tends to result in a shorter diaphragm life. Sizes up to 305 mm (12 in.) are available. The application of both patterns of valve are limited to about 1 MPa maximum (lower in larger valves) with temperatures and products limited by the diaphragm and optional body linings (typically 100°C maximum).

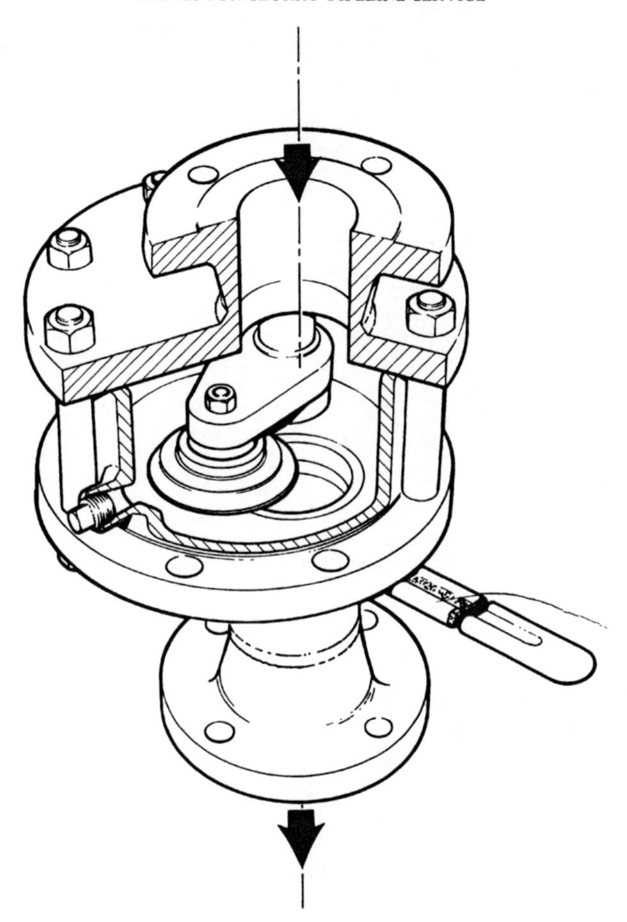

Fig. 22.4. Manually actuated rotating disc valve. (Courtesy of Everlasting Valve Co., UK.)

2.7. Pinch Valves

2.7.1. Design and Duty Considerations

In their simplest form these are essentially an elastomer tube with an external clamp to *pinch-off* the flow. On smaller sizes the elastomer tube is often contained within a steel body, adding support for the actuation and pinching mechanism, containment in the event of failure of the elastomer and facilitating line connection. An example of this design of valve is shown in Fig. 22.9. Figure 22.10 illustrates the method of flow control achieved using a double-acting mechanism as the tube

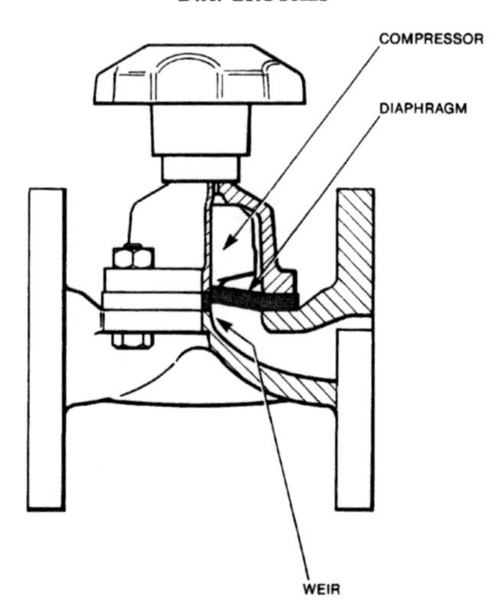

Fig. 22.5. Weir pattern diaphragm valve, flanged body style. (Courtesy of Saunders Valve Co. Ltd, UK.)

is *pinched off*; fully open valves are full-bore with no obstructions and very low pressure drops. This type of valve is attractive for larger size applications, for example, pulp and mineral slurry tailings pipelines.

Size, pressure and temperature ratings are limited by the elastomer and tube construction. Designs are available up to 914 mm (36 in) and pressures up to 1·5 MPa, but specials up to 7 MPa are claimed. Life is limited by the cycling and stretching of the elastomer tube, claims of four- to 10-times the life of special material ball valves are often made.

Designs are available where external pressure is applied to the sleeve to *neck* the tube (see iris valves in section 3.4).

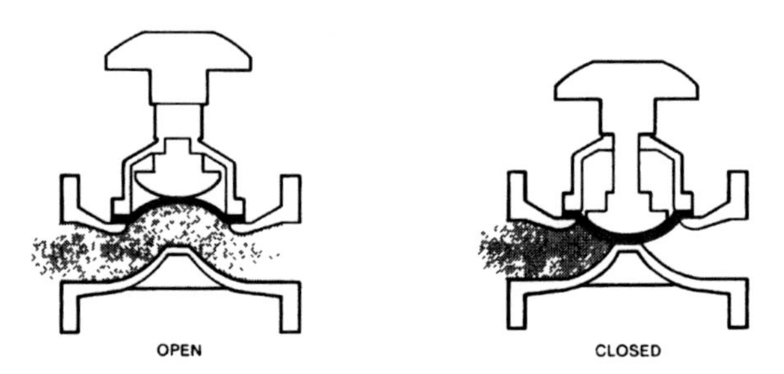

Fig. 22.6. Flow isolation is achieved in the weir pattern by *mushroom-shaped* compressor deforming the diaphragm. (Courtesy of Saunders Valve Co. Ltd, UK.)

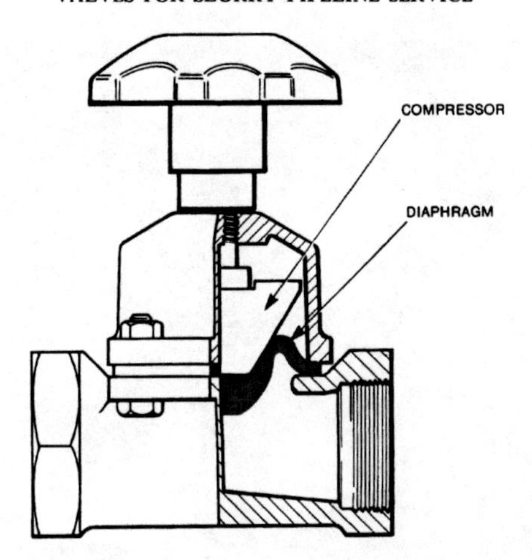

Fig. 22.7. Straight-through diaphragm valve, female screwed body design. (Courtesy of Saunders Valve Co. Ltd, UK.)

2.7.2. Advantage

For the larger, higher pressure applications a variety of power actuators are available, and, with a controller, these valves are good control valves.

2.7.3. Limitation

The valves tend to be expensive.

2.8. Plug Valves

This is one of the simplest forms of valve with origins in early hydraulic systems and consist of a tapered cylindrical plug with a bore that is rotated through 90° in a body. Its use in its simplest forms cannot be recommended for slurry service. However, innovative developments (such as lining the body cavity with PTFE)

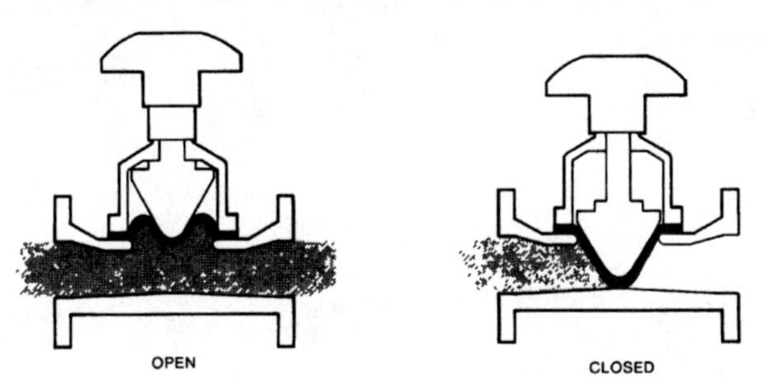

Fig. 22.8. Flow isolation is achieved in the straight-through pattern by *conical* compressor deforming the diaphragm. (Courtesy of Saunders Valve Co. Ltd, UK.)

Fig. 22.9. Manually operated pinch valve. (Courtesy of Red Valve Co. Inc, USA.)

have widened its application considerably and it is no longer the poor relation of the ball valve.

2.8.1. Tapered Plug

This design is available with external lubrication. However, they do tend to jam even on viscous clean fluids. Normal applications are for ANSI Class 300 (approx.

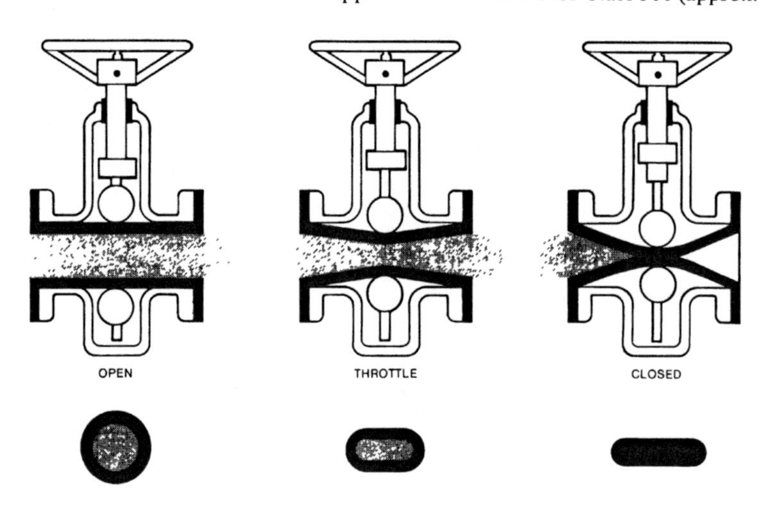

Fig. 22.10. Flow control achieved by *pinching* down on the elastomer tube. (Courtesy of Red Valve Co. Inc., USA.)

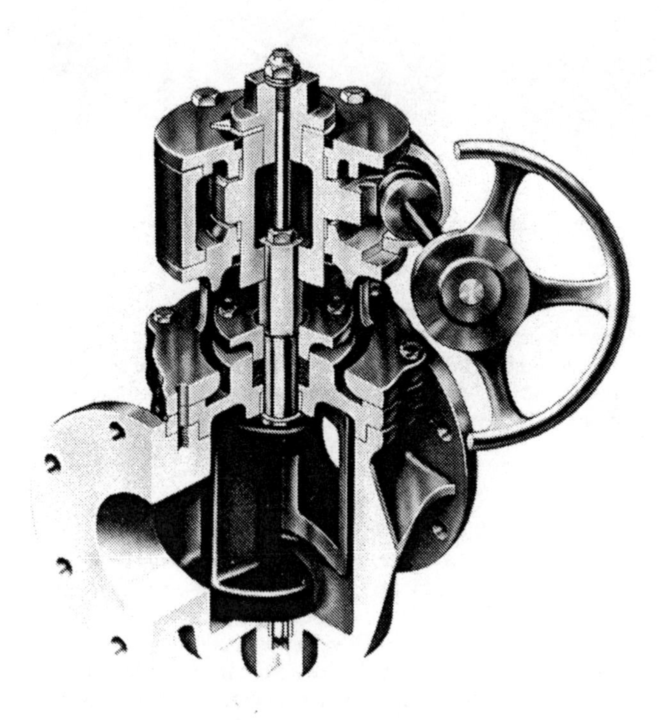

Fig. 22.11. Eccentric plug valve. (Courtesy of DeZurik, UK.)

5 MPa) in sizes up to 457 mm (18 in). Special high-pressure hard-faced designs were used on early long-distance coal and mineral slurry pipelines.

2.8.2. Lifting Tapered Plug

A special adaptation of a plug valve features a special cam action. This lifts the plug before it is rotated through 90° and lowers it after rotation. Metal–metal seals are used which automatically compensate for wear and thermal growth. Main applications are in high temperature (500°C), dirty or catalyst service in oil refining, e.g. petroleum coke, fluid catalytic cracking. Sizes up to 508 mm (20 in.) and pressures up to ANSI Class 900 (approx. 15 MPa). They are expensive.

2.8.3. Eccentric Plug

Another special version of the plug valve is the elastomer-lined eccentric plug, illustrated in Fig. 22.11. The method of sealing is illustrated in Fig. 22.12; an eccentrically mounted half-plug is rotating through 90° to seal against a single

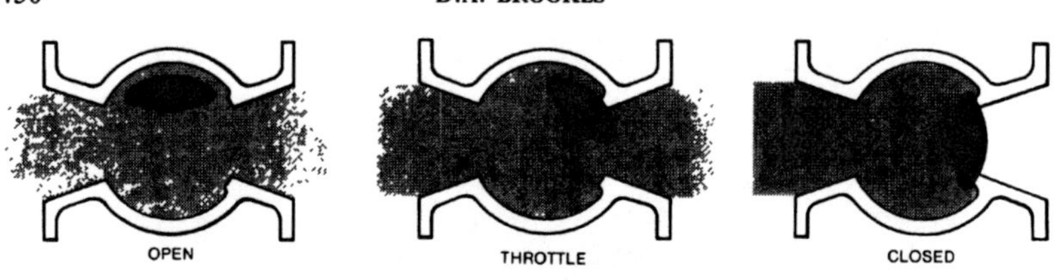

Fig. 22.12. Method of flow isolation using the eccentric plug valve. (Courtesy of DeZurik, UK.)

upstream seal. The whole of the inside of the valve body and plug are lined, usually with abrasion-resistant, soft rubber.

Maximum pressures are modest at 1·0–1·5 MPa but they are available in larger sizes up to 1·22 m (48 in.). They are much better than conventional plug valves. Limited application for flow control is also claimed.

2.9. Ball Valves

2.9.1. Design and Duty Considerations

Ball valves are not generally recommended for slurry service unless there are particular features of the application or special abrasion-resistant materials are used. Wide particle size distribution, relatively soft particle slurries (e.g. coal) in high-pressure service can be used in ball valves with conventional elastomer seals backed up by scraper rings. The valve ball should be trunnion-mounted (in preference to being supported by the seals), with external flushing to the top and bottom bearings and possibly to the body cavity. The matrix formed by a widely sized slurry will rapidly seal leaks. For the more abrasive arduous slurries all metal–metal seals, or, exceptionally, ceramic ball and seals, are available. Most designs use pressure energised sealing ring systems which, with appropriate design, can be bi-directional or occasionally double block and bleed. Encapsulation of the ball into the body of the valve can be an effective measure to prevent particles becoming jammed in dead spaces on closure.

2.9.2. Advantage

In high-pressure (ANSI Class 900, approx. 15 MPa) service, such as pipelines, they offer a light compact, piggable design available in all sizes.

2.10. Butterfly Valves

2.10.1. Design and Duty Considerations

The application of butterfly valves is usually limited to moderate pressure (1·0–1·5 MPa), dirty-water service with relatively soft particles in the fluid. They are very slim and are useful in similar situations to a knife gate. Metal disc with resilient seal versions are used for coal washery service and on fly ash. They are very economical and available in sizes to 610 mm (24 in.). They are not full bore or piggable.

3. THROTTLING VALVES

3.1. Introduction

The very action of throttling or restricting flow induces higher velocities with associated turbulence and erosive wear. With large particle slurries the action of reducing the pipe bore could lead to bridging, followed rapidly by pipe blockage. Effective control valves in slurry service are rare and are generally restricted to lower concentrations of finer particles. Control valves usually cause problems and their use should be avoided on coarser, high concentration slurries.

Control valves on slurry service will probably require regular maintenance. Duplication (with a stand-by manual, bypass valve) could be considered but the risk of blockage in the stagnant bypass line must be weighed against the disadvantages of plant shutdown.

Although initial cost may be higher, alternative means of flow control should be considered such as variable speed pumps using thyristor controls, multiple ratio V-belts or even fitting different size pump impellers. Careful system designs for simple tank transfer duties can often introduce an element of self-correction if flat pump performance curves are used with flat pipeline system head curves dominated by static head.

3.2. Quarter-Turn Rotary Ball Segment Valves

3.2.1. Design and Duty Considerations

There are now several of these types of valves; an example is illustrated in Fig. 22.13. They are developed from the ball valve principal and operate in a similar manner to the eccentric plug valve. A mushroom-shaped element rotates about an eccentric pivot point to form a single, continuous upstream seal in the closed position, retracting into the body cavity when open. The seat ring and pressure energising systems are simpler versions of the more complex arrangements used on ball valves. The bodies are steel or various corrosion resistant alloys, the ball segment or plugs are frequently hard faced using, for example, Stellite.

Sizes are up to 305 mm (12 in.) and pressure rating up to ANSI Class 600 (approx. 10 MPa). The size range can be limited for larger slurry pipeline application, e.g. tailings pipelines.

3.2.2. Advantages

The valves are compact and have good control characteristics. Suitable for high-temperature service.

3.3. Globe Valves

3.3.1. Design and Duty Considerations

Special high-pressure versions of this type of control valve have been developed for very arduous, high temperature and pressure applications on fine slurries, e.g.

Fig. 22.13. Quarter-turn rotary segment valve (one flange removed for clarity). (Courtesy of DeZurik, UK.)

Synfuel plants. The valves feature very durable, hard, wear-resistant alloys sometimes with ceramic inserts. Size is limited to 305 mm (12 in.) but pressure ratings go up to ANSI Class 2500 (approx. 40 MPa).

3.3.2. Advantage
They have excellent control characteristics on fine slurries.

3.3.3. Limitation
Like most special valves they are expensive.

3.4. Pinch Valves

3.4.1. Design and Duty Considerations
Pinch valves with controllers can be used to control flow providing the pressure drop across the elastomer tube is not excessive and the tube is not subjected to excessive flow induced vibrations. For larger sizes (above 457 mm, 18 in.), these are about the only type of control valve available. This design of valve has been used with considerable success as a secondary means of flow control, on a wide range of slurries in the Saskatchewan Research Council flow loops (see Chapter 10).

Iris valves are a type of pinch valve which use an external air or hydraulic oil pressure source to contract or *neck* the elastomer tube. Conceptually, this looks an ideal arrangement but, in practice, pressure control systems are required to reference the actuation pressure to the line pressure. Also pressure differences in excess of 25% of line pressure lead to valve instability.

3.5. Diaphragm Valves

These types of valves are sometimes used for control valve applications with suitable actuators on less arduous duties.

4. NON-RETURN VALVES

Non-return, or check, valves are rarely used on the more arduous in-plant duties due to the risk of blockage and premature failure (often with disastrous consequences). Systems should be designed to cater for run-back. Nearly all rotodynamic pumps can handle temporary reverse flow. There are few commercially available, slurry non-return valves for pipelines.

For fine-particle slurries at high pressures the best solution is to use multiple arrangements of the mushroom type, spring loaded non-return valves found in reciprocating pumps. Few commercial designs are available and they are expensive; the maximum particle size is limited to approximately 2 mm. Modified conventional non-return valves (e.g. swing checks or rubber flap valves) have been successfully used with low concentration, soft particle slurries. Low-pressure designs are available for slurries using membranes, often of a conical shape, that close against a metallic counter-face.

For coarse-particle slurries the only effective method is some arrangement of fast actuated slurry isolation valve to prevent reverse flow.

5. REFERENCES

Jennings, M.E. (1981). Samarco's 396 km pipeline—a major step in iron ore transportation. Mining Engng, **33**, 178–82.

O'Keefe, W. (1981). Coal conversion is a slurry valve challenge. *Power*, May, **125**(5), 91.

O'Keefe, W. (1984). How valve manufacturers and users solve tomorrow's problems. *Power*, Mar., **128**(3), 61–4.

O'Keefe, W. (1986). Valves for corrosives and slurries. *Power*, May, **130**(5), S1–S16.

Patton, A. (1981). Valve design for slurry systems. In *Proc. STA 6*. 184–8.*

Sabbagha, C.M. (1982). Practical experiences in pumping slurries at ERGO. In *Proc. HT 8*, paper A1, pp. 1–16.*

Sauermann, H.B. & Webber, C.E. (1980). High pressure pinch valves in hydraulic transport pipelines. In *Proc. HT 7*, paper A4, pp. 33–40.*

Schneider, L. (1982). Pinch valves fight clogging, leaking, and wear in FGD systems. *Power*, Dec., **126**(12), 49–52.

Sigrist, R. & Gehriger, C. (1987). Protection against abrasion. Plastic membrane valves with two parts contacted by the medium give longer service life. *Maschinemarkt*, 10 Mar., **93**(11), 26–81.

Subramanian, N. (1982). Applications of the butterfly valve—present concepts and trends. Chem. Age India, **33**(4), 222.

*Full details of the Hydrotransport (HT) and Slurry Transport Association (STA) series of conferences can be found on pp. 11–13 of Chapter 1.

23

Commercial Pipeline Instrumentation Techniques

Nigel P. Brown

The British Petroleum Company plc, London, UK

&

Nigel I. Heywood

Warren Spring Laboratory, Stevenage, UK

1. INTRODUCTION

This chapter reviews measuring techniques that are used in commercial pipeline instrumentation for both inventory evaluation and quality control. Emphasis is concentrated on the sensing element rather than on the transmitter and the signal processing. Commercial recommendations have been avoided; the suitability of a technique has been addressed through the identification of specific advantages and limitations. Instrumentation that can either be readily fabricated from standard components or that which is of a research nature is considered in Chapters 10 and 11. Cost comparisons are not considered since standard sizes are highly dependent upon optional features, materials of construction and pipe diameter—larger sizes are often one-offs, their cost being unrepresentative of scaling from smaller instruments.

2. GENERAL CONSIDERATIONS

2.1. General Selection Criteria

- Non-intrusive sensors reduce the potential for wear to occur and the opportunity for blockage of the process line.
- Non-restrictive sensors should be considered where head loss is of importance.
- Durability of instrumentation; slurry applications frequently involve harsh environments.
- Performance history on similar duties.
- Possibility of the long-term build-up of deposits or wear in the sensing element which will effect performance.

2.2. Installation Considerations

- Manufacturers usually calibrate large equipment with water; calibration under process conditions is desirable and essential for unfamiliar applications.
- Ability to bypass instrumentation.
- Orientation of the sensor; consideration of the presence of entrained air; flow regime; and erosive processes.
- The effect of downstream flow disturbance.

3. DENSITY

3.1. Introduction

Pipeline instrumentation measures the in-situ concentration of the solids (sometimes referred to as the *transport* concentration) which is likely to differ substantially from the delivered (or *bucket*) concentration in settling slurries containing large particles. The distinction between the two concentrations is discussed in Chapter 11.

3.2. Gravimetric

Operating principle. Continuous measurement of density is achieved by weighing an isolating section of pipework. The measuring element is either straight or U-shaped.

3.2.1. U-Shaped Measuring Element

System description. An instrument based on a U-shaped element is shown in Fig. 23.1. The measuring element is isolated from the process pipework by flexible connectors and free to pivot in the vertical plane. The mass of the measuring element is maintained in balance by weights and a force balance system. A change in density results in a change in the force necessary to keep the system in balance; the compensating force is converted to an output signal.

Advantages

- Direct measurement technique.
- Equipment is often designed for close control of relatively constant conditions.

Limitations

- Equipment tends to be relatively bulky.
- Maximum tube bore is around 50 mm.
- Small bore instruments tend to be sensitive to vibrations.
- Flexible connections pose a limit to the operating pressure, maximum operating pressure is typically 0·5 MPa.
- Additional bends are required for installation in straight sections of pipe work.
- Abrasive slurries and high flow rates could cause unacceptable wear on bends.

3.2.2. Straight Measuring Element

System description. In order to measure accurately the mass of a straight section of pipework a long length is generally required to overcome the stiffness of flexible couplings used for isolation. This has been overcome in the instrument shown in Fig. 23.2 by isolating the process pipe work using hydrostatic bearings.

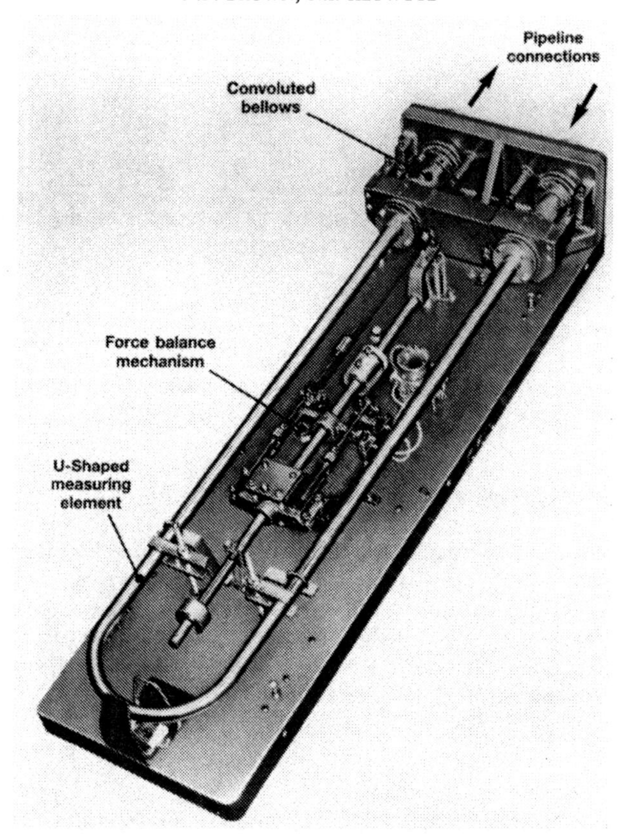

Fig. 23.1. Gravimetric density gauge based on a U-shaped measuring element with a 25 mm bore. (Courtesy of KDG Flowmeters, UK.)

These are supplied with carried fluid at a pressure slightly higher than the pipeline. Some of the sealing fluid necessarily flows into the pipe, the remainder is recirculated.

Advantages

- Direct measurement technique.
- Hydrostatic bearings provide compact design.
- Suitable for large diameter pipelines.

Limitations

- Hydrostatic bearings require auxiliary mechanical equipment to supply seal fluid; inevitable leakage will dilute slurry.

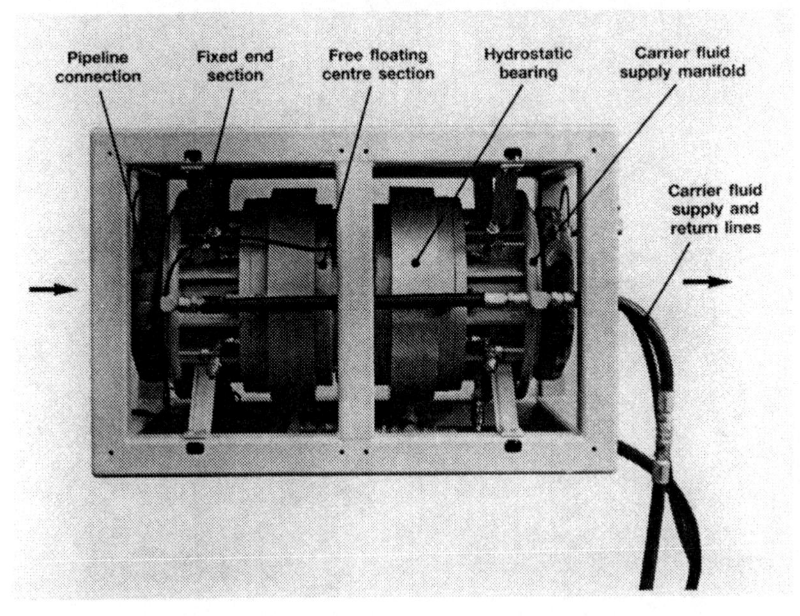

Fig. 23.2. Gravimetric density gauge based on a straight measuring element for a 300 mm bore pipe. (Courtesy of Riede Systems Inc, USA.)

- Instruments using flexible couplings are bulky; requirement of high operating pressure results in stiff couplings and poor resolution.

3.3. Radiometric Absorption

Operating principle. Gamma-photons (-rays), emitted from the decay of radioactive elements, are attenuated as they pass through matter. The attenuation is an exponential function of the path length through the medium and its density. An analytic relation (based on the Beer–Lambert law) enables the intensity of a beam of collimated gamma-photons transmitted through a pipe containing slurry to be related to the density of the slurry (see, for example, Brown, 1988). The most commonly used isotopes are 241-americium; 137-cesium and 60-cobalt. Radioactive decay is not a steady process and for this reason measurements are not instantaneous. In practice, a compromise has to be made between the error that is acceptable in the measurement and the frequency at which the concentration measurement is updated. The error and the time over which the measurement is made are reduced by increasing the activity size of the radioactive source.

System description. The radioactive source is securely contained in a shielded housing fitted with a lockable shutter designed to ensure that the device does not

Fig. 23.3. Radiometric density meter installed in a spool-piece of pipe of nominal bore 150 mm. (Courtesy of Krohne Measurement and Control Ltd, UK.)

contravene ionising radiation regulations. The source holder is usually designed to emit a narrowly collimated, fan-shaped beam to cover a large cross-section of the pipe. Gamma-photons transmitted across the pipe are detected by a material that scintillates which is optically coupled to a photo-multiplier tube or solid-state detector. Instruments are constructed either on a spool-piece of pipe or to be clamped-on to existing pipe work. An instrument constructed on a spool-piece is shown in Fig. 23.3; in the figure the detector is on the left, the source holder is diametrically opposite. If a path length through the medium is required to be longer than the diameter of the pipe in which the equipment is to be installed, special arrangements can be made that include, making measurements at an angle of 30 or 45° to the axis, or along the axis of an S-shaped spool-piece. The signal from the detector requires complex processing in order for a density, or concentration, to be indicated. This is increasingly carried out by a microprocessor.

Advantages

- Applicable to a wide range of slurries.
- Selection of the appropriate radionuclide enables large diameter pipes to be monitored.
- Commonly used density measuring instrument.

Limitations

- Accurate calibration is essential. Calibration usually involves a two-point determination with fluids of known density in the measuring section. This should preferably be carried out with the process slurry. In addition, clamp-on devices require calibration at the installed location.
- Attenuation of gamma-photons is not only dependent on the density of the slurry but also on its chemical composition. Re-calibration is necessary if the nature of the slurry changes.
- Air in the slurry has a pronounced effect on the measured density owing to its very low attenuating properties. Orientation of the beam at the 2 o'clock–8 o'clock position is likely to reduce such problems.
- Wear of the pipe wall affects calibration.
- Detection equipment is inherently sensitive to ambient temperature—equipment is available with sophisticated compensation.
- Radioactive isotopes are characterised by their half-life, a measure of the reduction in activity of the source with time. Automatic compensation is available; this is particularly important where the device is to be used over many years, or where isotopes with relatively short half-lives are employed (such as 60-cobalt).

3.4. Resonant Frequency

Operating principle. A section of pipe through which slurry is flowing is excited into vibration by an electromagnetic drive. The tube resonates at its natural frequency, dependent on the mass of the vibrating system. The vibrations are detected and maintained using a feedback signal. A change in density of the flowing mixture alters the mass of the vibrating system and hence its natural frequency. The shift in frequency is measured and converted to a density change.

System description. Instruments using this technique are available with flow tubes that tend to be either U-shaped or straight. Mass flowmeters based on the Coriolis principle usually have the facility to measure the process stream density.

Advantages

- Fast and continuous response.
- Typical maximum operating pressures in the order 15 MPa.
- Insensitive to external vibration.
- Accuracy typically better than 0·2% of the range used.

Limitations

- Maximum internal diameter of the flow element is of the order 25 mm.
- Application on large flows requires installation in a bypass loop. With slurries containing larger particles the attendant problems of sampling should be a consideration.

3.5. Venturi Meter

The venturi meter can be used to measure the density of the fluid provided that the flow rate can be measured using another instrument. This technique is considered here under its usual role of flowmetering.

4. VOLUMETRIC FLOW RATE

4.1. Electromagnetic Flowmeter

Operating principle. Electromagnetic flowmeters are based on the principle of Faraday's law of electromagnetic induction. A uniform magnetic field is generated across the pipe by a pair of electromagnets positioned on the outside of the flow element. Provided that the fluid flowing through the magnetic field is electrically conducting, a voltage is induced into the fluid in a direction at right angles to the magnetic field. The voltage signal is picked up by a pair of electrodes mounted opposite each other in the wall of the flow tube. The arrangement of the electromagnets and pick-up electrodes is shown in Fig. 23.4. The electrodes are usually in direct conductive contact with the fluid. However, indirect capacitive coupling can be used to extend the operating range of magnetic flowmeters to media with low conductivity (Hofmann, 1978).

System description. The electromagnets, electrodes and, increasingly, electronic circuitry are mounted on the flow tube. The flow tube is typically one to two pipe diameters in length. For smaller diameter lines the flow element is commonly a *wafer*, or *sandwich* design for installation between flanges (illustrated in Fig. 23.4); for larger pipe sizes flanged connections (shown in Fig. 23.5) are more common. The voltage signal picked up by the electrodes is conditioned and converted to a flow rate. Microprocessor-based signal conditioning is becoming increasingly more common.

Operational experience. In applications where there is a marked difference in velocity between the two phases (such as would be expected with coarse dense solids and water) there is debate as to what velocity is measured (see, for example, Wyatt, 1986). However, it is generally accepted that with such slurries electromagnetic flowmeters do indeed measure the average velocity of the mixture, rather than that of the fluid alone. Domnick *et al.* (1987) report results from an interesting study that assesses the size of errors that may be expected. Heywood and Mehta (1988) summarise published experience with a wide range of slurries.

Advantages

- Widely used.
- Measurement is independent of fluid rheology, density, temperature and pressure.

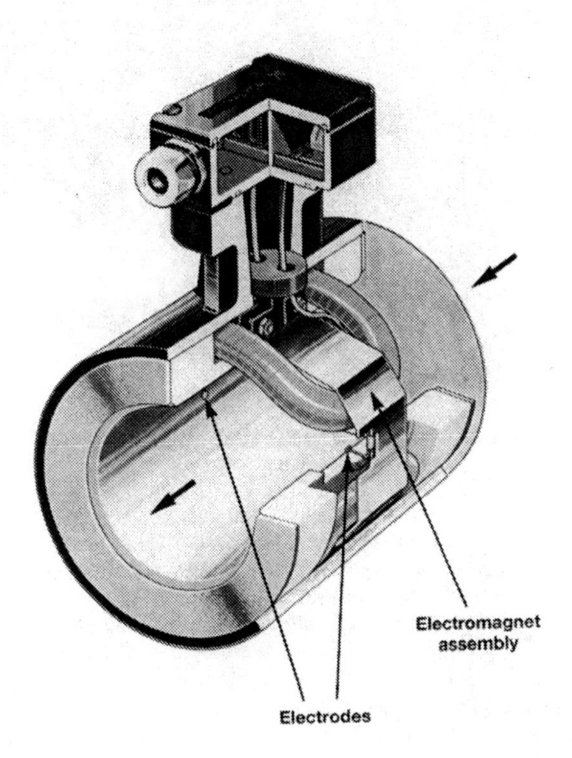

Fig. 23.4. Cut-away illustration of a wafer-type magnetic flowmeter with a fused aluminium oxide flow tube. (Courtesy of Krohne Measurement and Control Ltd, UK.)

- Increasingly compact design.
- Flow tubes available in standard inside diameters from 3–3000 mm.
- Accuracy usually claimed to be much better than 1% of the range of the instrument for flow rates in excess of 10% of the range.

Limitations

- A large error can occur in the indicated flow rate for flows below 10% of the range of the instrument.
- Even in clear fluids, swirling flows experienced downstream from bends have been observed to be responsible for systematic errors. The preferred vertical orientation may not always be ideal.

Fig. 23.5. Electromagnetic flowmeters with flanged connections. (Courtesy of Krohne Measurement and Control Ltd, UK.)

- The fluid must be electrically conducting, a minimum conductivity of $2\,\mu\text{S}\,\text{mm}^{-1}$ is usually specified; this is more than adequate for water-based slurries. Special flowmeters with capacitive signal pick-up are available that can be used with slurries with conductivities as low as $0{\cdot}005\,\mu\text{S}\,\text{mm}^{-1}$.
- Slurries that are capable of laying down insulating deposits on the walls of pipes tend to coat exposed electrodes; various techniques for electrode cleaning are available. The electrodes used with capacitive signal pick-up are located behind the measuring tube.
- Magnetic fields radiated from equipment such as large electric motors and transformers can adversely affect performance.
- Wear of exposed electrodes can affect long-term operation.
- Entrained air adversely affects measurements if the flow tube is installed horizontally and such that the electrodes are positioned at the top and bottom of the pipe.

4.2. Venturi Meter

Operating principle. Flowmeters that work on the venturi principle introduce a restriction in the flow area into the pipeline. The increase in the kinetic energy of the slurry in the throat section is achieved at the expense of pressure energy. The decrease in frictional pressure of the medium as it accelerates into the throat is proportional to the density of the medium and the square of the volumetric flow rate. Venturi meters can be used to measure flow rate if the density of the process medium is known or the density if the flow rate is known. The operation of venturi meters is usually characterised by a discharge coefficient—the ratio of the actual discharge through the device to that theoretically predicted. The discharge coefficient is close to unity for carefully designed and manufactured equipment operating on Newtonian fluids. The usual equation for clear fluids is used with the density term being that of the mixture (see Chapter 11).

System description. The venturi meter consists of an approach section with the same internal diameter as the process line, a conical nozzle section followed by a short parallel throat and a conical diverging section that returns the internal diameter to that of the process line. The pressure of the process medium is measured in the approach section, before the nozzle and in the throat. Venturi meters are usually supplied with flanged connections.

Operational experience. Experimental and theoretical studies reported by Shook and Masliyah (1974) and Shook (1982), indicate that venturi meters can be reliably used, without calibration using the slurry, for coarse-particle slurry flows that are not stratified. In other flow regimes the dependence of the discharge coefficient on pipeline conditions must be investigated. The results of Russian work with venturi meters (Kao & Kazanskij, 1979) with pipe sizes up to 300 mm indicate that venturi meters have been used successfully with sand–water and water–coal slurries where the top-size of the coal was 6 mm, although calibration with the slurry was required.

Advantages

- Reasonably priced.
- At velocities above those at which a stationary layer of solids is present in the pipe, the effects of heterogeneous flow are of secondary importance (Shook & Masliyah, 1974).
- The approach length for venturi meters mounted vertically is not of paramount importance (Shook & Masliyah, 1974).
- Wear resistance of the flow tube can be achieved by using a thermally deposited hardened lining material.
- Although the throat causes a flow restriction good pressure recovery is possible.

Limitations

- Calibration with the process medium should be considered for slurry applications in which the difference in velocities between the phases is large (large particle size and density difference between the solids and the liquid).
- Horizontal use with stratified flows requires knowledge of the quantity of suspended solids (Shook, 1982).

4.3. Ultrasonic

4.3.1. Introduction

Ultrasonic flowmeters utilise acoustic energy above the limits of human audibility generated by piezo-electric devices. The advent of relatively cheap microprocessors has seen the introduction of highly sophisticated instruments that use complex signal manipulation, such as correlation, to obtain a flow rate. Most units that are commercially available operate on one of two basic principles: either *frequency shift* or *time of flight*. Frequency shift, or Doppler flowmeters, require the acoustic signal to be reflected back from material in the flow while time of flight devices transmit the signal through the slurry. Flowmeters based on these principles are available in various physical configurations.

4.3.2. Operational Experience

Penetration of the ultrasonic beam in the process media is dependent on frequency and requires careful consideration for slurry applications. The frequency used by manufacturers varies greatly and is commonly between 16 kHz and 1 MHz. With single-phase liquids the flow rate is obtained from a knowledge of the Reynolds number that indicates the shape of the velocity profile. Especially with settling slurries, the possibility of a complex velocity profile introduces uncertainty into indicated values of flow rate.

A detailed systematic study into the response of ultrasonic flowmeters to water-based slurries composed of sand and polystyrene particles is reported by Colwell *et al.*, (1988). Heywood and Mehta (1988) summarise experience with Doppler flowmeters from other sources. From the experience that is reported it is clear that ultrasonic techniques are not suited to broad ranges of slurry flows. However, reasonable indication of flow rate can be obtained for certain flows.

4.3.3. Frequency Shift Flowmeters

Operating principle. The phenomenon that relative motion between a source of wave energy and a stationary receiver results in a shift in frequency was discovered by Christian Doppler. The effect can be observed as a shift in frequency as a moving sound source (such as a siren) first moves toward and then away from, a stationary observer. Flowmeters based on this principle are commonly referred to as *Doppler* flowmeters.

Two piezo-electric crystals are usually employed, one acting as a transmitter and one as a receiver. The transmitter sends a continuous signal into the process medium. The signal is scattered by discrete particles in the media. A proportion of the scattered signal is detected by the receiver. The difference in frequency between the outgoing and reflected signals is proportional to the velocity of the particle from which the outgoing signal was reflected.

System description. The transmitting and receiving crystals are usually encapsulated into a single transducer shown schematically in Fig. 23.6. The transducer

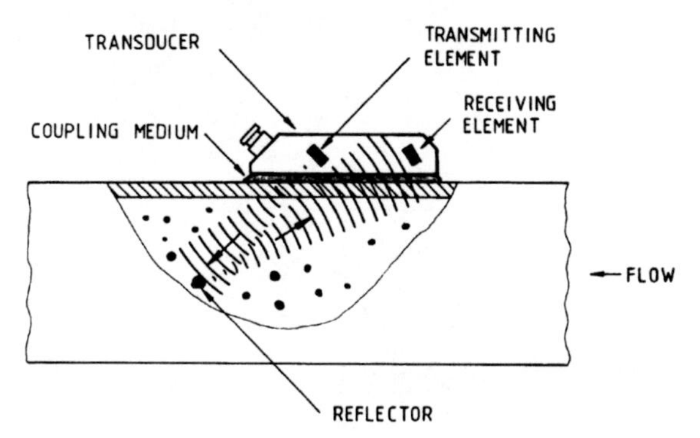

Fig. 23.6. Schematic representation of an ultrasonic flowmeter using the Doppler principle.

can be permanently bonded to the pipe using epoxy adhesive; temporary acoustic coupling can be achieved using grease or water-based gel. In either case, it is advisable to secure the transducer with straps. Units for permanent installation are commercially available in which the transducer is inserted through the wall of the pipe.

Advantage

- Portable and useful for trouble-shooting although misleading results can be obtained.

Limitations

- Absolute calibration requires the speed of sound in the slurry to be known.
- The depth of penetration of the signal will be dependent upon the acoustic attenuation of the coupling medium, the pipe wall material, debris adhering to both internal and external walls, and the nature of the slurry.
- The transducer should be mounted away from the top of the pipe to prevent misinterpretation from air bubbles.

4.3.4. Time of Flight Flowmeters

Operating principle. Time of flight flowmeters use the principle that the velocity of the acoustic signal is increased when the direction of propagation is with the flow and decreased when it is against. The time difference between the two signals is used to calculate a velocity of the flow. A large number of techniques utilise the transit time difference. The two most commonly encountered techniques based on this principle of operation are often referred to as *leading edge* and *sing-around.* The leading edge technique uses the time difference between two successive pulses in each direction. The average transit time is also continuously measured and used to compensate for variations in the speed of sound in the

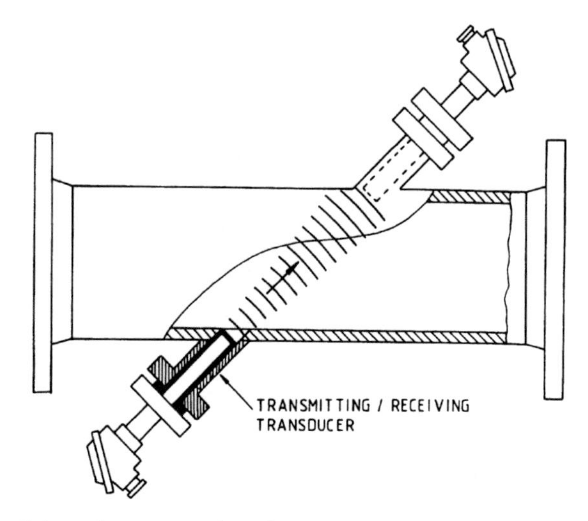

Fig. 23.7. Schematic representation of an ultrasonic flowmeter using the time of flight technique.

medium. The sing-around technique uses the frequency difference between two trains of pulses in each direction.

System description. Systems based on techniques related to the leading edge technique require a minimum of two transducers mounted diagonally across the pipe in line-of-sight (shown schematically in Fig. 23.7), while systems based on the sing-around technique require a minimum of four transducers. The angle between the beam and the axis of the pipe is usually between 20 and 60° depending upon the path length required for the application. The piezo-electric transducers are usually submerged in the medium to achieve maximum acoustic coupling. Some types of equipment enable the transducers to be removed from the pipeline while under pressure. Greater accuracy can be achieved (Lowell & Hirschfeld, 1979) by the use of multiple measuring paths. The transducers can be fitted to existing pipelines but they are more commonly available installed on a flanged spool-piece.

Advantages

- The measurements are independent of the speed of sound in the medium.
- Multiple-path instruments offer higher accuracy.

Limitations

- A chord-averaged rather than mean velocity is measured.
- Absolute calibration requires knowledge of the internal dimensions of the pipe and separation of the transducers.
- Transducers are usually in physical contact with the flowing medium.

- Attenuation of the ultrasound by the slurry may present problems where wide variations in concentration are encountered.

5. MASS FLOW RATE

5.1. Introduction

Mass flow rate can be measured indirectly by combining measurements of volumetric flow rate and density. Direct measurement is now feasible with equipment using the Coriolis effect.

5.2. Coriolis Flowmeters

Operating principle. The principle of operation is based on Newton's second law of motion that links the rate of change of momentum of a body to the force acting upon it. In a rotating environment, additional forces such as the Coriolis force are present. The effect of the Coriolis force is to deviate a body moving perpendicular to its velocity. For example, an object dropped from a high tower (that is perpendicular to the Earth's surface) will fall to a point slightly to the east of the base of the tower due to the rotation of the Earth. If a fluid were to flow down a pipe attached to the same tower, the falling fluid would carry with it the slightly larger eastward velocity experienced at the top of the tower. Because the fluid is constrained in the pipe a Coriolis force will be exerted on the pipe in an easterly direction. A more complete description of the theory on which this type of flowmeter is based is given by Young (1985).

System description. Flowmeters that utilise the Coriolis force require angular velocity to be imparted on the fluid. This is achieved by vibrating the flow sensor. The total flow is normally, but not always, divided into two loops. Each loop is excited to vibrate at its natural frequency using an electromagnetic drive. From measurements of the phase difference between the bending at the start of the loop and at the end of the loop the mass flow rate is calculated. The shape of the flow path used by manufacturers varies considerably (see, for example, Heywood & Mehta, 1988); straight tubes (see Fig. 23.8) are the exception rather than the rule. The earliest designs used B-shaped, U-shaped and single-turn spirals in order to minimise torsional stress in the tubes and produce high sensitivity for a given Coriolis force. Many designs incorporate a microprocessor and not only display mass flow rate, but also density and temperature of the process fluid.

Operational experience. Operational experience is limited by the relatively recent introduction of Coriolis-based flowmeters. Data published by Young (1985) indicate that they are potentially suitable for high-viscosity media, slurries and liquids containing small bubbles. Domnick *et al.* (1987) report comparative experimental results obtained with sand–water slurries at concentrations up to 9·6% (by

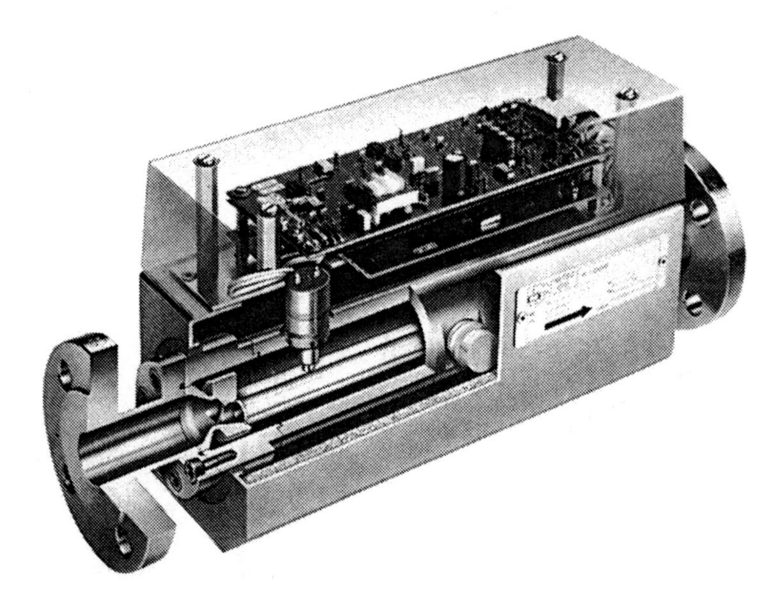

Fig. 23.8. Coriolis flowmeter. (Courtesy of Endress and Hauser, UK.)

volume). In these experiments the agreement was better than 1% between the mass flow rate measured with a Coriolis flowmeter and that obtained using a weigh-tank.

Advantages

- Calibration carried out by manufacturer.
- Compensation is incorporated to ensure that measurements are independent of pipeline pressure and temperature.
- Accuracy much better than 1% in the measured mass flow rate.
- Turn down of 100:1 is achievable.
- Instruments available for hygienic service.
- Early designs were for low flow rates and pressures—units suitable for line sizes up to 150 mm and capable of measuring 5500 kg min^{-1} are currently available. Units for smaller line sizes will handle pressures up to 15 MPa.

Limitations

- In keeping with all equipment using recently introduced techniques costs are high; however, a downward trend is apparent.
- Units for larger sized pipes are restricted in pressure capability.
- Designs in which the flow is required to make abrupt changes in direction

(especially designs incorporating B-shaped and U-shaped flow loops) could cause local wear with erosive slurries.

5.3. Thermal Sensing

Operating principle. The rate of heat transfer from the surface of an element by the flowing slurry is used to measure mass flow rate. The measurements are made either directly with separate temperature sensors or, indirectly using the heating element. The technique is similar to that used in hot-film anemometry.

System description. Equipment is available to cover flows from a few millilitres per minute to normal processing flow rates. The larger units are usually constructed around a flanged spool-piece of pipe. The heating element and temperature sensors are installed on the outside wall of the pipe. Intrusive probes are available for use in flumes.

Advantage

- Suitable for very low flow rates.

Limitations

- After calibration, a change in the thermal properties of the process medium will adversely affect measurements.
- Change in the nature of the flow in the vicinity of the pipe wall (or immersed sensor) will adversely affect calibration. A change in the flow regime (between laminar and turbulent) as well as a change of flow pattern will adversely affect the measured flow rate.
- Measurement very sensitive to scale build-up.
- Although compensation for process temperature and pressure is possible for clear liquids and gases, similar compensation for slurries may be difficult to realise in most applications.

6. PRESSURE SENSORS

6.1. Introduction

In the majority of slurry handling applications requiring the measurement of pressure, the selection of the type of transducer and signal transmitter are not usually of prime importance. However, in remote locations *smart* transmitters, that are showing increased presence, offer many advantages over more traditional instrumentation.

The most common method of measuring pressure in a transducer is to detect the distortion of a flexible element. The distortion of the element is measured using techniques that include strain gauges, capacitance, or magnetic reluctance. Clean,

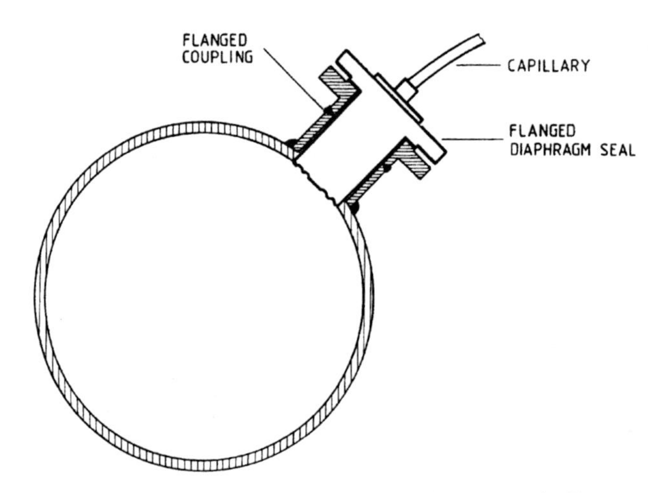

Fig. 23.9. Application of a diaphragm seal used to sense pipeline pressure.

non-corrosive fluids can often be permitted to be in direct contact with the sensing element. Slurry applications frequently require that the sensing element contained in the transducer is isolated from the slurry in order to reduce wear and corrosion and to prevent fouling with solids. Isolation is usually achieved using a secondary diaphragm seal, such as the one illustrated in Fig. 23.9, as a barrier between the slurry and a clean liquid. Liquids, such as mineral and silicone oils, are used to transmit the pressure signal to the sensing element. Over inter-plant distances this is achieved using a flexible capillary line.

Elastomeric sleeves are an alternative means of isolating the transmitting fluid from the slurry. The wafer-type, illustrated in Fig. 23.10, is installed in the pipeline between two flanges (as shown in Fig. 23.11). The transducer, or gauge, is usually close-coupled to the isolator. The isolator illustrated in Fig. 23.10 is fitted with a needle valve that can be used to attenuate mechanically noisy pressure signals and permits removal of the pressure instrument without shutting down the process flow.

6.2. Operational Experience

Direct contact of the transmission fluid with the slurry is common in flow loop applications. However, the tendency for blockage reduces the use of this type of tapping station in operational plants. Installation of single-point tappings on horizontal lines should be made at the 2 o'clock or 8 o'clock positions to avoid problems with entrained air in the slurry. Pressure signals from slurry applications are frequently notoriously noisy. Signal processing can be used to advantage to obtain averaged values.

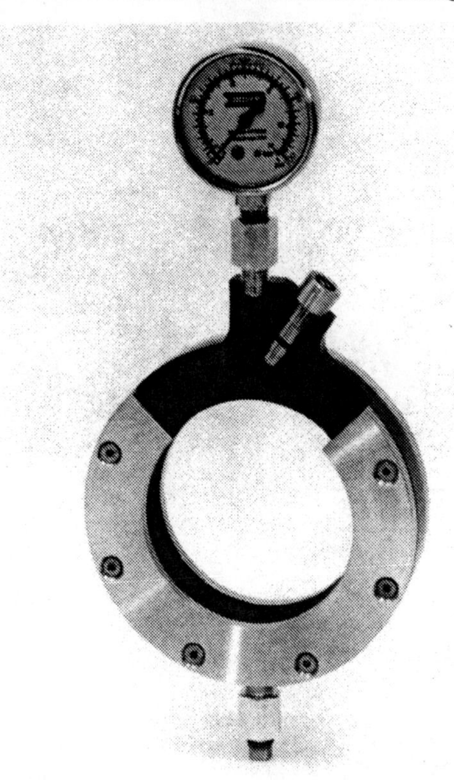

Fig. 23.10. Wafer-type of elastomeric isolator with integral needle valve. (Courtesy of Ronningen and Petter, Portage, USA.)

6.3. Direct Coupling

Advantages

- Direct contact with the medium permits fast response times up to 100 kHz using quartz transducers.
- Transducers tend to be physically small.

Limitations

- Cannot directly measure differential pressure.
- Under abrasive conditions life-expectancy can be short.
- Under harsh environmental conditions the small physical size of the transducer can reduce life-expectancy.
- Pipeline vibration will be transmitted directly to the transducer.

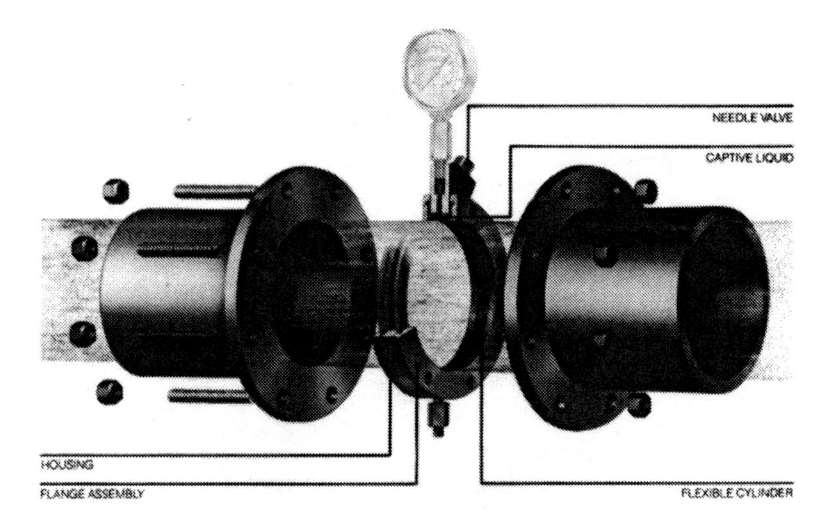

Fig. 23.11. Pipeline installation of a wafer-type of elastomeric isolator. (Courtesy of Ronningen and Petter, Portage, USA.)

6.4. Diaphragm Seals

Advantages

- Suitable for differential measurements.
- Environmentally sealed system from sensing element to transducer.
- Wide use in process industries provides cost advantages.
- Devices available with sophisticated microprocessor-based transmitters that can be remotely interrogated and reconfigured.
- Rugged construction.
- Remote location of the transducer can be advantageous where pipeline vibration is excessive.

Limitations

- The use of excessive capillary length can result in reduced response time unless the internal diameter of the capillary is increased.
- Large fill volumes of transmitting liquid tend to increase sensitivity to ambient temperature.
- In differential applications both seals should have the same environmental conditions. Contrary to popular belief, temperature effects will not cancel out; they are realised as a shift in the zero.
- Differential pressure transmitters should ideally be located at or below the mid-point between the pressure connections.

- The diameter of the diaphragm seal is typically 50 mm. Flush mounting with the inside wall of pipes cannot readily be achieved with pipe sizes below 250 mm.

6.5. Elastomeric Sleeves

Advantages

- Continuity of pipe bore maintained.
- Low tendency to clog or foul with solids.
- Available for pipe diameters above 25 mm.

Limitations

- Elastomer must be compatible with the temperature and chemical composition of the slurry.
- Not widely recommended for high-pressure applications.
- Large fill volume of transmitting fluid.
- Wafer designs require careful alignment to avoid potential wear.

7. PARTICLE SIZE

The analysis of particle size is a deservedly specialised discipline; it is common experience that different techniques produce different analyses of the same sample. A large number of techniques are available for determining particle size (see, for example, Allen, 1981). Considerable care has to be taken to ensure that the technique selected is compatible with the application and appropriate resources are available to carry out the analysis.

Considerable problems can arise where analysis is required for customer product acceptance; in such instances the sampling and measurement technique must be specified. Traditionally, particle sizing has been an off-line activity. However, in-line techniques are becoming established as the cost and complexity of the equipment is reduced. Sample collection and preparation is central to most size analyses; careful consideration is required to ensure that the sample is representative of the flow from which it was taken (Allen, 1981) (see also Chapter 11).

Although it is customary to report the results of analyses in the form of a distribution of particle size, often it is more appropriate for a simpler quantification to be provided, such as the amount of material retained on a particular screen in a sieve analysis. Complications exist in reporting a grading curve, on a common basis, when an analysis is required on a sample that contains a range of particle sizes spanning the capability of several analysis techniques. For control purposes comparative analyses are usually sufficient.

For samples containing particles above about 1 mm, the choice of techniques is limited to a form of screen analysis. Fully automatic batch sieving equipment is available, but this can only readily be used with relatively small quantities of dry sample. Analysers that contain hydrocyclones are used for sub-sieve analysis and for control of mineral grinding circuits where process conditions are not anticipated to vary over wide ranges. Cyclonic classifiers tend to be bulky, operate on batch or semi-batch basis and present results for a relatively small number of cuts. Although wet mechanical screening can be used to analyse samples down to around 5 μm, more sophisticated automated techniques are available that can analyse to smaller particle size.

Considerable interest is currently being shown in developing in-line instrumentation to analyse samples with particles in the size range 0·5 μm–2 mm. The equipment provides rapid analysis using sophisticated optical techniques (such as forward scattering, correlation spectroscopy, or diffraction). Optical techniques generally require a low concentration sample (typically less than 5% by volume). Sampling from streams at higher concentration necessarily requires dilution to be performed. In contrast to older equipment the latest instruments are reasonably priced, more rugged and frequently incorporate automatic sample preparation. A common concern with these techniques is obtaining a representative primary sample of the process medium and then ensuring the representativeness of the relatively small secondary sample used for analysis. Reproducible results have been achieved from coal slurries, containing particles below 800 μm, by concentrating the sample to a paste on a filter paper, sub-sampling the paste and diluting back to the required (low) concentration for analysis.

The development of equipment to perform size analysis directly on high-concentration slurries, without the need for either sampling or dilution, is a considerable challenge. Equipment with this capability has started to appear on the market but operational experience has yet to be accumulated.

8. VISCOSITY

8.1. Introduction

There are many commercially available designs of on-line instruments to measure viscosity or *consistency*, and a handful to measure other rheological properties, such as normal stress and dynamic properties (Cheng *et al.*, 1985). The term *consistency* tends to be used when an instrument is unable to give an absolute viscosity value. Viscosity or *consistency* are the most common quantities measured but sometimes other rheological properties may be more sensitive to changes in slurry composition.

These instruments are potentially useful for monitoring the viscosity of non-settling, pseudohomogeneous slurries only, and only a few are appropriate for slurries containing coarse particles (stabilised slurries). None of the instruments are applicable to coarse-particle settling slurries. Their purpose is to try to provide reliable on-line data to form part of a process flow loop and can be mounted in-line, in a side-stream through which slurry is sampled (on-line) or immersed in process equipment. However, in order to achieve this, viscosity (or other rheological property) has to be related to slurry parameters which can be adjusted, such as solids concentration, particle size and size distribution, slurry pH, etc. so that changes in one or more of these parameters can be monitored by change in the rheological property. There is, therefore, the need to carry out evaluation tests possibly in a pilot-scale facility before installation in process equipment or pipeline.

Important features of any instrument include its response time to changes in slurry properties and whether a continuous or intermittent signal is generated. In addition, it is sometimes important to know the shear rate at which viscosity is being measured. Because the bulk of commercial instruments were developed for Newtonian fluids, few instrument specifications include shear rate, although it is often possible in principle to estimate shear rate if details of the sensor geometry and operating conditions are known. Viscosity may have to be measured at more than one shear rate, particularly when viscosity is being maintained at one relevant shear rate but may be varying at another shear rate not employed on-line. As most instruments operate at a single speed, it is sometimes not possible to monitor changes in non-Newtonian slurry properties using a single shear rate.

On-line viscometers tend to be relatively expensive and output data from them over an extended time period are often not considered credible by production personnel. Part of the problem is the lack of knowledge regarding the long-term performance of an instrument. It is difficult for potential users to acquire this information as only short-term loans of instruments are generally offered by manufacturers. To overcome this, any drift in the measured quantity arising from long-term wear and fouling of the sensing element can be alleviated by regular programmed maintenance although this would appear to be rarely undertaken. Thus, there is not the widespread adoption of these instruments that manufacturers hope for and even less use is made of control loops in conjunction with on-line viscometers.

Dealy (1984) has summarised the main classes of on-line instrument available, and Cheng (1990) and Cheng *et al.* (1985) have considered the current and potential use of on-line viscometers in some detail. Sanders (1987) has demonstrated how precise on-line control of product viscosity from a continuous reactor is possible using a distributed control system with three alternative model-based control algorithms.

8.2. Tube or Pipe Viscometers

8.2.1. Use of Full Bore Pipe

Operating principle. A differential pressure transducer is connected across a suitable length of pipe carrying the slurry. Because the flow rate in the pipe is normally held constant, a single pressure loss value is obtained. This value can be related to viscosity if the flow in the pipe is laminar and will be more sensitive to changes in slurry parameters such as solids concentration or particle size in the laminar regime than in the turbulent regime. Additional information is possible if several differential pressure transducers are connected across two or more pipes of

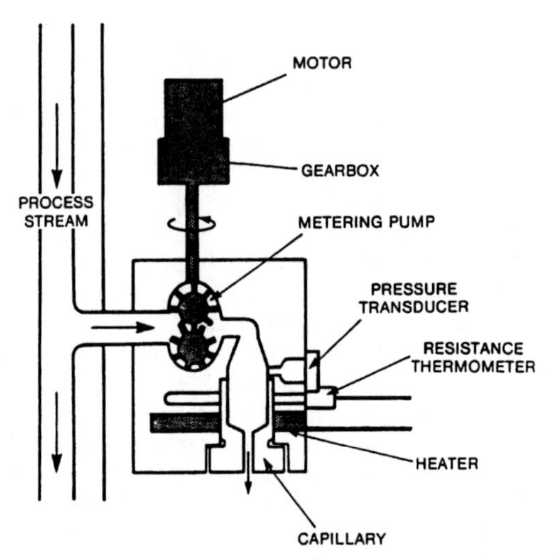

Fig. 23.12. Typical on-line capillary viscometer design. (Courtesy of Porpoise Viscometers Ltd, Skelmersdale, UK.)

different diameter carrying the same slurry. Then viscosity at different shear rates can be inferred.

Advantages

- The instrumentation is relatively inexpensive compared with purpose-built on-line instruments.
- There are no sampling problems as the pressure loss due to the whole flow of the slurry is monitored.

Limitation

- For turbulent flow in rough wall pipe, pressure loss is independent of viscosity and the system will essentially act as an on-line density meter.

Application areas

- Very wide application including stabilised slurries containing coarse particles.

8.2.2. Capillary Tube Viscometer

Operating principle. Material is sampled and pumped from a tank or pipeline using a positive displacement pump (frequently a precision gear pump) which can meter the flow through a capillary tube (Fig. 23.12). The pressure required to create the flow is measured using either a single flush-mounted pressure transducer at the entry to the capillary, or a differential pressure transducer.

Advantages

- They are suited to low-viscosity slurries.
- The use of a small diameter tube creates a measurable pressure drop while maintaining a sufficiently low Reynolds number for pipe flow.
- Sample stream can be readily brought to a pre-set temperature before viscosity measurement.

Limitations

- Most capillary tube viscometers are designed exclusively for side-stream installation. Therefore response time is slower than for in-line instruments.
- Care must be taken to ensure that the tube diameter is at least 10–20-times the maximum particle size in the slurry; otherwise either the tube will block or erroneously high pressure losses arising from partial blockage will occur.

Application areas

- Very wide application areas covering fine particle slurries with a wide range of viscosity.

8.3. Rotational Viscometers

8.3.1. Coaxial Viscometers

Operating principle. An inner cylinder is rotated inside an outer cylinder at a fixed speed and the resulting torque on the inner cylinder is measured as shown in Fig. 23.13. The principle is identical to laboratory co-axial cylinder viscometers.

Advantages

- Inner cylinder speed, and hence shear rate, can be easily varied to obtain a range of shear rates.
- Absolute viscosity values possible.

Limitation

- Must be mounted vertically in a process vessel or pipeline.

Application areas

- Wide range of non-settling fine-particle slurries.

8.3.2. Rotating Disc Consistency Transmitter

Operating principle. The disc sensor (Fig. 23.14) is immersed in a pipeline and rotated at a fixed speed; the resulting torque reading is monitored. This is a measure of *consistency*, which is defined in some industries as the ratio of solids

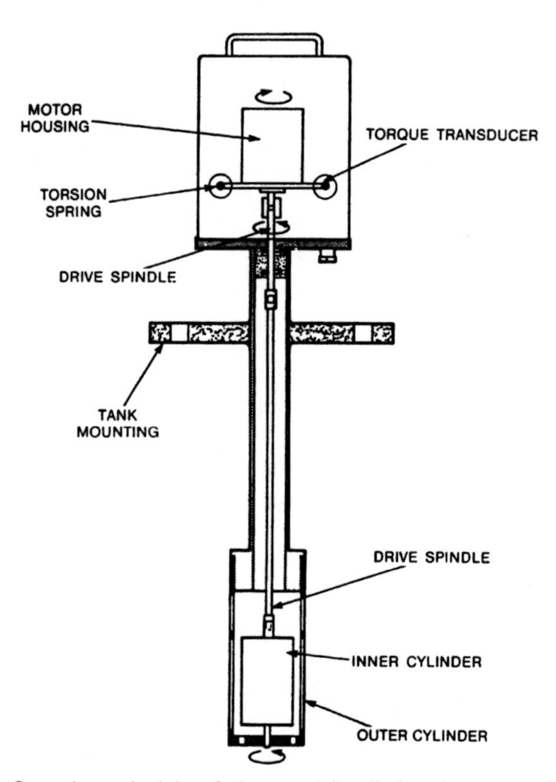

Fig. 23.13. Operating principle of the coaxial cylinder viscometer. (Courtesy of Brabender Messtechnik, Duisberg, Germany.)

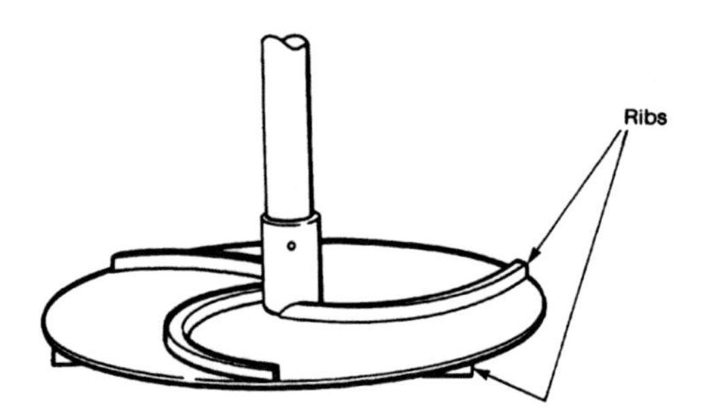

Fig. 23.14. Consistency disc viscometer sensor. (Courtesy of DeZurik International Ltd, Sevenoaks, UK.)

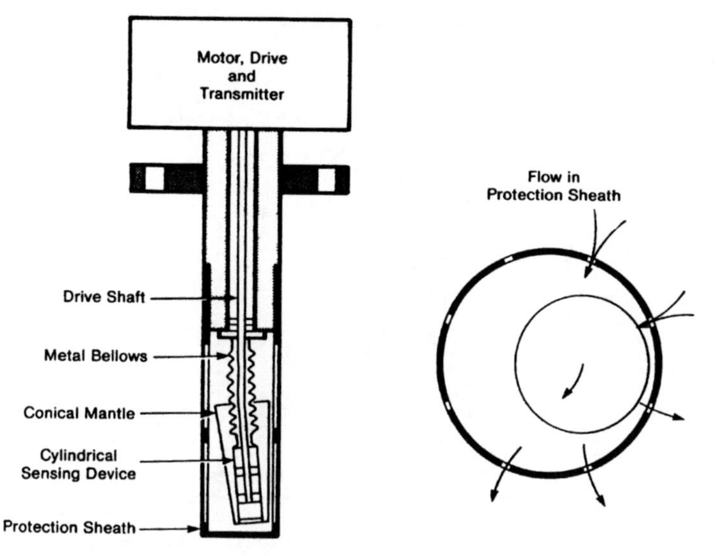

Fig. 23.15. Schematic of a rotational viscometer with gyratory bob motion. (Courtesy of Brabender Messtechnik, Duisberg, Germany.)

content to liquid in the slurry. The disc is typically fitted with involute ribs so that its output is unaffected by changes in flow rate in the pipeline and to increase sensitivity over a plain disc.

Advantage

- The sensor is designed to prevent fouling when monitoring the consistency of fibrous slurries.

Limitations

- Does not provide an absolute viscosity value and shear rate level is unknown.
- More useful for repeatability monitoring.

Application areas

- Commonly used to measure the consistency of paper stock fibres in suspension at concentrations typically in the range 0·7–6%.
- It has also been used for monitoring chalk slurry at a cement works.

8.3.3. Other Rotational Viscometers

Another design (Fig. 23.15) is based on a gyratory or nutating sensor caused by a bend in the lower end of the drive shaft. The sensor is cylindrical or fitted with a conical mantle for low to medium slurry viscosities. The sensor motion within the protective sheath causes a pumping effect that provides a fast exchange of slurry

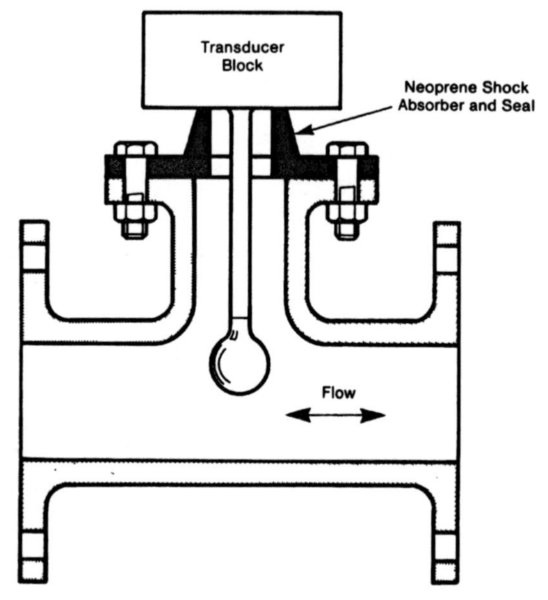

Fig. 23.16. Oscillating sphere viscometer. (Courtesy of Nametre, NJ, USA.)

sample. Hemmings and Boyes (1977) have used a nutating sensor for the control of a wet grinding circuit. Rapp *et al.* (1984) have used it to follow the degradation of cellulose from bacterialogical activity.

8.4. Vibrational Viscometers

8.4.1. Oscillating Sphere

Operating principle. A stainless steel sphere is immersed in-line in the slurry and oscillates about its polar axis (see Fig. 23.16) with a precisely controlled amplitude (typically 25 μm). The viscosity is computed from the force needed to maintain the predetermined amplitude of oscillation which tends to reduce as the sphere works against the slurry. This force is proportional to the product of viscosity and density.

Advantages

- Viscosity changes of 0·1% in a viscosity range of 0·1–100 mPa s can be monitored.
- Useage in temperatures up to 300°C and pressures to 20 mPa is possible.
- The measurements are claimed to be unaffected by the presence of solid particles or by rapid flow around the sphere.
- Suitable for low-viscosity slurries.

- Response time is normally fast; 10 s is typical but can be as low as 1 ms although there is considerable noise content in the output signal as the response time is reduced.
- Installation is simple as no pumps or motors are involved.

Limitation

- If slurry density varies as well as viscosity, an independent density measurement is required.

Application areas

- Can be used in the control of ceramic slip casting, mineral refining, transport of coal slurries, food processing, paint manufacture, and flocculation/deflocculation processes can be controlled (Matsuik & Scarna, 1981).

8.4.2. Vibrating Blade

Operating principle. The same as for the oscillating sphere but the sensor geometry is different.

Advantages

- Commercial designs cover a range of viscosities, temperatures and pressures.
- Response time normally fast when mounted in-line.

Limitations

- At high vibrational frequencies, the depth of penetration of the shear wave into the slurry sample is small and consequently the instrument may not, therefore, detect the presence of particles and give only the viscosity of the liquid component.
- If slurry density is varying with viscosity, an independent measurement of density is required.

Application area

- Picque and Corrieu (1988) have used a vibrating rod viscometer to follow a fermentation process by mounting the sensor on a bioreactor.

8.5. Falling Cylinder Viscometers

Operating principle. A piston is fitted inside a tube so that when it falls, the slurry is forced to flow through a cylindrical gap (Fig. 23.17). The time of fall is proportional to viscosity.

Advantages

- With variations in the clearance between piston and tube, viscosities can be measured in the range 0·1 mPa s–1000 Pa s.

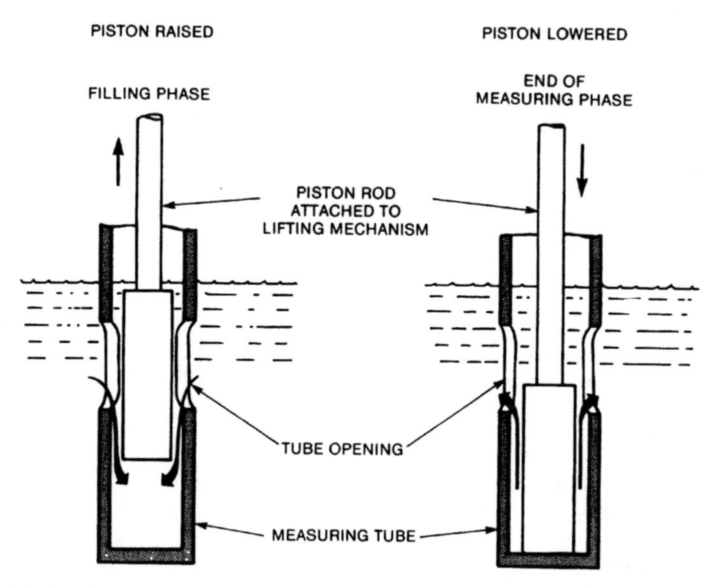

Fig. 23.17. Operating principle of the falling rod viscometer. (Courtesy of Norcross Corporation, MA, USA.)

- Claimed to be self-cleaning because the slurry sample is expelled during the down-stroke and on the up-stroke a fresh sample is drawn in.
- A large number of models are available for different types of installation, different process conditions and various viscosity levels (Dealy, 1984; Cheng *et al.*, 1985).

Limitations

- Must be mounted vertically and is designed primarily for mounting in process vessels.
- An intermittent rather than continuous measurement is obtained because time is required for the piston to be raised for a new measurement and for a new sample of material to enter the cylinder.

Application areas

- Various commercial designs are available for either in-line or tank mounting.

8.6. Drag Plate Consistency Transmitter

Operating principle. This uses the slurry drag on a sensing blade as slurry flows past (Fig. 23.18). A torque arm is connected to the blade and is attached to a torque transducer in the unit. As slurry consistency changes, the sensing blade and torque

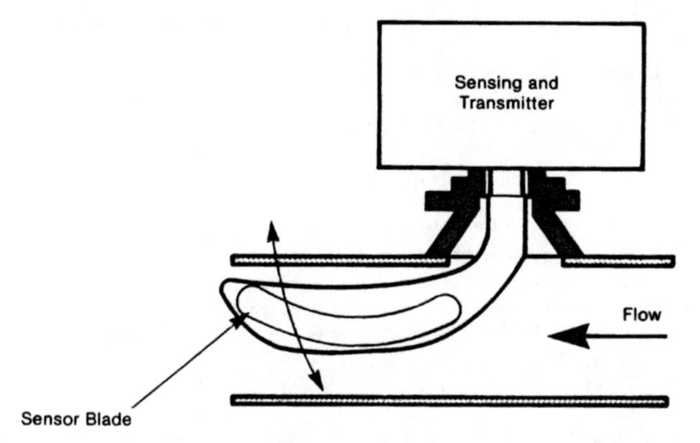

Fig. 23.18. Operating principle of the drag blade consistency transmitter.

arm pivot on the flexure mounts and a force balance feedback system is used to maintain the sensor blade position.

Advantage

- The instrument is designed for in-line use and is connected to a flanged coupling which can be welded into a pipeline at any angle.

Limitations

- The technique cannot measure viscosity directly, but rather *consistency*.
- In some designs, the slurry velocity should be maintained below about $1{\cdot}5\,\mathrm{m\,s^{-1}}$ to eliminate turbulence at the measuring point.

Application areas

- Typically used for measuring the consistency of paper stock, but would also be useful for other fibrous systems such as various foodstuffs, fermentation broths and sewage sludge.

9. REFERENCES

Allen, T. (1981). *Particle Size Measurement* (3rd edn). Chapman and Hall, London, UK.

Brown, N.P. (1988). An instrument to measure chord-averaged concentration profiles in two-phase pipeline flows. *J. Pipelines*, **7**, 177–89.

Cheng, D.C-H. (1990). A review of on-line rheological measurement. *Food Sci. Technol. Today*, **4**(4), 242–9.

Cheng, D.C.-H., Hunt, J.A. & Madhvi, P. (1985). Status report on process control viscometers: current applications and future needs. Warren Spring Laboratory, Stevenage, UK.

Colwell, J.M., Shook, C.A., Gillies, R.G. & Small, M. (1988). Ultrasonic flowmeter tests with slurries. *J. Pipelines*, **7**, 127–40.

Dealy, J.M. (1984). Viscometers for on-line measurement and control. *Chem. Engng*, 1 Oct., **91**(20), 62–70.
Domnick, J., Durst, F., Raszillier, H. & Zeisel, H. (1987). A method to measure mass and volume flowrates of two-phase flows. *Int. J. Multiphase Flow*, **13**, 685–98.
Hemmings, C.E. & Boyes, J.M. (1977). An on-line viscometry technique for improved operation and control of wet grinding circuits. In *Proceedings of the 12th International Mineral Processing Congress*. Brazilian Ministry of Mines and Energy, Sao Paulo, Brazil, pp. 46–64.
Heywood, N.I. & Mehta, K. (1988). A survey of non-invasive flowmeters for pipeline flow of high concentration non-settling slurries. In *Proc. HT 11*, paper C2, pp. 131–56.*
Hofmann, F. (1978). Magnetic flowmeter with capacitance signal pick-off. In *Flow Measurement of Fluids*, North-Holland Pub. Co., Amsterdam, The Netherlands, pp. 493–7.
Kao, D.T. & Kazanskij, I. (1979). On slurry flow velocity and solid concentration measuring techniques. In *Proc. STA 4*, pp. 102–20.*
Loughborough University. Current Awareness in Particle Size Technology. Loughborough University, UK.
Lowell, F. & Hirschfeld, F. (1979). Acoustic flowmeters for pipelines. *Mech. Eng.*, Oct., 29–35.
Matsuik, F.J. & Scarna, P.C. (1981). Latest instrument makes on-line viscosity control of slurries possible. *Control Engng*, Dec. 116–21.
Picque, D. & Corrieu, G. (1988). New instrument for on-line viscosity measurement of fermentation media. *Biotechnol. Bioeng.*, **31**, 19–23.
Rapp, P., Reng, H., Hempel, D.C. & Wagner, F. (1984). Cellulose degradation and monitoring of viscosity decrease in cultures of cellulomonas uda grown on printed newspaper. *Biotechnol. Bioeng.*, **26**, 1167–75.
Sanders, F.F. (1987). A comparison of model-based algorithms for on-line viscosity control. *Adv. Instrum.*, **42**, 753–66.
Shook, C.A. (1982). Flow of stratified slurries through horizontal venturi meters. *Can. J. Chem. Engng*, **60**, 342–5.
Shook, C.A. & Masliyah, J.H. (1974). Flows of a slurry through a venturi meter. *Can. J. Chem. Engng*, **52**, 228–33.
Wyatt, D.G. (1986). Electromagnetic flowmeter sensitivity with two-phase flow. *Int. J. Multiphase Flow*, **12**, 1009–17.
Young, A.M. (1985). Coriolis-based mass flow measurement. *Sensors*, Dec., 6–10.

*Full details of the Hydrotransport (HT) series and Slurry Transport Association (STA) of conferences can be found on pp. 11–13 of Chapter 1.

24

Control of Slurry Pipeline Systems

Ralph R. Buckwalter
Fluor Daniel Inc., Redwood City, USA

1. CONTROL VARIABLES

Transportation of slurry in a pipeline requires critical slurry parameters to be maintained or the transport system will not operate as designed. Slurry pipeline system control, as defined in the following section, was developed to permit slurry pipelines to be operated within narrow operating ranges inherent in all slurry pipeline systems and most inplant slurry piping design (Snoek *et al.*, 1978).

The following discussion relates primarily to slurries transporting minerals or other materials whose physical properties may vary due to in-situ conditions.

A large variety of bulk materials is being transported over long distances in slurry form. These materials include coal, copper concentrate, iron ore concentrate, limestone, phosphate, soda ash, and novel applications such as coarse coal in a fine magnetite slurry (Buckwalter, 1987). The physical properties, such as particle size distribution (PSD), solids relative density, particle shape and hardness of these materials affect the transport system hydraulic design. Only one of the physical properties, PSD, is a variable that can be controlled during slurry preparation for the system operation.

There are three other slurry properties that are controlled to allow reliable system operation. These are solids concentration, flow rate and pH. Of these three, flow rate is the primary controlled variable. However, both flow rate and solids concentration are frequently used to control transport velocity, the most critical of the slurry parameters for reliable pipeline operation. Solids concentration is used as a trimming type (cascade) control to adjust the throughput as required to maintain the slurry parameters within the slurry system's operation envelope.

To reiterate, the most critical slurry parameter for system design is transport velocity, which is controlled by flow rate and solids concentration. However, there are other related parameters which affect the velocity criteria to such an extent that they must also be monitored and maintained within design limits. These are listed and described briefly in the following sections.

2. SLURRY CHARACTERISTICS

2.1. Solids Relative Density

The relative density of mineral solids being processed will vary due to the physical variation of the ore body and the mining plan. The ore may or may not be blended to give a more uniform feed to the concentrator and/or the mineral beneficiation plant. Variations in solids relative density will affect the operating limits envelope of the slurry being transported. A shift in the density of the mineral solids will cause a shift in the slurry relative density which will in turn affect the transport velocity criteria.

2.2. Solids Size Consist

The size consist (PSD) of some materials being processed for slurry preparation will vary owing to the non-uniformity of the many physical characteristics of mineral ore bodies. The physical characteristics affect the grindability of the ore so the size consist of the material must be continuously monitored. If the top size of the ground material is out of slurry design limits, or the overall size consist is skewed from the design criteria, the pipeline control and operation may be detrimentally affected. The slurry system control cannot compensate for off-grind material and especially out of limits for top size. If the system is operated for extended periods of time with top size greater than the design basis, the pipe wear rate may be affected. Pipeline operation for slurries with excessive top size material should be suspended until specification size material is produced. This, of course, is more critical for long-distance systems than in-plant systems.

2.3. Slurry pH

The pH of slurry is also critical to the control of the system since it will normally affect the flow characteristics and therefore the pressure drop characteristics of the slurry. For instance, pH 7·5 to 10·0 or 11·0 is normal with many copper and iron concentrate systems. This high pH will normally reduce the slurry friction factor and will also minimise the internal corrosion rate of the pipeline. However, a pH above this range may increase the slurry friction factor. This characteristic should be checked during the initial design laboratory testing.

2.4. Slurry Concentration

The concentration of solids by weight or volume of the slurry being transported is very important since it affects the transport velocity (i.e. limit deposit velocity) criteria. This relationship between concentration and transport velocity has been studied in laboratory tests and correlated for many materials. However, laboratory pipe flow tests should be done on a representative sample of the material proposed to be transported for each system to be designed. This is again more critical for mineral-type slurries owing to the variations in ore bodies. Slurry concentration can be measured reliably and controlled directly by addition of water to the thickened slurry produced by the mineral beneficiation and ore concentration processes. In establishing the water addition requirements for density control during system design, care must be taken to include normal dilution water involved in the pump system operation. Centrifugal pump seal water or positive displacement plunger flush and seal water will provide an uncontrolled flow which can adversely affect dilution control at low flow rate.

Slurry concentration has a direct relationship with the slurry transport velocity criteria. Both slurry concentration and transport velocity criteria may be defined based on tests on the slurry rheological properties in the laboratory or with pipe loop testing, using actual material samples. These two parameters establish the system's pipe size, throughput and operation envelope limits.

Most materials can be transported at a lower velocity with higher concentrations since the larger particles are suspended by a high concentration of medium and smaller particles (see, for example, Buckwalter, 1987). Conversely, at lower concentrations the larger particles have a greater tendency to accumulate and be dragged along at the bottom of the pipe. This is detrimental to the pipe's useful life in that this tends to cause excessive wear in the bottom of the pipe. This excessive bottom wear can be tolerated in in-plant or short-distance, above-ground pipelines where the pipe wear can be more closely monitored and the pipe rotated to extend its useful life. However, excessive wear cannot be tolerated for long-distance and/or buried lines.

In long-distance slurry pipelines, pipe wear is monitored on a regular schedule but at long-distance intervals. If abnormal wear is detected and not mitigated then large sections of pipe may have to be dug up and replaced.

2.5. Corrosion Control

The control of corrosion is critical for slurry pipeline systems to enable the pipe to provide a useful design life and is considered in greater depth in Chapter 21. Corrosion is related to the oxygen entrained in the slurry and water used as the carrier fluid. Normally the slurry pH is maintained at an elevated level and no other inhibitors are used during slurry pumping.

However, to control corrosion during water pumping an oxygen scavenger is used. The type of inhibitor is dependent on the water quality. In some cases sodium sulphite has been used. A normal dosage would be 100 ppm of catalysed sodium sulphite added to the flushing water or any water batches.

3. CONTROL STRATEGIES

3.1. Operating Envelope

Figure 24.1 demonstrates a typical long-distance slurry pipeline operation diagram which shows the system's design operating envelope. This relates flow rate, solids concentration and solids throughput. The diagram is developed during the system's design phase. The upper and lower limits of solids concentration and transport velocities define the operation envelope. The upper throughput operation limit is sometimes limited by installed pump power as well as the velocity limit demonstrated in this example. Multiple operational envelopes should be developed for the upper and lower variations in solids relative density anticipated to be encountered during operation.

3.2. System Variations

At this point the differences in control strategies for in-plant piping systems/ short-distance slurry pipelines and long-distance slurry pipelines should be noted. Long-distance slurry pipelines will be discussed in more detail than in-plant systems in this control system section.

One basic difference in control strategies between in-plant/short-distance piping systems and long-distance pipelines involves the use of surge tanks on long distance pipelines (Aude, 1976).

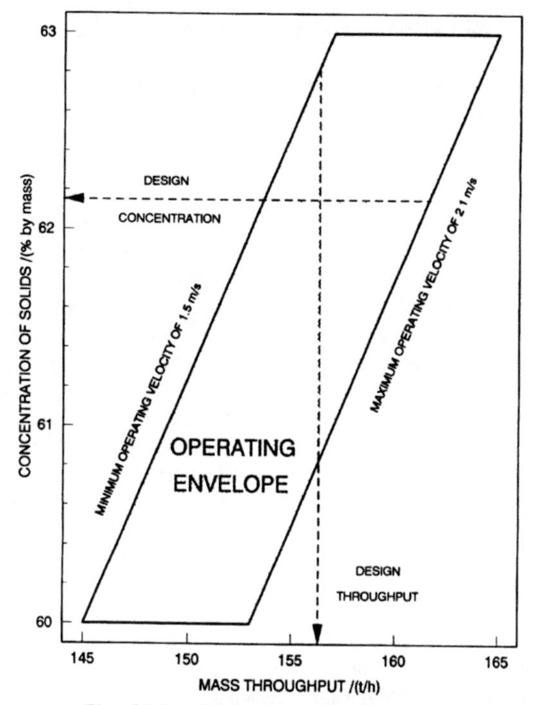

Fig. 24.1. Design operating envelope.

Normally, in-plant/short-distance piping systems operate as an integral part of the process plant operation. The throughput is based on the plant design criteria without allowance for operation during process shutdowns. Pumping operation uses sumps with minimum surge capabilities.

Long-distance slurry pipelines are designed to operate in a limited fashion independent of the processing plants at either end. This is accomplished by the use of surge tanks which provide 12–24 h of pipeline operation during process plant upsets or short-term shutdowns. These surge tanks are supplied with agitators to maintain the slurry solids in suspension. The agitator drive motors are provided with emergency power to minimise the possibility of sanding-up owing to the solids settling resulting from a commercial power outage.

The agitated slurry storage tanks also act as a buffer between the slurry processing plant at the head of the pipeline by normalising upsets in the slurry preparation process. The agitated slurry storage tanks at the pipeline terminal also serve to provide a uniform feed to the processing/dewatering plant. In this way, slurry concentration variations due to pipeline start and stop upsets and process upsets are reduced. Agitated tanks are covered in greater detail in Chapter 26.

3.3. In-Plant Piping System/Short-Distance Pipelines

When in-plant piping systems and short-distance pipelines use centrifugal pumps as the means for pressure augmentation, flow control is implemented by control of the pump speed which is accomplished by a variable speed drive. Either a variable speed drive motor, variable speed V-belt drive, or a fluid coupling between the drive motor and pump will allow the pump speed control to meet the flow variations required (Thompson *et al.*, 1972). (See also Chapter 14).

Another type of flow control for low head systems is a throttling type control valve. However, it is the throttling capability that makes this type of control a high-maintenance item and is a system not normally implemented. The increase in velocity due to reduction in flow area will cause high wear on the valve's internal components. A conventional in-line control valve is not recommended for normal slurry operation. However, a throttling device on a bypass line or recirculation line which will regulate only a portion of the total flow may be a more viable control scheme. The bypass scheme will also allow for maintenance without shutting down the system. Also, this bypass type installation eliminates the need to install additional valves to provide for a control valve bypass such as would be required in an in-line valve installation.

3.4. Long-Distance Pipelines

Control valves would not normally be used for long-distance system flow rate control due to the extremely high pressure drops and associated high velocities normally required to control mainline flow variations.

Long-distance slurry pipelines normally use positive displacement (PD) type pumps for pressure augmentation. PD pumps are perfectly suited for flow control by varying pump speed. The inherent design of the PD pump is such that volume flow rate is approximately directly proportional to pump speed. In fact, the well-maintained pump will provide an accurate measurement of flow, if the pump discharge has been calibrated as a function of differential pressure and slurry flow properties. The pump speed can be controlled by similar types of variable-speed devices such as those already mentioned for centrifugal pumps (except V-belt drives).

The basic long-distance slurry pipeline control scheme shown in Fig. 24.2 consists of four variables: flow rate, solids concentration, pump speed and pressure control. Pump speed is the primary control loop. The pump speed control set point is adjusted by a cascaded flow rate loop (Buckwalter, 1976). The pump speed control loop is also interfaced to a discharge pressure override control which must take over the pump speed control immediately on an increase in discharge pressure above the discharge pressure control set point.

The density control is an independent control loop and as such is not directly interfaced with the pump speed or flow control. However, the density control set point will affect the system envelope. This interface is left to the pipeline operator in some systems. However, this interface can be programmed directly into a

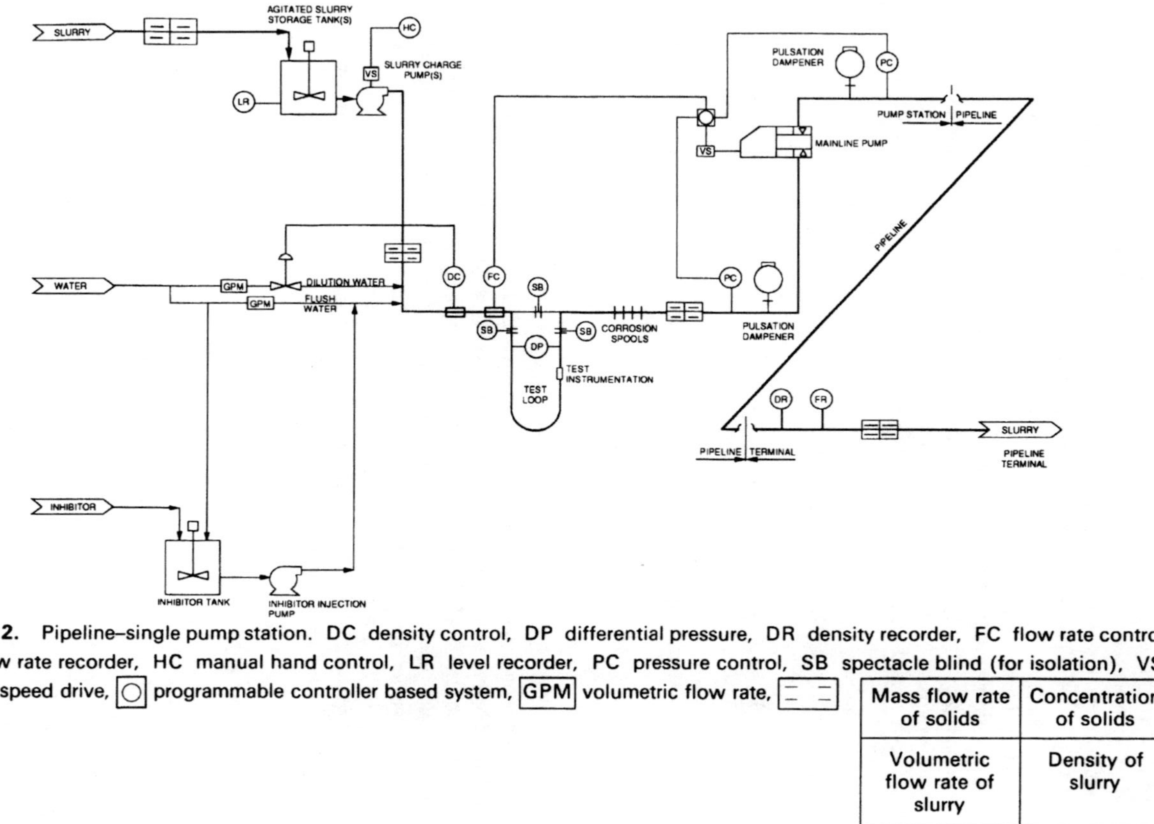

Fig. 24.2. Pipeline–single pump station. DC density control, DP differential pressure, DR density recorder, FC flow rate control, FR flow rate recorder, HC manual hand control, LR level recorder, PC pressure control, SB spectacle blind (for isolation), VS variable speed drive, ◯ programmable controller based system, GPM volumetric flow rate, ⊟

Mass flow rate of solids	Concentration of solids
Volumetric flow rate of slurry	Density of slurry

programmable-controller based system eliminating one additional manual interface problem area.

This control scheme also shows a test loop which is installed in many systems to enable the pipeline operator to run real-time on-line tests of the slurry properties. The test loop data should be used to verify the operating envelope limits that the system should be operated under based on the actual characteristics of the material being transported.

A new slurry parameter, not mentioned previously, which is monitored in the test loop is pressure differential. This is monitored as a method for determining the head losses to be expected in the pipeline. Adjustments to the operating envelope can also be based on actual pressure experienced at operating conditions.

4. START-UP AND SHUTDOWN

In addition to the control system design required for normal pipeline operation as mentioned above, consideration must be given to upset conditions and start-up and shutdown operations. Proper pipeline control during system start-up and shutdown is extremely important to minimise upset conditions. The system is more susceptible to over-pressure due to surge conditions during system start-up and shutdown. Over-pressure due to surge could cause high stress and ultimately pipe failure (for greater detail on the effects of surge the reader is directed to Chapter 9).

Start-up and shutdown control is primarily dependent on equipment timing cycles and proper valve sequencing. The start-up and shutdown sequencing can be performed by hard-wired relay logic, a microcomputer-based system or a programmable logic controller (PC) based system. The PC-based system can also be used in more complex pipelines where more than one pump station is involved or there are remote valve stations and pressure monitoring sites involved.

Systems involving remote sites and intermediate facilities should be interconnected via a dedicated communications link to a Supervisory Control And Data Acquisition (SCADA) system.

5. SLACK FLOW

Slack flow is a term used to describe a partially full pipe condition. This will occur in a pipeline or in-plant system where the friction losses are less than the head due to the ground profile. This slack flow condition is detrimental to the pipe because it will cause high pipe wear in the bottom segment of the pipe (Derammelaere & Chapman, 1979). This high pipe wear results from the high slurry velocities imposed by the partially full pipe. The same volume of slurry is passing through the pipe with a flowing cross-section that is less than a full pipe cross-section.

Remote pressure monitoring sites are used in a slurry system in which slack flow is anticipated. This slack flow condition exists in gravity systems in which the frictional head loss is less than head developed by the system's elevation profile. These systems will require back-pressure control at the terminal to control the flow rate within the operation envelope.

5.1. Slack Flow Control

Control of slack flow involves the addition of head dissipation devices to augment the system's friction head losses. This can be accomplished by reducing the pipe diameter in a section of the pipeline or installing fixed restricting orifices (chokes) in intermediate pipeline locations or at the pipeline terminal facility. These chokes can be installed in both a permanent in-line condition and a bypass condition. The combination of both permanent and bypass choke installations can be useful where a large pressure dissipation is required and the pressure dissipation requirement will vary over the operation envelope. The choke stations should be designed with in-line chokes required to augment the maximum system friction loss condition. Bypass chokes are designed to be added to cover the remainder of the operating range where the system head loss, including the in-line chokes, is below that required to eliminate slack flow. The choke station design must also take into consideration transient conditions which occur when the slurry pumping operation is switched to water and when a water pumping operation is switched back to slurry.

The choke design must also take the transient conditions during system start-up and shutdown into consideration.

6. BATCH OPERATION/THROUGHPUT CONTROL

Batch has at least three definitions in relation to slurry systems.

- *Batch* operation can be used as a means of controlling system throughput below the design rate by pumping a water *batch* between a slurry *batch* while maintaining a continuous pumping operation.
- *Batch* operation can be used as a means of controlling system throughput by pumping a slurry *batch* and then shutting down the system with slurry. A minimal water *batch* would be pumped at the end of the slurry *batch* to flush the pump station(s) piping. This is an intermittent type continuous slurry operation.
- *Batch* operation can also be used as a means of controlling system throughput, below the design rate, by pumping a slurry *batch* and then a water *batch* which completely flushes the system before the pipeline is shut down. The system is operated on an intermittent basis with all system shutdowns and restarts on water.

Batching should not be selected for the throughput design basis. However, if a batching mode is to be used for throughput control, the mode selected should be based on the operation intent and availability of water. Any intermittent batching operation will require an increase in water usage.

7. PRESSURE CONTROL

The subject of the direct control of system pressure has not been mentioned yet because it is not a primary control. Pressure control is a secondary control and a safety override control.

Pressure control is used primarily in a long-distance pipeline using PD pumps. Because of their inherent design, PD pumps can over-pressure the system if pressure override and safety shutdown controls are not provided. In addition, both discharge pressure and suction pressure may be used in pipeline control schemes to trim the primary flow rate control.

7.1. Station Discharge Pressure

Pump station discharge pressure can increase to a point where the pipeline will be stressed above the design limit. Therefore, the pump station discharge pressure is monitored and if the pressure reaches the upper design limit an automatic control override of the flow rate signal is energised. This will reduce the pump speed as required to maintain the pressure within the pipe stress limits. Note that the reduction in pump speed will continue and the pump speed would eventually be reduced to zero if the high-pressure condition continues. The pump speed rate of change will be dependent on the control algorithm and rate of change of system pressure. An alarm at low flow rates should be incorporated to alert the operator to a pressure override condition.

The station discharge pressure should also be monitored by use of pressure switches which will shut down the pump driver via a hard-wired shutdown circuit which is independent of the analogue process control system. This shutdown control should be set at a pressure above the pressure override control set-point but below the relief valve setting.

7.2. Station Suction Pressure

Suction pressure is also monitored in long-distance systems with intermediate pump stations and PD pumps. The suction pressure overrides the flow rate control to limit pump pulsation problems which are caused by low suction pressure on PD pumps. A low suction pressure condition is not as critical as a high discharge pressure condition and an alarm in lieu of override control could be used in pump stations with fewer than two operating pumps.

7.3. Pressure Relief Valves

An additional pressure safety relief device should be provided to protect the pump station piping from an inadvertent valve closure. A shear relief type valve should be installed on each pump discharge line and the station discharge header. The installation of the relief valve should include a flow monitoring device on the valve discharge line. Alternatively a position detector on the valve stem is used to monitor valve actuation and to shut down the pump drive using the same hard-wired pressure shutdown circuit described for over-pressure shutdowns.

A shear relief valve operates when the pressure in the system exerts a force that exceeds the pressure limit of the valve. The valve opens instantly and the flow discharges through a full port opening. Unlike spring-actuated relief valves, a shear relief valve will not reset when the pressure drops below the set-point. The valve action must be alarmed through the SCADA system so the operator can take corrective action. The shear relief valve will have to be manually reset and a new shear control limit installed before it can be returned to service.

The pressure switch and shear relief valve provide two independent safety circuits in the pump station to provide for personnel protection and equipment/piping protection.

8. SUMMARY AND COMMENTS

Short pipelines use the same control philosophies as long lines. The primary differences between systems will be related to the pipeline's profile and the number of intermediate pump stations and/or valve stations. Flow is the primary control variable used for mass throughput and slurry velocity regulation.

In-plant systems are primarily controlled to maintain sump or tank levels. These levels may be maintained by control of pump speed, control of discharge flow by an in-line control valve, or recirculating part of the discharge flow rate back to the tank or sump.

Long-distance lines are controlled to maintain throughput between two process plants. This throughput is controlled by monitoring process variables at the initial pump station. Any intermediate stations would be controlled using suction and discharge pressure override systems as discussed in section 7.

Other items that must be considered and/or controlled in long-distance pipelines are corrosion, slack flow, slurry pH and particle size.

9. REFERENCES

Aude, T.C. (1976). Slurry pipeline design—materials and equipment selection. In *Proc. STA 1*, paper 6.*

Buckwalter, R.K. (1976). Instrument design considerations for slurry pipelines utilising positive displacement pumps. In *Proc. HT 4*, paper H4, p. 53.*

Buckwalter, R.K. (1987). Freight pipelining coarse coal using a heavy medium slurry carrier. In *Proceedings of the 5th International Symposium on Freight Pipelines*.

Derammelaere, R.H. & Chapman, J.P. (1979). Slack flow in world's largest iron concentrate slurry pipeline. In *Proc. STA 4*, pp. 32–42.*

Snoek, P.E., Gandhi, R.L. & Weston, M.D. (1978). Design considerations for slurry pipeline systems. In *Proc. STA 3*, pp. 77–84.*

Thompson, T.L., Frey, R.J., Cowper, N.T. & Wasp, E.J. (1972). Slurry pumps—a review. In *Proc. HT 2*, paper H1, pp. 1–24.*

*Full details of the Hydrotransport (HT) and the Slurry Transport Association (STA) series of conferences can be found on pp. 11–13 of Chapter 1.

25

Solids Comminution and Grading

Anthony J. Elliott
Carnon Consultancy, Redruth, UK

1. INTRODUCTION

Within the general context of this book on solids in slurry form, it is pertinent to consider the methods available for producing particles at sizes suitable for materials handling as slurries and also the methods by which the particles can be classified into suitable sizes or size ranges.

Aspects of comminution can be subdivided into two general areas. First, size reduction from virtually any given top-size by means of mechanical compression or impact is usually carried out with the material in a dry state. The lower limit of this method is a top-size of about 25 mm. Secondly, comminution below this size is usually by grinding, the action being predominantly attrition and generally the particles are in slurry form. The most common equipment for the former range of reduction includes jaw, gyratory and cone crushers and, for grinding, rod and ball mills. The degree of comminution carried out by a particular machine is known as the reduction ratio and is the 80% passing size of the feed divided by the 80% passing size of the product from the device.

Methods of grading particles tend to correspond to the comminution areas. Coarse sizing is done by positive screening and fine sizing by classification in slurry form, exploiting the falling velocity differences between particle sizes. As examples of equipment, screens are fixed (vibrating or rotary) while fine sizing employs hydraulic classifiers and hydrocyclones.

It is not possible to deal technically with all aspects of these topics in one chapter. The reader is advised to consult the bibliography at the end for more detailed information and for more visual illustrations than can be included here.

2. COMMINUTION THEORY

Size reduction is the most expensive unit operation in mineral processing as energy requirements, capital outlay and operating costs are high.

The mechanical efficiency of comminution is extremely low; the bulk of the energy input appears as wasted heat. Consequently, theory has always been regarded as paramount in an effort to understand, and hence improve, the process. The earliest explanations of mechanisms of breakage were developed in the late 19th century by Rittinger and Kick (further details are to be found in the general works listed in the bibliography). More recently another comminution theory has been developed by Bond (1952).

A common feature of these approaches is that the work to break particles is proportional to particle diameter raised to a fixed exponent. Holmes (1957) proposed that the exponent on the diameter should be variable instead of fixed. While this approach has more wide applicability, and allows for variations in breakage resistance, particularly with size of particle, the practicality of determining the exponent renders the proposal tedious for fast answers to comminution problems.

In practice, the Bond proposition is the most used, particularly for wet grinding determinations. Here, the resistance to comminution is quantified by a grindability

Work Index. This can be used to calculate the required work input for any given size reduction. The grindability Work Index can be established in the laboratory for particular solids, although the test work is time consuming.

The basic equation proposed by Bond relates the energy required per tonne, W(kWh t^{-1}), to the grindability Work Index, Wi, from a feed size, F (μm), to a product size, P (μm) (where the sizes are the 80% passing sizes), by

$$W = 10Wi(P^{-0\cdot5} - F^{-0\cdot5}) \quad (25.1)$$

The grindability, Wi, is invariably quoted in terms of short tons and care should be taken when referencing such sources. Equation (25.1) can be used in the laboratory for a fast comparison of two ores to determine the grindability of an unknown mineral (Mular & Bhappu, 1980).

Some typical examples of grindability Work Indices are

Barite 5	Feldspar 12	Quartz 13
Bauxite 9	Fluorspar 9	Quartzite 10
Clay 7	Granite 15	Sandstone 7
Coal 11	Limestone 12	Shale 16
Dolomite 11	Magnetite 10	Taconite 23

The Work Index applies specifically to wet grinding in a standard size ball mill. To convert the power requirement to a realistic plant power input, various efficiency factors ($E_{1\text{-}8}$) should be applied. These are described briefly below, with a worked example. More details can be found in the general references, especially Mular and Bhappu (1980).

Dry grinding: $E_1 = 1\cdot3$.

Open circuit grinding: if the ball mill grinding is not in a conventional closed circuit with a classifier, the factor to be used is a function of the degree of control required on the product; for example, if the specification is 80% passing the product size, $E_2 = 1\cdot2$. If 95% quality is needed, $E_2 = 1\cdot57$.

Mill diameter: eqn (25.1) is based on a 2·44 m (96 in.) diameter mill. For other sizes,

$$E_3 = (2\cdot44/D)^{0\cdot2} \quad (25.2)$$

where D = the diameter of the mill (m).

Oversize feed: the optimum feed size, F_o, (μm) for rod milling is given by

$$F_o = 16\,000(13/Wi)^{0\cdot5} \quad (25.3)$$

And for ball milling by

$$F_o = 4\,000(13/Wi)^{0\cdot5} \quad (25.4)$$

for $F \leqslant F_o$, $E_4 = 1$;

for $F > F_o$ $\quad E_4 = ((F/P) + (Wi - 7)((F - F_o)/F_o))/(F/P) \quad (25.5)$

Fineness factor: if $P < 75\,\mu$m then

$$E_5 = (P + 10\cdot3)/1\cdot145P \quad (25.6)$$

Reduction ratio: this can be a circular calculation as the base size of the required mill must be established first; from this

$$R_{ro} = 8 + (5L/D) \quad (25.7)$$

where L and D are the length and diameter of the proposed mill; the material reduction ratio

$$R_r = F/P \quad (25.8)$$

if $R_r > (R_{ro} \pm 2)$, then

$$E_6 = 1 + ((R_r - R_{ro})^2/150) \quad (25.9)$$

Low reduction ratio for ball mills: if $R_r < 6$ (where R_r is given by eqn (25.8)), then

$$E_7 = (2(R_r - 1{\cdot}35) + 0{\cdot}26)/(2(R_r - 1{\cdot}35)) \quad (25.10)$$

Rod milling feed: if the material fed to a rod mill is produced in open circuit crushing, then $E_8 = 1{\cdot}4$; if closed circuit, then $E_8 = 1{\cdot}2$.

EXAMPLE

An ore has a grindability, Wi, of 14 and is to be ground in a rod mill from a feed produced in a closed crushing circuit, F of 19 mm to a product, P of 12 mm at a mass throughput of 400 th^{-1}. Estimate the power requirement and mill size.

From eqn (25.1), the work input per tonne, $W = 3{\cdot}03$ kWh t^{-1}.

Using the correction factors
Wet grinding: $E_1 = 1$
Closed circuit: $E_2 = 1$
Mill diameter: E_3 will be found after the power calculation.
Oversized feed: E_4
from eqn (25.3), $F_o = 15\,418$
from eqn (25.8), $R_r = 15{\cdot}8$
from eqn (25.5), $E_4 = 1{\cdot}103$
Fineness factor: $E_5 = 1$
Reduction ratio: E_6 will be found after sizing the mill.
Low reduction ratio for ball mills: $E_7 = 1$
Rod mill feed: $E_8 = 1{\cdot}2$.

The adjusted $W = 3{\cdot}03\ (E_4)(E_8) = 4{\cdot}01$ kWh t^{-1}. Converting this to horsepower, (in order to refer to manufacturer's data), the power is 5·38 hpt^{-1} and at a throughput of 400 th^{-1} is approximately 2 150 hp. Referring to catalogues, two mills will be needed, each at about 1 000 hp. The best diameter to match this power is 3·66 m (12 feet) with a range of lengths from which the accurate selection can be made.

Mill diameter: E_3—from eqn (25.2), $E_3 = 0{\cdot}922$. The power is thus,

$4{\cdot}01\ (E_3)\ (400) = 1{\cdot}48$ MW (or 1985 hp).

From the manufacturer's catalogue, these mills 4·88 m (16 feet) long draw 972 hp (725 kW)

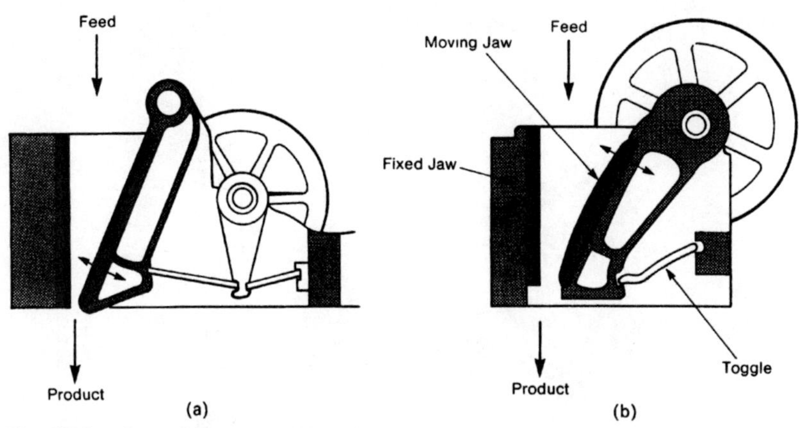

Fig. 25.1. Jaw crushers: (a) Blake jaw crusher, (b) overhead eccentric or single-toggle jaw pattern.

Reduction ratio: E_6—from eqn (25.7) $R_{ro} = 14{\cdot}67$

The value of R_r is within ± 2 of this; therefore no correction is needed. Two mills 3·66 m (12 feet) in diameter and 4·88 m (16 feet) long are satisfactory.

There are also methods of determining power needed for crushing by means of laboratory tests and then converting to plant equipment specifications. However, it is much more likely that the size of crusher or crushers is set by the maximum particle size that needs to be crushed. Thus, a 150 mm feed will need a minimum size opening in the crushing cavity and will therefore set the size of the machine. Calculations can then be made from manufacturers' data to check that the throughput rate will be sufficient for the individual need. In most cases the capacity of the machine will exceed the design tonnage. For this reason, crushers are often run for shorter periods than the matching milling stages. This also provides for the additional required time for maintenance.

3. CRUSHERS

3.1. Primary

Primary crushing can be regarded as the stage that accepts virtually any given top-size of material, dependent only on the gape of the machine, and produces a crushed material of approximately 50–150 mm. There are three main types of machine: jaw, gyratory and impact crushers, shown schematically in Figs 25.1–25.3.

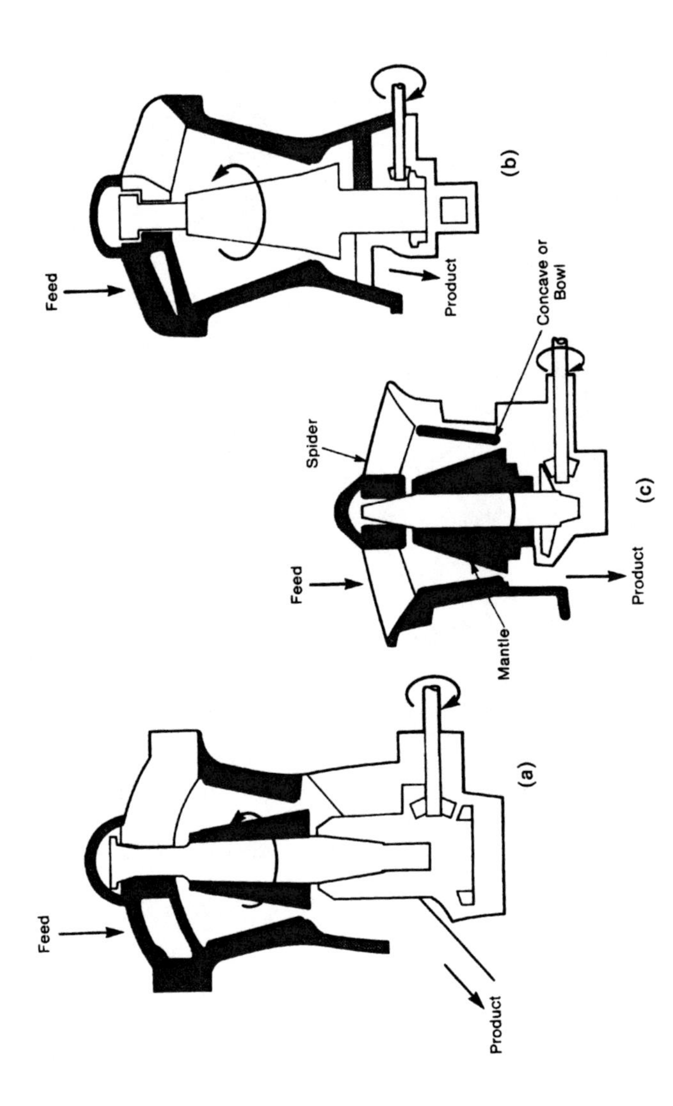

Fig. 25.2. Gyratory crushers: (a) suspended spindle pattern, (b) supported spindle pattern, (c) fixed spindle pattern.

3.1.1. Jaw Crushers

Jaw crushers operate by compressing rock between a fixed plate and a movable jaw. The crushing cavity is tapered vertically to allow successive compressions as rock falls down. The crushing zone is a straight line in the horizontal direction.

There are two principal types of jaw crusher (Fig. 25.1), with minor variations on each. The fundamental difference is in the method of operating the movable jaw. In the original type (the Blake crusher), the jaw is pivoted at the top and was oscillated at the bottom, whereas in more recent machines the jaw is pivoted at the bottom and driven at the top.

In the *overhead eccentric*, or *single toggle* crusher, the movement is imparted by the direct action of the eccentric driving the top of the movable jaw and also by the rocking action of the inclined toggle holding the lower end. This type tends to be smaller and lighter than the Blake type and is suitable for mobile plant and for materials that do not have severe abrasion problems that can cause wear on the jaws from the greater vertical motion. They are also lighter duty and should not be used where severe impact from the loading system is likely to occur.

The strength of the machines and the inertia imparted by the moving parts necessitate that protection is incorporated to prevent tramp material (such as hammerheads, bucket teeth, iron rail etc) wrecking the crusher. Protection is normally incorporated by a weak link in one of the toggles, which will snap if the moving jaw is hindered. It is also an important feature of a jaw crusher that feed is restricted on entry into the gape. Otherwise crowding of the crushing gap will occur as the rock shatters and expands, leading to a similar situation as tramp and restricting the movement of the swinging jaw. Cleaning out of the choked machine can be time consuming.

The shape of the jaws is important and various profiles can be chosen to allow for free fall of the rock being crushed between successive compressions; this reduces the likelihood of stalling. However, the angle between the jaws at any given point is also important, particularly with rock which exhibits elasticity. Too large an included angle will allow the rock to be squeezed out, often violently,

TABLE 25.1
Jaw Crusher Capacities

Size gape × length of feed opening		*Discharge gap (mm)*	*Capacity* $(t\,h^{-1})$
(in.)	*(m)*		
15 × 24	0·38 × 0·61	50	30
30 × 42	0·76 × 1·06	100	200
36 × 48	0·91 × 1·22	125	300
48 × 60	1·22 × 1·52	150	500
66 × 84	1·67 × 2·13	200	800

without being caught or nipped between the jaws. The angle is usually between 20 and 27°.

The jaws and side plates, or cheek plates, of the frame are always lined with replaceable sacrificial material and this is generally manganese steel. The wear rate is uneven across the jaws and liners are sometimes reversible to counteract this. The profiles of the liners vary from one manufacturer to another, and according to the material being crushed. They can be plain or ribbed vertically. The latter will produce less platey product than smooth liners if the ribs are offset with respect to each jaw.

Size of a jaw crusher is quoted in terms of the gape by the length of the feed opening. Selection of size of machine should be by ensuring that at least one dimension of the largest particle to be crushed is no more than 80% of the gape. This invariably leads to a machine with capacity in excess of the treatment rate required and the question should be asked as to whether a finer feed can be generated, by more efficient blasting, for example, in ore extraction. The closed setting of the discharge crusher, that is the minimum gap at the bottom of the jaws, will also affect capacity; the finer the product the lower the throughput. The reduction ratio (feed size/discharge size) should normally not exceed 8. Some reasonably typical capacities in tonnes per hour are given in Table 25.1 as a guide to performance. Manufacturers' machine sizes are given (with metric equivalents).

3.1.2. Gyratory Crushers

A gyratory crusher can be regarded as a jaw crusher wrapped round in a circle with the crushing surface gyrating. Various common designs are shown schematically in Fig. 25.2. All cone crushers are gyratory, but the larger machines used for coarse primary crushing are commonly called *gyratory*. *Cone* crushers are nearly always used in secondary or tertiary stages of crushing.

The gyratory crusher has a number of forms, with the method of mechanically arranging the moving crushing surface altering from type to type. In all cases, rock falls down the crushing cavity by gravity. The outer fixed surface takes the form of an inverted cone and is usually referred to as a concave, or bowl. The moving surface is an upright cone, called the mantle, and moves eccentrically within the

TABLE 25.2
Gyratory Crusher Capacities

Size opening × diameter of mantle (*in.*)	(*m*)	*Weight* *t*	*Power* (*kW*)	*Capacity* (th^{-1})	*Discharge gap* (*mm*)
30 × 60	0·76 × 1·55	100	150	500	75
42 × 66	1·06 × 1·67	150	190	850	100
48 × 70	1·22 × 1·78	250	260	1 400	125
54 × 74	1·37 × 1·88	265	370	2 000	150
72 × 90	1·83 × 2·29	490	520	3 300	200

concave. The distinguishing characteristic of the gyratory as opposed to a cone is that the mantle is invariably supported at the upper end. The mantle does not itself rotate, except sometimes when it is freewheeling with no feed passing through.

Owing to the large forces generated during crushing the massive shaft carrying the mantle is held at both ends. If the shaft and mantle weight is carried by a threaded bearing in the support frame (known as a spider) across the top of the receiving opening, the type is called a suspended spindle crusher. More commonly, the mass is held by a hydraulic piston at the base and is called a supported spindle, with the top still running in bearings in the spider. The third main type is where the shaft is carried at both ends and is known as a fixed spindle. In the first two types, the majority of the eccentricity of the mantle is at the lower end, whereas the fixed spindle tends to have the same throw all the way up the mantle.

Feed is often direct, for example, from trucks and for this reason the spider bearing is heavily protected. The manufacturers' size data refer to the feed opening and the size of the head. Thus a 48 × 80 has an opening measured radially across the crusher of 48 in. (1·22 m) and the maximum diameter of the mantle is 80 in. (2·13 m). Table 25.2 gives some indication of the sizes and capacities of gyratory crushers.

Liners are of similar materials to those in jaw crushers and there are similarities in the required angle of nip and the profiles of the liners to allow free fall of crushed rock through the crusher cavity. Gyratory crushers can be choke fed with suitable liner profiles, a major advantage with trucks discharge into the machine. This is because the crushing cavity is expanding circumferentially around the mantle, allowing room for rock swelling, unlike the jaw crusher.

3.1.3. Comparisons between Jaw and Gyratory Crushers

The jaw opening gape suits feed in block form whereas the gyratory favours slab-shaped rock. The gyratory has an advantage in that the curved crushing surface will tend to snap naturally slab-shaped rock, producing better quality pieces in, for example, aggregate production.

The jaw cycle of operation, by definition, gives crushing only for half the time. The gyratory operates continuously. Thus, for a similar gape, the gyratory has significantly larger capacity which may be excessive. Mechanically therefore, the jaw is more massive, but the simplicity of its construction makes maintenance easier. Liner wear on a jaw is uneven, requiring more changes. The gyratory wear is even, but the replacement of liners is more complex and the liners more expensive. The jaw drive and mechanism are more remote from the rock stream and this is an advantage if the material is wet or muddy.

Comparisons of cost are difficult as it is difficult to equate machines. However, where throughput is satisfactory for either machine with a given gape, then a jaw crusher will be cheaper, despite the fact that the foundations will be more massive to handle the shock loads. It is also important that material loses less vertical height in its passage through a jaw crusher and this is significant in terms of plant layout and conveyor lengths.

The conclusion is that, for smaller tonnage operations, a jaw crusher will be favoured for primary crushing, unless there is no means of feed control to give a non-choked feed. For large operations or where feed will be intermittent, by truck discharge for example, the gyratory will take precedence.

3.1.4. Impact Crushers

Impact crushers achieve breakage by the action of flailing or fixed hammers revolving on a shaft in the horizontal plane while rock falls through the hammer path. Generally, these are not primary crushers in the sense of accepting very large feed material, as with the jaw and gyratory types. Nevertheless, impact crushers are used in aggregate and industrial mineral applications as primary crushers, particularly with soft, platey material or with material that is wet. Owing to the severe wear on the rotating hammers, it is unusual to find these machines in applications where there is a high percentage of silica in the feed.

There are two types of machine, the impactor and the hammer mill (Fig. 25.3). The impactor uses the force of the hammer to shatter the material and, in addition, relies on the rock-on-rock impacts that are subsequently induced. The hammer mill uses the same initial impact, but also crushes by means of the attrition of the rock against a steel grid which retains the rock until it is fine enough to pass through. Obviously, the hammer mill gives a closer sized and more controlled size of product. A typical impact crusher is shown, cut-away, in Fig. 25.4.

Impactors can be subdivided into classes, but the main distinction is that some machines have twin rotors carrying the hammers. Both single and twin can be reversed in order to even out the wear on the leading edge of the hammers. The number of rows of hammers can vary. Three to six is common but the capacity is not dependent on the number of hammers.

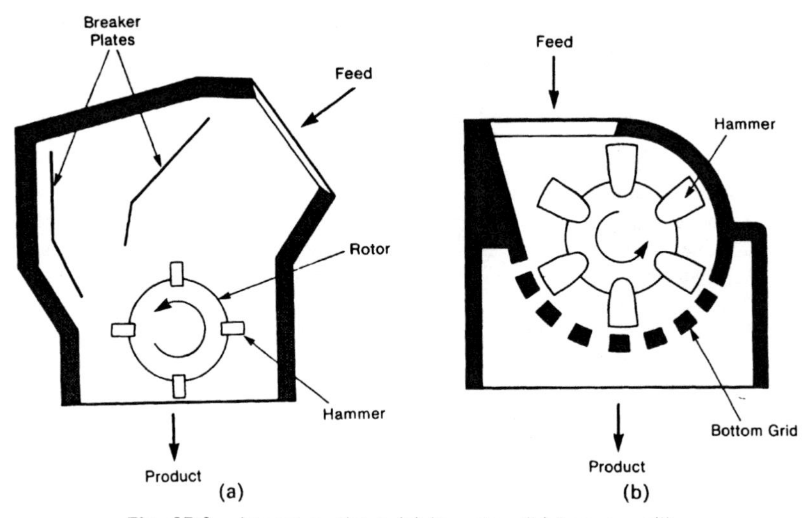

Fig. 25.3. Impact crushers: (a) impactor, (b) hammer mill.

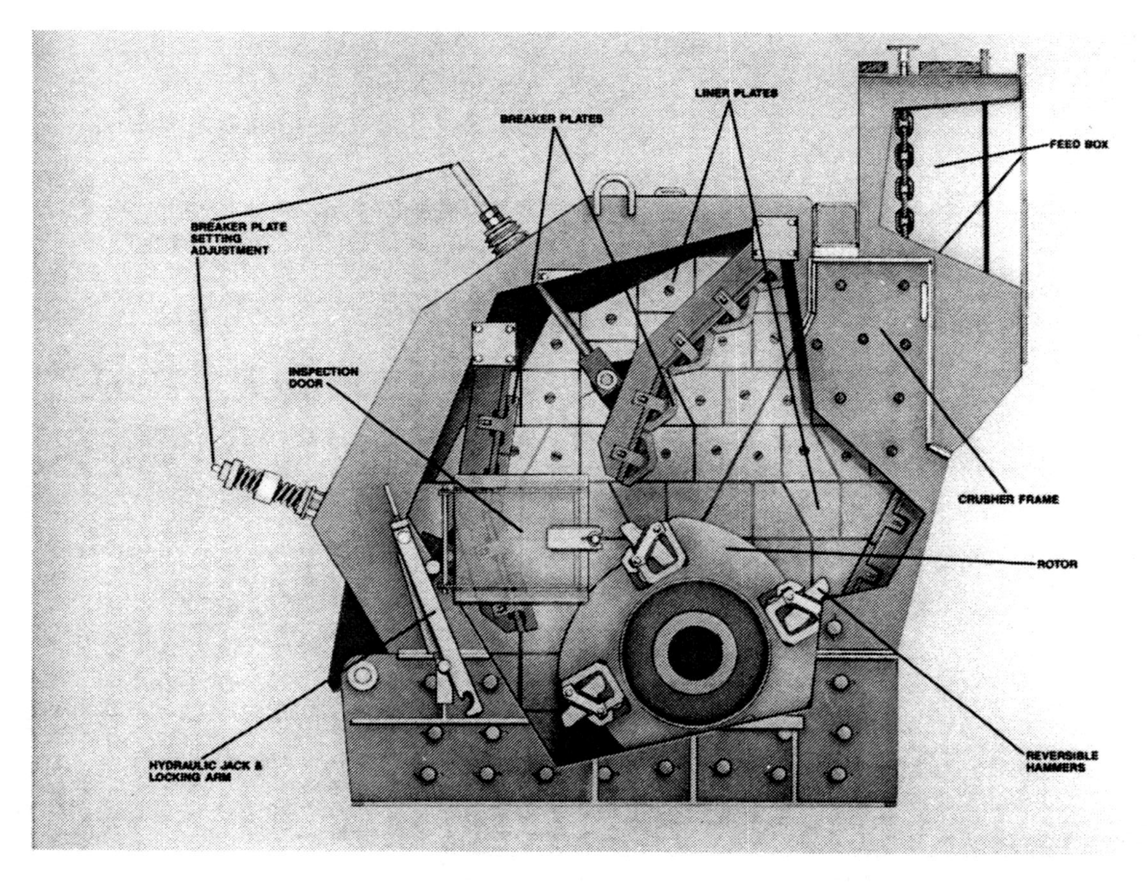

Fig. 25.4. An impact crusher. (Courtesy of Nordberg (UK), Ltd.)

TABLE 25.3
Primary Crusher Comparisons

Crusher type size	*Jaw* 48 × 60 in. (1·22 × 1·55 m)	*Gyratory* 54 × 74 in. (1·37 × 1·88 m)	*Impact* 46 × 54 in. (1·17 × 1·37 m)
Weight (t)	135	250	30
Power (kW)	150	375	300
Capacity (th^{-1})	500	2 300	600
Relative capital cost	240	750	100

The interior of the crushing chamber is lined, often with protruding breaker bars to provide additional point impact sites. Some sophisticated machines have spring-loaded breaker plates to avoid tramp chokage and some have pockets into which tramp can supposedly be diverted and held. Tramp material is, however, a problem.

The rotors operate at high speed, giving minimum hammer tip velocities of 17 ms^{-1} and often higher. This necessitates complex electrical start-up methods as the initial power may be higher than that needed to operate the crushing.

3.1.5. Comparisons between Impact and Compression Crushers

Table 25.3 shows types of machines that will accept approximately 1 m rocks to give a nominal 150 mm product.

The installation and capital cost of the impactor are significantly below the others for the required feed size, but the gyratory has a vastly higher capacity and is more efficient in terms of power consumption. The running cost of the machines is not shown as the wear and maintenance costs will be dependent on the material being crushed. Indeed, if the silica content is high, the impactor should not be considered at all. In any event, replacement costs will be high and, for a material such as limestone, the decision of initial machine choice may very well rest on the operators' own requirement to differentiate between capital and operating costs.

3.2. Secondary

No differentiation will be made here between secondary and tertiary crushing, although mention will be made of graduations in machine types that allow for finer crushing. Almost exclusively, secondary crushing is effected by cone crushers, these being the most developed form of machine. There is some use of crushing rolls.

The area between crushing and grinding is somewhat grey. Rod mills have been taken as the best way to achieve material degradation as the quality of the rods has improved. However, the waste of energy in heat and movement of the slurry has swung the fashion back to finer crushing in the dry state. To confuse the issue, autogenous grinding has improved dramatically with the ability to manufacture large diameter mills. There is an efficiency limit here as well and the top-size has probably been reached. Where the material is not suitable for autogenous grinding

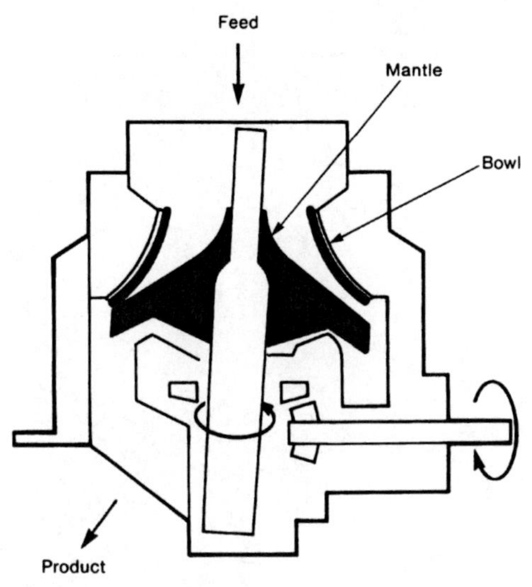

Fig. 25.5. Cone crusher.

and media has to be added to achieve the grinding required, then in all probability it would be more cost effective to revert to finer crushing.

3.2.1. Cone Crushers

There are similarities between cone and gyratory crushers as discussed earlier. The main distinction is that the fixed crushing surface (the bowl) is a cone in the same orientation as the mantle for a cone crusher. This is because it does not need to be designed to cater for large oversize feed. The crushing surface is inverted for the gyratory (compare the schematics in Figs 25.2 and 25.5). This means that the flare of the mantle can be much shallower, allowing greater room for expansion of the crushed rock and thus giving high capacities while still being choke fed. Other differences are that the shaft in a cone crusher is not supported at the upper end and the eccentric is operated at higher speed, being subjected to less stress, giving more impact than compressive crushing.

Classification of cone crushers falls into two types, the standard and the shorthead, the terms referring to the shape of the crushing chamber and coined by Symons, the main manufacturer. The standard tends to be used for secondary crushing and the shorthead for finer tertiary crushing. A cut-away view showing the detail of both types can be seen in Fig. 25.6. The difference between the standard and the shorthead is in the profile of the bowl, the standard having a much bigger size change from the feed to the discharge opening. The reduction ratio of the standard is in the range of 3–5 whereas the shorthead is of the order of 2.

Adjustment of the crusher is important to compensate for wear of the liners or

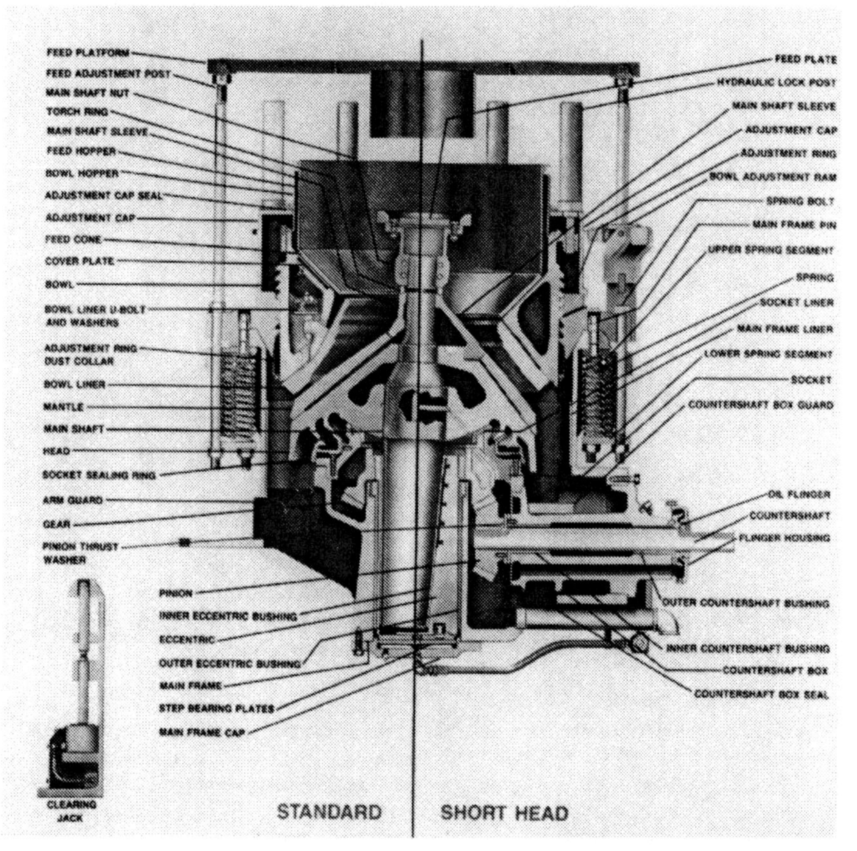

Fig. 25.6. Standard and shorthead pattern cone crushers. (Courtesy of Nordberg (UK), Ltd.)

for variations in the feed material. Automation of the clearance adjustment leads to complete process control possibilities in the crushing circuit. Power draw can control the feed and the percentage of return oversize can control the setting. It is usual for secondary crushers to operate in closed circuit with screens, and almost universal that tertiary crushers run in closed circuit.

Crusher size is specified by the diameter of the mantle. Selection of the correct size is often complicated. The prime consideration is the feed particle size. This will determine the crusher feed opening required. Too small an opening will restrict feed entry and reduce the effective capacity, whereas too large an opening will permit feed to fall too far through the crushing chamber before nipping and will also be detrimental to capacity. The presence of fines in the feed will also tend to clog the crushing action and it is common practice to have the closed circuit screens

TABLE 25.4
Cone Crusher Capacities

Crusher size (in.)	Crusher size (m)	Type	Circuit	Feed size (mm)	Product size (mm)	Capacity (th^{-1})
36	0·91	Standard	Open	125	12	50
				125	25	80
36	0·91	Shorthead	Closed	50	12	55
84	2·13	Standard	Open	300	25	600
				300	50	1 100
84	2·13	Shorthead	Open	150	12	300
				150	25	500

ahead of the crusher feed rather than on the crusher product. Selection of the crusher will also depend on whether it has to cope with a return circulating load.

Table 25.4 shows typical capacity data for cone crushers and typical product size distributions are given in Fig. 25.7.

3.2.2. Roll Crushers

Roll crushers are still in use, despite being somewhat old-fashioned and often having been replaced by cone crushers. Nevertheless, they find application in crushing soft, friable, less abrasive feeds and have the ability to produce a close-sized product that can be as fine as a rod mill discharge material.

The usual configuration is to have two horizontal cylinders rotating towards each other. The gap between the rolls can be adjusted by threaded bolts. In this case there is no relief of the spacing to protect against uncrushable tramp material and

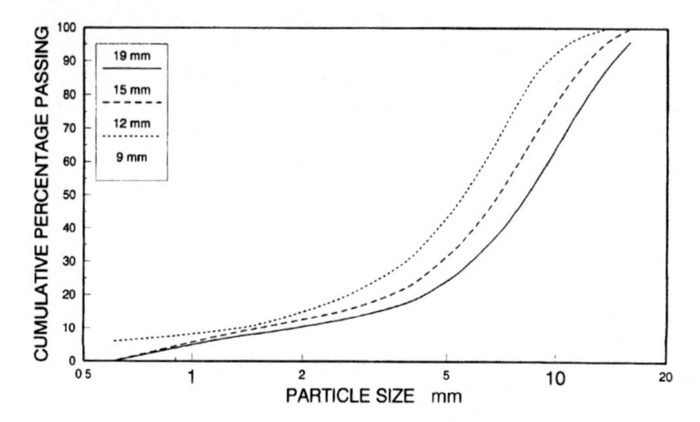

Fig. 25.7. Typical cone crusher particle size distributions—as a function of closed side setting.

the bolts at some point in the mechanical train rest against thin metal plates which shatter if the loading is too great. Alternatively, the rolls are held together by springs with the adjustment by shims. The disadvantage of this system is that if the feed causes the rolls to be forced apart, they will often stay spread until the feed stream ceases. This causes problems if the product is required to be accurately top-sized. The rolls are individually belt-driven.

The relationship between the top-size of the feed and the diameter of the rolls is critical. If the particle is too large it will not be nipped and will bounce out. This can be overcome to some extent by slowing the rolls and so increasing the coefficient of friction between the particle and the roll surface, but the latitude is restricted. Wear rate of the roll's surface is high for abrasive materials and replacement of the roll surface is not particularly easy.

The reduction ratio of rolls is small, rarely above 4, as there is no room for expansion of the material as it is crushed. Thus, the rolls must be fed gently and not allowed to choke. Feed must be evenly spread across the width of the rolls and a vibrating feeder is usually necessary.

Capacity can be calculated from the theoretical ribbon of crushed product by relating the peripheral speed of the rolls, the width of the rolls and the gap. Actual capacity is some 25% of this figure to allow for voids and slippage of the particles with respect to the rolls.

4. GRINDING

4.1. Introduction

Grinding is the final stage of comminution and will take material down to 10 μm top-size. Particles are reduced by a combination of impact and abrasion, either wet or dry, in rotating cylindrical, continuously fed, steel mills. The mechanism involves the action of the loose charge tumbling inside the mill. The media used to effect the grinding include rods, balls, hard selected rock, or the material itself, if suitable.

Particles are not reduced by positive action between surfaces as in crushing and the mechanism is much more random. Some particles are impacted by the media too often causing overgrinding and much energy is lost in media-to-media contact, friction, noise and heat. The power drawn by an operating mill will not change noticeably between the conditions of feed on or off. It is likely that grinding accounts for over 60% of the energy consumption of a full ore processing plant. A typical copper, zinc or similar concentrator will consume perhaps 12 $kWht^{-1}$ of ore in the grinding stage, compared with 2–3 $kWht^{-1}$ in crushing and the same for all other processes together. For this reason it is essential to grind the material only to the coarsest size permissible.

4.2. Grinding Circuits

There are two broad classifications of circuits used: open and closed (see Fig. 25.8). In the former, the mill and operating conditions must be selected to give the

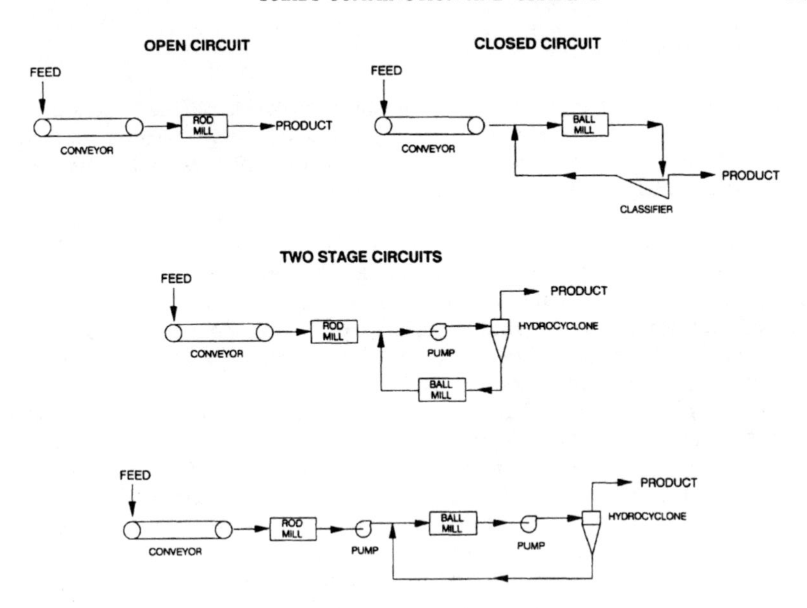

Fig. 25.8. Grinding circuits.

required product in one pass. This is not common practice as there is no control on top-size unless the circuit is only the first stage of grinding. Closed circuit means that the mill is in circuit with a classifier and the coarse oversize material is returned to the grinding unit. This gives control of oversize and tightens the product size distribution. It is not the intention to carry out all the grinding in one pass as this will produce overground material. Instead, the circuit is designed for a large return to the mill, called circulating load, and this is often in the range of 250% weight with respect to the new feed.

For a multi-stage milling circuit, it is usual to have a rod mill in open circuit followed by a ball mill in closed circuit. The classification stage may be at the rod mill discharge, removing fines before the second stage, or it may be at the ball mill discharge. The former is more common, involving only one sump and pump for the system. The classifying device can take many forms and these will be described in detail later in the chapter. Mechanical classifiers contain large volumes of water when handling fine products which tends to smooth fluctuations in the circuit. In contrast, the hydrocyclone contains a smaller volume resulting in a shorter residence time, allowing changes to be made to the circuit with faster response.

Control of grinding circuits is a very well documented subject and will not be described in detail here. The control strategies are usually to maximise throughput while maintaining a required product size, or to maximise throughput while maximising a down-stream recovery level (Lynch, 1977; Herbst & Rajamani, 1979; Wills, 1988).

4.3. Tumbling Mills

Horizontal, rotating grinding mills fall into three categories: rod, ball and autogenous. There are other types of fine grinding mills and these will be dealt with at the end of this section. However, tumbling mills can be described generally, and the uniqueness of each type will be summarised afterwards.

There are many small variations on the themes, but tumbling mills consist of horizontal steel shells, internally protected with replaceable liners. The shell is carried by hollow trunnions, running in bearings, attached to each end wall of the shell. Feed and discharge enters and leaves through these trunnions. The drive is normally transmitted to the shell by a large girth gear mounted around the circumference of the shell. Variable speed drive has been used and is a useful control if the capital expense can be justified.

The shell is fitted with replaceable liners, bolted through to the outside. The types of liners are many (Howat & Vermeulen, 1986), but most incorporate a wave form, step, rib, or are smooth with the intermediate liner being proud. They all promote the lifting action of rotation on the mill media. Rod mills have less flexibility, the lifting action must be equal along the length of the rod, and the mill end liners are usually smooth. Ball mills can exhibit greater variety—even liners that have a spiral lifter built in have been reported as being successful. The liners are of hardened steel, manganese steel, or even chrome steel, but Ni-hard is perhaps the most common. Replacement of liners is costly, not only in terms of the hardware, but also in the down-time incurred while the mill is out of production. It is probable that the most expensive liners are the cheapest overall. Rubber liners are in use, less so for rod mills. They offer ease of replacement, being lighter and faster to install, but are subject to greater wear. They do have the advantage of rendering the mill significantly quieter in operation.

Control of product size can be influenced by the amount of media charge, the feed rate of the material, the speed of rotation of the mill and the classification stage usually incorporated in the mill circuit. The media charge occupies between a third and half of the volume of the mill; there are many formulae for calculating the size of the media. An impression may be gained from a 2 m diameter mill which could contain 50 mm rods or 25 mm balls. Due to the rotation of the mill and the friction of the mill liners, the media is lifted up the rising side and then cascades down the mass of the charge. If the speed of the mill is set at this point, then all grinding would be by attrition. The results would be a fine product, but wear rates of media and the mill liners would be very high. At increased speed, the media free falls from a point nearer to the highest part of the mill shell. It then impacts on to the toe of the rotating media mass. Grinding then becomes by impact of the media on to particles. Increasing speed again will cause the media to begin to centrifuge in the mill. The speed at which this starts to occur is known as the critical speed. This can be easily calculated given the diameter of the mill and the media. Complete centrifuging would result in no grinding action. Mills are run at a speed around 75% of the critical speed.

4.3.1. Rod Mills

These are usually the first stage of the grinding process and are really fine crushers. They can take material down to 0·5 mm in size, but 1 mm is more common. They are preferred to a fine crusher when the material to be reduced is damp, wet or contains clay. The significance of the dimensions of a rod mill is that the length is between 1·5 and 2·5 times the diameter. This is critical in that a lower ratio will encourage the rods, which are marginally shorter than the internal length of the shell, to wedge across the shell causing tangling. Conversely, the length must not be too great, or worn rods may tend to snap, with the same result. The maximum mill length is about 6 m. The size of mill required is calculated by using the equations outlined at the beginning of this chapter.

Feed is introduced by a spout, or by rotating scoop, integral with the feed trunnion. Discharge is normally by overflow through the trunnion, although peripheral discharges, either at the discharge end or even in the centre of the shell, are in use.

The action of grinding is predominantly by impact of rod on rod, trapping the particles between them. This is significant in that it is evident that the largest trapped particle will be broken first, protecting the smaller particles in the gap. This leads to rod mills generating a closer size range than other tumbling mills, and with less production of undesirable fines in the discharge. It is unusual for the mills to have a classification stage incorporated; to some extent both top-size and bottom-size are controlled internally. There are, in fact, good reasons not to have a rod mill in closed circuit. If, for whatever reason, oversize is returned to the mill in large quantities, it will increase the rod–rod separation. This reduces the grinding action and also reduces the available space for feed, further reducing the grinding. The discharge will contain even more oversize which will be returned to the mill. The situation will escalate and the charge will swell blocking all feed entry. A major spillage will result, unless new feed can be quickly stopped when the situation first arises.

4.3.2. Ball Mills

Ball mills, by definition, use balls normally made of steel as the grinding media. These mills have a length that is less than twice their diameter, although longer ones are in use and should be called tube mills.

The balls are made of either forged steel, alloy steel or sometimes cast steel. They should be spherical initially, although in a well-operated mill they will wear to a pyramidal shape. High chrome balls may be an advantage (Meulendyke & Purdue, 1989). Unconventionally shaped media have been used, but the wear rate may be higher. Flashings on cast balls or imperfect spheres are not cost effective. Excess metal wears away quickly, and does no grinding, but has been paid for. The smallest ball should be used for charging the mill, providing it exerts enough energy in impacting to crush the largest particle, as smaller balls have a relatively larger surface area providing more point contacts for crushing. Media consumption is a significant operating cost, up to 1 kg of balls being used per tonne of material ground.

Particle breakage is random and consequently a much wider size distribution occurs in the discharge. This means that the mill should be operated in closed circuit with a classifier to return oversize to the mill and control the top-size of the product.

Efficiency of grinding in the mill can be controlled by several aspects. The slurry density is important and slurry concentration should be in the range 65–80% (by mass) of solids. The higher density slurry sticks the particles to the media, but this phenomenon is also affected by the particle size and lower densities are used for finer grinding. The mill should not be allowed to run at excessively high densities, because the complete charge mass may *freeze* with consequent spillage and time-consuming emptying of the mill. Density control by means of monitoring of the discharge related back to feed make-up water is essential, whether this is manual or completely automatic. Energy input to a mill increases with ball charge up to about 50% by volume and then reduces. Speed should be high as energy input is proportional to speed and should be set as near to critical as possible. The maximum is 80% of critical and will give most breakage by impact. This is useful for hard material or oversize feed.

The mill can be discharged through the trunnion. The alternative is by grate discharge, where there is an internal grate preventing the balls from emerging; the slurry flowing through is then elevated to the discharge by lifters. The extreme of this type is where the mill is mounted by means other than a trunnion bearing, as in a tyre-mounted mill, and the slurry can drop through the grate to the discharge launder. This results in a low level of slurry in the mill and gives a shorter residence time. This can prevent over-grinding of a normally softer, valuable mineral.

The maximum size of ball mills used to date is 6·5 m in diameter and 9·65 m long with 9·65 MW installed power and motors integral with the drum. In terms of efficiency, this size has been reported to give a 30% reduction in operating cost per tonne of ore treated over an equivalent set of smaller mills. However, selection of very large mills is not straightforward and there are problems of performance predictions at this level of scale-up (Harris & Arbiter, 1982). This seems to be associated with the effect of ball size.

4.3.3. Autogenous Mills

This type of mill has gained in popularity recently as the technology to build large mills has improved. Autogenous milling is defined as the use of the rock to effect grinding, hence saving grinding media costs. This may be by direct feed of the entire ore stream. In some cases, large rock when its proportion is small, may be screened out and added back to the ore stream after the remaining finer ore has been through a stage of conventional crushing. Semi-autogenous grinding (SAG) uses a small proportion of steel media, 6–10%, added to the mill charge to assist grinding. Either technique can be carried out in the dry state, although this can lead to severe dusting and noise problems and is not suitable for damp ore.

The characteristic of autogenous or SAG mills is their large diameter to length ratio, which is required to provide the necessary impact force from the falling rock of lower density than conventional media. Some of the largest are at the

Chuquicamata concentrator in Chile and are 9·75 m in diameter and 4·57 m in length. These have annular, gearless drive motors of 8·2 MW installed power and can treat approximately 1 000 th^{-1} using a small addition of 127 mm balls (Rosas, 1990). In the Republic of South Africa, 9·75 m diameter mills take 800 th^{-1} from 180 mm to less than 0.3 mm autogenously using 7 MW installed power.

In autogenous mills, particles tend to cleave along natural grain boundaries. This gives better liberation as the particles are not being impacted by indiscriminate media. This occurs for iron ore grinding and some industrial mineral applications, but it is less important in milling fine grained or homogeneous materials.

The smaller plant and reduction in conveyor systems is important in capital cost considerations. However, power consumption is probably higher than the equivalent conventional circuit. Selection of autogenous mills is best left to the manufacturers, who can carry out pilot-scale test work. The use of laboratory equipment to predict mill sizing is not practicable in view of the mechanism of the grinding action.

4.4. Tower Mills

A recent addition to the range of mills for fine grinding is the tower mill illustrated in Fig. 25.9. The device was developed in Japan for wet grinding of fine materials. Feed enters at the top along with water and is reduced by abrasion and attrition while the media is agitated by a vertical screw. The ground material is carried up by rising liquid and overflows to a classifier with the oversize returned to the mill base. The lack of impact grinding limits the top size of feed and 5 mm is the norm. The balls used have a maximum size of 32 mm and can be as small as 6 mm for very fine applications.

The advantages are a small floor-space area, low operating costs and reduced noise and vibration. Uses are predominantly in the industrial minerals field although they have been installed for re-grinding concentrates and are capable of throughputs of up to 100 th^{-1} (Russell, 1989; Kral, 1990).

4.5. High-Pressure Roller Mills

This new generation of roller mills works on the same principle and mechanical layout as roll crushers but are much more akin to briquetting presses or even steel-rolling mills in that the force exerted between the rolls is extremely high. The pressure produces an agglomerate of crushed material that requires light grinding to break up the particles. Even so, the required total energy consumption is claimed to be as little as 50% of the conventional equivalent milling stage because virtually all the energy is devoted to particle crushing, rather than the random action of a wet grinding mill. They are in use on industrial minerals, such as cement and limestone, but may find application in the metalliferrous minerals field and some work is being carried out on this (Brachthauser & Kellerwessel, 1988; Clarke & Wills, 1989).

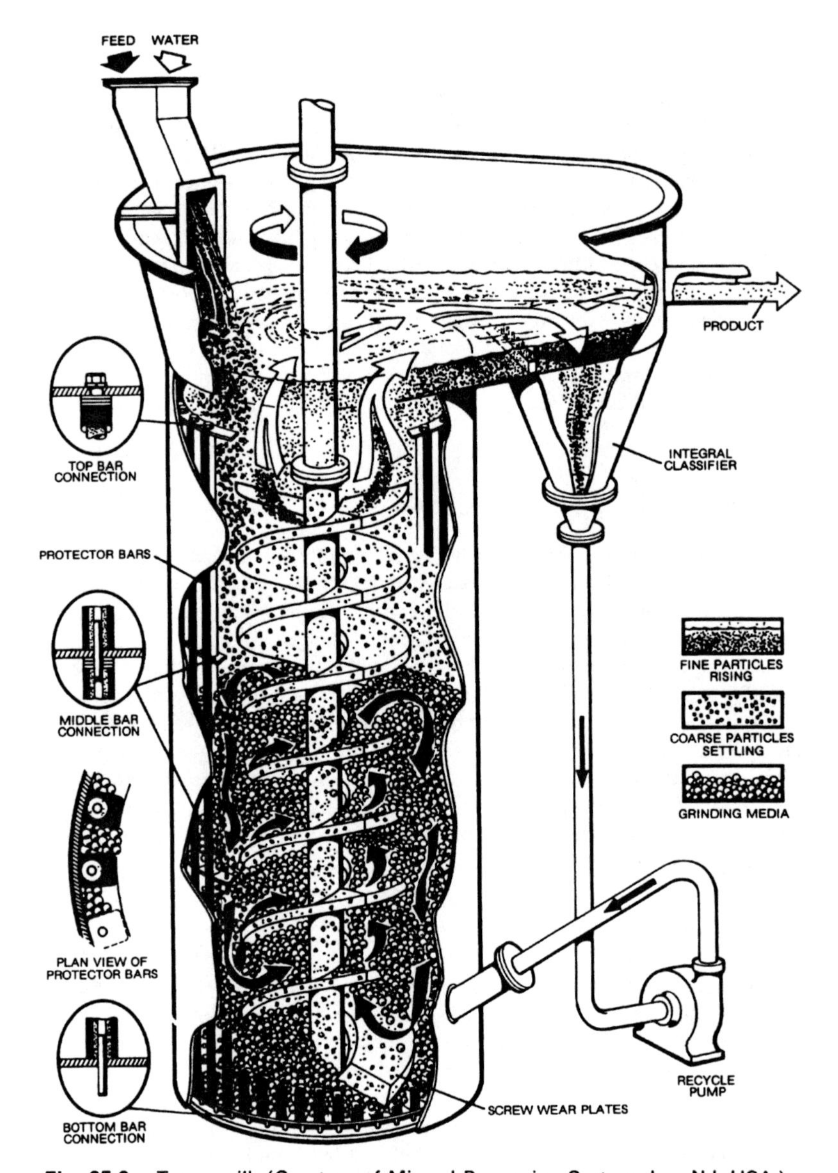

Fig. 25.9. Tower mill. (Courtesy of Mineral Processing Systems Inc, NJ, USA.)

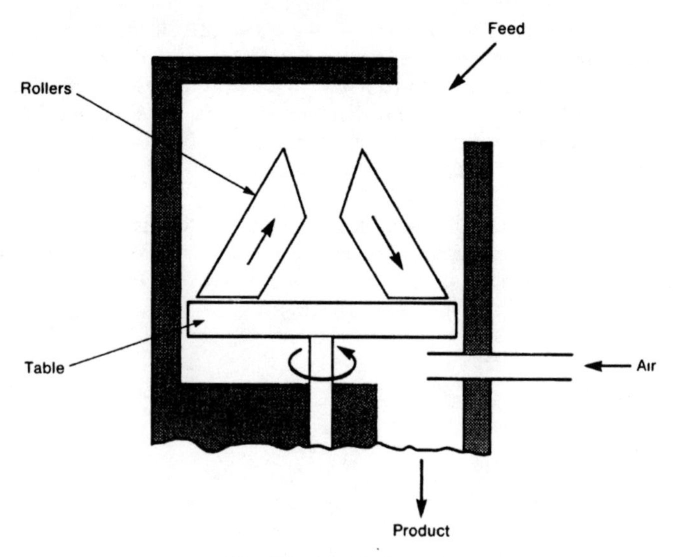

Fig. 25.10. Roller mill.

4.6. Roller Mills

These are used for the fine grinding of soft dry materials in the industrial minerals field. They consist of two or three rollers, held down by springs, running against a horizontal rotating table onto which the feed is distributed as shown schematically in Fig. 25.10. Ground material falling off the table is air swept into cyclonic classifiers with the oversize returned to the mill.

5. SIZING

In the same way that size reduction can be split into the fields of crushing and grinding, sizing can be split into screening and classification. The two methods almost correspond to the size reduction stages.

The grading of a material consisting of particles with a range of sizes involves separation of the material into a discrete number of samples in which the particles lie between known size limits. Various industries call this different names. The aggregates sector uses grading, which is confusing to the ore processors who use grade as a percentage measure of the valuable constituent of an ore. For the purposes of this chapter, sizing or size distribution will be used. It is normally expressed as a weight percentage passing a given size or screen aperture, e.g. 80% passing 10 mm. The alternative is the converse, 20% retained at 10 mm. A single size parameter is unreliable, as the slope of the size distribution could change below the defined size. Thus, it is important to quote at least two points, such as the 80 and 20% passing sizes. These passing sizes are also often defined as mesh sizes, originating from the number of apertures in a screen per inch of the sieve; 100 mesh implies 100 apertures in.^{-1} and corresponds to a size of 0·149 mm. Unfortunately, different countries still use different diameter wires in the sieve and hence the aperture can vary even though the mesh is the same. To circumvent this, sizing should be quoted in percentage passing a specific size, quoted either in millimetres or micrometres.

The easiest visualisation of a size distribution is to plot the cumulative weight percentage passing a given size against that size. If this is done on log–log or log–probability paper, the result often approaches a straight line or a gentle S-shaped curve which will define any required parameters.

6. SCREENING

6.1. Theory

During screening ideally all particles smaller than the screen aperture pass through and all particles larger report to the oversize. However, this is rarely achieved in practice and it is advantageous to measure the efficiency of a screening operation so that variables such as vibration frequency and quantity of spray water can be assessed.

The simplest formula to calculate efficiency, E, is given by

$$E = (c - f)/c(1 - f) \tag{25.11}$$

where c is the fraction of material greater than the cut size in the oversize product and f is the fraction of material greater than the cut size in the feed. The formula assumes that no oversize reports through the screen.

A more realistic measure uses the partition curve for a screen operation. This is most useful when a large proportion of material has a size close to the required cut size of the screen. The curve is constructed by plotting the percentage of a given size range reporting to the screen oversize with respect to the feed, against the geometric mean size of that range. The ideal cut point, that is the screen aperture, would appear as a vertical straight line at the screen size. In practice, the curve is an S-shaped one and the steepness of the curve measures the efficiency. The size corresponding to 50% probability is the actual cut point of the screen and is invariably smaller than the screen aperture.

The main factors affecting screen efficiency include the following.

Screen deck size. This is the most significant. Obviously a very long screen would allow all finer particles to pass through, but this is economically impractical. Reducing the area of the screen increases the depth of bed of material flowing across it and hence hinders all particles from having a chance of being presented to the openings. Generally, width influences capacity and length efficiency (Suttill, 1990). High capacity is in opposition to high efficiency and the result will always be a compromise. The screen dimensions should always be as large as the capital investment will permit.

Vibration. The majority of screens are vibrated, with the exception of fixed screens performing a very coarse tramp removal operation. The vibration helps to prevent blinding of the apertures, that is, clogging by near-size material. Vibration also helps to stratify the bed of material, allowing fine particles to pass down and through the screen, as well as increasing the number of times a particle strikes the screen, which increases the probability of particles passing through. However, too great a throw of the movement causes material to bounce and will reduce the efficiency. Too high a frequency may cause material to float along the screen, reducing the opportunity to pass through.

Presentation angle. The angle of presentation of the feed and the angle of the screen deck affect efficiency. Presentation should be at right angles to the surface and the screen should be horizontal, but this is also not possible in practice.

Screen material. For a given area of screen deck, the type of screen material chosen will have different available open area, that is the net area of aperture per area of screen. Rubber decks have significantly less open area than steel mesh decks, but may have other advantages; for example, the flexibility will reduce blinding.

Moisture content. Even bone-dry feeds will carry some fines across, damp feeds are almost impossible to screen and very wet feeds may screen well but are a disaster for the following machinery. However, recent developments show that flushing material through a protected crusher can increase capacity, reduce power and assist subsequent processing (Anon., 1989). Despite this, an appropriate combination of water sprays that improves efficiency but which is not deleterious to the rest of the materials handling system has to be determined in practice for the particular substance being screened.

6.2. Screening Devices

6.2.1. Moving Screens

The majority of screens move by some method. The simplest, and perhaps the oldest, is the rotating screen or trommel, which consists of a cylindrical drum made from screen mesh rotating in a plane slightly inclined from the horizontal. Feed is introduced at one end and the oversize passes through the drum emerging at the other. The screen surface can be made from a number of sizes of aperture, giving a range of sized products. However, the finest screen must be at the feed end and is subject to considerable wear from the total feed. It is possible to have a concentric series of cylindrical screens with the coarsest in the middle, thereby accepting the total feed, but maintenance is difficult, it is also hard to prevent the apertures blinding and is harder to clean them.

Shaking screens move reciprocally in the horizontal direction along the length of the screen. They are used for coarse separation but are subject to abrasion from hard materials. Reciprocating screens have a horizontal circular movement applied to the feed end. The discharge end is allowed to move in the direction of the screen surface. Therefore, the motion changes from circular, through eccentric, to straight line. They are mainly used for dry screening of light fine materials and are also subject to wear.

Vibrating screens are the most widely used and the vibration is in a direction at right angles to the screen surface, which is usually at a slope. The vibration is imparted by rotating out-of-balance weights carried on a shaft above the screen and attached to the screen frame. The alternative is to use a magnet, oscillated by alternating current electricity. The screens may be mounted on springs or on rubber pads. It is possible to fit multi-layered screen decks within the screen frame.

All vibrating screens are noisy, particularly for coarse applications. Levels in excess of 100 dB(A) are not uncommon. Vibration levels are high and the screen frames and carrier need to be substantially mounted.

A more recent screening device is the linear screen. In this, the screen surface is a coarse cloth that takes the form of a continuous loop carried around rollers and advancing by means of a driven head pulley. The slurry to be screened is fed onto the top of the cloth by a distributor and the fines pass through to a collection launder. Oversize is discharged at the head pulley into a chute. The cloth doubles around rollers where sprays remove any blinding material. To date, the main application is in the removal of small amounts of tramp oversize or trash, but the advantages of low power consumption and lack of noise or vibration may see this developing (Anon., 1986).

6.2.2. Fixed Screens

Stationary screens take two main forms. The fixed bar type, used for very coarse screening and known as a grizzly, is provided for tramp protection in feeding crushers. The bars, rods or rails are set at a steep angle to allow oversize to slide off. Frequent clogging is a common problem.

The exception to the statement that classifiers go with grinding is the sieve bend, originally termed the DSM (Dutch State Mines) screen, which has common use in grinding circuits. The sieve bend has a curved screen consisting of horizontal wedge wires with the top end of the screen almost vertical. Feed enters via a distribution box, falling vertically and following the curve of the screen. The layout is shown in Fig. 25.11. The particles pass successive openings with a portion of the water being removed through the slots. The size of particles passing through is not the size of the slots, but is a function of the apparent slot width presented to the slurry flow. In practice, the effective cut size is about half that of the slot width. This has a major advantage in that little blinding should occur. However, in most applications the flow of slurry decreases and the effectiveness of the action reduces, allowing coarser particles to attempt to pass through the slots and these can blind the screen. For this reason, the Dorr Rapifine depicted in the illustration has a rapper which is pneumatically actuated by a timer and periodically kicks the screen

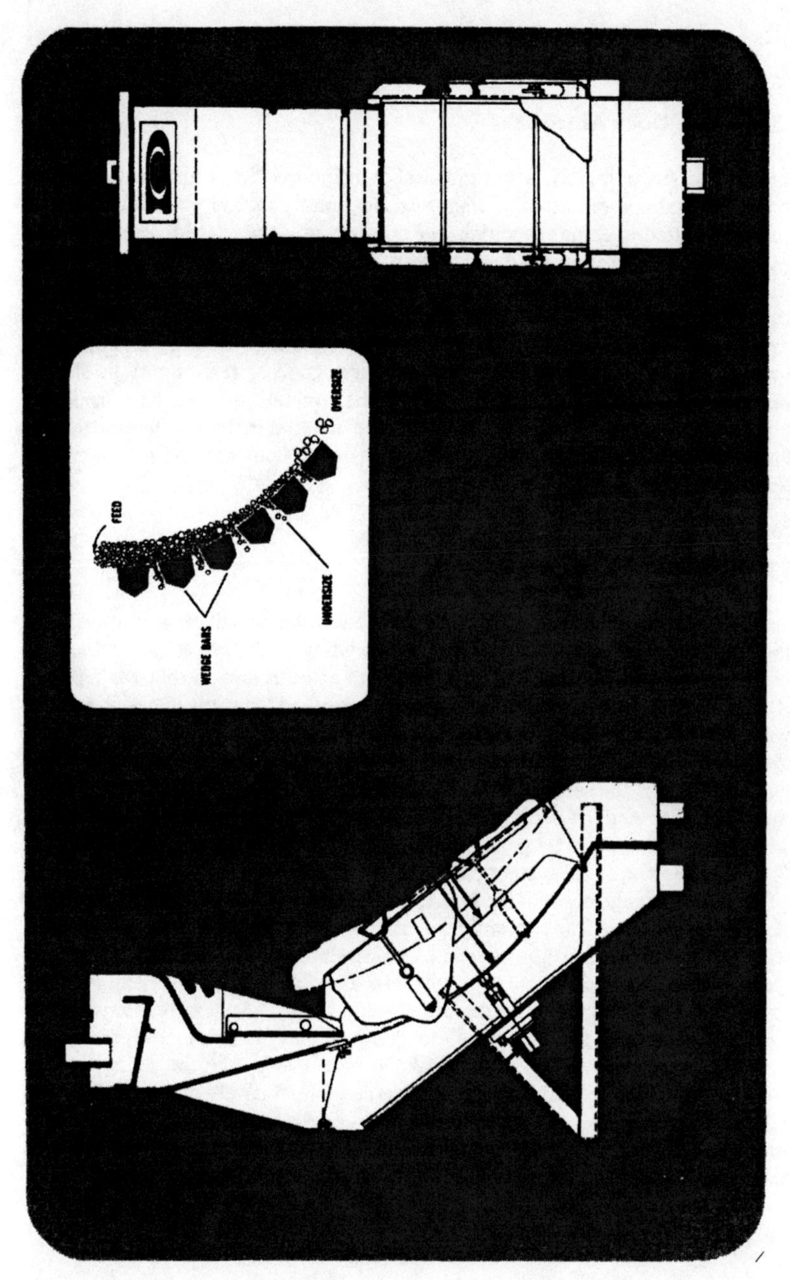

Fig. 25.11. Sieve bend. (Courtesy of Dorr-Oliver Co. Ltd, UK.)

jarring particles out of the slots. The size of effective separation can be as small as 50 μm.

6.3. Screen Deck Materials

There is a variety of materials that can be chosen and the screen surface will depend on the material and the use. For coarse applications, punched plate with circular, square or slotted openings is common. Rods held in place in the frame are useful in being replaceable in areas of high wear. Woven wire cloth is used for a wide range of aperture sizes, and the wire diameter can be selected to minimise wear and hence replacement time at the expense of open area. Wedge wire, often in stainless steel, is used for fine applications and has the advantage of being less prone to blinding. Rubber or polyurethane are extremely good at resisting abrasion and impact wear and are significantly quieter than metal surfaces. The flexibility reduces blinding, but to maintain strength, the material is thicker than metal and hence the open area is low. Polyurethane is better than rubber in this respect (Anderson, 1990).

7. CLASSIFICATION

The laws governing classification relate to the settling velocities of particles in a fluid medium. The theoretical treatment, including both Stokes' and Newton's laws, is covered in Chapter 2. The significance of both laws is that the terminal settling velocity is affected by both particle size and density; increasing either increases velocity. For two particles falling at the same speed, but with different sizes and densities, Stokes' or Newton's law applying to each can be ratioed. The resultant figure is the settling ratio. As an example, for two particles with specific gravities of 2·65 and 7·5 settling in water in the Newton regime the ratio is 3·94. This means that the lighter particles which are 3·94 times the size of the heavier particles will settle at the same rate.

There are two conditions of the settling environment which can affect the settling ratio. When the particle concentration is low, the condition is known as free settling. The calculation above refers to this state. If the number of particles increases above approximately 15% solids by mass, the suspending medium has a density effectively greater than the density of water. Turbulent resistance is the dominant effect and the settling ratio from Newton's law, incorporating the suspending medium density increase, would give a value for the same particles above of 5.2. Thus, denser particles will settle comparatively more quickly than under free settling. This gives rise to two types of classifiers; those which use free settling and are primarily size separators, and those which use hindered settling and can exploit the density separation, often to concentrate a desired mineral.

7.1. Hydraulic Classifiers

These use hindered settling regimes and normally consist of a column, or series of columns, in which water is introduced at the base to rise against the falling particles introduced in the feed stream at the top of the column. The overflow from the first stage flows on to the second stage, and so on. The products are removed from the base of the column through a restricted discharge which is throttled to balance the particles being removed while maintaining the upward liquid flow. The simplest device can be a single chamber with manually operated discharge. The most common type is the hydrosizer, consisting of a number of chambers. Typically, 4, 6, or 8 are in use. The discharge is controlled by a spigot valve that is set to adjust by means of the pressure in the chamber. As the amount of solids increases, the pressure increases and the valve will open to compensate. Under steady-state conditions, the operation is very well-controlled and efficient. The close sizings of the spigots are ideal for feed to subsequent, particularly gravity separation, processing stages. A change in feed quantity will be adequately accommodated by the device, but the solids in the spigot discharges will alter accordingly and this can be a problem for the subsequent processes.

7.2. Mechanical Classifiers

Free settling is the regime most used in this type which consists of a tank where the classification takes place and an inclined trough where the settled solids are removed by mechanical means. At the opposite side of the tank from the trough, the finer particles flow over an adjustable height weir. Feed is introduced from the side of the tank close to the end of the pool at the trough side in order to avoid turbulence near the weir.

The mechanisms used to remove solids are either a reciprocating rake or a spiral consisting of a series of flights carried on a rotating shaft. The elevating of the coarse solids allows draining of the material and sprays can be used to wash fines back down into the pool. The height gained is a prime reason for widespread use of these machines in grinding circuits. Mill discharge can flow into the classifier and the coarse material can be discharged into a chute returning oversize to the mill and eliminating the need for pumps. However, the floor space required is large and the capital cost is high so the hydrocyclone is the preferred modern option. Although operating costs are low, the wear rates on the rakes or spiral can be excessive and sacrificial polyurethane shoes are fitted.

Control of the classifier can be effected by a number of means. Initially, it is important that the speed of the rakes or spiral is sufficient to remove the settled solids. However, this must not be excessive otherwise agitation of the pool will occur and the cut point will change and efficiency of separation decrease. The adjustment of the weir height will change the pool area and, for a given flow into the pool, this will change the rising liquid velocity altering the cut point.

Water addition is the most critical variable. If the feed arises from a mill discharge at high slurry densities, this must be diluted for satisfactory classification. If the pool is in hindered settling conditions, increasing the water addition will increase the upward velocity, but, at some point, the mechanism will change to free settling and the cut point will actually occur at a finer particle size. A further increase in flow rate of water will increase the upward velocity again and, in this situation, the cut will occur at a coarser particle size. The point at which this occurs is known as the critical dilution. If the classifier is operating at this density, any change in water addition will coarsen the split.

In practice, a classifier will normally be operated below the critical dilution and water addition will be used to control the overflow size. This does give rise to a problem with this type of classifier in that the overflow will be fairly dilute and often needs to be thickened prior to the next processing stage.

If the classifier is operated on material that contains a high-density valuable mineral, fine mineral will return to the grinding circuit and this can be a disadvantage. This fault can be capitalised on by providing a recovery stage in the closed circuit. Mineral can be recovered immediately it is liberated and the liberation can be controlled to be in gentle stages if the mill is operated at high circulating loads.

7.3. Hydrocyclones

The hydrocyclone is probably the most widely used classifying device. It has a low capital cost, takes up little area, has low operating costs, requires very little operator control and can be used in a wide range of sizing applications as well as dewatering (see Chapter 27), de-sliming and a concentration device in coal washeries.

A hydrocyclone (often referred to misleadingly as just a cyclone) consists of a central cylindrical section with a conical lower piece open at its apex, which is the coarse fraction discharge, underflow, or spigot product. The upper part of the cylinder has a tangential feed inlet and the top is closed with a plate containing a tube, or vortex finder, which allows the fine fraction to overflow. A cross-section is shown in Fig. 25.12.

Materials of construction are steel or aluminium, often with replaceable liners which may be rubber or even ceramic. Moulded polyurethane for the entire cyclone can exhibit remarkable savings in wear rates and has the advantage of being light and easy to maintain.

Feed is pumped under pressure through the tangential inlet and swirls around the body, the vortex finder preventing any short-circuiting to the overflow. The centrifugal action generates an air core through the body of the cyclone. The forces acting on a particle are the centrifugal outward force and a radially inward drag caused by the flow of slurry from the inlet to the products (see Svarovsky, 1984). Rates of settling are enhanced by the higher than gravitational force. Coarse or dense particles are forced to the outer wall of the cyclone and then migrate down to the spigot. The smaller, lighter particles move to the air core and are then carried by the bulk liquid flow through the vortex finder to the overflow.

The performance of a cyclone can be calculated in the same way as for a screen by means of a partition curve. This will give a size corresponding to the 50%

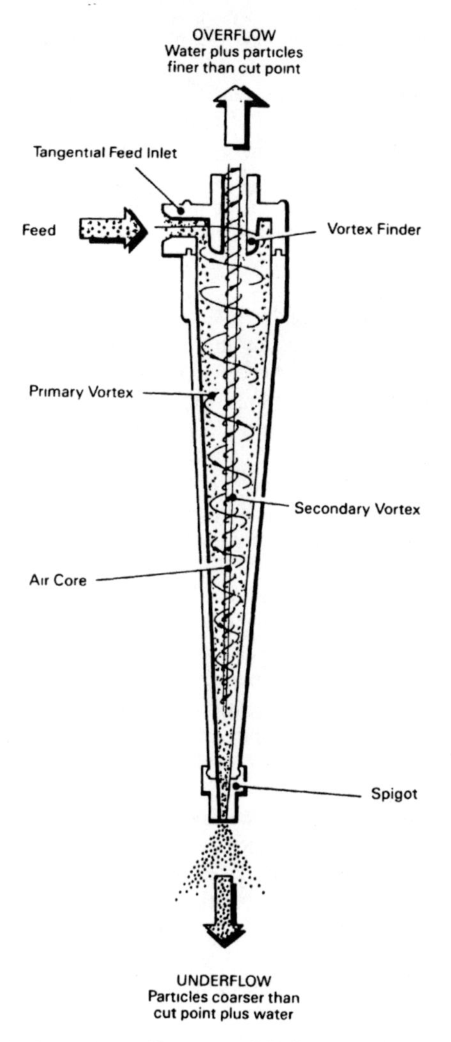

Fig. 25.12. Hydrocyclone. (Courtesy of Richard Mozley Ltd, Redruth, UK.)

probability point which is the cut point. Efficiencies are often calculated by taking the difference between the 75 and 25% cut points and dividing by twice the cut point. A proportion of feed water will report direct to the underflow taking that proportion of feed with it. The corresponding quantities of feed in each size range are sometimes deducted from the actual amounts used to generate the partition curve and the resulting new curve and efficiencies are often quoted as being *corrected* data and partitions. There are also reduced efficiency curves. All this is confusing and the real quantity that needs to be measured is the overflow or underflow size distribution. This will confirm that the product is that which is required and will indicate the amount of misplaced material.

An inordinate amount of time and energy has been devoted to deriving equations to predict and understand cyclone performance (see Bradley, 1965; Plitt, 1976). Mathematical modelling can be useful for predicting changes to performance in large-scale operations where plant test work would be difficult or time consuming. Otherwise, it is not impossible to change an operating parameter and observe the effect.

The main operating variables can be qualitatively described. It is important that the feed to the cyclone is kept constant and pump sump level control should be used; make-up water can be recycled cyclone overflow. Surging or air entrainment must be avoided, otherwise the cyclone will stop classifying. Feed pressure is important and the cyclone should have a pressure gauge on the inlet for monitoring purposes. An increase in pressure will increase the capacity of a cyclone and will reduce the cut point. As a result, the overflow becomes finer and the quantity of material reporting to the underflow will increase. The efficiency of the separation will also increase as the centrifugal force effect is enhanced. Pump power and wear rates will also increase. Increasing feed slurry density decreases the efficiency of the separation and also coarsens the cut point. Normal feed densities are around 30% (by mass) solids. Fine separations benefit from lower values and grinding circuits can be run at much higher levels if the inefficiency can be accepted. The area of the inlet can be changed for each cyclone and this will affect the flow rate and the pressure. Vortex finders are interchangeable; increasing the diameter will coarsen the overflow and increase capacity, given a constant pressure drop. Spigots are also made to a range of sizes; the opening must be large enough to allow the coarsest particles to pass and a guide that the opening is three times the size of the largest particle should prevent blocking of the apex. The highest practical underflow density should be chosen as this reduces the volume of water reporting to the underflow and hence reduces the amount of feed solids by-passing classification. Too small a spigot will cause *roping* of the discharge; this is sometimes done deliberately in dewatering applications but it leads to the breakdown of the air core and much reduced efficiency.

The diameter of the cyclone is the figure quoted by the manufacturer in specifying a cyclone. It is unclear how the diameter affects cut size, but as a rule, a larger cyclone will separate coarse sizes and a small diameter is better for fine separations. As a rough guide, a 500 mm cyclone could be used for 150 μm cut size, 250 mm for 75 μm, 100 mm for 40 μm, and 25 mm for down to 5 μm. The number of cyclones required for any particular operation is determined from the total slurry flow rate and knowing the capacity of any given size of cyclone for each manufacturer under normal set-up conditions. The installation is then fine-tuned for the required products using the replaceable spigots and vortex finders.

8. NOTATION

c Fraction of material greater than cut size in oversize product [1]
D Diameter of rod or ball mill (m)
E Screen efficiency [1]
$E_{1\text{-}8}$ Efficiency factors [1]
f Fraction of material greater than cut size in feed [1]
F Feed particle size, as 80% passing size (μm)
F_o Optimum particle feed size to mill (μm)
L Length of rod or ball mill (m)
P Product particle size, as 80% passing size (μm)
R_r Material reduction ratio (F/P) [1]
R_{ro} Ideal mill reduction ratio [1]
W Work input (energy) to grind a tonne of material (kWh t^{-1})
Wi Grindability Work Index (kWh t^{-1})

9. REFERENCES

Anderson, J. (1990). Screening problems: selecting solutions. *Pit and Quarry*, **83**(1), 24–6.
Anon. (1986). New linear screen offers wide applications. *E & MJ*, **187**(9), 79.
Anon. (1989). Nordberg introduces high-efficiency waterflush crushing. *E & MJ*, **190**(1), 106–7.
Bond, F.C. (1952). Third theory of comminution. *Trans. AIME*, **193**, 484.
Brachthäuser, M. & Kellerwessel, H. (1988). High-pressure comminution with roller presses in mineral processing. In *Proceedings XVIth International Mineral Processing Congress*. Elsevier, Amsterdam, The Netherlands, pp. 209–19.
Bradley, D. (1965). *The Hydrocyclone*. Pergamon Press, Oxford, UK.
Clarke, A.J. & Wills, B.A. (1989). Enhancement of cassiterite liberation by high-pressure roller comminution. *Minerals Engng*, **2**, 259–62.
Harris, C.C. & Arbiter, N. (1982). Grinding mill scale-up problems. *Mining Engng*, **34**, 43–6.
Herbst, J.A. & Rajamani, K. (1979). Control of grinding circuits. In *Computer Methods for the 80's in the Mineral Industry*, ed. A. Weiss. AIMME, New York, USA.
Holmes, J.A., (1957). A contribution to the study of comminution—a modified form of Kick's law. *Trans. Inst. Chem. Engrs*, **35**, 125–56.
Howat, D.D. & Vermeulen, L.A. (1986). The design of linings for rotary mills. *J. SA Inst. Mining and Metallurgy*, **87**(7), 251–9.
Kral, S. (1990). Grinding and particle size are critical to the profitability of industrial minerals. *Mining Engng*, **42**, 32–4.
Lynch, A.J. (1977). *Mineral Crushing and Grinding Circuits*. Elsevier, Amsterdam, The Netherlands.
Meulendyke, M.J. & Purdue, J.D. (1989). Wear of grinding media in the mineral processing industry. *Minerals and Metallurgical Processing*, **6**(4), 167–72.
Mular, A.L. & Bhappu, R.B. (1980). *Mineral Processing Plant Design*. Society of Mining Engineers, New York, USA, p. 149.
Plitt, L.R. (1976). A mathematical model of the hydrocyclone classifier. *CIM Bull.*, Dec., **69**(776), 114–123.
Rosas, G. (1990). The A-2 expansion of the concentrator at Chuquicamata. *Mining Magazine*, **163**(3), 200, 203.
Russell, A. (1989). Fine grinding—a review. *Ind. Minerals*, **259**, 57–70.
Suttill, K.R. (1990). Screens for all occasions. *E & MJ*, **191**(2), 18–22.

Svarovsky, L. (1984). *Hydrocyclones*. Holt Rinehart & Winston Ltd, Eastbourne, UK.
Wills, B.A. (1988). *Minerals Processing Technology*. Pergamon Press, Oxford, UK.

10. BIBLIOGRAPHY

Agricola, G. (1950). *De Re Metallica*, ed H.C. Hoover & L.H. Hoover. Dover Publications Inc., New York, USA.
Mular, A.L. & Bhappu, R.B. (1980). *Mineral Processing Plant Design*. Society of Mining Engineers, New York, USA.
Svarovsky, L. (1984). *Hydrocyclones*. Holt Rinehart & Winston Ltd, Eastbourne, UK.
Taggart, A.F. (1945). *Handbook of Mineral Dressing*. John Wiley & Sons Inc., New York, USA.
Weiss, N.L. (1985). *SME Mineral Processing Handbook*. Society of Mining Engineers, New York, USA.
Wills, B.A. (1988). *Minerals Processing Technology*. Pergamon Press, Oxford, UK.
Wills, B.A. (1989). *Proceedings of the International Symposium on Comminution*, Sept. *Minerals Engng*, **3**(1/2). (Published by Pergamon Press, Oxford, UK, 1990.)

International Conference on Hydrocyclones, BHRA Fluid Engineering, Cranfield, UK.
1st, Cambridge, UK, 1–3 Oct. 1980.
2nd, Bath, UK, 19–21 Sept. 1984.
3rd, Oxford, UK, 30 Sept.–2 Oct. 1987.
4th, Southampton, UK, 23–25 Sept. 1992.

26

Slurry Storage in Tanks

Nigel I. Heywood
Warren Spring Laboratory, Stevenage, UK

1. INTRODUCTION

Some of the main reasons for the tank storage of slurries are to

- convert from batch to continuous processing (or vice versa),
- provide a buffer against disruptions,

- even out fluctuations in product quality, and
- allow time for quality checks to be made.

In situations where storage alone is the requirement, with no additional processing involved (e.g. dissolution or heat transfer), the following need to be specified:

- slurry volume to be stored;
- storage vessel design, including inlet and outlet positions;
- whether slurry agitation to prevent particle settling under gravity is required and, if so, the type of agitation;
- the intensity of agitation to provide a specified degree of solids suspension; and
- agitation operational parameters, including power requirements.

Most slurry storage operations concern particles which are denser than the liquid and hence agitation is required to overcome gravity settling. However, occasionally either solids less dense than the liquid or slightly denser but poorly wetted solids will tend to float. This can occur, for instance, with some polymeric particles. Again, agitation is required to draw the solids down to achieve improved solids distribution in the slurry (Joosten *et al.*, 1977; Ohiaeri, 1980; Smith *et al.*, 1981; Hemrajani *et al.*, 1988; Bakker & Frijlink, 1989). These situations are comparatively rare and consequently this chapter focuses on slurries which settle under gravity.

2. SPECIFICATION OF STORAGE SYSTEM REQUIREMENTS

2.1. Slurry Volume to be Stored

The slurry volume to be stored is normally minimised while still achieving the objectives for storage listed above. This is so that the capital and operating costs of the storage system may be minimised. The slurry volume may be specified directly, such as when a reactor must be emptied rapidly, or it may be specified in terms of time derived from throughput. The latter may occur when upstream equipment is unreliable and a given volume must be held to enable the process to carry on for a prescribed time.

2.2. Degrees of Solid Suspension

Several different degrees of solids suspension (Fig. 26.1) can be identified depending on agitation intensity. When an agitator is used, the following progression occurs at progressively increasing agitator speed.

(1) *Non-progressive fillets.* At low agitator speeds, only a proportion of the solids are suspended. A significant amount remain unsuspended as fillets immediately below the eye of the agitator and in the bottom corners of the tank. These fillets remain the same size and may be of a different composition from the particle size distribution in a feed stream or the initial batch charge. They would typically contain the coarsest or most rapidly settling particles.

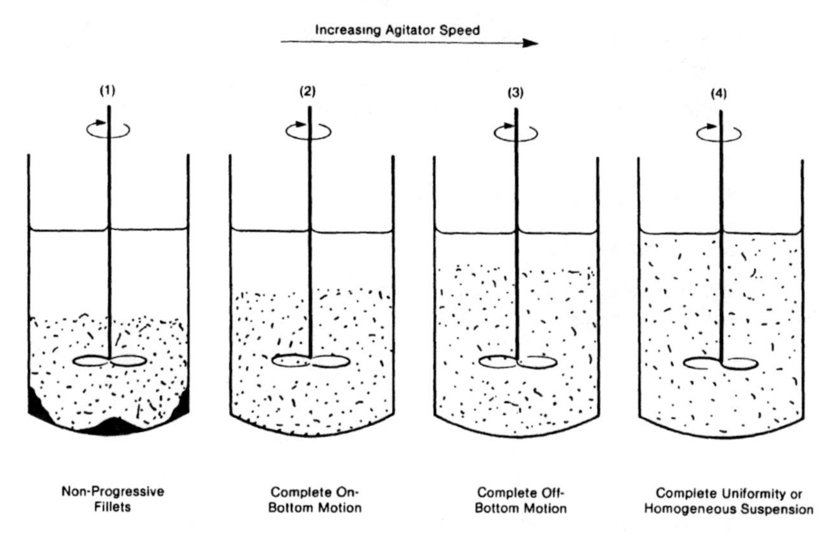

Fig. 26.1. Different degrees of solids suspension in an agitated tank.

(2) *Complete on-bottom motion.* All particles are in motion, either suspended or with some horizontal velocity on the tank bottom. This degree of suspension is often adequate for mass transfer where bulk diffusion is not the rate-limiting step.
(3) *Complete off-bottom motion.* All particles are moving with some vertical velocity and are therefore all in suspension. This degree of suspension would be specified when failure to suspend particles could result in the formation of a compacted sediment that is not easily resuspended.
(4) *Complete uniformity or homogeneous suspension.* All particles at all size fractions are uniformly suspended throughout the tank. This degree of suspension would typically be specified where solids have to be continuously fed into and withdrawn from the mixing tank in an application such as multi-stage counter-current leaching.

These four suspension levels can be summarised into two for many practical applications.

(a) *'Just-off-bottom' suspension* (*JOBS*). This is the condition when not all particles are lifted off the tank bottom at any one time but no particle remains on the tank bottom for longer than 1–2 s. The condition corresponds somewhere between (2) and (3) and has been used by many researchers who have developed correlations for the prediction of the critical agitator speed for this condition (N_{js}).
(b) *Height of a homogeneous layer or complete suspension.* This condition has again been used by researchers and is obtained at higher agitator speeds than

JOBS. Depending on the height of a homogeneous layer, the condition falls somewhere between (3) and (4).

Both agitation intensities (a) and (b) form the basis of several design methods described in section 4.

2.3. Batch or Continuous Operation

Downstream processing determines

(1) the degree of solids suspension necessary; and
(2) whether the tank is emptied on

   (a) a continuous basis (the slurry level remains constant on discharge),
   (b) a semi-continuous basis (the slurry level is allowed to fall to a specified height, e.g. above the agitator), or
   (c) on a batch basis (the tank is emptied completely).

EXAMPLE

> A continuously operating filter or centrifuge is fed with slurry from a storage vessel. The feed must have a near-constant solids concentration. If the slurry tank is filled and emptied batchwise, high-intensity agitation is required to maintain near-homogeneous suspension throughout the cycle. The agitator will probably have to be located near to the tank bottom to keep the solids in suspension when the tank is emptied and two storage tanks may need to be operated alternately. If, on the other hand, feed and discharge to and from the tank are continuous, the agitation intensity may be less and only that required to obtain a known degree of suspension at the point in the tank where slurry is drawn off.

A lower agitation intensity than for the example above may only be required if the downstream process is a batch operation and the batch size is equal to the slurry volume stored in the tank. In this situation, the degree of suspension will be that necessary to allow complete removal of solids when the tank is emptied. Intermittent agitation may be acceptable in these circumstances.

The upstream process can also affect the slurry tank operation. For instance, variations in solids concentration in the feed to a slurry storage tank which is discharging continuously can affect the solids concentration in the tank discharge. Thus, it may be preferable to operate two slurry tanks alternately so that variations in the feed composition can be evened out before each tank is emptied.

3. OPTIONS FOR EQUIPMENT SELECTION

3.1. Methods for Suspending Solids in Storage Vessels

The most common method of achieving the suspension of settling solids in a liquid is the use of one or more rotating agitators. These are usually mounted with their

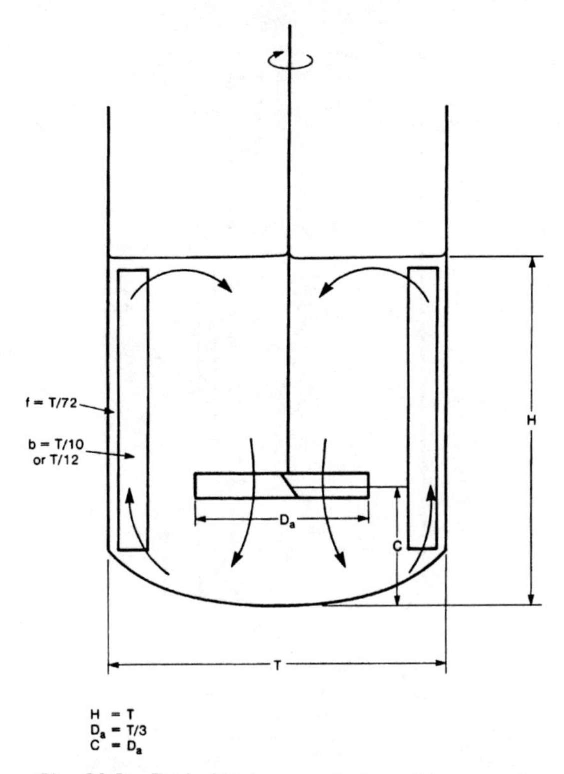

Fig. 26.2. Typical tank geometry for solids suspension.

shafts vertical in the storage vessel. The tank will typically contain four equi-spaced baffles (see Fig. 26.2) or other inserts (occasionally a draught tube, see Fig. 26.5) to aid the liquid circulation pattern and the production of turbulent eddies, both necessary for effective solids suspension. Various tank designs can and have been used to achieve this (section 3.2). Much industrial and academic interest is directed towards solids suspension in relatively low viscosity liquids (often water), although high slurry concentrations will significantly raise the slurry viscosity, particularly if fines are present.

A range of agitator designs is available for both low and high viscosity slurries (section 3.3). These include several proprietary designs which, manufacturers claim, can achieve a specified degree of solids suspension at a reduced power input compared with more conventional agitator types. For large diameter tanks containing slowly settling slurries, side-entry agitators (Kipke, 1984) are a useful option (section 3.4).

Other methods for imparting kinetic energy to the suspending liquid and appropriate turbulence intensities include the use of

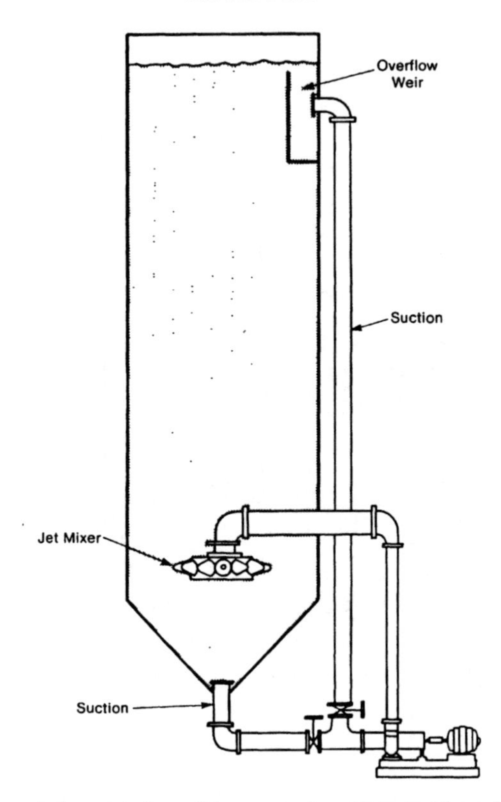

Fig. 26.3. Use of slurry jets for solids suspension. (Adapted from Bathija (1982).)

- liquid or slurry jets issuing from nozzles mounted in the storage tank, as shown in Fig. 26.3 (Racz *et al.*, 1977; Bathija, 1982; Butcher *et al.*, 1982; Hamm *et al.*, 1989; Shamlou & Zolfagharian, 1990);
- a central, single vertical air jet as used in the Pachuca tank (see Fig. 26.4) in the minerals industry (Lamont, 1958) or several vertical air jets (Eyler, 1984);
- vibrating plates, used primarily in relatively small storage vessels (Murakami *et al.*, 1980; Tojo *et al.*, 1981; Ramsey *et al.*, 1989); and
- air bubbles created by single or multiple orifices in a distributor plate, via a sparge ring or via a porous sinter (Roy *et al.*, 1964; Narayanan *et al.*, 1969*a*).

3.2. Tank Design

Tanks which discharge under gravity are usually cylindrical with either flat, dished or occasionally conical bottoms (Musil & Vlk, 1978). In order to minimise capital cost, the tank height-to-diameter ratio (H/T) is usually set close to one, but the ratio may be significantly less than one if large slurry volumes are to be stored.

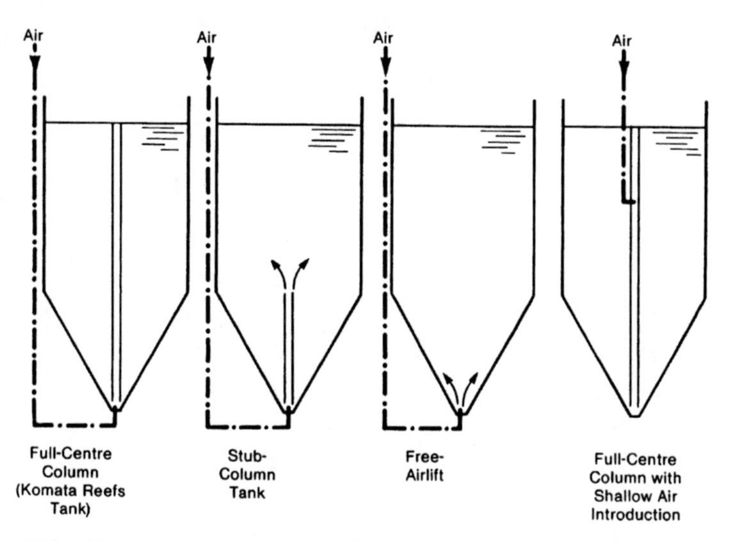

Fig. 26.4. Various designs of Pachuca Tank using air jetting. (Adapted from Lamont (1958).)

Occasionally H/T can be significantly greater than one. The use of H/T ratios greater or less than one often requires the use of several agitators mounted on a single shaft ($H/T > 1$) or on several shafts ($H/T < 1$). This is discussed further in section 3.4.

Square or rectangular tanks are sometimes used for slurry viscosities above about 5 Pa s where the square corners provide sufficient baffling, but for lower viscosities, baffles centred on each of the four walls may be necessary.

3.2.1. Bourne bottom

The geometry of the tank bottom internally determines the energy requirement to suspend solids. The *Bourne bottom* (Aeschbach & Bourne, 1972; Bourne & Sharma, 1974*a*, *b*) is an attempt to optimise tank design to obtain homogeneous suspension with least agitator power and is essentially a streamlining of the tank base to avoid dead zones in the tank corners and immediately below the eye of the agitator. The design (Fig. 26.5) was suggested through visual observation of the distribution of settled solids in slurries agitated at speeds below those corresponding to *just-off-bottom* suspension. The aim is to provide a gentle circulation of the whole suspension, characterised by a constant cross-sectional area for the recirculating flow and therefore a constant recirculation speed, which must be slightly greater than the settling rate of the largest particle.

A draught tube avoids dead zones which would otherwise form a horizontal ring at the agitator level (Bourne & Sharma, 1974*b*). The agitator is placed at the base of the draught tube and is therefore lower than the frequently recommended position of $T/3$ or $T/4$ off the base, although there is now substantial evidence

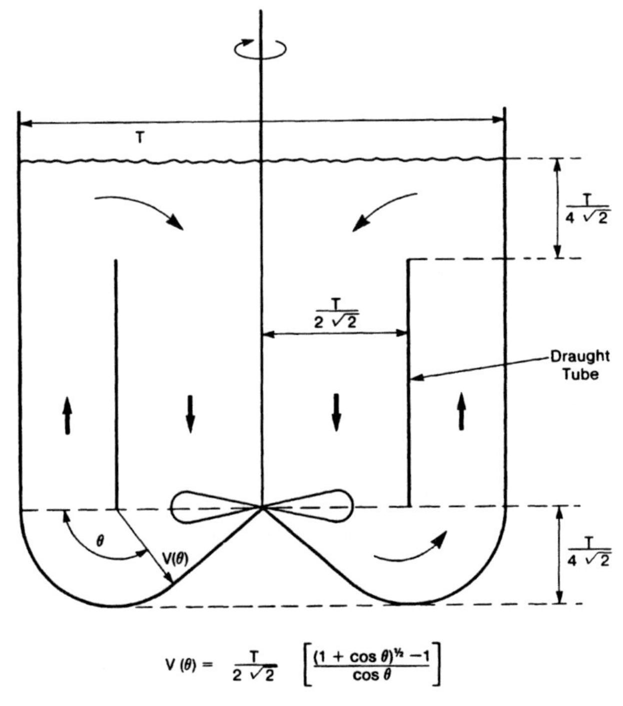

$$V(\theta) = \frac{T}{2\sqrt{2}} \left[\frac{(1 + \cos\theta)^{1/2} - 1}{\cos\theta} \right]$$

Fig. 26.5. Idealised tank geometry to minimise agitator power consumption during solids suspension. (Adapted from Bourne and Sharma (1974*b*).)

(Conti *et al.*, 1981; Rieger & Ditl, 1982) suggesting that this is too large a clearance for solids suspension applications and significant power savings can be made using smaller clearances.

It is clear that the Bourne design presents construction problems which would significantly increase the capital cost above that of more conventional designs. Thus, there is a trade-off between initial cost and long-term operating costs.

3.2.2. Chudacek simplification of the Bourne bottom

Chudacek (1982, 1983, 1984, 1985*a*, *b*, 1986) has rationalised the Bourne tank bottom design in an attempt to reduce the construction cost while at the same time retaining much of its benefits. The modified design (Fig. 26.6) incorporates a cone beneath the eye of the agitator and a 45° fillet to remove the tank bottom's sharp corners. With this design, it is claimed that power savings of over 50% can be made for *just-off-bottom* suspension compared with the power required in similar sized flat-bottomed tanks.

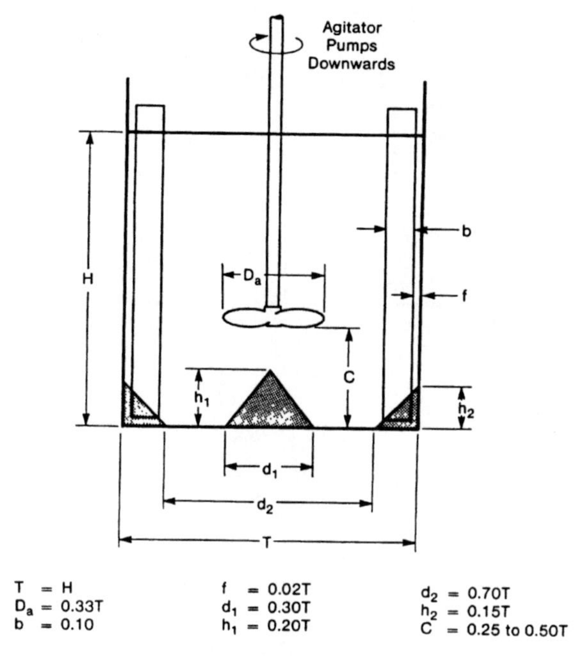

Fig. 26.6. Practical tank geometry using modified *Bourne bottom*. (Adapted from Chudacek (1986).)

3.2.3. Design of vessel discharge

The design of the tank discharge arrangement is often of crucial importance. A bottom outlet line located centrally can accumulate solids above the valve (see Fig. 26.7(a)). This can cause blockages in the valve and can be a particular problem when the agitation is barely adequate because solids will tend to accumulate beneath the eye of the agitator if a Chudacek cone (see above) is not fitted. Even a sample valve which lifts the solids out of the outlet line when it is opened tends to restrict the flow and blockages can still occur. A better option, if the bottom outlet is necessary, is to locate the outlet beneath the periphery of the agitator where solids are less likely to settle out at the bottom of the tank.

The preferred take-off system is a downward-directed outlet pipe where solids cannot accumulate. Two possible arrangements are shown in Fig. 26.7(b) and (c). Such systems are not possible when large diameter agitators are used, such as gates, anchors or helical ribbons. In addition, it is not possible to achieve complete emptying of the tank.

When continuous feed and discharge are required, an overflow system is useful. However, the discharge slurry concentration (and also the particle size distribution) will usually be less than the bulk values, unless homogeneous suspension has been achieved (Bourne & Sharma, 1974*a*).

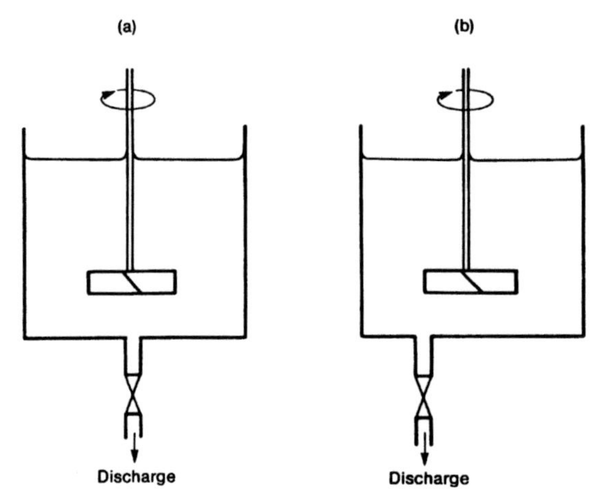

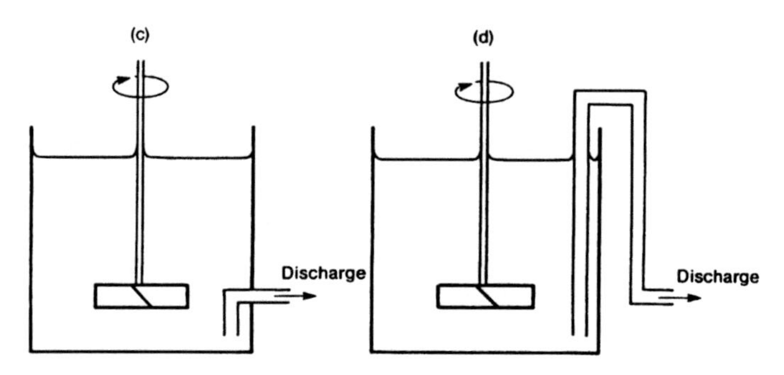

Fig. 26.7. Alternative take-off points for a slurry tank.

3.3. Types of Agitator

3.3.1. Non-proximity agitators

Many alternative agitator designs are specified either by manufacturers or users. Some are much better suited to developing flow patterns appropriate for suspending solids than others. Normally an agitator having a diameter approximately one-third the vessel diameter (a *non-proximity* agitator) would be selected to operate in the transitional or turbulent flow regimes (see section 5). Typical agitator speeds fall in the range 2–25 rev s^{-1}.

One of the most common agitators and least expensive to fabricate is the four-blade, 45° pitched turbine which produces both axial and radial flows and which is employed in the large slurry tanks at the Mohave power station at the discharge of the Black Mesa coal slurry pipeline (Cobb *et al.*, 1978). Several other *standard* designs (Fig. 26.8) are also employed in the process industries, including the marine propeller (three-blade), six-blade, pitched agitator and the Rushton disc turbine. The marine propeller produces mostly axial flow, i.e. directly downwards towards the tank bottom, whereas agitators with flat, vertical blades tend to produce radial flow. As might be expected, the pitched blade turbine produces flow patterns intermediate between the propeller and the flat, vertical blade turbine. Because of the predominantly axial flow produced, the marine propeller is more efficient than the Rushton turbine for solids suspension and dissipates energy mainly through bulk fluid movement rather than through local small-scale recirculation arising from turbulence.

Relatively novel, non-proximity agitator designs have been developed by several manufacturers (Fig. 26.9) with a view to reducing power requirements. These designs include the following.

- Chemineer HE-3
- Denver 'MIL' propeller
- Ekato 'Intermig' (see Muller & Pysall, 1986)
- Lightnin A310 (see Chervenic & Coyle, 1987; Pharamond & Olderstein, 1980; Salzman *et al.*, 1983)
- Scaba 'SHP'

In addition, the CBMF (convex bladed mixed flow) agitator shows promise for solids suspension duties (Pandit *et al.*, 1987, 1989), although this agitator is not yet available commercially.

There has been almost no independent research to compare the performance of these agitators. However, Heywood *et al.* (1991) have compared the power efficiencies of three proprietary designs, a marine propeller and a four-blade, 45° pitched turbine using two alternative measures of suspension efficiency for sand slurries at concentrations in the range 50–65% (by mass). It was concluded that the Denver and Ekato designs gave the highest power efficiencies for intermediate degrees of suspension, but the Scaba design and the marine propeller were the only two out of the five designs tested that could achieve near-homogeneous suspension. The four-blade, 45° pitched turbine compared poorly with the other four designs under a variety of test conditions.

3.3.2. Proximity agitators

These agitators are characterised by a diameter which is typically around 90% of the tank diameter (Fig. 26.10). They are used for agitating higher viscosity slurries, where the use of a *non-proximity* agitator would create insufficient flow for solids suspension, including insufficient motion near the tank wall. Typical speeds are much lower than *non-proximity* agitators at around 0·2 to 0·5 rev s^{-1}. However, even at these low speeds, torque levels can be high and substantial motors and drive units are often necessary.

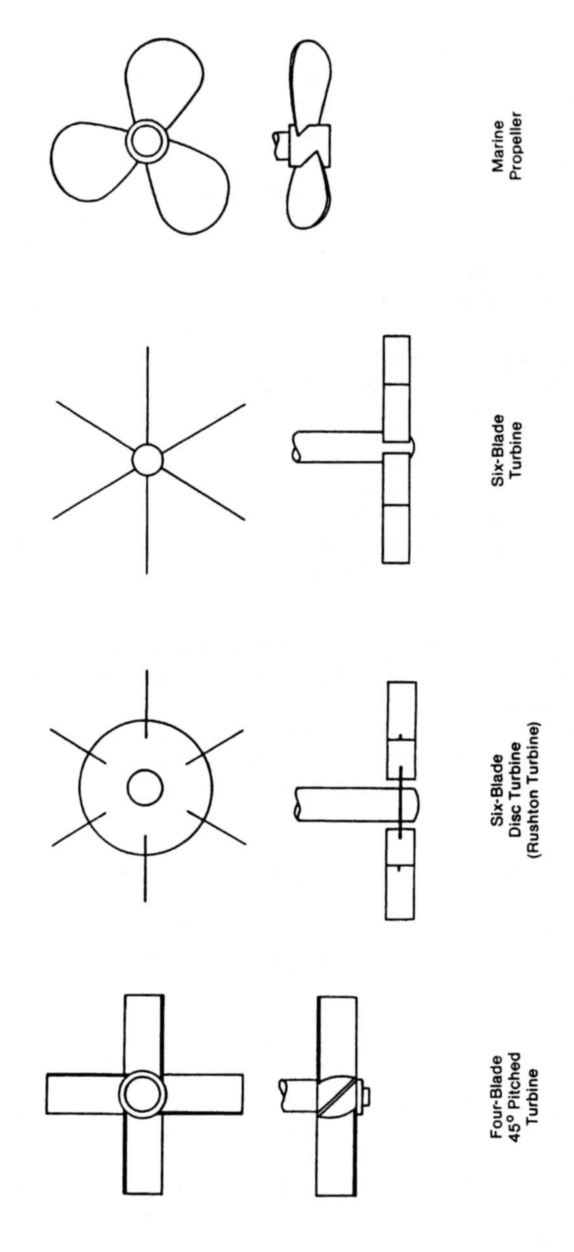

Fig. 26.8. Types of generic proximity agitator for low viscosity or coarse particle slurry.

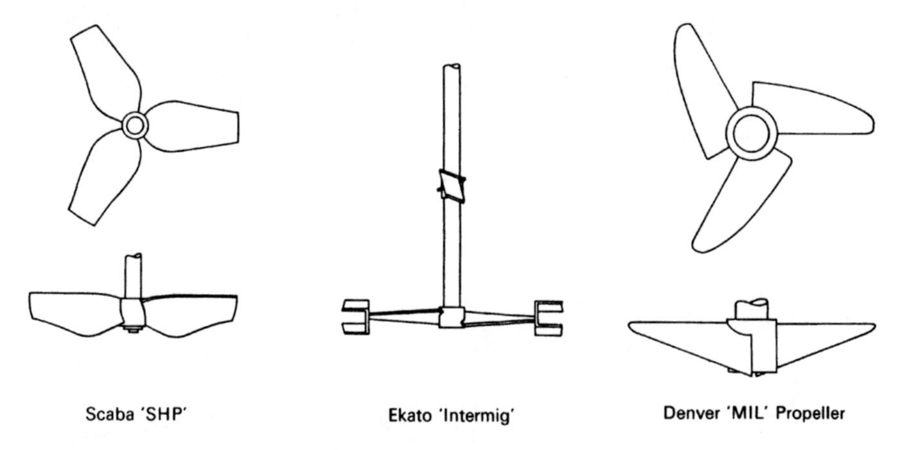

Fig. 26.9. Various proprietary agitator designs used for solids suspension duties.

3.4. Agitator/Tank Configuration

3.4.1. Top-entry agitators

When the tank height-to-diameter ratio is approximately one, a single agitator is normally specified. The tank to agitator diameter ratio (T/D_a) and the agitator clearance, the distance between agitator and tank floor, will then need specifying. The T/D_a ratio typically lies between 2 and 4. Manufacturers may base the value on experience with similar slurries or duties or it may be predetermined from bench-scale test work. Lyons (1967) has produced curves (Fig. 26.11) which relate the T/D_a ratio to particle size and particle density. He suggests that the curves may

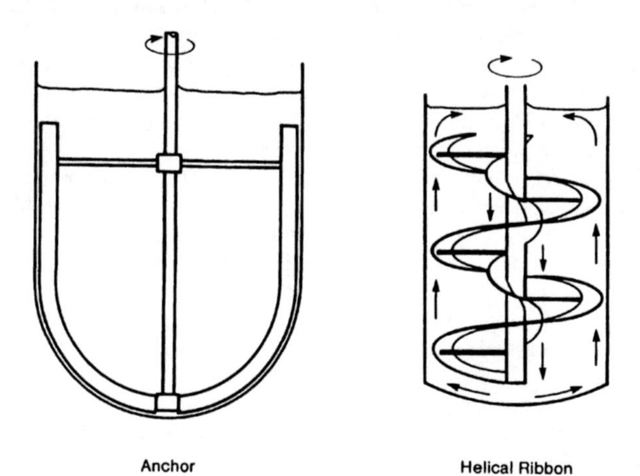

Fig. 26.10. Examples of proximity agitators for high viscosity slurries.

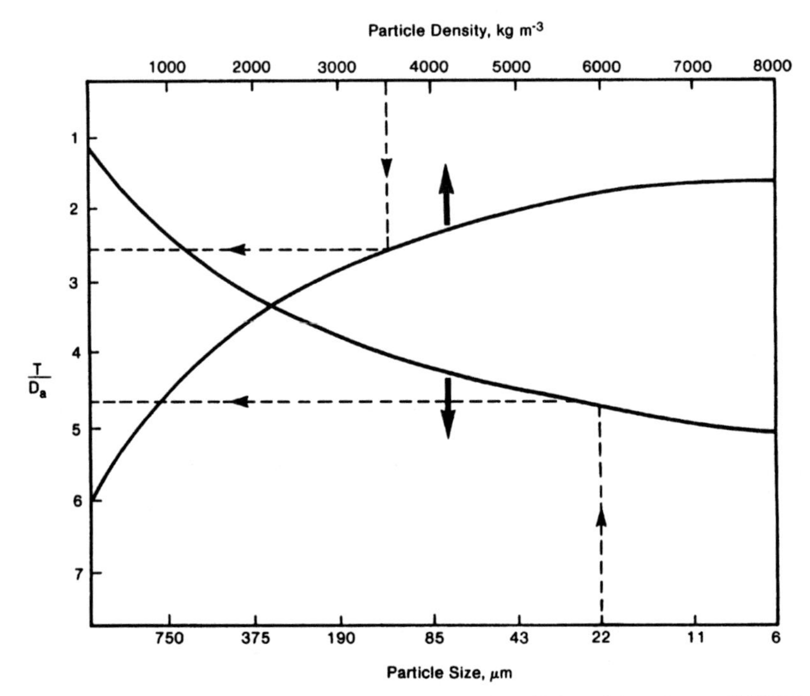

Fig. 26.11. Optimum tank to agitator diameter ratios. (Adapted from Lyons (1967).)

be used for *order of magnitude* sizing of an agitator. The T/D_a ratio could then be used in bench-scale tests on which large-scale sizing is then based for non-fibrous slurry concentrations between 20 and 40% (by weight). Figure 26.11 shows an example of a slurry with 22 μm particles of density 3500 kg m^{-3} in water. The curves indicate a 2·6 T/D_a ratio based on density and a 4·65 ratio based on particle size. The arithmetic mean of 3·62 is taken for design.

For slurry concentrations below 10%, Lyons (1967) proposes that the curves may be used directly for full-scale sizing. At concentrations above 40%, the effects of hindered settling and high viscosities will suggest smaller T/D_a ratios.

Agitator clearance off bottom has traditionally been set at around one-third of the tank diameter. However, some studies (e.g. Conti *et al.*, 1981) suggest that significant agitator power savings can be made if the clearance is set at between 0·1 and 0·2 T. Bench-scale tests for a given slurry will assist in an appropriate clearance specification.

When the tank diameter is large (i.e. H/T in the range 0·2–0·6), top-entry agitators can require an elaborate support structure and jet agitation (Bathija, 1982) may be preferable. Alternatively, multiple agitators located at different positions in the tank (see Fig. 26.12) may be needed. Considerations for arriving

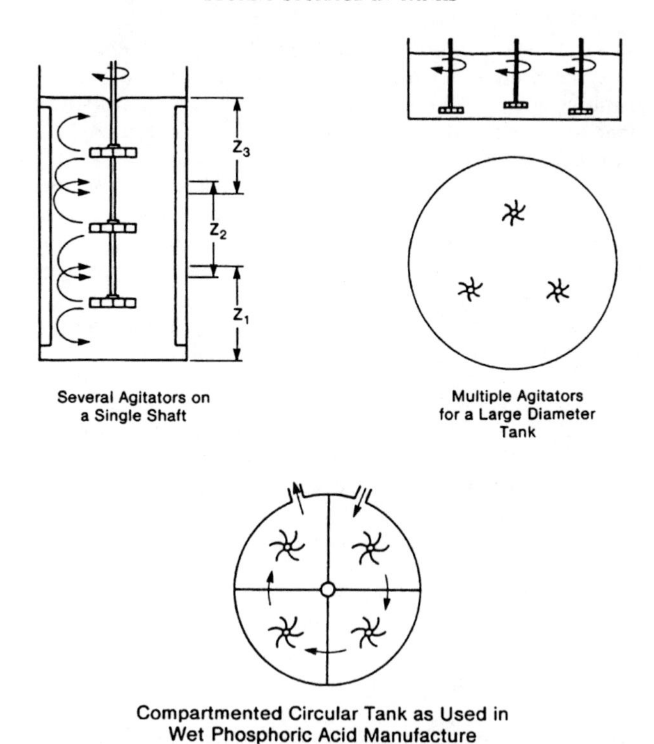

Fig. 26.12. Multiple agitators for tanks with large or small H/T. (Adapted from Lyons (1967).)

at an appropriate T/D_a for each individual agitator are similar to those for a single agitator. Baffles, if used, will be located on the wall adjacent to each to provide a vertical direction to the agitator discharge and assist in a more uniform suspension of solids.

EXAMPLE

> In the manufacture of wet process phosphoric acid, it is common practice to create continuous, staged-mixer compartments by partitioning a large, circular tank into sections (see Fig. 26.12). Each section is usually sufficiently angular to obviate the need for baffling. In estimating the appropriate T/D_a ratio, the tank *diameter* is normally taken as twice the average distance between the vertical shaft and the farthest corner and the distance from the shaft to the nearest side wall.

Relatively tall, narrow tanks ($H/T > 1$), or large diameter tanks with H/T approximately one, usually require multiple agitators on a single shaft (Fajner *et al.*, 1985; Magelli *et al.*, 1990). T/D_a ratio considerations remain the same as those for a single agitator. The number and location of agitators on the shaft are determined as the result of a simple geometric evaluation based on an overlapping of the vertical heights of the circulation patterns of the agitators. Lyons (1967) suggests that to ensure that the slurry circulated to the top of the circulation pattern of the bottom agitator will be picked up by the bottom half pattern of the next agitator in vertical sequence, this overlap should be equal to approximately one-quarter of the circulation depth on the side of each agitator adjacent to the interchanging streams, as indicated by Z_1, Z_2 and Z_3 in Fig. 26.12. Fibrous slurries require one-third overlap. These guidelines apply to agitator types giving mainly radial flow discharge.

Harrah (1974) provides some useful details of several applications where large tank diameters are required.

3.4.2. Side-entry agitators

Side-entry agitators are sometimes a more economic alternative to centrally mounted, vertical shaft agitators for large tanks where the solids suspension duty is not arduous. They are used in the pulp and paper industry (Oldshue, 1983), and Kipke (1984) has discussed the use of three, equally-spaced agitators in tanks having a height-to-diameter ratio of about 0·5.

Kipke (1984) also provides some general installation guidelines. Two angles, α and β, define the configuration of the shaft of a side-entry agitator. The agitator points downwards and the angle α is the angle the shaft makes with a horizontal plane. Kipke varied this angle between 5 and 15° in his experiments and found that the degree of suspension achieved for a given agitator speed was unaffected by the angle α used.

The angle β, the angle that the agitator shaft makes with a line normal to the tank wall in the horizontal plane, was found to have an important influence. When using a 700 mm tank size, an optimum value for β of 14° was found. This fell to 10° when a 1400 mm tank was used. The agitator should rotate so that fluid is pumped towards the tank centre. Thus, when viewed from behind, the agitator rotates in a clockwise direction and should be angled to the left by approximately 10–14° depending on the tank size.

3.4.3. Submersible agitators

These operate on a similar principle to side-entry agitators. However, because they are mounted on a vertical shaft, or guide-bar system, and are fully submerged in the slurry (Karlsen, 1987), there is no longer the problem of sealing the agitator shaft in the tank wall. In addition, the vertical shaft introduces a greater flexibility in varying both the height of the agitator in the tank and the angles of the agitator shaft in the horizontal and vertical planes.

4. DESIGN METHODS FOR OPERATING PARAMETERS

Whether particles settle rapidly or slowly under gravity largely determines the type of agitation employed, the operational procedure for agitation (for instance, whether the agitation is continuous or intermittent) and the degree of suspension that can be achieved at realistic power inputs. Coarse particle, rapidly settling slurries normally require top-entry agitators whereas slowly settling slurries can be effectively agitated by side-entry agitators or through jet agitation.

Flocculated fine particles will often settle more rapidly than non-flocculated particles, even though flocculated particles create higher viscosity, non-Newtonian slurries. If settling rates are low, intermittent agitation may be possible provided any sediment can be readily resuspended without excessive agitator power requirement (see section 5.3). Flocculated slurries typically create non-cohesive, easily resuspendable sediments, whereas non-flocculated slurries give rise to hard, compacted sediments which may be sometimes almost impossible to resuspend. Coarse particle, rapidly settling slurries usually develop a resuspendable sediment provided few fines are present.

Slowly settling flocculated slurries are often shear-thinning and may, in addition, exhibit a yield stress. This yield stress may assist in the support of coarser particles to such an extent that no agitation is required. For a known yield stress, it is possible to determine the maximum particle size which can be supported. Alternatively, for slurries with an indistinct yield stress, the settling rate for the maximum particle size in the slurry can be determined by measuring the slurry's zero shear viscosity (see Chapter 4).

4.1. Design for Rapidly Settling Slurries

4.1.1. Use of Just-Off-Bottom Suspension Agitator Speed

The minimum agitator speed (and therefore agitator power consumption) to achieve *just-off-bottom* suspension, N_{js}, has been the subject of much experimental, and some theoretical, investigation since the late 1950s. The aim has been to develop general correlations relating this minimum agitator speed to both equipment and slurry parameters. However, just as in the case of the various critical slurry velocities for horizontal pipe flow (see Chapter 6), each correlation is usually reasonably reliable only for the range of experimental variables for which it was developed.

Pronounced vertical solids concentration profiles are still present at the *just-off-bottom* suspension condition, but the specification of this condition is nevertheless useful for various duties. For instance, it ensures that no compacted bed of particles will develop on the tank bottom and optimises power input for mass transfer operations (Ovsenik, 1982).

One of the most used correlations for the critical agitator speed, N_{js}, was developed by Zwietering (1958). He carried out over 1000 experiments with sand–water and other systems, using several agitator designs, tank geometries and tank sizes (0·003–0·17 m^3), and obtained

$$N_{js} = \frac{S d^{0\cdot 2} \eta_f^{0\cdot 1} [g(\rho_p - \rho_f)]^{0\cdot 45} B^{0\cdot 13}}{\rho_f^{0\cdot 55} D_a^{0\cdot 85}} \tag{26.1}$$

where B is the total solids concentration in the tank expressed as a percent by

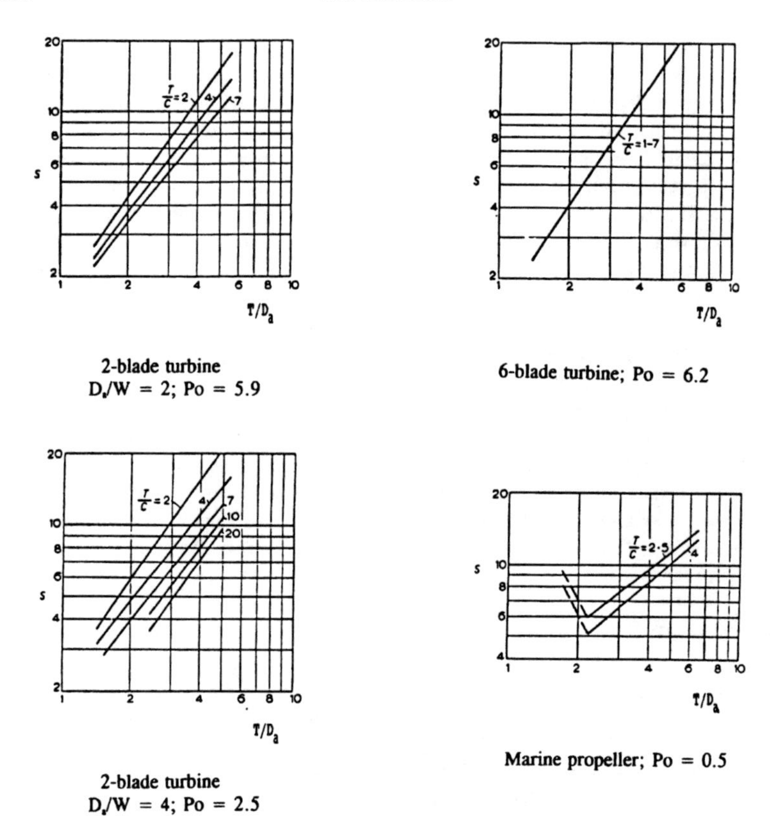

Fig. 26.13. Correlations for S-value in Zwietering correlation for minimum agitator speed for 'just-off-bottom' suspension. (Adapted from Zwietering (1958).)

weight solids to liquid ratio. S is a constant for a given system geometry and also depends on the tank diameter to agitator diameter ratio (T/D_a) ratio (see Fig. 26.13),

$$S = S'\left(\frac{T}{D_a}\right)^a \tag{26.2}$$

in which the exponent a is approximately 1·5.

The correlation has been verified by other studies (e.g. Nienow, 1968), and is useful for an initial estimate of N_{js}, provided it is noted that the correlation was developed using slurry concentrations below 20% (by weight) and for tank diameters less than 0·6 m. For more accurate N_{js} estimation, predictions from a number of correlations should be made using, for instance, those by Narayanan *et al.* (1969*b*),

Baldi *et al.* (1978), Herringe (1979), Ohiaeri (1980), Cliff *et al.* (1981), Ovsenik (1982), Rieger and Ditl (1982), Musil *et al.* (1984), and Raghava Rao *et al.* (1988). Look at the spread and mean of the N_{js} predictions. If the spread is unacceptably wide, it may be worth the expense of performing small-scale experiments on the specific slurry and scaling up the results according to the considerations outlined below.

4.1.2. Scale-up from Small-Scale Data for Just-Off-Bottom Suspension

In principle, a correlation which predicts the agitator speed for the *just-off-bottom* suspension criterion could be used directly for design. However, many of these correlations were developed for specific ranges of variables. An alternative approach is to perform some small-scale tests using the specific slurry. For one or more agitator types, the effects of variations in parameters such as agitator diameter and clearance may be investigated. Ideally, more than one scale should be used involving near geometrically similar agitation systems so that a scale-up rule can be developed. The nearer the largest scale used to the full-scale, the greater the confidence in the final design, but obviously this becomes progressively more expensive.

The agitator speed, and hence the power requirement, is the main design parameter to be scaled up. However, each correlation for N_{js} implies a different scale-up basis. The power per unit volume, P/V, is often quoted as a suitable scaling parameter but it is likely that the use of a single scaling parameter is too simple an approach. Nevertheless, the P/V approach still provides a useful comparison between the various correlations and is still widely used. Most correlations for N_{js} imply that P/V is inversely proportional to agitator power raised to some power.

The Zwietering (1958) correlation for N_{js}, eqn (26.1), implies that

$$N_{js} D_a^{0.85} = \text{constant} \tag{26.3}$$

If the power number ($Po = P/\rho_m N^3 D_a^5$) is assumed constant in the fully turbulent regime (see section 5), the power per unit volume may be derived from eqn (26.3),

$$P/V \propto D_a^{-0.55} \tag{26.4}$$

However, Nienow (1968) suggests that the exponent on D_a in eqn (26.3) is 0·71, giving an alternative scale-up rule:

$$P/V \propto D_a^{-0.13} \tag{26.5}$$

This, therefore, implies a much smaller reduction in P/V on scale-up than Zwietering's work. The wide range of exponents on D_a when using P/V as the scaling parameter is illustrated by the spread of lines in Fig. 26.14; these lines are frequently referred to in the literature as a "harp". It is often considered that a reasonable interpretation of the current disagreement between the N_{js} correlations is to keep the P/V ratio constant when scaling up for *just-off-bottom* suspension.

Further work needs to be undertaken to determine whether P/V is the relevant scaling parameter. The following effects require further study.

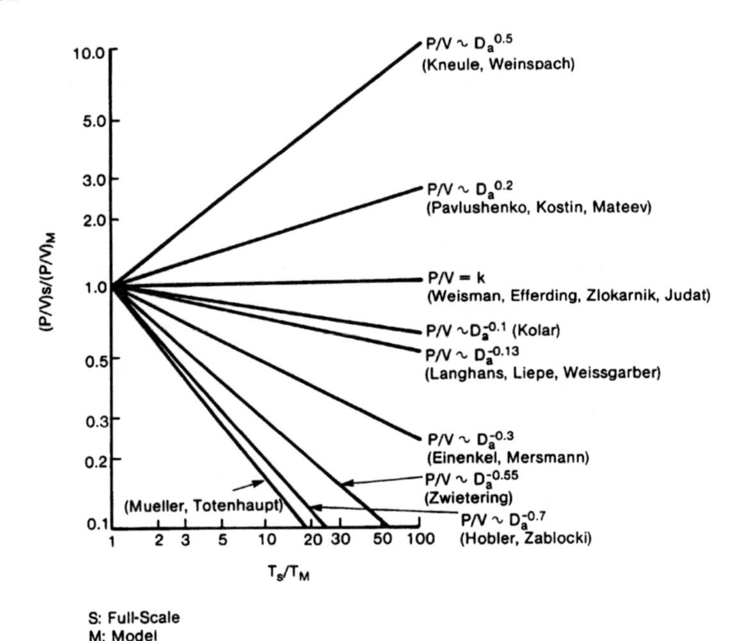

Fig. 26.14. Comparison of scale-up correlations for 'just-off-bottom' suspension criterion based on agitator power consumption. After Mersmann *et al.* (1985).

(1) Particle size stays constant when equipment is scaled up (Herringe, 1979; Ditl & Rieger, 1985; Mersmann & Laufhütte, 1985).
(2) The slurry density immediately in the vicinity of the agitator varies according to the degree of suspension and on scale-up (Herringe, 1979; Mersmann & Laufhütte, 1985).
(3) Frequent lack of complete geometrical similarity of the agitator on scale-up (Buurman *et al.*, 1985), which can result in a non-constant power number on scale-up.

4.1.3. Gates et al. (1976) *Approach*

A design procedure which attempts to make use of a specified degree of suspension at different points between *just-off-bottom* and homogeneous suspension is described by Gates *et al.* (1976) of Chemineer Inc. (Dayton, OH, USA) for a pitched blade turbine. To determine an appropriate agitator speed, the procedure is as follows.

First, a degree of suspension on an arbitrary scale of 0 to 10 is selected together with a specified agitator-to-tank diameter ratio. Next, the value of a parameter ϕ is obtained using Fig. 26.15. This parameter is defined by

$$\phi = \frac{N^{3.75} D_a^{2.81}}{V_d} \tag{26.6}$$

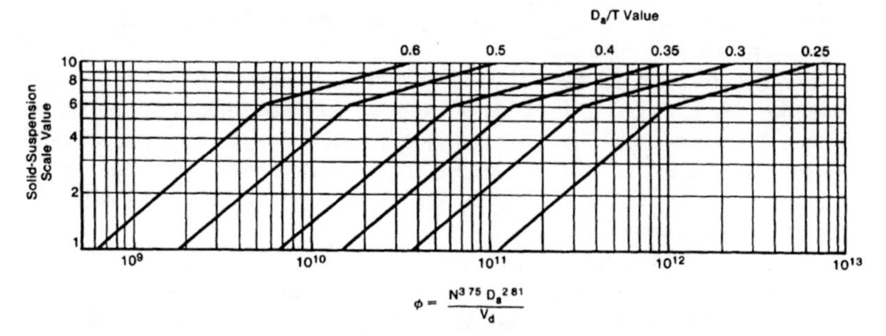

Fig. 26.15. The Gates *et al.* (1976) solid suspension scale values. (Adapted from Gates *et al.* (1976).)

where V_d is the *design settling velocity* which is related to the particle terminal settling velocity through a parameter f_w:

$$V_d = f_w u_\infty \tag{26.7}$$

The parameter f_w is tabulated in the reference as a function of solids concentration. Values range from 0·8 at 2% solids to 1·85 at 50% solids. Thus, if u_∞ is calculated or measured, V_d is estimated from eqn (26.7) and the agitator speed N calculated from eqn (26.6).

It is of interest to infer the scale-up dependency from eqn (26.6). For a specified slurry, eqn (26.6) reduces to

$$N^{3.75} D_a^{2.81} = \text{constant} \tag{26.8}$$

which, for a constant power number, implies

$$P/V \propto D_a^{-0.25} \tag{26.9}$$

The exponent on D_a therefore falls between the two values given in eqns (26.4) and (26.5) and indicates that the method applicable to a number of different degrees of suspension is consistent with correlations for the *just-off-bottom* suspension agitator speed.

One practical aspect to be considered in scale-up is that motors and gearboxes are generally standardised, and even when the size and power have been predicted by a particular scale-up equation, it may not be possible to meet these requirements closely. Gates *et al.* (1976) give extracts from detailed tables for various scales of agitation using various tank sizes with suitable stirring systems which are indicated in terms of standard motor powers and standard gearbox shaft speeds.

4.1.4. Design of Near-Homogeneous Suspension

The criterion of *just-off-bottom* suspension is of limited use in slurry storage design because prediction of the agitator speed at this condition gives no information regarding the distribution of the solids throughout the tank contents. Data on solids distribution can be useful in designing the slurry discharge from the tank so that a specified slurry concentration on discharge (and perhaps also particle size distribution) may be obtained.

Considerable effort has been directed towards measurement of either the one-dimensional (e.g. Einenkel, 1980; Heywood *et al.*, 1991) and the two-dimensional solids concentration profiles. However, for practical design purposes, a single parameter such as the height of a defined slurry interface is sometimes all that is necessary, for instance, when the slurry discharges continuously over a weir at a specified concentration.

Several studies (Pavlushenko *et al.*, 1957; Weisman & Efferding, 1960; Buurman *et al.*, 1985) have resulted in correlations for agitator speed giving a defined interface height as a function of slurry properties and tank/diameter parameters.

The Pavlushenko *et al.* (1957) correlation for the agitator speed giving *near-homogeneous suspension* throughout the vessel is

$$N_h = \frac{0{\cdot}415\rho_p^{0{\cdot}8} d^{0{\cdot}4} T^{1{\cdot}9}}{\rho_f^{0{\cdot}6} \eta_f^{0{\cdot}2} D_a^{2{\cdot}5}} \tag{26.10}$$

The correlation was derived using the following variable ranges

$$730 < \mathrm{Re}_a < 3{\cdot}79 \times 10^5$$

$$2 < T/D_a < 3$$

$$2{\cdot}3 \times 10^{-4} < d/D_a < 8{\cdot}25 \times 10^{-3}$$

$$3{\cdot}48 \times 10^6 < \mathrm{Ga} < 7 \times 10^{10}$$

where the Galileo number, Ga, is defined by

$$\mathrm{Ga} = D_a^3 \rho_f^2 g / \eta_f^2 \tag{26.11}$$

Buurman *et al.* (1985) correlated the height of a homogeneous zone, Z, with a Froude number as shown in Fig. 26.16. They suggested that for homogeneous suspension

$$N_h^2 D_a^{1{\cdot}55} = \text{constant} \tag{26.12}$$

On scaling-up using a constant power number, this leads to the following expression for geometrically similar systems:

$$P/V \propto D_a^{-0{\cdot}33} \propto T^{-0{\cdot}33} \tag{26.13}$$

Thus, on a larger scale, homogeneity is obtained at a lower specific power input than on a smaller scale. The data shown in Fig. 26.16 were obtained for when $H/T = 1$. By way of example, if the criterion that homogeneity should be achieved throughout the lower 90% of the tank ($Z/T = 0{\cdot}9\,T$), then Fig. 26.16 suggests that

$$\frac{\rho_f N_h^2 D_a^2}{g \Delta\rho d}\left(\frac{d}{D_a}\right)^{0{\cdot}45} > 20 \tag{26.14}$$

in which $\Delta\rho$ is $\rho_m - \rho_f$.

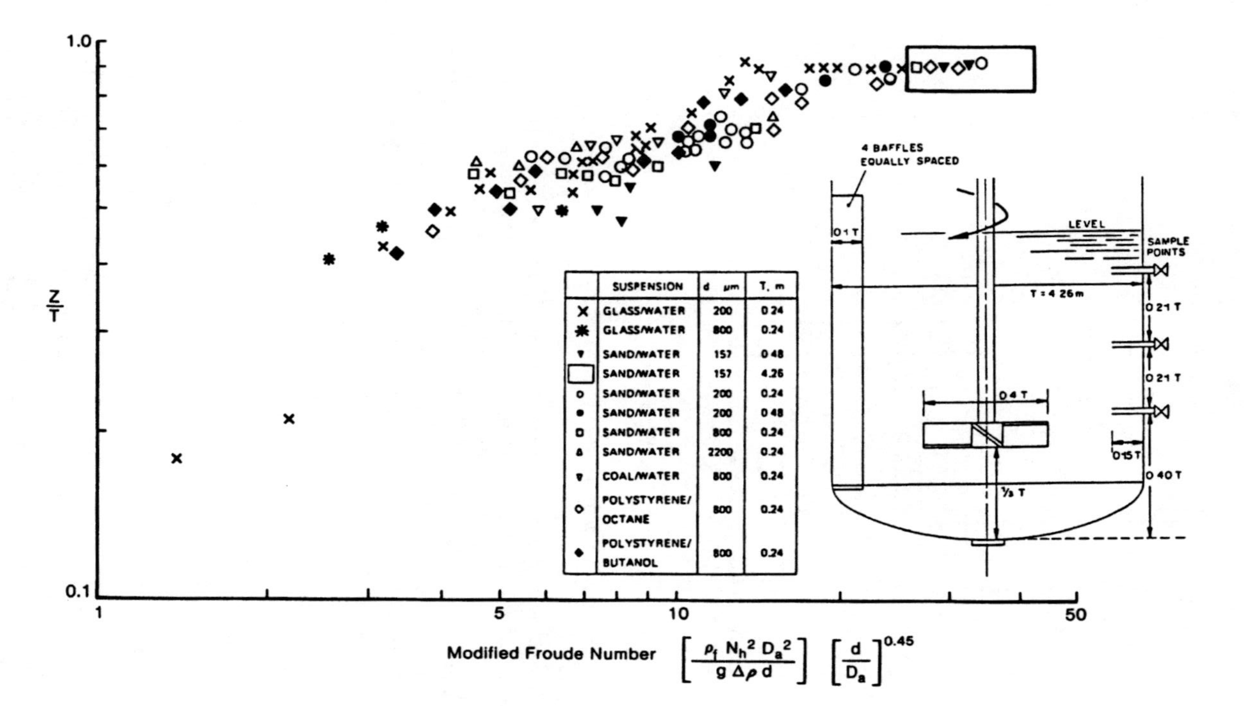

	SUSPENSION	d μm	T, m
×	GLASS/WATER	200	0.24
✱	GLASS/WATER	800	0.24
▼	SAND/WATER	157	0.48
□ (box)	SAND/WATER	157	4.26
○	SAND/WATER	200	0.24
●	SAND/WATER	200	0.48
□	SAND/WATER	800	0.24
△	SAND/WATER	2200	0.24
▽	COAL/WATER	800	0.24
◇	POLYSTYRENE/ OCTANE	800	0.24
◆	POLYSTYRENE/ BUTANOL	800	0.24

Fig. 26.16. Tank geometry, test slurries and correlation for homogeneous suspension. (After Buurman *et al.* (1985).)

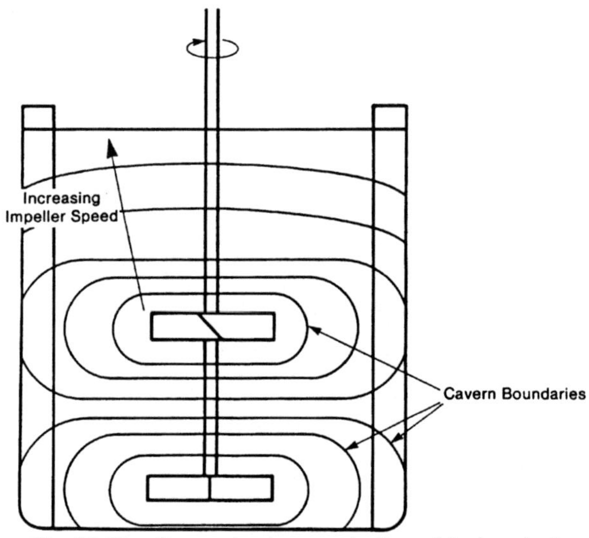

Fig. 26.17. Cavern development in shear-thinning slurries.

4.2. Design for Slowly Settling Slurries

4.2.1. Agitator Speed for Shear-Thinning Slurries with or without a Yield Stress

Slurries of fine particles in a flocculated state (see Chapter 4) will often exhibit shear-thinning, non-Newtonian flow properties. This may result in a discernible yield stress which may be sufficiently large to support coarser particles present in the slurry without the need for agitation. In fact, agitation may inhibit the development of a sufficiently large yield stress for coarse particle support if the slurry is thixotropic. Alternatively, if the yield stress is low or if there is no distinct yield stress, agitation may prove necessary for particle suspension.

On many occasions, the relatively high viscosity of many shear-thinning slurries requires the use of a proximity agitator. However, sometimes a non-proximity agitator is specified because of availability or from cost considerations. It is important to operate the agitator at a speed which produces sufficient agitation throughout the tank. Too low an agitator speed and the result is stagnant slurry adjacent to the tank wall and bottom. For slurry with a yield stress, no motion at all can occur in these tank regions, and even for highly shear-thinning slurries not exhibiting a distinct yield stress, the motion generated can be insufficient for coarse particle support.

When a non-proximity agitator is used, a *cavern* is generated around it (Elson, 1988, 1990; Elson *et al.*, 1986; Nienow & Elson, 1988). This *cavern* represents the volume of slurry in motion in the agitator region (Fig. 26.17). To ensure slurry motion from the agitator tip to the tank wall, an agitator speed must be selected such that the cavern diameter equals the tank diameter. Higher agitator speeds are required to ensure motion near the slurry surface. Correlations which allow the prediction of these agitator speeds are available in the literature. Etchells *et al.* (1987) have applied these correlations with some success.

4.2.2. Jet Agitation

Jet agitation can become economic for large tanks storing slowly settling solids in low viscosity liquids over relatively long time periods. Rather little research has been carried out on solids suspension using jets and, in addition, there is little published on operational experience. However, Bathija (1982) gives a design procedure for the number of nozzles required, the volumetric flow rate through the jets, and the power requirement. The procedure is based on a typical jet system shown in Fig. 26.3 and is claimed to be valid within the following limits: $3 < T < 12{\cdot}2$; $0{\cdot}6 < H < 9{\cdot}1$; $V < 1\,370\,\mathrm{m}^3$; $\eta_m < 30\,\mathrm{Pa\,s}$.

Butcher *et al.* (1982) have provided design recommendations when using horizontally aimed jets across the bottom of a rectangular tank. Hamm *et al.* (1989) have developed a model for the resuspension of sludge settled on a tank bottom using two jets issuing from a submerged centrifugal pump. Shamlou and Zolfagharian (1990) have correlated the jet velocity for the *just-off-bottom* suspension condition with slurry and equipment parameters for a single jet pointing vertically downwards in the centre of a tank.

5. PREDICTION OF POWER CONSUMPTION IN AGITATED VESSELS

Estimation of agitator power consumption is necessary for sizing the agitator motor and drive, and to use the scale-up procedures described in section 4.

5.1. Non-Proximity Agitators

5.1.1. Newtonian Slurries

For fully baffled tanks, agitator power consumption can be correlated with agitator Reynolds number using the power number

$$\mathrm{Po} = \frac{P}{\rho_m N^3 D_a^5} \tag{26.15}$$

The agitator Reynolds number is defined for a Newtonian slurry as

$$\mathrm{Re}_a = \frac{\rho_m N D_a^2}{\eta_m} \tag{26.16}$$

Specific agitator geometries lead to different forms of the $\mathrm{Po}/\mathrm{Re}_a$ relationship; typical power curves are shown in Fig. 26.18 and standard texts on mixing (see bibliography) illustrate power curves for many other agitator types. All power curves are composed of three regions which correspond to the flow regimes: laminar, transitional, and turbulent.

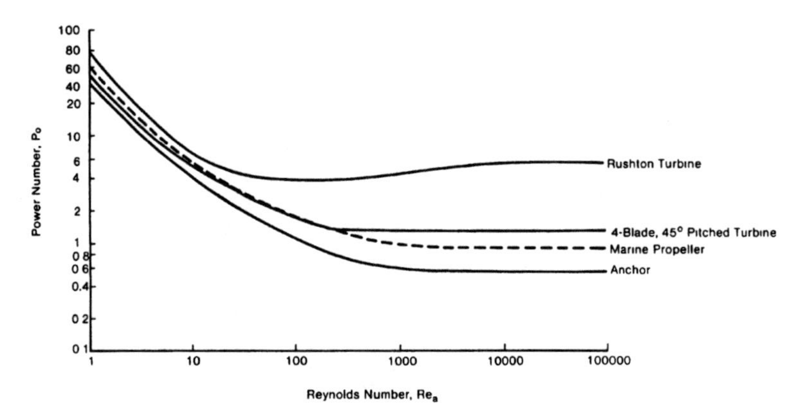

Fig. 26.18. Power number/agitator Reynolds number relationships for five agitator designs. (Adapted from Oldshue (1983).)

Laminar regime. At low Re_a (e.g. $Re_a < 10$ for disc turbines), the laminar regime is characterised by a straight-line relationship between log Po and log Re_a with slope -1, i.e.

$$\mathrm{PoRe_a} = K_L \tag{26.17}$$

The constant K_L depends on agitator/tank geometry. From eqns (26.15)–(26.17) agitator power is given by

$$P = K_L N^2 D_a^3 \eta_m \tag{26.18}$$

Turbulent regime. At high Re_a values, fully turbulent flow results in near constant values for the power number. Table 26.1 lists values of Po for various agitator types for the turbulent flow regime.

TABLE 26.1
Power Number Values in the Turbulent Regime for Some Commonly Used Agitators

Agitator	*Po*
4-blade disc turbine	4·0
6-blade (Rushton) disc turbine	5·6
12-blade disc turbine	8·7
18-blade disc turbine	9·5
6-concave-blade disc turbine	4·0
6-blade, 45° pitched turbine	1·2
16-blade vaned disc	4·0
Marine propeller	0·5–0·8
Ekato 'Intermig' (1 stage)	0·25
4-blade CBMF	0·85
6-blade CBMF	1·16
12-blade CBMF	1·45

Transitional regime. At intermediate Re_a values, the dependence of power number on agitator Reynolds number is complex. No mathematical relations have been widely described which will reproduce the entire transitional regime power curve, although for a limited range of data a power law relation is sometimes used:

$$Po \propto Re_a^b \tag{26.19}$$

Sometimes when coarse particles are being agitated, centrifugal effects are important and the slurry density to be used in power calculations is often less than the bulk slurry density. Herringe (1979) has discussed this in some detail.

5.1.2. Non-Newtonian slurries

Power correlations for non-Newtonian slurries have usually been based on the power law flow model (see Chapter 4). The power number is then a function of both agitator Reynolds number and the slurry flow behaviour index, n.

$$Po = fn\,(Re_a', n) \tag{26.20}$$

The definition of Re_a' can vary depending on the approach adopted, but for any given agitator/tank geometry, all slurries with the same power law index, n, will have a single power curve across the three flow regimes. The curve would be similar in shape to the Newtonian curve for that agitator/tank combination, but the exact location and shape of the transitional section would be a function of n.

Laminar regime. The constant K_L depends on the flow behaviour index,

$$PoRe_a' = K_L(n) \tag{26.21}$$

and decreases with decreasing n for a given agitator/tank combination (Bourne & Butler, 1969). In general, a plot of Po against Re_a' results in a family of lines rather than a single line in the case of Newtonian slurries. However, several different ways have been developed to collapse these curves on to one by appropriate definition of the Reynolds number (Foresti & Liu, 1959; Calderbank & Moo-Young, 1961). These have been discussed at length by Skelland (1967). This is still not an entirely satisfactory route and a useful alternative approach is the method developed by Metzner and Otto (1957).

This method has already been outlined in Chapter 4 for the measurement of slurry viscosity using an agitator. Using the proportionality constant between shear rate and agitator speed (k_s), a Reynolds number can be defined which brings the laminar regime power data onto a single line for a given agitator:

$$Re_{MO} = \frac{\rho_m N^{2-n} D_a^2}{K k_s^{n-1}} \tag{26.22}$$

Turbulent regime. As for Newtonian slurries, Po is taken to be independent of Reynolds number. It might be expected that Po would be a function of flow behaviour index, n, but this appears not to be the case for practical purposes.

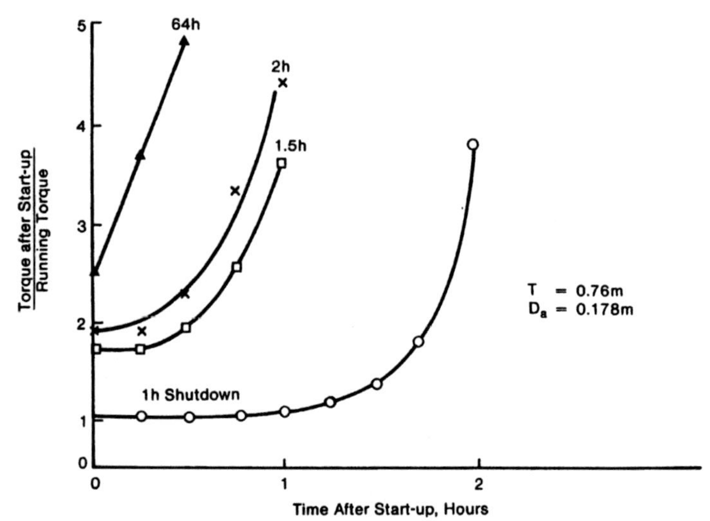

Fig. 26.19. Increasing agitator torque on start-up in a settled bed. (Taconite: 70% by mass.) (Adapted from Oldshue (1983).)

5.2. Proximity Agitators

5.2.1. Newtonian Slurries

Agitation generally occurs in the laminar regime because slow agitator speeds are used and slurry viscosities are high. There is considerably less information on power consumption for proximity agitators than for non-proximity turbines partly because there is less conformity towards a standard design.

Because proximity agitators operate close to the vessel wall, geometric ratios such as T/D_a influence the predictive eqn (26.18) for power consumption in the laminar flow regime. Some attempt has been made to correct for the effect of agitator design variables in the laminar flow regime. For instance, Bourne and Butler (1969), and Hall and Godfrey (1970) give expressions for power number for a helical ribbon which includes blade width and pitch.

5.2.2. Non-Newtonian Slurries

As a first approximation, power number may be estimated from predictive equations such as that given by Hall and Godfrey (1970) by defining the agitator Reynolds number according to the non-Newtonian flow model adopted. Alternatively, the Metzner–Otto method can be used (see Chapter 4, section 3.5) to obtain calibration constants for the specific agitator.

5.3. Start-Up Torque for Agitator in a Settled Bed

Allowance must be made for the possibility of the agitator primary power supply failing. In critical situations it is usual to have stand-by power supplies or strategies to alleviate settling of the slurry to form a cohesive bed. For less critical situations,

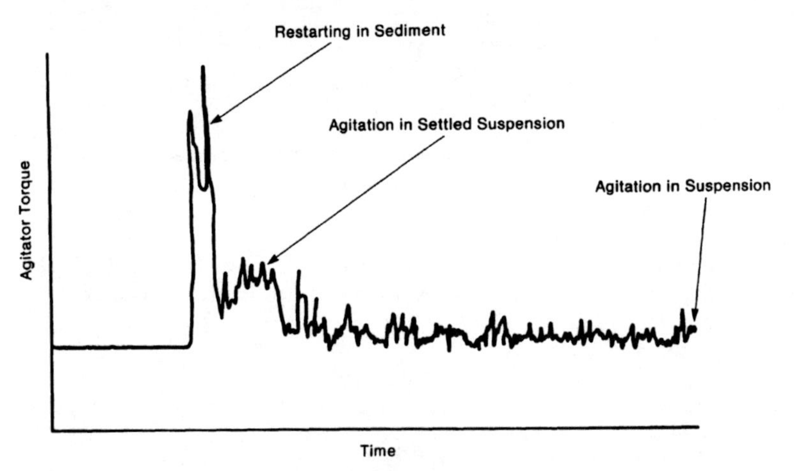

Fig. 26.20. Agitator torque development on start-up in a settled bed. (Adapted from Kipke (1983).)

typically when relatively small tanks are involved, such contingency planning would not normally occur in which case significant settling can result if the power supply is out for an extended period.

In this case, two situations normally arise: either the settled solids interface lies below or above the agitator; occasionally the agitator may be only partially submerged. Problems do not normally arise regarding torque requirements if the interface is below the agitator and the settled solids are in a flocculated, non-cohesive state or the particles have a low density. For dispersed (i.e. non-flocculated) solids which have a highly angular particle shape, restarting presents the most difficult problem when the agitator is submerged.

Often agitators must be restarted with liquid or air probes, or lances to loosen the material surrounding the agitator. In order to protect the agitator drive, there are usually clutch couplings installed between the motor and drive, but they may not be adequate to start the agitator if the torque required exceeds the agitator drive capacity. Figure 26.19 shows the high start-up torques when solids are allowed to settle for various lengths of time (Oldshue, 1983).

It is possible that even if start-up may not have been difficult, compaction of settled solids over time may result causing *sanding-in* of the agitator. In such situations, Kipke (1983) observes that the laws of fluid mechanics no longer apply and granular property may predominate. Kipke has attempted to model the complicated flow in a simplified way to predict the critical stress level at which the solids will be disturbed.

Figure 26.20 shows a typical torque–time plot (Kipke, 1983) during start-up with a fully submerged agitator with a covering layer of 220 mm. In this case, the agitator motion was transmitted to the sediment surface from where the solids are slowly distributed throughout the vessel. The torque then drops to the value prevailing at normal operating conditions. Sidorich and Shebatin (1985) have also looked at this problem.

6. NOTATION

a	Exponent in eqn (26.2) [1]
b	Width of baffle [L]
B	Total solids concentration in tank, by weight, percentage solids to liquid ratio, as used in Zwietering eqn (26.1) [1]
C	Agitator clearance off tank bottom [L]
d	Particle diameter [L]
d_1	Diameter of central cone fillet in Chudacek tank design, Fig. 26.6 [L]
d_2	Diameter of inner tank area excluding corner fillets in Chudacek design, Fig. 26.6 [L]
D_a	Agitator diameter [L]
f	Separation between baffle and tank wall [L]
f_w	Gates *et al.* (1976) parameter relating *design settling velocity* to particle terminal settling velocity [1]
g	Gravitational acceleration [$L\,T^{-2}$]
Ga	Galileo number, defined by eqn (26.11) [1]
h_1	Height of central fillet in Chudacek tank design, Fig. 26.6 [L]
h_2	Height of corner fillets in Chudacek tank design, Fig. 26.6 [L]
H	Height of slurry in tank [L]
k_s	Constant of proportionality between average shear rate in vessel and agitator speed [1]
K	Consistency coefficient in power law flow model [$M\,L^{-1}T^{-n}$]
K_L	Product of power number and agitator Reynolds number in laminar regime [1]
n	Flow behaviour index in power law flow model [1]
N	Agitator speed, ($rev\,s^{-1}$) [T^{-1}]
N_h	Agitator speed to obtain near-homogeneous suspension [T^{-1}]
N_{js}	Agitator speed for *just-off-bottom* suspension [T^{-1}]
P	Agitator power consumption [$M\,L^2T^{-3}$]
Po	Agitator power number [1]
Re_a	Agitator Reynolds number for Newtonian slurry defined by eqn (26.16) [1]
Re'_a	Agitator Reynolds number for non-Newtonian, power law slurry [1]
Re_{MO}	Agitator Reynolds number defined by eqn (26.22) [1]
S	Parameter in Zwietering eqn (26.1) for N_{js} [1]
S'	Constant in eqn (26.2)
T	Diameter of circular storage tank [L]
u_∞	Particle terminal settling velocity [$L\,T^{-1}$]
V	Total volume of slurry in tank [L^3]
V_d	*Design settling velocity* used in Gates *et al.* (1976) design method [$L\,T^{-1}$]
$V(\theta)$	Geometric length, see Fig. 26.5 [L]
W	Agitator blade width [L]
Z	Height of *homogeneous zone* in Buurman *et al.* (1985) correlation [L]
α	Angle of dip from horizontal of shaft of side-entering agitator [1]

β Angle in horizontal plane of shaft of side-entering agitator [1]
$\Delta\rho$ $\rho_m - \rho_f$ [M L^{-3}]
η_f Suspending liquid viscosity [M L^{-1}T^{-1}]
η_m Newtonian slurry viscosity [M L^{-1}T^{-1}]
ρ_f Liquid density [M L^{-3}]
ρ_m Slurry density [M L^{-3}]
ρ_p Particle density [M L^{-3}]
ϕ Gates *et al.* (1976) design parameter, defined by eqn (26.6)
θ Angle, see Fig. 26.5 [1]

7. REFERENCES

Aeschbach, S. & Bourne, J.R. (1972). The attainment of homogeneous suspension in a continuous stirred tank. *Chem. Egn. J.*, **4**, 234–42.

Baldi, G., Conti, R. & Alaria, I.N. (1978). Complete suspension of particles in mechanically agitated vessels. *Chem. Eng. Sci.*, **33**, 21-5.

Bakker, A. & Frijlink, J.J. (1989). The drawdown and dispersion of floating solids in aerated and unaerated stirred vessels. *Chem. Eng. Res. Des.*, **67**, 208–10.

Bathija, P.R. (1982). Jet mixing design and applications. *Chem. Engng*, 13 Dec., **89**(25), 89–94.

Bourne, J.R. & Butler, H. (1969). Power consumption of helical ribbon impellers in viscous liquids. *Trans. Inst. Chem. Engrs*, **47**, 263–70.

Bourne, J.R. & Sharma, R.N. (1974*a*). Homogeneous particle suspension in propeller-agitated flat-bottomed tanks. *Chem. Eng. J.*, **8**, 243–50.

Bourne, J.R. & Sharma, R.N. (1974*b*). Suspension characteristics of solid particles in propeller-agitated tanks. In *Proceedings of the 1st European Mixing Conference*, ed. N.G. Coles. BHRA Fluid Engineering, Cranfield, UK, paper B3.

Butcher, T., Krishna, C. & Saunders, J. (1982). Recirculation mixing of coal slurry storage tanks. In *Proceedings of the 4th International Symposium on Coal Slurry Combustion.*

Buurman, C., Resoort, G. & Plaschkes, A. (1985). Scaling-up rules for solids suspension in stirred vessels. In *Proceedings of the 5th European Conference on Mixing*, ed. J. Stanbury. BHRA Fluid Engineering, Cranfield, UK, paper 5.

Calderbank, P.H. & Moo-Young, M.B. (1961). The power characteristics for the mixing of Newtonian and non-Newtonian fluids. *Trans. Inst. Chem. Engrs*, **39**, 337–47.

Chervenic, M.W. & Coyle, C.K. (1987). Recent developments affecting coal–water fuel preparation and storage. In *Proc. STA 12*, pp. 185–9.*

Chudacek, M.W. (1982). Formation of unsuspended solid profile in a slurry mixing vessel. In *Proceedings of the 4th European Mixing Conference*, ed. H.S. Stephens & D.H. Goodes. BHRA Fluid Engineering, Cranfield, UK, paper H2.

Chudacek, M.W. (1983). New efficient geometry for slurry holding tanks. In *Proc. STA 8*, pp. 185–92.*

Chudacek, M.W. (1984). Does your tank bottom have the right shape? *Chem. Engng*, 1 Oct., **91**(20), 79–83.

Chudacek, M.W. (1985*a*). Solids suspension behaviour in profiled and flat-bottomed mixing tanks. *Chem. Eng. Sci.*, **40**, 385–92.

Chudacek, M.W. (1985*b*). Impeller power numbers and impeller flow numbers in profiled bottom tanks. *Ind. Eng. Chem. PDD*, **24**, 858–67.

Chudacek, M.W. (1986). Relationships between solids suspension criteria, mechanism of suspension tank geometry and scale-up parameters in stirred tanks. *Ind. Eng. Chem. Fundam.*, **25**(3), 391–401.

Cliff, M.J., Edwards, M.F. & Ohaeri, I.N. (1981). The suspension of settling solids in

agitated vessels. In *Proceedings of Fluid Mixing II* (Symp. Ser. no. 64). Institution of Chemical Engineers, Rugby, UK, pp. M1–M11.
Cobb, D.B., Giles, C.O., Hornbuckle, J.D. & Leavitt, F.O. (1978). Coal slurry storage and reclaim facility for Mohave generating station. *In Proc. STA 3*, pp. 58–68.*
Conti, R., Sicardi, S. & Specchia, V. (1981). Effect of stirrer clearance on particle suspension in agitated tanks. *Chem. Eng. J.*, **22**, 247–9.
Ditl, P. & Rieger, F. (1985). Suspension of solid particles—relative velocity of particles. In *Proceedings of the 5th European Mixing Conference*, ed. J. Stanbury. BHRA Fluid Engineering, Cranfield, UK, paper 8.
Einenkel, W.D. (1980). Influence of physical properties and equipment design on the homogeneity of suspensions in agitated vessels. *Ger. Chem. Eng.*, **3**, 118–24.
Elson, T.P. (1988). Mixing of fluids possessing a yield stress. In *Proceedings of the 6th European Mixing Conference*. The Italian Association of Chemical Engineering/EFCE Working Party on Mixing, pp. 485–92.
Elson, T.P. (1990). The growth of caverns formed around rotating impellers during the mixing of a yield stress fluid. *Chem. Eng. Commun.*, **96**, 303–19.
Elson, T.P., Cheeseman, D.J. & Nienow, A.W. (1986). X-ray studies of cavern sizes and mixing performance with fluids possessing a yield stress. *Chem. Eng. Sci.*, **41**, 2555–62.
Etchells, A.W., Ford, W.N. & Short, D.G.R. (1987). Mixing of Bingham plastics. In *Proceedings of Fluid Mixing III* (Symp. Ser. No. 108). Institution of Chemical Engineers, Rugby, UK.
Eyler, L.L. (1984). Three-dimensional numerical simulation of settling and resuspension of solids in storage tanks with air injection circulators. In *Proc. STA 9*, pp. 71–6.*
Fajner, D., Magelli, F., Nocentini, M. & Pasquali, G. (1985). Solids concentration profiles in a mechanically stirred and staged column slurry reactor. *Chem. Eng. Res. Des.*, **63**, 235–40.
Foresti, R. & Liu, T. (1959). How to measure power requirements for agitation of non-Newtonian liquids in the laminar regime. *Ind. Eng. Chem.*, **51**, 860–4.
Gates, L.E. Morton, J.R. & Fondy, P.L. (1976). Selecting agitator systems to suspend solids in liquids. *Chem. Engng*, 24 May, 144–50.
Hall, K.R. & Godfrey, J.C. (1970). Power consumption by helical ribbon impellers. *Trans. Inst. Chem. Engrs*, **48**, T201–8.
Hamm, B.A., Wesr, W.L. & Tatterson, G.B. (1989). Sludge suspension in waste storage tanks. *AIChE J.*, **35**, 1391–4.
Harrah, H.W. (1974). Slurry storage agitators for slurry pumping. In *Proc. HT 3*, paper B4, pp. 43–52.*
Hemrajani, R.R. (1987). Designing mixers for suspension of solids in liquid filled tanks. *Chem. Proc.*, July, 22–6.
Hemrajani, R.R., Smith, D.L., Koros, R.M. & Tarmy, B.L. (1988). Suspending floating solids in stirred tanks—mixer design, scale-up and optimisation. In *Proceedings of the 6th European Mixing Conference*. Italian Association of Chemical Engineering/EFCE Working Party on Mixing, pp. 259–65.
Herringe, W.D. (1979). The behaviour of mono-size particle slurries in a fully baffled turbulent mixer. In *Proceedings of the 3rd European Conference on Mixing*, ed. H.S. Stephens & C.A. Stapleton. BHRA Fluid Engineering, Cranfield, UK, paper D1.
Heywood, N.I., Rehman, S. & Whittemore, R.G. (1991). A comparison of the efficiency of five agitator designs for solids suspension duties at high solids concentrations. In *Proceedings of the 7th European Congress on Mixing*, ed. M. Bruyelmane & G. Froment. Bruges, Belgium, 18–20 Sept.
Joosten, G.E.H., Schilder, J.G.M. & Broere, A.M. (1977). The suspension of floating solids in stirred vessels. *Trans. Inst. Chem. Engrs*, **55**, 220–2.
Karlsen, H. (1987). A new concept in suspending slurries using a submersible mixer. In *Proc. STA 12*, pp. 201–5.*
Kipke, K. (1983). Restarting of agitators in settled solids. *Ger. Chem. Eng.*, **6**, 264–8.

Kipke, K. (1984). Suspension by side-entering agitators. *Chem. Eng. Process*, **18**, 233–8.
Lamont, A.G.W. (1958). Air agitation and Pachuca tanks. *Can. J. Chem. Engng*, **37**, 153–60.
Lyons, E.J. (1967). Suspension of solids. Chapter 9 in *Mixing Theory and Practice* (Vol. 2), ed. V.W. Uhl & J.B. Gray. Academic Press, New York, USA.
Magelli, F., Fajner, D., Nocentini, M. & Pasquali, G. (1990). Solid distribution in vessels stirred with multiple impellers. *Chem. Eng. Sci.*, **45**, 615–25.
Mersmann, A. & Laufhütte, H.D. (1985). Scale-up of agitated vessels for different mixing processes. In *Proceedings of the 5th European Conference on Mixing*, ed. J. Stanbury. BHRA Fluid Engineering, Cranfield, UK, paper 28.
Metzner, A.B. & Otto, R.E. (1957). Agitation of non-Newtonian fluids. *AIChE J.*, **3**, 3–10.
Muller, W. & Pysall, D. (1986). Das Suspendieren von Feststoff mittels Balkenrühren in Stufenanordnung. *Chem-Ing. Tech.*, **58**, 508–9.
Murakami, Y., Hirose, T. & Ohshima, M. (1980). Mixing with an up and down impeller. *Chem. Eng. Prog.*, May, **76**(5), 78–82.
Musil, L. & Vlk, J. (1978). Suspending solid particles in an agitated conical-bottomed tank. *Chem. Eng. Sci.*, **33**, 1123–31.
Musil, L., Vlk, J. & Jiroudkova, H. (1984). Suspending solid particles in an agitated tank with axial-type agitators. *Chem. Eng. Sci.*, **39**, 621–8.
Narayanan, S., Bhatia, V.K. & Guha, D.K. (1969*a*). Suspension of solids by bubble agitation. *Can. J. Chem. Engng*, **47**, 360–4.
Narayanan, S., Bhatia, V.K., Guha, D.K. & Rao, M.N. (1969*b*). Suspension of solids by mechanical agitation. *Chem. Eng. Sci.*, **24**, 223–30.
Nienow, A.W. (1968). Suspension of solid particles in turbine agitated vessels. *Chem. Eng. Sci.*, **23**, 1453–9.
Nienow, A.W. & Elson, T.P. (1988). Aspects of mixing in rheologically complex fluids. *Chem. Eng. Res. Des.*, **66**, 5–15.
Ohiaeri, I.N. (1980). The suspension of solids in mixing vessels. PhD thesis, University of Bradford, UK.
Oldshue, J.Y. (1983). Solids suspension. Chapter 5 in *Fluid Mixing Technology*. McGraw-Hill, New York, USA.
Ovsenik, A. (1982). Optimisation of solid particles suspending. In *Proceedings of the 4th European Conference on Mixing*, ed. H.S. Stephens & D.H. Goodes. BHRA Fluid Engineering, Cranfield, UK, pp. 463–70.
Pandit, A.B., Niranjan, K. & Davidson, J.F. (1987). British Patent Application No. 8716870.
Pandit, A.B., Rielly, C.D., Niranjan, K. & Davidson, J.F. (1989). The convex bladed mixed flow impeller: a multipurpose agitator. *Chem. Eng. Sci.*, **44**, 2463–74.
Pavlushenko, I.S., Kostin, N.M. & Matveer, S.F. (1957). Stirrer speeds in the stirring of suspensions. *J. Appl. Chem. USSR*, **30**, 1235–43.
Pharamond, J.C. & Olderstein, A.J. (1980). New concepts in slurry storage. In *Proc. STA 5*, pp. 184–7.*
Racz, I., Dees, P. & Wassinck, J.G. (1977). Suspension of solid particles in axial jet agitators. *Chem-Ing. Tech.*, **49**(10), 841.
Raghava Rao, K.S.M.S., Rewatkar, V.B. & Joshi, J.B. (1988). Critical impeller speed for solid suspension in mechanically agitated contactors. *AIChE J.*, **34**, 1332–40.
Ramsey, C.J., Kyser, E.A. & Tatterson, G.B. (1989). Mixing and solids suspension by up-down agitators in a slab tank. *AIChE J*, **35**, 1219–23.
Rieger, F. & Ditl, P. (1982). Suspension of solid particles in agitated vessels. In *Proceedings of the 4th European Conference on Mixing*, ed. H.S. Stephens & D.H. Goodes. BHRA Fluid Engineering, Cranfield, UK, paper H1.
Roy, A.N., Guha, D.K. & Rao, M.N. (1964). Suspension of solids in a bubbling liquid. Critical gas flow rates for complete suspension. *Chem. Eng. Sci.*, **19**, 215–25.
Salzman, R.N., Coyle, C.K., Weetman, R.J. & Pharamond, J.C. (1983). High efficiency impeller for slurry storage. *In Proc. STA 8*, pp. 305–10.*

Shamlou, P.A. & Zolfagharian, A. (1990). Suspension of solids in liquid-jet stirred vessels. In *Proceedings of Fluid Mixing IV* (Symp. Ser. no. 121). Institution of Chemical Engineers, Rugby, UK, pp. 365–78.
Sidorovich, A.I.M. & Shebatin, V.C. (1985). Start-up loads on the stirrer when suspending sediments. *Chem. Petr. Engng*, **21**, 115–17.
Skelland, A.H.P. (1967). *Non-Newtonian Flow and Heat Transfer*. John Wiley & Sons, New York, USA.
Smith, D.L., Hemrajani, R.R., Koros, R.M. & Tarmy, B.L. (1981). Mixing technology for homogeneous suspension of buoyant solids and liquid drops. Paper presented at AIChE Annual Meeting, 8–12 November, New Orleans, USA.
Tojo, K., Miyanami, K. & Mitsui, H. (1981). Vibratory agitation in solids–liquid mixing. *Chem. Eng. Sci.*, **36**, 271–84.
Weisman, J. & Efferding, L.E. (1960). Suspension of slurries by mechanical stirrers. *AIChE J*, **6**, 419–26.
Zwietering, T.N. (1958). Suspending of solids in liquids by agitators. *Chem. Eng. Sci.*, **8**, 244–53.

* Full details of the Hydrotransport (HT) and Slurry Transport Association (STA) series of conferences can be found on pp. 11–13 of Chapter 1.

8. BIBLIOGRAPHY

Harnby, N., Edwards, M.F. & Nienow, A.W. (1985). *Mixing in the Process Industries*. Butterworths, London, UK.
Nagata, S. (1975). *Mixing Principles and Applications*. John Wiley & Sons, New York, USA.
Nienow, A.W. (1985). The dispersion of solids in liquids. Chapter 8 in *Mixing of Liquids by Mechanical Agitation*, ed. J.J. Ulbrecht & G.K. Patterson. Gordon & Breach, USA.
Oldshue, J.Y. (1983). *Fluid Mixing Technology*. McGraw-Hill, New York, USA.
Skelland, A.H.P. (1967). *Non-Newtonian Flow and Heat Transfer*. John Wiley & Sons, New York, USA.
Sterbacek, Z. & Tausk, P. (1965). *Mixing in the Chemical Industry*. Pergamon Press, Oxford, UK.
Uhl, V.W. & Gray, J.B. (eds) (1967). *Mixing Theory and Practice* (Vol. 2). Academic Press, New York, USA.

Proceedings of Fluid Mixing Conferences, Institution of Chemical Engineers, Rugby, UK. All held in Bradford, UK.
1st (Symposium Series no. 64), Mar. 1981.
2nd (Symposium Series no. 89), 3–5 Apr. 1984.
3rd (Symposium Series no. 108) 8–10 Sept. 1987.
4th (Symposium Series no. 121) 11–13 Sept. 1990.

Proceedings of European Mixing Conferences.
The proceedings of the 1st–5th conferences were published by BHRA Fluid Engineering, Cranfield, UK. The 6th was published by the Italian Association of Chemical Engineering and the EFCE Working Party on Mixing. The 7th was published jointly by the Royal Flemish Society of Engineers (K. VIV), the Belgian Branch of the Societe de Chemie Industrielle and the EFCE Working Party on Mixing.
1st, Cambridge, UK, 9–11 Sept. 1974.
2nd, Cambridge, UK, 30 Mar.–1 Apr. 1977.
3rd, York, UK, 4–6 Apr. 1979.
4th, Noordwijkerhout, Holland, 15–17 Sept. 1982.
5th, Würzburg, Germany, 10–12 June 1985.
6th, Pavia, Italy, 24–26 May 1988.
7th, Bruges, Belgium, 18–20 Sept. 1991.

27

Recovery of Solids from Slurries

Alan J. Carleton
Warren Spring Laboratory, Stevenage, UK

1. INTRODUCTION

At the end of most slurry pipelines there is at least one and possibly more solid–liquid separation stages. However, there is a wide range of equipment to choose from (see Figs 27.2–27.5) so selection of the best systems may not be an easy task. The aim of this chapter is to help the reader to select the best equipment for the application and to show how multiple equipment can be integrated into a flow sheet.

The chapter does not claim to give definitive answers for every situation. Instead, it gives a rationale in how to proceed, check lists of factors to consider, and examples of successful flow sheets. The aim is to give the reader information to help him conduct a fruitful dialogue with equipment vendors.

Figure 27.1 shows the steps required to develop a flow sheet of separation equipment downstream of a slurry pipeline, and shows the appropriate section numbers in this chapter. It will be seen that some of the steps are interactive so that some of the early decisions may need to be reappraised in the light of later evidence.

The various solid–liquid separations are classified in section 2. Check lists of factors which need to be considered are given in section 3. Methods of assessing the approximate size of equipment from simple tests are given in section 4. A check list of points to consider when integrating separators into a flow sheet is given in section 5 which is illustrated with examples of typical flow sheets.

2. CLASSES OF SEPARATOR

Equipment can be classified by the mechanism used to separate the solid from the liquid.

- Evaporation (section 2.1.)
- Sedimentation (section 2.2., Fig. 27.2)
- Filtration (section 2.3., Figs 27.3, 27.4)
- Other mechanisms (section 2.4., Fig. 27.5)

In Figs 27.2–27.5, *batch* refers to discharge of the solids. Most equipment has continuous discharge of the liquid. Equipment commonly used for recovery of solids from industrial scale pipelines is underlined. The emphasis is on this type of equipment but other types are included because of the frequent need to achieve other objectives, e.g. to produce environmentally acceptable liquid, free from solids.

The separators are illustrated in Figs 27.6–27.15. In these figures, *F* refers to the feed, *L* to the liquid product and *S* to the solid product which may be a thickened slurry or a cake. More detailed descriptions of equipment can be found in Cheremisinoff and Azbel (1983), Matteson and Orr (1987), Perry and Green (1984), Purchas (1981), Svarovsky (1981, 1985), and Warring (1981).

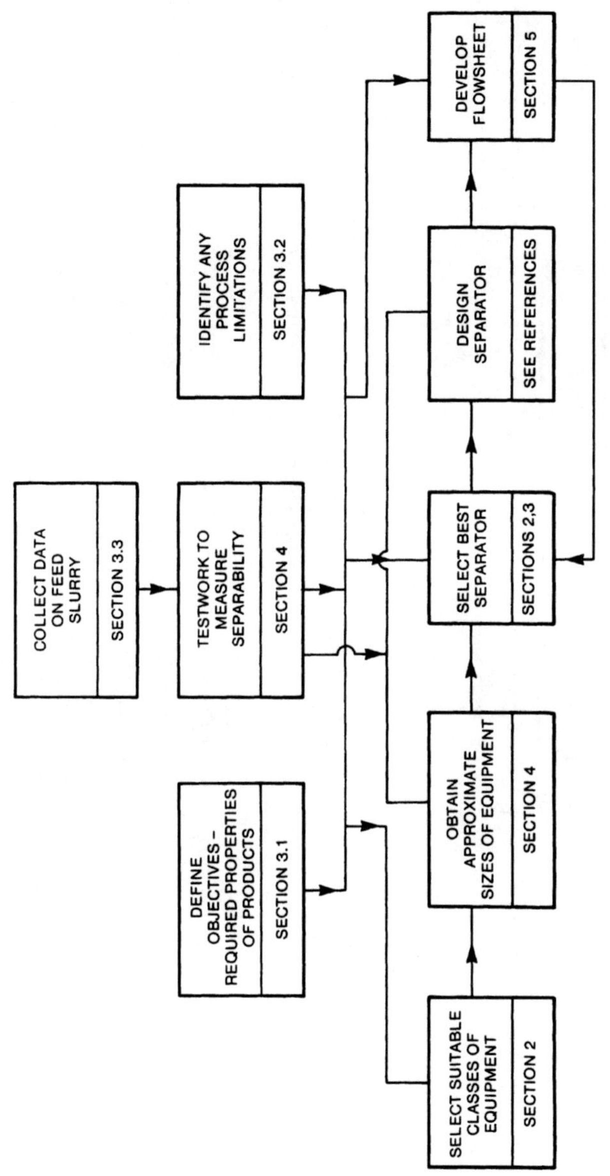

Fig. 27.1. Procedure for selecting solid–liquid separation equipment.

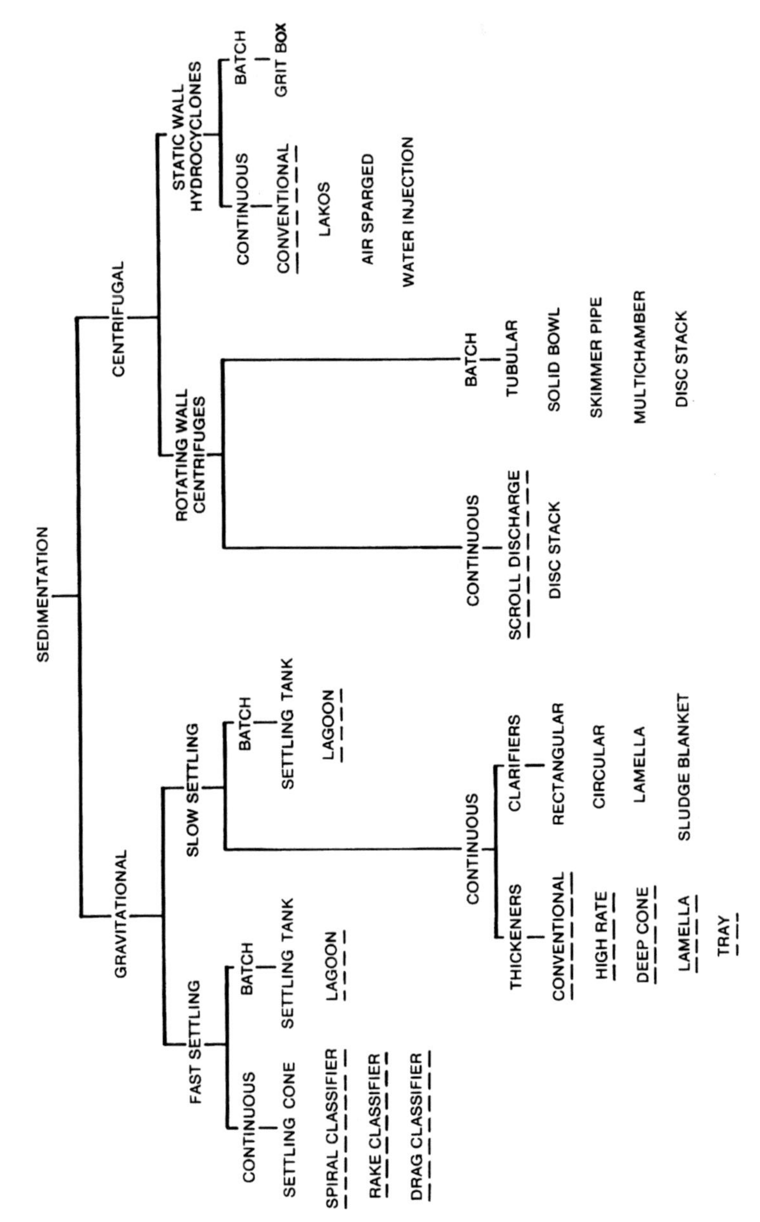

Fig. 27.2. Sedimentation equipment.

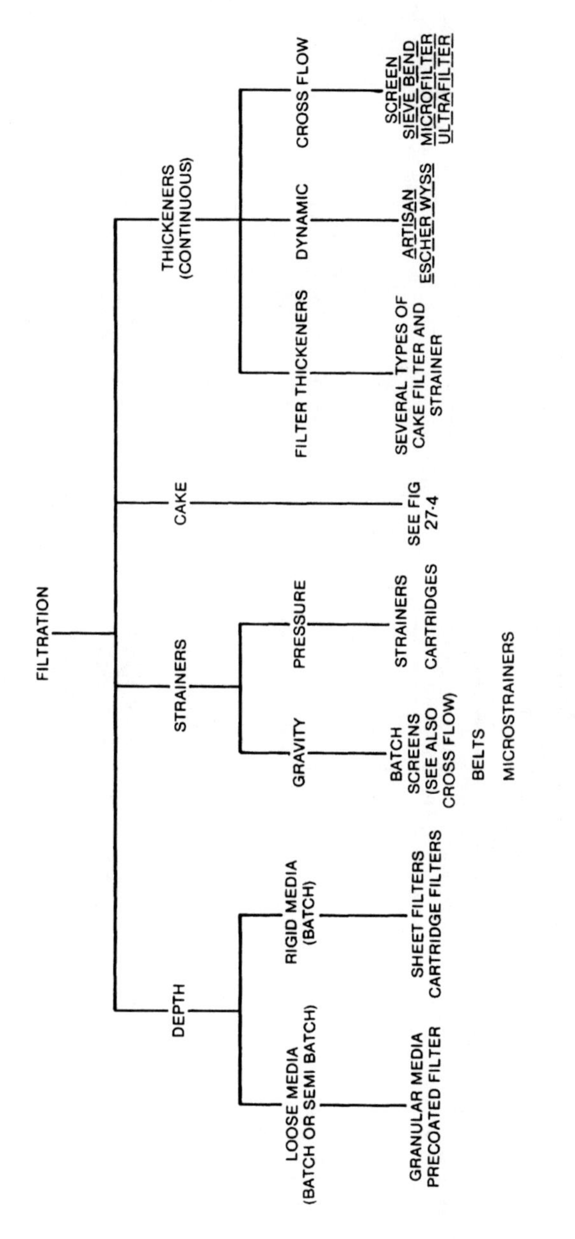

Fig. 27.3. Filtration equipment.

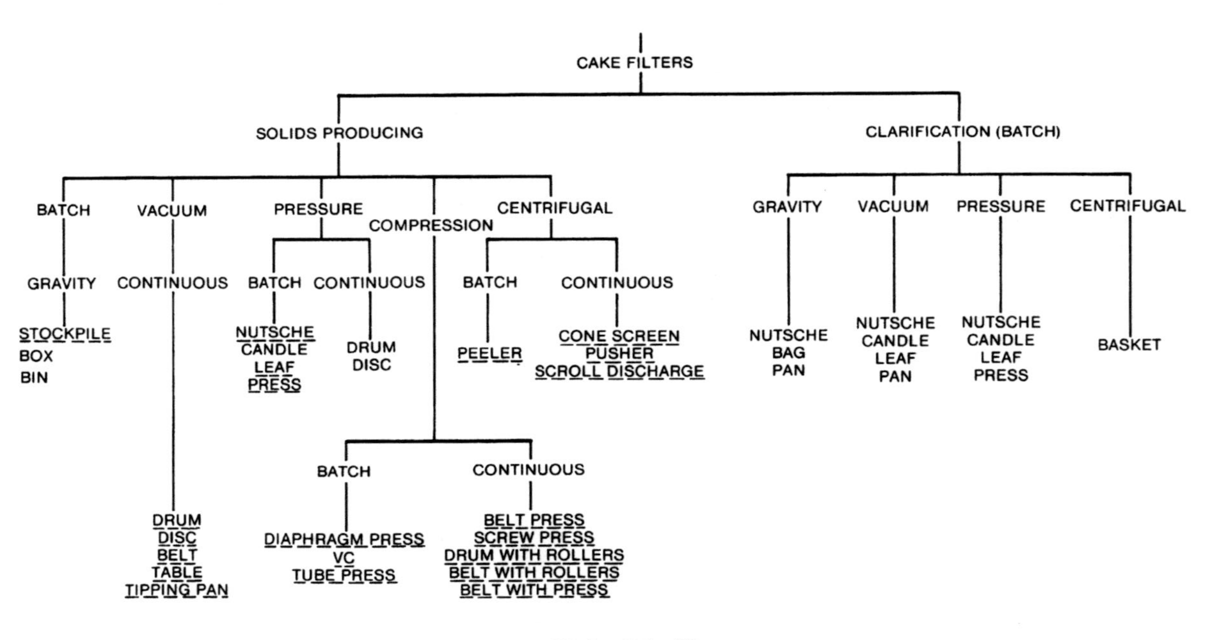

Fig. 27.4. Cake filters.

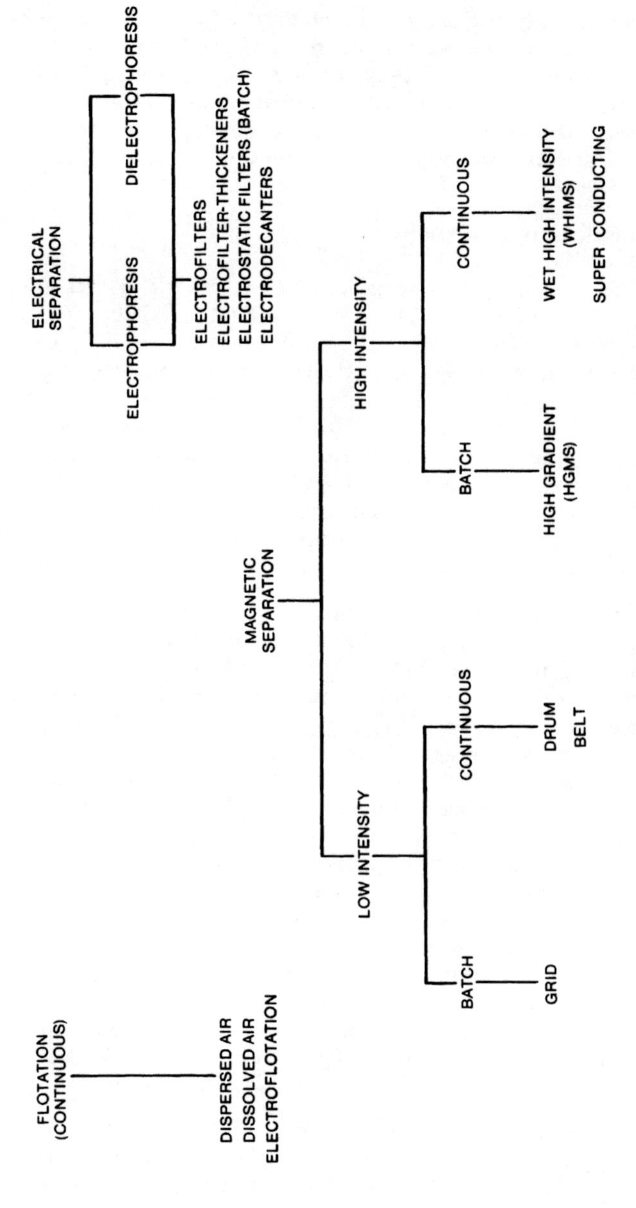

Fig. 27.5. Less common methods of separation.

2.1. Evaporation

Evaporation is the only mechanism which produces a bone-dry solid product so, if this is the ultimate objective, the flow sheet will probably contain a dryer. However, evaporation is a very energy-intensive way of removing liquid so it is generally wise to use mechanical methods to remove as much of the liquid as possible. The remainder of this section and section 3 are concerned only with mechanical separators. The integration of these with the dryer is discussed in section 5.

2.2. Sedimentation Equipment

Figure 27.2 lists sedimentation equipment which utilises a difference in density between the solid and the liquid. This is subdivided into gravitational equipment and centrifugal equipment.

The effectiveness of this separation mechanism depends on settling rate which is a function of particle size and density difference (see section 4). If settling rate is very slow, the effective particle size can be increased by adding coagulants or flocculants. In principle, any density difference can be used to effect a separation although most commercial equipment is designed for solids which are denser than the liquid. The objective may be to produce a clear liquid (clarification) or to produce a thickened slurry from a feed containing a moderate concentration of solids (thickening). The equipment can be designed to achieve one or both of these objectives.

2.2.1. Gravity Sedimentation Equipment

Gravity settling of fine particles is slow so that very large equipment may be required. However, this is offset by low operating and maintenance costs. Landfill *lagoons* may cover several acres and are used when the solids are not required. Conventional raked *thickeners* (Fig. 27.6(a)) may have diameters up to 200 m. Raked *clarifiers* are similar in appearance to thickeners but have a lighter construction. The so-called *high rate clarifiers* are also similar but added attention is paid to mixing of the flocculant and the position of the feed injection which may be below the level of the thickened slurry. Clarification performance can also be improved by the addition of inclined plates which reduce the distance the particles have to settle before being collected. This design is known as a *lamella separator* (Fig. 27.6(b)). *Deep cone thickeners* (Fig. 27.6(c)) are designed to produce thickened sludges with a high solids concentration.

Coarse particles ($> 50\ \mu m$) should not be fed to thickeners because they settle out rapidly and form beds which cannot be raked or pumped. *Settling cones* (Fig. 27.6(d)) have been used for this type of material but are difficult to control. More often the solids are removed with a screw (*spiral classifier*, Fig. 27.6(e)), a conveyor (*drag classifier*, Fig. 27.6(f)) or rake (*rake classifier*, Fig. 27.6(g)).

2.2.2. Centrifugal Sedimentation Equipment

Settling rates can be increased in a centrifugal field so reducing equipment size. This can be done in equipment with static walls (hydrocyclones) or with rotating

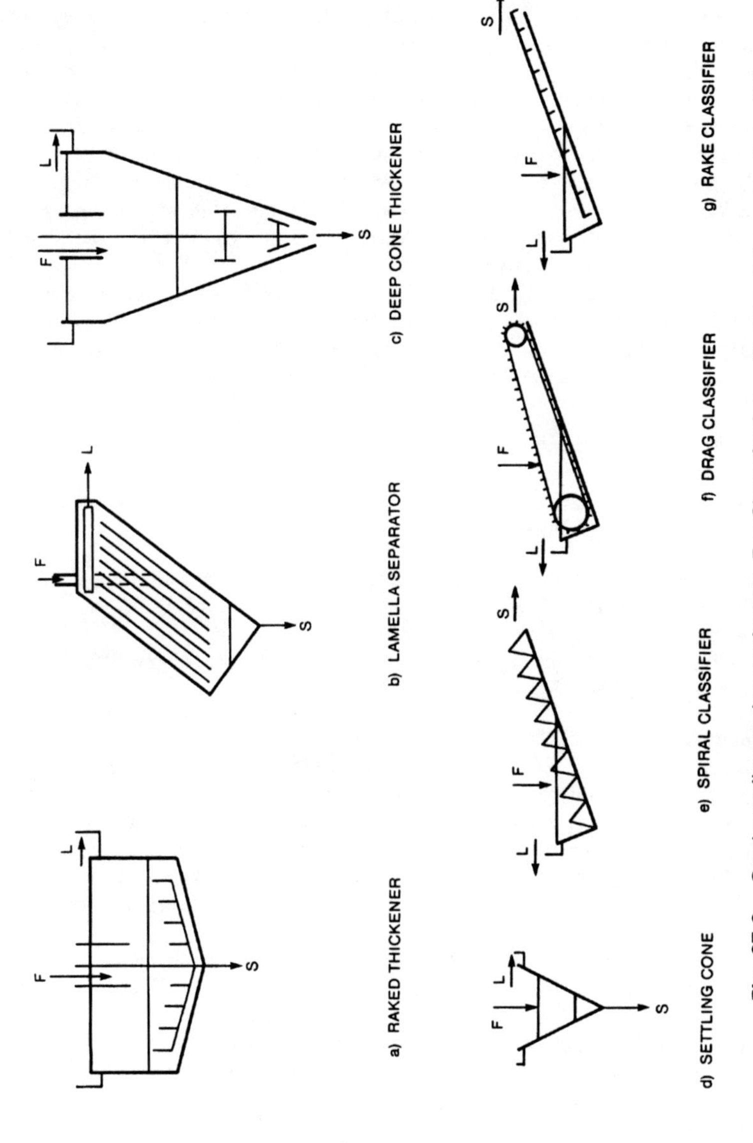

Fig. 27.6. Gravity sedimentation equipment. F = Slurry feed, L = liquid and S = separated solids.

walls (solid-bowl centrifuges). The use of flocculants to increase rates is generally less effective than with gravity equipment because the flocs get broken down in the high shear conditions, particularly in hydrocyclones.

In *hydrocyclones* (Fig. 27.7(a)) the centrifugal field is achieved by introducing the feed slurry tangentially at high velocity. Hydrocyclones can be designed to operate as clarifiers, thickeners or classifiers. They are simple but versatile pieces of equipment which can operate with solids which are denser or lighter than the liquid.

In *sedimenting* (solid bowl) *centrifuges* the heavier solids are thrown towards the wall while the clarified liquid overflows a central weir. *Batch* machines (Fig. 27.7(b)) are used mainly for clarification with the solids often removed manually. In *scroll discharge decanter centrifuges* (Fig. 27.7(c)), the solids are continuously scrolled to one end of the machine while the liquid overflows at the other end. These are versatile high throughput machines.

Disc-stack centrifuges (Fig. 27.7(d)) make use of the lamella principle. Solids settle in the discs and slide-off them to the outside of the bowl where they can be discharged continuously through nozzles or intermittently by opening the bowl. These are high-performance machines used for fine particles.

2.3. Filtration Equipment

Filters present a medium to the slurry which retains the solids while allowing the liquid to pass through. Filters can be classified by

- mechanism of filtration—e.g. depth filtration, straining, cake filtration or filter thickening (see Fig. 27.8);
- batch or continuous discharge of solids; and
- driving force—e.g. gravity, vacuum, pressure, compression, centrifugal (see section 2.3.3, Fig. 27.4).

A summary is given in Table 27.1.

In *depth filters* (Fig. 27.8(a)) the particles penetrate the medium and are trapped within it. They can trap very fine particles but have limited capacity. Solids are often not recoverable so these filters are used exclusively for clarification.

Particles are deposited on the surface of *strainers* to a depth of one or two layers (Fig. 27.8(b)). They are often used for clarification but can also be used to classify the solids, i.e. by using a coarse screen so that fine particles pass through while the coarser particles are retained.

Most filters for pipeline slurries are *cake filters*. Particles deposit on the surface of the medium (e.g. a cloth) to form a cake having a depth of several hundred layers (Fig. 27.8(c)). In the early stages they act as strainers when some bleed through of fines may occur. This filtrate may need to be recycled. However, when the cake builds up this acts as the medium and very fine particles can be trapped, possibly by a depth filtration mechanism. Thus, cake filters can be used for clarification or for solids recovery.

A problem with cake filtration of fine particles is that resistance to flow can be

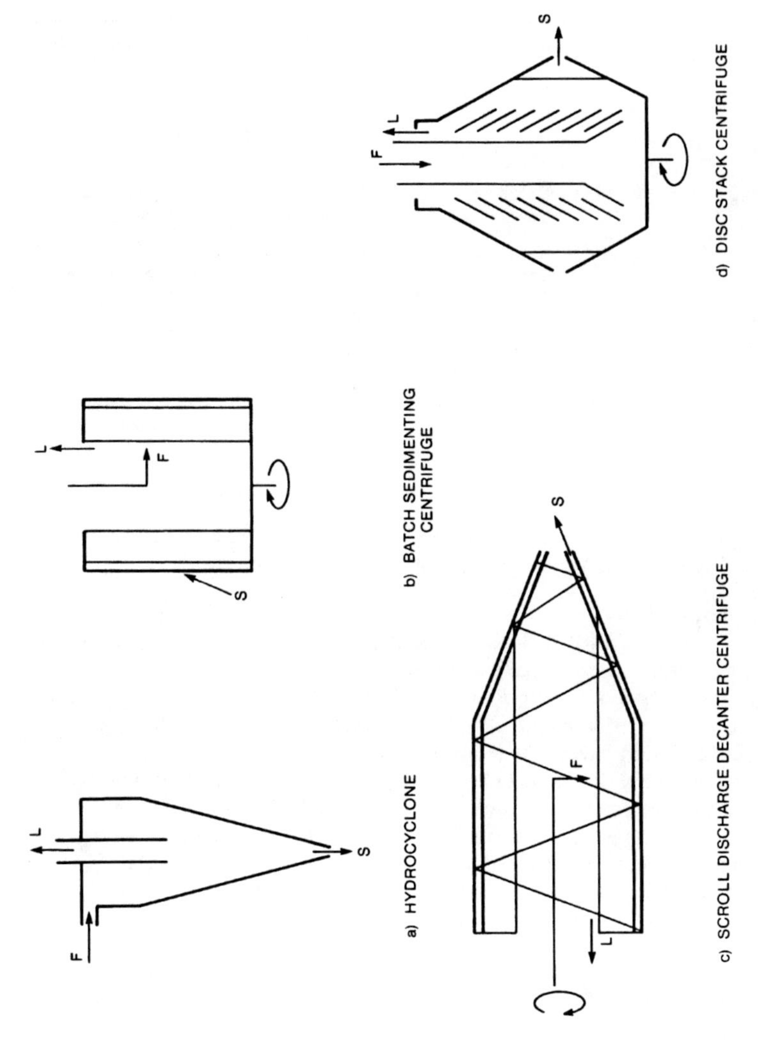

Fig. 27.7. Centrifugal separation equipment. F = Slurry feed, L = liquid and S = separated solids.

TABLE 27.1
Summary of Filter Classifications

Mechanism	*Fig.*	*Batch/ continuous*	*Driving force*	*Objective*
Depth	27.8(a)	Solids often not recoverable, batch recovery if they are	Granular medium filters—gravity or pressure. Sheet filters—pressure. Precoats—any driving force	Clarification only
Straining	27.8(b)	Batch or continuous	Gravity, vacuum or pressure	Clarification, classification
Cake	27.8(c)	Batch or continuous. Gravity, pressure, usually batch; vacuum usually continuous	Any driving force	Clarification, solids recovery
Filter thickening	27.8(d)	Continuous	Generally pressure	Clarification, recovery of solids as thickened slurry

very high. A solution to this is to prevent cake from building up on the filtering surface (Fig. 27.8(d)). Ways of doing this are given in section 2.3.3.

2.3.1. Depth Filters

Depth filters can consist of loose particles packed into a bed or rigid medium consisting of thick cloth or paper, or sintered polymers or metals.

Granular media filters comprise a bed of loose coarse media like sand or coke. The slurry can be fed in by gravity or under pressure. The bed can be regenerated by back-flushing to release the retained particles.

Filter aids such as diatomite or perlite can be formed into a *precoat* in most types of cake filter. The precoat can then be used as a depth filter. Finer particles can be removed than with granular media filters but the precoat cannot be regenerated.

Rigid media depth filters are typically loaded into filter presses (Fig. 27.12(a)) known as *sheet filters* or into pressure leaf filters (Fig. 27.12(b)) and known as *pad filters*.

2.3.2. Strainers

Strainers can consist of screens for coarse separations or cloth or paper for finer solids.

Gravity fed *screens* may be batch but are often made continuous by vibrating the screen and/or giving it a slope (Figs 27.9(a) and (b)). Continuous screens are often used for classifying the solids into coarse and fine fractions. They are solids producing filters and can be classified as filter thickeners (see section 2.3.5). All

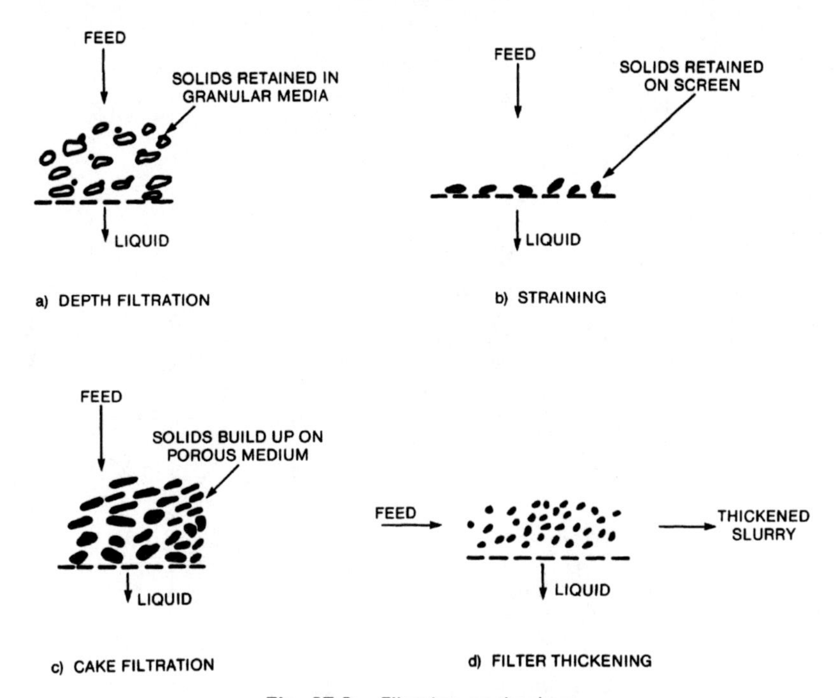

Fig. 27.8. Filtration mechanisms.

other strainers are used for clarification. *Pipeline strainers* are pressure-fed screens fitted in pipes. They may be cleaned *in situ* (Fig. 27.9(c)) or removed from the pipe for cleaning.

Belt filters used as strainers consist of a paper (or possibly cloth) belt and may operate by gravity or under a slight vacuum. Belt movement can be continuous but is often intermittent (Fig. 27.9(d)). In this type, build-up of slurry triggers belt movement and exposes clean cloth to the slurry.

Micro-strainers consist of a cloth covered rotating drum (Fig. 27.9(e)). Solids are picked up inside the drum at the bottom and are discharged by back-flushing at the top.

Cartridge filters (Fig. 27.9(f)) are made of cloth or paper and fitted into pipes. When full of solids they have to be removed for disposal.

2.3.3. Solids Producing Cake Filters

Cake filters can be classified by the driving force and whether they are batch or continuous. Continuous filters are generally preferred when the primary objective is cake production. Cake filters for clarification are discussed in section 2.3.4. The driving forces available are as follows.

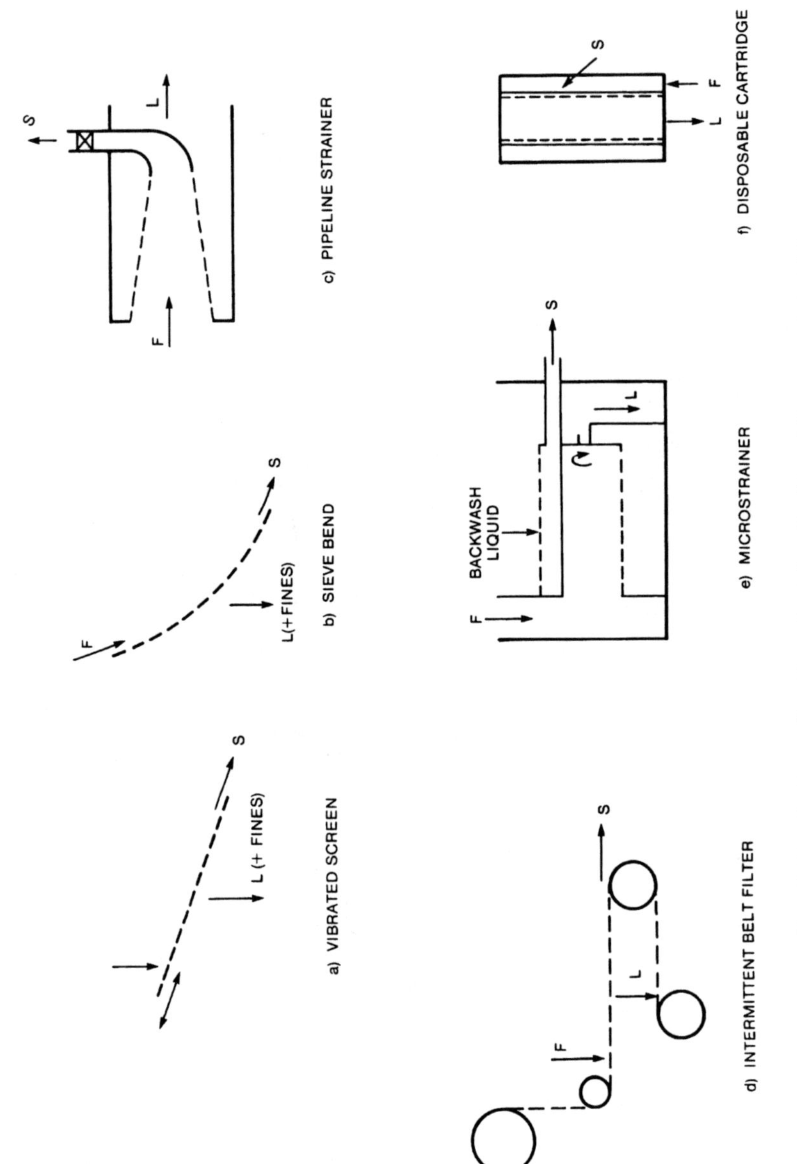

Fig. 27.9. Strainers. **F** = Slurry feed, **L** = liquid and **S** = separated solids.

- *Gravity*. Here, the driving force is the head of slurry. These are used mainly for coarse solids.
- *Vacuum*. Here, a vacuum is applied to the inside of the filter so that cake is picked up on the outside where it can be discharged. Because of the ease of discharge, these machines are generally continuous.
- *Pressure*. Here, slurry is pumped into the filter under pressure. Cake is on the inside of the filter so discharge tends to be batchwise. However, because the driving force is greater than for vacuum filters, they give faster filtration rates and drier cakes, particularly when the particles are fine.
- *Compression*. In this type of filter the cake is formed and is then compressed mechanically to reduce its moisture content.
- *Centrifugal*. Here, the driving force is the head of slurry enhanced by a centrifugal field.

Gravity filters. If the particles are very coarse and fast settling they can be pumped into any retaining vessel and allowed to drain by gravity. A porous base is preferable but any vessel with an outlet for the liquid can be made to work. Such systems are classified here as filters because the liquid drains through the bottom of the bed of solids (i.e. cake). Most systems require batchwise recovery of the solids. Examples are a *stockpile* surrounded by a bund (Fig. 27.10(a)) where solids recovery may be by a grab crane; a *box* (Fig. 27.10(b)) where recovery may be by lowering a hinged side and sending in a front loader; or a *bin* (Fig. 27.10(c)) with recovery through a gate at the bottom. Continuous recovery of the solids can be achieved by pumping the slurry onto a *conveyor belt* (Fig. 27.10(d)) which may have drainage holes.

Vacuum filters. Continuous vacuum filters are commonly used for dewatering pipeline slurries. *Drum filters* (Fig. 27.11(a)) are popular and can be supplied with a number of discharge options. The *disc filter* (Fig. 27.11(b)) can be made in large sizes but is more restricted in the number of discharge options. Top-fed filters are more suitable if the particles tend to settle or if extensive washing is required. *Belt filters* (Fig. 27.11(c)) are commonly used and are available in a number of different designs. *Table filters* (Fig. 27.11(d)) can also be used but most designs do not discharge the cake completely; they leave a cake heel. Rotary *tipping pan filters* (Fig. 27.11(e)) are another alternative.

Pressure filters. Most pressure filters are batch so require holding tanks if they are at the end of continuous pipelines. *Filter presses* (Fig. 27.12(a)) are commonly used. They have a reputation for being dirty and labour intensive but modern machines can be highly automated and in some cases can be run without supervision. *Pressure vessel filters* consist of a pressure vessel which contains a number of filter elements which can be flat (*leaf filters*) or tubular (*candle filters*) and may be horizontal (Fig. 27.12(b)) or vertical. The *pressure Nutsche* (Fig. 27.12(c)) has a single horizontal filtering surface and is suitable for small batches. Modern designs are automated and allow crystallisation, filtration, re-slurrying, washing

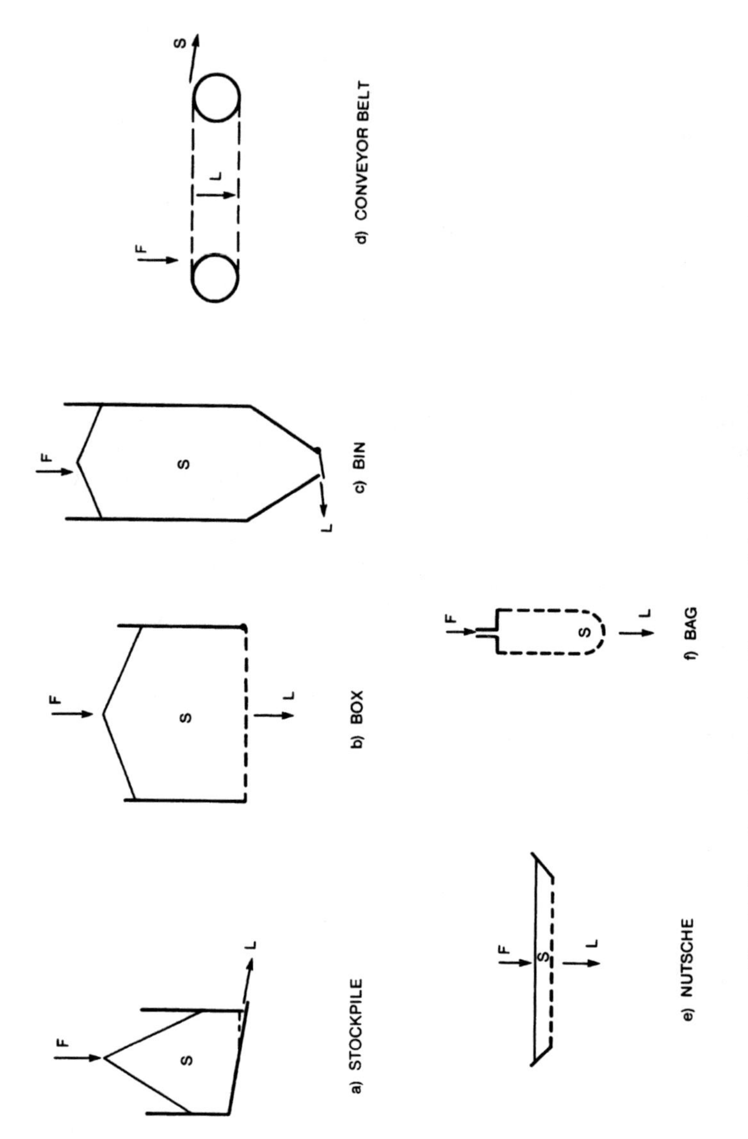

Fig. 27.10. Gravity filters. F = Slurry feed, L = liquid and S = separated solids.

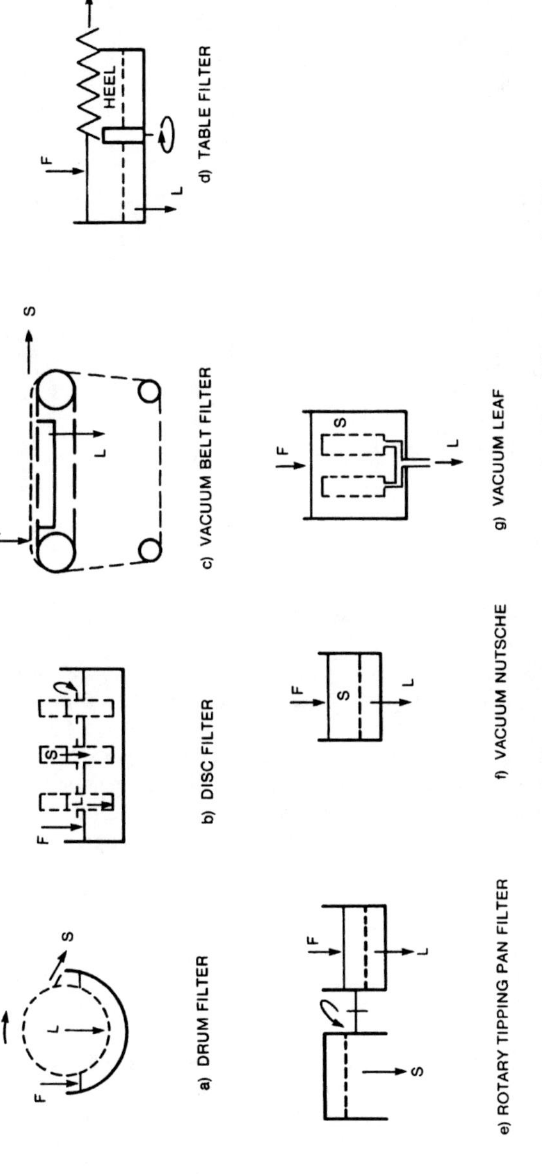

Fig. 27.11. Vacuum filters. F = Slurry feed, L = liquid and S = separated solids.

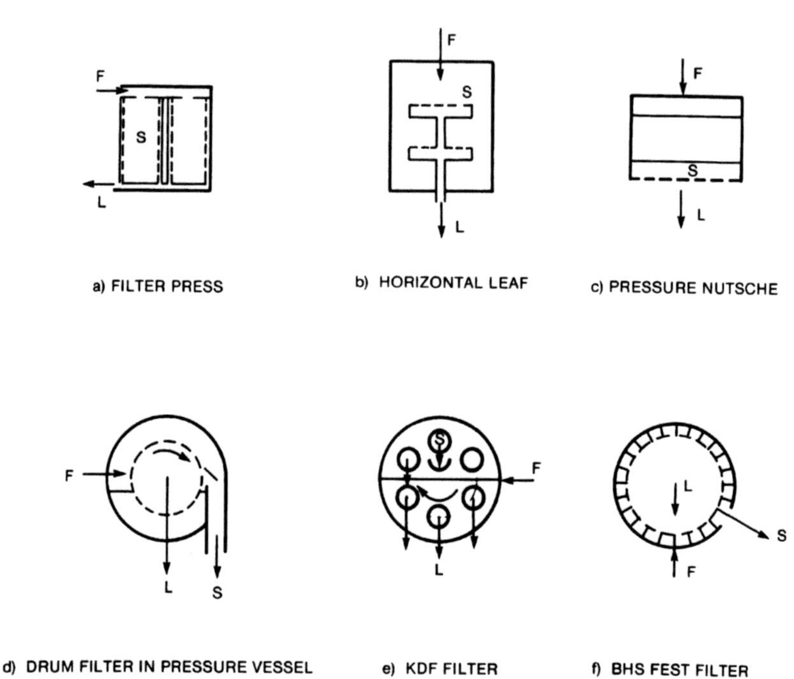

Fig. 27.12. Pressure filters. F = Slurry feed, L = liquid and S = separated solids.

and drying to be carried out in a single vessel. This can eliminate the need for slurry pipelines between separate pieces of equipment. They are sometimes known as *filter dryers*.

Attempts have been made to combine the advantages of pressure filtration (faster filtration, drier cakes) with continuous operation. However, continuous pressure filters are not commonly used because the equipment tends to be expensive and cake discharge can be a problem. The most common type consists of a conventional *drum filter* enclosed within a *pressure vessel* (Fig. 27.12(d)). The *KDF filter* (Fig. 27.12(e)) consists of a number of discs which discharge into a central conveyor. In both these designs the cake is discharged internally and must be removed from the vessel intermittently through a solids valve. The *BHS filter* (Fig. 27.12(f)) has a drum which fits closely into the outer shell. There is a long seal between the two and the cake is discharged externally.

Compression filters. Some designs of batch pressure filter can be fitted with rubber diaphragms which can be inflated to compress the cake and squeeze out additional liquid. Diaphragms can be fitted to filter presses which can have vertical or horizontal chambers. Rubber sleeves can also be used to compress cakes formed on cylindrical filtering surfaces which can be vertical or horizontal (Fig. 27.13(a)).

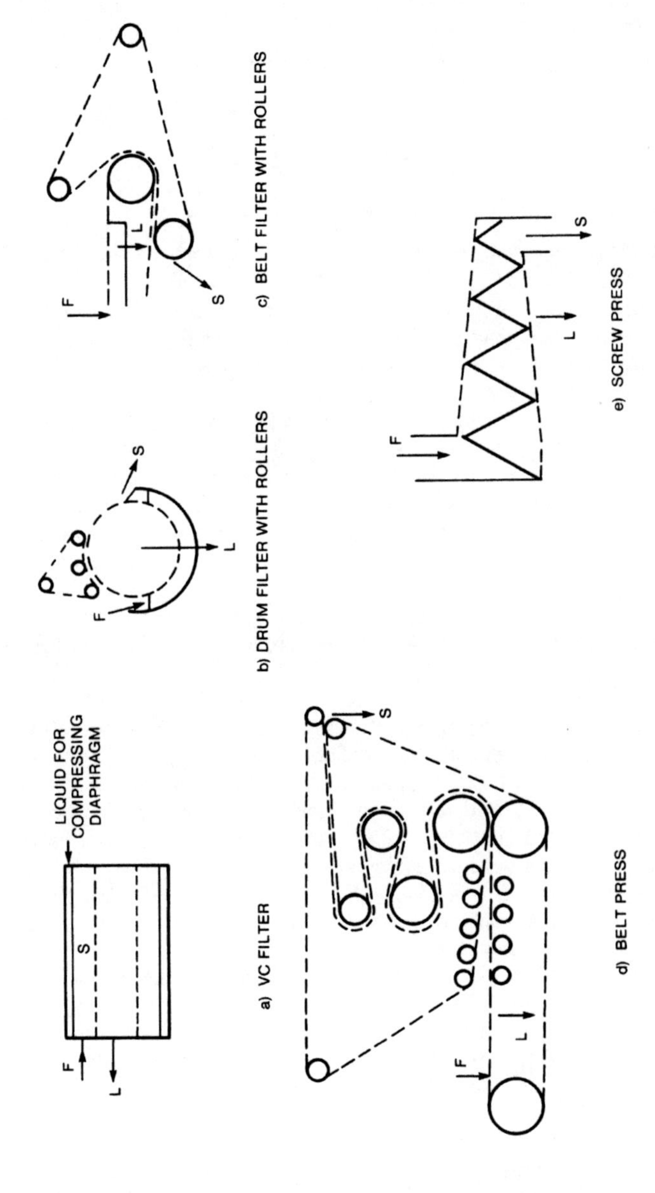

Fig. 27.13. Compression filters. F = Slurry feed, L = liquid and S = separated solids.

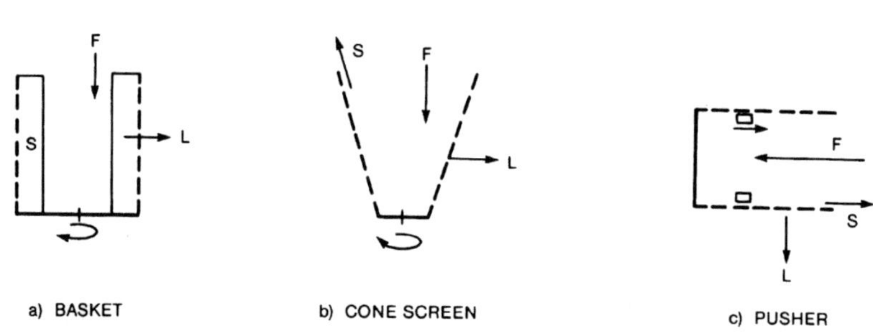

Fig. 27.14. Filtering centrifuges. F = Slurry feed, L = liquid and S = separated solids.

Continuous compression filters overcome the need for large-scale storage and recovery equipment between batch filters and continuous dryers. However, continuous compression filters usually operate at lower pressures so produce wetter cakes.

Cakes can be compressed by passing them under *rollers* on *drum filters* (Fig. 27.13(b)) or at the end of *belt filters* (Fig. 27.13(c)). Low-pressure rollers can be fitted to existing filters or special filters can be built to withstand higher roller pressure.

Another configuration is the belt press (Fig. 27.13(d)). This can have a gravity drainage section, a low-pressure tapering section, and a high-pressure section where the cake is trapped between two cloths which pass around rollers of gradually decreasing size.

The screw press (Fig. 27.13(e)) consists of a screw in a perforated cage. The pitch or diameter of the screw decreases so that pressure builds up which dewaters the slurry.

Filtering centrifuges. Filtering centrifuges have perforated bowls (screens) and should not be confused with solid-bowl (sedimenting) centrifuges. The solids are retained on the screen while the liquid passes through. *Batch machines* (Fig. 27.14(a)) are often fitted with cloths to retain fine solids and can be highly automated. Slurry is fed in at slow speed to minimise crystal damage. Filtration and dewatering are then done at high speed before the machine is slowed down again for cake discharge. This is done with a knife and leaves a heel on the cloth.

Continuous discharge machines are fitted with metal screens to withstand the discharge device so are suitable only for coarse ($> 50\,\mu m$) solids.

In *cone screen* machines the solid moves from the apex of the cone to its base. Transport of the solids may be by simple slippage (Fig. 27.14(b)), by applying vibration or a gyratory motion to the cone, or by means of a screw.

In *pusher centrifuges* (Fig. 27.14(c)) the cake is pushed intermittently along the bowl by an annular pusher ring.

2.3.4. Cake Filters for Clarification

Clarification of dilute suspensions is often carried out in batch cake filters. If the solids concentration is low enough, the filter can be run with a long cycle time.

Fig. 27.15. Dynamic filters. F = Slurry feed, L = liquid and S = separated solids.

Simple *Nutsche filters* with manual discharge can be used and are available in gravity (Fig. 27.10(c)), vacuum (Fig. 27.11(f)) or pressure versions (Fig. 27.12(c)). *Pan filters* are similar but tip for discharge; they are available in gravity or vacuum versions. *Candle* and *leaf filters* are available in vacuum (Fig. 27.11(g)) or pressure versions (Fig. 27.12(b)). *Filter presses* (Fig. 27.12(a)) are often used for clarification. Batch basket centrifuges (Fig. 27.14(a)) fitted with cloths can also be used.

2.3.5. Filter Thickeners. Filter thickening is used with fine solids which would form high-resistance cakes. It can be done in conventional filters, in dynamic filters or in cross-flow filters.

Filter thickening can be achieved in many types of filter by discharging the cake continuously or at frequent intervals. Examples are *vibrating screens* (Fig. 27.9(a)) where vibration is used to prevent cake formation, frequent sluicing the cake off *pressure vessel filters*, or knifing a cake off a *drum filter* so that it drops back into the trough. Ultrasonics and electrophoretic forces have also been used to prevent cake formation. The solids may be discharged as a thickened slurry or a cake may eventually be formed—this is known as *delayed cake formation.*

In *dynamic filters* cake formation is prevented by applying shear to the cake. This can be done by rotating discs close to a flat filtering surface (Fig. 27.15(a)) or by rotating a tubular filtering surface (Fig. 27.15(b)).

Cake formation can also be prevented by feeding the slurry tangentially to the filtering surface at high velocity. This is known as *cross-flow filtration* and is the principle of *microfiltration* (particle sizes 0·02–10 μm) and ultrafiltration (particle sizes 0·001–0·02 μm). Continuous screens like *sieve bends* (Fig. 27.9(b)) also operate on this principle.

2.4. Other Separation Methods

Sedimentation, filtration and centrifugation are the options most commonly used for solid–liquid separation of pipeline slurries. Other methods are generally used only for separating out very difficult (usually very fine) solids which cannot be separated by conventional methods.

Flotation requires the solid to attach itself to gas bubbles which bring it to the

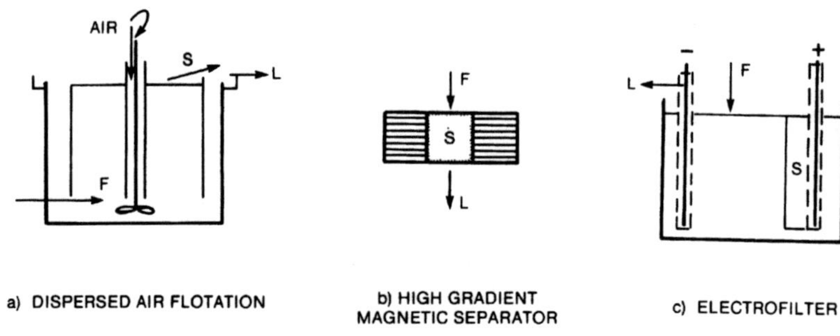

Fig. 27.16. Other separators. F = Slurry feed, L = liquid and S = separated solids.

surface (Fig. 27.16(a)). Flotation can be used for clarification or for separating solids with different surface properties.

Magnetic filters can be used to clarify liquids containing magnetic particles. Permanent magnets can be used to remove ferromagnetic materials batchwise or continuously. High-gradient magnetic separators (Fig. 27.16(b)) using electromagnets can be used for paramagnetic materials.

Electrophoretic forces act when a DC field is applied to a slurry containing charged particles. Dielectrophoresis is a phenomenon which depends on the polarisability of particles in non uniform AC or DC fields in liquids with low conductivity. These forces can be used to improve separations in a number of ways. *Electrofilters* can operate so that the electrical forces compress the cake and dewater it (Fig. 27.16(c)). Alternatively the forces can be applied in the opposite direction to prevent cake formation, i.e. achieve filter thickening. In the absence of a filter medium, settling rates can be enhanced as in *electrodecanters.*

Other forces such as *ultrasonics* have also been used to enhance separations.

3. SELECTION FACTORS

This section gives a check list of factors to consider when selecting the most suitable equipment from the options described in section 2. Selection depends on the objectives of the separation, i.e. on the required properties of the products (section 3.1), process limitations (section 3.2) and the properties of the feed slurry (section 3.3). The data collected here may be supplemented with quantitative data from section 4 on the likely size of alternative pieces of equipment. With these data it should be possible to identify suitable pieces of equipment. The final choice will involve discussions with equipment vendors and may require pilot-scale testing to compare different options.

3.1. Objectives of Separation

Possible objectives of the separation are summarised in Table 27.2.

TABLE 27.2
Objectives of Separation

Objective	*Section*	*Required phase*	*Phase not required*	*Equipment*
Thickening	3.1.1.	Often intermediate stage		Sedimentation equipment. Filter thickeners
Cake formation	3.1.2.	Solid	Liquid	Filter or centrifuge. Drainage, gas displacement or compression may be used to reduce liquid content.
Washing	3.1.3.	Solid	Mother liquor	Re-slurry washing in multi-stage sedimentation equipment. Displacement washing in filter with washing capability—usually horizontal surface.
Washing	3.1.4.	Liquid	Solid, wash liquid	Filter with counter-current washing capabilities, e.g. top-fed vacuum filter.
Clarification	3.1.5.	Liquid	Solid	Gravity clarifier. Sedimenting centrifuge, depth filter, batch cake filter
Classification	3.1.6.	Solid	Other solids, by size	Classifiers, hydrocyclones, continuous screens

3.1.1. Recovery of Solids as a Thickened Slurry

If the slurry is dilute it can be thickened up to a solids concentration which is just pumpable. Pipeline slurries may already be at this concentration in which case thickening is not an option. Thickening is not usually the final objective but offers a cheap way of removing liquid before feeding to a more expensive cake filter.

Thickening is an option for slow settling solids and can be carried out in gravity thickeners, sedimenting centrifuges, hydrocyclones (Fig. 27.2) or filter thickeners (Fig. 27.3).

3.1.2. Recovery of Dry Solids

A final objective is often the recovery of the solids in a bone-dry form. A dryer is usually required to remove the last traces of liquid but the bulk of the liquid can be removed more economically in a cake filter or centrifuge. Suitable equipment is shown in Fig. 27.4.

When a cake is first formed it will have a relatively high moisture content (Fig. 27.17). However, this can be reduced by drainage, gas displacement or compression. Drainage and gas displacement require the liquid in the pores of the cake to

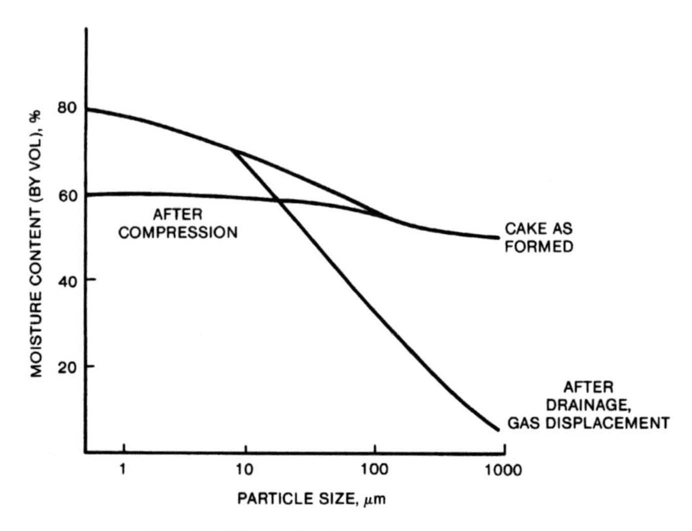

Fig. 27.17. Cake dewatering mechanisms.

be replaced by gas. Compression reduces the porosity of the cake. The range of application of different methods of dewatering is summarised in Fig. 27.18.

For the liquid in the pores to be replaced by gas it is necessary to overcome capillary (surface tension) forces and these increase as the pore size decreases (see section 4.5). If the particles are coarse enough, gravitational forces will be sufficient to drain the cake, e.g. in a stockpile. Finer particle cakes (with higher capillary

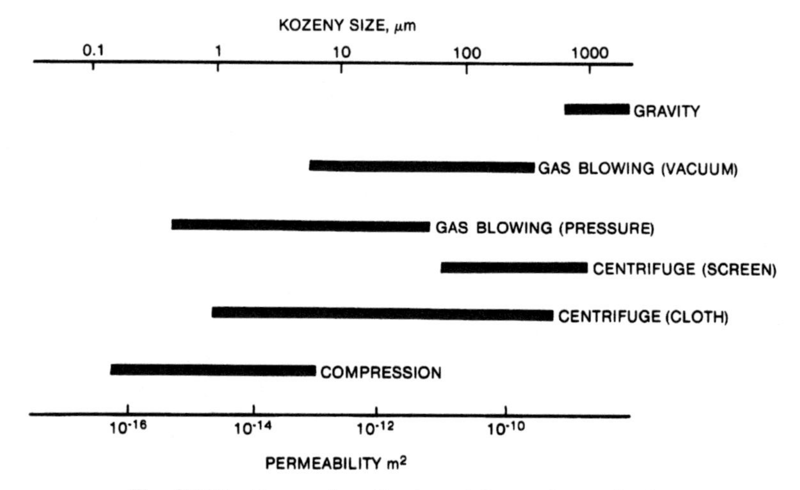

Fig. 27.18. Range of application of dewatering methods.

forces) can be dewatered at the higher *g* forces available in filtering centrifuges. However, continuous filtering centrifuges are fitted with screens and so cannot be used for very fine particles.

Cakes formed in filters can be desaturated by gas displacement. Gas can be sucked through the cake in a vacuum filter, but finer particle cakes can be desaturated by blowing gas through the cake in a pressure filter. Gas displacement is most effective when the liquid is initially hot or volatile, or when hot gas or steam are used. Under some conditions bone-dry cakes can be produced on the filter (Carleton & Salway, 1987).

If the particles are very fine the capillary pressures are very high and the cake cannot be desaturated. However, this type of cake is often voluminous and can be compressed in a compression filter (see section 4.6).

3.1.3. Recovery of Washed Solids

It is sometimes necessary to recover the solids free from impurities contained in the liquid. This can be done by displacement washing and/or re-slurry washing (see section 4.7).

In displacement washing a cake is formed and wash liquid passed through it to displace the mother liquor. It can be carried out in most cake filters but those with horizontal filtering surfaces are usually preferred when high efficiency is required.

In re-slurry washing the feed slurry is concentrated up to form a thickened slurry or a cake. Wash liquid is then added to reform the slurry which is then reconcentrated. The efficiency of each stage depends on the amount of concentration achieved at the stage. High efficiency requires multiple stages. Re-slurry washing can be carried out in gravity thickeners or hydrocyclones. Higher efficiency per stage can be achieved in sedimenting centrifuges or cake filters, but each stage is more expensive.

A single stage of displacement washing uses less liquid and gives higher efficiency than a single stage of re-slurry washing. However, very high efficiency may be difficult to achieve by displacement washing alone as some mother liquor is trapped in inaccessible pores. A solution to this is to have a re-slurry stage between displacement washes. With multiple stages of either type of washing the amount of wash liquid can be minimised by counter-current flow between stages.

3.1.4. Recovery of Maximum Liquid

When the liquid is a valuable product it may be necessary to recover as much of it as possible. This can be done by displacement washing of a cake. If it is also necessary to avoid dilution with wash liquid, this can be done by multistage countercurrent washing on a top-fed vacuum filter, e.g. belt, tipping pan or table filter.

3.1.5. Recovery of Liquid Free from Solids

With pipeline slurries the primary objective is usually recovery of the solids. However, clarification of liquid may be an important secondary objective if the liquids from the primary separators still contain solids so cannot be disposed of without further treatment.

Clarification can be carried out in gravity clarifiers, sedimenting centrifuges, hydrocyclones (Fig. 27.2), depth filters (Fig. 27.3), some cake filters (Fig. 27.4), float cells, magnetic filters and electrical separators (Fig. 27.5).

3.1.6. Classification of Solids

The solids may be classified by size,

- so that the fine particles can be treated in separate equipment from the coarse particles;
- to remove tramp material from a slurry product, e.g. removal of coarse lumps from paint; and
- to remove coarse particles from a carrier fluid, e.g. removal of cuttings from a drilling mud.

Classification can be done in gravity or centrifugal sedimentation equipment designed with a suitable cut size. Hydrocyclones are often used. Alternatively continuous strainers (e.g. screens) can be used.

3.2. Process Requirements

Process factors which affect the choice of equipment include the following.

- *Throughput.* Some equipment (e.g. gravity sedimentation equipment, continuous vacuum filters, presses) can be designed for high solids throughputs whereas other equipment (e.g. much batch equipment) is limited to small operations.
- *Compatability with downstream processes.* For example, dryers. Examples of this are given in section 5.
- *Space requirements.* For example, when floor area, headroom or weight limitations affect the choice of equipment.
- *Reliability required.* In some cases reliability is of vital importance, e.g. when breakdown would be very expensive or equipment would be hazardous to repair. In other cases, cheaper, less reliable equipment could be selected, e.g. when there is parallel equipment.
- *Use of additives.* Separation performance can be improved by the use of coagulants or flocculants to increase settling or filtration rates, surfactants to increase dewatering rates, filter aid precoats to improve liquid clarity, or filter aid body filters to increase filtration rates. However, these additives may be incompatible with the final liquid or solid product.

3.3. Properties of the Feed Slurry

This section gives a check list of factors associated with the properties of the feed slurry which should be considered in equipment design. The use of measured data on the separation properties of the slurry is discussed in section 4.

3.3.1. Particle Size Distribution

Particle size distribution is an important parameter in selection. In some cases (e.g. with screens) it is important in its own right. More often it is important in as far as it affects settling rate or filtration rate (see section 4). If the size range is very wide it may be necessary to classify the solids and to treat the fines in a different separator from the coarse (see section 5).

Both settling rates and filtration rates increase with the square of particle diameter so separation costs increase rapidly as the particle size decreases. However, pipeline pumping costs generally decrease as the particle size decreases. So if particle size is a variable which can be selected there may be an optimum size for minimum overall costs. For long pipes the optimum size will be small to minimise pumping costs but for short pipes the particle size should be kept large to minimise separation costs.

Colloidal solids ($\ll 1\ \mu m$) are very difficult to recover from the liquid. They can be thickened in a filter thickener or fed directly to a suitable dryer (e.g. spray dryer). Fine solids (1–50 μm) can be categorised as *non-settling* in the turbulent conditions of the pipeline but they will settle in sedimentation equipment. They can also be filtered. Vacuum filters are used for the coarser sizes. For finer particles, pressure and compression filters and filter thickeners are used. Coarse solids ($> 50\ \mu m$) are best handled in classifiers, gravity filters, top-fed vacuum filters and continuous centrifugal filters. Table 27.3 is a guide to the operating ranges of common separation equipment.

3.3.2. Specific Gravity Difference

Difference in density between the solids and the liquid affects the settling rate (see section 4). If this is low then sedimentation methods can be ruled out. Most sedimentation methods require the solids to be denser than the liquid.

3.3.3. Solids Concentration

Table 27.3 shows the range of application of various separators. Liquids containing $< 0.1\%$ solids require clarifying. Slurries containing 1–10% solids can be thickened. Slurries containing over 20% solids are usually fed to filters. Some shear-thinning pastes may have the consistency of cakes when they leave the pipeline and the shear is released. Such pastes cannot be filtered but additional liquid can be removed by expression in a compression filter.

3.3.4. Particle Strength

Fragile crystals may be damaged in high shear equipment like centrifuges or hydrocyclones or by some discharge devices, e.g. knives, although damage may be no greater than in the pumps and pipe. However, flocs which have been formed after the pipe should not be fed into high shear equipment.

3.3.5. Hazardous Properties

Toxic, reactive, volatile, flammable and explosive materials should be handled in enclosed equipment (e.g. in pressure filters). Some materials require special pre-

TABLE 27.3
Operating Ranges of Separation Equipment

	Particle size (μm)	*Feed slurry concentration (% by mass)*
Classifier	50–1000	20–50
Thickener	1–50	5–50
Clarifier	1–50	1–5
Scroll discharge decanter	2–50	2–50
Continuous disc stack	0·2–20	0·1–20
Batch disc stack	0·2–20	0·005–0·1
Solid bowl centrifuge	2–5000	0·1–5
Hydrocyclone	5–200	0·1–40
Granular media filter	0·1–50	0·001–0·1
Precoated filter	0·1–50	0·001–0·1
Sheet filter	0·1–50	0·001–0·1
Drum filter	1–50	5–50
Disc filter	1–50	5–50
Belt filter	5–100 000	10–50
Table filter	50–500	10–50
Tipping pan filter	20–50 000	10–50
Pressure vessel filter	1–100	0·01–5
Filter press	1–10	0·001–30
Strainer	5–500	0·001–0·01
Cartridge filter	0·5–50	0·001–0·1
Edge filter	1–200	0·001–0·1
Pressure drum	5–200	0·5–10
Diaphragm press	1–10	5–30
Belt press	1–100	1–50
Screw press	50–500	10–50
Basket centrifuge	2–20 000	10–50
Peeler centrifuge	20–500	10–50
Cone screen centrifuge	50–20 000	10–50
Pusher centrifuge	50–5000	10–50
Dynamic filter	0·1–10	10–50
Sieve bend	50–2000	10–50
Microfilter	0·02–10	0·1–10
Ultrafilter	0·001–0·02	0·1–10
Flotation	20–200	0·1–5
Magnetic separation	1–500	0·001–10
Electrofilter	0·1–10	0·1–20

cautions like nitrogen blanketing. Materials which can be contaminated should be handled in enclosed equipment which can be easily cleaned or sterilised.

3.3.6. Abrasivity and Corrosiveness

Abrasive solids can cause problems in high shear equipment (e.g. centrifuges, hydrocyclones). Abrasion and corrosion can also be a problem in equipment with vulnerable parts (e.g. seals, screens, discharge knives or scrolls).

3.3.7. Temperature

High or low slurry temperatures can affect equipment selection in a number of ways. It may affect the choice of materials of construction including those of any filter media. If jacketing is required, it is an option on some types of equipment (e.g. pressure vessel filters) but not others. Liquids near their boiling point tend to flash-off in vacuum filters so need to be handled in other equipment. If the liquid is saturated with dissolved solids then any cooling will result in precipitation and this needs to be taken into account in the design of the separator.

4. ROUGH SIZING OF EQUIPMENT

This section describes test work and calculations for estimating the approximate size of particular pieces of equipment and deciding whether they are feasible for the application. This section does not go into detailed design of separators which is usually left to the equipment supplier. The reader who wants to be armed with more design details during his discussions with the supplier should refer to Purchas and Wakeman (1986).

4.1. Sedimentation Equipment

The settling rate of fine particles ($< 50\,\mu$m) at low concentrations is given by Stokes law:

$$u_\infty = \frac{d^2(\rho_p - \rho_f)x}{18\eta_f} \tag{27.1}$$

where $x = g$ for gravity sedimentation, $x = \Omega^2 r$ for centrifugal sedimentation, and $x = 4P_H/D\rho_m$ for hydrocyclones.

For higher concentrations the hindered settling velocity is given by

$$u_H = u_\infty(1 - C)^{4.6} \tag{27.2}$$

Equations (27.1) or (27.2) can be used for all centrifugal sedimentations but Stokes law does not apply to gravity sedimentation of fine particles unless the slurry is very dilute. These particles will be flocculated and will all settle at the same velocity irrespective of their size. This velocity (u_∞) can be measured in a jar test (Fig. 27.19(a)).

Under ideal conditions in a clarification operation the percentage of particles size, d, which are removed from the liquid is given by

$$G(d) = u_\infty A/Q \tag{27.3}$$

For gravity sedimentation this gives the area, A, required to achieve the required removal of particles size, d (settling velocity u_∞) at flowrate, Q. For sedimenting centrifuges the effective area depends on the geometry so the capacity of centrifuges is often expressed in terms of a Σ factor where

$$\Sigma = Q/2u_g \tag{27.4}$$

i.e. the area of a gravity clarifier required for 50% recovery.

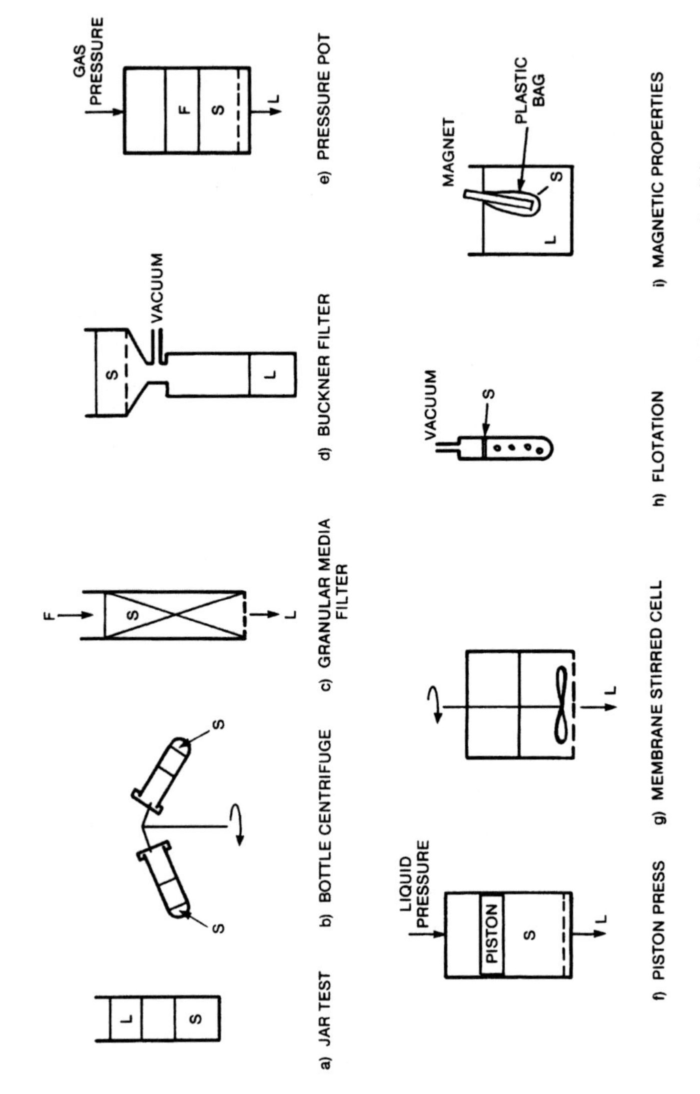

Fig. 27.19. Small-scale tests. F = Slurry feed, L = liquid and S = separated solids.

For hydrocyclones, the effective area can be related to equipment size by

$$A = \pi D_0 l \tag{27.5}$$

The area and depth of a gravity thickener required to achieve a given concentration of thickened slurry can be assessed by jar tests or by low speed centrifuge tests. However, the procedures are complex and are not covered here (see Purchas and Wakeman, 1986).

An idea of the solids residence time necessary to achieve a given concentration of solids in a sedimenting centrifuge can be obtained by tests in a laboratory bottle centrifuge (Fig. 27.19(b)). The consistency of the thickened slurry (e.g. fluid, pasty, firm) affects the choice of centrifuge.

4.2. Depth Filtration

The effectiveness of clarification filters is best assessed with tests on a small area of the filter medium which preferably has the full depth (Fig. 27.19(c)). The test should be done under the same conditions as will be used in the full scale. A sample of the contaminated liquid, if available, should be used in the test and if the contamination does not vary too much. If a sample is not available, a fixed concentration of particles of known size distribution in the right range should be used. Capacity is determined by the time taken for the solids to break through so that the effluent clarity no longer meets the required specification. Capture can be characterised by

$$\frac{C}{C_0} = \exp(-\lambda L) \tag{27.6}$$

4.3. Strainers

When screens are used for classification, the size of aperture chosen depends on the cut size required. Square apertures should be close to the cut size. Slotted screens with the slots parallel to the direction of flow will retain particles half the size of the slots. The proportion of fines retained on the screens depends on the length of the screen and figures are available for assessing this (Wills, 1981). The width of screen required for a given throughput is given in tables supplied by the manufacturers.

Strainers used for clarification are limited by the build-up of pressure caused by cake formation. The rate of cake formation can be estimated from the concentration of solids in the contaminated liquid and its throughput. The cake resistance can be estimated by the methods given in section 4.4. This then gives the rate of build-up of pressure and allows decisions to be made about the size of strainer required and the time between cleaning which will affect the choice of type.

4.4. Cake Formation

The rate of cake formation can be assessed by tests on a small filter leaf. For design purposes the orientation of the leaf and the applied pressure should simulate full-scale practice, but for an initial assessment tests on a vacuum Buchner are satisfactory (Fig. 27.19(d)). Data should fit the following equation (see Fig. 27.20(a)):

$$\frac{V^2}{A^2} = \frac{2Pt_F}{\eta_f\left(\frac{v}{B} + \frac{RA}{V}\right)} \tag{27.7}$$

The build up of cake is given by

$$L = \frac{vV}{A} \tag{27.8}$$

If the time taken to build up 10 mm of cake is less than 2 min a continuous filter is indicated (< 30 s indicates a top-fed filter). Times longer than 5 min indicate a batch filter. Equation (27.7) can then be used to estimate the filtration area required for cake formation. If the cake is incompressible (coarse particles) eqn (27.7) can be used to extrapolate the data from the vacuum tests to higher pressures. However, many fine-particle cakes are compressible and permeability, B, varies with pressure, P. The relationship between B and P can be determined by tests at different pressures covering the range of interest. A pressure pot (Fig. 27.19(e)) can be used to simulate pressure vessel filters.

If a sample of material is not available, a value of B can be estimated from

$$B = \frac{e^3 d_K^2}{150(1 - e)^2} \tag{27.9}$$

where d_K is a surface averaged particle size. If a particle size distribution by volume is available then an estimate of d_K is given by the size of the smallest 10% of the particles.

4.5. Gas Displacement

The moisture content which can be achieved by a long period of gas blowing is a function of the applied pressure P_D as shown in Fig. 27.20(b).

This curve can be obtained from small scale tests in appropriate equipment (Figs 27.19(d) and (e)). Carleton and Mackay (1988) gives method of predicting the curve when experimental data are not available. No dewatering will occur unless

$$P_D > \frac{0{\cdot}275\gamma}{B^{0{\cdot}5}} \tag{27.10}$$

The rate at which the equilibrium value shown in Fig. 20(b) is approached is

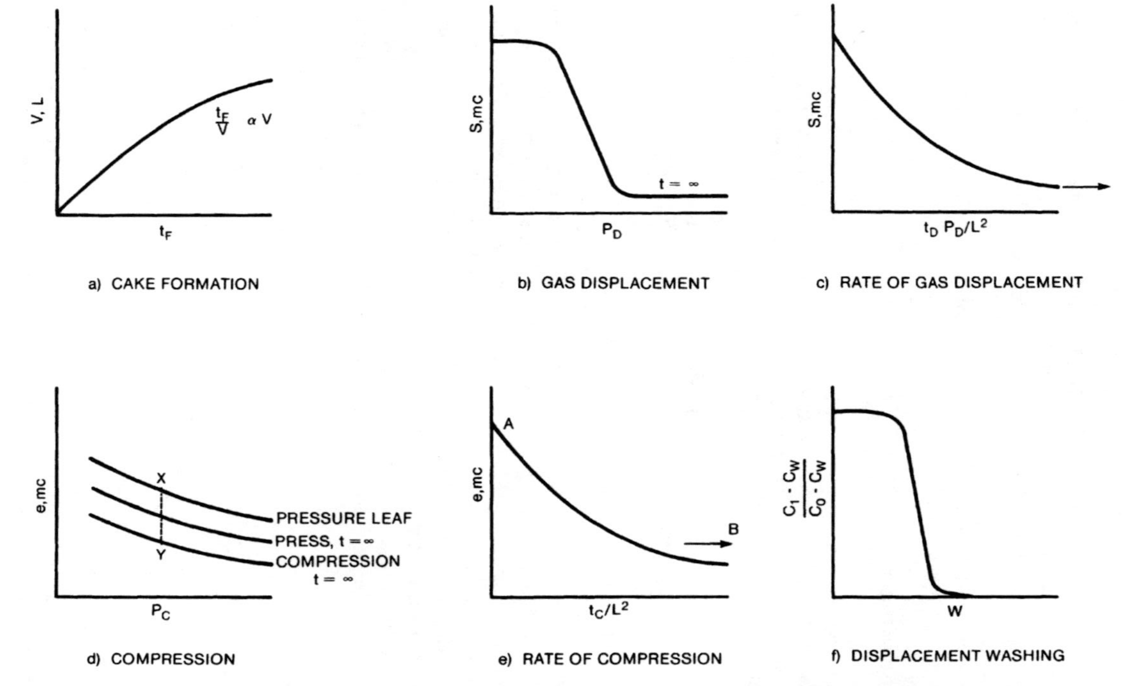

Fig. 27.20. Assessment of data for cake filters. (a) From eqns (27.7) and (27.8); (b) from eqn (27.10); (c) S from Fig. 27.20(b).

found from the small scale tests. If the data are plotted as moisture content against

$$\frac{Bt_D P_D}{\eta_f e L^2}$$

(see Fig. 27.20(c)) they can be extrapolated to different values of L and, for incompressible cakes, to different values of P_D. This gives the time required to achieve a required moisture content. Optimisation of dewatering filters is described by Carleton (1982).

Similar methods apply to gravity, or centrifugal drainage with P_D replaced by $\rho_f g L$ and $\rho_f \Omega^2 r L$ respectively.

4.6. Compression

The compressibility of a cake can be assessed by measuring its porosity (i.e. moisture content) as a function of pressure in a small-scale test. A steep curve (Fig. 27.20(d)) indicates that the higher capital costs of working at high pressure can be justified by the energy savings achieved by producing a cake with a low moisture content.

Tests in a pressure pot (Fig. 27.19(e)) will show what moisture content can be achieved at different pressures in a pressure vessel filter. Form time will be given by eqn (27.7).

Tests in a compression cell (Fig. 27.19(f)) will show what moisture content can be achieved in a batch compression filter. After the cake has formed, the moisture content will be at point X (Fig. 27.20(d)) but this can be reduced by the compression stage to point Y. The rate at which this is achieved is found by plotting the data as moisture content against t_C/L^2 (Fig. 27.20(e)). This then gives the time required to achieve any required moisture content. Compression may be continued until the cost of removing any further liquid exceeds the cost of removing the liquid thermally in a dryer (Carleton, 1985).

The moisture content achieved at a given pressure in a filter press will be between the values obtained in the pressure pot tests and those in the compression cell tests (Fig. 27.20(d)).

4.7. Washing

Figure 27.21 shows a single stage of re-slurry washing. In the absence of adsorption,

$$\frac{C_1 - C_W}{C_0 - C_W} = \frac{1}{1 + W} \tag{27.11}$$

where

$$W = V_W / V_L \tag{27.12}$$

The efficiency of a single stage of displacement washing is found from small-scale tests in suitable equipment (Figs 27.19(d) and (e)). The data should be plotted as

$$\frac{C_1 - C_W}{C_0 - C_W}$$

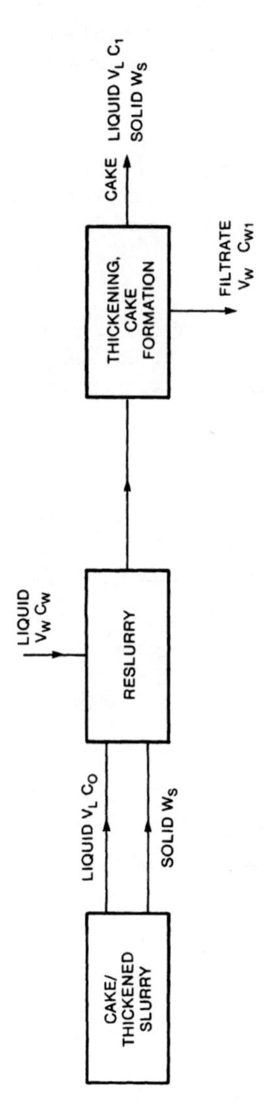

Fig. 27.21. Single reslurry washing stage.

against W (Fig. 27.20(f)), where W is the volume of wash liquid added in terms of the void volume of the cake. This is analogous to the definition in eqn (27.12).

The efficiency of multiple stage washing (re-slurry or displacement) can be found from mass balances. The amount of wash liquid and number of stages needed to achieve a given reduction in the concentration of impurity can be optimised.

4.8. Cross-Flow Filtration

Many cross-flow filters are available in modular form and their performance can be assessed from tests with a single module. If the membrane is available in sheet form its performance can be assessed in a stirred cell (Fig. 27.19(g)).

4.9. Flotation

The feasibility of flotation as a method of clarification can be assessed by putting a sample of the contaminated slurry in a test tube and shaking for half a minute to entrain air. The tube is then fitted to a vacuum line and observed to see if the solids float (Fig. 27.19(h)).

4.10. Magnetic Filtration

The suitability of low-intensity magnetic filters for removing ferromagnetic materials can be assessed with a hand magnet (Fig. 27.19(i)). Assessment of high-gradient separators for paramagnetic materials requires specialised equipment.

5. FLOW SHEETS

Points to bear in mind when designing a flow sheet downstream of a pipeline include the need:

- for multiple separators (section 5.1.);
- to make separators and dryer compatible (section 5.2.), and
- to avoid handling problems (section 5.3.).

Figure 27.22 shows some typical flow sheets which illustrate the points below.

5.1. Multiple Separators

Multiple separators may be required for the following:

- To achieve multiple objectives, e.g. using a cake filter and dryer to produce dry solids and a clarifier to clean up the dirty filtrate (Figs 27.22(a) and (b)).
- To handle slurry containing a wide range of particle sizes, e.g. using a screen to classify the solids; then using a thickener and filter for the fine fraction and a filtering centrifuge for the coarse fraction (Fig. 27.22(a)). If the coarse material were fed to a thickener it would form an immovable deposit. If the fines were fed to the centrifuge they would bleed through the screens.

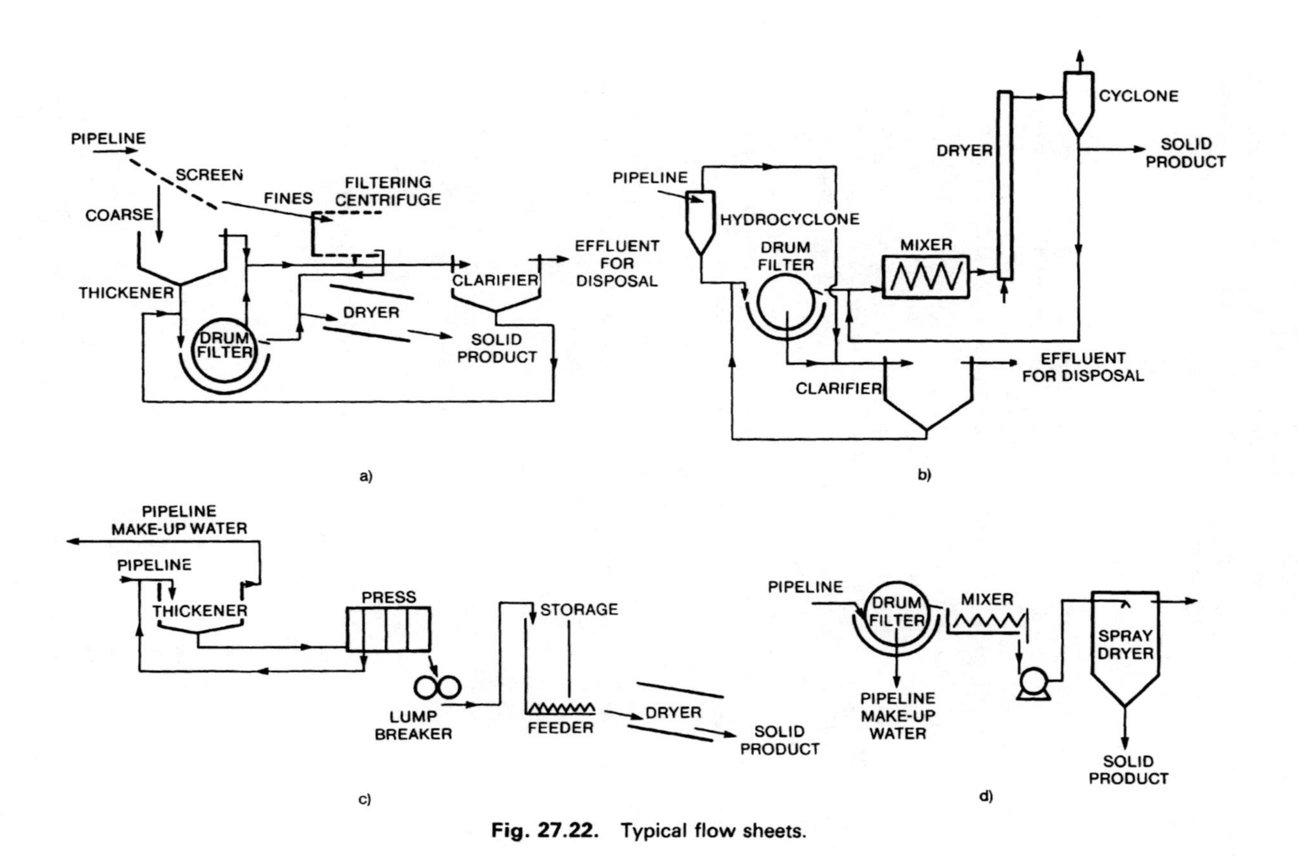

Fig. 27.22. Typical flow sheets.

- To minimise overall costs by using the most economic equipment at each stage, e.g. using a hydrocyclone to thicken up a dilute slurry, then using a filter to form a cake, before using a dryer to remove the last traces of liquid (Fig. 27.22(b)).

5.2. Matching Separator with Dryer

Ideally, the separator and the dryer should be selected so that the material from the separator is in a form which can be accepted by the dryer. If this is not done then some treatment stage will be required between the separator and the dryer. Some examples of mismatches and treatment stages are as follows.

- The dryer requires a pumpable material (e.g. spray dryer) but the filter produces a cake. If the cake contains fine particles and is produced on a vacuum filter it may be saturated and shear-thinning. If so it can be *worked* (e.g. in a paddle mixer) into a paste and then pumped to the dryer (Fig. 27.22(d)).
- The cake entering the dryer must be dry to prevent fouling (e.g. pneumatic conveyor dryer) but the cake from the filter is sticky (e.g. from a vacuum filter). One solution is to dry partially the cake between the filter and the dryer. Another is to mix the cake with recycled dry product (Fig. 27.22(b)).
- The dryer cannot handle large lumps (e.g. fluidised bed dryer) but the filter produces lumps. Brittle cakes from a press can be broken in a lump breaker (Fig. 27.22(c)). Lumps of sticky cake can be extruded into spaghetti-like pieces.
- The dryer requires fines to be removed (e.g. to allow circulation in through circulation dryers or to avoid gas cleaning problems in direct fired dryers). Here, it may be necessary to agglomerate the cake.

5.3. Avoidance of Handling Problems

Cakes can be difficult to handle and conveying equipment can be unreliable so, if possible, cake handling should be minimised. Ways of doing this are:

- to combine equipment to avoid intermediate conveying, e.g. use of spray dryers or filter dryers;
- to allow cake to fall directly from a continuous filter into a continuous dryer without an intermediate conveyor;
- to use gravity transfer;
- to minimise conveying distances;
- to avoid changes in direction; and
- to use continuous filters to feed continuous dryers (a way of matching the rates is to use a measurement of dryer performance, e.g. outlet gas temperature, to control the speed of the filter).

Pressure filters may be preferred because they give drier cakes which put less load on the dryer. However, they are normally batch so need cake storage facilities before continuous dryers. Ways of avoiding the need for intermediate cake storage and recovery are:

- to use continuous equipment, e.g. a belt press;
- to use batch filters and batch dryers with the cycle times and capacities matched; and
- to use multiple batch filters or ones with short cycle times to feed a long residence time dryer.

A thickener between the pipeline and the batch filter can be used for slurry storage (Fig. 27.22(c)). The thickener is discharged only when the press needs filling.

6. NOTATION

A	Area of separator [L^2]
B	Cake permeability [L^2]
C	Volumetric concentration of solids in slurry [1]
C_0	Value of C at surface [1]
C_W	Concentration of impurity in wash liquid [$M L^{-3}$]
C_{W1}	Concentration of impurity in filtrate from stage 1 [$M L^{-3}$]
C_0	Concentration of impurity in liquid in cake before washing [$M L^{-3}$]
C_1	Concentration of impurity in liquid in cake after washing [$M L^{-3}$]
d	Particle diameter [L]
d_K	Kozeny diameter of particle [L]
D	Diameter of hydrocyclone [L]
D_0	Diameter of vortex finder [L]
e	Cake porosity [1]
g	Acceleration due to gravity [$L T^{-2}$]
$G(d)$	Grade efficiency at size d defined by eqn (27.3) [1]
l	Vortex finder clearance length [L]
L	Cake thickness [L]
mc	Moisture content [$M L^{-3}$]
P	Pressure drop during cake formation [$M L^{-1} T^{-2}$]
P_C	Pressure drop during compression [$M L^{-1} T^{-2}$]
P_D	Pressure drop during dewatering [$M L^{-1} T^{-2}$]
P_H	Pressure drop across hydrocyclone [$M L^{-1} T^{-2}$]
Q	Flow rate [$L^3 T^{-1}$]
r	Radius of centrifuge bowl [L]
R	Cloth resistance [L^{-1}]
S	Saturation, i.e. fraction of cake voids filled with liquid [1]
t_C	Time for cake compression [T]
t_D	Time for cake dewatering [T]
t_F	Time for cake formation [T]
u_g	Gravitational settling velocity [$L T^{-1}$]
u_H	Hindered settling velocity [$L T^{-1}$]
u_∞	Settling velocity [$L T^{-1}$]
v	Volume of cake formed per unit volume of filtrate [1]

V Volume of filtrate [L^3]
V_L Volume of liquid in cake [L^3]
V_W Volume of wash liquid [L^3]
W Volume of wash liquid in terms of void volumes of cake [1]
W_S Mass of solids [M]
x Parameter in eqn (27.1) [$L T^{-2}$]

γ Surface tension [$M T^{-2}$]
η_f Viscosity of liquid [$M L^{-1} T^{-1}$]
λ Efficiency factor for depth filter [L^{-1}]
π 3·14159 . . . [1]
ρ_f Liquid density [$M L^{-3}$]
ρ_p Solid density [$M L^{-3}$]
ρ_m Slurry density [$M L^{-3}$]
Σ Sigma factor defined by eqn. (27.4) [1]
Ω Rotational speed of centrifuge, rad s^{-1} [T^{-1}]

7. REFERENCES

Carleton, A.J. (1982). Optimisation of dewatering filters. *J. Sepn Proc. Technol.*, **3**(4), 14–20.
Carleton, A.J. (1985). Dewatering Filter Cakes by Compression. *Inst. Chem. Engrs (Symp. Ser.)*, **91**, 175–86.
Carleton, A.J. & Mackay, D.J. (1988). Assessment of models for predicting the dewatering of filter cakes by gas blowing. *Filtn Sepn*, **25**, 187–91.
Carleton, A.J. & Salway, A.G. (1987). Evaporation effects in gas blowing. In *Proceedings of FILTECH '87 Conference*. Filtration Society, Loughborough, UK, pp. 8–17.
Cheremisinoff, N.P. & Azbel, D.S. (1983). *Liquid Filtration*. Ann Arbor Science, Woburn, MA, USA.
Matteson, M.J. & Orr, C. (eds) (1987). *Filtration Principles and Practices*. Marcel Dekker, New York, USA.
Perry, R.H. & Green, D.W. (eds) (1984). *Chemical Engineers' Handbook*. McGraw-Hill, New York, USA.
Purchas, D.B. (1981). *Solid–Liquid Separation Technology*. Uplands Press, Croydon, UK.
Purchas, D.B. & Wakeman, R.J. (eds) (1986). *Solid–Liquid Separation Scale-Up*. Uplands Press, London, UK.
Svarovsky, L. (ed.) (1981). *Solid–Liquid Separation*. Butterworths, London, UK.
Svarovsky, L. (1985). *Solid–Liquid Separation Processes and Technology*. Elsevier, Amsterdam, The Netherlands.
Warring, R.H. (1981). *Filters and Filtration Handbook*. Trade and Technical Press, Morden, Surrey, UK.
Wills, B.A. (1981). *Mineral Processing Technology*. Pergamon Press, Oxford, UK.

28

Economic Evaluation of Slurry Transport Projects

Marvin H. Muenzler
International Development Planners, San Francisco, USA

1. INTRODUCTION

Overland transport of coal comprises a major portion of the overall cost of coal as an energy source. In many cases in the US the transport costs represent up to two-thirds of the total cost of coal to the ultimate customer. It is, therefore, not surprising that a large degree of effort has been devoted to developing more efficient and less costly means of coal transport.

At the present state-of-the-art, there are two technologies that have been applied commercially.

- *Conventional coal slurry system* employed by Black Mesa (see Chapters 15 and 29) and Ohio coal; and
- *Coal–water mixture (CWM)* recently installed in a pipeline in Siberia (see, for example, Ercolani, 1986).

Other technologies are still in the pilot-scale development stage and have not been proven commercially. Included in this category are

- *Coal/carbon dioxide (CO_2)* in which liquid CO_2 is the carrier, and
- *stabilised flow (Stabflow)* in which a fine coal is used as carrier for coarse coal so that it does not settle (Brookes & Dodwell, 1985).

This chapter presents the bases, major parameters and methodology to be used by an engineer to make a first appraisal of the merits of alternative slurry transport methods for the application under consideration. Each application has its own set of unique conditions so that after this first appraisal more refined estimates of costs can then be applied to the selected technology. The methodology described in this chapter is applied to the two systems in commercial operation, the conventional coal and the CWM technologies. The other technologies (coal/CO_2 and Stabflow) are reviewed briefly giving some of the specific conditions in which these technologies may be applied.

Although there are a number of slurry pipeline systems currently in operation the capital investment and operating costs of these systems are not readily available. The data used as a basis for the capital investment and operating cost estimates in this chapter have been derived primarily from the estimates of Energy Transport Systems, Inc. (ETSI). This was a large-scale proposed project extending from Wyoming/Montana to power plants in the south-western US. Although the pipeline was never laid, detailed estimates have been prepared on both capital investment operating costs, and since the pipeline system was the subject of litigation in a US District Court, the estimates are now in the public domain. This chapter considers some of the aspects of deriving market-based and cost-based tariffs for the proposed ETSI coal system. Methods for adjusting the costs and tariffs for inflation are also discussed.

2. CASE DESCRIPTION

Conventional. The conventional slurry system as used in Black Mesa includes slurry preparation, pipeline and pumping stations, and dewatering (Aude &

Thompson, 1985). The slurry, consisting of approx. 50% coal by mass is transported through a buried, conventionally constructed, steel pipeline with intermediate pump stations with positive displacement piston pumps. At the terminus the dewatered coal is reduced to approx. 10% surface moisture, ready for use in a conventional coal fired power plant.

Coal–water mixture (CWM). In this technology, the slurry is composed of finely ground coal in water. Through the use of chemical additives, a slurry mixture composed of approx. 70% coal by mass can be obtained (Aude & Thompson, 1985). The slurry, formulated to remain stable over a long period of time, is transported under laminar flow conditions. The high concentration of coal in the slurry mixture allows it to be fired directly in modified oil burners without further processing. The costs of preparing the slurry, including the costs of the chemical additives, are estimated to be in the range of US\$15–25 t^{-1}.

In both cases the pipeline would be designed to have a 20 year life.

2.1. Capital Investment and Operating Costs

In each of the cases, a computer model was used to estimate the following.

Optimum pipe diameter. Optimum pipe diameter was selected with the objective of achieving the lowest transport costs over the project life, staying within the minimum and maximum slurry flow velocity constraints. The economic trade-off was between the incremental costs of increasing pipe diameter and corresponding savings in friction loss in the pipeline. The model predicted pressure drops in a similar manner to those described in Chapters 6 and 7 for specified values of hourly throughput, particle size and solids concentration.

Size of the major facilities. The major facilities sized by the model include the size and number of

- pump stations,
- slurry storage tanks,
- coal preparation, and
- slurry dewatering plants.

The maximum capacity of each coal preparation and dewatering plant was set at 5 Mt y^{-1}. The number of these plants required for systems having greater than 5 Mt y^{-1} are in increments of 5 Mt y^{-1}. Slurry storage tank requirements assume 8 h of equivalent storage at preparation plants and 24 h at dewatering plants.

Estimates of capital investments. The estimates include

- preparation plants with required storage;
- land and right-of-way costs;
- pipe, coating materials, other pipeline material;

- pipeline installation;
- pump stations;
- dewatering plant with storage;
- sales tax;
- engineering, procurement, construction management (EPCM);
- an allowance for contingencies and omissions; and
- owners' costs (including research and development, start-up expenses, training, owners' staff costs prior to start-up, but excluding interest during construction).

The model developed the capital investment costs of each system analysed by applying the cost factors shown in Table 28.1 to the selected facilities. The cost factors are based primarily on the capital investment estimates developed by ETSI, adjusted to 1990 prices using appropriate inflation indices. Regression analysis techniques have been used in a manner so that a broad range of systems designs can be estimated. The inflation indices used are explained in sections 6.3 and 6.4. By using these indices, capital investment estimates can be kept current with prices beyond 1990.

Estimates of operating costs. The operating costs include

- power,
- labour,
- supplies, and
- water.

The ETSI system included its own water supply facility transporting water from an existing reservoir. In this analysis, an allowance for water at US\$0·80 m^{-3} was included.

2.2. Capital Charges

A factor of 20% has been used to compute the annual capital charges. This factor yields an internal rate of return (IRR) of 15% over a 20 year operating life. The 15% IRR is commonly used as a *hurdle rate* in evaluating investments in energy projects. A three-year construction period is assumed and property taxes are included at 1·5%. It is further assumed that the project would begin operation at full capacity during the first operating year. All costs are based on 1990 prices and are not escalated. The cost of transport is thus obtained by factoring the capital costs by 20%, adding the operating costs and dividing the sum by the annual volume.

3. BASE CASE

A base case has been selected to demonstrate the application of the model to the two technologies. The basic characteristics of the two systems are shown in Table 28.2.

TABLE 28.1
1990 Bases for Unit Capital and Operating Costs

Unit capital costs		
Preparation plant base facilities	7 200 000	US$ plant^{-1}
Preparation plant equipment	0·059	US$ t^{-1}
Right-of-way and damages	10 000	US$ km^{-1}
Land in fee	39 000	US$ site^{-1}
Mainline pipe	700	US$ t^{-1}
Pipe coating material	3·70	US$ m^{-2}
Other pipeline material	10·0	% of pipe
Pipeline construction		
Non-variable	39·40	US$ m^{-1}
Variable	2·00	US$ inch diameter^{-1} m^{-1}
Slurry pump station base facilities	1 600 000	US$ station^{-1}
Slurry pump station pump units	1 700 000	US$ unit^{-1}
Slurry storage tanks	100	US$ m^{-3}
Tank manifolds, transfer pumps	890 000	US$ tank^{-1}
Tank manifolds, pumps	360	US$ kW^{-1}
Water ponds	13	US$ m^{-3}
Dewatering plant base facilities	38 600 000	US$ plant^{-1}
Dewatering plant equipment	0·079	US$ t^{-1}
Steam plants base facilities	10 800 000	US$ station^{-1}
Steam plants variable costs	1·70	US$ t^{-1} coal
Power supply	887 000	US$ site^{-1}
Communications	15 000	US$ km^{-1}
Controls	38 000	US$ station^{-1}
Maintenance bases	6 400 000	US$ base^{-1}
Spare parts	2·0	% above-ground facilities
Sales and use taxes	2·3	% of all facilities
Engineering, procurement, construction and management	10·0	% of all facilities
Contingencies	10·0	% of subtotal
Owners' costs	5·0	% of total costs
Ocean freight	56	US$ t^{-1}
Inland freight	28	US$ t^{-1}
Unit operating costs		
Water supply	0·80	US$ m^{-3}
Labour		
Preparation plants	1 430 000	US$ station^{-1}
Pump stations	150 000	US$ station^{-1}
Dewatering plants	3 060 000	US$ station^{-1}
Maintenance base	1 430 000	US$ station^{-1}
Supplies		
Preparation plants	2·6	% capital costs
Pump stations	4·6	% capital costs
Dewatering plants	6·6	% capital costs
Power	0·05	US$ kWh^{-1}
Administration	40·0	% of labour

TABLE 28.2
Base Case Characteristics

	Slurry technology	
	Conventional	CWM
Pipeline capacity ($Mt\,y^{-1}$)	5	5
Length of pipeline (km)	400	400
Pipeline diameter (mm)	457	457
Pump stations	5	5
Mass of steel in pipeline (t)	33 000	33 000
Coal concentration (% by mass)	54	70
Slurry flow rate ($m^3\,h^{-1}$)	970	700
Flow velocity ($m\,s^{-1}$)	1·7	1·3

3.1. Capital Investment and Operating Costs

Table 28.3 summarises the investment and operating costs of the two technologies by facility. The slurry preparation includes the plant, associated storage, piping and power supply. The transport includes the pipeline, pump stations, land and right-of-way and communications. The dewatering facilities include the dewatering plant, steam supply and associated storage. All indirect costs including engineering, contingencies and owners' costs have been allocated to the facilities in proportion to the direct costs of each facility.

TABLE 28.3
Summary of Investment and Operating Costs[a]

	Slurry technology	
	Conventional	CWM
Investment costs		
Slurry preparation	51	142
Slurry transport	177	169
Dewatering	135	26
Total	362	336
Operating costs		
Slurry preparation	5	81
Slurry transport	17	13
Dewatering	12	5
Total	34	100

[a] Measured in US$ millions for 5 $Mt\,y^{-1}$, 400 km pipeline.

TABLE 28.4
Economic Comparisons[a]

	Slurry technology	
	Conventional	*CWM*
Coal preparation (US$ t^{-1})	3·10	22·00
Slurry transport (US$ t^{-1})	10·40	9·60
Dewatering (US$ t^{-1})	7·90	2·00
Total (US$ t^{-1})	21·20	33·60
Total (cents t^{-1} km^{-1})	5·30	8·40

[a]5·0 Mt y^{-1}, 400 km.

3.2. Economic Comparisons

The results of the economics of the two systems are shown in Table 28.4. As expected the high cost of slurry preparation for the CWM system makes it much more expensive than conventional coal slurry. The slurry transport is slightly lower for the CWM system, but not sufficiently low to offset the CWM preparation cost differential.

It should be pointed out here that the economics are not directly comparable since a grinding credit can be taken on the CWM system, which is estimated to be in the range of US$12 t^{-1}. This would make the total costs of the two systems similar. However, there are other adjustments that should be made. For example, the surface moisture on the coal reduces the boiler efficiency by 10% in the case of CWM and 3% for conventional coal slurry (Aude & Thompson, 1985). The costs and availability of water at the coal source may be a prime consideration. A shortage or high cost of water would tend to favour the CWM system.

4. SENSITIVITY ANALYSIS

The following paragraphs describe the impact on costs resulting from changes in some of the key variables.

4.1. Impact of Distance

Figure 28.1 shows the transport costs of a system with a capacity of 5 Mt y^{-1} for varying lengths ranging from 100 to 1 200 km.

As expected, the transport costs increase almost directly as a straight-line function of distance with a zero intercept of approximately US$12 t^{-1} for the conventional slurry and US$25 t^{-1} for the CWM. The zero-intercept represents the non-variable costs of coal preparation and dewatering. Although the pipeline transport cost of CWM is slightly less than that of conventional coal slurry (2·4 cents t^{-1} km^{-1} for CWM as compared with 2·6 cents t^{-1} km^{-1} for conventional),

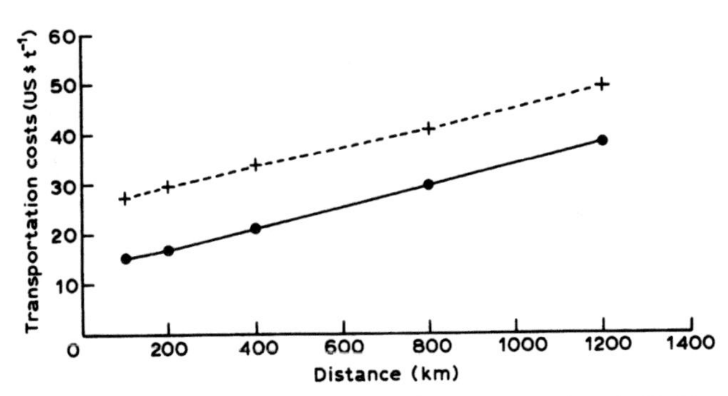

Fig. 28.1. Effect of length. (——) Conventional, (----) CWM.

the difference is not great enough to offset the added cost of slurry preparation for CWM.

4.2. Impact of Throughput

Figure 28.2 shows the impact on transport costs resulting from increases in throughput. The two systems exhibit some economy of scale; this is slightly more apparent in the conventional system. This is because there is some economy of scale in conventional slurry preparation/dewatering, whereas with CWM the preparation costs are essentially linear with volume. This is demonstrated in Fig. 28.3, for a 400 km pipeline. This figure shows that for both the conventional and CWM systems the economy of scale is largely due to decreases in pipeline costs. Although not shown here, it is obvious that longer pipelines will demonstrate a greater economy of scale where the pipeline costs represent a larger portion of the total transport costs.

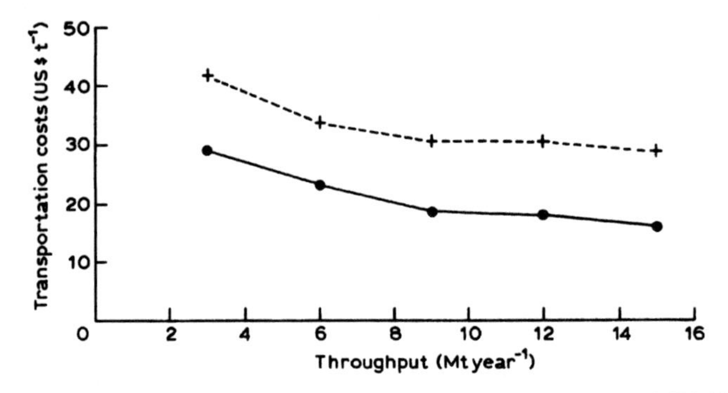

Fig. 28.2. Impact of throughput. (——) Conventional, (----) CWM.

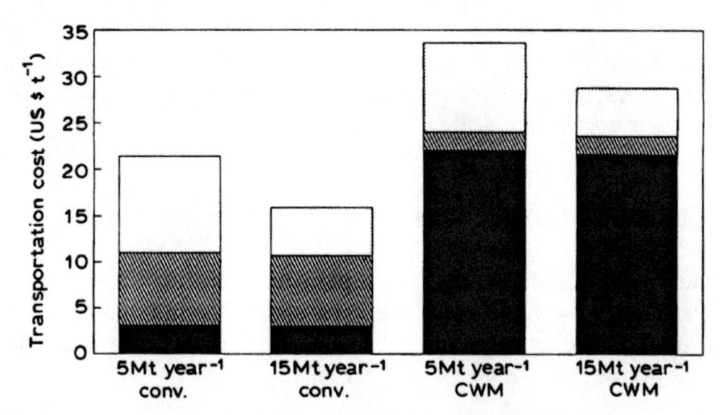

Fig. 28.3. Transport costs by facility. (■) Slurry preparation, (▨) slurry dewatering, (□) pipeline.

4.3. Impact of Financial Factors

As mentioned previously, a capital charge factor of 20%, derived from an implied IRR of 15%, has been used in computing transport costs. Considering the inherent risks of long-term energy projects and the relatively new technology employed in coal slurry pipelines, the investors may require a rate of return higher than the 15%. Assuming that the requirement is a 20% IRR, the resulting capital charge factor is 23%. The transport costs of the conventional system is increased from US\$21·20 t^{-1} to US\$23·40 t^{-1} and the CWM system is increased from US\$33·60 t^{-1} to US\$35·50 t^{-1} for the Base Case 5 Mt y^{-1}, 400 km pipelines.

4.4. Impact of Construction Duration

A further analysis can be made regarding the assumption of a three-year construction period. The duration of the construction period has a direct effect on the capital charge factor since imputed interest on the construction expenditures has to be recovered as part of the transport costs. If a three-year construction period is considered reasonable for a 5 Mt y^{-1} project, it might also be considered reasonable to assume a four-year construction period for a 15 Mt y^{-1} project. For this mega-project, more than one coal source and several coal delivery terminals would probably be required. The capital investment for a 15 Mt y^{-1} project is over twice that for a 5 Mt y^{-1} project. An additional one-year added to the construction period therefore seems reasonable when the difference in the magnitude of the projects is considered. It is also unlikely that the ultimate throughput capacity of 15 Mt y^{-1} would be attained in the first year of operation, but that a three-year build-up of throughput may be more reasonable.

Using these assumptions together with a four-year construction period and a three-year build-up in volume, a new capital charge factor of 24·8% was derived that results in a transport cost of US\$18·50 t^{-1} for a 400 km, 15 Mt y^{-1} system. This compares with a transport cost of US$16·00 t^{-1} shown in Fig. 28.2. The analysis indicates that when delays in construction completion and the attainment of ultimate project capacity are factored into the economics, the economy of scale of mega-projects is greatly reduced.

5. OTHER COAL SLURRY TECHNOLOGIES

Research and pilot-plant work have been performed on two other coal transport technologies that may have merit for very special conditions. The following describes these technologies in general terms and discusses the situations where they may be applied.

5.1. Stabilised Flow

This system has been tested by, amongst others, a British Petroleum/Bechtel test facility (Brookes & Dodwell, 1985). The results for *Stabflow* slurries indicate that stable slurries of up to 70% coal by mass can be obtained by using fine coal as a carrier for coarse coal. These slurries were successfully pumped through a 305 mm (12 in.) pipe at a test facility, but scaled-up results in larger diameter pipes have never been published. The tests indicated that the slurries have significantly higher pipeline friction losses than conventional slurry or CWM. However, the higher pumping costs may be offset by savings in coal preparation and dewatering costs. Stabflow may be considered as a transport option for short distances where no railroad exists. Economic analyses conducted using the test facility data (Brookes & Snoek, 1988) showed that a *window* may exist for this technology for distances of 50–300 km. For distances less than this, mechanical systems, such as cable or conveyor belt, would probably be more economic. For greater distances, other slurry technologies are more economic.

The joint venture has also suggested this method for loading coal tankers by pumping coal directly from the pipeline terminus through single buoy moorings (Brookes & Dodwell, 1985); however, this again has not been demonstrated.

5.2. Coal/CO_2

In this system, liquid CO_2 replaces water as the carrier in the slurry. Liquid CO_2 has a very low viscosity and as coal slurry viscosities are a function of the liquid phase viscosity and the coal concentration, high coal concentrations of coal can be pipelined at high velocities and low pressure drops. This system has not been demonstrated to be commercially viable but some pilot-scale tests have been done (Santhanam *et al.*, 1984).

Based on the optimistic assumption that this system is technologically feasible, it still would be limited to very special cases where a source of CO_2 is available at or near the coal source. It would also require a market for CO_2 at the destination. Some economic analyses conducted on the system (Aude & Thompson, 1985) have shown that in order for the system to be economic, the CO_2 would have to be sold at the terminus and the sales price would have to bear some of the costs of the transport system. The analysis indicated that a transport charge of US$8·8 per 1 000 m^3 of CO_2 would be required for the coal/CO_2 system to break-even with other technologies. In the absence of this market for the carrier, a return line carrying the CO_2 back to the coal source would be required.

6. TRANSPORT TARIFFS

The discussion up to now has centred on how to estimate capital and operating costs, and then how to determine the transport costs of a slurry pipeline project given the capital and operating cost estimates. The owner is usually faced with some competing mode of transport and must set the pipeline rates at a level to be competitive. These competitive rates must be high enough to provide the owner with a reasonable profit margin. If they are not, the project will never be financed.

6.1. Setting Initial Tariffs

In the US, coal slurry pipelines are viewed as an alternative to either shipment by rail or by barge. Because of the large investment required for coal pipelines, the electric utility companies or coal shippers must commit to long-term *take-or-pay* contracts in order for the pipeline to obtain financing. This commitment by the shippers coupled with the fact that coal slurry pipelines are relatively uncommon in the US, forces the coal pipeline owner to propose an initial tariff substantially below the alternative railroad cost in order to be accepted by the shipper. ETSI's initial tariffs were about 10% below rail costs. (The rail costs included the transport cost paid to the railroad plus the cost of the rail cars which were owned by the shipper.)

6.2. Tariff Escalation

After the initial tariff is set, it must then be adjusted periodically to protect the owners from cost inflation. For this purpose, two escalation indices were derived.

- Construction cost index (CCI)
- Operating cost index (OCI)

The derivation of the indices is explained in the next sections. The ETSI contract agreement with coal shippers divided the transport tariff into the following three components with the approximate allocation of the tariff assigned to each.

- Operating cost component (34%)
- Property tax component (5%)
- Balance-of-payments (61%)

The ETSI contract called for monthly adjustments of the tariff using the CCI and

OCI as the escalators. After the initial tariff was set, it was then adjusted for inflation in two phases. The first phase was between the setting of the initial tariff and the beginning of operation of the pipeline. The second phase began with start-up of the pipeline and continued during the operating phase.

During the first phase, the property tax and balance-of-payments components were escalated in accordance with the CCI. The operating component was escalated in accordance with the OCI.

In the second phase, the property tax component was escalated at one-half the monthly change in the producer price index (PPI), published monthly by the US Department of Labor. Half of the balance-of-payments component was escalated in accordance with the PPI, while the remaining 50% was held constant. The operating cost component continued to be escalated in accordance with the OCI. This method of escalation protected the owners from inflation of operating and construction costs and provided an increasing profitability resulting from the escalation of 50% of the balance-of-payments portion.

6.3. Construction Cost Index

This is a weighted index derived by allocating the estimated costs of the major facilities to various commodity price indices. The derivation of the index is shown in Table 28.5. The commodity indices shown in this table are published monthly by the US Bureau of Labor Statistics as the *Producer Price Index* (PPI), or by Chemical Engineering (published by McGraw Hill) as the *CE Plant Cost Index*. The PPI is a composite index of 12 major industrial commodity groups, which are in turn composites of many sub-groups. As noted in Table 28.5, 9% of the total capital investment for the base case is allocated to the PPI. The remaining US Bureau of Labor Statistics indices in Table 28.5 of industrial commodity sub-groups have been selected as being closely related to the capital investment of the corresponding facility.

By applying the changes in the indices, factored by the weighted percentages shown in Table 28.5, the CCI can be kept current. For example, to determine the CCI at a date other than that at which the capital investment has been prepared, the percentage change in each of the indices from those prevailing at the base date to those at the required date is computed. The percentage changes are multiplied by the corresponding weights for each index. The percent change in the CCI from the base date to the required date is the sum of these products.

6.4. Operating Cost Index

This index was derived in a similar manner. The main categories of operating expenses (labour, material and supplies, and power) were correlated with the following indices and weights.

- *Labour*: average hourly earnings pipeline transport (046)—26%
- *Materials and supplies*—37%
- *Power*: producer price index for electric power (054)—37%

TABLE 28.5

Weighted Construction Cost Indices[a]

Capital investment	*Cost for base case (US$ million)*	*PPI*	*Storage tanks 1072–0147*[b]	*General mechanical equipment 114*[b]	*Piping 1017–06*[b]	*Electrical equipment 117*[b]	*Buildings*[c]	*Pumps and drivers 1141–02*[b]	*Pipe*[d]	*Equipment rental 112*[b]	*Fuel 2911–413*[b]	*Construction labour*[c]
Coal preparation plants	36 7			51%	8%	8%	8%					25%
Lands and right-of-way	4 3	100%										
Pipeline construction	30·2	11%								14%	6%	69%
Mainline pipe	25·6								100%			
Other pipeline material	4 4				94%	6%						
Slurry pump stations	43·0		6%	3%	18%	7%	8%	42%				16%
Slurry tank farms	10 2		32%	18%	7%	8%	2%	7%				27%
Dewatering plants	78 1		8%	45%	16%	6%	5%					20%
Steam plants	19 3		69%			6%						25%
Power supplies	6 2					60%						40%
Communications and controls	6·3					100%						
Maintenance bases	12 8			50%			29%					21%
Spare parts	3 1			50%				50%				
Sales tax	6·3	100%										
EPCM[e]	28·4	50%										50%
Contingencies	31 5	9%	8%	21%	9%	7%	5%	6%	8%	1%	1%	25%
Owners costs	15 7	9%	8%	21%	9%	7%	5%	6%	8%	1%	1%	25%
Total capital investment	362 1											
Weighted percentage[f]		9%	8%	21%	9%	7%	5%	6%	8%	1%	1%	25%

[a] For base case on conventional slurry 5 $Mt y^{-1}$, 400 km

[b] Producer Price Index (PPI) commodity number.

[c] Based on Chemical Engineering Cost Indices.

[d] Based on pipe quotes from manufacturer

[e] Engineering, procurement, construction and management

[f] Example of derivation for PPI—((4 3)(100%) + (30 2)(11%) + (6·3)(100%) + (28 4)(50%) + (31 5)(9%) + (15 7)(9%))/362 1 = 9%

As in the case of the CCI components, these indices can be read monthly to derive an OCI.

7. PROJECTS IN DEVELOPING COUNTRIES

The previous sections of this chapter have been devoted to the financial and economic analysis of slurry transportation projects assumed to be located in industrialised countries, where the owners and financial institutions would be guided in investment decisions by standard profitability analysis of the project. In developing countries the owner of the project will be a state institution and the project will likely be financed by either the World Bank or a regional development bank. Although many of the same general economic concepts will apply, some of the inputs to the estimates of costs and benefits may have to be developed differently.

The following sections describe briefly these differences or divergences in the application of the methods to projects in developing countries as compared to projects in industrialised countries.

7.1. Economic Rates of Return

The conventional method of measuring the economic viability of a project is by computing the net present values of the projected cash flows generated by the project during its life. This method has the advantage of placing a value on the timing of the cash flows, that is, it gives more value to cash flows received sooner rather than later. The discount rates used in the calculations are a matter of judgement, typically ranging from 10 to 15%, and are influenced by the perception of risk in the project. The discount rate at which the sum of the annual discounted net cash flows equals zero is the internal rate of return and is one of the standard criteria of economic viability. The net cash flows used in the net present value calculations exclude consideration of income and sales taxes, duties and other fees paid to the state.

7.2. Shadow Pricing

One of the fundamental innovations in the economic analysis of developing country projects is the replacement of actual prices with shadow prices, or accounting prices, since actual prices may be a poor guide to economic benefits and costs. Typically, in developing countries, the prices of essential goods are set low in comparison to costs, while the prices of less essential goods are set high (Wood, 1984). The prices of many raw materials and energy prices may be irrationally low while the prices of some other goods may be excessive. The result is that actual prices may not accurately state the true economics of a project. Some projects may appear very profitable when the costs and benefits are valued at actual prices, whereas in fact they may be very unattractive from the standpoint of the overall economy of the country. Care, therefore, must be taken in estimating appropriate shadow prices to use in the economic analysis for both the costs and benefits.

7.3. Project Costs

The costs of a project can be divided into the following broad categories.

Traded goods. A traded good is one in which the production or use of the good has a direct impact on the country's import or export of the good (Wood, 1984). If, for example, the country is a net importer or exporter of steel, then the steel that goes into the manufacture of pipe, or the pipe itself, would be treated as a traded good. The real price of a traded good is the world price of that good, excluding any duties or taxes, adjusted for international and domestic freight. All capital equipment that is purchased internationally would be classified as traded goods and would be priced based upon the world market for those goods. As a general rule, the major items of equipment for a slurry pipeline constructed in a developing country should be treated as traded goods and priced accordingly.

Non-traded goods. Some of the significant items included in this category are buildings, communications facilities and utilities used in operation of the pipeline. Shadow prices for these items can be derived by breaking down the items into their input components (Duvigneau & Prasad, 1984). As an example, the real costs of electric power can be broken down into its components of capital investment in the generating and transmission equipment, installation of the equipment (discussed below) and fuel requirements. A major portion of the generating and transmission equipment can be categorised as traded goods and priced using international prices for these commodities. The fuel component can also be priced on an international price basis since developing countries either import or export some form of energy. Although the generating plant may use domestic coal as its energy source, the shadow price for the coal would be its international price (less freight to a reference port such as Singapore or Rotterdam if the country is a net exporter of coal, or plus freight if an importer).

Installation costs. These costs will also need to be broken down into their input categories of labour, installation equipment, consumable material and indirect costs so shadow prices can be derived separately for each component. Based upon a weighting given to each category, the total installation costs can then be estimated. Shadow prices for installation equipment can be derived from international prices for this category. Fuel costs are a major component of consumable material and the shadow price can also be derived from international prices using diesel and other fuel commodities as the proxy.

Shadow wage rates used for labour costs are, in theory, a measure of the amount that a worker would contribute to national production had he not been employed on this project (Duvigneau & Prasad, 1984). This is the cost to the economy of employing the worker on the project. For highly skilled workers, the actual wages paid may be substantially less than the shadow wage since many developing countries follow an egalitarian wage policy (Wood, 1984). For these workers, who generally do not make up a large part of the overall project cost, the shadow wage rates can be based on the costs of employing foreign workers with similar skills. For unskilled or semi-skilled workers, the contribution to total project costs may be much greater. In determining the shadow wages for these workers, the estimator has to consider the fact that in developing countries, unskilled or semi-skilled workers would likely be underemployed and contributing very little to the overall production of the country were it not for this project. The shadow wage rates of these workers should be set at a level somewhat below the actual wage rates (including benefits and welfare expenditures). In cases where the shadow wage rate of these workers is critical to the economic profitability of the project, a level of 50% may be used. In other cases, actual wages may be used as shadow wages (Duvigneau & Prasad, 1984).

Indirect costs related to installation costs are composed of overheads and profits, contractors' field and home office expenses. These costs should be estimated as a percentage of total costs, similar to the method used in projects in industrialised countries.

7.3.1. Interest Rates

A shadow interest rate is needed in economic analysis of developing country projects to calculate the discounted cash flows and as a yardstick against which to measure internal rates of return. The shadow interest rate should be the cost of diverting funds from some other project to this project (Wood, 1984). It can be taken as the cost of securing fully convertible funds from the most expensive foreign source, usually a commercial bank, deflated by the expected rate of world inflation. The shadow interest rate used by the World Bank for projects in many developing countries ranges from 10 to 15%.

7.3.2. Direct Operating Expenses

The main components of direct operating expenses are material and supplies, power, and labour. Previous sections described the methodologies for shadow pricing power and labour. The material and supplies component has to be divided into categories of traded or non-traded goods, and shadow prices for these categories developed separately as discussed previously.

7.4. Benefits

The economic benefit of a coal slurry pipeline is equal to the avoided cost of the service that it displaces, which in most developing countries would be rail transportation of coal. In order to make this evaluation, the cost of service of the rail transport has to be evaluated by applying the techniques described in the previous sections.

Once the costs of the displaced service have been estimated, the economic rate of return, or other economic indicators, of the pipeline system can then be determined. This can be done incrementally, that is, by taking the yearly differences in the economic costs of the pipeline and the displaced service. The rate of return of these differences can be derived and measured against the yardstick as established by the bank for that country.

8. REFERENCES

Aude, T.C. & Thompson, T.L. (1985). Alternative coal systems (CWM, CO_2 or Black Mesa technology). In *Proc. STA 10*, pp. 281–90.*

Brookes, D.A. & Dodwell, C.H. (1985). The economic and technical evaluation of slurry pipeline transport techniques in the international coal trade. In *Proc. STA 10*, pp. 67–81.*

Brookes, D.A. & Snoek, P.E. (1988). The potential for Stabflow coal slurry pipelines—an economic study. In *Proc. HT 11*, pp. 33–48.*

Duvigneau, J.C. & Prasad, R.N. (1984). *Guidelines for Calculating Financial and Economic Rates of Return for DFC Projects.* World Bank Technical Paper, No. 33.

Ercolani, D. (1986). Production plants and pipeline systems for Snamprogetti's coal water slurries, recent experience and current projects in Italy and USSR. In *Proc. HT 10*, Paper A3, pp. 19–30.*

Santhanam, C.J., Dale, S.E., Peirson, J.F., Burke, W. & Hanks, R.W. (1984). Coal–liquid CO_2 slurry pipeline technology. In *Proc. STA 9*, pp. 147–53.*

Wood, A. (1984). *Economic Evaluation of Investment Projects. Possibilities and Problems of Applying Western Methods in China.* World Bank Staff Working Paper, No. 631.

*Full details of the Hydrotransport (HT) and the Slurry Transport Association (STA) series of conferences can be found on pp. 11–13 of Chapter 1.

29

Case Studies of Some Major Projects

Norman T. Cowper
Slurry Systems Pty Ltd, Northbridge, NSW, Australia

1. INTRODUCTION

The transfer of solids in water by pipeline over relatively short distance has been practised for over 125 years. Historically, the major application was for hydraulic dredging and later in mine tailings disposal. However, it is only in comparatively recent times that transportation of solids cross-country by slurry pipelines was established as a viable and proven technology.

Long-distance slurry pipeline transportation systems have their beginnings in the 1950s with the 174 km, 250 mm, Consolidation Coal Pipeline in Ohio, USA, and the 116 km, 150 mm, American Gilsonite Pipeline in Nevada, USA. Both of these systems were highly successful. It took another 20 years to pass before long-distance pipeline technology was commercialised by Edward J. Wasp (Wasp *et al.*, 1977) to transport products other than coal and gilsonite. The commercialisation era commenced with the commissioning of the world's first long-distance iron ore slurry pipeline at Savage River Mines, Tasmania, Australia, in 1967 (McDermott *et al.*, 1968). The World's longest system is the 440 km, 457 mm, Black Mesa coal slurry pipeline, commissioned in 1970.

This chapter on practical applications reviews some of the practical aspects common to long-distance pipelines, reviews major operating systems in the Western World, briefly explores other areas of application, and finally presents in detail three recent projects with varying unique features.

2. PRACTICAL ASPECTS OF LONG-DISTANCE SLURRY PIPELINING

2.1. Overview

This section discusses some of the practical aspects common to the operation of most long-distance slurry pipelines including commissioning, operation and maintenance. (Aude *et al.*, 1971; Cowper *et al.*, 1972).

2.2. System Commissioning

Commissioning commences after completion of the normal mechanical, electrical instrumentation and computer control checks on water. These pre-slurry start-up activities are followed by commissioning on slurry. The commissioning programme is planned to demonstrate and prove system performance under operating conditions, shutdown and restart, pipeline flushing and pigging operations. The programme is also used to train operators.

2.3. Pipeline Operation

Most long-distance slurry pipelines operate in a batching mode. Batch operation involves pumping accumulated slurry produced and stored at the mine site followed by on-line switching to pumping flush-water to maintain a normal pipeline flow rate. This is done while continuing to transport the slurry to the terminal or next mainline booster station.

Earlier systems such as Savage River and Black Mesa operated in this mode.

However, with confidence in the capability to shutdown and restart the pipelines with slurry in the line, the batching mode has been superseded with a shutdown–restart mode of operation to minimise energy and flush-water usage. Shutdown–restart operation involves pumping slurry until the slurry storage at the mine site is depleted and then to shut the pipeline down until the next batch of slurry is ready for transfer.

Control of slurry batches in the pipeline does not present a problem. Flowmeters or monitoring of positive displacement pump strokes provide an accurate means to locate and control batch interfaces by volume displacement calculation. Although slurry density is higher than water, slurry batches are transported with only minor solids trail-out occurring even over 80–90 km transport distance. The trail-out from a batch is picked-up by the head of the following batch of slurry.

All long-distance slurry pipelines, unless they are automatically flushed or self-draining, must be capable of shutdown and restart with slurry in the pipeline. The most common cause of pipeline shutdown is electrical power failure. The pumps shut down, the flow in the pipeline drops and the slurry is locked in by closure of a mainline block valve. Blocking the slurry eliminates flow below the deposition velocity, minimising slurry break-up which could create problems with restart and, in extreme cases, pipeline plugging. Pipelines are normally restarted within 24 h. Restart involves opening mainline block valves and commencing pumping flush-water. Operating flow rate is achieved as soon as practical. Restart characteristics depend on the individual pipeline. Increased energy is required to accelerate the slurry (including solids) to operating velocity.

2.4. Pipeline Control

Slurry pipelines are normally automatically and remotely controlled from a central control room. The operation of the pipeline requires only a minimum of direct control and can be performed as part of the duties of a mine site operator. Pump station instrumentation and control systems monitor key operating variables.

Prior to pumping, slurry batches are pre-tested to check if slurry is within pipeline slurry specification. Batch composite samples are gathered during production and tested in the laboratory for particle size consist (particularly particle top-size), percent solids, pH and rheology. If test results are within specification, the batch is acceptable for pumping. Off-specification slurry is a danger to pipeline operation. Chapter 24 considered control of pipeline systems in greater detail.

2.5. Corrosion Control

Monitoring and control of pipeline internal corrosion is required to ensure the life of the pipeline. Corrosion control involves a programme of controlling corrosivity of the slurry for day-to-day operations and to monitor the long-term corrosion in the pipeline by corrosion spool weighing or ultrasonic pipe wall thickness measurement at selected pipeline locations. Fortunately, most freshly ground ore concentrates have an affinity for dissolved oxygen. The lack of dissolved oxygen in the slurry allows corrosion control by adding lime to adjust the pH. Slurry pH in the range 9–10 is normally required.

Every six months it is necessary to check the cathodic system protecting the outside wall of the pipe measuring ground to pipeline potential at electrical test points. The test points, installed during construction, are regularly spaced along the pipeline route at approximately 1·5 km intervals.

Both the internal and external corrosion of slurry pipelines is discussed in Chapter 21.

2.6. Pipeline Pigging

Pipeline pigging is the internal cleaning of a pipeline using a special scraper (called a *pig*). Pigs come equipped with rubber cups or can be made of polyurethane with a diameter slightly larger than the internal diameter of the pipeline. They are driven along by the pipeline flow.

Some concentrate slurry pipelines require pigging at regular intervals. If small corrosion pits develop on the pipe wall the pump station discharge pressure will, in time, gradually rise through increase in pipeline roughness. When the discharge pressure has increased by 5–10% a wire brush or scraper pig is launched into the flow in a small water batch. The pig removes pipe corrosion products and the pump station discharge pressure reduces to normal. Slurry pipelines for iron ore concentrates are normally pigged at 10–12 week intervals.

2.7. Pump Station Maintenance

Pump station operating equipment is regularly inspected and pump wear components are replaced prior to failure on a preventative maintenance schedule. Normally, each pump station contains one additional pump as a stand-by spare. An effective and efficient preventative maintenance schedule increases total system availability to in excess of 99%.

Pump station discharge pressures are high, normally in excess of 8 MPa, and any slight leak at this pressure causes erosion by release of abrasive slurry at high velocity. On one occasion a leak of magnetite slurry at 11 MPa caused by fatigue failure in a drain valve weld opened up from a pinhole leak to a major opening in a matter of 10 min, and the resulting jet of slurry *sand-blasted* the surrounding buildings walls up to a distance of 3–4 m.

2.8. Pipeline Maintenance

The pipeline right-of-way is regularly inspected at approximately monthly intervals. This inspection may be conducted by either aerial or ground reconnaissance. A permanent all-weather access road along the pipeline route is not required for reconnaissance or maintenance purposes. Right-of-way damage by washerways or land slips and particularly from third party construction activities are the major concerns.

3. SLURRY PIPELINES—REVIEW OF OPERATING SYSTEMS

3.1. Overview

The following section briefly discusses applications of slurry pipelines to transport limestone, iron ore, mineral concentrations and coal slurry, including a review of developments in design and operational advancements of these systems.

3.2. Cement Industry

Slurry pipelines in the cement industry are installed to transport limestone or cement kiln feed slurries from the limestone quarry to the manufacturing plant. Pumping of limestone slurries over long distances is a proven technology with a number of successful operations around the World (Cowper & Venton, 1978). The World's longest system is the 92 km Rugby limestone–chalk pipeline in the UK, which has operated since 1964.

The major raw materials are drawn from natural deposits of limestone and clay or shale. They are mined and individually pre-blended in stockpiles to minimise fluctuations in the chemical composition. The raw mix is wet ground in grinding mills, which discharge a slurry (or powder) of the desired size consist (usually, $< 250\,\mu m$) to homogenising basins. This raw mix is fed into a direct fired counter-current kiln and is discharged as cement clinker. The wet process for cement manufacture is ideally suited for the application of slurry pipelines; both the particle size distribution and the slurry concentration required by the process are suitable for pipelining.

Limestone slurry design aspects. Both limestone and cement kiln fed slurries are transported at high solids contents (in the order of 60–70% solids by mass) to minimise the energy required to dry these components during manufacture. Both slurries exhibit Bingham plastic non-Newtonian rheological characteristics.

Some existing long-distance limestone slurry pipelines are operated continuously in the laminar flow regime. Systems under laminar flow conditions are susceptible to upset from changes in slurry rheology and coarse particle deposition. Design in the turbulent flow regime is a more stable condition with head loss less dependent upon slurry yield stress. Coarse particles are suspended in the flow by turbulent forces.

To achieve high-concentration pumping at reasonable head losses, viscosity modifiers are added to the slurry. Depending on the type of thinner used, the results can be dramatic. Figure 29.1 shows the reduction in pipeline friction losses for a 67·7% concentration (by mass) limestone slurry, resulting from the addition of caustic soda at a rate of 0·4 kg of 50% caustic soda solution per tonne of slurry solids. The addition of excessive quantities of modifier can cause flocculation with an adverse effect on the rheological properties.

Mainline pumps. Slurry is produced and transported near the limit of pump-

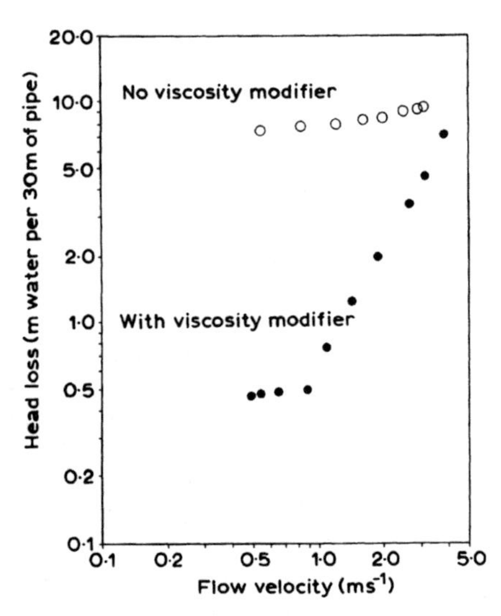

Fig. 29.1. Effect of viscosity modifier on limestone slurry head loss. (Pipe diameter, 100 mm; solids density, 2690 kg m^{-3}; solids concentration 67·7% by mass.)

able concentration since any additional water has to be removed thermally, reducing kiln throughput and increasing operating costs. Thus, the use of flushed gland pumps, both centrifugal and plunger, is discouraged and, depending on the abrasivity of the slurry, piston pumps and piston diaphragm positive displacement pumps are preferred.

Pipeline. Limestone slurries are essentially non-corrosive and exhibit settled-bed characteristics which permit shutdown and restart of the system with slurry in the pipeline. Slope restrictions are not critical. Kiln feed slurries, which contain a high proportion of clays, may form a semi-rigid gel after prolonged settling and this does place a limit upon the maximum period of shutdown for some formulations.

Economics. A new wet process plant with slurry pipeline transportation can provide an economic alternative to dry process with conventional transportation. Transportation costs are sufficiently high in relation to processing costs that the wet process–slurry pipeline system was found, for Australian conditions, to be viable when the distance between the raw materials mine and the market exceeds 25 km. Figure 29.2 shows the conclusions from a study for a facility producing 1 Mt of cement annually.

The economic study considered a raw materials mine remote from the market, as is often the case in this industry. Because the cement process uses about 1·6 t of raw material to produce 1 t of cement clinker, the alternatives chosen to represent the extreme conditions for plant location were

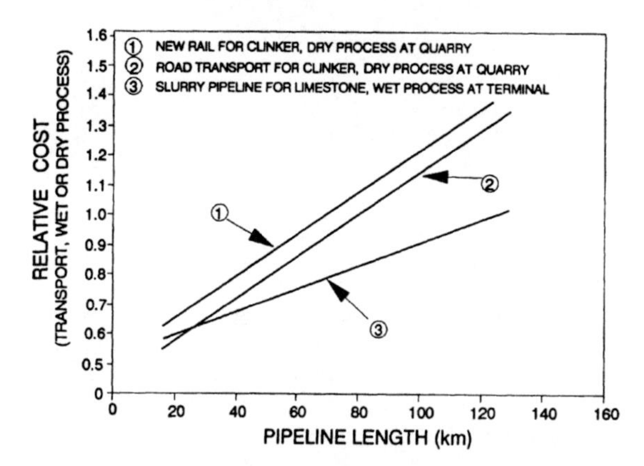

Fig. 29.2. Cement industry application of pipelines—relative economics and system alternatives.

- a dry process cement plant located at the raw materials mine with road or rail transport of product to the market or to a central distribution point; and
- a wet process cement plant located near the market or central distribution point, with slurry pipeline transporting raw materials from the remote mine.

These alternative schemes are shown in Fig. 29.3. Costs established were escalated to those currently applying in Australia. They include the cost of viscosity modifiers and average operating costs for processing components for dry and wet cement plants (i.e. kiln, grinding mills, homogenising, etc.). Rail transportation charges are those applying for short haul non-unit train transportation. Energy costs are a factor in any new facility. Wet process plants require approximately 70% more energy than dry process plants. In high energy cost areas, mechanical dewatering of slurry may be used to reduce the impact of energy on the economic comparison.

In the US, and many other countries of the Western World there are many existing wet process plants in operation. These facilities can utilise slurry trans-

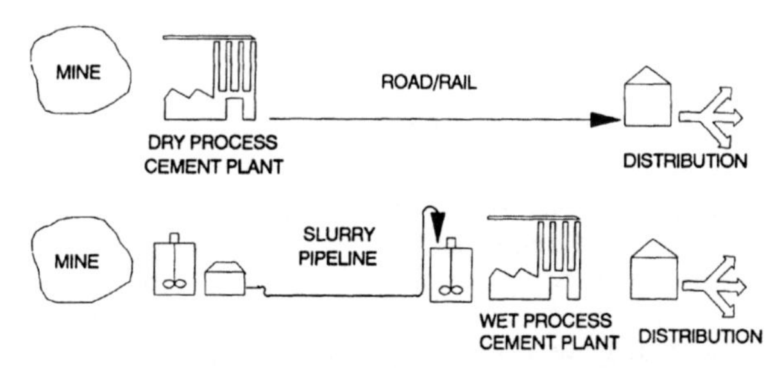

Fig. 29.3. Alternative system for economic study of cement industry pipelining.

TABLE 29.1
Typical Properties of Iron Ore Slurries

Yield stress (Pa)	3–9
Plastic viscosity (mPa s)	5–10
Concentration (% by mass)	55–65
Particle size	
<100 μm (%)	100
<44 μm (%)	80

portation of raw materials as a means of reducing operating costs. The cost of new grass roots cement plants has rapidly inflated in recent years. This, combined with environmental restrictions on new facilities, can be expected to favour the future development of remote mines, rather than construction of new facilities at the new mine, following depletion of existing sources.

3.3. Iron Ore Pipelining

3.3.1. Suitability for Pipelining

Distributed throughout the World are a number of pipelines transporting iron ore. The lengths of these pipelines vary up to 390 km; annual tonnages range from 2 to 12 Mt y^{-1}.

An iron ore slurry pipeline is an element in the total processing of the ore from minerals extraction at the mine site to agglomeration at the pipeline terminal. Mineral extraction generally involves fine grinding of the ore for efficient extraction of the magnetite or hematite concentrate. These concentrates are produced in slurry form, in a size consist suited to pumping. The rheological properties of the slurry are typically within the range shown in Table 29.1 and the iron ore slurry is *soft-settling*.

In general, iron concentrate slurries are highly abrasive requiring installation of either positive displacement plunger pumps or piston–diaphragm pumps to ensure long-life pump wearing parts. At the pipeline terminal the slurry is dewatered, generally by filters, and the filter cake is agglomerated to form iron ore pellets. The slurry corrosivity can be controlled to allow use of bare steel pipe.

3.3.2. Savage River Pipeline

Storage and pump station feed facilities. The earlier design of slurry storage and pump station feed facilities reflected a degree of conservatism due to the pioneering aspect of the project (Cowper & Wasp, 1973). The Savage River pump station includes two 8 h agitated storage tanks plus a 1 h pump feed tank. The Savage River flow circuit is illustrated in Fig. 29.4.

The original concept behind a 1 h slurry feed tank was threefold. First, a constant level could be maintained in the feed tank which resulted in a steady pressure in the pump station suction header. Secondly, the feed tank acted as a mixing tank for dilution of the slurry. Thirdly, the test loop could be run in parallel

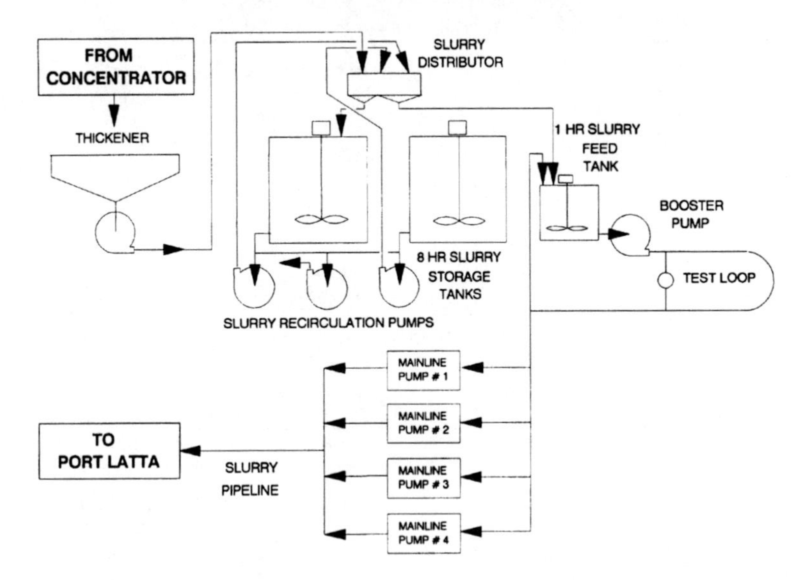

Fig. 29.4. Savage River iron ore slurry pipeline—flow diagram.

to the mainline suction header. During initial operations it was determined that these functions can be maintained without the need for a 1 h feed tank.

The magnetite in slurry storage tanks is maintained in suspension by downward pumping turbine agitators. Since Savage River the design and sizing of agitators has become highly competitive with a choice of good equipment from several manufacturers. It was earlier thought that mixing required assistance by recirculation via a centrifugal slurry pump. Recirculation is an inefficient and inadequate means of maintaining particle suspension.

Today, the amount of agitated storage capacity for a slurry pipeline system can be better defined. The required slurry storage for new slurry pipelines can be determined from the availability of the slurry preparation facility, the transit time for the pipeline and the availability of receiving facilities at the terminal of the pipeline.

Earlier pump stations incorporated a horizontal test loop in the suction piping to test a slurry batch prior to pumping in the main pipeline. Slurry technology has advanced to a point where testing in flow loops has been replaced partly by laboratory tests.

Mainline pumps. Pump units were basically oil-field mud pumps upgraded for long-life continuous operation in the mining industry. The Wilson–Snyder 560 PT pumps are triplex reciprocating plunger pumps operating at 169 strokes min^{-1} driven by 448 kW electric motors. A total of four units are installed, two fixed speed and two variable speed driven via a fluid drive. These units have operated for nearly

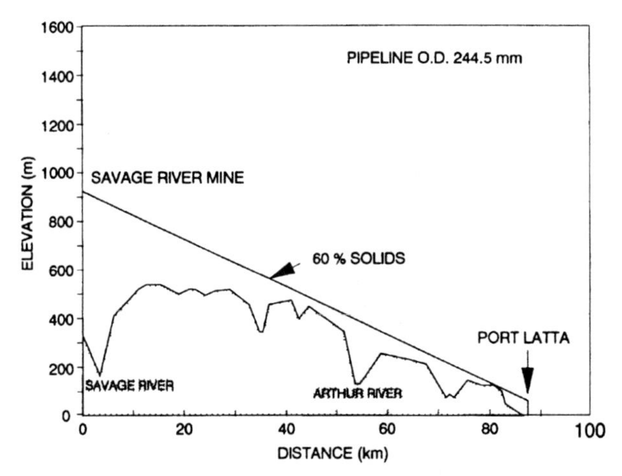

Fig. 29.5. Savage River iron ore slurry pipeline—ground profile and hydraulic gradient.

23 years and are still in operation. The pump station discharge pressure is of the order of 11 MPa. The major pump unit improvement includes replacing the single pump head block with three separate individual heads, one per plunger. This change isolates any head failures to an individual plunger, significantly reducing the repair labour. Wear parts experience relatively short lives because of the high pump speed. For instance, valves are changed on average every 350 h whereas those at Samarco, Brazil, have an average life of about 1200 h (Oliveiro, 1986).

Selection of a suitable valve to isolate pumps for maintenance or repair was a problem. High-pressure lubricated-plug valves did not provide satisfactory life in this service. In early operation, the life of these valves was about 2–3 months. Savage River developed a swing-type check valve based on the material selected for pump poppet valves. Samarco developed and installed a similar check-isolation valve some years later to replace their lubricated-plug valves.

Pipeline. Quickly settling iron concentrate slurries are transported in heterogeneous flow. Pipeline flow rate must be maintained above particle deposition velocity to avoid segregation. Lime is added to maintain a high slurry pH to control pipeline corrosion. Shutdown and restart operation with slurry in the pipeline as an operating mode is possible. However, the system must be designed with slope restriction. The 10% slope restriction is considered conservative and in future systems this may be increased to, say, 16%. However, the choice of slope limit for a new installation needs to be fully analysed by performing laboratory slope and slurry settling tests, analysing the pipeline profile and predicting restart pressure.

Pipeline operation. The Savage River pipeline profile (Fig. 29.5) commences at the mine site pump station at an elevation of 305 m, a peak of 518 m is crossed at 19·3 km and the discharge at the pellet plant is 5 m above sea level. The Savage

TABLE 29.2
Typical Mineral Concentrate Slurry Properties

Property (units)	*Magnetite*	*Copper concentrate*	
	Savage River	*Bougainville*	*West Irian*
Yield stress (Pa)	5·0	7·0	8·0
Plastic viscosity (mPa s)	7·0	9·5	9·5
Concentration (% by mass)	61·0	70·0	62·0
Density of solids ($kg\,m^{-3}$)	5000	4100	4100

River pipeline restart requires time to build up the flow rate without over-pressuring at the pump station. The pipeline is initially restarted by operating one variable speed unit at slow speed. The pump speed is increased in increments until full speed is reached and pressure stabilises. The second pump is then started and pressure stability is achieved in a similar manner. The final pump is then started. Normal flow is established in about 15–20 min.

3.4. Mineral Concentrate Pipelines

3.4.1. Overview

Mineral concentrates include copper, lead and zinc. In every case the ore is finely ground to increase the efficiency of extraction and the final concentrate is in slurry form with properties suited to long-distance pipelining. Ore bodies may contain a number of minerals that are produced as separate concentrates at the mine site. Slurry transportation of the products without contamination is achieved by batching the products separately down the pipeline. Each batch is followed by a short water slug to eliminate contamination of one concentrate into the next during transportation. Typical properties of mineral concentrates are shown in Table 29.2.

Mineral concentrate pump stations are similar to iron ore slurry pump stations. Both Bougainville and West Irian have single pump stations utilising Ingersoll–Rand/Aldrich triplex vertical plunger pumps. Each has two pumps installed (one acts as a stand-by spare). The pump drive motor at Bougainville is 520 kW while in West Irian it is only 224 kW.

3.4.2. Bougainville Copper Concentrate

The Bougainville Copper Ltd (Papua New Guinea) copper concentrate pipeline commenced operating in April 1972 and has successfully operated for over 17 years. The hydraulic design of the pipeline was a unique step in long-distance pipeline technology. The design was based on laboratory tests of a representative sample of the slurry to be pumped. The hydraulics were computer predicted using the laboratory test data. Full-scale loop tests were not required, eliminating the need for costly extraction and preparation of large test samples. Up to January 1981, the pipeline probably transported the greatest value of goods of any slurry pipeline: 5·3 Mt of concentrates containing 1·5 Mt of copper, 172 t of gold and 390 t of silver.

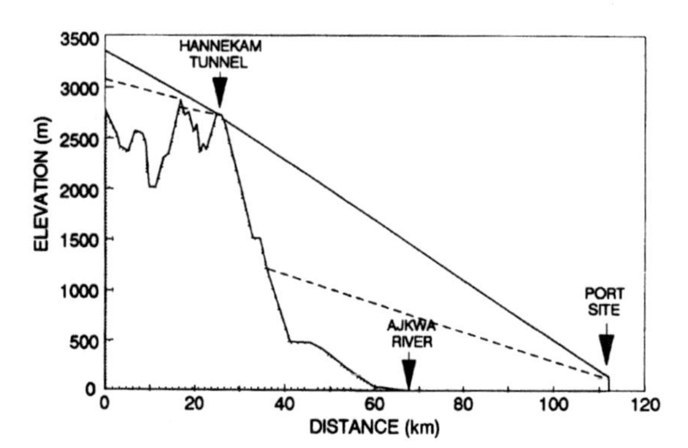

Fig. 29.6. West Irian copper concentrate slurry pipeline—ground profile and hydraulic gradient.

Benefits of the pipeline. Factors affecting the decision to use a pipeline are summarised elsewhere (Piercy & Cowper, 1981). These factors are common to most mineral pipeline systems. The reliability of any transport system in a continuous production operation is essential. High availabilities were being achieved with similar pipeline systems around the World. The relative location of the concentrator and the port and the intervening mountainous terrain necessitated road transport of all materials to the mine site. Pipeline systems proved to have better economics than trucking and consequently a pipeline system was accepted for Bougainville. Significant losses of concentrate could be expected to occur from road transportation whereas security of product was certain with a pipeline. The pipeline system is capital intensive and, as such, the cost of transport of concentrate to the port is virtually unaffected by inflationary effects—a most important advantage during periods of low metal prices. In the case of Bougainville, the pipeline is buried and is therefore completely unobtrusive, making it a most environmentally acceptable transportation system.

3.4.3. West Irian Copper Concentrate

The Bougainville pipeline was closely followed by another long-distance copper concentrate pipeline in the remote area of West Irian (Indonesia). This 108 mm diameter, 111 km pipeline commenced operation in November 1972.

Pipeline. The pipeline and access road traverse some of the remotest and roughest terrain in the World as illustrated by the ground profile and hydraulic gradient shown in Fig. 29.6. The pump station is located at the concentrator site at an elevation of 2800 m and the terminal is just above sea level. The initial portion of the line traverses a sawtooth ridge line for 24 km. Ground elevation varies from 1980 m up to 2830 m along the ridge with pipeline slopes as steep as 25% between these extremes. The pipeline follows the access road for 14·5 km down an escarpment from elevation 2650 m down to 610 m and thence a relatively gentle slope to flat swampy terrain approximately 72·5 km to the coast.

The pipeline was designed for an expected mine life of 13 years; it has been successfully operated for over 18 years. It will be replaced with a larger diameter pipeline in the near future to transport increased production from a new ore body.

After one year's operation a unique problem was experienced. Serious leaks developed in the steep pipe section downstream of the Hannekam Tunnel (Fig. 29.6). The pipeline flow in this section changes from *slack flow* to *packed flow*, where the hydraulic gradient at 60% solids intersects the ground profile. The hydraulic gradient of the slurry traversing the high point at Hannekam Tunnel is less than the steeply falling ground profile. The velocity increases and flow changes to open channel. A reduced pressure is created above the slurry surface. Hence the term *slack flow*. Slack flow in itself is not a problem provided the wall thickness of the pipeline is increased to allow for increase wear resulting from the higher slurry velocity. This increased wear was recognised. However, the significant increase in erosion occurring at the lower end of the slack flow section where the flow resumed *packed* conditions was unexpected. In this zone excess energy is dissipated in a hydraulic jump. In combination with high slurry velocity and cavitation, rapid loss of pipe wall steel was experienced. Extended sections up to 4 m long comprising over 9 mm of pipe wall on an arc of 60–80° were completely removed leaving only the external pipe coating to contain the slurry. The remainder of the pipe wall was in new condition with pipe mill scale unaffected.

Slack flow can be eliminated by installing a smaller diameter pipeline downstream of the slack flow area which increases the hydraulic gradient. Other options are to install pressure-reducing stations containing a number of orifices or *chokes* or a pipeline *harp*. Harps incorporate a number of smaller diameter pipelines installed in parallel. The velocity in each leg of the harp is low enough to prevent erosion.

3.5. Black Mesa Coal Pipeline

The longest and largest slurry pipeline in operation to date is the 440 km Black Mesa coal pipeline (USA). The pipeline traverses the state of Arizona and was commissioned in the autumn of 1970. This 457 mm diameter pipeline transports over 4·5 Mt y^{-1} of coal from a mine site in the Navajo–Hopi Indian reservation, Arizona, to the Mohave Power plant site on the Nevada side of the Colorado River.

Benefits of coal slurry pipelines. Pipelines can transport large tonnages (up to 30 Mt y^{-1}) requiring pipe diameters in the range 300–850 mm. The benefits of coal slurry pipelines and the success of the Black Mesa system are well documented in the literature (see, for example, Wasp & McCamish, 1978). Coal slurry pipelines are not large users of water when compared to that consumed in power generation. About 7 t of water is needed for each tonne of coal burned for electrical purposes. This compares with 1 t of water to each tonne of coal transported in a coal slurry

TABLE 29.3
Typical Properties of Black Mesa Slurry

Concentration (% by mass)	46–48
Particle size	
>1 168 μm (%)	1–2
<44 μm (%)	18 (minimum)

pipeline. Slurry pipelines do require a supply of water. However, at the pipeline terminal this water can be recovered for industrial use or in the power station for cooling.

The unit cost of transporting coal by slurry pipeline is of the order 70% of a new railway system. Railways have high labour requirements and 75–80% of the operating cost varies with inflation. A fully automated pipeline requires only a minimum of labour for operation. Only 30% of its operating cost varies with inflation. The unit cost (capital charges plus operating costs) of pipeline transportation inflates at less than half of the inflation in railway unit costs.

Slurry preparation. For the Black Mesa pipeline, coal with a top size of 50 mm is reduced to 8 mm by impactors and then ground to pass 1·2 mm in three parallel rod-mill lines. The slurry is stored in three agitated slurry tanks that feed the initial pump station. The properties of the slurry are shown in Table 29.3.

The optimum velocity for long-distance pipelining is in the range 1·5–1·8 $m\,s^{-1}$. Pipeline operation is sensitive to pumping off-specification size consist slurries. Coal slurry of the above size consist is pumped at the optimum velocity with only a 0·2 $m\,s^{-1}$ margin on the deposition velocity. Increasing the coarse content to 5% > 1168 μm with only 13–14% < 44 μm will deposit in the pipeline forming a bed of coarse solids.

The improved preparation circuit consisting of two impact mills in series has been tested as a viable alternative to the primary impact crusher/rod-mill grinding circuit developed for Black Mesa. Coal slurry produced by two-stage cage impact crushing is equal to or better than the alternative rod-mill product. The capital cost of the circuit is reduced at the expense of a higher maintenance cost for an impact crusher versus rod usage in a rod-mill.

Pump stations. Three pump stations are required along the route. Pumps are electric powered 1700 PT Wilson–Snyder double-acting duplex piston pumps, the largest of this type in existence. Each station has three pumps in parallel except station 2 which has four.

Pipeline. Coal pipelines are constructed to specifications similar to mineral slurry pipelines. The slope of the pipeline is restricted to about 16% grade.

Separation facilities. At the power plant, the coal discharges into one of three 24 h agitated holding tanks and is then pumped to centrifuges for dewatering. There are 20 centrifuges for each of the two 750 MW generating units. The dewatered coal, at about 25% total moisture, then goes through a pulverising mill and is blown into the boiler with heated air.

Major efforts have been directed to improve slurry dewatering with the aim to dewater the coal to a surface moisture suited for handling and bunker storage. Developments include slurry pre-heating, and the use of screen bowl centrifuges and belt filters. The required moisture content cannot be achieved by mechanical dewatering alone, and it is necessary to provide some thermal assistance. Tests have confirmed that pre-heating the slurry prior to centrifuging or filtering reduces the resultant cake moisture. The coal slurry is pre-heated in shell and tube heat exchangers using waste heat from the power plant. The reduced viscosity of the water in the slurry at the elevated temperature allows improved flow during separation. In addition, the coal cake is discharged at an elevated temperature and releases further moisture during adiabatic cooling.

3.6. Other Areas of Practical Application

The previous sections concentrated on transportation of solids cross-country over a long distance where the design must be highly refined to achieve a practical system. This section provides a summary of more common applications requiring less-sophisticated design procedures.

3.6.1. Mineral Processing

Mineral processing plants make extensive use of pumping and pressure piping for slurry handling for both in-plant and waste disposal application (Cowper & Venton, 1976). In-plant pressure pipelines are used in combination with unpressurised open launders. This is due partly to the need for pressure feeding of some process equipment (e.g. cyclones), and partly because pumps enable a more flexible design and a more economic use of plant floor area.

The optimum service from slurry pipes is obtained using the shortest pipe runs between process equipment, with a minimum number of 90° turns. Any near-horizontal runs are normally sloped to ensure drainage of both solids and liquids from the pipes during shutdown. Generally, abrasion becomes a consideration at velocities above about $2{\cdot}5\,m\,s^{-1}$ and is a major problem with velocities above $4{\cdot}5\,m\,s^{-1}$. Extra-long radius elbows or rubber hose are often used to minimise wear at changes of direction. Couplings that require grooving the pipe ends introduce weak points. Valves should preferably provide a full, straight-through opening, should not depend on machined surfaces for closure and should not have dead pockets that fill with solids and restrict operation.

For pumping system design, the major power requirement is often to overcome a static lift to an elevated process unit. Energy losses from slurry friction are often only a minor consideration. Consequently, a highly refined design procedure to establish slurry head loss is not always critical for proper sizing of pumping units. The other important design factor is to ensure that the flow velocity in near-horizontal piping runs is adequate to prevent deposition of solids.

Centrifugal slurry pumps are the work horses of slurry transfer systems. When sizing and selecting a centrifugal pumping unit, it is good practice to ensure the motor has extra capacity and the drive system has a simple arrangement for adjusting pump speed. These allow for adjustment during start-up of units and for possible change in slurry properties not accounted for in the design.

3.6.2. Tailings Disposal

In typical mineral plants the recovered mineral is less than 3% of the total ore being treated, with 97% reporting to tailings. Consequently, tailings pipelines transport the greatest tonnages of all slurry pipelines. For economic reasons tailings dams are sited as close to the process plant as feasible depending on terrain. However, tailings pipeline length can vary from 200 m to 30 km. Depending on the difference in elevation between the process plant and the tailing dam, the tailings may require pumping. However, it frequently occurs that the tailing is transferred under gravity by flume.

Pumped tailing systems use centrifugal slurry pumps in series to provide the required slurry head. Most tailings have relatively coarse size consist requiring a high velocity in the tailings pipeline to prevent deposition.

Pipe wear is a problem that requires careful selection of pipeline materials to ensure a long life. Pipes are laid above ground on a prepared bench for ease of maintenance. Pipes are turned to balance pipe wear and to increase the pipeline life. The careful selection of pipe sizes is necessary to ensure safe operation for the varying throughputs of the process plant. Multiple parallel pipelines may be required to provide for the full operating envelope.

Where terrain will allow transfer of tailings by gravity, unpressurised pipe flumes offer an ideal solution. The slope of the flume needs to be carefully selected to ensure that the velocity in the flume is only slightly above deposition velocity. In steep terrain it is necessary to provide carefully designed, vertical drop boxes between pipe sections.

3.6.3. Dredging

The recovery and pumping of solids by dredging is the oldest of all slurry transfer systems. Dredging technology evolved as operator art, without a high degree of technical understanding of the slurry hydraulics. Dredge operators are highly skilled and the operation is controlled by monitoring pump suction pressure, engine exhaust temperature and pump discharge pressure.

A cutter suction dredge consists of a floating, pontoon-mounted pumping unit. These large centrifugal pumps are driven via diesel engines. The suction pipe to the centrifugal pump is mounted on a ladder that can be lowered or raised in the water. Generally, a hydraulically driven cutter head is mounted at the end of the ladder. The cutter head provides the shear necessary to break up the material being excavated. If the excavating depth is in excess of 5 m below the water level, the flow in the suction pipe is assisted by jet pumps. Dredges are capable of transferring a wide range of solids from fine silt to coarse gravel. The throughput of the dredge system is significantly affected by the size consist of the material to be pumped.

3.6.4. Sewage Sludge

The long-distance transfer of sewage sludge by pipeline is technically feasible and a recent example is the sludge pipeline constructed for North-West Water (UK) from Manchester to Liverpool. Digested sludge may be pumped from cities to areas for land restoration and fertilisation. The rheology of sewage sludge limits the solids content to 4–6%. The sewage sludge from a city of 2 million population can be transported in a pipeline with a diameter in the range 200–250 mm. The technology used for digested sewage sludge pumping is similar to long-distance mineral slurry pipelines.

3.6.5. Offshore Loading of Slurries

The Waipipi ironsands project, New Zealand, started in 1972, was the World's first system to load ships in oil-tanker fashion via a submarine slurry pipeline.

Ironsands are separated from the dredged ore and pumped to the port site where the ironsands are recovered from the slurry by dewatering cyclones mounted above the stockpile. The offshore loading system consists of gravity feed onto a reclaim conveyor located in a gallery under the stockpile. The conveyor discharges over a trash screen directly into a conical bottom sump. The sump concentrates the solids and feeds them into the suction of the ship-loading pump station which contains six centrifugal pumps in series. Each pump is sized at 400 mm × 350 mm driven by a 597 kW motor. Three units are fixed speed and three are variable speed driven via eddy current couplings.

The submarine pipeline is 300 mm diameter and approx. 2·6 km long. The pipe material is bare steel. The pipeline terminates under a mono-mooring with a riser hose between the end of the submarine pipeline and the mooring. The ship loading hose is connected to the central 360° rotating pipe of the mono-mooring.

The holds of the ship are specifically designed to receive the ironsands. The excess water in the slurry is decanted over-board during loading. An under-hold drainage system filters the remaining water in the ironsands during the voyage. At the destination the ironsands are unloaded by conventional grabs. The ships are loaded with up to 1500 $t\,h^{-1}$ of ironsands in all weather conditions except gale-force. The offshore loading systems cost less than 35% of alternative mechanical offshore loading systems.

4. RECENT SLURRY SYSTEMS—CASE STUDIES

4.1. Overview

This section describes three recent practical applications of slurry technology.

- Ok Tedi copper concentrate pipeline
- New Zealand Steel ironsand pipeline
- Nerang River sand bypass system

TABLE 29.4
Ok Tedi Design Parameters

Pipeline length	156 km
Nominal pipeline diameter	150 mm
Wall thickness	5·56–10·97 mm
Pipe steel	API 5L Grade X52
Pressure rating	ANSI Class 900
Maximum solids throughput	97 t h^{-1}
Normal solids throughput	91 t h^{-1}
Available static head	1500 m
Maximum pressure	14·7 MPa

4.2. Ok Tedi Copper Concentrate Pipeline

General description of system. The 156 km, 150 mm, Ok Tedi Copper Concentrate pipeline in Papua New Guinea is the World's longest copper concentrate slurry pipeline pumping a fine mineral slurry (Cowper & Venton, 1988). It was constructed in less than six months in one of the World's most remote and inaccessible locations. It has been in operation since June 1987. Major design parameters are summarised in Table 29.4.

Slurry properties. Slurry properties were determined from a 40 litre sample. This fine-particle slurry settles homogeneously with little particle size segregation. As settling proceeds, a clear layer of water forms at the top. Rheological measurements were made using a rotational viscometer. The data were analysed using the Bingham plastic model. Slurry properties are summarised in Table 29.5.

Pipeline hydraulics. At design conditions, turbulent flow was required. Laminar flow was to be avoided for two reasons: first, deposition is likely to occur, and secondly, under laminar flow conditions the pressure gradient is much more sensitive to changes in concentration.

The pressure gradient and the critical deposit velocity were computer predicted. The turbulent flow pressure gradient was predicted using a proprietary computer model that takes into account all relevant phenomena including non-Newtonian

TABLE 29.5
Ok Tedi Slurry Properties

Property (units)	*Design*	*Operating range*
Yield stress (Pa)	4·0	1–4
Plastic viscosity (mPa s)	15·5	6–15
Concentration (% by mass)	60	56–62
Particle size		
>212 μm (%)	0	0–0·2
>106 μm (%)	15	1–5
Density of solids (kg m^{-3})	3560	3500–3900

TABLE 29.6
Ok Tedi Hydraulic Parameters

Parameters (units)	*Design*	*Operating range*
Velocity ($m\,s^{-1}$)	1·25	1·18–1·5
Pressure gradient ($Pa\,m^{-1}$)		
km 155–59	259	230 (maximum)
km 59–0[a]	247	180 (minimum)

[a] Reduced pressure gradient due to pump flush water dilution.

rheology (Wilson & Thomas, 1985), heterogeneous effects (Wilson & Judge, 1978), and viscous sub-layer effects (Thomas, 1978). The design and operating hydraulic parameters given in Table 29.6 include a factor of safety.

System design. Figure 29.7 shows the pipeline route profile and the hydraulic gradient. The pipeline starts about 1 km from the concentrator. Slurry is accumulated in a 500 m^3 agitated tank which discharges directly into the pipeline under gravity head. Provision is made to control flow into the line, but this is necessary only when the line is being initially filled with slurry. The route profile makes it necessary to provide additional pressure when the pipeline is being flushed because the flushing water density does not provide sufficient energy to overcome the slurry friction losses. To ensure the pipeline flow remains above the minimum deposit velocity, a piston pump is installed in the flush-water line. The pump station contains a Wilson–Snyder triplex plunger pump driven by a 468 kW diesel. Corrosion inhibitor is automatically dosed into the flush-water. A small programmable logic controller is used to sequence the equipment operation, monitor and annunciate alarms, and to initiate alarm handling routines.

At Kiunga, slurry is received into an agitated storage tank and transferred to the filtration plant as required. On-line nuclear density metering facilitates the transfer

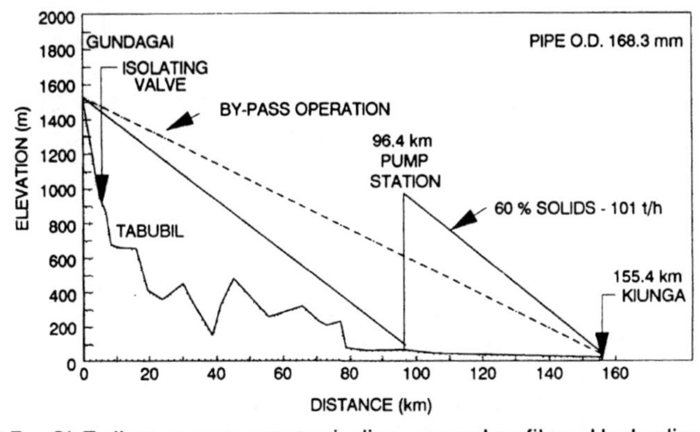

Fig. 29.7. Ok Tedi copper concentrate pipeline—ground profile and hydraulic gradient.

of flush-water slugs to the plant thickener as they are detected to minimise dilution of the slurry batches.

Pipeline design. The required pipe wall thickness was calculated from the pressure requirements together with an allowance for corrosion over the 20 year design life of the pipeline. Based on laboratory corrosion tests and experience with similar pipelines, a 0·076 mm y^{-1} corrosion allowance was added to the pipe wall thickness for most of the pipeline length. A higher wear margin was provided for the first 5 km of pipeline. Over this length the pipeline falls 690 m (see Fig. 29.7) raising the probability of slack flow with consequent high velocities and erosive pipe wear.

Externally the pipeline was protected with a 0·05 mm Polyken coating and 0·05–1·0 mm thick outer wrap. The pipeline was hydrostatically tested in five sections at maximum section pressure of 20 MPa, in accordance with Australian standards.

Pipeline route and construction. The terrain between the mine and Kiunga consists of jungle-covered, steep ridges. There is little natural rock and the high annual rainfall makes it impossible for tracked construction equipment to operate reliably except on a prepared road base of river gravel. The existing access road offered the only feasible route for the pipeline and it was installed as close as possible to the edge of the road. The road follows the relatively dry ridges and consequently there are up to 25 long-radius bends per kilometre and many vertical bends.

Pipeline operation. Historically, the fitness of a batch of slurry for pumping in long-distance slurry pipelines has been assessed by a combination of laboratory analysis and loop testing. No test loop is provided in the Ok Tedi pipeline and so a decision on slurry acceptability is based solely on laboratory tests. Slurry is rejected if the calculated deposit velocity, transition velocity or pressure gradient exceed preset limits.

To minimise slack flow the pipeline is operated without flow control at the pipeline inlet. The positive displacement pump is used to control the flow. When the available head exceeds the pipeline friction loss, such as when the pipeline is being refilled with slurry, flow increases beyond design velocity and by-passes the pump. As the pipeline loss increases as a result of slurry hydraulics and the position of a slurry batch in the pipeline, the velocity falls and at the appropriate time the pump is started.

The pipeline is designed for shutdown–restart operation, and, to limit the shutdown head on the pump station, an isolating valve station is provided approximately 5 km from the pipeline start. This valve is always closed before the pump station is stopped and is opened after the pump station has commenced acceleration during restart.

Following commissioning, concentrate production increased rapidly to 1000 t day^{-1} and then to 1500 t day^{-1}. Initially, with production below design the pipeline was operated on a shutdown–restart basis to limit flush-water use. Because static differential pressures at the isolating valve stations can be as high as 14 MPa, the pipeline is operated batchwise with the aim of stopping the pipeline with slugs of water at both isolating valve stations to reduce the risk of premature failure of the hard-faced plug valves. This requires careful monitoring by the pipeline operator to track the position of successive batches of slurry and water.

Laboratory studies in the design phase showed that some increases in concentration at the low points would occur at 12% gradient, but the settled solids remained removable under hydraulic pressure. It was concluded that slippage of particles to a low point would not significantly hinder pipeline shutdown–restart operation. In practice, dribble flow, caused by delays in completely isolating the pipeline with sectioning valves, could be responsible for considerable solids movement. However, shutdown–restart operation is extremely smooth, provided the sequences are performed with minimum delay.

4.3. New Zealand Steel Ironsand Pipeline

General description of system. This 18 km pipeline transporting titanomagnetite ironsand was commissioned in 1986 (Venton & Cowper, 1986). It is the first high-pressure pipeline in the World to use a combination of positive displacement pumps and a buried, polyurethane-lined pipeline to transport abrasive granular solids. The pipeline was selected in preference to other transport options because of environmental concerns.

Project development. An earlier investigation by the client into the slurry pipeline option had considered a conventional centrifugal pump system with bare steel pipe. However, the forecast replacement life of the pipeline made this option unacceptable. Earlier experience with the Waipipi ironsand ship-loading project convinced the designers that positive displacement pumps and polyurethane-lined pipe were essential to satisfy the design requirements. Two system designs were considered.

- A single-pump station system using a 250 mm diameter pipe operating at 30% solids and a maximum pressure of 14 MPa.
- A two-pump station system using a 200 mm diameter pipe operating at 50% solids and a maximum pressure of 10 MPa.

The design study concluded that the second option was the most economic. However, three technical concerns remained.

- Were positive displacement pumps capable of reliably pumping high density, coarse granular material, without excessive maintenance or operator attention?
- Was polyurethane-lined pipe capable of withstanding 25 years of operation at high and cyclic pressures?
- Was it possible to use welded pipeline joints with polyurethane-lined pipes?

To answer these questions an extensive investigation was undertaken.

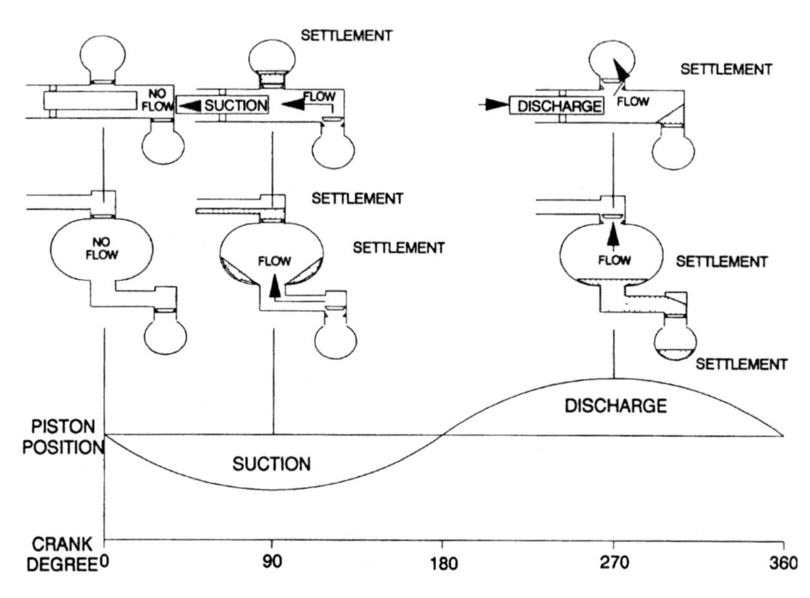

Fig. 29.8. Positive displacement pump cycle—ironsands slurry.

Pumps. Two types of positive displacement pump were identified as potentially suitable: a diaphragm and a flushed plunger pump. Tests were conducted by potential pump suppliers in Holland and the US to determine suitability for duty on ironsand slurry. Figure 29.8 shows the fluid end used in both pumps and indicates the areas of concern through the stroke. Tests showed there was sufficient turbulence during the stroke to prevent deposition, although settlement was evident during the low velocity portions of the cycle. Also, there was no problem restarting a pump after shutdown with slurry at normal concentrations. The pump cylinder on the suction stroke progressively fluidised the settled ironsand, while ironsand in the cylinder on the discharge stroke was displaced sufficiently for the stroke, without evidence of jamming. The diaphragm pump was assisted by the flexibility of the diaphragm, while the flush-water used in the plunger pump partially fluidised the settled solids ahead of the plunger.

Pipeline. Model pumps were operated through a range of speeds to assess whether suction and discharge pipes could become plugged. These tests showed that when the flowing velocity drops below the deposit velocity, ironsand deposits until the superficial velocity equals the deposit velocity. Immediately the velocity is increased the settled bed erodes to maintain the superficial velocity at the deposit velocity until full suspension is obtained. Detailed study (and correlation with other experience) confirmed the maintenance requirements offered by suppliers were realistic.

The pipe challenge was more difficult. Although polyurethane lining has been

TABLE 29.7
New Zealand Steel Ironsand Slurry Properties

Mean particle size (μm)	120
<212 μm × >75 μm (%)	94
Density of solids ($kg\,m^{-3}$)	4760

used in slurry pipelines since the 1970s there was little published history, and certainly no history of service in long-term high-pressure ironsand pipelines. Good experience with rubber-lined pipe coupled with the limited information available for polyurethane pipe linings in the UK, US and Australia provided confidence in polyurethane lining. A special coupling and welding technique was developed to weld the steel pipe without damage to the polyurethane (see Chapter 19).

Slurry properties. Typical properties of the ironsand concentrate are given in Table 29.7. During transport up to 1% of the ironsand solids are released as clay slimes.

Pipeline hydraulics. The particle size analysis and density of the ironsand (shown in Table 29.7) dictate a heterogeneous flow regime in the slurry pipeline. Care was required to select an operating velocity sufficiently above the deposit velocity to provide safe operation with minimum hold-up while not penalising the operating cost. These parameters were restricted by the available pipe sizes, the weekly operating cycle of 120 h, and system operation as the pipe lining wears. The projected operating hydraulic gradient was calculated from the known solids properties. Full advantage was taken of the smooth finish provided by centrifugally cast polyurethane-lined pipe in determining the design hydraulic gradient. The final hydraulic design is summarised in Table 29.8.

Pipeline route. The pipeline route is across undulating farmland and along country roads. The pipeline crosses approximately 6 km of reclaimed swamp where the water table lies to within 0·6 m of ground level. This region is crossed with numerous drainage ditches which are machine cleaned annually. The minimum pipeline cover in this region is 0·75 m with 1·0 m cover provided below the invert of drains. The pipeline travels 9·2 km across undulating country to pump station 2. The pipeline is constructed at a maximum slope of 15% uphill in the direction of flow, and 12% downhill, to limit accumulation at low points on shutdown.

System design. The design parameters for the system are given in Table 29.9,

TABLE 29.8
New Zealand Steel Ironsand Hydraulic Parameters

Concentration by mass dry solids (%)	48·5
Minimum operating velocity ($m\,s^{-1}$)	3·6
Design operating velocity ($m\,s^{-1}$)	3·9
Design hydraulic gradient ($Pa\,m^{-1}$)	1024
Specific energy consumption ($kWh\,t^{-1}$)	7·82

TABLE 29.9
New Zealand Steel Ironsand Design Parameters

Pipeline length[a] (km)	
Section 1	9·2
Section 2	8·8
Pipe inside diameter (mm)	187·4
Pressure rating	ANSI Class 600
Solids throughput (based on a 120 h week) ($t\,h^{-1}$)	250
Maximum pressure (MPa)	10

[a] System schematic shown in Fig. 29.9.

the schematic of the system is shown in Fig. 29.9. The slurry pipeline system accepts ironsand either direct from the concentration plant product cyclones, or reclaimed from the plant stockpile (or both), and delivers the ironsand to a belt filter dewatering plant or an emergency cyclone at the pipeline terminal.

One of the benefits of positive displacement pumps is the guarantee of constant volumetric throughput. Ironsand is delivered at a controlled rate over a safety screen to the constant level pipeline feed tank. The ironsand settles to the tank outlet, and is delivered to the pump at the correct concentration. This enables pipeline operation to be switched between ironsand slurry and flush-water by simply starting and stopping the ironsand feed. No valves or flush pumps are required.

Each pump station contains two Geho piston diaphragm pumps driven through a common drive train. This not only provides a capital cost saving, but more importantly, allows the pump crankshafts to be aligned for minimum pulsation. Each pump set is fitted with a scoop controlled fluid coupling. Pump station no. 1 is driven at 100% speed and pump station no. 2 is driven at controlled speed to maintain an adequate suction pressure. The terminal elevation is approximately 10 m above pump station 1, allowing the system to be operated without blocking

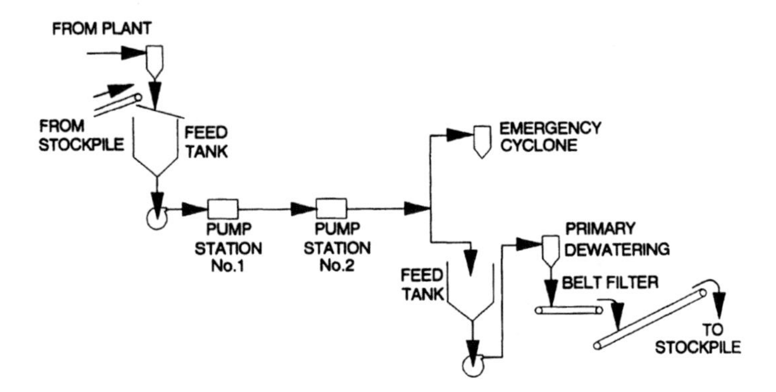

Fig. 29.9. New Zealand Steel ironsands pipeline—system schematic.

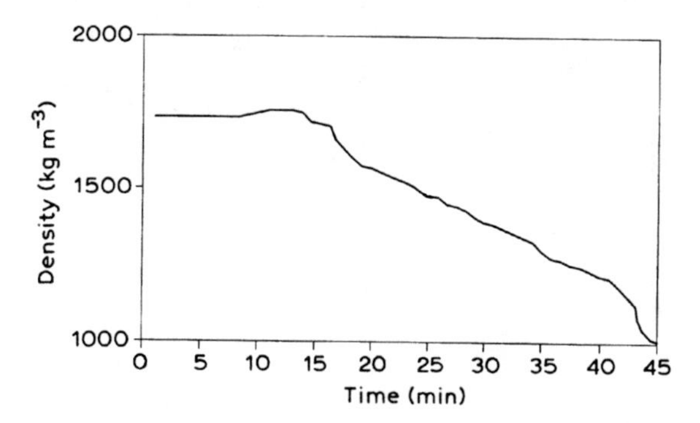

Fig. 29.10. New Zealand Steel ironsands pipeline—terminal density trail-out.

valves. Rising stem-gate valves fitted with replaceable liners are provided at each pump station to allow isolation for maintenance only. The pipeline system is fully automated, and operated by the mine site concentration plant operator. Pump station no. 2 is unattended, except for an inspection each shift.

System operation. Figure 29.10 illustrates the tail-out which occurs when the pipeline is flushed. There is no clear slurry–water interface. The gradual reduction in concentration as solids are flushed from the line is a consequence of the heterogeneous flow.

Restart of the pipeline system, when filled with ironsand at 50% concentration, is normally accomplished by the control system using the normal start sequence to accelerate the flow to full speed in 3 min. Acceleration is smooth, and the discharge pressure increases with flow, without significant over-pressure. During restart the pump station discharge pressure is controlled. It will override the acceleration ramp if necessary to maintain safe discharge pressures.

4.4. Nerang River Sand Bypass System

General description of system. A sand bypass system is used to combat the effects of drifting coastal sand blocking river access. The system provides a fixed sand recovery and transfer system from a sand trap located on the south side of the Nerang River entrance, to a disposal beach on the north side of the entrance on South Stradbroke Island, Queensland, Australia. The flow diagram of the system is shown in Fig. 29.11.

Design criteria. The client required the facility to operate on a 40 h week day-shift basis (including maintenance time) for average sand-drift conditions, with extended operation at peak conditions. Under normal operating conditions the system is designed to operate at approximately 60% of the design capacity, but is capable of operating below 60% at a reduced efficiency. The sand-trap design is based on installation of sufficient recovery jets to excavate the sand trap to 11·0 m

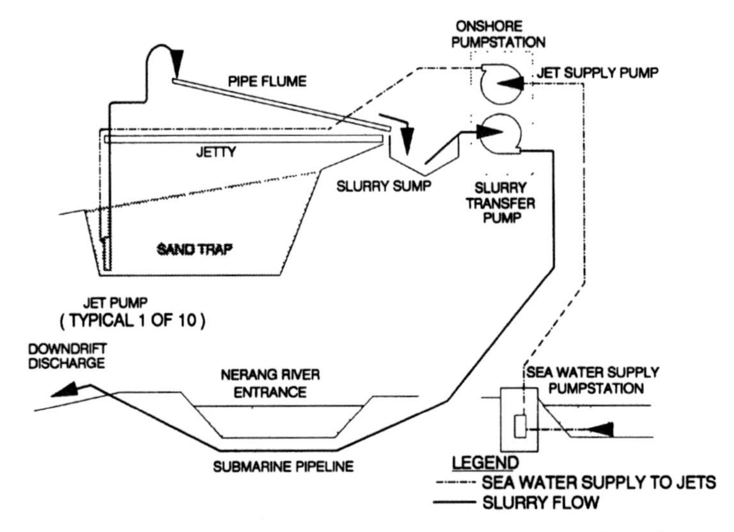

Fig. 29.11. Nerang River sand bypass system—system schematic.

below mean sea-level for the full length of the jetty proposed by the client. The design criteria are summarised in Table 29.10.

Design philosophy. The Nerang River entrance is a remote site, and the operating duties specified require the system to operate day and night during periods of peak sand drift. The design philosophy adopted is based on the following:

- operation from a shore-based control room using day-shift labour;
- sufficient automation for unattended operation, minimising the problems of manning a facility required to operate on a one, two or three shift basis depending on the sand loading cycle;

TABLE 29.10
Nerang River Sand Bypass System Design Criteria

Parameter (units)	*Peak conditions*	*Normal conditions*[a]
Sand throughput ($m^3 h^{-1}$)	585	335
Design concentration of sand (% by mass solids)	40	31
Transfer pipeline flow rate ($m^3 h^{-1}$)	1 760	1 420
Weekly operating hours (h)		30
Weekly sand throughput (m^3)		10 000

[a] Based on average weekly throughput.

- a system designed to trap the peak five-day sand drift thereby preventing training wall or entrance channel sedimentation from drift under all conditions;
- capacity for future extension;
- use of proven manufacturers' standard sand dredging equipment using electric power supplied from the local grid;
- minimum maintenance, based on periodic replacement of expendable components;
- a non-plugging transfer pipeline;
- modern control concepts to simplify operation, enhance reliability and eliminate the need for a skilled operator; and
- an environmentally acceptable and aesthetically pleasing installation.

System design. The sand bypass system (shown in Fig. 29.11) consists of a jet pump sand recovery system, a flume transfer pipe, slurry pump and sand slurry discharge pipeline.

Two low-pressure, 150 kW vertical turbine, sea-water supply pumps are installed in a remote pump station. They supply clear water from the basin to the suction of two high-pressure jet water supply pumps. These 560 kW centrifugal pumps are located in the main onshore pump station. The jet water supply pumps feed seven of the 10 jet lift pumps with motive water through the jetty header. The pumps operate singularly or in parallel depending on four or seven jet operation. High-pressure sea-water can also be supplied to the slurry transfer line to assist in slurry restart in an emergency.

Ten, 90 mm (3·5 in.) Genflo *Sandbug* jet pumps are installed on the jetty, to service the full length of the sand trap. The jet pumps are designed specifically for sand dredging operations and are non-clogging under fully buried conditions. Each pump includes integral fluidising jets which expand and fluidise the sand bed, enabling sand to be freely entrained by the jet pump at a controlled concentration. As sand is excavated from the region around the pump, the sand bed collapses to maintain a fluid bed adjacent to the pump. The trap continues to expand until the walls stabilise at the prevailing angle of repose.

Each of the jet pumps discharges into an elevated 375 m long pipe flume. The flume provides a non-blocking transfer system which has capacity to handle a wide range of flow rates and solids concentrations which enables the jet pumps to be properly balanced for equal performance. The jet pump discharge and the flume are lined with polyurethane for extended life.

The pipe flume discharges into a cone-bottom slurry pit which supplies the transfer pipeline main pump. This centrifugal pump is driven by a 710 kW electric motor. During peak flow operation, excess water will overflow the pipeline feed sump and the sand concentration automatically increase to the maximum design value in the transfer pipeline.

A single, polyurethane-lined steel pipeline (400 mm nominal bore) is provided to transfer the slurry 1482 m to the other side of the river entrance where it is discharged onto the beach. Three discharge off-takes, each controlled by manually operated full-bore knife-gate valves, are arranged to prevent sand build-up and blockage during shutoff. The pipeline slope is limited to minimise plug formation during an emergency shutdown on sand. The pipeline is insulated from the pump station and protected against external corrosion by an impressed current cathodic protection system.

System operation. Under normal operation, one low-pressure sea-water pump and one high-pressure jet-water supply pump operates to supply four jet pumps. When peak transfer rates are required the second sea-water supply and jet-water supply pumps and remaining three jet pumps are operated. Any combination of seven of the ten jet pumps covering the trap can be operated.

The pipeline system includes pressure, density and flow measurement instrumentation to monitor operation. The system is controlled by an integrated, microprocessor-based system.

5. REFERENCES

Aude, T.C., Cowper, N.T., Thompson, T.L. & Wasp, E.J. (1971). Slurry pipeline systems, trends, design methods, guideline. *Chem. Engng*, 28 June, **78**, 74–90.

Cowper, N.T. & Venton, P.B. (1976). Long distance slurry pipelines. In *Proceedings of the Annual Conference of the Australian Pipeline Industry Association.*

Cowper, N.T. & Venton, P.B. (1978). Slurry pipelines in the cement industry. In *Proc. STA 3*, pp. 118–24.*

Cowper, N.T. & Venton, P.B. (1988). The Ok Tedi copper concentrate pipeline—156 km fast track development. In *Proc. STA 13* pp. 577–88.*

Cowper, N.T. & Wasp, E.J. (1973). Slurry pipelines—recent developments. Paper presented at IMM Transportation Symposium, Melbourne, Australia.

Cowper, N.T., Thompson, T.L., Aude, T.C. & Wasp, E.J. (1972). Processing steps: key to successful slurry-pipeline systems. *Chem. Engng*, 7 Feb., **79**, 58–67.

McDermott, W.F., Cowper, N.T., Davis, R.A. & Wasp, E.J. (1968). Savage River Mines —The World's first long-distance iron ore slurry pipeline. Paper presented at AIME Society of Mining Engineers Fall Meeting, Minneapolis, MN, USA, 18–20 Sept.

Oliveiro, M.C. (1986). Samarco's iron concentrate pipeline. In *Proc. STA 11*, pp. 231–42.*

Piercy, P. & Cowper, N.T. (1981). Bougainville Copper Limited, concentrate pipeline—nine years of successful operation. *J. Pipelines*, **1**, 127–38.

Thomas, A.D. (1978). Coarse particles in a heavy medium—turbulent pressure drop reduction and deposition under laminar flow. In *Proc. HT 5*, paper D5.*

Venton, P.B. & Cowper, N.T. (1986). New Zealand Steel ironsands slurry pipeline. In *Proc. HT 10*, paper G3, pp. 237–46.*

Wasp, E.J., Kenny, J.P. & Gandhi, R.L. (1977). *Solid–Liquid Flow Slurry Pipeline Transportation.* Trans. Tech. Publications, Clausthal-Zellerfeld, Germany.

Wasp, E.J. & McCamish, H.M. (1978). A perspective on coal slurry pipelines for the next decade. Paper presented at AIM Petroleum Institute Conference, Houston, TX, USA, 17–18 Apr.

Wilson, K.C. & Judge, D.G. (1978). Analytically based nomographic charts for sand water flow. In *Proc. HT 5*, paper A1.*

Wilson, K.C. & Thomas, A.D. (1985). A new analysis of the turbulent flow of non-Newtonian fluids. *Can. J. Chem. Engng*, **63**, 539–46.

*Full details of the Hydrotransport (HT) and the Slurry Transport Association (STA) series of conferences can be found on pp. 11–13 of Chapter 1.

Appendix 1: Conversion between SI and Other Commonly Used Symbols

Nigel P. Brown
The British Petroleum Company plc, London, UK

TABLE A1.1: LENGTH

	metre *m*	*millimetre* *mm*	*yard* *yd*	*foot* *ft*	*inch* *in*
1 m	1	1·000E+03	1·094E+00	3·281E+00	3·937E+01
1 mm	1·000E−03	1	1·094E−03	1·094E−03	1·094E−03
1 yd	9·144E−01	9·144E+02	1	3·000E+00	3·600E+01
1 ft	3·048E−01	3·048E+02	3·333E−01	1	1·200E+01
1 in	2·540E−02	2·540E+01	2·778E−02	8·333E−02	1

(SI unit, metre, m)

TABLE A1.2: MASS

	kilogram *kg*	*tonne* *t*	*pound* *lb*	*long ton* *UKton*	*short ton* *USton*
1 kg	1	1·000E−03	2·205E+00	9·842E−04	1·102E−03
1 t	1·000E+03	1	2·205E+03	9·842E−01	1·102E+00
1 lb	4·536E−01	4·536E−04	1	4·464E−04	5·000E−04
1 UKton	1·016E+03	1·016E+00	2·240E+03	1	1·120E+00
1 USton	9·072E+02	9·072E−01	2·000E+03	8·929E−01	1

(SI unit, kilogram, kg)

TABLE A1.3: TIME

	second *s*	*minute* *min*	*hour* *h*	*day* *d*
1 s	1	1·667E−02	2·778E−04	1·157E−05
1 min	6·000E+01	1	1·667E−02	6·944E−04
1 h	3·600E+03	6·000E+01	1	4·167E−02
1 d	8·640E+04	1·440E+03	2·400E+01	1

(SI unit, second, s)

TABLE A1.4: AREA

	mm^2	*m*^{2}	*hectare* *ha*	*in*^{2}	*ft*^{2}	*yd*^{2}	*acre*
1 mm^2	1	1·000E−06	1·000E−10	1·550E−03	1·076E−05	1·196E−06	2·471E−10
1 m^2	1·000E+06	1	1·000E−04	1·550E+03	1·076E+01	1·196E+00	2·471E+04
1 ha	1·000E+10	1·000E+04	1	1·550E+07	1·076E+05	1·196E+04	2·471E+00
1 in^2	6·452E+02	6·452E−04	6·452E−08	1	6·944E−03	7·716E−04	1·594E−07
1 ft^2	9·290E+04	9·290E−02	9·290E−06	1·440E+02	1	1·111E−01	2·296E−05
1 yd^2	8·361E+05	8·361E−01	8·361E−05	1·296E+03	9·000E+00	1	2·066E−04
1 acre	4·047E+09	4·047E+03	4·047E−01	6·273E+06	4·356E+04	4·840E+03	1

TABLE A1.5: VOLUME

	cubic metre *m*^{3}	*litre* *l*	*gallon (UK)* *UKgal*	*gallon (US)* *USgal*
1 m^3	1	1·000E+03	2·200E+02	2·642E+02
1 l	1·000E−03	1	2·200E−01	2·642E−01
1 UKgal	4·546E−03	4·546E+00	1	1·201E+00
1 USgal	3·785E−03	3·785E−00	8·327E−01	1

TABLE A1.6: FORCE

	newton *N*	*kilogram-force* *kg-f or kp*	*pound-force* *lb-f*	*poundal* *pdl*
1 N	1	1·020E−01	2·248E−01	7·233E+00
1 kg-f	9·807E+00	1	2·205E+00	7·093E+01
1 lb-f	4·448E+00	4·536E−01	1	3·217E+01
1 pdl	1·383E−01	1·410E−02	3·108E−02	1

(SI derived unit, newton, N)

TABLE A1.7: POWER

	watt W	kW	horsepower hp	ft lb-f s^{-1}
1 W	1	1·000E−03	1·341E−03	7·376E−01
1 kW	1·000E+03	1	1·341E+00	7·376E+02
1 hp	7·457E+02	7·457E−01	1	5·500E+02
1 ft lb-f s^{-1}	1·356E+00	1·356E−03	1·818E−03	1

(SI derived unit, watt, W)

TABLE A1.8: ENERGY, WORK, HEAT

	joule J	British thermal unit Btu
1 J	1	9·478E−04
1 Btu	1·055E+03	1

(SI derived unit, joule, J)

TABLE A1.9: PRESSURE, STRESS

	pascal Pa	kPa	MPa	kg-f cm^{-2}	bar	psi lb-f in^{-2}	metre water m water
1 Pa	1	1·000E−03	1·000E−06	1·020E−05	1·000E−05	1·450E−04	1·020E−04
1 kPa	1·000E+03	1	1·000E−03	1·020E−02	1·000E−02	1·450E−01	1·020E−01
1 MPa	1·000E+06	1·000E+03	1	1·020E+01	1·000E+01	1·450E+02	1·020E+02
1 kg-f cm^{-2}	9·807E+04	9·807E+01	9·807E−02	1	9·807E−01	1·422E+01	1·000E+01
1 bar	1·000E+05	1·000E+02	1·000E−01	1·020E+00	1	1·450E+01	1·020E+01
1 lb-f in^{-2}	6·895E+03	6·895E+00	6·895E−03	7·031E−02	6·895E−02	1	7·031E−01
1 m water[a]	9·807E+03	9·807E+00	9·807E−03	1·000E−01	9·807E−02	1·422E+00	1

(SI derived unit, pascal, Pa equivalent to N m^2)
[a]density of water taken as 1000 kg m^{-3}

TABLE A1.10: PRESSURE GRADIENT

	kPa m^{-1}	*bar km*$^{-1}$	*m water (m pipe)*$^{-1}$	*m water (100 m pipe)*$^{-1}$	*lb-f in*$^{-2}$ *(1000 ft)*$^{-1}$
1 kPa m^{-1}	1	1·000E−01	1·020E−01	1·020E+01	4·421E+01
1 bar km^{-1}	1·000E−01	1	1·020E−02	1·020E+00	4·421E+00
1 m water (m pipe)$^{-1}$ [a]	9·807E+00	9·807E+01	1	1·000E+02	4·335E+02
1 m water (100 m pipe)$^{-1}$ [a]	9·807E−02	9·807E−01	1·000E−02	1	4·335E+00
1 lb-f in^{-2} (1000 ft)$^{-1}$	2·262E−02	2·262E−01	2·307E−03	2·307E−01	1

[a] density of water taken as 1000 kg m^{-3}

TABLE A1.11: VOLUMETRIC FLOW RATE

	m^3 *s*$^{-1}$	*l s*$^{-1}$	*m*^{3} *h*$^{-1}$	*m*^{3} *d*$^{-1}$	*UKgal h*$^{-1}$	*UKgal d*$^{-1}$	*USgal h*$^{-1}$	*USgal d*$^{-1}$
1 m^3 s^{-1}	1	1·000E+03	3·600E+03	8·640E+04	7·919E+05	1·901E+07	9·510E+05	2·282E+07
1 l s^{-1}	1·000E−03	1	3·600E+00	8·640E+01	7·919E+02	1·901E+04	9·510E+02	2·282E+04
1 m^3 h^{-1}	2·778E−04	2·778E−01	1	2·400E+01	2·200E+02	5·279E+03	2·642E+02	6·340E+03
1 m^3 d^{-1}	1·157E−05	1·157E−02	4·167E−02	1	9·165E+00	2·200E+02	1·101E+02	2·642E+02
1 UKgal h^{-1}	1·263E−06	1·263E−03	4·546E−03	1·091E−01	1	2·400E+01	1·201E+00	2·882E+01
1 UKgal d^{-1}	5·262E−08	5·262E−05	1·894E−04	4·546E−03	4·167E−02	1	5·004E−02	1·201E+00
1 USgal h^{-1}	1·052E−06	1·051E−03	3·785E−03	9·085E−02	8·327E−01	1·998E+01	1	2·400E+01
1 USgal d^{-1}	4·381E−08	4·381E−05	1·577E−04	3·785E−03	3·469E−02	8·327E−01	4·167E−02	1

TABLE A1.12: DENSITY

	kg m^{-3}	*relative density unit* [a]	*specific gravity unit*	*lb UKgal*$^{-1}$	*lb USgal*$^{-1}$
1 kg m^{-3}	1	1·002E−03	1·000E+03	1·002E−02	8·345E−03
1 relative density unit	9·982E+02	1	9·982E+05	1·000E+01	8·331E+00
1 specific gravity unit	1·000E−03	1·002E−06	1	1·002E−05	8·345E−06
1 lb UKgal^{-1}	9·978E+01	9·995E−02	9·978E+04	1	8·327E−01
1 lb USgal^{-1}	1·198E+02	1·200E−01	1·198E+05	1·201E+00	1

[a] ratio of the mass of a body to that of an equal volume of water at the triple point.

Appendix 2: Units and Dimensions

Nigel P. Brown
The British Petroleum Company plc, London, UK

The Système International d'Unités (abbreviated to SI in all languages) has been used in this handbook. It is internationally accepted as a system of units. The system is based on seven *base units*, one for each of the seven *base quantities* that are dimensionally independent; these are shown in Table A2.1.

Variables used in equations are defined using these base dimensions. Amongst other uses, this enables other coherent, absolute, length–mass–time (LMT) systems to be used. Care must be exercised when non-coherent systems (such as the US Customary Units) are used. In a non-coherent system of units, constants appear (such as g_c) when units are derived from base units. This is not the case in the SI system as illustrated in the example below.

EXAMPLE
The definitions of some commonly encountered SI *derived units* and their dimensions are given below.

Newton (N)
Unit of force. 1 N is the force required to impart an acceleration of $1\,m\,s^{-2}$ to a mass of 1 kg. The dimensions are $[M]\,[LT^{-2}] = [MLT^{-2}]$.

Pascal (Pa)
Unit of pressure, or stress which equals $1\,N\,m^{-2}$. The dimensions are $[MLT^{-2}]\,[L^{-2}] = [ML^{-1}T^{-2}]$.

Joule (J)
Unit of work, energy and heat. 1 J is the work done when a force of 1 N moves its point of application 1 m in the direction of the force. The dimensions are $[MLT^{-2}]\,[L] = [ML^{2}T^{-2}]$.

Watt (W)
Unit of power, equal to $1\,J\,s^{-1}$. The dimensions are $[ML^{2}T^{-2}]$ $[T^{-1}] = [ML^{2}T^{-3}]$.

TABLE A2.1
The Seven Base Quantities in the SI System

Base quantity	*Base SI Unit*	*Symbol*	*Dimension*
Length	metre	m	L
Mass	kilogram	kg	M
Time	second	s	T
Thermodynamic temperature	kelvin	K	Θ
Electric current	ampere	A	I
Amount of substance	mole	mol	N
Luminous intensity	candela	cd	J

Throughout this handbook dimensions are enclosed in square brackets and round brackets are used, where appropriate, to enclose abbreviations of units.

The use of gravimetric systems of units is the most frequent cause of confusion. For example, the pound-mass (lb) is often written and used in calculation schemes implying pound-force (lb–f). In SI, the kilogram is restricted to the unit of mass; the newton is the only unit of force.

The SI unit of temperature is the kelvin. This unit of temperature is not particularly useful for most engineering applications since a temperature of 20°C is 293·16 K. The difference is unimportant for temperature intervals since the Celsius (formally Centigrade) temperature scale is directly related to the thermodynamic temperature scale; 1°C is exactly equal to 1 K.

Interpreted details of systems of units, source documentation, conversion factors and much other useful information is given by Drazil (1983).

REFERENCE

Drazil, J.V. (1983). *Quantities and Units of Measurement—A Dictionary and Handbook.* Mansell Pub Co., London, UK.

List of contributors

David A. Brookes
British Petroleum Engineering, London, UK

Dave Brookes is a Chartered Mechanical Engineer with a background in pipeline and utilities engineering. Before joining British Petroleum he worked for EXXON as a pipeline contractor and with Seltrust Engineering where he was actively involved in mining and minerals processing. With British Petroleum he has gained experience in several sub-sea projects and managed the Stabflow R&D project on behalf of the joint venture with Bechtel Inc. He holds a BSc degree in engineering from City University, UK.

Nigel P. Brown
The British Petroleum Company plc, London, UK

Nigel Brown has been active in the field of two-phase solid–liquid fluid mechanics since 1976. He has a background in technical consultancy and is currently working for BP Exploration Co., London, in the drilling technical department. He holds BSc(Eng) and PhD degrees from Imperial College, London and is a Chartered Chemical Engineer.

Ralph K. Buckwalter
Fluor Daniel Inc., Redwood City, USA

Ralph Buckwalter is a registered professional engineer in the state of California in both mechanical and control systems engineering. He has over 27 years of experience in conceptual and detail engineering, construction and start-up phases of major pipeline projects. He has been involved in many pipeline projects since the Black Mesa coal slurry system in 1968. He is a graduate of Healds College of

Engineering and a member of the board of the Coal and Slurry Technology Association.

Alan J. Carleton
Warren Spring Laboratory, Stevenage, UK

Alan Carleton is a Senior Principal Scientific Officer at Warren Spring Laboratory. He is a Chemical Engineer with over 25 years of experience in contract research and development in topic areas that include paste handling, hydraulic transport and solid–liquid separation. Current responsibility is for the solid–liquid separation work of the Separation Processes Service, an international co-operative project. He holds a BSc in Chemical Engineering for Leeds University, UK.

Norman T. Cowper
Slurry Systems Pty Ltd, Northbridge, NSW, Australia

Norm Cowper formed Slurry Systems in 1978. His company has been responsible for engineering and commissioning many systems including New Zealand Steel ironsands pipeline, Fixed Sands By-pass system in Australia and the Ok Tedi mining pipeline in Papua New Guinea. He has in excess of 20 years experience in practical application of slurry pipelining technology.

Denis Delaroute
Denver Process Equipment Ltd, Leatherhead, UK

Denis Delaroute has held the position of Product Manager for Pumps at Denver Process Equipment since 1984. He has a background in the minerals industries, working for BP Minerals, Sykes Pumps and Warman Int in the UK. He is currently interested in furthering understanding of the handling of non-Newtonian fluids by centrifugal pumps. He holds a BSc from Salford University, UK.

Anthony J. Elliott
Carnon Consultancy, Redruth, UK

Tony Elliott manages the Carnon Consultancy which is involved in laboratory and plant test work in the minerals area for world-wide clients. Previously, he has been involved in the copper mining industry in Zambia and tin mining in the UK. He graduated from the University of Birmingham, UK, in Minerals Engineering and is a member of the Society for Mining, Metallurgy and Exploration and a fellow of the Minerals Engineering Society.

Ramesh L. Gandhi
Bechtel Inc., San Francisco, USA

Ramesh Gandhi is Chief Slurry Engineer for the Bechtel Corporation. He has been involved in process design, start-up and technical research and development for slurry pipelines for over 20 years. He is a registered Professional Engineer in the State of California, a member of the ASCE and the International Freight Pipeline

Society. He holds BE and ME degrees from Indian Universities and an MBA from Golden Gate University, USA. He has completed studies for a PhD.

Randall G. Gillies
Saskatchewan Research Council, Saskatoon, Canada

Randy Gillies has been employed at the Saskatchewan Research Council since 1975 and is currently manager of the Pipeline Technology Programme. He has been involved in many pipelining research projects and is currently working on the development of improved modelling techniques for the flow of settling slurries. He holds BSc and PhD degrees from the University of Saskatchewan.

Nigel I. Heywood
Warren Spring Laboratory, Stevenage, UK

Nigel Heywood is a Chartered Chemical Engineer and fellow of the IChemE. He has worked as an industrial researcher and consultant in slurry handling and rheology for almost 20 years. After graduating with BSc (Eng) and MSc degrees from Imperial College, London he gained a PhD for research into the co-current flow of gas and non-Newtonian slurries at the University of Wales. This research was continued at the University of Toronto, Canada. He currently runs the Wet Solids Handling research consortium at Warren Spring Laboratory.

Alan G. Huggett
British Petroleum Engineering, London, UK

Alan Huggett is a Chartered Chemical Engineer, specialising in fluid flow. He has been involved in many surge-related projects on both major pipelines and smaller liquid transfer systems. He holds a BSc degree.

Melissa J. McKibben
University of Saskatchewan, Saskatoon, Canada

Melissa McKibben is a graduate chemical engineer who has been investigating slurry pipeline erosion as her doctoral research project since 1986.

Marvin H. Muenzler
International Development Planners, San Francisco, USA

Marvin Muenzler has over 30 years experience in the energy transportation sector. Currently he is an independent consultant. Previously he worked for Bechtel Inc for over 20 years. While with Bechtel he was involved in economic and planning activities for most of the major pipelines. He holds a degree in Natural Gas and Petroleum Engineering and an MBA from Golden Gate University, USA.

Bryan Poulson
NEI International Research and Development Ltd, Newcastle-upon-Tyne, UK

Bryan Poulson currently manages the Corrosion Technology Group at NEI, where he has worked since 1974. Previously he worked for the Danish Corrosion Centre. He holds BSc and PhD degrees and has published extensively in the field of stress corrosion cracking and the effects of fluid flow on corrosion.

Clifton A. Shook
University of Saskatchewan, Saskatoon, Canada

Cliff Shook is Professor of Chemical Engineering at the University of Saskatchewan, Canada. Since 1965 he has been a major contributor to the understanding of slurry flows and has inspired and led research into a wide range of topics concerned with fluid-solid systems. He conducted research under Professors Newitt and Richardson at Imperial College, London. Currently, he actively collaborates with the Saskatchewan Research Council in pipeline studies. His current research activities are concerned with wear and modelling the flow behaviour of slurries.

Peter E. Snoek
Bechtel Inc., San Francisco, USA

Peter Snoek is the Manager of Slurry Technology for Bechtel Corporation. He has been engaged in the design, construction and operation of pipeline facilities for over 25 years. He holds BSc and MSc engineering degrees and an MBA. He is a registered Professional Engineer in the State of California, member of the ASCE and the International Freight Pipeline Society.

Michael Streat
Loughborough University of Technology, UK

Mike Streat currently holds the Chair of Chemical Engineering at Loughborough University of Technology, UK. Previously he was Reader in Process Engineering in the Department of Chemical Engineering, Imperial College, London, UK. He has been involved in the hydraulic transport of slurries since 1965 and he devoted much effort to conveying high concentration mixtures. He is a Chartered Engineer and a fellow of the Institution of Chemical Engineers.

Philip Venton
Williams Brothers – CMPS Engineers, Chatswood, NSW, Australia

Phil Venton currently works for Williams Brothers – CMPS Engineers. Previously he has worked for Slurry Systems, Associated Pipelines and Campbell Brothers Ltd in Australia. He has been involved in the design and operation of many major slurry pipeline engineering projects throughout the world. He holds a BSc degree in Chemical Engineering from the University of Queensland, Australia.

Anthony W. Wakefield
A.W. Wakefield, Stamford, UK

Tony Wakefield is a consulting engineer who has specialised since 1965 in the design of jet pumps and their application to handling solids. Jet pumps that he has designed find world-wide application in the aggregate and food industries. He is a Chartered Engineer and holds a BSc (Eng) degree. Currently he is actively pursuing a research degree.

Kenneth C. Wilson
Queen's University, Kingston, Canada

Ken Wilson is Professor, Department of Civil Engineering, at Queen's University, Canada. Since 1970 he has been a leading personality responsible for furthering understanding of the behaviour of settling slurries. His physically-based approach has provided the inspiration for many developments. Prior to his academic career, he gained design-office experience in hydraulic engineering. He holds a BSc in Civil Engineering from the University of British Colombia, an MSc from Imperial College, London and a PhD from his present University.

Index

Breinigsville, PA USA
27 April 2010
236883BV00004B/39/A

9 781851 666454